Integral Transforms and Operational Calculus

Special Issue Editor

H. M. Srivastava

MDPI • Basel • Beijing • Wuhan • Barcelona • Belgrade

I0049394

Special Issue Editor
H. M. Srivastava
University of Victoria
Canada

Editorial Office
MDPI
St. Alban-Anlage 66
4052 Basel, Switzerland

This is a reprint of articles from the Special Issue published online in the open access journal *Symmetry* (ISSN 2073-8994) from 2018 to 2019 (available at: https://www.mdpi.com/journal/symmetry/special_issues/Integral_Transforms_Operational_Calculus)

For citation purposes, cite each article independently as indicated on the article page online and as indicated below:

LastName, A.A.; LastName, B.B.; LastName, C.C. Article Title. *Journal Name* **Year**, *Article Number*, Page Range.

ISBN 978-3-03921-618-5 (Pbk)
ISBN 978-3-03921-619-2 (PDF)

© 2019 by the authors. Articles in this book are Open Access and distributed under the Creative Commons Attribution (CC BY) license, which allows users to download, copy and build upon published articles, as long as the author and publisher are properly credited, which ensures maximum dissemination and a wider impact of our publications.
The book as a whole is distributed by MDPI under the terms and conditions of the Creative Commons license CC BY-NC-ND.

Integral Transforms and Operational Calculus

Contents

About the Special Issue Editor

H. M. Srivastava has held the position of Professor Emeritus in the Department of Mathematics and Statistics at the University of Victoria in Canada since 2006, having joined the faculty there in 1969, first as an Associate Professor (1969–1974) and then as a Full Professor (1974–2006). He began his university-level teaching career right after having received his M.Sc. degree in 1959 at the age of 19 years from the University of Allahabad in India. He earned his Ph.D. degree in 1965 while he was a full-time member of the teaching faculty at the J. N. V. University of Jodhpur in India. He has held numerous visiting research and honorary chair positions at many universities and research institutes in different parts of the world. Having received several D.Sc. (infoTitlehonoris causa) degrees as well as honorary memberships and honorary fellowships of many scientific academies and learned societies around the world, he is also actively associated editorially with numerous international scientific research journals. His current research interests include several areas of Pure and Applied Mathematical Sciences such as (for example) Real and Complex Analysis, Fractional Calculus and Its Applications, Integral Equations and Transforms, Higher Transcendental Functions and Their Applications, q-Series and q-Polynomials, Analytic Number Theory, Analytic and Geometric Inequalities, Probability and Statistics, and Inventory Modeling and Optimization. He has published 33 books, monographs and edited volumes, 33 book (and encyclopedia) chapters, 48 papers in international conference proceedings, and **more than 1,200** scientific research articles in peer-reviewed international journals, as well as Forewords and Prefaces to many books and journals, and so on. He is a Clarivate Analytics [Thomson-Reuters] (Web of Science) Highly Cited Researcher. For further details about his other professional achievements and scholarly accomplishments, as well as honors, awards and distinctions, including the lists of his most recent publications such as Journal Articles, Books, Monographs and Edited Volumes, Book Chapters, Encyclopedia Chapters, Papers in Conference Proceedings, Forewords to Books and Journals), the interested reader should look into the following Web Site: http://www.math.uvic.ca/~harimsri/.

Preface to "Integral Transforms and Operational Calculus"

This volume contains a total of 36 accepted submissions (including several invited feature articles) to the Special Issue of the MDPI's journal, ***Symmetry*** on the general subject-area of "Integral Transforms and Operational Calculus" from all over the world.

Investigations involving the theory and applications of integral transforms and operational calculus are remarkably wide-spread in many diverse areas of the mathematical, physical, chemical, engineering and statistical sciences. In this Special Issue, we invited and welcome review, expository and original research articles dealing with the recent advances on the topics of integral transforms and operational calculus as well as their multidisciplinary applications.

The suggested topics of interest for the call of papers for this Special Issue included, but were not limited to, the following keywords:

•Integral Transforms and Integral as well as Other Related Operators
•Applications Involving Mathematical (or Higher Transcendental) Functions
•Applications Involving Fractional-Order Differential and Differintegral Equations
•Applications Involving q-Series and q-Polynomials
•Applications Involving Analytic Number Theory
•Applications Involving Special Functions of Mathematical Physics and Applied Mathematics
•Applications Involving Geometric Function Theory of Complex Analysis

Several well-established scientific research journals, which are published by such publishers as (for example) Elsevier Science Publishers, John Wiley and Sons, Hindawi Publishing Corporation, Springer, De Gruyter, MDPI, and other publishing houses, have published and continue to publish a number of infoTitleSpecial Issues of many of their journals on recent advances on different aspects, especially of the subject of one of the above-mentioned keywords, "Applications Involving Fractional-Order Differential and Differintegral Equations." Many widely-attended international conferences, too, continue to be successfully organized and held world-wide ever since the very first one on this particular subject-area in U.S.A. in the year 1974.

Finally, it gives me enormous pleasure in thanking all of the participants in this Special Issue as well as the editorial personnel in the MDPI Editorial Office for Symmetry for their contributions toward the success of this Special Issue. The wholehearted support and dedication of one and all are indeed greatly appreciated.

H. M. Srivastava
Special Issue Editor

symmetry

MDPI

Article

Some Symmetric Identities for the Multiple (p,q)-Hurwitz-Euler eta Function

Kyung-Won Hwang [1] and Cheon Seoung Ryoo [2,*]

1 Department of Mathematics, Dong-A University, Busan 604-714, Korea; khwang@dau.ac.kr
2 Department of Mathematics, Hannam University, Daejeon 34430, Korea
* Correspondence: ryoocs@hnu.kr

Received: 28 March 2019; Accepted: 1 May 2019; Published: 8 May 2019

Abstract: The main purpose of this paper is to find some interesting symmetric identities for the (p,q)-Hurwitz-Euler eta function in a complex field. Firstly, we define the multiple (p,q)-Hurwitz-Euler eta function by generalizing the Carlitz's form (p,q)-Euler numbers and polynomials. We find some formulas and properties involved in Carlitz's form (p,q)-Euler numbers and polynomials with higher order. We find new symmetric identities for multiple (p,q)-Hurwitz-Euler eta functions. We also obtain symmetric identities for Carlitz's form (p,q)-Euler numbers and polynomials with higher order by using symmetry about multiple (p,q)-Hurwitz-Euler eta functions. Finally, we study the distribution and symmetric properties of the zero of Carlitz's form (p,q)-Euler numbers and polynomials with higher order.

Keywords: Euler numbers and polynomials; q-Euler numbers and polynomials; Hurwitz-Euler eta function; multiple Hurwitz-Euler eta function; higher order q-Euler numbers and polynomials; (p,q)-Euler numbers and polynomials of higher order; symmetric identities; symmetry of the zero

MSC: 11B68; 11S40; 11S80

---

1. Introduction

The area of the specific functions like the gamma and beta functions, the hypergeometric functions, special polynomials, the zeta functions and the area of series such as q-series, and series representations are a rapidly developing area in advanced mathematics (see [1–15]). Many q-extensions of specific functions and polynomials have been studied (see [1,3,6–10,13,16]). Srivastava [15] discussed some properties and q-extensions of the Bernoulli polynomials, Euler polynomials, and Genocchi polynomials. Choi, Anderson and Srivastava have developed the q-extension of the Riemann zeta function and functions related to the Riemann zeta function (see [5]). Choi and Srivastava presented a generalized Hurwitz formula and Hurwitz-Euler eta function (see [4]). Recently, many authors have developed (p,q)-extensions of the special functions, Riemann zeta function and related functions (see [1,13,17–19]). The symmetry of special polynomials is also actively studied (see [8,9,19]).

We use this

$$\sum_{m_1=0}^{n}\cdots\sum_{m_r=0}^{n}=\sum_{m_1,\cdots,m_r=0}^{n}.$$

We know the binomial formula as

$$(1-a)^n=\sum_{i=0}^{n}\binom{n}{i}(-a)^i,\ \text{where}\ \binom{n}{i}=\frac{n(n-1)\dots(n-i+1)}{i!},$$

and

$$\frac{1}{(1-a)^n}=(1-a)^{-n}=\sum_{i=0}^{\infty}\binom{-n}{i}(-a)^i=\sum_{i=0}^{\infty}\binom{n+i-1}{i}a^i.$$

Choi and Srivastava [4] constructed and made formulas about the multiple Hurwitz-Euler eta function $\eta_r(s,a)$ defined by following r-ple series:

$$\eta_r(s,a) = \sum_{k_1,\cdots,k_r=0}^{\infty} \frac{(-1)^{k_1+\cdots+k_r}}{(k_1+\cdots+k_r+a)^s}, \quad (Re(s)>0; a>0; r\in\mathbb{N}),$$

where $\mathbb{N}$ is the set of natural numbers. It is known that $\eta_r(s,a)$ can be analytically continued to be all complex s-plane (see [4]). The (p,q)-number was defined as

$$[n]_{p,q} = \frac{p^n-q^n}{p-q} = p^{n-1}+p^{n-2}q+p^{n-3}q^2+\cdots+p^2q^{n-3}+pq^{n-2}+q^{n-1}.$$

It can be seen that the (p,q)-number contains a symmetric property, and this number is q-number when $p=1$. In particular, we can see $\lim_{q\to 1}[n]_{p,q}=n$ with $p=1$. Since $[n]_{p,q}=p^{n-1}[n]_{\frac{q}{p}}$, we observe that p-numbers and (p,q)-numbers are different. In other words, by substituting q by $\frac{q}{p}$ in the q-number, we could not obtain a (p,q)-number. Therefore, much research has been conducted in the area of special functions by using (p,q)-number (see [1,13,18,19]). In this article, the (p,q)-extension of the multiple form of Hurwitz-Euler eta function can be defined as follows: For $s,x\in\mathbb{C}$ with $Re(x)>0$, the multiple (p,q)-Hurwitz-Euler eta function $\eta_{p,q}^{(r)}(s,x)$ is defined by

$$\eta_{p,q}^{(r)}(s,x) = [2]_q^r \sum_{m_1,\ldots,m_r=0}^{\infty} \frac{(-1)^{m_1+\cdots+m_r} q^{m_1+\cdots+m_r}}{[m_1+\cdots+m_r+x]_{p,q}^s}.$$

The aim of this paper is to introduce and study a new some generalizations of the Carlitz's form higher order q-Euler numbers and polynomials, the multiple q-Euler zeta function, and the multiple Hurwitz q-Euler zeta function. We call them Carlitz's type higher-order (p,q)-Euler numbers and polynomials, the multiple (p,q)-Euler zeta function, and the multiple (p,q)-Hurwitz-Euler eta function. The paper is structured as follows. In Section 2 we define Carlitz's type higher-order (p,q)-Euler numbers and (p,q)-Euler polynomials and induce some of their properties involving elementary properties, distribution relation, property of complement, and so on. In Section 3, by using the Carlitz's type higher-order (p,q)-Euler numbers and polynomials, the multiple (p,q)-Euler zeta function and the multiple (p,q)-Hurwitz-Euler eta function are defined. We also present some connection formulae between the Carlitz's type higher-order (p,q)-Euler numbers and polynomials, the multiple (p,q)-Euler zeta function, and the multiple (p,q)-Hurwitz-Euler eta function. In Section 4 we give several symmetric identities about the multiple (p,q)-Hurwitz-Euler eta function and Carlitz's type higher-order (p,q)-Euler numbers and polynomials. In Section 5, we investigate the distribution and symmetry of the zero of Carlitz's type higher-order (p,q)-Euler polynomials using a computer. Our paper ends with Section 6, where the conclusions and future developments of this work are presented.

Definition 1. *The classical higher-order Euler numbers denoted by $E_n^{(r)}$ and Euler polynomials denoted by $E_n^{(r)}(x)$ are defined as the below generating functions*

$$\left(\frac{2}{e^t+1}\right)^r = \sum_{n=0}^{\infty} E_n^{(r)} \frac{t^n}{n!}, \quad (|t|<\pi),$$

and

$$\left(\frac{2}{e^t+1}\right)^r e^{xt} = \sum_{n=0}^{\infty} E_n^{(r)}(x) \frac{t^n}{n!}, \quad (|t|<\pi),$$

respectively (see [15]).

Definition 2. *For $0 < q < p \leq 1$, the Carlitz's type (p,q)-Euler polynomials denoted by $E_{n,p,q}(x)$ are defined as the below generating function (see [13])*

$$\sum_{n=0}^{\infty} E_{n,p,q}(x)\frac{t^n}{n!} = [2]_q \sum_{m=0}^{\infty}(-1)^m q^m e^{[m+x]_{p,q}t}.$$

2. Carlitz's Form Higher-Order (p,q)-Euler Numbers and Polynomials

First, we think the Carlitz's form with high-order (p,q)-Euler numbers and polynomials as follows:

Definition 3. *For $r \in \mathbb{N}$, the high-order (p,q)-Euler polynomials denoted by $E^{(r)}_{n,p,q}(x)$ are defined like the generating function:*

$$\sum_{n=0}^{\infty} E^{(r)}_{n,p,q}(x)\frac{t^n}{n!} = [2]_q^r \sum_{m_1,\cdots,m_r=0}^{\infty}(-1)^{m_1+\cdots+m_r} q^{m_1+\cdots+m_r} e^{[m_1+\cdots+m_r+x]_{p,q}t}. \tag{1}$$

If $x = 0$, $E^{(r)}_{n,p,q} = E^{(r)}_{n,p,q}(0)$ are called the higher-order (p,q)-Euler numbers $E^{(r)}_{n,p,q}$. Note that if $r = 1$, then $E^{(r)}_{n,p,q} = E_{n,p,q}$ and $E^{(r)}_{n,p,q}(x) = E_{n,p,q}(x)$. Observe that if $p = 1, q \to 1$, then $E^{(r)}_{n,p,q} \to E^{(r)}_n$ and $E^{(r)}_{n,p,q}(x) \to E^{(r)}_n(x)$.

Definition 4. *For $r \in \mathbb{N}$, the (h,p,q)-Euler polynomials with high-order denoted by $E^{(r,h)}_{n,p,q}(x)$ are defined as the below generating function:*

$$\sum_{n=0}^{\infty} E^{(r,h)}_{n,p,q}(x)\frac{t^n}{n!} = [2]_q^r \sum_{m_1,\cdots,m_r=0}^{\infty}(-q)^{m_1+\cdots+m_r} p^{h(m_1+\cdots+m_r)} e^{[m_1+\cdots+m_r+x]_{p,q}t}. \tag{2}$$

If $x = 0$, $E^{(r,h)}_{n,p,q} = E^{(r,h)}_{n,p,q}(0)$ is called (h,p,q)-Euler numbers with higher-order denoted by $E^{(r)}_{n,p,q}$. Remark that if $h = 0$, then $E^{(r,h)}_{n,p,q} = E^{(r)}_{n,p,q}$ and $E^{(r,h)}_{n,p,q}(x) = E^{(r)}_{n,p,q}(x)$. We see that if $r = 1$, then $E^{(r,h)}_{n,p,q} = E^{(h)}_{n,p,q}$ and $E^{(r,h)}_{n,p,q}(x) = E^{(h)}_{n,p,q}(x)$ (see [13]). Observe that if $p = 1, q \to 1$, then $E^{(r,h)}_{n,p,q} \to E^{(r)}_n$ and $E^{(r,h)}_{n,p,q}(x) \to E^{(r)}_n(x)$.

By (1) and (2), we know that

$$\begin{aligned} E^{(r)}_{n,p,q}(x+y) &= \sum_{i=0}^{n}\binom{n}{i} p^{(n-i)x} q^{yi} E^{(r,n-i)}_{i,p,q}(x)[y]^{n-i}_{p,q}, \\ E^{(r)}_{n,p,q}(x) &= \sum_{i=0}^{n}\binom{n}{i} q^{xi}[x]^{n-i}_{p,q} E^{(r,n-i)}_{i,p,q}. \end{aligned} \tag{3}$$

Theorem 1. *For $r \in \mathbb{N}$, we have*

$$\begin{aligned} E^{(r)}_{n,p,q}(x) &= [2]_q^r \sum_{m_1,\cdots,m_r=0}^{\infty}(-1)^{m_1+\cdots+m_r} q^{m_1+\cdots+m_r}[m_1+\cdots+m_r+x]^n_{p,q} \\ &= \frac{[2]_q^r}{(p-q)^n}\sum_{l=0}^{n}\binom{n}{l}(-1)^l q^{xl} p^{(n-l)x}\left(\frac{1}{1+q^{l+1}p^{n-l}}\right)^r. \end{aligned}$$

Proof. When we use the Taylor series expansion of $e^{[x]_{p,q}t}$, we can get

$$\begin{aligned}\sum_{l=0}^{\infty} E_{l,p,q}^{(r)}(x)\frac{t^l}{l!} &= [2]_q^r \sum_{m_1,\cdots,m_r=0}^{\infty} (-1)^{m_1+\cdots+m_r} q^{m_1+\cdots+m_r} e^{[m_1+\cdots+m_r+x]_{p,q}t} \\ &= \sum_{l=0}^{\infty}\left([2]_q^r \sum_{m_1,\cdots,m_r=0}^{\infty} (-1)^{m_1+\cdots+m_r} q^{m_1+\cdots+m_r}[m_1+\cdots+m_r+x]_{p,q}^l\right)\frac{t^l}{l!}.\end{aligned}$$

The first part of the theorem follows when we compare the coefficients of $\frac{t^l}{l!}$ in the above equation. By (p,q)-numbers and binomial expansion, we also note that

$$\begin{aligned}E_{n,p,q}^{(r)}(x) &= [2]_q^r \sum_{m_1,\cdots,m_r=0}^{\infty} (-1)^{m_1+\cdots+m_r} q^{m_1+\cdots+m_r}[m_1+\cdots+m_r+x]_{p,q}^n \\ &= [2]_q^r \sum_{m_1,\cdots,m_r=0}^{\infty} (-1)^{m_1+\cdots+m_r} q^{m_1+\cdots+m_r}\left(\frac{p^{m_1+\cdots+m_r+x}-q^{m_1+\cdots+m_r+x}}{p-q}\right)^n \\ &= \frac{[2]_q^r}{(p-q)^n}\sum_{l=0}^{n}\binom{n}{l}(-1)^l q^{xl}p^{(n-l)x} \\ &\quad\times \sum_{m_1,\cdots,m_r=0}^{\infty}(-1)^{m_1+\cdots+m_r} q^{(l+1)(m_1+\cdots+m_r)}p^{(n-l)(m_1+\cdots+m_r)} \\ &= \frac{[2]_q^r}{(p-q)^n}\sum_{l=0}^{n}\binom{n}{l}(-1)^l q^{xl}p^{(n-l)x}\left(\frac{1}{1+q^{l+1}p^{n-l}}\right)^r.\end{aligned}$$

We finish the proof of Theorem 1. □

Theorem 2. *For $r \in \mathbb{N}$, we get*

$$E_{n,p,q}^{(r)}(x) = [2]_q^r \sum_{m=0}^{\infty}\binom{r+m-1}{m}(-1)^m q^m [m+x]_{p,q}^n. \tag{4}$$

Proof. By Taylor-Maclaurin series expansion of $(1-a)^{-n}$, we have

$$\left(\frac{1}{1+q^{l+1}p^{n-l}}\right)^r = \sum_{m=0}^{\infty}\binom{m+r-1}{m}(-1)^m (q^{l+1}p^{n-l})^m.$$

Also, by Theorem 1 and binomial expansion, one can obtain the desired result immediately. □

For $d \in \mathbb{N}$ with $d \equiv 1(\bmod 2)$, by Theorem 1 we can show

$$\begin{aligned}E_{n,p,q}^{(r)}(x) &= \frac{[2]_q^r}{(p-q)^n}\sum_{l=0}^{n}\binom{n}{l}(-1)^l q^{xl}p^{(n-l)x}\sum_{a_1,\cdots,a_r=0}^{d-1}\sum_{m_1,\cdots,m_r=0}^{\infty}(-1)^{a_1+\cdots+a_r} \\ &\quad\times(-1)^{m_1+\cdots+m_r}q^{(l+1)(a_1+dm_1+\cdots+a_r+dm_r)}p^{(n-l)(a_1+dm_1+\cdots+a_r+dm_r)}.\end{aligned}$$

Theorem 3. *(Distribution relation of (p,q)-Euler polynomials with higher-order). For $d \in \mathbb{N}$ with $d \equiv 1(\bmod 2)$, we have*

$$E_{n,p,q}^{(r)}(x) = \frac{[2]_q^r}{[2]_{q^d}^r}[d]_{p,q}^n \sum_{a_1,\cdots,a_r=0}^{d-1}(-q)^{a_1+\cdots+a_r}E_{n,p^d,q^d}^{(r)}\left(\frac{a_1+\cdots+a_r+x}{d}\right).$$

Proof. Since

$$E^{(r)}_{n,p^d,q^d}\left(\frac{a_1+\cdots+a_r+x}{d}\right)$$
$$=\frac{[2]^r_{q^d}}{(p^d-q^d)^n}\sum_{l=0}^{n}\binom{n}{l}(-1)^l q^{l(a_1+\cdots+a_r+x)}p^{(n-l)(a_1+\cdots+a_r+x)}\left(\frac{1}{1+q^{d(l+1)}p^{d(n-l)}}\right)^r,$$

we have

$$\sum_{a_1,\cdots,a_r=0}^{d-1}(-q)^{a_1+\cdots+a_r}E^{(r)}_{n,p^d,q^d}\left(\frac{a_1+\cdots+a_r+x}{d}\right)$$
$$=\frac{[2]^r_{q^d}}{(p^d-q^d)^n}\sum_{l=0}^{n}\binom{n}{l}(-1)^l q^{lx}p^{(n-l)x}$$
$$\times\sum_{a_1,\cdots,a_r=0}^{d-1}(-1)^{a_1+\cdots+a_r}q^{a_1+\cdots+a_r}q^{l(a_1+\cdots+a_r)}p^{(n-l)(a_1+\cdots+a_r)}\left(\frac{1}{1+q^{d(l+1)}p^{d(n-l)}}\right)^r.$$

Hence, we derive

$$\frac{[2]^r_q}{[2]^r_{q^d}}[d]^n_{p,q}\sum_{a_1,\cdots,a_r=0}^{d-1}(-q)^{a_1+\cdots+a_r}E^{(r)}_{n,p^d,q^d}\left(\frac{a_1+\cdots+a_r+x}{d}\right)$$
$$=\frac{[2]^r_q}{(p-q)^n}\sum_{l=0}^{n}\binom{n}{l}(-1)^l q^{xl}p^{(n-l)x}\left(\frac{1}{1+q^{l+1}p^{n-l}}\right)^r.$$

We prove Theorem 3. □

3. Multiple (p,q)-Hurwitz-Euler eta Function

We define multiple (p,q)-Hurwitz-Euler eta function. This function makes (p,q)-Euler polynomials at negative integers with higher-order. Choi and Srivastava [4] defined $\eta_r(s,a)$ by means of

$$\eta_r(s,a)=\sum_{k_1,\cdots,k_r=0}^{\infty}\frac{(-1)^{k_1+\cdots+k_r}}{(k_1+\cdots+k_r+a)^s},\quad (Re(s)>0; a>0; r\in\mathbb{N}).$$

It is known that $\eta_r(s,a)$ can be continued analytically to be all complex s-plane (see [4]). The (p,q)-extension of $\eta_r(s,a)$ can be defined as follows:

Definition 5. *For $s,x\in\mathbb{C}$ with $Re(x)>0$, the multiple (p,q)-Hurwitz-Euler eta function $\eta^{(r)}_{p,q}(s,x)$ is defined as*

$$\eta^{(r)}_{p,q}(s,x)=[2]^r_q\sum_{m_1,\ldots,m_r=0}^{\infty}\frac{(-1)^{m_1+\cdots+m_r}q^{m_1+\cdots+m_r}}{[m_1+\cdots+m_r+x]^s_{p,q}}.$$

Observe that when $p=1, q\to 1$, then $2^r\eta^{(r)}_{p,q}(s,a)=\eta_r(s,a)$.
Let

$$F^{(r)}_{p,q}(t,x)=\sum_{n=0}^{\infty}E^{(r)}_{n,p,q}(x)\frac{t^n}{n!}$$
$$=[2]^r_q\sum_{m_1,\ldots,m_r=0}^{\infty}(-1)^{m_1+\cdots+m_r}q^{m_1+\cdots+m_r}e^{[m_1+\cdots+m_r+x]_{p,q}t}. \tag{5}$$

Theorem 4. *For $r\in\mathbb{N}$, we get*

$$\eta^{(r)}_{p,q}(s,x)=\frac{1}{\Gamma(s)}\int_0^{\infty}F^{(r)}_{p,q}(x,-t)t^{s-1}dt, \tag{6}$$

where $\Gamma(s) = \int_0^\infty z^{s-1}e^{-z}dz$.

Proof. From (5) and Definition 5, we get

$$\begin{aligned}
\eta_{p,q}^{(r)}(s,x) &= [2]_q^r \sum_{m_1,\cdots,m_r=0}^{\infty} \frac{(-1)^{m_1+\cdots+m_r} q^{m_1+\cdots+m_r}}{[m_1+\cdots+m_r+x]_{p,q}^s} \\
&= [2]_q^r \frac{1}{\Gamma(s)} \sum_{m_1,\cdots,m_r=0}^{\infty} \frac{(-1)^{m_1+\cdots+m_r} q^{m_1+\cdots+m_r}}{[m_1+\cdots+m_r+x]_{p,q}^s} \int_0^\infty z^{s-1}e^{-z}dz \\
&= \frac{[2]_q^r}{\Gamma(s)} \sum_{m_1,\cdots,m_r=0}^{\infty} (-1)^{m_1+\cdots+m_r} q^{m_1+\cdots+m_r} \int_0^\infty e^{[m_1+\cdots+m_r+x]_{p,q}t} t^{s-1}dt \\
&= \frac{1}{\Gamma(s)} \int_0^\infty F_{p,q}^{(r)}(x,-t)t^{s-1}dt.
\end{aligned}$$

We are finished Theorem 4. □

The value of multiple (p,q)-Hurwitz-Euler eta function $\eta_{p,q}^{(r)}(s,x)$ at negative integers is given explicitly by the following theorem:

Theorem 5. *Let $n \in \mathbb{N}$. Then we obtain*

$$\eta_{p,q}^{(r)}(-n,x) = E_{n,p,q}^{(r)}(x).$$

Proof. Again, by (5) and (6), we have

$$\eta_{p,q}^{(r)}(s,x) = \frac{1}{\Gamma(s)} \int_0^\infty F_{p,q}^{(r)}(x,-t)t^{s-1}dt = \frac{1}{\Gamma(s)} \sum_{m=0}^{\infty} E_{m,p,q}^{(r)}(x) \frac{(-1)^m}{m!} \int_0^\infty t^{m+s-1}dt. \tag{7}$$

We note that

$$\Gamma(-n) = \int_0^\infty e^{-z}z^{-n-1}dz = \lim_{z\to 0} 2\pi i \frac{1}{n!} \left(\frac{d}{dz}\right)^n (z^{n+1}e^{-z}z^{-n-1}) = 2\pi i \frac{(-1)^n}{n!}. \tag{8}$$

For $n \in \mathbb{N}$, let us take $s = -n$ in (7). Then, by (7), (8), and Cauchy residue theorem, we have

$$\begin{aligned}
\eta_{p,q}^{(r)}(-n,x) &= \lim_{s\to -n} \frac{1}{\Gamma(s)} \sum_{m=0}^{\infty} E_{m,p,q}^{(r)}(x) \frac{(-1)^m}{m!} \int_0^\infty t^{m-n-1}dt \\
&= 2\pi i \left(\lim_{s\to -n} \frac{1}{\Gamma(s)} \right) \left(E_{n,p,q}^{(r)}(x) \frac{(-1)^n}{n!} \right) \\
&= 2\pi i \left(\frac{1}{2\pi i \frac{(-1)^n}{n!}} \right) \left(E_{n,p,q}^{(r)}(x) \frac{(-1)^n}{n!} \right) = E_{n,p,q}^{(r)}(x).
\end{aligned}$$

The proof of Theorem 5 is finished. □

By (4), we have

$$\sum_{n=0}^{\infty} E_{n,p,q}^{(r)} \frac{t^n}{n!} = [2]_q^r \sum_{m=0}^{\infty} \binom{m+r-1}{m} (-1)^m q^m e^{[m]_{p,q}t}.$$

From Taylor series of $e^{[m]_{p,q}t}$ in the above formula, we can get

$$\sum_{n=0}^{\infty} E_{n,p,q}^{(r)} \frac{t^n}{n!} = \sum_{n=0}^{\infty} \left([2]_q^r \sum_{m=0}^{\infty} \binom{m+r-1}{m} (-1)^m q^m [m]_{p,q}^n \right) \frac{t^n}{n!}.$$

If we compare coefficients $\frac{t^n}{n!}$, then we know

$$E^{(r)}_{n,p,q} = [2]^r_q \sum_{m=0}^{\infty} \binom{m+k-1}{m} (-1)^m q^m [m]^n_{p,q}. \tag{9}$$

By using (9), we define multiple (p,q)-Euler zeta function like below formula:

Definition 6. *For $s \in \mathbb{C}$, we define*

$$\zeta^{(r)}_{p,q}(s) = [2]^r_q \sum_{m=1}^{\infty} \binom{m+r-1}{m} \frac{(-1)^m q^m}{[m]^s_{p,q}}. \tag{10}$$

The function $\zeta^{(r)}_{p,q}(s)$ makes the number $E^{(r)}_{n,p,q}$ in negative integers. Instead of s, $s=-n$ for $n \in \mathbb{N}$ into (10), and using (9), we can obtain the below theorem:

Theorem 6. *Let $n \in \mathbb{N}$, We have*

$$\zeta^{(r)}_{p,q}(-n) = E^{(r)}_{n,p,q}.$$

4. Symmetric Identities for the Multiple (p,q)-Hurwitz-Euler eta Function

Let $w_1, w_2 \in \mathbb{N}$ where, $w_1 \equiv 1 \pmod 2$, $w_2 \equiv 1 \pmod 2$. For $r \in \mathbb{N}$ and $n \in \mathbb{Z}_+$, we get symmetry identities about the multiple (p,q)-Hurwitz-Euler eta function.

Theorem 7. *Let w_1, w_2 be natural numbers, where $w_1 \equiv 1 \pmod 2$, $w_2 \equiv 1 \pmod 2$. Then we obtain*

$$\begin{aligned}
&[w_2]^s_{p,q}[2]^r_{q^{w_2}} \sum_{j_1,\cdots,j_r=0}^{w_1-1} (-1)^{\sum_{l=1}^r j_l} q^{w_2 \sum_{l=1}^r j_l} \\
&\quad \times \eta^{(r)}_{p^{w_1}q^{w_1}}(s, w_2 x + \frac{w_2}{w_1}(j_1+\cdots+j_r)) \\
&= [w_1]^s_{p,q}[2]^r_{q^{w_1}} \sum_{j_1,\cdots,j_r=0}^{w_2-1} (-1)^{\sum_{l=1}^r j_l} q^{w_1 \sum_{l=1}^r j_l} \\
&\quad \times \eta^{(r)}_{p^{w_2},q^{w_2}}(s, w_1 x + \frac{w_1}{w_2}(j_1+\cdots+j_r)).
\end{aligned} \tag{11}$$

Proof. We know that $[xy]_q = [x]_{q^y}[y]_q$ for any $x,y \in \mathbb{C}$. Hence, using $w_2 x + \frac{w_2}{w_1}(j_1+\cdots+j_r)$ instead of x and replacing by q^{w_1} and p^{w_1} instead of q and p in (11), respectively, we induce the next result

$$\begin{aligned}
&\frac{1}{[2]^r_{q^{w_1}}} \eta^{(r)}_{p^{w_1}q^{w_1}}(s, w_2 x + \frac{w_2}{w_1}(j_1+\cdots+j_r)) \\
&= \sum_{m_1,\cdots,m_r=0}^{\infty} \frac{(-1)^{m_1+\cdots+m_r} q^{w_1 m_1+\cdots+w_1 m_r}}{[m_1+\cdots+m_r+w_2 x+\frac{w_2}{w_1}(j_1+\cdots+j_r)]^s_{p^{w_1},q^{w_1}}} \\
&= \sum_{m_1,\cdots,m_k=0}^{\infty} \frac{(-1)^{m_1+\cdots+m_r} q^{w_1 m_1+\cdots+w_1 m_r}}{\left[\frac{w_1(m_1+\cdots+m_r)+w_1 w_2 x+w_2(j_1+\cdots+j_r)}{w_1}\right]^s_{p^{w_1},q^{w_1}}} \\
&= \sum_{m_1,\cdots,m_r=0}^{\infty} \frac{(-1)^{m_1+\cdots+m_r} q^{w_1 m_1+\cdots+w_1 m_r}}{\frac{[w_1(m_1+\cdots+m_k)+w_1 w_2 x+w_2(j_1+\cdots+j_k)]^s_{p,q}}{[w_1]^s_{p,q}}} \\
&= [w_1]^s_{p,q} \sum_{m_1,\cdots,m_k=0}^{\infty} \frac{(-1)^{m_1+\cdots+m_r} q^{w_1 m_1+\cdots+w_1 m_r}}{[w_1(m_1+\cdots+m_r)+w_1 w_2 x+w_2(j_1+\cdots+j_r)]^s_{p,q}}
\end{aligned}$$

$$
\begin{aligned}
&= [w_1]_{p,q}^s \sum_{m_1,\cdots,m_k=0}^{\infty} \sum_{i_1,\cdots,i_k=0}^{w_2-1} \frac{(-1)^{m_1+\cdots+m_r} q^{w_1 m_1+\cdots+w_1 m_r}}{[w_1(m_1+\cdots+m_r)+w_1 w_2 x+w_2(j_1+\cdots+j_r)]_{p,q}^s} \\
&= [w_1]_{p,q}^s \sum_{m_1,\cdots,m_r=0}^{\infty} \sum_{i_1,\cdots,i_r=0}^{w_2-1} (-1)^{\sum_{j=1}^r (w_2 m_j+i_j)} q^{w_1 \sum_{j=1}^r (w_2 m_j+i_j)} \\
&\quad \times \left([w_1(w_2 m_1+i_1)+\cdots+w_1(w_2 m_r+i_r)+w_1 w_2 x+w_2(j_1+\cdots+j_r)]_{p,q}^s\right)^{-1} \\
&= [w_1]_{p,q}^s \sum_{m_1,\cdots,m_r=0}^{\infty} \sum_{i_1,\cdots,i_r=0}^{w_2-1} (-1)^{\sum_{j=1}^r m_j} (-1)^{\sum_{j=1}^r i_j} q^{w_1 w_2 \sum_{j=1}^r m_j} q^{w_1 \sum_{j=1}^r i_j} \\
&\quad \times \left([w_1 w_2(x+m_1+\cdots+m_r)+w_1(i_1+\cdots+i_r)+w_2(j_1+\cdots+j_r)]_{p,q}^s\right)^{-1}.
\end{aligned} \tag{12}
$$

Thus, from (12), we see the following equation.

$$
\begin{aligned}
&\frac{[w_2]_{p,q}^s}{[2]_{q^{w_1}}^r} \sum_{j_1,\cdots,j_r=0}^{w_1-1} (-1)^{j_1+\cdots+j_r} q^{w_2(j_1+\cdots+j_r)} \eta_{p^{w_1},q^{w_1}}^{(r)}(s, w_2 x+\frac{w_2}{w_1}(j_1+\cdots+j_r)) \\
&= [w_1]_{p,q}^s [w_2]_{p,q}^s \sum_{m_1,\cdots,m_r=0}^{\infty} \sum_{i_1,\cdots,i_r=0}^{w_2-1} \sum_{j_1,\cdots,j_r=0}^{w_1-1} (-1)^{\sum_{l=1}^r (j_l+i_l+m_l)} q^{w_1 w_2 \sum_{l=1}^r m_l} \\
&\quad \times q^{w_1 \sum_{l=1}^r i_l} q^{w_2 \sum_{l=1}^r j_l} \\
&\quad \times \left([w_1 w_2(x+m_1+\cdots+m_r)+w_1(i_1+\cdots+i_r)+w_2(j_1+\cdots+j_r)]_{p,q}^s\right)^{-1}
\end{aligned} \tag{13}
$$

By using the same method as (13), we have

$$
\begin{aligned}
&\frac{[w_1]_{p,q}^s}{[2]_{q^{w_2}}^r} \sum_{j_1,\cdots,j_r=0}^{w_2-1} (-1)^{j_1+\cdots+j_r} q^{w_1(j_1+\cdots+j_r)} \eta_{p^{w_2},q^{w_2}}^{(r)}(s, w_1 x+\frac{w_1}{w_2}(j_1+\cdots+j_r)) \\
&= [w_1]_{p,q}^s [w_2]_{p,q}^s \sum_{m_1,\cdots,m_k=0}^{\infty} \sum_{j_1,\cdots,j_r=0}^{w_2-1} \sum_{i_1,\cdots,i_r=0}^{w_1-1} (-1)^{\sum_{l=1}^r (j_l+i_l+m_l)} \\
&\quad \times q^{w_1 w_2 \sum_{l=1}^r m_l} q^{w_2 \sum_{l=1}^r i_l} q^{w_1 \sum_{l=1}^r j_l} \\
&\quad \times \left([w_1 w_2(x+m_1+\cdots+m_r)+w_1(j_1+\cdots+j_r)+w_2(i_1+\cdots+i_r)]_{p,q}^s\right)^{-1}
\end{aligned} \tag{14}
$$

Therefore, by (13) and (14), we complete the proof Theorem 7. □

Taking $w_2 = 1$ in Theorem 7, we obtain the below corollary.

Corollary 1. *Let w_1 be natural numbers, where $w_1 \equiv 1 \pmod 2$. For $r \in \mathbb{N}$ and $n \in \mathbb{Z}_+$, we obtain*

$$
\begin{aligned}
\eta_{n,p,q}^{(r)}(s, w_1 x) = &\frac{[2]_q^r}{[2]_{q^{w_1}}^r [w_1]_{p,q}^s} \sum_{j_1,\cdots,j_r=0}^{w_1-1} (-1)^{\sum_{l=1}^r j_l} q^{w_2 \sum_{l=1}^r j_l} \\
&\times \eta_{n,p^{w_1},q^{w_1}}^{(r)}\left(s, x+\frac{j_1+\cdots+j_r}{w_1}\right).
\end{aligned} \tag{15}
$$

If $p = 1, q \to 1$ in above Corollary 1, then we can see the below corollary.

Corollary 2. *Let $m \in \mathbb{N}$. $m \equiv 1 \pmod 2$. For $r \in \mathbb{N}$ and $n \in \mathbb{Z}_+$, we obtain*

$$
\eta_r(s, x) = \frac{1}{m^s} \sum_{j_1,\cdots,j_r=0}^{m-1} (-1)^{j_1+\cdots+j_r} \eta_r\left(s, \frac{x+j_1+\cdots+j_r}{m}\right). \tag{16}
$$

For $r \in \mathbb{N}$ and $n \in \mathbb{Z}_+$, we see symmetry identities about higher-order (p,q)-Euler polynomials.

Theorem 8. *Let w_1, w_2 be natural numbers with $w_1 \equiv 1 \pmod 2$, $w_2 \equiv 1 \pmod 2$. For $r \in \mathbb{N}$ and $n \in \mathbb{Z}_+$, we obtain*

$$\begin{aligned}
&[w_1]_{p,q}^n [2]_{q^{w_2}}^r \sum_{j_1,\cdots,j_r=0}^{w_1-1} (-1)^{\sum_{l=1}^r j_l} q^{w_2 \sum_{l=1}^r j_l} \\
&\quad \times E_{n,p^{w_1},q^{w_1}}^{(r)} \left(w_2 x + \frac{w_2}{w_1}(j_1 + \cdots + j_r) \right) \\
&= [w_2]_{p,q}^n [2]_{q^{w_1}}^r \sum_{j_1,\cdots,j_r=0}^{w_2-1} (-1)^{\sum_{l=1}^r j_l} q^{w_1 \sum_{l=1}^r j_l} \\
&\quad \times E_{n,p^{w_2},q^{w_2}}^{(r)} \left(w_1 x + \frac{w_1}{w_2}(j_1 + \cdots + j_r) \right).
\end{aligned} \tag{17}$$

Proof. Using Theorems 5 and 7, we see easily the Theorem 8. □

Taking $w_2 = 1$ in Theorem 8, we have the below corollary.

Corollary 3. *Let w_1 be the natural number with $w_1 \equiv 1 \pmod 2$. For $r \in \mathbb{N}$ and $n \in \mathbb{Z}_+$, we obtain*

$$\begin{aligned}
E_{n,p^{w_1},q^{w_1}}^{(r)}(w_1 x) = \frac{[2]_q^r}{[2]_{q^{w_1}}^r} [w_1]_{p,q}^n \sum_{j_1,\cdots,j_r=0}^{w_1-1} (-1)^{\sum_{l=1}^r j_l} q^{w_2 \sum_{l=1}^r j_l} \\
\times E_{n,p^{w_1},q^{w_1}}^{(r)} \left(s, x + \frac{j_1 + \cdots + j_r}{w_1} \right).
\end{aligned} \tag{18}$$

If $p = 1, q \to 1$ in the above Corollary, then we get the another Corollary.

Corollary 4. *Let m be the natural number, where $m \equiv 1 \pmod 2$. Let $r \in \mathbb{N}$ and $n \in \mathbb{Z}_+$, we see*

$$E_n^{(r)}(x) = m^n \sum_{j_1,\cdots,j_r=0}^{m-1} (-1)^{j_1+\cdots+j_r} E_n^{(r)} \left(\frac{x + j_1 + \cdots + j_r}{m} \right). \tag{19}$$

By (3), we have

$$\begin{aligned}
&\sum_{j_1,\cdots,j_r=0}^{w_1-1} (-1)^{\sum_{l=1}^r j_l} q^{w_2 \sum_{l=1}^r j_l} \\
&\qquad \times E_{n,p^{w_1},q^{w_1}}^{(r)} \left(w_2 x + \frac{w_2}{w_1}(j_1 + \cdots + j_k) \right) \\
&= \sum_{j_1,\cdots,j_r=0}^{w_1-1} (-1)^{\sum_{l=1}^r j_l} q^{w_2 \sum_{l=1}^r j_l} \\
&\times \sum_{i=0}^{n} \binom{n}{i} q^{w_2(n-i)(j_1+\cdots+j_r)} p^{w_1 w_2 x i} E_{n-i,p^{w_1},q^{w_1}}^{(r,i)}(w_2 x) \left[\frac{w_2}{w_1}(j_1 + \cdots + j_r) \right]_{p^{w_1},q^{w_1}}^i \\
&= \sum_{j_1,\cdots,j_r=0}^{w_1-1} (-1)^{\sum_{l=1}^r j_l} p^{w_2 \sum_{l=1}^r j_l} \\
&\times \sum_{i=0}^{n} \binom{n}{i} q^{w_2(n-i) \sum_{l=1}^r j_l} p^{w_1 w_2 x i} E_{n-i,p^{w_1},q^{w_1}}^{(r,i)}(w_2 x) \left(\frac{[w_2]_{p,q}}{[w_1]_{p,q}} \right)^i [j_1 + \cdots + j_r]_{p^{w_1},q^{w_1}}^i
\end{aligned} \tag{20}$$

therefore, we can see the below theorem.

Theorem 9. *Let $w_1, w_2 \in \mathbb{N}$. Let $w_1 \equiv 1 \pmod 2$, $w_2 \equiv 1 \pmod 2$. Let $r \in \mathbb{N}$ and $n \in \mathbb{Z}_+$, we get*

$$\begin{aligned}
&\sum_{j_1,\cdots,j_r=0}^{w_1-1} (-1)^{\sum_{l=1}^r j_l} q^{w_2 \sum_{l=1}^r j_l} \\
&\qquad \times E_{n,p^{w_1},q^{w_1}}^{(r)}\left(w_2 x + \frac{w_2}{w_1}(j_1 + \cdots + j_r)\right) \\
&= \sum_{i=0}^{n} \binom{n}{i} [w_2]_{p,q}^i [w_1]_{p,q}^{-i} p^{w_1 w_2 x i} E_{n-i,p^{w_1},q^{w_1}}^{(r,i)}(w_2 x) \\
&\times \sum_{j_1,\cdots,j_r=0}^{w_1-1} (-1)^{\sum_{l=1}^r j_l} q^{w_2(n-i+1)\sum_{l=1}^r j_l} [j_1 \cdots + j_r]_{p^{w_2},q^{w_2}}^i .
\end{aligned}$$

For all different integers $n \geq 0$, let

$$\mathcal{S}_{n,i,p,q}^{(r)}(w) = \sum_{j_1,\cdots,j_r=0}^{w-1} (-1)^{\sum_{l=1}^r j_l} q^{(n-i+1)\sum_{l=1}^r j_l} [j_1 \cdots + j_k]_{p,q}^i .$$

This sum $\mathcal{S}_{n,i,p,q}^{(k)}(w)$ is called the alternating (p,q)-power sums.

By above Theorem 9, we get the result

$$\begin{aligned}
&[2]_{q^{w_2}}^r [w_1]_{p,q}^n \sum_{j_1,\cdots,j_r=0}^{w_1-1} (-1)^{\sum_{l=1}^r j_l} q^{w_2 \sum_{l=1}^r j_l} \\
&\qquad \times E_{n,p^{w_1},q^{w_1}}^{(r)}\left(w_2 x + \frac{w_2}{w_1}(j_1 + \cdots + j_r)\right) \\
&= [2]_{q^{w_2}}^r \sum_{i=0}^{n} \binom{n}{i} [w_2]_{p,q}^i [w_1]_{p,q}^{n-i} p^{w_1 w_2 x i} E_{n-i,p^{w_1},q^{w_1}}^{(r,i)}(w_2 x) \mathcal{S}_{n,i,p^{w_2},q^{w_2}}^{(r)}(w_1).
\end{aligned} \tag{21}$$

By using the same method as in (21), we have

$$\begin{aligned}
&[2]_{q^{w_1}}^r [w_2]_{p,q}^n \sum_{j_1,\cdots,j_r=0}^{w_2-1} (-1)^{\sum_{l=1}^r j_l} q^{w_1 \sum_{l=1}^k j_l} \\
&\qquad \times E_{n,p^{w_2},q^{w_2}}^{(r)}\left(w_1 x + \frac{w_1}{w_2}(j_1 + \cdots + j_r)\right) \\
&= [2]_{q^{w_1}}^r \sum_{i=0}^{n} \binom{n}{i} [w_1]_{p,q}^i [w_2]_{p,q}^{n-i} p^{w_1 w_2 x i} E_{n-i,p^{w_2},q^{w_2}}^{(r,i)}(w_1 x) \mathcal{S}_{n,i,p^{w_1},q^{w_1}}^{(r)}(w_2).
\end{aligned} \tag{22}$$

So we see the following result using (21) and (22) and Theorem 3.

Theorem 10. *Let w_1, w_2 be the natural numbers, where $w_1 \equiv 1 \pmod 2$, $w_2 \equiv 1 \pmod 2$. Let $r \in \mathbb{N}$ and $n \in \mathbb{Z}_+$, we can see*

$$\begin{aligned}
&[2]_{q^{w_1}}^r \sum_{i=0}^{n} \binom{n}{i} [w_1]_{p,q}^i [w_2]_{p,q}^{n-i} p^{w_1 w_2 x i} E_{n-i,p^{w_2},q^{w_2}}^{(r,i)}(w_1 x) \mathcal{S}_{n,i,p^{w_1},q^{w_1}}^{(r)}(w_2) \\
&= [2]_{q^{w_2}}^r \sum_{i=0}^{n} \binom{n}{i} [w_2]_{p,q}^i [w_1]_{p,q}^{n-i} p^{w_1 w_2 x i} E_{n-i,p^{w_1},q^{w_1}}^{(r,i)}(w_2 x) \mathcal{S}_{n,i,p^{w_2},q^{w_2}}^{(r)}(w_1).
\end{aligned}$$

Using Theorem 10, we induce the symmetric identity (p,q)-Euler numbers $E_{n,p,q}^{(r)}$ for the higher-order in complex field.

Corollary 5. *Let w_1, w_2 be the natural numbers which have $w_1 \equiv 1 \pmod 2$, $w_2 \equiv 1 \pmod 2$. For $k \in \mathbb{N}$ and $n \in \mathbb{Z}_+$, we get*

$$[2]^r_{q^{w_1}} \sum_{i=0}^{n} \binom{n}{i} [w_1]^i_{p,q} [w_2]^{n-i}_{p,q} p^{w_1 w_2 x i} S^{(r)}_{n,i,p^{w_1},q^{w_1}}(w_2) E^{(r,i)}_{n-i,p^{w_2},q^{w_2}}$$
$$= [2]^r_{q^{w_2}} \sum_{i=0}^{n} \binom{n}{i} [w_2]^i_{p,q} [w_1]^{n-i}_{p,q} p^{w_1 w_2 x i} S^{(r)}_{n,i,p^{w_2},q^{w_2}}(w_1) E^{(r,i)}_{n-i,p^{w_1},q^{w_1}}.$$

5. Zeros of the Higher-Order (p,q)-Euler Polynomials $E^{(r)}_{n,p,q}(x) = 0$

If it is difficult to find solutions of equations, visualizing distributions of solutions using a computer can help to find regular patterns of solutions. These are particularly interesting because it is hard to approach theoretically. Therefore, the work of the last section is of interest to us. Based on these results, we suggest a few unsolved problems.

The values of the $E^{(r)}_{n,p,q}(x)$ are given by

$$E^{(r)}_{0,p,q}(x) = 1,$$
$$E^{(r)}_{1,p,q}(x) = \frac{[2]^r_q \left(p^x \left(\frac{1}{1+pq}\right)^r - q^x \left(\frac{1}{1+q^2}\right)^r\right)}{p-q},$$
$$E^{(r)}_{2,p,q}(x) = \frac{[2]^r_q \left(p^{2x} \left(\frac{1}{1+p^2q}\right)^r - 2p^x q^x \left(\frac{1}{1+pq^2}\right)^r + q^{2x} \left(\frac{1}{1+q^3}\right)^r\right)}{(p-q)^2},$$
$$E^{(r)}_{3,p,q}(x) = \frac{[2]^r_q \left(p^{3x} \left(\frac{1}{1+p^3q}\right)^r - 3p^{2x} q^x \left(\frac{1}{1+p^2q^2}\right)^r + 3p^x q^{2x} \left(\frac{1}{1+pq^3}\right)^r - q^{3x} \left(\frac{1}{1+q^4}\right)^r\right)}{(p-q)^3}.$$

We see that the numerical results about approximate solutions of zeros of $E^{(r)}_{n,p,q}(x) = 0$ are in Tables 1 and 2. In Table 1, the numbers of zeros of $E^{(r)}_{n,p,q}(x) = 0$ are listed about a fixed $p = \frac{1}{2}$ and $q = \frac{1}{10}$.

Table 1. Numbers of real and complex zeros of $E^{(r)}_{n,p,q}(x)$.

	$r = 1, p = \frac{1}{2}, q = \frac{1}{10}$		$r = 3, p = \frac{1}{2}, q = \frac{1}{10}$	
Degree n	**Real Zeros**	**Complex Zeros**	**Real Zeros**	**Complex Zeros**
1	1	0	0	1
2	2	0	*	*
3	1	2	1	2
4	2	2	*	*
5	1	4	1	4
6	2	4	2	4
7	1	6	1	6
8	*	*	*	*
9	1	8	1	8
10	2	8	2	8
11	1	10	1	10
12	2	10	2	10
13	1	12	1	12
14	*	*	2	12
15	1	14	1	14
16	*	*	*	*
17	1	16	1	16

The $*$ mark in inside of Table 1 means that there is no solution of $E^{(r)}_{n,p,q}(x) = 0$. It is possible to visualize the zeros of $E^{(r)}_{n,p,q}(x) = 0$ using computer graphics. The zeros of $E^{(r)}_{n,p,q}(x) = 0$, where $x \in \mathbb{C}$ are visualized in Figure 1.

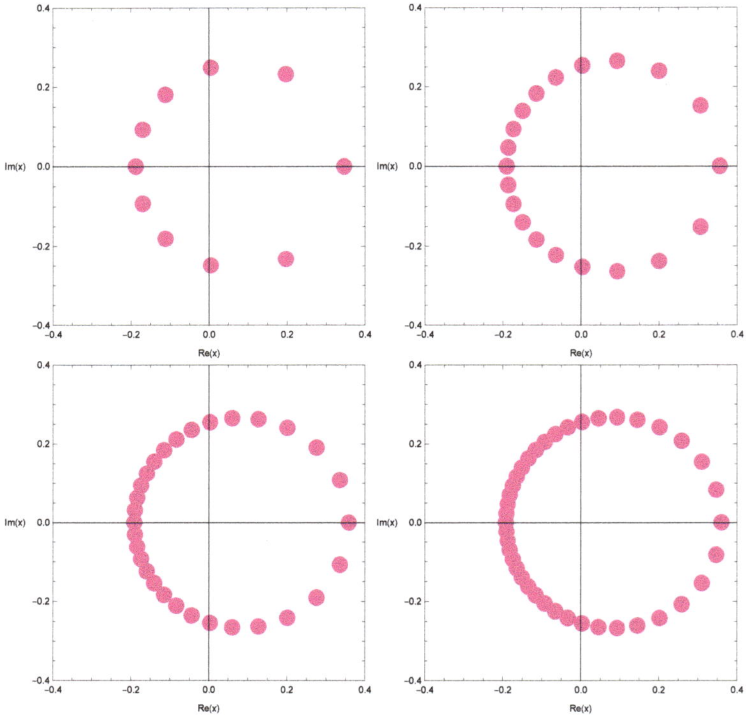

Figure 1. Zeros of $E^{(r)}_{n,p,q}(x) = 0$.

In Figure 1 (top-left), we chose $r = 7, n = 10, p = 1/2$ and $q = 1/10$. In Figure 1 (top-right), we chose $r = 7, n = 20, p = 1/2$ and $q = 1/10$. In Figure 1 (bottom-left), we chose $r = 7, n = 30, p = 1/2$ and $q = 1/10$. In Figure 1 (bottom-right), we chose $r = 7, n = 40, p = 1/2$ and $q = 1/10$. We can see that distribution of zeroes of $E^{(r)}_{n,p,q}(x) = 0$ is very regular. Therefore, the theoretical prediction of the regularity of distributions of the zeros of $E^{(r)}_{n,p,q}(x) = 0$ will remain as future research problems (Table 1).

Now, we have the numerical solution satisfying higher-order Euler polynomials $E^{(r)}_{n,p,q}(x) = 0$ for $x \in \mathbb{R}$. The numerical solutions of the higher-order Euler polynomials $E^{(r)}_{n,p,q}(x) = 0$ are listed in Table 2 about a fixed $r = 3, p = \frac{1}{2}$, and $q = \frac{1}{10}$ and different value of n.

Table 2. Numerical solutions of $E^{(3)}_{n,p,q}(x) = 0, p = \frac{1}{2}, q = \frac{1}{10}$.

Degree n	x
1	0.0723976
2	$*$
3	0.206956
4	$*$
5	0.258552
6	-0.163912, 0.273465

The $*$ mark in Table 2 means that there is no solution of $E^{(r)}_{n,p,q}(x) = 0$.

6. Conclusions and Future Developments

This paper introduced the Carlitz's form higher-order Euler numbers and polynomials. We have induced some formulas about the Carlitz's form Euler numbers and polynomials with high-order. Symmetric identities about Carlitz's form Euler numbers and polynomials with high-order are also gained. In addition, the result of [19] is a special case of $r = 1$, which can be induced from our paper. We make the following conjectures by numerical experiments:

Conjecture 1. *Prove or disprove that* $E^{(r)}_{n,p,q}(x), x \in \mathbb{C}$, *has* $Im(x) = 0$ *reflection symmetry analytic complex functions. Furthermore,* $E^{(r)}_{n,p,q}(x)$ *has* $Re(x) = a$ *reflection symmetry for* $a \in \mathbb{R}$.

It have been checked about many values of n. It is still unknown when the conjecture 1 is true or false about each value n (see Figure 1).

In Table 1, there is no solution of that the Carlitz's form (p,q)-Euler polynomials with higher-order is 0. Find such n so that there is no solution. If the Carlitz's form (p,q)-Euler polynomials with higher-order has solutions, it is doubtful whether it has distinct solutions.

Conjecture 2. *Prove or disprove that* $E^{(r)}_{n,p,q}(x) = 0$ *has* n *distinct solutions.*

We use the following symbols. $R_{E^{(r)}_{n,p,q}(x)}$ denotes the number of real zeros of $E^{(r)}_{n,p,q}(x) = 0$ on the real plane $Im(x) = 0$ and $C_{E^{(r)}_{n,p,q}(x)}$ denotes the number of complex zeros of $E^{(r)}_{n,p,q}(x) = 0$. We can check $R_{E^{(r)}_{n,p,q}(x)} = n - C_{E^{(r)}_{n,p,q}(x)}$ (see Tables 1 and 2) because n is the degree of the polynomial $E^{(r)}_{n,p,q}(x)$.

Also, when the Carlitz's form higher-order (p,q)-Euler polynomials is 0, if the equation has solutions, we have the following question:

Conjecture 3. *Prove or disprove that*

$$R_{E^{(r)}_{n,p,q}(x)} = \begin{cases} 1, & \text{if } n = \text{ odd}, \\ 2, & \text{if } n = \text{ even}. \end{cases}$$

We expect that the research in this direction will be a new approach using numerical methods for the study of Carlitz's form Euler polynomials $E^{(r)}_{n,p,q}(x) = 0$ (See [13,17,19,20]).

Author Contributions: All authors contributed equally in writing this article. All authors read and approved the final manuscript.

Funding: This work was supported by the Dong-A university research fund.

Conflicts of Interest: The authors declare no conflict of interest.

References

1. Araci, S.; Duran, U.; Acikgoz, M.; Srivastava, H.M. A certain (p,q)-derivative operato rand associated divided differences. *J. Ineq. Appl.* **2016**, *2016*. [CrossRef]
2. Andrews,G.E.; Askey, R.; Roy, R. Special Functions. In *Encyclopedia of Mathematics and Its Applications 71*; Cambridge University Press: Cambridge, UK, 1999.
3. Carlitz, L. Expansion of q-Bernoulli numbers and polynomials. *Duke Math. J.* **1958**, *25*, 355–364. [CrossRef]
4. Choi, J.; Srivastava, H.M. The Multiple Hurwitz Zeta Function and the Multiple Hurwitz-Euler Eta Function. *Taiwan. J. Math.* **2011**, *15*, 501–522. [CrossRef]
5. Choi, J.; Anderson, P.J.; Srivastava, H.M. Carlitz's q-Bernoulli and q-Euler numbers and polynomials and a class of generalized q-Hurwiz zeta functions. *Appl. Math. Comput.* **2009**, *215*, 1185–1208.
6. Guariglia, E.; Silvestrov, S. A functional equation for the Riemann zeta fractional derivative. *AIP Conf. Proc.* **2017**, *1798*, 020063.
7. Guariglia, E. Fractional derivative of the Riemann zeta function. In *Fractional Dynamics*; De Gruyter: Berlin, Germany, 2015; pp. 357–368.
8. He, Y. Symmetric identities for Carlitz's q-Bernoulli numbers and polynomials. *Adv. Diff. Equ.* **2013**, *246*, 10. [CrossRef]
9. Kim, D.; Kim, T.; Seo, J.-J. Identities of symmetric for (h,q)-extension of higher-order Euler polynomials. *Appl. Math. Sci.* **2014**, *8*, 3799–3808.
10. Kim T. Barnes type multiple q-zeta function and q-Euler polynomials. *J. Phys. A Math. Theor.* **2010**, *43*, 255201. [CrossRef]
11. Li, C.; Dao, X.; Guo, P. Fractional derivatives in complex planes. *Nonlinear Anal.* **2009**, *71*, 1857–1869. [CrossRef]
12. Ortigueira, M.D. A coherent approach to non-integer order derivatives. *Signal Process.* **2006**, *86*, 2505–2515. [CrossRef]
13. Ryoo, C.S. (p,q)-analogue of Euler zeta function. *J. Appl. Math. Inform.* **2017**, *35*, 113–120. [CrossRef]
14. Simsek, Y. Twisted (h,q)-Bernoulli numbers and polynomials related to twisted (h,q)-zeta function and L-function. *J. Math. Anal. Appl.* **2006**, *324*, 790–804. [CrossRef]
15. Srivastava, H.M. Some generalizations and basic (or q-) extensions of the Bernoulli, Euler and Genocchi Polynomials. *Appl. Math. Inform. Sci.* **2011**, *5*, 390–444.
16. Kurt, V. A further symmetric relation on the analogue of the Apostol-Bernoulli and the analogue of the Apostol-Genocchi polynomials. *Appl. Math. Sci.* **2009**, *3*, 53–56.
17. Agarwal, R.P.; Kang, J.Y.; Ryoo, C.S. Some properties of (p,q)-tangent polynomials. *J. Comput. Anal. Appl.* **2018**, *24*, 1439–1454.
18. Duran, U.; Acikgoz, M.; Araci, S. On (p,q)-Bernoulli, (p,q)-Euler and (p,q)-Genocchi polynomials. *J. Comput. Theor. Nanosci.* **2016**, *13*, 7833–7846. [CrossRef]
19. Ryoo, C.S. Some symmetric identities for (p,q)-Euler zeta function. *J. Comput. Anal. Appl.* **2019**, *27*, 361–366.
20. Ryoo, C.S. On the generalized Barnes type multiple q-Euler polynomials twisted by ramified roots of unity. *Proc. Jangjeon Math. Soc.* **2010**, *13*, 255–263.

© 2019 by the authors. Licensee MDPI, Basel, Switzerland. This article is an open access article distributed under the terms and conditions of the Creative Commons Attribution (CC BY) license (http://creativecommons.org/licenses/by/4.0/).

symmetry

MDPI

Article

A Note on the Truncated-Exponential Based Apostol-Type Polynomials

H. M. Srivastava [1,2,*], Serkan Araci [3], Waseem A. Khan [4] and Mehmet Acikgöz [5]

1 Department of Mathematics and Statistics, University of Victoria, Victoria, BC V8W 3R4, Canada
2 Department of Medical Research, China Medical University Hospital, China Medical University, Taichung 40402, Taiwan
3 Department of Economics, Faculty of Economics, Administrative and Social Science, Hasan Kalyoncu University, TR-27410 Gaziantep, Turkey; mtsrkn@hotmail.com
4 Department of Mathematics, Integral University, Lucknow 226026, Uttar Pradesh, India; waseem08_khan@rediffmail.com
5 Department of Mathematics, Faculty of Science and Arts, Gaziantep University, TR-27310 Gaziantep, Turkey; acikgoz@gantep.edu.tr
* Correspondence: harimsri@math.uvic.ca

Received: 3 April 2019; Accepted: 12 April 2019; Published: 15 April 2019

Abstract: In this paper, we propose to investigate the truncated-exponential-based Apostol-type polynomials and derive their various properties. In particular, we establish the operational correspondence between this new family of polynomials and the familiar Apostol-type polynomials. We also obtain some implicit summation formulas and symmetric identities by using their generating functions. The results, which we have derived here, provide generalizations of the corresponding known formulas including identities involving generalized Hermite-Bernoulli polynomials.

Keywords: truncated-exponential polynomials; monomiality principle; generating functions; Apostol-type polynomials and Apostol-type numbers; Bernoulli, Euler and Genocchi polynomials; Bernoulli, Euler, and Genocchi numbers; operational methods; summation formulas; symmetric identities

PACS: Primary 11B68; Secondary 33C05

1. Introduction

Operational techniques involving differential operators, which is a consequence of the monomiality principle, provide efficient tools in the theory of conventional polynomial systems and their various generalizations. Steffensen [1] suggested the concept of poweroid, which happens to be behind the idea of monomiality. The principle of monomiality was subsequently reformulated and developed by Dattoli [2]. The strategy underlining this viewpoint is apparently simple, but the outcomes are remarkably deep.

In the theory of the monomiality principle, a polynomial set $p_n(x)$ $(n \in \mathbb{N};\ x \in \mathbb{C})$ is quasi-monomial if there exist two operators $\widehat{M}$ and $\widehat{P}$, which are named the *multiplicative* and the *derivative* operators, respectively, are defined as follows:

$$\widehat{M}\{p_n(x)\} = p_{n+1}(x) \qquad \text{and} \qquad \widehat{P}\{p_n(x)\} = np_{n-1}(x),$$

together with the initial condition given by

$$p_0(x) = 1. \tag{1}$$

The operators $\widehat{M}$ and $\widehat{P}$ satisfy the following commutation relation:

$$[\widehat{M}, \widehat{P}] = \widehat{1}. \tag{2}$$

Thus, clearly, these operators display a Weyl group structure.

The properties of the polynomials $p_n(x)$ can be deduced from those of the operators $\widehat{M}$ and $\widehat{P}$. If $\widehat{M}$ and $\widehat{P}$ possess a differential character, then the polynomials $p_n(x)$ satisfy the following differential equation:

$$\widehat{M}\widehat{P}\{p_n(x)\} = np_n(x). \tag{3}$$

The polynomial family $p_n(x)$ can be explicitly constructed through the action of $\widehat{M^n}$ on $p_0(x)$ as follows:

$$p_n(x) = \widehat{M}^n\{p_0(x)\}. \tag{4}$$

Just as in (1), we shall always assume that $p_0(x) = 1$. In view of the above identity (4), the exponential generating function of $p_n(x)$ can be written in the form:

$$\exp(t\widehat{M})\{1\} = \sum_{n=0}^{\infty} p_n(x)\,\frac{t^n}{n!} \quad (|t| < \infty)\,. \tag{5}$$

We now introduce the truncated-exponential polynomials $e_n(x)$ (see [3]) defined by the following series:

$$e_n(x) = \sum_{k=0}^{n} \frac{x^k}{k!}, \tag{6}$$

that is, by the first $n+1$ terms of the Taylor-Maclaurin series for the exponential function e^x. These truncated-exponential polynomials play an important rôle in many problems in optics and quantum mechanics. However, their properties are apparently as widespread as they should be. The truncated-exponential polynomials $e_n(x)$ have been used to evaluate several overlapping integrals associated with the optical mode evolution or for characterizing the structure of the flattened beams. Their usefulness has led to the possibility of appropriately extending their definition. Actually, Dattoli et al. [4] systematically studied the properties of these polynomials.

The definition (6) does lead us to most (if not all) of the properties of the polynomials $e_n(x)$. We note the following representation:

$$e_n(x) = \frac{1}{n!}\int_0^{\infty} e^{-\xi}\,(x+\xi)^n\,\mathrm{d}\xi, \tag{7}$$

which follows readily from the classical gamma-function representation (see, for details, [3]). Consequently, we have the following generating function for the truncated-exponential polynomials $e_n(x)$ (see [4]):

$$\frac{e^{xt}}{1-t} = \sum_{n=0}^{\infty} e_n(x)\,t^n. \tag{8}$$

The definition (6) of $e_n(x)$ can thus be extended to a family of potentially useful truncated-exponential polynomials as follows (see [4]):

$$[2]e_n(x) = \sum_{k=0}^{[\frac{n}{2}]} \frac{x^{n-2k}}{(n-2k)!}, \tag{9}$$

which obviously possesses a generating function in the form (see [4]):

$$\frac{e^{xt}}{1-t^2} = \sum_{n=0}^{\infty} [2]e_n(x)t^n. \tag{10}$$

We also recall the higher-order truncated-exponential polynomials $[r]e_n(x)$, which are defined by the following series (see [4]):

$$[r]e_n(x) = \sum_{k=0}^{[\frac{n}{2}]} \frac{x^{n-rk}}{(n-rk)!} \tag{11}$$

and specified by the following generating function (see [4]):

$$\frac{e^{xt}}{1-t^r} = \sum_{n=0}^{\infty} [r]e_n(x)t^n. \tag{12}$$

The special two-variable case of the polynomials in (11) (that is, the case when $r = 2$) are important for applications. Moreover, these polynomials help us derive several potentially useful identities in a simple way and in investigating other novel families of polynomial systems. Actually, Equation (12) enables us to give a new family of polynomials as has been given in Theorem 1.

A 2-variable extension of the truncated-exponential polynomials is given by (see [4])

$$[2]e_n(x,y) = \sum_{k=0}^{[\frac{n}{2}]} \frac{y^k x^{n-2k}}{(n-2k)!} \tag{13}$$

and possesses the following generating function (see [4]):

$$\frac{e^{xt}}{1-yt^2} = \sum_{n=0}^{\infty} [2]e_n(x,y)t^n. \tag{14}$$

With a view to introducing a mixed family of polynomials related to the familiar Sheffer sequence, we first consider the 2-variable truncated-exponential polynomials (2VTEP) $e_n^{(r)}(x,y)$ of order r, which are expressed explicitly by (see [5])

$$e_n^{(r)}(x,y) = \sum_{k=0}^{[\frac{n}{2}]} \frac{y^k x^{n-rk}}{(n-rk)!} \tag{15}$$

and which are generated by

$$\frac{e^{xt}}{1-yt^r} = \sum_{n=0}^{\infty} e_n^{(r)}(x,y)\,\frac{t^n}{n!}. \tag{16}$$

From (8), (10), (12), (14) and (16), we can deduce several special cases of the 2VTEP $e_n^{(r)}(x,y)$, For example, we have

$$e_n^{(2)}(x,y) = [2]e_n(x,y) \quad e_n^{(1)}(x,1) = [r]e_n(x) \quad e_n^{(2)}(x,1) = [2]e_n(x) \quad \text{and} \quad e_n^{(1)}(x,1) = e_n(x). \tag{17}$$

As it is shown in [6,7], the 2VTEP $e_n^{(r)}(x,y)$ are quasi-monomial (see also [1,2]) with respect to multiplicative and derivative operators given by

$$\widehat{M}_{e^{(r)}} = (x + ry\partial_y y \partial_x^{r-1}) \tag{18}$$

and

$$\widehat{P}_{e^{(r)}} = \partial_x, \tag{19}$$

where

$$\partial_x = \frac{\partial}{\partial x} \quad \text{and} \quad \partial_y = \frac{\partial}{\partial y}.$$

Thus, if we apply the monomiality principle as well as the Equations (18) and (19), we have

$$\widehat{M}_{e^{(r)}}\{e_n^{(r)}(x,y)\} = e_{n+1}^{(r)}(x,y) \tag{20}$$

and

$$\widehat{P}_{e^{(r)}}\{e_n^{(r)}(x,y)\} = ne_{n-1}^{(r)}(x,y), \tag{21}$$

respectively.

The 2VTEP $e_n^{(r)}(x,y)$ are quasi-monomial, so their properties can be derived from those of the multiplicative and derivative operators $\widehat{M}_{e^{(r)}}$ and $\widehat{P}_{e^{(r)}}$, respectively. We thus find that

$$\widehat{M}_{e^{(r)}}\widehat{P}_{e^{(r)}}\{e_n^{(r)}(x,y)\} = ne_n^{(r)}(x,y), \tag{22}$$

which satisfies a differential equation for $e_n^{(r)}(x,y)$ as follows:

$$(r\partial_x + ry\partial_y y\partial_x^r - n)e_n^{(r)}(x,y) = 0. \tag{23}$$

Again, since $e_0^{(r)}(x,y) = 1$, the 2VTEP $e_n^{(r)}(x,y)$ can be explicitly constructed as follows:

$$e_n^{(r)}(x,y) = \widehat{M}^n_{e^{(r)}}\{e_0^{(r)}(x,y)\} = \widehat{M}^n_{e^{(r)}}\{1\}. \tag{24}$$

Equation (24) yields the following generating function of the 2VTEP $e_n^{(r)}(x,y)$:

$$\exp(\widehat{M}_{e^{(r)}}t)\{1\} = \sum_{n=0}^{\infty} e_n^{(r)}(x,y)\frac{t^n}{n!} \quad (|t| < \infty). \tag{25}$$

We can easily verify the following relation between $\widehat{M}_{e^{(r)}}$ and $\widehat{P}_{e^{(r)}}$:

$$[\widehat{P}_{e^{(r)}}, \widehat{M}_{e^{(r)}}] = \widehat{1}. \tag{26}$$

Denoting the classical Bernoulli, Euler and Genocchi polynomials by $B_n(x)$, $E_n(x)$ and $G_n(x)$, respectively, we now recall their familiar generalizations $B_n^{(\alpha)}(x)$, $E_n^{(\alpha)}(x)$ and $G_n^{(\alpha)}(x)$ of order α, which are generated by (see, for details, [8–14]; see also [15] as well as the references cited therein):

$$\left(\frac{t}{e^t-1}\right)^{\alpha} e^{xt} = \sum_{n=0}^{\infty} B_n^{(\alpha)}(x)\frac{t^n}{n!} \qquad (|t| < 2\pi;\ 1^{\alpha} := 1), \tag{27}$$

$$\left(\frac{2}{e^t+1}\right)^{\alpha} e^{xt} = \sum_{n=0}^{\infty} E_n^{(\alpha)}(x)\frac{t^n}{n!} \qquad (|t| < \pi;\ 1^{\alpha} := 1) \tag{28}$$

and

$$\left(\frac{2t}{e^t+1}\right)^{\alpha} e^{xt} = \sum_{n=0}^{\infty} G_n^{(\alpha)}(x)\frac{t^n}{n!} \quad (|t| < \pi;\ \alpha \in \mathbb{N}_0). \tag{29}$$

Obviously, we have

$$B_n^{(1)}(x) =: B_n(x), \quad E_n^{(1)}(x) =: E_n(x) \quad \text{and} \quad G_n^{(1)}(x) =: G_n(x). \tag{30}$$

It is also known that

$$B_n^{(1)}(0) =: B_n, \quad E_n^{(1)}(0) =: E_n \quad \text{and} \quad G_n^{(1)}(0) =: G_n \tag{31}$$

for the Bernoulli, Euler, and Genocchi numbers B_n, E_n and G_n, respectively.

The Apostol-Bernoulli polynomials $B_n^{(\alpha)}(x;\lambda)$ of order α was introduced by Luo and Srivastava (see [16,17]). Subsequently, the Apostol-Euler polynomials $E_n^{(\alpha)}(x;\lambda)$ and the Apostol-Genocchi polynomials $G_n^{(\alpha)}(x;\lambda)$ of order α were analogously studied by Luo (see [18–20]; see also [21–27]).

Definition 1. *The Apostol-Bernoulli polynomials $B_n^{(\alpha)}(x)$ of order α are defined by*

$$\left(\frac{t}{\lambda e^t-1}\right)^{\alpha}=\sum_{n=0}^{\infty}B_n^{(\alpha)}(x;\lambda)\,\frac{t^n}{n!} \tag{32}$$

$$\left(|t|<2\pi \text{ when } \lambda=1;\ |t|<|\log\lambda| \text{ when } \lambda\neq 1;\ 1^{\alpha}:=1\right)$$

with

$$B_n^{(\alpha)}(x)=B_n^{(\alpha)}(x;1) \quad \text{and} \quad B_n^{(\alpha)}(\lambda)=B_n^{(\alpha)}(0;\lambda), \tag{33}$$

where $B_n^{(\alpha)}(\lambda)$ denotes the Apostol-Bernoulli numbers of order α.

Definition 2. *The Apostol-Euler polynomials $E_n^{(\alpha)}(x)$ of order α are defined by*

$$\left(\frac{2}{\lambda e^t+1}\right)^{\alpha}=\sum_{n=0}^{\infty}E_n^{(\alpha)}(x;\lambda)\,\frac{t^n}{n!} \tag{34}$$

$$\left(|t|<\pi \text{ when } \lambda=1;\ |t|<|\log(-\lambda)|<\pi \text{ when } \lambda\neq 1;\ 1^{\alpha}:=1\right)$$

with

$$E_n^{(\alpha)}(x)=E_n^{(\alpha)}(x;1) \quad \text{and} \quad E_n^{(\alpha)}(\lambda)=E_n^{(\alpha)}(0;\lambda), \tag{35}$$

where $E_n^{(\alpha)}(\lambda)$ denotes the Apostol-Euler numbers of order α.

Definition 3. *The Apostol-Genocchi polynomials $G_n^{(\alpha)}(x)$ of order α are defined by*

$$\left(\frac{2t}{\lambda e^t+1}\right)^{\alpha}=\sum_{n=0}^{\infty}G_n^{(\alpha)}(x;\lambda)\,\frac{t^n}{n!} \tag{36}$$

$$\left(|t|<\pi \text{ when } \lambda=1;\ |t|<|\log(-\lambda)| \text{ when } \lambda\neq 1;\ 1^{\alpha}:=1\right) \tag{37}$$

with

$$G_n^{(\alpha)}(x)=G_n^{(\alpha)}(x;1) \quad \text{and} \quad G_n^{(\alpha)}(\lambda)=G_n^{(\alpha)}(0;\lambda), \tag{38}$$

where $G_n^{(\alpha)}(\lambda)$ denotes the Apostol-Genocchi numbers of order α.

Remark 1. *Whenever $\lambda=1$ in (32) and $\lambda=-1$ in (36), the order α of the Apostol-Bernoulli polynomials $B_n^{(\alpha)}(x;\lambda)$ and the order α of the Apostol-Genocchi polynomials $G_n^{(\alpha)}(x;\lambda)$ should obviously be constrained to take on nonnegative integer values (see, for details, [14]). A similar remark would apply also to the order α in all other analogous situations considered in this paper.*

Among other authors, Özden (see [28,29]), Özden et al. ([30]) and Özarslan (see [31,32]) introduced and studied the unification of the above-defined Apostol-type polynomials. In particular, Özden ([29]) defined the unified polynomials $Y_{n,\beta}^{(\alpha)}(x;k,a,b)$ of higher order by

$$\left(\frac{2^{1-k}t^k}{\beta^b e^t - a^b}\right)^{\alpha} e^{xt} = \sum_{n=0}^{\infty} Y_{n,\beta}^{(\alpha)}(x;k,a,b)\,\frac{t^n}{n!} \tag{39}$$

$$\left(|t| < 2\pi \text{ when } \beta = a;\ |t| < \left|b\log\left(\frac{\beta}{a}\right)\right| \text{ when } \beta \neq a;\ 1^{\alpha} := 1;\ k \in \mathbb{N}_0;\ a,b \in \mathbb{R}\setminus\{0\};\ \alpha,\beta \in \mathbb{C}\right).$$

By putting $x = 0$ in (39), we can readily obtain the corresponding unification $Y_{n,\beta}^{(\alpha)}(k,a,b)$ of the Apostol-type polynomials, which is generated by

$$\left(\frac{2^{1-k}t^k}{\beta^b e^t - a^b}\right)^{\alpha} = \sum_{n=0}^{\infty} Y_{n,\beta}^{(\alpha)}(k,a,b)\,\frac{t^n}{n!}. \tag{40}$$

In fact, from Equations (32), (34), (36) and (39), we have

$$Y_{n,\lambda}^{(\alpha)}(x;1,1,1) = B_n^{(\alpha)}(x;\lambda), \tag{41}$$

$$Y_{n,\lambda}^{(\alpha)}(x;0,-1,1) = E_n^{(\alpha)}(x;\lambda) \tag{42}$$

and

$$Y_{n,\lambda}^{(\alpha)}\left(x;1,-\frac{1}{2},1\right) = G_n^{(\alpha)}(x;\lambda). \tag{43}$$

Definition 4. *For an arbitrary real or complex parameter λ, the number $S_k(n,\lambda)$ is given by Zhang and Yang (see [19])*

$$\sum_{k=0}^{\infty} S_k(n,\lambda)\,\frac{t^k}{k!} = \frac{\lambda e^{(n+1)t} - 1}{\lambda e^t - 1}, \tag{44}$$

which, for $\lambda = 1$, yields

$$S_k(n,1) =: S_k(n).$$

Our main objective in this article is to first appropriately combine the 2-variable truncated-exponential polynomials and the Apostol-type polynomials by means of operational techniques. This leads us to the truncated-exponential-based Apostol-type polynomials. By framing these polynomials within the context of the monomiality principle, we then establish their potentially useful properties. We also derive some other properties and investigate several implicit summation formulas for this general family of polynomials by making use of several different analytical techniques on their generating functions. We choose to point out some relevant connections between the truncated-exponential polynomials and the Apostol-type polynomials and thereby derive extensions of several symmetric identities.

2. Two-Variable Truncated-Exponential-Based Apostol-Type Polynomials

We now start with the following theorem arising from the generating functions for the truncated-exponential-based Apostol-type polynomials (TEATP), which are denoted by ${}_{e^{(r)}}Y_{n,\beta}^{(\alpha)}(x,y;k,a,b)$.

Theorem 1. *The generating function for the 2-variable truncated-exponential-based Apostol-type polynomials ${}_{e^{(r)}}Y_{n,\beta}^{(\alpha)}(x,y;k,a,b)$ is given by*

$$\sum_{n=0}^{\infty}\left({}_{e^{(r)}}Y_{n,\beta}^{(\alpha)}(x,y;k,a,b)\right)\frac{t^n}{n!} = \left(\frac{2^{1-k}t^k}{\beta^b e^t - a^b}\right)^{\alpha} e^{xt}\left(\frac{1}{1-yt^r}\right). \tag{45}$$

Proof. Replacing x in the left-hand side and the right-hand side of (39) by the multiplicative operator $\widehat{M}_e^{(r)}$ of the 2VTEATP $_{e^{(r)}}Y_{n,\beta}^{(\alpha)}(x,y;k,a,b)$, we have

$$\left(\frac{2^{1-k}t^k}{\beta^b e^t - a^b}\right)^{\alpha} \exp(\widehat{M}_e^{(r)}t)\{1\} = \sum_{n=0}^{\infty} Y_{n,\beta}^{(\alpha)}(\widehat{M}_e^{(r)};k,a,b)\,\frac{t^n}{n!} \qquad \left(|t| < \left|b\,\log\left(\frac{\beta}{a}\right)\right|\right). \tag{46}$$

Using Equation (25) in the left-hand side and Equation (18) in the right-hand side of Equation (46), we see that

$$\left(\frac{2^{1-k}t^k}{\beta^b e^t - a^b}\right)^{\alpha} \sum_{n=0}^{\infty} e_n^{(r)}(x,y)\frac{t^n}{n!} = \sum_{n=0}^{\infty} Y_{n,\beta}^{(\alpha)}\left(x + \frac{\phi'(y,\partial_x)}{\phi(y,\partial_x)};k,a,b\right)\frac{t^n}{n!}. \tag{47}$$

Now, using Equation (16) in the left-hand side and denoting the resulting 2-variable truncated-exponential-based Apostol-type polynomials (2VTEATP) in the right-hand side by $_{e^{(r)}}Y_{n,\beta}^{(\alpha)}(x,y;k,a,b)$, we have

$$_{e^{(r)}}Y_{n,\beta}^{(\alpha)}(x,y;k,a,b) = Y_{n,\beta}^{(\alpha)}(\widehat{M}_e^{(r)};k,a,b) = Y_{n,\beta}^{(\alpha)}\left(x + \frac{\phi'(y,\partial_x)}{\phi(y,\partial_x)};k,a,b\right), \tag{48}$$

which yields the assertion (45) of Theorem 1. □

Remark 2. *Equation (48) gives the operational representation involving the unified Apostol-type polynomials* $Y_{n,\beta}^{(\alpha)}(x,y;k,a,b)$ *and 2VTEATP* $_{e^{(r)}}Y_{n,\beta}^{(\alpha)}(x,y;k,a,b)$.

To frame the 2VTEATP $_{e^{(r)}}Y_{n,\beta}^{(\alpha)}(x,y;k,a,b)$ within the context of monomiality principle, we state the following result.

Theorem 2. *The 2VTEATP* $_{e^{(r)}}Y_{n,\beta}^{(\alpha)}(x,y;k,a,b)$ *are quasi-monomial with respect to the following multiplicative and derivative operators:*

$$\widehat{M}_{e^{(r)}Y} = x + ry\partial_y y\partial_x^{r-1} + \frac{\alpha k(\beta^b e^t - a^b) - \alpha\beta^b\partial_x e^{\partial_x}}{\partial_x(\beta^b e^t - a^b)} \tag{49}$$

and

$$\widehat{P}_{e^{(r)}Y} = \partial_x. \tag{50}$$

Proof. Let us consider the following expression:

$$\partial_x\left\{e^{xt}\frac{1}{1-yt^r}\right\} = t\left\{e^{xt}\frac{1}{1-yt^r}\right\}. \tag{51}$$

Differentiating both sides of Equation (45) partially with respect to t, we see that

$$\begin{aligned}\left(x + ry\partial_y y\partial_x^{r-1} + \frac{\alpha k(\beta^b e^t - a^b) - \alpha\beta^b t e^t}{t(\beta^b e^t - a^b)}\right)\left(\frac{2^{1-k}t^k}{\beta^b e^t - a^b}\right)^{\alpha}\frac{e^{xt}}{1-yt^r}\\ = \sum_{n=0}^{\infty} {}_{e^{(r)}}Y_{n+1,\beta}^{(\alpha)}(x,y;k,a,b)\,\frac{t^n}{n!}.\end{aligned} \tag{52}$$

Since

$$\phi(y,t) = \frac{1}{1-yt^r}$$

is an invertible series of t, therefore,

$$\frac{\phi'(y, \partial_x)}{\phi(y, \partial_x)}$$

possesses a power-series expansion in t. Thus, using (51), Equation (52) becomes

$$\left(x + ry\partial_y y\partial_x^{r-1} + \frac{\alpha k(\beta^b e^{\partial_x} - a^b) - \alpha \beta^b \partial_x e^{\partial_x}}{\partial_x(\beta^b e^t - a^b)} \right) \left(\frac{2^{1-k} t^k}{\beta^b e^t - a^b} \right)^\alpha \frac{e^{xt}}{1 - yt^r}$$

$$= \sum_{n=0}^{\infty} {}_{e^{(r)}} Y_{n+1,\beta}^{(\alpha)}(x, y; k, a, b) \frac{t^n}{n!}. \tag{53}$$

Again, by using the generating function (45) in left-hand side of Equation (53) and rearranging the resulting summation, we have

$$\sum_{n=0}^{\infty} \left(x + ry\partial_y y\partial_x^{r-1} + \frac{\alpha k(\beta^b e^{\partial_x} - a^b) - \alpha \beta^b \partial_x e^{\partial_x}}{\partial_x(\beta^b e^t - a^b)} \right) \left\{ {}_{e^{(r)}} Y_{n,\beta}^{(\alpha)}(x, y; k, a, b) \right\} \frac{t^n}{n!}$$

$$= \sum_{n=0}^{\infty} {}_{e^{(r)}} Y_{n+1,\beta}^{(\alpha)}(x, y; k, a, b) \frac{t^n}{n!}. \tag{54}$$

Comparing the coefficients of $\frac{t^n}{n!}$ in the Equation (54), we get

$$\left(x + ry\partial_y y\partial_x^{r-1} + \frac{\alpha k(\beta^b e^{\partial_x} - a^b) - \alpha \beta^b \partial_x e^{\partial_x}}{\partial_x(\beta^b e^t - a^b)} \right) \left\{ {}_{e^{(r)}} Y_{n,\beta}^{(\alpha)}(x, y; k, a, b) \right\}$$

$$= {}_{e^{(r)}} Y_{n+1,\beta}^{(\alpha)}(x, y; k, a, b), \tag{55}$$

which, in view of the monomiality principle exhibited in Equation (20) for ${}_{e^{(r)}} Y_{n,\beta}^{(\alpha)}(x, y; k, a, b)$, yields the assertion (49) of Theorem 2.

We now prove the assertion (50) of Theorem 2. For this purpose, we start with the following identity arising from Equations (45) and (51):

$$\partial_x \left\{ \sum_{n=0}^{\infty} {}_{e^{(r)}} Y_{n,\beta}^{(\alpha)}(x, y; k, a, b) \frac{t^n}{n!} \right\} = \sum_{n=1}^{\infty} {}_{e^{(r)}} Y_{n-1,\beta}^{(\alpha)}(x, y; k, a, b) \frac{t^n}{(n-1)!}. \tag{56}$$

Rearranging the summation in the left-hand side of Equation (56), and then equating the coefficients of the same powers of t in both sides of the resulting equation, we find that

$$\partial_x \left\{ {}_{e^{(r)}} Y_{n,\beta}^{(\alpha)}(x, y; k, a, b) \right\} = {}_{e^{(r)}} Y_{n-1,\beta}^{(\alpha)}(x, y; k, a, b) \qquad (n \in \mathbb{N}), \tag{57}$$

which, in view of the monomiality principle exhibited in Equation (21) for ${}_{e^{(r)}} Y_{n,\beta}^{(\alpha)}(x, y; k, a, b))$, yields the assertion (50) of Theorem 2. Our demonstration of Theorem 2 is thus completed. $\square$

We note that the properties of quasi-monomials can be derived by means of the actions of the multiplicative and derivative operators. We derive the differential equation for the 2VTEATP ${}_{e^{(r)}} Y_{n,\beta}^{(\alpha)}(x, y; k, a, b)$ in the following theorem.

Theorem 3. *The 2VTEATP* ${}_{e^{(r)}} Y_{n,\beta}^{(\alpha)}(x, y; k, a, b)$ *satisfies the following differential equation:*

$$\left(x\partial_x + ry\partial_y y\partial_x^r + \frac{\alpha k(\beta^b e^t - a^b) - \alpha \beta^b \partial_x e^{\partial_x}}{(\beta^b e^t - a^b)} - n \right) \left\{ {}_{e^{(r)}} Y_{n,\beta}^{(\alpha)}(x, y; k, a, b) \right\} = 0, \tag{58}$$

Proof. Theorem 3 can be easily proved by combining (49) and (50) with the monomiality principle exhibited in (22). □

Remark 3. *When $r = 2$, the 2VTEP $e^{(r)}(x,y)$ of order r reduces to the 2VTEP ${}_{[2]}e_n(x,y)$. Therefore, if we set $r = 2$ in Equation (45), we get the following generating function for the 2-variable truncated-exponential Apostol-type polynomials (2VTEATP) ${}_{[2]e^{(r)}}Y^{(\alpha)}_{n,\beta}(x,y;k,a,b)$:*

$$\left(\frac{2^{1-k}t^k}{\beta^b e^t - a^b}\right)^{\alpha} e^{xt}\left(\frac{1}{1-yt^2}\right) = \sum_{n=0}^{\infty} {}_{[2]e^{(r)}}Y^{(\alpha)}_{n,\beta}(x,y;k,a,b)\,\frac{t^n}{n!}. \tag{59}$$

The series definition and other results for the 2VTEATP ${}_{[2]e^{(r)}}Y^{(\alpha)}_{n,\beta}(x,y;k,a,b)$ can be obtained by taking $r = 2$ in Theorems 1 and 2. Table 1 shown the special cases of the 2VTEATP ${}_{\cdot e^{(r)}}Y_n(x,y;k,a,b)$.

Remark 4. *For the case $y = 1$, the polynomials ${}_{[2]}e_n(x,1)$ reduce to the truncated-exponential polynomials ${}_{[2]}e_n(x)$. Therefore, by taking $y = 1$ in Equation (59), we get the following generating function for the truncated-exponential Apostol-type polynomials (TEATP) ${}_{[2]e^{(r)}}Y^{(\alpha)}_{n,\beta}(x;k,a,b)$:*

$$\left(\frac{2^{1-k}t^k}{\beta^b e^t - a^b}\right)^{\alpha} e^{xt}\left(\frac{1}{1-t^2}\right) = \sum_{n=0}^{\infty} {}_{[2]e^{(r)}}Y^{(\alpha)}_{n,\beta}(x;k,a,b)\,\frac{t^n}{n!}. \tag{60}$$

Table 1. Some special cases of the 2VTEATP ${}_{\cdot e^{(r)}}Y_n(x,y;k,a,b)$.

S. No.	Values of the Parameter	Relation between the 2VTEATP ${}_{e^{(r)}}Y_n(x,y;k,a,b)$ and Its Special Case	Name of the Resultant Special Polynomials	Generating Functions and the Resultant of Special Polynomials
I.	$k=a=b=1, \beta=\lambda$	${}_{e^{(r)}}Y_n(x,y;1,1,\lambda) = {}_{e^{(r)}}B^{(\alpha)}_n(x,y;\lambda)$	2-variable truncated-exponential-based Apostol-Bernoulli polynomial	$\left(\frac{t}{\lambda e^t-1}\right)^{\alpha} e^{xt}\left(\frac{1}{1-yt^r}\right) = \sum_{n=0}^{\infty} {}_{e^{(r)}}B^{(\alpha)}_n(x,y;\lambda)\frac{t^n}{n!}$
II.	$k+1=-a=b=1, \beta=\lambda$	${}_{e^{(r)}}Y_n(x,y;0,-1,1,\lambda) = {}_{e^{(r)}}E^{(\alpha)}_n(x,y;\lambda)$	2-variable truncated-exponential-based Apostol-Euler polynomial	$\left(\frac{2}{\lambda e^t+1}\right)^{\alpha} e^{xt}\left(\frac{1}{1-yt^r}\right) = \sum_{n=0}^{\infty} {}_{e^{(r)}}E^{(\alpha)}_n(x,y;\lambda)\frac{t^n}{n!}$
III.	$k=-2a=b=1, 2\beta=\lambda$	${}_{e^{(r)}}Y_n(x,y;1,-\frac{1}{2},1,\lambda) = {}_{e^{(r)}}G^{(\alpha)}_n(x,y;\lambda)$	2-variable truncated-exponential-based Apostol-Genocchi polynomial	$\left(\frac{2t}{\lambda e^t+1}\right)^{\alpha} e^{xt}\left(\frac{1}{1-yt^r}\right) = \sum_{n=0}^{\infty} {}_{e^{(r)}}G^{(\alpha)}_n(x,y;\lambda)\frac{t^n}{n!}$

In the case when $\lambda = 1$, the results obtained above for the 2VTEABP ${}_{e^{(r)}}B^{(\alpha)}_n(x,y;\lambda)$, 2VTEAEP ${}_{e^{(r)}}E^{(\alpha)}_n(x,y;\lambda)$ and 2VTEAGP ${}_{e^{(r)}}G^{(\alpha)}_n(x,y;\lambda)$ give the corresponding results for the 2-variable truncated-exponential Bernoulli polynomials (2VTEBP) (of order α) ${}_{e^{(r)}}B^{(\alpha)}_n(x,y)$, 2-variable truncated-exponential Euler polynomials (2VTEBP) (of order α) ${}_{e^{(r)}}E^{(\alpha)}_n(x,y)$ and 2-variable truncated-exponential Genocchi polynomials (2VTGBP) (of order α) ${}_{e^{(r)}}G^{(\alpha)}_n(x,y)$ [6]. Again for $\alpha = 1$, we get the corresponding results for the 2-variable truncated-exponential Bernoulli polynomials (2VTEBP) ${}_{e^{(r)}}B_n(x,y)$, 2-variable truncated-exponential Euler polynomials (2VTEEP) ${}_{e^{(r)}}E_n(x,y)$ and 2-variable truncated-exponential Genocchi polynomials (2VTEGP) ${}_{e^{(r)}}G_n(x,y)$.

3. Implicit Formulas Involving the 2-Variable Truncated-Exponential Based Apostol-Type Polynomials

In this section, we employ the definition of the 2-variable truncated-exponential-based Apostol-type polynomials ${}_{e^{(r)}}Y^{(\alpha)}_{n,\beta}(x,y;k,a,b)$ that help in proving the generalizations of the previous works of Khan et al. [33] and Pathan and Khan (see [34–36]). For the derivation of implicit formulas involving the 2-variable truncated-exponential-based Apostol-type polynomials ${}_{e^{(r)}}Y^{(\alpha)}_{n,\beta}(x,y;k,a,b)$, the same considerations as developed for the ordinary Hermite and related polynomials in the works by Khan et al. [33] and Pathan et al. (see [34–36]) apply as well. We first prove the following results involving the 2-variable truncated-exponential-based Apostol-type polynomials ${}_{e^{(r)}}Y_{n,\beta}^{(\alpha)}(x,y;k,a,b)$.

Theorem 4. *The following implicit summation formulas for the 2-variable truncated-exponential-based Apostol-type polynomials* ${}_{e^{(r)}}Y_{n,\beta}^{(\alpha)}(x,y;k,a,b)$ *holds true:*

$$ {}_{e^{(r)}}Y_{q+l,\beta}^{(\alpha)}(z,y;k,a,b) = \sum_{n=0}^{q}\sum_{p=0}^{l}\binom{q}{n}\binom{l}{p}(z-x)^{n+p}\; {}_{e^{(r)}}Y_{q+l-n-p,\beta}^{(\alpha)}(x,y;k,a,b). \tag{61}$$

Proof. We replace t by $t+u$ and rewrite (45) as follows:

$$\left(\frac{2^{1-k}(t+u)^k}{\beta^b e^{t+u}-a^b}\right)^{\alpha}\left(\frac{1}{1-y(t+u)^r}\right) = e^{-x(t+u)}\sum_{q,l=0}^{\infty}{}_{e^{(r)}}Y_{q+l,\beta}^{(\alpha)}(x,y;k,a,b)\,\frac{t^q}{q!}\,\frac{u^l}{l!}. \tag{62}$$

Replacing x by z in the Equation (62) and equating the resulting equation to the above equation, we get

$$e^{(z-x)(t+u)}\sum_{q,l=0}^{\infty}{}_{e^{(r)}}Y_{q+l,\beta}^{(\alpha)}(x,y;k,a,b)\frac{t^q}{q!}\frac{u^l}{l!} = \sum_{q,l=0}^{\infty}{}_{e^{(r)}}Y_{n,\beta}^{(\alpha)}(z,y;k,a,b)\,\frac{t^q}{q!}\,\frac{u^l}{l!}. \tag{63}$$

Upon expanding the exponential function (63), we get

$$\sum_{N=0}^{\infty}\frac{[(z-x)(t+u)]^N}{N!}\sum_{q,l=0}^{\infty}{}_{e^{(r)}}Y_{q+l,\beta}^{(\alpha)}(x,y;k,a,b)\,\frac{t^q}{q!}\,\frac{u^l}{l!} = \sum_{q,l=0}^{\infty}{}_{e^{(r)}}Y_{q+l,\beta}^{(\alpha)}(z,y;k,a,b)\,\frac{t^q}{q!}\,\frac{u^l}{l!}, \tag{64}$$

which, by appealing to the following series manipulation formula:

$$\sum_{N=0}^{\infty}f(N)\,\frac{(x+y)^N}{N!} = \sum_{m,n=0}^{\infty}f(m+n)\,\frac{x^m}{m!}\,\frac{y^n}{n!} \tag{65}$$

in the left-hand side of (64), becomes

$$\sum_{n,p=0}^{\infty}\frac{(z-x)^{n+p}t^n u^p}{n!\,p!}\sum_{q,l=0}^{\infty}{}_{e^{(r)}}Y_{q+l,\beta}^{(\alpha)}(x,y;k,a,b)\,\frac{t^q}{q!}\,\frac{u^l}{l!} = \sum_{q,l=0}^{\infty}{}_{e^{(r)}}Y_{q+l,\beta}^{(\alpha)}(z,y;k,a,b)\,\frac{t^q}{q!}\,\frac{u^l}{l!}. \tag{66}$$

Now, replacing q by $q-n$ and l by $l-p$, and using a lemma in [37] in the left-hand side of (66), we get

$$\begin{aligned}\sum_{q,l=0}^{\infty}\sum_{n=0}^{q}\sum_{p=0}^{l}\frac{(z-x)^{n+p}}{n!\,p!}{}_{e^{(r)}}Y_{q+l-n-p,\beta}^{(\alpha)}(x,y;k,a,b)\,\frac{t^q}{(q-n)!}\,\frac{u^l}{(l-p)!}\\ = \sum_{q,l=0}^{\infty}{}_{e^{(r)}}Y_{q+l,\beta}^{(\alpha)}(z,y;k,a,b)\,\frac{t^q}{q!}\,\frac{u^l}{l!}.\end{aligned} \tag{67}$$

Finally, on equating the coefficients of the like powers of t and u in the equation (67), we get the required result (61) asserted by Theorem 4. □

If we set

$$k=a=b=1 \qquad \text{and} \qquad \beta=\lambda$$

in Theorem 4, we get the following corollary.

Corollary 1. *The following implicit summation formula for the truncated-exponential-based Bernoulli polynomials* ${}_{e^{(r)}}B_n^{(\alpha)}(x,y;\lambda)$ *holds true:*

$$
{}_{e^{(r)}}B_{q+l}^{(\alpha)}(z,y;\lambda)=\sum_{n=0}^{q}\sum_{p=0}^{l}\binom{q}{n}\binom{l}{p}(z-x)^{n+p}\ {}_{e^{(r)}}B_{q+l-p-n}^{(\alpha)}(x,y;\lambda). \tag{68}
$$

For

$$
k+1=-a=b=1 \qquad \text{and} \qquad \beta=\lambda
$$

in Theorem 4, we get the following corollary.

Corollary 2. *The following implicit summation formula for the truncated-exponential-based Euler polynomials* ${}_{e^{(r)}}E_n^{(\alpha)}(x,y;\lambda)$ *holds true:*

$$
{}_{e^{(r)}}E_{q+l}^{(\alpha)}(z,y;\lambda)=\sum_{n=0}^{q}\sum_{p=0}^{l}\binom{q}{n}\binom{l}{p}(z-x)^{n+p}\,{}_{e^{(r)}}E_{q+l-p-n}^{(\alpha)}(x,y;\lambda). \tag{69}
$$

Letting

$$
k=-2a=b=1 \qquad \text{and} \qquad 2\beta=\lambda
$$

in Theorem 4, we get the following corollary.

Corollary 3. *The following implicit summation formulas for the truncated-exponential-based Genocchi polynomials* ${}_{e^{(r)}}G_n^{(\alpha)}(x,y;\lambda)$ *holds true:*

$$
{}_{e^{(r)}}G_{q+l}^{(\alpha)}(z,y;\lambda)=\sum_{n=0}^{q}\sum_{p=0}^{l}\binom{q}{n}\binom{l}{p}(z-x)^{n+p}\,{}_{e^{(r)}}G_{q+l-p-n}^{(\alpha)}(x,y;\lambda). \tag{70}
$$

Theorem 5. *The following implicit summation formula involving the 2-variable truncated-exponential-based Apostol-type polynomials* ${}_{e^{(r)}}Y_{n,\beta}^{(\alpha)}(x,y;k,a,b)$ *holds true:*

$$
{}_{e^{(r)}}Y_{n,\beta}^{(\alpha)}(x,y;k,a,b)=\sum_{s=0}^{n}\binom{n}{s}Y_{n-s,\beta}^{(\alpha)}(k,a,b)e_s^{(r)}(x,y). \tag{71}
$$

Proof. By the definition (45), we have

$$
\left(\frac{2^{1-k}t^k}{\beta^b e^t-a^b}\right)^{\alpha}e^{xt}\left(\frac{1}{1-yt^r}\right)=\sum_{n=0}^{\infty}Y_{n,\beta}^{(\alpha)}(k,a,b)\,\frac{t^n}{n!}\sum_{s=0}^{\infty}e_s^{(r)}(x,y)\,\frac{t^s}{s!}. \tag{72}
$$

Now, replacing n by $n-s$ in the right-hand side of the Equation (72) and comparing the coefficients of t, we get the result (71) asserted by Theorem 5. □

If we set

$$
k=a=b=1 \qquad \text{and} \qquad \beta=\lambda
$$

in Theorem 5, we get the following corollary.

Corollary 4. *The following implicit summation formula for the 2-variable truncated-exponential-based Bernoulli polynomials* ${}_{e^{(r)}}B_n^{(\alpha)}(x,y;\lambda)$ *holds true:*

$$
{}_{e^{(r)}}B_n^{(\alpha)}(x+z,y+u;\lambda)=\sum_{s=0}^{n}\binom{n}{s}B_{n-s}^{(\alpha)}(\lambda)e_s^{(r)}(x,y). \tag{73}
$$

For

$$k + 1 = -a = b = 1 \quad \text{and} \quad \beta = \lambda$$

in Theorem 5, we get the following corollary.

Corollary 5. *The following implicit summation formula for the 2-variable truncated-exponential-based Euler polynomials* ${}_{e^{(r)}}E_n^{(\alpha)}(x, y; \lambda)$ *holds true:*

$$ {}_{e^{(r)}}E_n^{(\alpha)}(x + z, y + u; \lambda) = \sum_{s=0}^{n} \binom{n}{s} E_{n-s}^{(\alpha)}(\lambda) e_s^{(r)}(x, y). \tag{74}$$

Letting

$$k = -2a = b = 1 \quad \text{and} \quad 2\beta = \lambda$$

in Theorem 5, we get the following corollary.

Corollary 6. *The following implicit summation formula for the 2-variable truncated-exponential-based Genocchi polynomials* ${}_{e^{(r)}}G_n^{(\alpha)}(x, y; \lambda)$ *holds true:*

$$ {}_{e^{(r)}}G_n^{(\alpha)}(x + z, y + u; \lambda) = \sum_{s=0}^{n} \binom{n}{s} G_{n-s}^{(\alpha)}(\lambda) e_s^{(r)}(x, y). \tag{75}$$

Theorem 6. *The following implicit summation formula involving the 2-variable truncated-exponential-based Apostol-type polynomials* ${}_{e^{(r)}}Y_{n,\beta}^{(\alpha)}(x, y; k, a, b)$ *holds true:*

$$ {}_{e^{(r)}}Y_{n,\beta}^{(\alpha)}(x + z, y; k, a, b) = \sum_{s=0}^{n} \binom{n}{s} {}_{e^{(r)}}Y_{n-s,\beta}^{(\alpha)}(x, y; k, a, b) z^s. \tag{76}$$

Proof. We first replace x by $x + z$ in (45). Then, by using (16), we rewrite the generating function (45) as follows:

$$\begin{aligned} \left(\frac{2^{1-k}t^k}{\beta^b e^t - a^b}\right)^{\alpha} e^{(x+z)t} \left(\frac{1}{1 - yt^r}\right) &= \sum_{n=0}^{\infty} {}_{e^{(r)}}Y_{n,\beta}^{(\alpha)}(x, y; k, a, b) \frac{t^n}{n!} \sum_{s=0}^{\infty} \frac{(zt)^s}{s!} \\ &= \sum_{n=0}^{\infty} {}_{e^{(r)}}Y_{n,\beta}^{(\alpha)}(x + z, y; k, a, b) \frac{t^n}{n!}. \end{aligned} \tag{77}$$

Furthermore, upon replacing n by $n - s$ in *l.h.s* and comparing the coefficients of t^n, we complete the proof of Theorem 6. □

For

$$k = a = b = 1 \quad \text{and} \quad \beta = \lambda$$

in Theorem 6, we get the following corollary.

Corollary 7. *The following implicit summation formula for the 2-variable truncated-exponential-based Bernoulli polynomials* ${}_{e^{(r)}}B_n^{(\alpha)}(x, y; \lambda)$ *holds true:*

$$ {}_{e^{(r)}}B_n^{(\alpha)}(x + z, y + u; \lambda) = \sum_{s=0}^{n} \binom{n}{s} {}_{e^{(r)}}B_{n-s}^{(\alpha)}(x, y; \lambda) H_s(z, u). \tag{78}$$

Upon setting

$$k + 1 = -a = b = 1 \quad \text{and} \quad \beta = \lambda$$

in Theorem 6, we get the following corollary.

Corollary 8. *The following implicit summation formula for the 2-variable truncated-exponential-based Euler polynomials* ${}_{e^{(r)}}E_n^{(\alpha)}(x,y;\lambda)$ *holds true:*

$$ {}_{e^{(r)}}E_n^{(\alpha)}(x+z,y+u;\lambda) = \sum_{s=0}^{n} \binom{n}{s} {}_{e^{(r)}}E_{n-s}^{(\alpha)}(x,y;\lambda) H_s(z,u). \tag{79} $$

Letting

$$ k = -2a = b = 1 \qquad \text{and} \qquad 2\beta = \lambda $$

in Theorem 6, we get the following corollary.

Corollary 9. *The following implicit summation formula for the 2-variable truncated-exponential-based Genocchi polynomials* ${}_{e^{(r)}}G_n^{(\alpha)}(x,y;\lambda)$ *holds true:*

$$ {}_{e^{(r)}}G_n^{(\alpha)}(x+z,y+u;\lambda) = \sum_{s=0}^{n} \binom{n}{s} {}_{e^{(r)}}G_{n-s}^{(\alpha)}(x,y;\lambda) H_s(z,u). \tag{80} $$

Theorem 7. *The following implicit summation formula for the 2-variable truncated-exponential-based Apostol-type polynomials* ${}_{e^{(r)}}Y_{n,\beta}^{(\alpha)}(x,y;k,a,b)$ *holds true:*

$$ {}_{e^{(r)}}Y_{n,\beta}^{(\alpha)}(x,y;k,a,b) = \sum_{r=0}^{n} \binom{n}{r} Y_{n-r,\beta}^{(\alpha)}(x-z;k,a,b) e^{(r)}(z,y). \tag{81} $$

Proof. Let us rewrite Equation (45) as follows:

$$ \left(\frac{2^{1-k}t^k}{\beta^b e^t - a^b} \right)^{\alpha} e^{(x-z+z)t} \left(\frac{1}{1-yt^r} \right) = \sum_{n=0}^{\infty} Y_{n,\beta}^{(\alpha)}(x-z;k,a,b) \frac{t^n}{n!} \sum_{r=0}^{\infty} e^{(r)}(z,y) \frac{t^r}{r!}. \tag{82} $$

Replacing n by $n-r$ and using (45), and then equating the coefficients of the of t^n, we complete the proof of Theorem 7. □

For

$$ k = a = b = 1 \qquad \text{and} \qquad \beta = \lambda $$

in Theorem 7, we get the following corollary.

Corollary 10. *The following implicit summation formula for the 2-variable truncated-exponential-based Apostol-type Bernoulli polynomials* ${}_{e^{(r)}}B_n^{(\alpha)}(x,y;\lambda)$ *holds true:*

$$ {}_{e^{(r)}}B_n^{(\alpha)}(x,y;\lambda) = \sum_{r=0}^{n} \binom{n}{r} B_{n-r}^{(\alpha)}(x-z;\lambda) e^{(r)}(z,y). \tag{83} $$

Letting

$$ k+1 = -a = b = 1 \qquad \text{and} \qquad \beta = \lambda $$

in Theorem 7, we get the following corollary.

Corollary 11. *The following implicit summation formula for the 2-variable truncated-exponential-based Apostol-type Euler polynomials* ${}_{e^{(r)}}E_n^{(\alpha)}(x,y;\lambda)$ *holds true:*

$$ {}_{e^{(r)}}E_n^{(\alpha)}(x,y;\lambda) = \sum_{r=0}^{n} \binom{n}{r} E_{n-r}^{(\alpha)}(x-z;\lambda) e^{(r)}(z,y). \tag{84} $$

If we set

$$k = -2a = b = 1 \qquad \text{and} \qquad 2\beta = \lambda$$

in Theorem 7, we get the following corollary.

Corollary 12. *The following implicit summation formula for the 2-variable truncated-exponential-based Apostol-type Genocchi polynomials* ${}_{e^{(r)}}G_n^{(\alpha)}(x,y;\lambda)$ *holds true*:

$$ {}_{e^{(r)}}G_n^{(\alpha)}(x,y;\lambda) = \sum_{r=0}^{n} \binom{n}{r} G_{n-r}^{(\alpha)}(x-z;\lambda) e^{(r)}(z,y). \tag{85}$$

Theorem 8. *The following implicit summation formula for the 2-variable truncated-exponential-based Apostol-type polynomials* ${}_{e^{(r)}}Y_{n,\beta}^{(\alpha)}(x,y;k,a,b)$ *holds true*:

$$ {}_{e^{(r)}}Y_{n,\beta}^{(\alpha)}(x+1,y;k,a,b) = \sum_{m=0}^{n} \binom{n}{m} {}_{e^{(r)}}Y_{n-m,\beta}^{(\alpha)}(x,y;k,a,b). \tag{86}$$

Proof. Using the generating function (45), we find that

$$\begin{aligned}
&\sum_{n=0}^{\infty} \left({}_{e^{(r)}}Y_{n,\beta}^{(\alpha)}(x+1,y;k,a,b) - {}_{e^{(r)}}Y_{n,\beta}^{(\alpha)}(x,y;k,a,b) \right) \frac{t^n}{n!} \\
&= \left(\frac{2^{1-k}t^k}{\beta^b e^t - a^b} \right)^{\alpha} \left(\frac{1}{1-yt^r} \right) (e^t - 1) \\
&= \sum_{n=0}^{\infty} {}_{e^{(r)}}Y_{n,\beta}^{(\alpha)}(x,y;k,a,b) \frac{t^n}{n!} \left(\sum_{r=0}^{\infty} \frac{t^m}{m!} - 1 \right) \\
&= \sum_{n=0}^{\infty} {}_{e^{(r)}}Y_{n,\beta}^{(\alpha)}(x,y;k,a,b) \frac{t^n}{n!} \sum_{r=0}^{\infty} \frac{t^m}{m!} - \sum_{n=0}^{\infty} {}_{e^{(r)}}Y_{n,\beta}^{(\alpha)}(x,y;k,a,b) \frac{t^n}{n!} \\
&= \sum_{n=0}^{\infty} \left[\sum_{r=0}^{n} \binom{n}{r} {}_{e^{(r)}}Y_{n-m,\beta}^{(\alpha)}(x,y;k,a,b) - {}_{e^{(r)}}Y_{n,\beta}^{(\alpha)}(x,y;k,a,b) \right] \frac{t^n}{n!}.
\end{aligned}$$

which, upon equating the coefficients of t^n, yields the assertion (86) of Theorem 8. □

Remark 5. *Several corollaries and consequences of Theorem 11 can be deduced by using many of the aforementioned specializations of the various parameters involved in Theorem 8.*

4. General Symmetry Identities

In this section, we give general symmetry identities for the 2-variable truncated-exponential-based Apostol-type polynomials ${}_{e^{(r)}}Y_{n,\beta}^{(\alpha)}(x,y;k,a,b)$ by applying the generating functions (39) and (45). The results extend some known identities of Özarslan (see [31,32]), Khan [38], and Pathan and Khan (see [34–36]).

Theorem 9. *Let* $\alpha, k \in \mathbb{N}_0$, $a, b \in \mathbb{R} \setminus \{0\}$, $\beta \in \mathbb{C}$, $x, y \in \mathbb{R}$ *and* $n \in \mathbb{N}_0$. *Then the following symmetry identity holds true*:

$$\begin{aligned}
&\sum_{m=0}^{n} \binom{n}{m} d^m c^{n-m} {}_{e^{(r)}}Y_{n-m,\beta}^{(\alpha)}(dx, d^r y;k,a,b) \, {}_{e^{(r)}}Y_{m,\beta}^{(\alpha)}(cX, c^r Y;k,a,b) \\
&\quad = \sum_{m=0}^{n} \binom{n}{m} c^m d^{n-m} {}_{e^{(r)}}Y_{n-m,\beta}^{(\alpha)}(cx, c^r y;k,a,b) {}_{e^{(r)}}Y_{m,\beta}^{(\alpha)}(dX, d^r Y;k,a,b).
\end{aligned} \tag{87}$$

Proof. Let us first consider the following expression:

$$g(t) = \left(\frac{c^k d^k 2^{2(1-k)} t^{2k}}{(\beta^b e^{ct} - a^b)(\beta^b e^{dt} - a^b)} \right)^{\alpha} e^{cdxt} \left(\frac{1}{1 - y(cdt)^r} \right) e^{cdXt} \left(\frac{1}{1 - Y(cdt)^r} \right),$$

which shows that the function $g(t)$ is symmetric in the parameters a and b. Then, by expanding $g(t)$ into series in two different ways, we get

$$\begin{aligned} g(t) &= \sum_{n=0}^{\infty} {}_{e(r)}Y_{n,\beta}^{(\alpha)}(dx, d^r y; k, a, b) \frac{(ct)^n}{n!} \sum_{m=0}^{\infty} {}_{e(r)}Y_{m,\beta}^{(\alpha)}(cX, c^r Y; k, a, b) \frac{(dt)^m}{m!} \\ &= \sum_{n=0}^{\infty} \sum_{m=0}^{n} \binom{n}{m} d^m c^{n-m} {}_{e(r)}Y_{n-m,\beta}^{(\alpha)}(dx, d^r y; k, a, b) {}_{e(r)}Y_{m,\beta}^{(\alpha)}(cX, c^r Y; k, a, b) t^n \end{aligned} \tag{88}$$

and

$$\begin{aligned} g(t) &= \sum_{n=0}^{\infty} {}_{e(r)}Y_{n,\beta}^{(\alpha)}(cx, c^r y; k, a, b) \frac{(dt)^n}{n!} \sum_{m=0}^{\infty} {}_{e(r)}Y_{m,\beta}^{(\alpha)}(dX, d^r Y; k, a, b) \frac{(ct)^m}{m!} \\ &= \sum_{n=0}^{\infty} \sum_{m=0}^{n} \binom{n}{m} c^m d^{n-m} {}_{e(r)}Y_{n-m,\beta}^{(\alpha)}(cx, c^r y; k, a, b) {}_{e(r)}Y_{m,\beta}^{(\alpha)}(dX, d^r Y; k, a, b) t^n. \end{aligned} \tag{89}$$

Comparing the coefficients of t^n on the right-hand sides of Equations (88) and (89), we arrive at the desired result (87). □

For

$$k = a = b = 1 \quad \text{and} \quad \beta = \lambda$$

in Theorem 9, we get the following corollary.

Corollary 13. *For all $c, d, r \in \mathbb{N}$, $n \in \mathbb{N}_0$ and $\lambda \in \mathbb{C}$, the following symmetry identity for the 2-variable truncated-exponential-based Apostol-type Bernoulli polynomials holds true:*

$$\begin{aligned} &\sum_{m=0}^{n} \binom{n}{m} d^m c^{n-m} {}_{e(r)}B_{n-m}^{(\alpha)}(dx, d^r y; \lambda) {}_{e(r)}B_{m}^{(\alpha)}(cX, c^r Y; \lambda) \\ &\quad = \sum_{m=0}^{n} \binom{n}{m} c^m d^{n-m} {}_{e(r)}B_{n-m}^{(\alpha)}(cx, c^r y; \lambda) {}_{e(r)}B_{m}^{(\alpha)}(dX, d^r Y; \lambda). \end{aligned} \tag{90}$$

Putting

$$k + 1 = -a = b = 1 \quad \text{and} \quad \beta = \lambda$$

in Theorem 9, we get the following corollary.

Corollary 14. *For all $r \in \mathbb{N}$, $n \in \mathbb{N}_0$ and $\lambda \in \mathbb{C}$, the following symmetry identity for the 2-variable truncated-exponential-based Apostol-type Euler polynomials holds true:*

$$\begin{aligned} &\sum_{m=0}^{n} \binom{n}{m} d^m c^{n-m} {}_{e(r)}E_{n-m}^{(\alpha)}(dx, d^r y; \lambda) {}_{e(r)}E_{m}^{(\alpha)}(cX, c^r Y; \lambda) \\ &\quad = \sum_{m=0}^{n} \binom{n}{m} c^m d^{n-m} {}_{e(r)}E_{n-m}^{(\alpha)}(cx, c^r y; \lambda) {}_{e(r)}E_{m}^{(\alpha)}(dX, d^r Y; \lambda). \end{aligned} \tag{91}$$

If we set

$$k = -2a = b = 1 \quad \text{and} \quad 2\beta = \lambda$$

in Theorem 9, we get the following corollary.

Corollary 15. *For all $r \in \mathbb{N}$, $n \in \mathbb{N}_0$ and $\lambda \in \mathbb{C}$, the following symmetry identity for the 2-variable truncated-exponential-based Apostol-type Genocchi polynomials holds true:*

$$\sum_{m=0}^{n} \binom{n}{m} d^m c^{n-m} \, {}_{e^{(r)}}G_{n-m}^{(\alpha)}(dx, d^r y; \lambda) \, {}_{e^{(r)}}G_m^{(\alpha)}(cX, c^r Y; \lambda)$$
$$= \sum_{m=0}^{n} \binom{n}{m} c^m d^{n-m} \, {}_{e^{(r)}}G_{n-m}^{(\alpha)}(cx, c^r y; \lambda) \, {}_{e^{(r)}}G_m^{(\alpha)}(dX, d^r Y; \lambda). \tag{92}$$

Theorem 10. *Let $\alpha, k \in \mathbb{N}_0$, $a, b \in \mathbb{R} \setminus \{0\}$, $\beta \in \mathbb{C}$, $x, y \in \mathbb{R}$ and $n \in \mathbb{N}_0$. Then the following symmetry identity holds true:*

$$\sum_{m=0}^{n} \binom{n}{m} \sum_{i=0}^{c-1} \sum_{j=0}^{d-1} c^{n-m} d^m \, {}_{e^{(r)}}Y_{n-m,\beta}^{(\alpha)}\left(dx + \frac{d}{c}i + j, d^r y; k, a, b\right) \, {}_{e^{(r)}}Y_{m,\beta}^{(\alpha)}(cX, c^r Y; k, a, b)$$
$$= \sum_{m=0}^{n} \binom{n}{m} \sum_{i=0}^{d-1} \sum_{j=0}^{c-1} d^{n-m} c^m \, {}_{e^{(r)}}Y_{n-m,\beta}^{(\alpha)}\left(cx + \frac{c}{d}i + j, c^r y; k, a, b\right) \, {}_{e^{(r)}}Y_{m,\beta}^{(\alpha)}(dX, d^r Y; k, a, b). \tag{93}$$

Proof. Let us first consider the following application:

$$g(t) = \left(\frac{c^k d^k 2^{2(1-k)} t^{2k}}{(\beta^b e^{ct} - a^b)(\beta^b e^{dt} - a^b)}\right)^{\alpha} e^{cdxt} \left(\frac{1}{1 - y(cdt)^r}\right) \frac{(e^{cdt} - 1)^2}{(e^{ct} - 1)(e^{dt} - 1)} e^{cdXt} \left(\frac{1}{1 - Y(cdt)^r}\right)$$
$$= \left(\frac{2^{(1-k)} c^k t^k}{\beta^b e^{ct} - a^b}\right)^{\alpha} e^{cdxt} \left(\frac{1}{1 - y(cdt)^r}\right) \left(\frac{e^{cdt} - 1}{e^{ct} - 1}\right) \left(\frac{2^{(1-k)} d^k t^k}{\beta^b e^{dt} - a^b}\right)^{\alpha}$$
$$\cdot e^{cdXt} \left(\frac{1}{1 - Y(cdt)^r}\right) \left(\frac{1}{e^{cdt} - 1} e^{dt} - 1\right)$$
$$= \left(\frac{2^{(1-k)} c^k t^k}{(\beta^b e^{ct} - a^b)}\right)^{\alpha} e^{cdxt} \left(\frac{1}{1 - y(cdt)^r}\right) \sum_{i=0}^{c-1} e^{dti} \left(\frac{2^{(1-k)} d^k t^k}{\beta^b e^{dt} - a^b}\right)^{\alpha}$$
$$\cdot e^{cdXt} \left(\frac{1}{1 - Y(cdt)^r}\right) e^{cdyt} \sum_{j=0}^{d-1} e^{ctj}$$
$$= \sum_{n=0}^{\infty} \left[\sum_{m=0}^{n} \binom{n}{m} \sum_{i=0}^{c-1} \sum_{j=0}^{d-1} c^{n-m} d^m \, {}_{e^{(r)}} \right.$$
$$\left. \cdot Y_{n-m,\beta}^{(\alpha)}\left(dx + \frac{d}{c}i + j, d^r y; k, a, b\right) \, {}_{e^{(r)}}Y_{m,\beta}^{(\alpha)}(cX, c^r Y; k, a, b) \right] t^n. \tag{94}$$

On the other hand, we have

$$g(t) = \sum_{n=0}^{\infty} \left(\sum_{m=0}^{n} \binom{n}{m} \sum_{i=0}^{d-1} \sum_{j=0}^{c-1} d^{n-m} c^m \right.$$
$$\left. \cdot \, {}_{e^{(r)}}Y_{n-m,\beta}^{(\alpha)}\left(cx + \frac{c}{d}i + j, c^r y; k, a, b\right) \, {}_{e^{(r)}}Y_{m,\beta}^{(\alpha)}(dX, d^r Y; k, a, b) \right) t^n. \tag{95}$$

By comparing the coefficients of t^n on the right-hand sides of (94) and (95), we arrive at the desired result (93) asserted by Theorem 10. □

Remark 6. *Several corollaries and consequences of Theorem 11 can be derived by making use of many of the aforementioned specializations of the various parameters involved in Theorem 10.*

Theorem 11. *For each pair of integers a and b and all integers $n \in \mathbb{N}_0$, the following identity holds true:*

$$
\sum_{m=0}^{n} \binom{n}{m} \sum_{i=0}^{c-1} \sum_{j=0}^{d-1} c^{n-m} d^m \,_{e^{(r)}} Y_{n-m,\beta}^{(\alpha)} \left(dx + \frac{d}{c}\, i, d^r y; k, a, b \right) \,_{e^{(r)}} Y_{m,\beta}^{(\alpha)}(cX + \frac{c}{d}\, j, c^r Y; k, a, b)
$$

$$
= \sum_{m=0}^{n} \binom{n}{m} \sum_{i=0}^{d-1} \sum_{j=0}^{c-1} d^{n-m} c^m \,_{e^{(r)}} Y_{n-m,\beta}^{(\alpha)} \left(cx + \frac{c}{d}\, i, c^r y; k, a, b \right)
$$

$$
\cdot \,_{e^{(r)}} Y_{m,\beta}^{(\alpha)}(dX + \frac{d}{c}\, j, d^r Y; k, a, b). \tag{96}
$$

Proof. The proof of Theorem 11 is analogous to that of Theorem 10, so we omit the details involved in the proof of Theorem 11. □

Remark 7. *Several corollaries and consequences of Theorem 11 can be derived by applying many of the aforementioned specializations of the various parameters involved in Theorem 11.*

We conclude our present investigation by proving the following symmetric identity involving the number $S_k(n, \lambda)$, which is defined by (44).

Theorem 12. *For all positive integers a and b, and for $n \in \mathbb{N}_0$, the following symmetric identity holds true:*

$$
\sum_{m=0}^{n} \binom{n}{m} c^{n-m} d^m \,_{e^{(r)}} Y_{n-m,\beta}^{(\alpha)} (dx, d^r y; k, a, b) \sum_{i=0}^{m} \binom{m}{i} S_i \left(c-1; \left(\frac{\beta}{a}\right)^b \right) \,_{e^{(r)}} Y_{m-i,\beta}^{(\alpha)}(cX, c^r Y; k, a, b)
$$

$$
= \sum_{m=0}^{n} \binom{n}{m} c^m d^{n-m} \,_{e^{(r)}} Y_{n-m,\beta}^{(\alpha)} (cx, c^r y; k, a, b) \sum_{i=0}^{m} \binom{m}{i} S_i \left(d-1; \left(\frac{\beta}{a}\right)^b \right)
$$

$$
\cdot \,_{e^{(r)}} Y_{m-i,\beta}^{(\alpha)}(dX, d^r Y; k, a, b). \tag{97}
$$

Proof. We first consider the function $g(t)$ given by

$$
g(t) = \frac{(2^{2(1-k)} c^k d^k t^{2k})^{\alpha} (\beta^b e^{cdt} - a^b)}{(\beta^b e^{ct} - a^b)^{\alpha} (\beta^b e^{dt} - a^b)^{\alpha+1}} e^{cdxt} \left(\frac{1}{1 - y(cdt)^r} \right) e^{cdXt} \left(\frac{1}{1 - Y(cdt)^r} \right)
$$

$$
= \left(\frac{2^{(1-k)} c^k t^k}{\beta^b e^{ct} - a^b} \right)^{\alpha} e^{cdxt} \left(\frac{1}{1 - y(cdt)^r} \right) \left(\frac{\beta^b e^{cdt} - a^b}{\beta^b e^{dt} - a^b} \right) \left(\frac{2^{(1-k)} d^k t^k}{\beta^b e^{dt} - a^b} \right)^{\alpha} e^{cdXt} \left(\frac{1}{1 - Y(cdt)^r} \right)
$$

$$
= \left(\sum_{n=0}^{\infty} \,_{e^{(r)}} Y_{n,\beta}^{(\alpha)} (dx, d^r y; k, a, b) \frac{(ct)^n}{n!} \right) \left[\sum_{n=0}^{\infty} S_n \left(c-1; \left(\frac{\beta}{a}\right)^b \right) \frac{(dt)^n}{n!} \right]
$$

$$
\cdot \left(\sum_{n=0}^{\infty} \,_{e^{(r)}} Y_{n,\beta}^{(\alpha)} (cX, c^r Y; k, a, b) \frac{(dt)^n}{n!} \right).
$$

Using similar arguments as above, we get

$$
g(t) = \left(\sum_{n=0}^{\infty} \,_{e^{(r)}} Y_{n,\beta}^{(\alpha)} (cx, c^r y; k, a, b) \frac{(dt)^n}{n!} \right) \left[\sum_{n=0}^{\infty} S_n \left(d-1; \left(\frac{\beta}{a}\right)^b \right) \frac{(ct)^n}{n!} \right]
$$

$$
\cdot \left(\sum_{n=0}^{\infty} \,_{e^{(r)}} Y_{n,\beta}^{(\alpha)} (dX, d^r Y; k, a, b) \frac{(ct)^n}{n!} \right). \tag{98}
$$

Finally, after a suitable manipulation with the summation index in (98) followed by a comparison of the coefficients of t^n, the proof of Theorem 12 is completed. □

5. Conclusions

Özden ([29]) defined the unified polynomials $Y_{n,\beta}^{(\alpha)}(x;k,a,b)$ of order α by means of the following generating function (see also Remark 1 above):

$$\left(\frac{2^{1-k}t^k}{\beta^b e^t - a^b}\right)^{\alpha} e^{xt} = \sum_{n=0}^{\infty} Y_{n,\beta}^{(\alpha)}(x;k,a,b)\,\frac{t^n}{n!}$$

$$\left(|t| < 2\pi \text{ when } \beta = a;\ |t| < \left|b\log\left(\frac{\beta}{a}\right)\right| \text{ when } \beta \neq a;\ 1^{\alpha} := 1;\ k \in \mathbb{N}_0;\ a,b \in \mathbb{R}\setminus\{0\};\ \alpha,\beta \in \mathbb{C}\right).$$

Basing our investigation upon this generating function, we have introduced generating function for the 2-variable truncated-exponential-based Apostol-type polynomials denoted by ${}_{e^{(r)}}Y_{n,\beta}^{(\alpha)}(x,y;k,a,b)$ as follows:

$$\sum_{n=0}^{\infty} {}_{e^{(r)}}Y_{n,\beta}^{(\alpha)}(x,y;k,a,b)\,\frac{t^n}{n!} = \left(\frac{2^{1-k}t^k}{\beta^b e^t - a^b}\right)^{\alpha} e^{xt}\left(\frac{1}{1-yt^r}\right),$$

which we have found to be instrumental in deriving quasi-monomiality with respect to the following multiplicative and derivative operators:

$$\widehat{M}_{e^{(r)}Y} = x + ry\partial_y y\partial_x^{r-1} + \frac{\alpha k(\beta^b e^t - a^b) - \alpha\beta^b\partial_x e^{\partial_x}}{\partial_x(\beta^b e^t - a^b)}$$

and

$$\widehat{P}_{e^{(r)}Y} = \partial_x.$$

We have also presented a further investigation to obtain some implicit summation formulas and symmetric identities by means of their generating functions.

In our next investigation, we propose to study an appropriate combination of the operational approach with that involving integral transforms with a view to studying integral representations related to the truncated-exponential-based Apostol-type polynomials which we have introduced and studied in this article.

Author Contributions: All authors contributed equally to this investigation.

Funding: This research received no external funding.

Conflicts of Interest: The authors declare no conflicts of interest.

References

1. Steffensen, J.F. The poweroid, an extension of the mathematical notion of power. *Acta Math.* **1941**, *73*, 333–366. [CrossRef]
2. Dattoli, G. Hermite-Bessel and Laguerre-Bessel-functions: A by-product of the monomiality principle. In *Advanced Special Functions and Applications, Proceedings of the First Melfi School on Advanced Topics in Mathematics and Physics, Melfi, Italy, 9–12 May 1999*; Cocolicchio, D., Dattoli, G., Srivastava, H.M., Eds.; Aracne Editrice: Rome, Italy, 2000; pp. 147–164.
3. Andrews, L.C. *Special Functions for Engineers and Mathematicians*; Macmillan Company: New York, NY, USA, 1985.
4. Dattoli, G.; Cesarano, C.; Sacchetti, D. A note on truncated polynomials. *Appl. Math. Comput.* **2003**, *134*, 595–605. [CrossRef]
5. Dattoli, G.; Migliorati, M.; Srivastava, H.M. A class of Bessel summation formulas and associated operational methods. *Fract. Calc. Appl. Anal.* **2004**, *7*, 169–176.
6. Khan, S.; Yasmin, G.; Ahmad, N. On a new family related to truncated exponential and Sheffer polynomials. *J. Math. Anal. Appl.* **2014**, *418*, 921–937. [CrossRef]

7. Yasmin, G.; Khan, S.; Ahmad, N. Operational methods and truncated exponential-based Mittag-Leffler polynomials. *Mediterr. J. Math.* **2016**, *13*, 1555–1569. [CrossRef]
8. Apostol, T.M. On the Lerch zeta function. *Pac. J. Math.* **1951**, *1*, 161–167. [CrossRef]
9. Sándor, J.; Crsci, B. *Handbook of Number Theory*; Kluwer Academic Publishers: Dordrecht, The Netherlands; Boston, MA, USA; London, UK, 2004; Volume II.
10. Srivastava, H.M.; Choi, J. *Series Associated with the Zeta and Related Functions*; Kluwer Academic Publishers: Dordrecht, The Netherlands; Boston, MA, USA; London, UK, 2001.
11. Guariglia, E. Fractional Derivative of the Riemann Zeta function. In *Fractional Dynamics*; Cattani, C., Srivastava, H.M., Yang, X.-J., Eds.; Emerging Science Publishers (De Gruyter Open): Berlin, Germany; Warsaw, Poland, 2015; pp. 357–368.
12. Gaboury, S. Some relations involving generalized Hurwitz-Lerch zeta function obtained by means of fractional derivatives with applications to Apostol-type polynomials. *Adv. Differ. Equ.* **2013**, *2013*, 1–13. [CrossRef]
13. Lin, S.-D.; Srivastava, H.M. Some families of the Hurwitz-Lerch zeta functions and associated fractional derivative and other integral representations. *Appl. Math. Comput.* **2004**, *154*, 725–733. [CrossRef]
14. Srivastava, H.M. Some formulas for the Bernoulli and Euler polynomials at rational arguments. *Math. Proc. Camb. Philos. Soc.* **2000**, *129*, 77–84. [CrossRef]
15. Guariglia, E.; Silvestrov, S. A functional equation for the Riemann zeta fractional derivative. *AIP Conf.* **2017**, *1798*, 020063; doi:10.1063/1.4972738. [CrossRef]
16. Luo, Q.-M.; Srivastava, H.M. Some relationships between the Apostol-Bernoulli and Apostol-Euler polynomials. *Comput. Math. Appl.* **2006**, *51*, 631–642. [CrossRef]
17. Luo, Q.-M.; Srivastava, H.M. Some generalizations of the Apostol-Bernoulli and Apostol-Euler polynomials. *J. Math. Anal. Appl.* **2005**, *308*, 290–302. [CrossRef]
18. Luo, Q.-M. Fourier expansions and integral representations for the Apostol-Bernoulli and Apostol-Euler polynomials. *Math. Comput.* **2009**, *78*, 2193–2208. [CrossRef]
19. Zhang, Z.; Yang, H. Several identities for the generalized Apostol-Bernoulli polynomials. *Comput. Math. Appl.* **2008**, *56*, 2993–2999. [CrossRef]
20. Luo, Q.-M. Some formulas for the Apostol-Euler polynomials associated with Hurwitz zeta function at rational arguments. *Appl. Anal. Discret. Math.* **2009**, *3*, 336–346. [CrossRef]
21. He, Y.; Araci, S. Sums of products of Apostol-Bernoulli and Apostol-Euler polynomials. *Adv. Differ. Equ.* **2014**, *2014*, 1–13. [CrossRef]
22. He, Y.; Araci, S.; Srivastava, H.M. Some new formulas for the products of the Apostol type polynomials. *Adv. Differ. Equ.* **2016**, *2016*, 1–18. [CrossRef]
23. He, Y.; Araci, S.; Srivastava, H.M.; Acikgöz, M. Some new identities for the Apostol-Bernoulli polynomials and the Apostol-Genocchi polynomials. *Appl. Math. Comput.* **2015**, *262*, 31–41. [CrossRef]
24. Luo, Q.-M. Fourier expansions and integral representations for the Genocchi polynomials. *J. Integer Seq.* **2009**, *12*, 1–9.
25. Luo, Q.-M. *q*-Extension for the Apostol-Genocchi polynomials. *Gen. Math.* **2009**, *17*, 113–125.
26. Luo, Q.-M. Extensions for the Genocchi polynomials and their Fourier expansions and integral representations. *Osaka J. Math.* **2011**, *48*, 291–310.
27. Luo, Q.-M.; Srivastava, H.M. Some generalizations of the Apostol-Genocchi polynomials and the Stirling numbers of the second kind. *Appl. Math. Comput.* **2011**, *217*, 5702–5728. [CrossRef]
28. Özden, H. Unification of generating functions of the Bernoulli, Euler and Genocchi numbers and polynomials. *AIP Conf. Proc.* **2010**. [CrossRef]
29. Özden, H. Generating function of the unified representation of the Bernoulli, Euler and Genocchi polynomials of higher order. *AIP Conf. Proc.* **2011**, *1389*, 349. [CrossRef]
30. Özden, H.; Simsek, Y.; Srivastava, H.M. A unified presentation of the generating functions of the generalized Bernoulli, Euler and Genocchi polynomials. *Comput. Math. Appl.* **2010**, *60*, 2779–2287. [CrossRef]
31. Özarslan, M.A. Hermite-Based unified Apostol-Bernoulli, Euler and Genocchi polynomials. *Adv. Differ. Equ.* **2013**, *2013*, 1–13. [CrossRef]
32. Özarslan, M.A. Unified Apostol-Bernoulli, Euler and Genocchi polynomials. *Comput. Math. Appl.* **2011**, *6*, 2452–2462. [CrossRef]
33. Khan, S.; Pathan, M.A.; Makhboul, H.N.A.; Yasmin, G. Implicit summation formula for Hermite and related polynomials. *J. Math. Anal. Appl.* **2008**, *344*, 408–416. [CrossRef]

34. Pathan, M.A.; Khan, W.A. Some implicit summation formulas and symmetric identities for the generalized Hermite-based polynomials. *Acta Univ. Apulensis.* **2014**, *39*, 113–136. [CrossRef]
35. Pathan, M.A.; Khan, W.A. Some implicit summation formulas and symmetric identities for the generalized Hermite-Bernoulli polynomials. *Mediterr. J. Math.* **2015**, *12*, 679–695. [CrossRef]
36. Pathan, M.A.; Khan, W.A. A new class of generalized polynomials associated with Hermite and Euler polynomials. *Mediterr. J. Math.* **2016**, *13*, 913–928. [CrossRef]
37. Srivastava, H.M.; Manocha, H.L. *A Treatise on Generating Functions*; Halsted Press: New York, NY, USA; Ellis Horwood Limited: New York, NY, USA; John Wiley and Sons; New York, NY, USA, 1984.
38. Khan, W.A. Some properties of the generalized Apostol type Hermite-Based polynomials. *Kyungpook Math. J.* **2015**, *55*, 597–614. [CrossRef]

© 2019 by the authors. Licensee MDPI, Basel, Switzerland. This article is an open access article distributed under the terms and conditions of the Creative Commons Attribution (CC BY) license (http://creativecommons.org/licenses/by/4.0/).

symmetry

MDPI

Article

A Certain Family of Integral Operators Associated with the Struve Functions

Shahid Mahmood [1], H. M. Srivastava [2,3], Sarfraz Nawaz Malik [4,*], Mohsan Raza [5], Neelam Shahzadi [4] and Saira Zainab [6]

1 Department of Mechanical Engineering, Sarhad University of Science and I.T, Ring Road, Peshawar 25000, Pakistan; shahidmahmood757@gmail.com
2 Department of Mathematics and Statistics, University of Victoria, Victoria, BC V8W 3R4, Canada; harimsri@math.uvic.ca
3 Department of Medical Research, China Medical University Hospital, China Medical University, Taichung 40402, Taiwan
4 Department of Mathematics, COMSATS University Islamabad, Wah Campus 47040, Pakistan; nshahzadi356@gmail.com
5 Department of Mathematics, Government College University Faisalabad, Faisalabad 38000, Pakistan; mohsan976@yahoo.com
6 Department of Mathematics, University of Wah, Wah Cantt 47040, Pakistan; sairazainab07@yahoo.com
* Correspondence: snmalik110@yahoo.com

Received: 31 January 2019; Accepted: 21 March 2019; Published: 2 April 2019

Abstract: This article presents the study of Struve functions and certain integral operators associated with the Struve functions. It contains the investigation of certain geometric properties like the strong starlikeness and strong convexity of the Struve functions. It also includes the criteria of univalence for a family of certain integral operators associated with the generalized Struve functions. The starlikeness and uniform convexity of the said integral operators are also part of this research.

Keywords: analytic functions; convex functions; starlike functions; strongly convex functions; strongly starlike functions; uniformly convex functions; Struve functions

MSC: 30C45; 30C50

---

1. Introduction

We denote by $\mathcal{A}$ the class of functions f that are analytic in the open unit disc $\mathcal{D} = \{z : |z| < 1\}$ and of the form:

$$f(z) = z + \sum_{n=2}^{\infty} a_n z^n. \tag{1}$$

Let $\mathcal{S}$ denote the class of all functions in $\mathcal{A}$, which are univalent in $\mathcal{D}$. Let $\mathcal{S}^*(\alpha)$, $\widetilde{\mathcal{S}}^*(\alpha)$ and $\widetilde{\mathcal{C}}(\alpha)$ denote the classes of starlike, strongly starlike and strongly convex functions of order α, respectively, and defined as:

$$\begin{aligned} \mathcal{S}^*(\alpha) &= \left\{ f : f \in \mathcal{A} \text{ and } \Re\left(\frac{zf'(z)}{f(z)}\right) > \alpha,\ z \in \mathcal{U}, \alpha \in [0,1) \right\}, \\ \widetilde{\mathcal{S}}^*(\alpha) &= \left\{ f : f \in \mathcal{A} \text{ and } \left|\arg\left(\frac{zf'(z)}{f(z)}\right)\right| < \frac{\alpha\pi}{2},\ z \in \mathcal{U}, \alpha \in [0,1) \right\}, \end{aligned}$$

and:

$$\widetilde{\mathcal{C}}(\alpha) = \left\{ f : f \in \mathcal{A} \text{ and } \left|\arg\left(1 + \frac{zf''(z)}{f'(z)}\right)\right| < \frac{\alpha\pi}{2},\ z \in \mathcal{U}, \alpha \in [0,1) \right\}.$$

It is clear that:

$$\widetilde{\mathcal{S}^*}(1) = \mathcal{S}^*(0) = \mathcal{S}^*,\ \widetilde{\mathcal{C}}(1) = \mathcal{C}(0) = \mathcal{C}.$$

The class $\mathcal{UCV}$ of uniformly convex functions is defined as:

$$\mathcal{UCV} = \left\{ f \in \mathcal{A} : \Re\left(1 + \frac{zf''(z)}{f'(z)}\right) > \left|\frac{zf''(z)}{f'(z)}\right|,\ z \in \mathcal{D} \right\}.$$

For more detail, see [1]. If f and g are analytic functions, then the function f is said to be subordinate to g, written as $f(z) \prec g(z)$, if there exists a Schwarz function w with $w(0) = 0$ and $|w| < 1$ such that $f(z) = g(w(z))$. Furthermore, if the function g is univalent in $\mathcal{U}$, then we have the following equivalent relation:

$$f(z) \prec g(z) \Longleftrightarrow f(0) = g(0) \quad \text{and} \quad f(\mathcal{U}) \subset g(\mathcal{U}).$$

Now, we consider the second order inhomogeneous differential equation:

$$z^2 w''(z) + zw'(z) + \left(z^2 - L^2\right) w(z) = \frac{4\left(\frac{z}{2}\right)^{L+1}}{\sqrt{\pi}\Gamma\left(L + \frac{1}{2}\right)}. \tag{2}$$

The solution of the homogeneous part is Bessel functions of order L, where L is a real or complex number. For more details about Bessel functions, we refer to [2–8]. The particular solution of the inhomogeneous equation defined in Equation (2) is called the Struve function of order L; see [9]. It is defined as:

$$X_L(z) = \sum_{n=0}^{\infty} \frac{(-1)^n (z/2)^{2n+L+1}}{\Gamma(n+3/2)\Gamma(L+n+3/2)}. \tag{3}$$

Now, we consider the differential equation:

$$z^2 w''(z) + zw'(z) - \left(z^2 + L^2\right) w(z) = \frac{4\left(\frac{z}{2}\right)^{L+1}}{\sqrt{\pi}\Gamma\left(L + \frac{1}{2}\right)}. \tag{4}$$

The Equation (4) differs from the Equation (2) in the coefficients of w. Its particular solution is called the modified Struve functions of order L and is given as:

$$Y_L(z) = -ie^{-ip\pi/2} X_L(iz) = \sum_{n=0}^{\infty} \frac{\left(\frac{z}{2}\right)^{2n+L+1}}{\Gamma(n+3/2)\Gamma\left(L+n+\frac{3}{2}\right)}.$$

Again, consider the second order inhomogeneous differential equation:

$$z^2 w''(z) + bzw'(z) + \left[cz^2 - L^2 + (1-b)L\right] w(z) = \frac{4\left(\frac{z}{2}\right)^{L+1}}{\sqrt{\pi}\Gamma\left(L + \frac{b}{2}\right)}, \tag{5}$$

where $b,\ c,\ L \in \mathbb{C}$. The Equation (5) generalizes the Equations (2) and (4). In particular, for $b = 1$, $c = 1$, we obtain Equation (2), and for $b = 1$, $c = -1$, we obtain Equation (4). Its particular solution has the series form:

$$w_{L,b,c}(z) = \sum_{n=0}^{\infty} \frac{(-1)^n c^n (z/2)^{2n+L+1}}{\Gamma(n+3/2)\Gamma(L+n+(b+2)/2)} \tag{6}$$

and is called the generalized Struve function of order L. This series is convergent everywhere. We take the transformation:

$$u_{L,b,c}(z) = 2^{L}\sqrt{\pi}\Gamma\left(L+(b+2)/2\right) z^{(-L-1)/2} w_{L,b,c}\left(\sqrt{z}\right) = \sum_{n=0}^{\infty} \frac{(-c/4)^{n} z^{n}}{(3/2)_{n}(q)_{n}}, \tag{7}$$

where $q = L + (b+2)/2 \neq 0, -1, -2, \ldots$ and $(\gamma)_n = \frac{\Gamma(\gamma+n)}{\Gamma(\gamma)} = \gamma(\gamma+1)\ldots(\gamma+n-1)$. This function is analytic in the whole complex plane and satisfies the differential equation:

$$4z^{2}w''(z) + 2(2p+b+3)zw'(z) + [cz+2p+b]\,w(z) = 2p+b,$$

where $\Gamma(.)$ denotes the Gamma function. The function $u_{L,b,c}$ unifies the Struve functions and modified Struve functions. The function $u_{L,b,c}$ is not in the class $\mathcal{A}$ of analytic functions; therefore, we consider the following normalized form of the Struve function as:

$$v_{L,b,c}(z) = zu_{L,b,c} = z + \sum_{n=1}^{\infty} \frac{(-c/4)^{n} z^{n+1}}{(3/2)_{n}(q)_{n}}. \tag{8}$$

Special cases:

(i) For $b=1,\ c=1$, we have the normalized Struve function $\mathcal{X}_L : \mathcal{A} \to \mathcal{A}$ of order L. It is given as:

$$\begin{aligned} \mathcal{X}_L(z) &= 2^{L}\sqrt{\pi}\Gamma\left(L+\frac{3}{2}\right) z^{\frac{(-L+1)}{2}} X_L\left(\sqrt{z}\right) \\ &= z + \sum_{n=1}^{\infty} \frac{(-1/4)^{n} z^{n+1}}{(3/2)_{n}(q)_{n}}. \end{aligned} \tag{9}$$

(ii) For $b=1,\ c=-1$, we have the normalized Struve function $\mathcal{Y}_L : \mathcal{A} \to \mathcal{A}$ of order L. It is given as:

$$\begin{aligned} \mathcal{Y}_L(z) &= 2^{L}\sqrt{\pi}\Gamma\left(L+\frac{3}{2}\right) z^{\frac{(-L+1)}{2}} Y_L\left(\sqrt{z}\right) \\ &= z + \sum_{n=1}^{\infty} \frac{(1/4)^{n} z^{n+1}}{(3/2)_{n}(q)_{n}}. \end{aligned} \tag{10}$$

The functions $u_{L,b,c}$ and $v_{L,b,c}$ were introduced and studied by Orhan and Yugmur [10] and further investigated by other authors [11–13]. In the last few years, many mathematicians have set the univalence criteria of several of those integral operators that preserve the class $\mathcal{S}$. By using a variety of different analytic techniques, operators and special functions, several authors have studied the univalence criterion. Recently Din et al. [14] studied the univalence of integral operators involving generalized Struve functions. These operators are defined as follows:

$$\mathcal{F}_{\alpha_1,\ldots,\alpha_n,\beta}(z) = \left[\beta \int_0^z t^{\beta-1} \prod_{i=1}^{n} \left(\frac{v_{L_i,b,c}(t)}{t}\right)^{\frac{1}{\alpha_i}} dt\right]^{\frac{1}{\beta}}, \tag{11}$$

$$\mathcal{M}_{n,\gamma}(z) = \left[(n\gamma+1)\int_0^z \prod_{i=1}^{n} \left\{v_{L_i,b,c}(t)\right\}^{\gamma} dt\right]^{\frac{1}{n\gamma+1}}, \tag{12}$$

and:

$$\mathcal{Z}_{\lambda}(z) = \left[\lambda \int_{0}^{z} t^{\lambda-1} \left(e^{v_{L_i,b,c}(t)} \right)^{\lambda} dt \right]^{1/\lambda}. \tag{13}$$

Now, we introduce the following integral operators $H_{Li,b,c,\gamma_1,..,\gamma_n,\beta}$, $I_{Li,b,c,\gamma_1,...,\gamma_n,\delta,\beta} : \mathcal{A} \to \mathcal{A}$ involving the generalized Struve functions as:

$$H_{L_i,b,c,\gamma_i,\beta}(z) = \left\{ \beta \int_{0}^{z} t^{\beta-1} \prod_{i=1}^{n} \left(\frac{v_{L_i,b,c}(t)}{g_i(t)} \right)^{\gamma_i} dt \right\}^{\frac{1}{\beta}}, \tag{14}$$

$$I_{Li,b,c,\gamma_i,\delta_i,\beta}(z) = \left\{ \beta \int_{0}^{z} t^{\beta-1} \prod_{i=1}^{n} \left(\frac{v'_{L_i,b,c}(t)}{t} \right)^{\gamma_i} \left(g'_i(t) \right)^{\delta_i} dt \right\}^{\frac{1}{\beta}}, \tag{15}$$

where γ_i, δ_i, β are nonzero complex numbers, $L_i \in \mathbb{R}$ for all $i = 1, 2, \cdots, n$ and $g_i \in \mathcal{A}$.

In this paper, our aim is to study certain geometric properties like the strong starlikeness and strong convexity of the Struve functions and univalence for the integral operators $H_{L_i,b,c,\gamma_i,\beta}$ and $I_{L_i,b,c,\gamma_i,\delta_i,\beta}$ associated with the generalized Struve functions. The starlikeness and uniform convexity of the said integral operators are also part of this research.

2. Preliminary Results

We need the following lemmas to prove our main results.

Lemma 1 ([15]). *Let $G(z)$ be convex and univalent in the open unit disc with condition $G(0) = 1$. Let $F(z)$ be analytic in the open unit disc with condition $F(0) = 1$ and $F \prec G$ in the open unit disc. Then, $\forall\, n \in \mathbb{N} \cup \{0\}$, we obtain:*

$$(n+1)z^{-1-n} \int_{0}^{z} t^{n} F(t) dt \prec (n+1)z^{-1-n} \int_{0}^{z} t^{n} G(t) dt.$$

Lemma 2 ([16]). *If $g \in \mathcal{A}$ satisfies:*

$$\left| 1 + \frac{zg''(z)}{g'(z)} \right| < 2, \quad \text{then } g \in \mathcal{S}^*.$$

Lemma 3 ([17]). *If $g \in \mathcal{A}$ satisfies:*

$$\left| \frac{zg''(z)}{g'(z)} \right| < \frac{1}{2}, \quad \text{then } g \in \mathcal{UCV}.$$

Lemma 4 ([10]). *If $b, L \in \mathbb{R}$ and $c \in \mathbb{C}$, $q = L + \frac{b+2}{2}$ are so constrained that $q > \max\left\{0, \frac{7|c|}{24}\right\}$, then the function $v_{L,b,c} : \mathcal{D} \longrightarrow \mathbb{C}$ satisfies the following inequalities.*

(i) $\left| \frac{zv'_{L,b,c}(z)}{v_{L,b,c}(z)} - 1 \right| \leq \frac{|c|(6q-|c|)}{3(4q-|c|)(3q-|c|)}$,

(ii) $\left| \frac{zv''_{L,b,c}(z)}{v'_{L,b,c}(z)} \right| \leq \frac{6|c|}{(12q-7|c|)}$.

Lemma 5 ([18]). *If $g \in \mathcal{A}$ satisfies the following inequality:*

$$\frac{1-|z|^{2\Re(\alpha)}}{\Re(\alpha)} \left| \frac{zg''(z)}{g'(z)} \right| \leq 1, \Re(\alpha) > 0,$$

then for every complex number $\beta, \Re\beta \geq \Re(\alpha)$, *the function:*

$$G_\beta(z) = \left(\beta \int_0^z t^{\beta-1} g'(t)\, dt\right)^{\frac{1}{\beta}} \in \mathcal{S}.$$

Lemma 6 ([19]). *Let* $g(z) = z + a_2 z^2 + \cdots$ *be the analytic function in* $\mathcal{D}$. *If:*

$$\left|\frac{g''(z)}{g'(z)}\right| \leq K, \;\; z \in \mathcal{D},$$

where $K \simeq 3.05$, *then* g *is univalent in* $\mathcal{D}$.

Remark 1. *The constant* K *is the solution of the equation* $8\left[x(x-2)^3\right]^{\frac{1}{2}} - 3(4-x)^2 = 12$. *An approximation by using the computer programs suggest the value* 3.03902118847875. *Kudriasov used the approximated value equal to* 3.05.

3. Geometric Properties of Generalized Struve Functions

Theorem 1. *If* $q \geq \frac{7|c|}{12}$, *then* $v_{L,b,c} \in \widetilde{\mathcal{S}^*}(\alpha)$, *where:*

$$\alpha = \frac{2}{\pi} \arcsin\left(\psi\sqrt{1-\frac{\psi^2}{4}} + \frac{\psi}{2}\sqrt{1-\psi^2}\right) \tag{16}$$

and $\psi = \frac{4|c|}{3(4q-|c|)}$ *is such that* $\arcsin\frac{\psi}{2} + \arcsin\psi \in \left[-\frac{\pi}{2}, \frac{\pi}{2}\right]$.

Proof. By using Equation (8) with the triangle inequality, we have:

$$\left|v'_{L,b,c}(z) - 1\right| \leq \sum_{n=1}^{\infty} \frac{|c|^n (n+1)}{(3/2)_n 4^n (q)_n}$$

By the help of the inequalities:

$$(3/2)_n \geq \frac{3}{4}(n+1), \; (q)_n \geq q^n, \; \forall\, n \geq 1,$$

we obtain:

$$\begin{aligned} \left|v'_{L,b,c}(z) - 1\right| &\leq \frac{|c|}{3q} \sum_{n=1}^{\infty} \left(\frac{|c|}{4q}\right)^{n-1} \\ &= \frac{4|c|}{3(4q-|c|)} = \psi, \quad q > \frac{|c|}{4}. \end{aligned} \tag{17}$$

For $q \geq \frac{7|c|}{12}$, it is clear that $0 < \psi \leq 1$. Furthermore, from expression (17), we concluded that:

$$v'_{L,b,c}(z) \prec 1 + \psi z \;\; \Rightarrow \;\; \left|\arg\left(v'_{L,b,c}(z)\right)\right| < \arcsin\psi. \tag{18}$$

With the help of Lemma 1, take $n = 0$ with $F(z) = v'_{L,b,c}(z)$ and $G(z) = 1 + \psi z$, and we get:

$$\frac{v_{L,b,c}(z)}{z} \prec 1 + \frac{\psi}{2} z. \tag{19}$$

As a result:

$$\left|\arg\left(\frac{v_{L,b,c}(z)}{z}\right)\right| < \arcsin\frac{\psi}{2}. \tag{20}$$

By using relations (18) and (19) , we obtain:

$$\begin{aligned}\left|\arg\left(\frac{zv'_{L,b,c}(z)}{v_{L,b,c}(z)}\right)\right| &= \left|\arg\left(\frac{z}{v_{L,b,c}(z)}\right) - \arg\left(v'_{L,b,c}(z)\right)\right| \\ &\leq \left|\arg\left(\frac{z}{v_{L,b,c}(z)}\right)\right| + \left|\arg\left(v'_{L,b,c}(z)\right)\right| \\ &< \arcsin\frac{\psi}{2} + \arcsin\psi.\end{aligned}$$

As $0 < \psi \leq 1$, thus one can write the above last expression as:

$$\left|\arg\left(\frac{zv'_{L,b,c}(z)}{v_{L,b,c}(z)}\right)\right| < \arcsin\left(\psi\sqrt{1-\frac{\psi^2}{4}} + \frac{\psi}{2}\sqrt{1-\psi^2}\right),$$

which shows that $v_{L,b,c} \in \widetilde{\mathcal{S}^*}(\alpha)$ for $\alpha = \frac{2}{\pi}\arcsin\left(\psi\sqrt{1-\frac{\psi^2}{4}} + \frac{\psi}{2}\sqrt{1-\psi^2}\right)$. □

Theorem 2. *If $q \geq \frac{4|c|}{3}$, then $v_{L,b,c} \in \widetilde{\mathcal{C}}(\alpha)$, where:*

$$\alpha = \frac{2}{\pi}\arcsin\left(\varphi\sqrt{1-\frac{\varphi^2}{4}} + \frac{\varphi}{2}\sqrt{1-\varphi^2}\right), \tag{21}$$

and $\varphi = \frac{2|c|}{3q-2|c|}$ is such that $\arcsin\frac{\varphi}{2} + \arcsin\varphi \in \left[-\frac{\pi}{2}, \frac{\pi}{2}\right]$.

Proof. By using the well-known triangle inequality:

$$|z_1 + z_2| \leq |z_1| + |z_2|,$$

with the inequalities:

$$(n+1)^2 \leq 4^n,\ (q)_n \geq q^n \quad \forall\, n \in \mathbb{N},$$

we obtain:

$$\begin{aligned}\left|\left(zv'_{L,b,c}(z)\right)' - 1\right| &\leq \sum_{n=1}^{\infty} \frac{|c|^n (n+1)^2}{(3/2)_n 4^n (q)_n} \\ &\leq \frac{2|c|}{3q}\sum_{n=1}^{\infty}\left(\frac{2|c|}{3q}\right)^{n-1} \\ &= \frac{2|c|}{3q-2|c|} = \varphi.\end{aligned} \tag{22}$$

It is clear that $0 < \varphi \leq 1$ for $q \geq \frac{4|c|}{3}$, and from the expression (22), we conclude that:

$$\left(zv'_{L,b,c}(z)\right)' \prec 1 + \varphi z \quad \Rightarrow \quad \left|\arg\left(zv'_{L,b,c}(z)\right)'\right| < \arcsin\varphi. \tag{23}$$

With the help of Lemma 1, take $n = 0$ with $F(z) = \left(zv'_{L,b,c}(z)\right)'$ and $G(z) = 1 + \varphi z$, and we get:

$$\frac{zv'_{L,b,c}(z)}{z} \prec 1 + \frac{\varphi}{2}z. \tag{24}$$

This implies that:

$$v'_{L,b,c}(z) \prec 1 + \frac{\varphi}{2} z.$$

As a result:

$$\left|\arg v'_{L,b,c}(z)\right| < \arcsin \frac{\varphi}{2}. \tag{25}$$

By using relations (23) and (25), we obtain:

$$\begin{aligned}
\left|\arg\left(\frac{\left(zv'_{L,b,c}(z)\right)'}{v'_{L,b,c}(z)}\right)\right| &= \left|\arg\left(zv'_{L,b,c}(z)\right)' - \arg v'_{L,b,c}(z)\right| \\
&\leq \left|\arg\left(zv'_{L,b,c}(z)\right)'\right| + \left|\arg\left(v'_{L,b,c}(z)\right)\right| \\
&< \arcsin \frac{\varphi}{2} + \arcsin \varphi.
\end{aligned}$$

As $0 < \varphi \leq 1$, thus one can write the above last expression as:

$$\left|\arg\left(\frac{\left(zv'_{L,b,c}(z)\right)'}{v'_{L,b,c}(z)}\right)\right| < \arcsin\left(\varphi\sqrt{1-\frac{\varphi^2}{4}} + \frac{\varphi}{2}\sqrt{1-\varphi^2}\right),$$

which shows that $v_{L,b,c} \in \widetilde{\mathcal{C}}(\alpha)$ for $\alpha = \frac{2}{\pi}\arcsin\left(\varphi\sqrt{1-\frac{\varphi^2}{4}} + \frac{\varphi}{2}\sqrt{1-\varphi^2}\right)$. □

Theorem 3. *Let $q > \frac{19|c|}{12}$, then $v_{L,b,c} \in \mathcal{UCV}$.*

Proof. Since:

$$\left|\frac{zv''_{L,b,c}(z)}{v'_{L,b,c}(z)}\right| \leq \frac{6|c|}{(12q - 7|c|)}.$$

By using Lemma 3, we have the required result. □

4. Univalence Criteria for Integral Operators

In this section, we find the univalence of these integral operators defined by generalized Struve functions, by using the above lemmas.

Theorem 4. *Let $L_1, \ldots, L_n$, $b \in \mathbb{R}$, $c \in \mathbb{C}$ and $q_i > \frac{7|c|}{24}$ with $q_i = L_i + \frac{b+2}{2}$, $i = 1, ..., n$. Let $v_{L_i,b,c} : \mathcal{D} \longrightarrow \mathbb{C}$ be defined in the Equation (8). Suppose $q = \min(q_1, q_2, \ldots, q_n)$, γ_i are non-zero complex numbers and if $g_i \in \mathcal{A}$ with:*

$$\left|\frac{g_i''(z)}{g_i'(z)}\right| \leq K, \ z \in \mathcal{D},$$

where $K \simeq 3.05$, these numbers satisfying the relations:

$$\frac{1}{\Re(\alpha)}\left(1 + \frac{|c|(6q - |c|)}{3(4q - |c|)(3q - |c|)}\right)\sum_{i=1}^{n}|\gamma_i| + \frac{4}{\Re(\alpha)}\sum_{i=1}^{n}|\gamma_i| < 1, \tag{26}$$

when $0 < \Re(\alpha) < 1$ and for $\Re(\alpha) \geq 1$:

$$\frac{1}{\Re(\alpha)}\left(1 + \frac{|c|(6q - |c|)}{3(4q - |c|)(3q - |c|)}\right)\sum_{i=1}^{n}|\gamma_i| + 4\sum_{i=1}^{n}|\gamma_i| < 1, \tag{27}$$

then for every complex number β, $\Re(\beta) \geq \Re(\alpha) > 0$, the function $H_{L_i,b,c,\gamma_i,\beta}$ defined in (14) is univalent.

Proof. Consider the function:

$$H_{L_i,b,c,\gamma_i}(z) = \int_0^z \prod_{i=1}^n \left(\frac{v_{L_i,b,c}(t)}{g_i(t)} \right)^{\gamma_i} dt. \tag{28}$$

By taking the derivative of Equation (28), we get:

$$H'_{L_i,b,c,\gamma_i}(z) = \prod_{i=1}^n \left(\frac{v_{L_i,b,c}(z)}{g_i(z)} \right)^{\gamma_i}. \tag{29}$$

It is clear that $H_{L_i,b,c,\gamma_i}(0) = H'_{L_i,b,c,\gamma_i}(0) - 1 = 0$. It follows easily that:

$$\frac{zH''_{L_i,b,c,\gamma_i}(z)}{H'_{L_i,b,c,\gamma_i}(z)} = \sum_{i=1}^n \gamma_i \left\{ \left(\frac{zv'_{L_i,b,c}(z)}{v_{L_i,b,c}(z)} \right) - \left(\frac{zg'_i(z)}{g_i(z)} \right) \right\}$$

and:

$$\frac{1-|z|^{2\Re(\alpha)}}{\Re(\alpha)} \left| \frac{zH''_{L_i,b,c,\gamma_i}(z)}{H'_{L_i,b,c,\gamma_i}(z)} \right| \le \frac{1-|z|^{2\Re(\alpha)}}{\Re(\alpha)} \left\{ \sum_{i=1}^n |\gamma_i| \left| \frac{zv'_{L_i,b,c}(z)}{v_{L_i,b,c}(z)} \right| + \sum_{i=1}^n |\gamma_i| \left| \frac{zg'_i(z)}{g_i(z)} \right| \right\}.$$

Now, using the Lemma 6, we have $g_i \in \mathcal{S}, i = 1, ..., n$, and:

$$\left| \frac{zg'_i(z)}{g_i(z)} \right| \le \frac{1+|z|}{1-|z|}. \tag{30}$$

By virtue of the above inequality (30), we get:

$$\begin{aligned} \frac{1-|z|^{2\Re(\alpha)}}{\Re(\alpha)} \left| \frac{zH''_{L_i,b,c,\gamma_i}(z)}{H'_{L_i,b,c,\gamma_i}(z)} \right| &\le \frac{1-|z|^{2\Re(\alpha)}}{\Re(\alpha)} \left\{ \sum_{i=1}^n |\gamma_i| \left| \frac{zv'_{L_i,b,c}(z)}{v_{L_i,b,c}(z)} \right| + \sum_{i=1}^n |\gamma_i| \frac{1+|z|}{1-|z|} \right\} \\ &\le \frac{1-|z|^{2\Re(\alpha)}}{\Re(\alpha)} \sum_{i=1}^n |\gamma_i| \left| \frac{zv'_{L_i,b,c}(z)}{v_{L_i,b,c}(z)} \right| + \frac{1-|z|^{2\Re(\alpha)}}{\Re(\alpha)} \frac{2}{1-|z|} \sum_{i=1}^n |\gamma_i|. \end{aligned}$$

First, we consider the part:

$$\frac{1-|z|^{2\Re(\alpha)}}{\Re(\alpha)} \sum_{i=1}^n |\gamma_i| \left| \frac{zv'_{L_i,b,c}(z)}{v_{L_i,b,c}(z)} \right|.$$

This implies that:

$$\frac{1-|z|^{2\Re(\alpha)}}{\Re(\alpha)} \sum_{i=1}^n |\gamma_i| \left| \frac{zv'_{L_i,b,c}(z)}{v_{L_i,b,c}(z)} \right| \le \frac{1}{\Re(\alpha)} \sum_{i=1}^n |\gamma_i| \left| \frac{zv'_{L_i,b,c}(z)}{v_{L_i,b,c}(z)} \right|.$$

Using Lemma 5, we have:

$$\frac{1-|z|^{2\Re(\alpha)}}{\Re(\alpha)} \sum_{i=1}^n |\gamma_i| \left| \frac{zv'_{L_i,b,c}(z)}{v_{L_i,b,c}(z)} \right| \le \frac{1}{\Re(\alpha)} \sum_{i=1}^n |\gamma_i| \left\{ 1 + \frac{|c|(6q_i - |c|)}{3(4q_i - |c|)(3q_i - |c|)} \right\}.$$

We define the function $\tau : \left(\frac{7|c|}{24}, \infty \right) \longrightarrow \mathbb{R}$, $\tau(x) = \frac{|c|(6x-|c|)}{3(4x-|c|)(3x-|c|)}$. It is a decreasing function; therefore:

$$\frac{|c|(6q_i - |c|)}{3(4q_i - |c|)(3q_i - |c|)} \le \frac{|c|(6q - |c|)}{3(4q - |c|)(3q - |c|)},$$

hence:

$$\frac{1-|z|^{2\Re(\alpha)}}{\Re(\alpha)}\sum_{i=1}^{n}|\gamma_i|\left|\frac{zv'_{L_i,b,c}(z)}{v_{L_i,b,c}(z)}\right| \le \frac{1}{\Re(\alpha)}\sum_{i=1}^{n}|\gamma_i|\left\{1+\frac{|c|(6q-|c|)}{3(4q-|c|)(3q-|c|)}\right\}. \quad (31)$$

Now, we consider the part:

$$\frac{1-|z|^{2\Re(\alpha)}}{\Re(\alpha)}\frac{2}{1-|z|}\sum_{i=1}^{n}|\gamma_i|.$$

For this, we have the following cases:

(1) For $0<\Re(\alpha)<1$, then the function $v:(0,1)\longrightarrow\mathbb{R}$, $v(x)=1-a^{2x}$, $x=\Re(\alpha)$ and $|z|=a$ is increasing and:

$$1-|z|^{2\Re(\alpha)}\le 1-|z|^2;$$

therefore:

$$\frac{1-|z|^{2\Re(\alpha)}}{\Re(\alpha)}\frac{2}{1-|z|}\sum_{i=1}^{n}|\gamma_i| \le \frac{4}{\Re(\alpha)}\sum_{i=1}^{n}|\gamma_i|. \quad (32)$$

From the inequalities (31) and (32), for $0<\Re(\alpha)<1$, we have:

$$\frac{1-|z|^{2\Re(\alpha)}}{\Re(\alpha)}\left|\frac{zH''_{L_i,b,c,\gamma_i}(z)}{H'_{L_i,b,c,\gamma_i}(z)}\right| \le \frac{1}{\Re(\alpha)}\left(1+\frac{|c|(6q-|c|)}{3(4q-|c|)(3q-|c|)}\right)\sum_{i=1}^{n}|\gamma_i|+\frac{4}{\Re(\alpha)}\sum_{i=1}^{n}|\gamma_i|. \quad (33)$$

(2) For $\Re(\alpha)\ge 1$, consider the function $w:[1,\infty)\longrightarrow\mathbb{R}$, $w(x)=\frac{1-a^{2x}}{x}$, $x=\Re(\alpha)$ and $|z|=a$ is a decreasing function and:

$$\frac{1-|z|^{2\Re(\alpha)}}{\Re(\alpha)}\le 1-|z|^2;$$

therefore:

$$\frac{1-|z|^{2\Re(\alpha)}}{\Re(\alpha)}\frac{2}{1-|z|}\sum_{i=1}^{n}|\gamma_i| \le 4\sum_{i=1}^{n}|\gamma_i|. \quad (34)$$

By combining the inequalities (31) and (34) for $\Re(\alpha)\ge 1$, we get:

$$\left(\frac{1-|z|^{2\Re(\alpha)}}{\Re(\alpha)}\right)\left|\frac{zH''_{L_i,b,c,\gamma_i}(z)}{H'_{L_i,b,c,\gamma_i}(z)}\right| \le \frac{1}{\Re(\alpha)}\left(1+\frac{|c|(6q-|c|)}{3(4q-|c|)(3q-|c|)}\right)\sum_{i=1}^{n}|\gamma_i|+4\sum_{i=1}^{n}|\gamma_i|. \quad (35)$$

From the inequalities (26), (27), (33) and (35), we obtain:

$$\frac{1-|z|^{2\Re(\alpha)}}{\Re(\alpha)}\left|\frac{zH''_{L_i,b,c,\gamma_i}(z)}{H'_{L_i,b,c,\gamma_i}(z)}\right|<1.$$

Therefore, using Lemma 5, we get the required result. □

Theorem 5. *Let $L_1,\dots L_n$, $b\in\mathbb{R}$, $c\in\mathbb{C}$ and $q_i>\frac{7|c|}{24}$ with $q_i=L_i+\frac{(b+2)}{2}$, $i=1,\dots,n$. Let $v_{L_i,b,c}:\mathcal{D}\longrightarrow\mathbb{C}$ be defined in the Equation (8). Suppose $q=min(q_1,q_2,......q_n)$, γ_i, δ_i are non-zero complex numbers and if $g_i\in\mathcal{A}$ with*

$$\left|\frac{g''_i(z)}{g'_i(z)}\right|\le K,\ z\in\mathcal{D},$$

where $K\simeq 3.05$, and these numbers satisfy the relation:

$$\frac{1}{\Re(\alpha)}\frac{|c|(6q-|c|)}{3(4q-|c|)(3q-|c|)}\sum_{i=1}^{n}|\gamma_i|+\frac{2K}{(2\Re(\alpha)+1)^{\frac{(2\Re(\alpha)+1)}{2\Re(\alpha)}}}\sum_{i=1}^{n}|\delta_i|<1. \quad (36)$$

Then, for every complex number β, $\Re(\beta) \geq \Re(\alpha) > 0$, the function $I_{L_i,b,c,\gamma_i,\delta_i,\beta}$ defined in Equation (15) is univalent.

Proof. Consider the function:

$$I_{L_i,b,c,\gamma_i,\delta_i}(z) = \int_0^z \prod_{i=1}^{n} \left(\frac{v_{L_i,b,c}(t)}{t} \right)^{\gamma_i} \left(g_i'(t) \right)^{\delta_i} dt. \tag{37}$$

By taking the derivative of Equation (37), we get:

$$I'_{L_i,b,c,\gamma_i,\delta_i}(z) = \prod_{i=1}^{n} \left(\frac{v_{L_i,b,c}(z)}{z} \right)^{\gamma_i} \left(g_i'(z) \right)^{\delta_i}. \tag{38}$$

It is clear that $I_{L_i,b,c,\gamma_i,\delta_i} \in \mathcal{A}$. It follows easily that:

$$\frac{zI''_{L_i,b,c,\gamma_i,\delta_i}(z)}{I'_{L_i,b,c,\gamma_i,\delta_i}(z)} = \sum_{i=1}^{n} \gamma_i \left(\frac{zv'_{L_i,b,c}(z)}{v_{L_i,b,c}(z)} - 1 \right) + \sum_{i=1}^{n} \delta_i \left\{ \frac{zg_i''(z)}{g_i'(z)} \right\}.$$

Therefore, we obtain:

$$\begin{aligned} &\frac{1-|z|^{2\Re(\alpha)}}{\Re(\alpha)} \left| \frac{zI''_{L_i,b,c,\gamma_i,\delta_i}(z)}{I'_{L_i,b,c,\gamma_i,\delta_i}(z)} \right| \\ \leq\ & \frac{1-|z|^{2\Re(\alpha)}}{\Re(\alpha)} \left\{ |\gamma_i| \left| \frac{zv'_{L_i,b,c}(z)}{v_{L_i,b,c}(z)} - 1 \right| + |z| \, |\delta_i| \left| \frac{g_i''(z)}{g_i'(z)} \right| \right\}. \end{aligned}$$

This implies that:

$$\begin{aligned} \frac{1-|z|^{2\Re(\alpha)}}{\Re(\alpha)} \left| \frac{zI''_{L_i,b,c,\gamma_i,\delta i}(z)}{I'_{L_i,b,c,\gamma i,\delta_i}(z)} \right| \leq\ & \left\{ \frac{1-|z|^{2\Re(\alpha)}}{\Re(\alpha)} \sum_{i=1}^{n} |\gamma_i| \left| \frac{zv'_{L_i,b,c}(z)}{v_{L_i,b,c}(z)} - 1 \right| \right. \\ & \left. + \frac{1-|z|^{2\Re(\alpha)}}{\Re(\alpha)} |z| \sum_{i=1}^{n} |\delta_i| \left| \frac{g_i''(z)}{g_i'(z)} \right| \right\}. \end{aligned} \tag{39}$$

Using Lemmas 4 and 6, we get:

$$\begin{aligned} \frac{1-|z|^{2\Re(\alpha)}}{\Re(\alpha)} \left| \frac{zI''_{L_i,b,c,\gamma_i,\delta_i}(z)}{I'_{L_i,b,c,\gamma_i,\delta_i}(z)} \right| \leq\ & \left\{ \frac{1-|z|^{2\Re(\alpha)}}{\Re(\alpha)} \sum_{i=1}^{n} |\gamma_i| \frac{|c|(6q_i - |c|)}{3(4q_i - |c|)(3q_i - |c|)} \right. \\ & \left. + \frac{1-|z|^{2\Re(\alpha)}}{\Re(\alpha)} |z| K \sum_{i=1}^{n} \right\} |\delta_i|. \end{aligned}$$

As was mentioned before:

$$\frac{|c|(6q_i - |c|)}{3(4q_i - |c|)(3q_i - |c|)} \leq \frac{|c|(6q - |c|)}{3(4q - |c|)(3q - |c|)};$$

therefore:

$$\begin{aligned} \frac{1-|z|^{2\Re(\alpha)}}{\Re(\alpha)} \left| \frac{zI''_{L_i,b,c,\gamma_i,\delta_i}(z)}{I'_{L_i,b,c,\gamma_i,\delta_i}(z)} \right| \leq\ & \left\{ \frac{1-|z|^{2\Re(\alpha)}}{\Re(\alpha)} \frac{|c|(6q - |c|)}{3(4q - |c|)(3q - |c|)} \sum_{i=1}^{n} |\gamma_i| \right. \\ & \left. + \frac{1-|z|^{2\Re(\alpha)}}{\Re(\alpha)} |z| K \sum_{i=1}^{n} |\delta_i| \right\}. \end{aligned}$$

Consider the function $h : [0,1] \longrightarrow \mathbb{R}, h(x) = \frac{x(1-x^{2a})}{a}, x = |z|, a = \Re(\alpha)$. Then:

$$\max h(x) = \frac{2}{(2a+1)^{\frac{(2a+1)}{2a}}}, x \in [0,1].$$

This implies that:

$$\begin{aligned}\frac{1-|z|^{2\Re(\alpha)}}{\Re(\alpha)}\left|\frac{zI''_{L_i,b,c,\gamma_i,\delta_i}(z)}{I'_{L_i,b,c,\gamma_i,\delta_i}(z)}\right| &\le \left\{\frac{1-|z|^{2\Re(\alpha)}}{\Re(\alpha)}\frac{|c|(6q-|c|)}{3(4q-|c|)(3q-|c|)}\sum_{i=1}^{n}|\gamma_i|\right.\\ &\left.+\frac{2K}{(2\Re(\alpha)+1)^{\frac{(2\Re(\alpha)+1)}{2\Re(\alpha)}}}\sum_{i=1}^{n}|\delta_i|\right\}.\end{aligned}$$

Using the inequalities (36) and (39), we get:

$$\frac{1-|z|^{2\Re(\alpha)}}{\Re(\alpha)}\left|\frac{zI''_{L_i,b,c,\gamma_i,\delta_i}(z)}{I'_{L_i,b,c,\gamma_i,\delta_i}(z)}\right| < 1.$$

Therefore, by using Lemma 5, we get the required result. □

Corollary 1. *Consider the function $\mathcal{X}_{L_i}(z) : \mathcal{D} \longrightarrow \mathbb{C}$ defined in the Equation (9). Let $L_1, \ldots, L_n > -1.75$ ($n \in \mathbb{N}$) and $L = min\{L_1, \ldots, L_n\}$. Furthermore, let the parameter γ_i be non-zero complex numbers with $\{i = 1, 2, 3, \ldots, n\}$ and if $g_i \in \mathcal{A}$ with:*

$$\left|\frac{g_i''(z)}{g_i'(z)}\right| \le K, \quad z \in \mathcal{D},$$

where $K \simeq 3.05$, and these numbers satisfy the relations:

$$\frac{1}{\Re(\alpha)}\left(1+\frac{4(3L+4)}{3(24L^2+58L+35)}\right)\sum_{i=1}^{n}|\gamma_i| + \frac{4}{\Re(\alpha)}\sum_{i=1}^{n}|\gamma_i| < 1,$$

when $0 < \Re(\alpha) < 1$ and for $\Re(\alpha) \ge 1$:

$$\left[\frac{1}{\Re(\alpha)}\left(1+\frac{4(3L+4)}{3(24L^2+58L+35)}\right)\sum_{i=1}^{n}|\gamma_i| + 4\sum_{i=1}^{n}|\gamma_i| < 1\right];$$

then for every complex number β, $\Re(\beta) \ge \Re(\alpha) > 0$, the function $H_{Li,b,c,\gamma_i,\beta}$ is univalent.

Corollary 2. *Consider the function $\mathcal{X}_{L_i}$ defined in the Equation (9). Let $L_1, \ldots, L_n > -1.75$ ($n \in \mathbb{N}$) and $L = min\{L_1, \ldots, L_n\}$. Furthermore, let the parameter γ_i, δ_i be non-zero complex numbers and if $g_i \in \mathcal{A}$ with:*

$$\left|\frac{g_i''(z)}{g_i'(z)}\right| \le K, \quad z \in \mathcal{D},$$

where $K \simeq 3.05$, and these numbers satisfy the relation:

$$\frac{1}{\Re(\alpha)}\frac{4(3L+4)}{3(24L^2+58L+35)}\sum_{i=1}^{n}|\gamma_i| + \frac{2K}{(2\Re(\alpha)\Re(\gamma)+1)^{\frac{(2\Re(\alpha)+1)}{2\Re(\alpha)}}}\sum_{i=1}^{n}|\delta_i| < 1;$$

then for every complex number β, $\Re(\beta) \ge \Re(\alpha) > 0$, the function $I_{L_i,b,c,\gamma_i,\delta_i,\beta}$ is univalent.

Corollary 3. *Consider the function* $\mathcal{Y}_{L_i}(z) : \mathcal{D} \longrightarrow \mathbb{C}$ *defined in the Equation* (10). *Let* $L_1, \ldots, L_n > -1.75$ $(n \in \mathbb{N})$ *and* $L = \min\{L_1, L_2, ..., L_n\}$. *Furthermore, let the parameters* γ_i *be non-zero complex numbers with* $\{i = 1, \ldots, n\}$ *and if* $g_i \in \mathcal{A}$ *with:*

$$\left|\frac{g_i''(z)}{g_i'(z)}\right| \leq K, \ z \in \mathcal{D},$$

where $K \simeq 3.05$, *and these numbers satisfy the relations:*

$$\frac{1}{\Re(\alpha)}\left(1 + \frac{4(3L+4)}{3(24L^2+58L+35)}\right)\sum_{i=1}^{n}|\gamma_i| + \frac{4}{\Re(\alpha)}\sum_{i=1}^{n}|\gamma_i| < 1,$$

when $0 < \Re(\alpha) < 1$ *and for* $\Re(\alpha) \geq 1$:

$$\left[\frac{1}{\Re(\alpha)}\left(1 + \frac{4(3L+4)}{3(24L^2+58L+35)}\right)\sum_{i=1}^{n}|\gamma_i| + 4\sum_{i=1}^{n}|\gamma_i| < 1\right];$$

then for every complex number β, $\Re(\beta) \geq \Re(\alpha) > 0$, *the function* $H_{L_i,b,c,\gamma_i,\beta}$ *is univalent.*

Corollary 4. *Consider the function* $\mathcal{Y}_{L_i}(z) : \mathcal{D} \longrightarrow \mathbb{C}$ *defined in the Equation* (10). *Let* $L_1, \ldots, L_n > -1.75$ $(n \in \mathbb{N})$ *and* $L = \min\{L_1, L_2, ..., L_n\}$. *Furthermore, let the parameter* γ_i, δ_i *be non-zero complex numbers and if* $g_i \in \mathcal{A}$ *with:*

$$\left|\frac{g_i''(z)}{g_i'(z)}\right| \leq K, \ z \in \mathcal{D},$$

where $K \simeq 3.05$, *and these numbers satisfy the relation:*

$$\frac{1}{\Re(\alpha)\Re(\gamma)}\frac{4(3L+4)}{3(24L^2+58L+35)}\sum_{i=1}^{n}|\gamma_i| + \frac{2K}{(2\Re(\alpha)+1)^{\frac{(2\Re(\alpha)+1)}{2\Re(\alpha)}}}\sum_{i=1}^{n}|\delta_i| < 1;$$

then for every complex number β, $\Re(\beta) \geq \Re(\alpha) > 0$, *the function* $I_{L_i,b,c,\gamma_i,\delta_i,\beta}$ *is univalent.*

5. Starlikeness and Uniform Convexity Criteria for the Integral Operator

In this section, we find the starlikeness and uniform convexity of these integral operators defined by generalized Struve functions.

Theorem 6. *Let* $L_1, \ldots, L_n$, $b \in \mathbb{R}$, $c \in \mathbb{C}$ *and* $q_i > \frac{7|c|}{24}$ *with* $q_i = L_i + \frac{(b+2)}{2}$, $i = 1, \ldots, n$. *Let* $v_{L_i,b,c} : \mathcal{D} \longrightarrow \mathbb{C}$ *be defined in the Equation* (8). *Let the function* g_i *satisfy the condition* $\left|\frac{zg_i'(z)}{g_i(z)}\right| \leq M$, *where* M *is a positive integer. Suppose* $q = \min(q_1, \ldots, q_n)$ *and* γ_i *are non-zero complex numbers and these numbers satisfy the relation:*

$$\sum_{i=1}^{n}\left\{|\gamma_i|\left(\frac{|c|(6q-|c|)}{3(4q-|c|)(3q-|c|)} + 1\right) + |\gamma_i| M\right\} < 1,$$

then the function $H_{L_i,b,c,\gamma_i,1}$ *defined in the Equation* (14) *is in class* $\mathcal{S}^*$.

Proof. Consider the function:

$$H_{L_i,b,c,\gamma_i,1}(z) = \int_0^z \prod_{i=1}^{n}\left(\frac{v_{L_i,b,c}(t)}{g_i(t)}\right)^{\gamma_i} dt. \tag{40}$$

Hence:

$$1 + \frac{zH''_{L_i,b,c,\gamma_i,1}(z)}{H'_{L_i,b,c,\gamma_i,1}(z)} = \sum_{i=1}^{n}\left[\gamma_i\left(\frac{zv'_{L_i,b,c}(z)}{v_{L_i,b,c}(z)}\right) - \gamma_i\left(\frac{zg_i'(z)}{g_i'(z)}\right)\right] + 1.$$

This implies that:

$$\left|1+\frac{zH''_{L_i,b,c,\gamma_i,1}(z)}{H'_{L_i,b,c,\gamma_i,1}(z)}\right| \leq \sum_{i=1}^{n}|\gamma_i|\left|\frac{zv'_{L_i,b,c}(z)}{v_{L_i,b,c}(z)}\right| + \sum_{i=1}^{n}|\gamma_i|\left|\frac{zg'_i(z)}{g'_i(z)}\right| + 1. \tag{41}$$

Using Lemma 4(i),

$$\left|\frac{zv'_{L,b,c}(z)}{v_{L,b,c}(z)} - 1\right| \leq \frac{|c|(6q-|c|)}{3(4q-|c|)(3q-|c|)}$$

and:

$$\left|\frac{zg'_i(z)}{g_i(z)}\right| \leq M,$$

we have:

$$\left|1+\frac{zH''_{L_i,b,c,\gamma_i,1}(z)}{H'_{L_i,b,c,\gamma_i,1}(z)}\right| \leq \sum_{i=1}^{n}\left\{\gamma_i\left(\frac{|c|(6q_i-|c|)}{3(4q_i-|c|)(3q_i-|c|)}+1\right)+\gamma_i M\right\}+1.$$

Since:

$$\frac{|c|(6q_i-|c|)}{3(4q_i-|c|)(3q_i-|c|)} \leq \frac{|c|(6q-|c|)}{3(4q-|c|)(3q-|c|)},$$

therefore:

$$\left|1+\frac{zH''_{L_i,b,c,\gamma_i,1}(z)}{H'_{L_i,b,c,\gamma_i,1}(z)}\right| \leq \sum_{i=1}^{n}\left\{|\gamma_i|\left(\frac{|c|(6q-|c|)}{3(4q-|c|)(3q-|c|)}+1\right)+|\gamma_i| M\right\}+1.$$

Furthermore,

$$\sum_{i=1}^{n}\left\{|\gamma_i|\left(\frac{|c|(6q-|c|)}{3(4q-|c|)(3q-|c|)}+1\right)+|\gamma_i| M\right\} < 1,$$

implies that:

$$\left|1+\frac{zH''_{L_i,b,c,\gamma_i,1}(z)}{H'_{L_i,b,c,\gamma_i,1}(z)}\right| < 2.$$

By using Lemma 2, the function $H_{L_i,b,c,\gamma_i,1} \in \mathcal{S}^*$. □

Theorem 7. *Let $L_1,\ldots,L_n$, $b \in \mathbb{R}$, $c \in \mathbb{C}$ and $q_i > \frac{7|c|}{24}$ with $q_i = L_i + \frac{b+2}{2}$, $i = 1,\ldots,n$. Let $v_{L_i,b,c} : \mathcal{D} \longrightarrow \mathbb{C}$ be defined in the Equation (8). Let the function g_i satisfy the condition $\left|\frac{zg'_i(z)}{g_i(z)}\right| \leq M$, where M is a positive integer. Suppose $q = min(q_1, q_2, \ldots, q_n)$ and γ_i are non-zero complex numbers and these numbers satisfy the relation:*

$$\sum_{i=1}^{n}\left\{|\gamma_i|\left(\frac{|c|(6q-|c|)}{3(4q-|c|)(3q-|c|)}+1\right)+|\gamma_i| M\right\} < \frac{1}{2},$$

then the function $H_{L_i,b,c,\gamma_i,1} \in \mathcal{UCV}$.

Proof. Consider the function:

$$H_{L_i,b,c,\gamma_i,1}(z) = \int_0^z \prod_{i=1}^{n}\left(\frac{v_{L_i,b,c}(t)}{g_i(t)}\right)^{\gamma_i} dt. \tag{42}$$

This implies that:

$$\frac{zH''_{L_i,b,c,\gamma_i,1}(z)}{H'_{L_i,b,c,\gamma_i,1}(z)} = \sum_{i=1}^{n}\left[\gamma_i\left(\frac{zv'_{L_i,b,c}(z)}{v_{L_i,b,c}(z)}\right) - \gamma_i\left(\frac{zg'_i(z)}{g'_i(z)}\right)\right].$$

Therefore:

$$\left|\frac{zH''_{L_i,b,c,\gamma_i,1}(z)}{H'_{L_i,b,c,\gamma_i,1}(z)}\right| \leq \sum_{i=1}^{n} |\gamma_i| \left|\frac{zv'_{L_i,b,c}(z)}{v_{L_i,b,c}(z)}\right| + \sum_{i=1}^{n} |\gamma_i| \left|\frac{zg'_i(z)}{g'_i(z)}\right|, \tag{43}$$

Using Lemma 4(i):

$$\left|\frac{zv'_{L,b,c}(z)}{v_{L,b,c}(z)} - 1\right| \leq \frac{|c|(6q - |c|)}{3(4q - |c|)(3q - |c|)}$$

and:

$$\left|\frac{zg'_i(z)}{g_i(z)}\right| \leq M,$$

we have:

$$\left|\frac{zH''_{L_i,b,c,\gamma_i,1}(z)}{H'_{L_i,b,c,\gamma_i,1}(z)}\right| \leq \sum_{i=1}^{n} \left\{ |\gamma_i| \left(\frac{|c|(6q_i - |c|)}{3(4q_i - |c|)(3q_i - |c|)} + 1 \right) + |\gamma_i| M \right\}.$$

Since:

$$\frac{|c|(6q_i - |c|)}{3(4q_i - |c|)(3q_i - |c|)} \leq \frac{|c|(6q - |c|)}{3(4q - |c|)(3q - |c|)};$$

therefore:

$$\left|\frac{zH''_{L_i,b,c,\gamma_i,1}(z)}{H'_{L_i,b,c,\gamma_i,1}(z)}\right| \leq \sum_{i=1}^{n} \left\{ |\gamma_i| \left(\frac{|c|(6q - |c|)}{3(4q - |c|)(3q - |c|)} + 1 \right) + |\gamma_i| M \right\}.$$

Using:

$$\sum_{i=1}^{n} \left\{ |\gamma_i| \left(\frac{|c|(6q - |c|)}{3(4q - |c|)(3q - |c|)} + 1 \right) + |\gamma_i| M \right\} < \frac{1}{2},$$

then:

$$\left|\frac{zH''_{L_i,b,c,\gamma_i,1}(z)}{H'_{L_i,b,c,\gamma_i,1}(z)}\right| < \frac{1}{2}.$$

Hence, by using Lemma 3, $H_{L_i,b,c,\gamma_i,1} \in \mathcal{UCV}$. □

Corollary 5. (1) *Consider the function $\mathcal{X}_{L_i}$ defined in the Equation (9). Let $L_1, \ldots, L_n > -1.75$ $(n \in \mathbb{N})$, $L = min\{L_1, L_2, ..., L_n\}$ and the function g_i satisfy the condition $\left|\frac{zg'_i(z)}{g_i(z)}\right| \leq M$, where M is a positive integer. Suppose $q = min(q_1, q_2, ...q_n)$ and γ_i are non-zero complex numbers and these numbers satisfy the inequality:*

$$\sum_{i=1}^{n} \left\{ |\gamma_i| \left(\frac{4(3L+4)}{3(24L^2 + 58L + 35)} + 1 \right) + |\gamma_i| M \right\} < 1,$$

then the function $H_{L_i,b,c,\gamma_i,1} \in \mathcal{S}^*$.

(2) *Consider the function $\mathcal{X}_{L_i}$ defined as the Equation (9). Let $L_1, \ldots, L_n > -1.75$ $(n \in \mathbb{N})$, $L = min\{L_1, L_2, \ldots, L_n\}$ and the function g_i satisfy the condition $\left|\frac{zg'_i(z)}{g_i(z)}\right| \leq M$, where M is a positive integer. Suppose $q = min(q_1, q_2, \ldots, q_n)$ and γ_i are non-zero complex numbers and these numbers satisfy the inequality:*

$$\sum_{i=1}^{n} \left\{ |\gamma_i| \left(\frac{8(3L+4)}{3(24L^2 + 58L + 35)} + 1 \right) + |\gamma_i| M \right\} < \frac{1}{2},$$

then the function $H_{L_i,b,c,\gamma_i,1} \in \mathcal{UCV}$.

Corollary 6. (1) *Consider the function* $\mathcal{Y}_{L_i}$ *defined as the Equation* (10). *Let* $L_1, \ldots, L_n > -1.75$ $(n \in \mathbb{N})$, $L = min\,\{L_1, L_2, ..., L_n\}$ *and the function* g_i *satisfy the condition* $\left|\frac{zg_i'(z)}{g_i(z)}\right| \leq M$, *where M is a positive integer. Suppose* $q = min\,(q_1, q_2, ...q_n)$ *and* γ_i *are non-zero complex numbers and these numbers satisfy the inequality:*

$$\sum_{i=1}^{n}\left\{|\gamma_i|\left(\frac{4\,(3L+4)}{3\,(24L^2+58L+35)}+1\right)+|\gamma_i|\,M\right\}<1,$$

then the function $H_{L_i,b,c,\gamma_i,1} \in \mathcal{S}^*$.

(2) *Consider the function* $\mathcal{Y}_{L_i}$ *defined as the Equation* (10). *Let* $L_1, \ldots, L_n > -1.75$ $(n \in \mathbb{N})$, $L = min\,\{L_1, L_2, \ldots, L_n\}$ *and the function* g_i *satisfy the condition* $\left|\frac{zg_i'(z)}{g_i(z)}\right| \leq M$, *where M is a positive integer. Suppose* $q = min\,(q_1, q_2, \ldots, q_n)$ *and* γ_i *are non-zero complex numbers and these numbers satisfy the inequality:*

$$\sum_{i=1}^{n}\left\{|\gamma_i|\left(\frac{8\,(3L+4)}{3\,(24L^2+58L+35)}+1\right)+|\gamma_i|\,M\right\}<\frac{1}{2},$$

then the function $H_{L_i,b,c,\gamma_i,1} \in \mathcal{UCV}$.

Author Contributions: Conceptualization, H.M.S. and M.R.; Formal analysis, H.M.S. and S.N.M.; Funding acquisition, S.M.; Investigation, M.R. and N.S.; Methodology, M.R. and N.S.; Supervision, S.N.M.; Validation, S.N.M.; Visualization, S.Z.; Writing—original draft, S.Z.; Writing—review & editing, S.Z.

Funding: This work is partially supported by Sarhad University of Science and I.T., Peshawar, Pakistan.

Acknowledgments: The authors are grateful to the referees for their valuable comments, which improved the quality of the work and the presentation of the paper.

Conflicts of Interest: The authors declare no conflict of interest.

References

1. Goodman, A.W. On uniformly convex functions. *Ann. Pol. Math.* **1991**, *56*, 87–92. [CrossRef]
2. Arif, M.; Raza, M. Some properties of an integral operator defined by Bessel function. *Acta Univ. Apul.* **2011**, *26*, 69–74.
3. Baricz, A.; Ponnusamy, S. Starlikeness and convexity of generalized Bessel functions. *Integral Transf. Spec. Funct.* **2010**, *21*, 641–653. [CrossRef]
4. Baricz, A. Some inequalities involving generalized Bessel functions. *Math. Inequal. Appl.* **2007**, *10*, 827–842. [CrossRef]
5. Baricz, A.; Frasin, B.A. Univalence of integral operators involving Bessel functions. *Appl. Math. Lett.* **2010**, *23*, 371–376. [CrossRef]
6. Deniz, E.; Orhan, H.; Srivastava, H.M. Some sufficient conditions for univalence of certain families of integral operators involving generalized Bessel functions. *Taiwan J. Math.* **2011**, *15*, 883–917.
7. Frasin, B.A. Sufficient conditions for integral operators defined by Bessel functions. *J. Math. Inequal.* **2010**, *4*, 301–306. [CrossRef]
8. Zhang, S.; Jin, J.-M. *Computation of Special Functions*; Wiley Interscience Publication: New York, NY, USA, 1996.
9. Struve, H. Beitrag zur Theorie der Diffraction an Fernrohren. *Ann. Phys. Chem.* **1882**, *17*, 1008–1016. [CrossRef]
10. Orhan, H.; Yagmur, N. Geometric properties of generalized Struve function. *Analele ştiinţifice ale Universităţii "Al. I. Cuza" din Iaşi. Matematică* **2017**, *63*, 229–244. [CrossRef]
11. Baricz, A.; Dimitrov, D.K.; Orhan, H.; Yagmur, N. Radii of starlikeness of some special functions. *Proc. Am. Math. Soc.* **2016**, *144*, 3355–3367. [CrossRef]
12. Orhan, H.; Yagmur, N. Starlikeness and convexity of generalized Struve functions. *Abstr. Appl. Anal.* **2013**, *2013*, 954513.
13. Raza, M.; Yagmur, N. Some properties of a class of analytic functions defined by generalized Struve functions. *Turk. J. Math.* **2015**, *39*, 931–944. [CrossRef]

14. Din, M.U.; Srivastava, H.M.; Raza, M. Univalence of certain integral operators involving generalized Struve functions. *Hacet. J. Math. Stat.* **2018**, *47*, 821–833.
15. Hallenbeck, D.J.; Ruscheweyh, S. Subordination by convex functions. *Proc. Am. Math. Soc.* **1975**, *52*, 191–195. [CrossRef]
16. Miller, S.S.; Mocanu, P.T. Univalence of Gaussian and confluent hypergeometric functions. *Proc. Am. Math. Soc.* **1990**, *110*, 333–342 [CrossRef]
17. Ravichandran, V. On uniformly convex functions. *Ganita* **2002**, *53*, 117–124.
18. Pescar, V. A new generalization of Ahfors and Beckers criterion of univalence. *Bull. Malays. Math. Soc.* **1996**, *19*, 53–54.
19. Kudriasov, N.S. Onekotorih priznakah odnolistnosti analiticesschihfunktii. *Mathmaticskie Zametki* **1973**, *13*, 359–366.

© 2019 by the authors. Licensee MDPI, Basel, Switzerland. This article is an open access article distributed under the terms and conditions of the Creative Commons Attribution (CC BY) license (http://creativecommons.org/licenses/by/4.0/).

symmetry

MDPI

Article

Exploiting the Symmetry of Integral Transforms for Featuring Anuran Calls

Amalia Luque [1,*], Jesús Gómez-Bellido [1], Alejandro Carrasco [2] and Julio Barbancho [3]

[1] Ingeniería del Diseño, Escuela Politécnica Superior, Universidad de Sevilla, 41004 Sevilla, Spain; jesgombel@outlook.es

[2] Tecnología Electrónica, Escuela Ingeniería Informática, Universidad de Sevilla, 41004 Sevilla, Spain; acarrasco@us.es

[3] Tecnología Electrónica, Escuela Politécnica Superior, Universidad de Sevilla, 41004 Sevilla, Spain; jbarbancho@us.es

* Correspondence: amalialuque@us.es; Tel.: +34-955-420-187

Received: 23 February 2019; Accepted: 18 March 2019; Published: 20 March 2019

Abstract: The application of machine learning techniques to sound signals requires the previous characterization of said signals. In many cases, their description is made using cepstral coefficients that represent the sound spectra. In this paper, the performance in obtaining cepstral coefficients by two integral transforms, Discrete Fourier Transform (DFT) and Discrete Cosine Transform (DCT), are compared in the context of processing anuran calls. Due to the symmetry of sound spectra, it is shown that DCT clearly outperforms DFT, and decreases the error representing the spectrum by more than 30%. Additionally, it is demonstrated that DCT-based cepstral coefficients are less correlated than their DFT-based counterparts, which leads to a significant advantage for DCT-based cepstral coefficients if these features are later used in classification algorithms. Since the DCT superiority is based on the symmetry of sound spectra and not on any intrinsic advantage of the algorithm, the conclusions of this research can definitely be extrapolated to include any sound signal.

Keywords: spectrum symmetry; DCT; MFCC; audio features; anuran calls

1. Introduction

Automatic processing of sound signals is a very active topic in many fields of science and engineering which find applications in multiple areas, such as speech recognition [1], speaker identification [2,3], emotion recognition [4], music classification [5], outlier detection [6], classification of animal species [7–9], detection of biomedical disease [10], and design of medical devices [11]. Sound processing is also applied in urban and industrial contexts, such as environmental noise control [12], mining [13], and transportation [14,15].

These applications typically include, among their first steps, the characterization of the sound: a process which is commonly known as feature extraction [16]. A recent survey of techniques employed in sound feature extraction can be found in [17], of which Spectrum-Temporal Parameters (STPs) [18], Linear Prediction Coding (LPC) coefficients [19], Linear Frequency Cepstral Coefficients (LFCC) [20], Pseudo Wigner-Ville Transform (PWVT) [21], and entropy coefficients [22] are of note.

Nevertheless, the Mel-Frequency Cepstral Coefficients (MFCC) [23] are probably the most widely employed set of features in sound characterization and the majority of the sound processing applications mentioned above are based on their use. Additionally, these features have also been successfully employed in other fields, such as analysis of electrocardiogram (ECG) signals [24], gait analysis [25,26], and disturbance interpretation in power grids [27].

On the other hand, the processing and classification of anuran calls have attracted the attention of the scientific community for biological studies and as indicators of climate change. This taxonomic group is regarded as an outstanding gauge of biodiversity. Nevertheless, frog populations have suffered a significant decrease in the last years due to habitat loss, climate change and invasive species [28]. So, the continual monitoring of frog populations is becoming increasingly important to develop adequate conservation policies [29].

It should be mentioned that the system of sound production in ectotherms is strongly affected by the ambient temperature. Therefore, the temperature can significantly influence the patterns of calling songs by modifying the beginning, duration, and intensity of calling episodes and, thus, the anuran reproductive activity. The presence or absence of certain anuran calls in a certain territory, and their evolution over time, can therefore be used as an indicator of climate change.

In our previous work, several classifiers for anuran calls are proposed that use non-sequential procedures [30] or temporally-aware algorithms [31], or that consider score series [32], mainly using a set of MPEG-7 features [33]. MPEG-7 is an ISO/IEC standard developed by MPEG (Moving Picture Experts Group). In [34], the comparison of MPEG-7 and MFCC are undertaken both in terms of classification performance and computational cost. Finally, the optimal values of MFCC options for the classification of anuran calls are derived in [35].

State of the art classification of sound relies on Convolutional Neural Networks (CNN) that take input from some form of the spectrogram [36] or even the raw waveform [37]. Moreover, CNN deep learning approaches have also been used in the identification of anuran sound [38]. In spite of that, studying and optimizing the process of extracting MFCC features is of great interest at least for three reasons. First, because sound processing goes beyond the classification task, including procedures such as compression, segmentation, semantic description, sound database retrieval, etc. Secondly, because the spectrograms that feed the state-of-the-art deep CNN classifiers can be constructed using MFCC [39]. And finally due to the fact that CNN classifiers based on spectrograms or raw waveforms require intensive computing resources which makes them unsuitable for implementation in low-cost low-power-consumption distributed nodes, as is the usual case in environmental monitoring networks [35].

As presented in greater detail later, the MFCC features are a representation of the sounds in the cepstral domain. They are derived after a first integral transform (from time to frequency domain), which obtains the sound spectrum, and then a second integral transform is carried out (from frequency to cepstral domain). In this paper, we will show that, by exploiting the symmetry of the sound spectra, it is possible to obtain a more accurate representation of the anuran calls and the derived features will therefore more precisely reflect the sound.

The main contribution of the paper is to offer a better understanding of the reason (symmetry) that justify and quantify why Discrete Cosine Transform (DCT) has been extensively used to compute MFCC. In more detail, the paper will show that DCT-based sound features yielded to a significantly lower error representing spectra, which is a very convenient result for several applications such as sound compression. Additionally, through the paper it will be demonstrated that symmetry-based features (DCT) are less correlated, which is an advantage to be exploited in later classification algorithms.

2. Materials and Methods

2.1. Extracting MFCC

The process of extracting the MFCC features from the n samples of a certain sound requires 7 steps in 3 different domains, which are depicted in Figure 1, and can be summarized as follows:

1. Pre-emphasis (time domain): The sound's high frequencies are increased to compensate for the fact that the Signal-to-Noise Ratio (SNR) is usually lower at these frequencies.
2. Framing (time domain): The n samples of the full-length sound segment are split into frames of short duration (N samples, $N \ll n$). These frames are commonly obtained using non-rectangular

overlapping windows (for instance, Hamming windows [40]). The subsequent steps are executed on the N samples of each frame.

3. Log-energy spectral density (spectral domain): Using the Discrete Fourier Transform (DFT) or its faster version, the Fast Fourier Transform (FFT), the N samples of each frame are converted into the N samples of an energy spectral density, which are usually represented in a log-scale.
4. Mel bank filtering (spectral domain): The N samples of each frame's spectrum are grouped into M banks of frequencies, using M triangular filters centred according to the mel scale [41] and the mel Filter Bank Energy (mel-FBE) is obtained.
5. Integral transform (cepstral domain): The M samples of the mel-FBE (in the spectral domain) are converted into M samples in the cepstral domain using an integral transform. In this article, it will be shown that the exploitation of the symmetry of the DFT integral transform obtained in step 3 yields a cepstral integral transform with a better performance.
6. Reduction of cepstral coefficients (cepstral domain): The M samples of the cepstrum are reduced to C coefficients by discarding the least significant coefficients.
7. Liftering (cepstral domain): The C coefficients of the cepstrum are finally liftered to compensate for the fact that high quefrency coefficients are usually much smaller than their low quefrency counterparts.

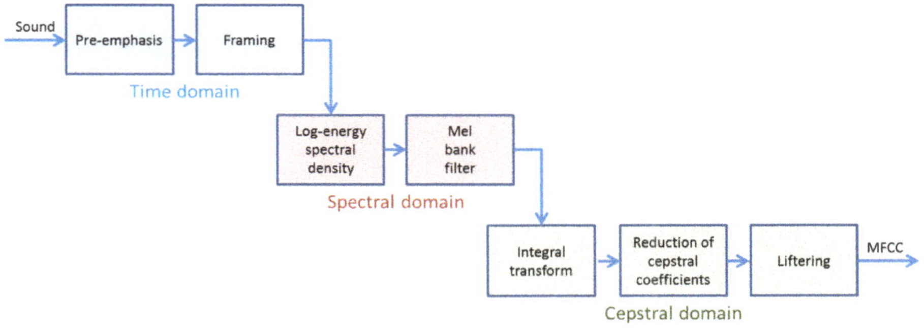

Figure 1. The process of extracting the Mel-Frequency Cepstral Coefficients (MFCC) features from a certain sound.

In this process, integral transforms are used twice: in step 3 to move from the time domain into the spectral domain; and in step 5 to move forward into the cepstral domain. In this paper, the symmetric properties of the DFT integral transform in step 3 will be exploited for the selection of the most appropriate integral transform required in step 5.

2.2. Integral Transforms of Non-Symmetric Functions

As detailed in the previous subsection, a sound spectrum is featured in order to obtain the MFCC of a sound, specifically by characterizing the logarithm of its energy spectral density. In short, this would be a particular case of the characterization of a function $f(x)$ by means of a reduced set of values where, in this case, $f(x)$ is the spectrum of a sound. To address this problem, which is none other than that of the compression of information, several techniques have been proposed, from among which the frequency representation of the function stands out. In effect, the idea underlying this type of technique is to consider the original signal, expand it in Fourier series, and then approximate the function by means of a few terms of its expansion. Thus, instead of having to supply the values of the function corresponding to each value of x, only the amplitude values (and eventually also the phase) of a reduced number of harmonics are provided.

Let us consider an arbitrary example function $f(x)$, such as that shown in Figure 2, of which we know only one fragment in the interval $[x_0, x_0 + P]$ (dashed line). Now let us consider that this function is sampled, and the values only at specific points for $x = x_n$, separated at intervals Δx, are known. By denoting N as the total number of points (samples) in a period, we know that $\Delta x = P/N$. The sampled function will be called $\hat{f}(x_n) = f_n$ where the hat (^) above f represents a sampled function.

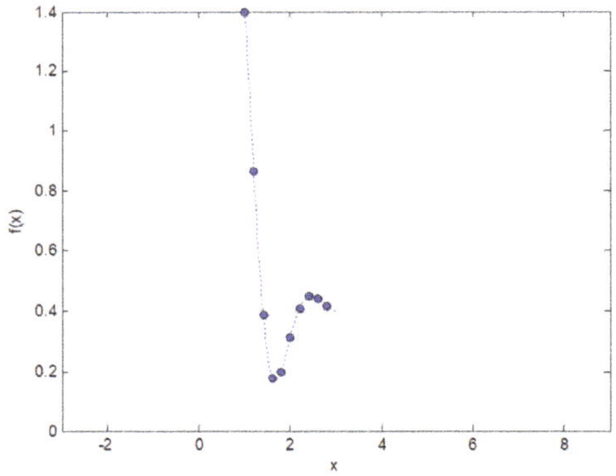

Figure 2. Known fragment of an example function $f(x)$ (dashed line) and its corresponding sampled function $\hat{f}(x_n)$ (dots).

The usual way to obtain the spectrum of that function is to define a periodic function $f_p(x)$ of period P that coincides with the previous function in the known interval (see Figure 3), and to proceed to compute the spectrum of that new function. The spectral representation of the function $f_p(x)$ is composed of the complex coefficients of the Fourier series expansion given by [42].

$$c_k = \frac{1}{P} \int_{x_0}^{x_0+P} f(x) e^{-j\frac{2\pi k x}{P}} dx. \tag{1}$$

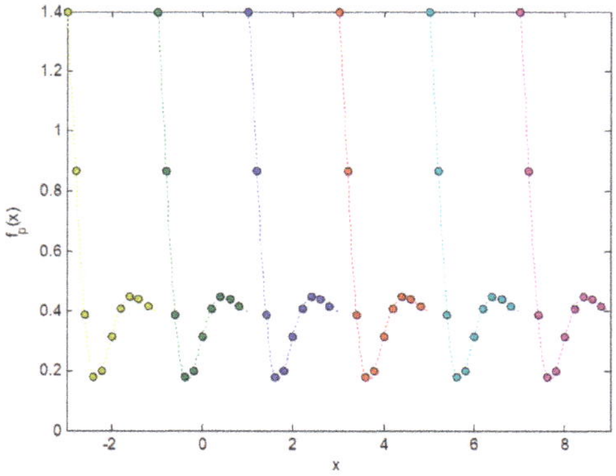

Figure 3. Periodic function $f_p(x)$ obtained by repetition of the known fragment of $f(x)$.

On the other hand, the sampled function, $\hat{f}(x_n) = f_n$, will have a spectral representation $\hat{c}_k$ that corresponds to c_k, when the sampling of the variable x is taken into account. Now let us call $I(x)$ the integrand of Equation (1), i.e.,

$$I(x) = f(x) e^{-j\frac{2\pi k x}{P}}, \tag{2}$$

and hence the spectral representation of the non-sampled function $f_p(x)$ is featured by the coefficients

$$c_k = \frac{1}{P}\int_{x_0}^{x_0+P} I(x)dx. \tag{3}$$

in order to obtain the values $\hat{c}_k$ that take into account the sampling of the variable x, the continuous calculation of the area that supposes the integral of the previous expression is substituted with the sum of the rectangles corresponding to the discrete values (sum of Riemann). In Figure 4, the calculation of the real part of $\hat{c}_1$ is depicted for the example function $f_p(x)$.

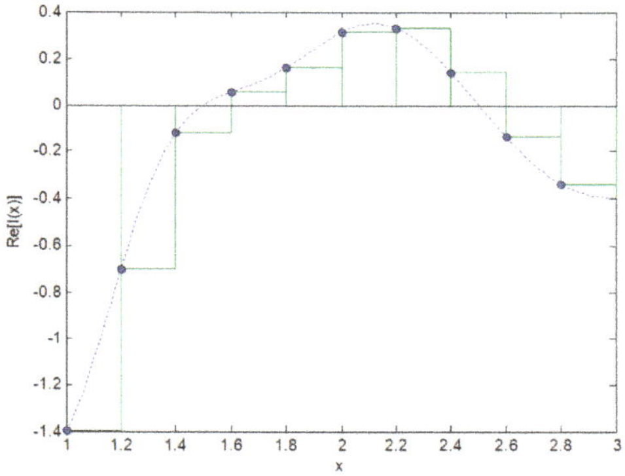

Figure 4. Integration of sampled functions (sum of Riemann).

Therefore,

$$\hat{c}_k \equiv [c_k]_{x=x_n} = \left[\frac{1}{P}\int_{x_0}^{x_0+P} I(x)dx\right]_{x=x_n}. \tag{4}$$

From this equation it can be derived (see supplementary material) that

$$\hat{c}_k = \frac{1}{N}e^{-j\frac{2\pi k x_0}{N\Delta x}}\sum_{n=0}^{N-1} f_n\, e^{-j\frac{2\pi kn}{N}}. \tag{5}$$

It can be observed that the spectral representation $\hat{c}_k$ depends on the point x_0 selected as the origin of coordinates, due to the factor $e^{-j\frac{2\pi k x_0}{N\Delta x}}$. This factor does not affect the amplitude spectrum (since its modulus is 1), but it does affect the phase spectrum corresponding to the known time-shift property of the Fourier Transform. For practical purposes, the origin of coordinates is usually considered to be the starting point of the sequence, that is, at $x_0 = 0$, and hence the spectral representation finally becomes

$$\hat{c}_k = \frac{1}{N}\sum_{n=0}^{N-1} f_n\, e^{-j\frac{2\pi kn}{N}}. \tag{6}$$

This expression coincides with the usual definition of the Discrete Fourier Transform (DFT) [43]. In other words: The Discrete Fourier Transform of a known fragment of a function presupposes the periodic repetition of that fragment.

2.3. Integral Transforms of Symmetric Functions

Let us now again consider the function $f(x)$ of which we know only sampled values of a fragment f_n in the interval $[x_0, x_0 + P]$, as shown in Figure 2. An alternative way of representing its spectrum to that of periodically repeating the values f_n as in Figure 3, lies in defining a sequence of values g_n

of length $2P$ that coincides with f_n in the interval $[x_0, x_0 + P]$, which is its symmetric in the interval $[x_0 - P, x_0]$, as depicted in Figure 5.

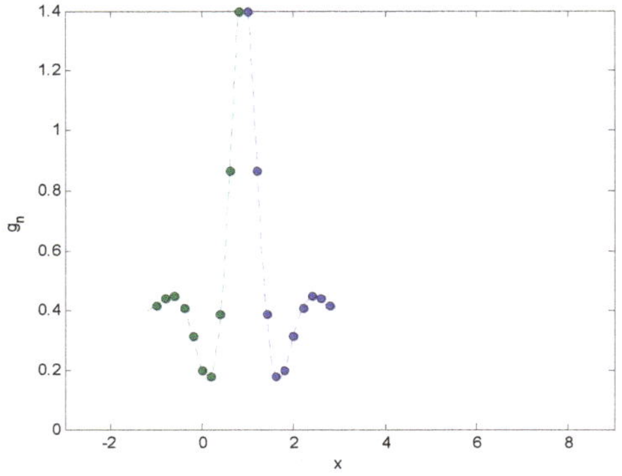

Figure 5. Known fragment of a symmetric example function $g(x)$ (dashed line) and its corresponding sampled function $\hat{g}(x_n)$ (dots). These functions are obtained by considering the original fragment of the example function $f(x)$ (blue) and its symmetric (green).

It can be observed that

$$\begin{array}{c} g_n = f_n \; \forall n \in [0, N-1] \\ g_n = f_{-n-1} \; \forall n \in [-N, -1] \end{array}. \tag{7}$$

Subsequently, a sequence of periodic values h_n of period $P' = 2P$ is defined that coincides with g_n in the interval $[x_0 - P, x_0 + P]$, as shown in Figure 6.

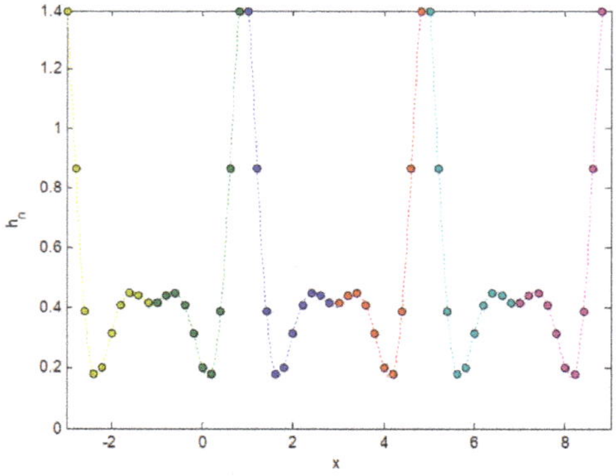

Figure 6. Periodic function h_n obtained by repetition of the known fragment of g_n.

In order to obtain the spectrum of the sequence of values h_n it can be written that

$$\hat{c}_k = \frac{1}{P\prime} \sum_{x_n = x_0 - P}^{x_n = x_0 + P - \Delta x} h_n \, e^{-j\frac{2\pi k x_n}{P\prime}} \, \Delta x. \tag{8}$$

From this equation it can be derived (see supplementary material) that

$$\hat{c}_k = \frac{1}{2N} e^{-j\frac{\pi k x_0}{N \Delta x}} \left[e^{j\frac{\pi k}{N}} \sum_{n=0}^{N-1} f_n \, e^{j\frac{\pi k n}{N}} + \sum_{n=0}^{N-1} f_n \, e^{-j\frac{\pi k n}{N}} \right]. \tag{9}$$

As can be observed, due to the factor $e^{-j\frac{2\pi k x_0}{N\Delta x}}$, the spectral representation $\hat{c}_k$ depends on the point x_0 where the origin of coordinates is defined. This factor does not affect the amplitude spectrum (since its modulus is 1), but it does affect the phase spectrum, which corresponds to the known time-shifting property of the Fourier transform. For practical purposes, the origin of coordinates is usually considered to be located the midpoint of the symmetric sequence g_n, that is, $x_0 = \Delta x/2$, as shown in Figure 7.

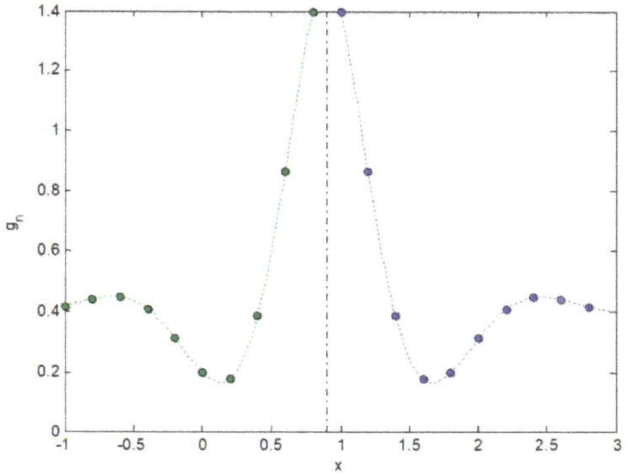

Figure 7. Defining the origin of coordinates.

Finally, the spectral representation becomes (see supplementary material)

$$\hat{c}_k = \frac{1}{N}\sum_{n=0}^{N-1} f_n \cos\left[\frac{\pi k}{N}\left(n+\frac{1}{2}\right)\right]. \tag{10}$$

This expression coincides with the usual definition of the Discrete Cosine Transform (DCT) [44]. In other words, the Discrete Cosine Transform of a known fragment of a function presupposes the periodic repetition of that fragment and its symmetric.

2.4. Representing Anuran Call Spectra

With this digression, we can now address the question posed at the beginning of Section 2.2 concerning the best way to characterize the spectrum of a sound by using the sum of its harmonics. Note that it is necessary to compute the spectrum (step 5) of a spectrum (step 4), that is, the trans-spectrum or the cepstrum, as previously discussed. The decision regarding whether this trans-spectrum (cepstrum) should be derived using either the Fourier transform, or the cosine transform, is based on the form of the fragment f_n (in this case the spectral values of the sound). That is, it should be considered whether the best approximation to the spectrum is either a periodic repetition of f_n or, in contrast, a periodic repetition of f_n and its symmetric.

Although this is a general question, we have addressed it in the context of a specific application by featuring anuran calls for their further classification. The dataset employed contains 1 hour and 13 minutes of sounds which have been recorded at five different locations (four in Spain, and one in Portugal) [32] and they were subsequently sampled at 44.1 kHz. The recordings include 4 types of anuran calls and, since they have been taken in their natural habitat, are affected by highly significant surrounding environmental noise (such as that of wind, water, rain, traffic, and voices).

In this paper, the duration of the frames (step 2) was set to 10 ms, such that each frame has $N = 441$ data points and a total of $W = 434,313$ frames are considered. The log-energy spectral density (step 3) is obtained using a standard FFT algorithm, which obtains a spectrum with $N = 441$ values. The mel-scaling (step 4) employs a set of $M = 23$ filters, and hence the mel-FBE spectrum is characterised by this number of values ($M = 23$). In step 5, two different approaches for obtaining the cepstrum are used and compared: DFT and DCT. The results are then analysed for a different number of cepstral coefficients ($1 \leq C \leq M$).

In order to carry out a more systematic study of the spectrum approximation error, let us call $E_i(n)$ the original mel-FBE spectrum of the i-th frame (the result of step 4), where n is the filter index (equivalent to the frequency in mel scale). Let us also call $H_i(m)$ the spectrum of $E_i(n)$, that is, the cepstrum as obtained in step 5, where m is the cepstral index (equivalent to the quefrency in mel scale). It can be written that $H_i(m) = \mathcal{F}[E_i(n)]$, where $\mathcal{F}$ represents either the DFT or the DCT Fourier expansions.

After reducing the number of cepstral coefficients to a value of $C \leq M$, the resulting approximate cepstrum (step 6) will be called $\widetilde{H}_i(m)$, where the tilde ($\widetilde{\ }$) above the H represents an approximation. Using these C values in the corresponding Fourier expansion leads to an approximation of the mel-FBE, that is, $\widetilde{E}_i(n) = \mathcal{F}^{-1}\left[\widetilde{H}_i(m)\right]$. The approximation error for the i-th frame is therefore $\varepsilon_i(n) = E_i(n) - \widetilde{E}_i(n)$, that is, a different error for each value of n, the filter index (or frequency in mel-scale). An error measure for the overall spectrum of the i-th frame can be obtained using the Root Mean Square Error ($RMSE_i$) defined as:

$$RMSE_i \equiv \sqrt{\frac{1}{M}\sum_{n=0}^{M-1}[\varepsilon_i(n)]^2} = \sqrt{\frac{1}{M}\sum_{n=0}^{M-1}\left[E_i(n) - \widetilde{E}_i(n)\right]^2}. \tag{11}$$

In this paper, an arbitrary selected single frame is first considered, mainly for illustration purposes. Its time-domain representation is depicted in Figure 8A while its spectrum is plotted in Figure 8B. Some other examples can be found in [32].

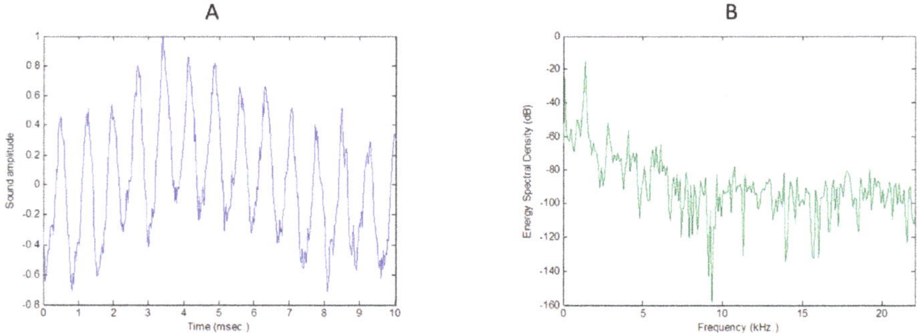

Figure 8. Sound amplitude for an arbitrarily selected frame of an anuran call (**A**); and its log-scale Energy Spectral Density (**B**).

Additionally, in order to compare the performance of the 2 competing algorithms obtaining the cepstrum, an overall metric for the whole dataset is considered and defined as the mean RMSE for every frame, that is,

$$RMSE \equiv \frac{1}{W}\sum_{i=1}^{W} RMSE_i = \frac{1}{W}\sum_{i=1}^{W}\sqrt{\frac{1}{M}\sum_{n=0}^{M-1}[\varepsilon_i(n)]^2}. \tag{12}$$

3. Results

Let us first consider a single frame, arbitrarily selected from the whole sound dataset. Although these results are limited to that specific sound frame, very similar results are obtained if a different frame is selected. Moreover, at the end of this section, the overall sound dataset is considered.

For the case of the single frame, the mel-FBE spectrum obtained in step 4 is depicted in Figure 9. This is the $f(x)$ function whose spectrum (cepstrum in this case) must be computed in step 5.

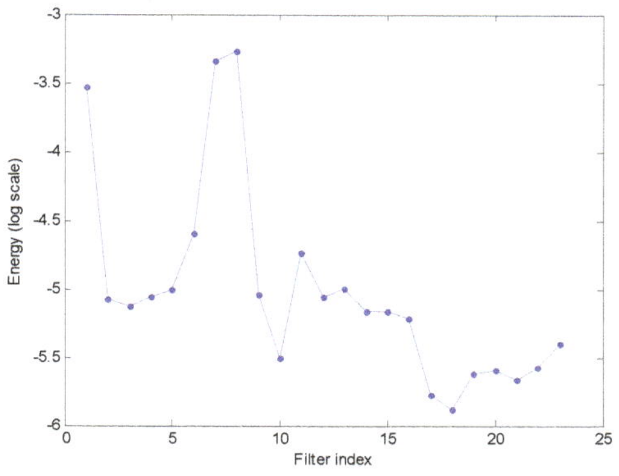

Figure 9. Mel Filter Bank Energy (mel-FBE) spectrum for an arbitrarily selected frame of an anuran call.

For this frame, let us consider whether it is better to use either a DFT or a DCT. The decision depends on whether the function $f(x)$ can be considered as a fragment of a periodic repetition of: (A) the fragment, as shown in Figure 10A, or (B) the function and its symmetric, as shown in Figure 10B. In the first case, the DFT should be more appropriate, while in the second case the DCT would obtain better results.

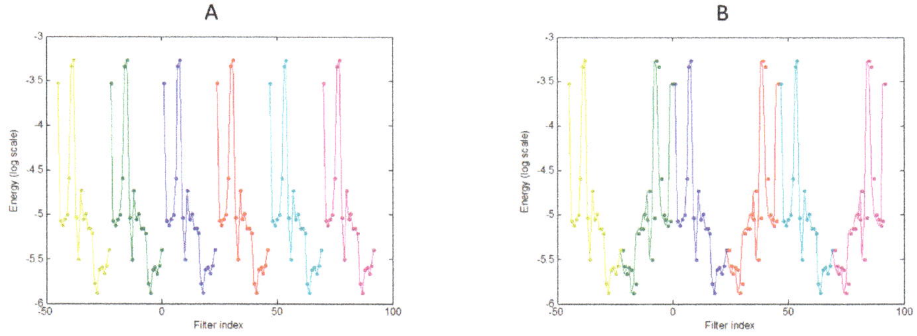

Figure 10. Periodic repetition of the mel-FBE spectrum (**A**); and the mel-FBE spectrum and its symmetric (**B**).

However, the mel-FBE is nothing but a rescaled and compressed way of presenting a spectrum. On the other hand, it is a well-known fact that the spectrum of a real signal is symmetric with respect to the vertical axis [43]. And finally, it is also known that the spectrum of a sampled signal is periodic [45]. For this reason, the repetition of the fragment of Figure 9 corresponds to Figure 10B and, therefore, using the DCT to compute its trans-spectrum (or cepstrum) should obtain better results. This hypothesis is verified in the following paragraphs for the selected frame, and, later in this section, it is verified for the whole dataset.

The number of coefficients obtained by applying either DCT or DFT is $M = 23$, that is, they have the same number of values that define the mel-FBE. The resulting cepstrum for the selected frame is shown in Figure 11.

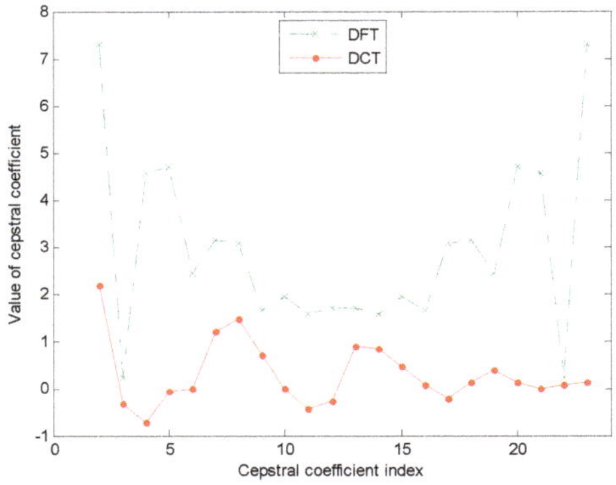

Figure 11. Cepstral representation of the mel-FBE spectrum (cepstrum).

The ability to compress information of the Fourier transforms (either in the DFT or DCT version) lies in the fact that it is not necessary to consider the full set of the M coefficients of the Fourier expansion to obtain a good approximation of the original function. In Figure 12, the original mel-FBE spectrum is depicted for the example frame, and those spectra recovered using $C \leq M$ cepstral coefficients obtained using DCT.

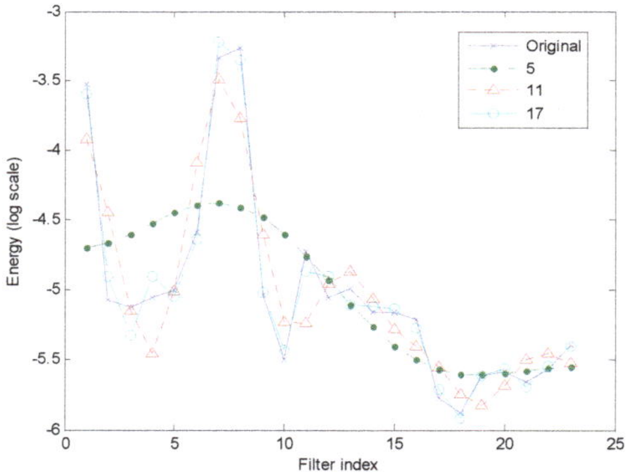

Figure 12. Mel-FBE spectrum for an arbitrarily selected frame of an anuran call. Original spectrum and recovered spectra using a different number of Discrete Cosine Transform (DCT) cepstral coefficients.

Additionally, as expected, the DCT achieves approximations to the original spectrum that are, in general, significantly better than those obtained for the DFT with the same number of coefficients. In Figure 13, the original mel-FBE spectrum is depicted for the example frame, and those spectra recovered using $C = 11$ cepstral coefficients obtained using DFT and DCT.

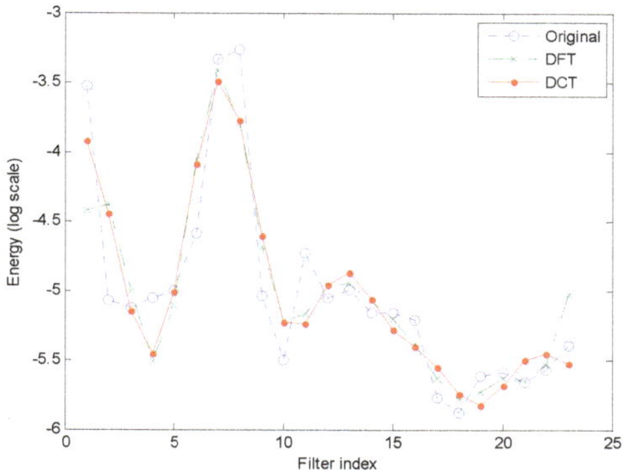

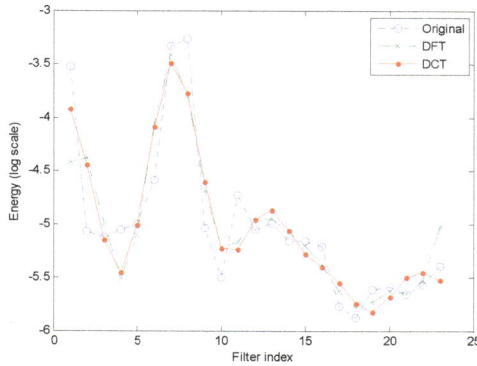

Figure 13. Mel-FBE spectrum for an arbitrarily selected frame of an anuran call. Original spectrum and recovered spectrum using $C = 11$ coefficients obtained using Discrete Fourier Transform (DFT) and DCT.

In order to quantify the error of recovering the selected mel-FBE spectrum using $C \leq M$ cepstral coefficients, the Root Mean Square Error (RMSE) is computed in accordance with Equation (11). The value of RMSE as a function of the number C of cepstral coefficients used for the recovery of the spectrum is depicted in Figure 14, both for DFT and DCT.

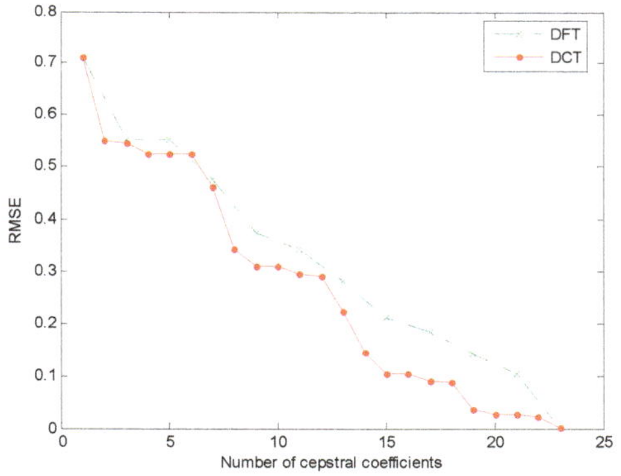

Figure 14. Root Mean Square Error recovering the original mel-FBE spectrum when a different number of C cepstral coefficients are used. The cepstral coefficients are obtained applying either DFT or DCT.

This analysis can be extended to include the computation of the RMSE for the whole dataset in accordance with Equation (12). The value of RMSE as a function of the number C of cepstral coefficients used for the recovery of the spectrum is depicted in Figure 15 for DFT and DCT separately.

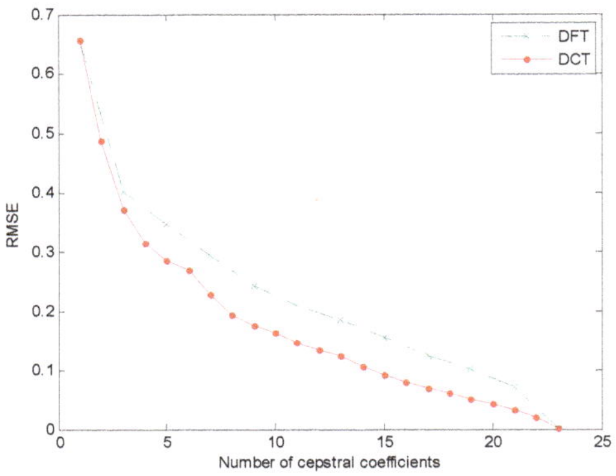

Figure 15. Root Mean Square Error for the whole dataset when either DFT or DCT is employed.

4. Discussion

Let us first consider the $RMSE_i$ for a single frame as depicted in Figure 14. Let us now regard the case where, for instance, the number of values required to describe the mel-FBE spectrum ($M = 23$) is halved, and hence the number of cepstral coefficients used for the recovering an approximation of the spectrum is $C = 11$ (in accordance with Equations (6) and (10)).

In this case, it can be observed that $RMSE_i$ is 0.34 for DFT, and 0.30 for DCT. On the other hand, as depicted in Figure 9, the values of the mel-FBE spectrum lie within the range $[-6, -3]$, with a mean value of -5.02. This means that the relative error of the spectrum representation is only 6.84% for DFT (5.36% for DCT) when the number of values employed for that representation are halved.

Let us now focus on the RMSE when the DFT is used (green line), either for a single frame (Figure 14) or for the whole dataset (Figure 15). In both cases, it can be observed that RMSE has values only for an odd number of cepstral coefficients. This fact can be explained by recalling that, according to Equation (6), every DFT cepstral coefficient $\hat{c}_k$ is a complex number for $1 \leq k \leq M-1$ and a real number for $k = 0$. On the other hand, according to Equation (10), the DCT cepstral coefficients $\hat{c}_k$ are real numbers for every value of k. Additionally, it has to be considered that DFT cepstrum is symmetric (green line in Figure 11). Therefore, for $k > 0$, it can be written that $\hat{c}_k = \hat{c}_{M-k+1}$ and, therefore, only one of these 2 terms have to be kept for recovery purposes. These circumstances jointly explain the odd number of DFT cepstral coefficients.

To clarify this idea, let us consider an example where $M = 23$ and $C = 5$. The DCT cepstrum is then described using $\hat{c}_0$, $\hat{c}_1$, $\hat{c}_2$, $\hat{c}_3$ and $\hat{c}_4$, that is, 5 real numbers which can be employed to approximately recover the mel-FBE spectrum. On the other hand, the DFT cepstrum is described using $\hat{c}_0$, which is a real number, and $\hat{c}_1$ and $\hat{c}_2$, which are complex numbers, that is, although 3 terms are used, a total of 5 values (coefficients) are required. However, to approximately recover the mel-FBE spectrum, the terms $\hat{c}_0$, $\hat{c}_1$, $\hat{c}_2$, $\hat{c}_{23}$ and $\hat{c}_{22}$ can be used since $\hat{c}_1 = \hat{c}_{23}$ and $\hat{c}_2 = \hat{c}_{22}$.

As regards the results obtained for the whole dataset (Figure 15), it can be seen that DCT is better at describing the mel-FBE spectra than is its DFT counterpart. This improvement (decrease of the RMSE), can be measured by defining $\Delta RMSE \equiv RMSE_{DFT} - RMSE_{DCT}$ (Figure 16A) or its relative value $\Delta RMSE(\%) \equiv 100 \cdot \Delta RMSE / RMSE_{DFT}$ (Figure 16B). For example, for $C = 11$, the RMSE is reduced from 0.209 (DFT) to 0.146, which involves an improvement of approximately 30%. For the degenerated cases where $C = 1$ and $C = M$, there is no improvement. In the first case, only $\hat{c}_0$ is used which, according to Equations (6) and (10), is the mean value of the mel-FBE spectrum, that is, the DFT and DCT recovering methods have the same error. On the other hand, if $C = M$ then no reduction on the number of coefficients is achieved, and both equations exactly recover the original spectrum (no error).

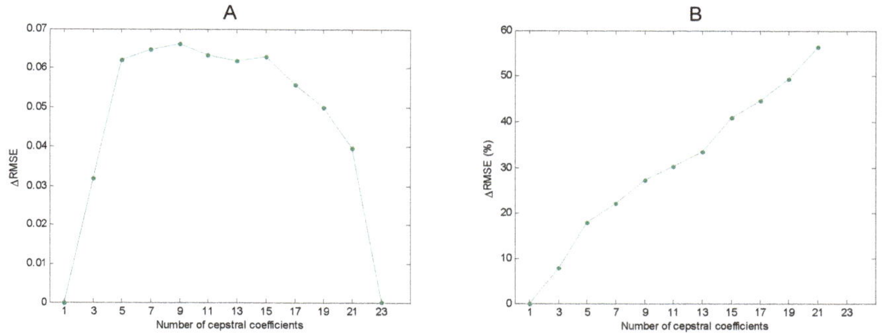

Figure 16. Improvement of DCT over DFT describing mel-FBE spectra. (**A**): $\Delta RMSE$. (**B**): $\Delta RMSE(\%)$.

The above results concern the mean improvement of DCT over DFT for every frame in the dataset. In a more in-depth analysis, let us also compute its probability density function (pdf). The results are depicted in Figure 17. In panel A, the pdf is shown for several values of the number of cepstral coefficients (C). In panel B, the value of the pdf is colour-coded as a function of the improvement ($\Delta Error$) and of the number of cepstral coefficients (C). It can be observed that only a negligible number of the frames present a significant negative improvement, thereby demonstrating that DCT is superior to DFT.

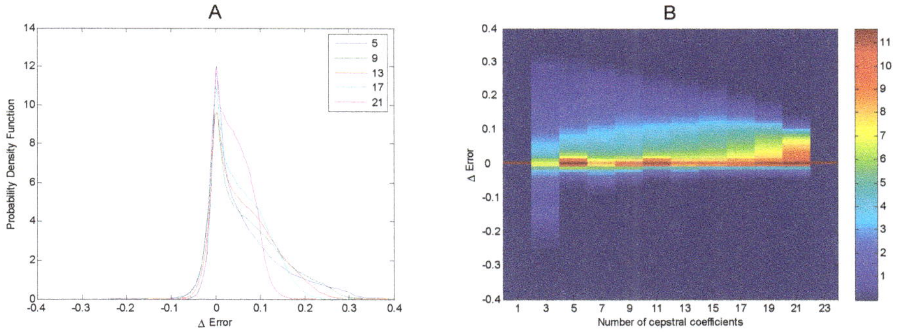

Figure 17. Improvement of DCT over DFT in describing mel-FBE spectra. (**A**): Probability density function for several values of the number of cepstral coefficients. (**B**): Probability density function for each value of the number of cepstral coefficients.

The higher performance of DCT over DFT is due to the fact that the mel-FBE spectra are a special type of function derived from symmetric sound spectra. Consequently, if DCT and DFT were compared in the task of recovering arbitrary functions, they would each present equal performance. To demonstrate this claim, one million M-value arbitrary functions are randomly generated ($M = 23$), and DFT and DCT are then employed to recover the original function with a reduced set of C coefficients to measure the errors of that recovery. Finally, the improvement of DCT over DFT is computed. The results are depicted in Figure 18 where it can be observed that positive and negative improvements are symmetrically distributed around a zero-mean improvement. Therefore, it can be concluded that DCT and DFT have similar performance in describing arbitrary functions.

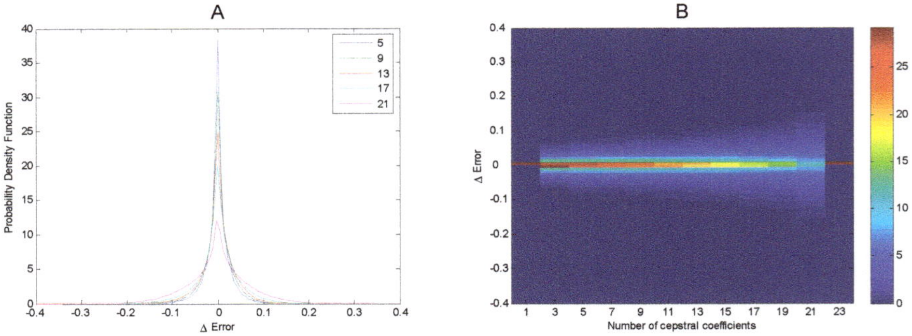

Figure 18. Improvement of DCT over DFT in describing arbitrary function. (**A**): Probability density function for several values of the number of cepstral coefficients. (**B**): Probability density function for each value of the number of cepstral coefficients.

From the above results, it is clear that DCT offers superior performance featuring mel-FBE spectra and, therefore offers superior performance featuring sounds. When the purpose of these features is to be used as input to some kind of classifier, then DCT offers an additional advantage. It is a well-established result that classifiers obtain better results if their input features are low-correlated. The reason is clear: a classification algorithm that includes a new feature that is highly correlated with previous features adds almost no new information and, therefore, almost no classification improvement should be expected. Let us therefore examine the correlation between coefficients obtained by DFT and those by DCT.

Let us call μ_u the mean value of the u-th coefficient $\hat{c}_{ui}$ describing the i-th frame, obtained by

$$\mu_u = \frac{1}{W}\sum_{i=1}^{W}\hat{c}_{ui}, \tag{13}$$

where W is the total number of frames in the dataset. The variance σ_u^2 of the u-th coefficient can be obtained by

$$\sigma_u^2 = \frac{1}{W-1}\sum_{i=1}^{W}(\hat{c}_{ui} - \mu_u)^2. \tag{14}$$

The correlation ρ_{uv} between the u-th and the v-th coefficient for the whole dataset is therefore given by

$$\rho_{uv} = \frac{1}{W-1}\sum_{i=1}^{W}\frac{\hat{c}_{ui} - \mu_u}{\sigma_u}\cdot\frac{\hat{c}_{vi} - \mu_v}{\sigma_v}. \tag{15}$$

In Figure 19, the absolute values of the correlation are shown, whereby the values for the case $M = 23$ are colour-coded. The correlations corresponding to the DFT are shown in panel A and those corresponding to DCT in panel B. In the DFT case, each $\hat{c}_{ui}$ factor is a complex number, and hence the total number of values is 46, whereby the first 23 coefficients represent the real parts and the last 23 the imaginary parts. By simply considering the colours in that figure, it is clear that DCT coefficients are less correlated.

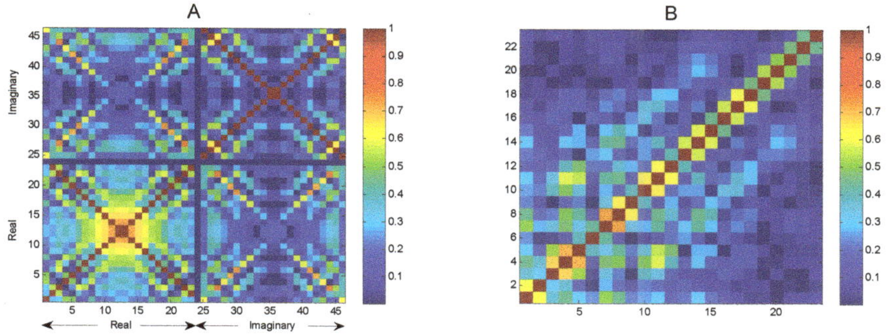

Figure 19. Correlation between cepstral coefficients describing mel-FBE spectra for DFT (panel **A**) and DCT (panel **B**).

An alternative way to present this result is by using a histogram of the values of the correlation coefficients, as depicted in Figure 20. Those corresponding to DCT are more frequent for the low values of correlation, that is, DCT-obtained features are less correlated than those obtained using DFT. Hence, classifiers of a more efficient nature should be expected from using DCT.

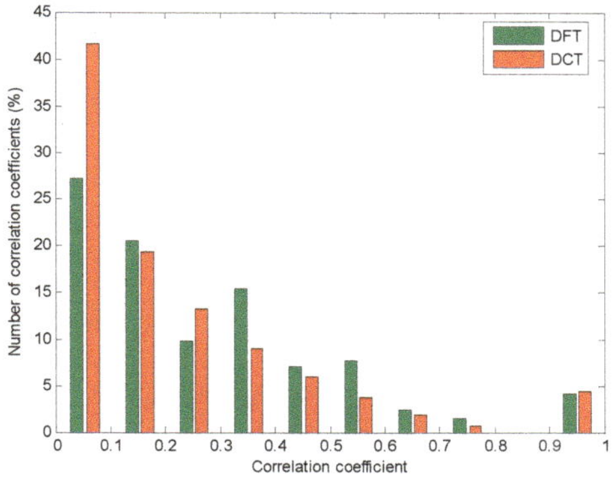

Figure 20. Histogram of the correlation among cepstral coefficients describing mel-FBE spectra for DFT and DCT.

When the MFCC features are used as input of a later classification algorithm, the lower correlation of DCT-obtained features should yield to a better classification performance. The results obtained classifying anuran calls [35] do confirm a slight advantage for the DCT as it is reflected in Table 1. This table has been produced taking the best result (geometric mean of sensitivity and specificity) obtained through a set of ten classification procedures: minimum distance, maximum likelihood, decision trees, k-nearest neighbors, support vector machine, logistic regression, neural networks, discriminant function, Bayesian classifiers and hidden Markov models.

Table 1. Classification performance metrics for DCT and DFT.

Cepstral Transform	ACC	PRC	F_1
DFT	94.27%	74.46%	77.67%
DCT	94.85%	76.76%	78.93%

Let us finally consider the computing efforts required for these two algorithms which mainly depend on the number of samples defining the mel-FBE spectra. Fast versions of DFT and DCT algorithms have been tested on a conventional desktop personal computer. The results are depicted in Figure 21. It can be seen that DCT is about one order of magnitude slower than DFT. Although this fact is certainly a drawback of DCT it has a limited impact on conventional MFCC extraction process because the number of values describing the mel-FBE spectra is usually very low (about 20). Additional studies on processing times for anuran sounds classification can be found in [34].

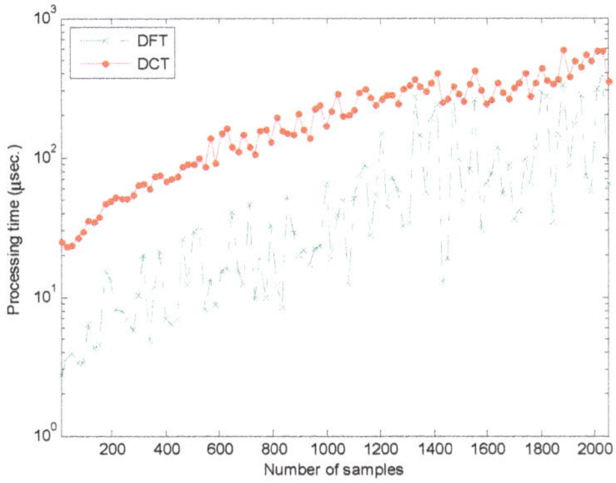

Figure 21. Processing time required to compute the DFT and DCT vs. the number of samples describing mel-FBE spectra.

5. Conclusions

In this article, it has been shown that DCT outperforms DFT in the task of representing sound spectra. It has also been shown that this improvement is due to the symmetry of the spectrum and not to any intrinsic advantage of DCT.

In representing the mel-FBE spectra required to obtain the MFCC features of anuran calls, DCT errors are approximately 30% lower than DFT errors. This type of spectra is therefore much better represented using DCT.

Additionally, it has been shown than MFCC features obtained using DCT are remarkably less correlated than those obtained using DFT. This result will make DCT-based MFCC features more powerful in later classification algorithms.

Although only one specific dataset has been analysed herein, the advantage of DCT can easily be extrapolated to include any sound since this advantage is based on the symmetry of the spectrum of the sound

Supplementary Materials: The following are available online at http://www.mdpi.com/2073-8994/11/3/405/s1, supplementary material: Derivation of integral transforms expressions.

Author Contributions: Conceptualization, A.L.; investigation, A.L., J.G.-B., A.C. and J.B.; writing—original draft, A.L., J.G.-B., A.C. and J.B.

Funding: This research received no external funding.

Acknowledgments: The authors would like to thank Rafael Ignacio Marquez Martinez de Orense (Museo Nacional de Ciencias Naturales) and Juan Francisco Beltrán Gala (Faculty of Biology, University of Seville) for their collaboration and support.

Conflicts of Interest: The authors declare there to be no conflict of interest.

References

1. Haridas, A.V.; Marimuthu, R.; Sivakumar, V.G. A critical review and analysis on techniques of speech recognition: The road ahead. *Int. J. Knowl.-Based Intell. Eng. Syst.* **2018**, *22*, 39–57. [CrossRef]
2. Gómez-García, J.A.; Moro-Velázquez, L.; Godino-Llorente, J.I. On the design of automatic voice condition analysis systems. Part II: Review of speaker recognition techniques and study on the effects of different variability factors. *Biomed. Signal Process. Control* **2019**, *48*, 128–143. [CrossRef]
3. Vo, T.; Nguyen, T.; Le, C. Race Recognition Using Deep Convolutional Neural Networks. *Symmetry* **2018**, *10*, 564. [CrossRef]
4. Dahake, P.P.; Shaw, K.; Malathi, P. Speaker dependent speech emotion recognition using MFCC and Support Vector Machine. In Proceedings of the 2016 International Conference on Automatic Control and Dynamic Optimization Techniques (ICACDOT), Pune, India, 9–10 September 2016; pp. 1080–1084.
5. Chakraborty, S.S.; Parekh, R. Improved Musical Instrument Classification Using Cepstral Coefficients and Neural Networks. In *Methodologies and Application Issues of Contemporary Computing Framework*; Springer: Singapore, 2018; pp. 123–138.
6. Panteli, M.; Benetos, E.; Dixon, S. A computational study on outliers in world music. *PLoS ONE* **2017**, *12*, e0189399. [CrossRef] [PubMed]
7. Noda, J.J.; Sánchez-Rodríguez, D.; Travieso-González, C.M. A Methodology Based on Bioacoustic Information for Automatic Identification of Reptiles and Anurans. In *Reptiles and Amphibians*; IntechOpen: London, UK, 2018.
8. Desai, N.P.; Lehman, C.; Munson, B.; Wilson, M. Supervised and unsupervised machine learning approaches to classifying chimpanzee vocalizations. *J. Acoust. Soc. Am.* **2018**, *143*, 1786. [CrossRef]
9. Malfante, M.; Mars, J.I.; Dalla Mura, M.; Gervaise, C. Automatic fish sounds classification. *J. Acoust. Soc. Am.* **2018**, *143*, 2834–2846. [CrossRef]
10. Wang, Y.; Sun, B.; Yang, X.; Meng, Q. Heart sound identification based on MFCC and short-term energy. In Proceedings of the 2017 Chinese Automation Congress (CAC), Jinan, China, 20–22 October 2017; pp. 7411–7415.
11. Usman, M.; Zubair, M.; Shiblee, M.; Rodrigues, P.; Jaffar, S. Probabilistic Modeling of Speech in Spectral Domain using Maximum Likelihood Estimation. *Symmetry* **2018**, *10*, 750. [CrossRef]
12. Cao, J.; Cao, M.; Wang, J.; Yin, C.; Wang, D.; Vidal, P.P. Urban noise recognition with convolutional neural network. *Multimed. Tools Appl.* **2018**. [CrossRef]
13. Xu, J.; Wang, Z.; Tan, C.; Lu, D.; Wu, B.; Su, Z.; Tang, Y. Cutting Pattern Identification for Coal Mining Shearer through Sound Signals Based on a Convolutional Neural Network. *Symmetry* **2018**, *10*, 736. [CrossRef]
14. Lee, J.; Choi, H.; Park, D.; Chung, Y.; Kim, H.Y.; Yoon, S. Fault detection and diagnosis of railway point machines by sound analysis. *Sensors* **2016**, *16*, 549. [CrossRef]
15. Choi, Y.; Atif, O.; Lee, J.; Park, D.; Chung, Y. Noise-Robust Sound-Event Classification System with Texture Analysis. *Symmetry* **2018**, *10*, 402. [CrossRef]
16. Guyon, I.; Elisseeff, A. An introduction to feature extraction. In *Feature Extraction*; Springer: Berlin/Heidelberg, Germany, 2006; pp. 1–25.
17. Alías, F.; Socoró, J.; Sevillano, X. A review of physical and perceptual feature extraction techniques for speech, music and environmental sounds. *Appl. Sci.* **2016**, *6*, 143. [CrossRef]
18. Zhang, H.; McLoughlin, I.; Song, Y. Robust sound event recognition using convolutional neural networks. In Proceedings of the 2015 IEEE International Conference on Acoustics, Speech and Signal Processing (ICASSP), Brisbane, Australia, 19–24 April 2015; pp. 559–563.
19. Dave, N. Feature extraction methods LPC, PLP and MFCC in speech recognition. *Int. J. Adv. Res. Eng. Technol.* **2013**, *1*, 1–4.
20. Paul, D.; Pal, M.; Saha, G. Spectral features for synthetic speech detection. *IEEE J. Sel. Top. Signal Process.* **2017**, *11*, 605–617. [CrossRef]
21. Taebi, A.; Mansy, H.A. Analysis of seismocardiographic signals using polynomial chirplet transform and smoothed pseudo Wigner-Ville distribution. In Proceedings of the 2017 IEEE Signal Processing in Medicine and Biology Symposium (SPMB), Philadelphia, PA, USA, 2 December 2017; pp. 1–6.

22. Dayou, J.; Han, N.C.; Mun, H.C.; Ahmad, A.H.; Muniandy, S.V.; Dalimin, M.N. Classification and identification of frog sound based on entropy approach. In Proceedings of the 2011 International Conference on Life Science and Technology, Mumbai, India, 7–9 January 2011; Volume 3, pp. 184–187.
23. Zheng, F.; Zhang, G.; Song, Z. Comparison of different implementations of MFCC. *J. Comput. Sci. Technol.* **2001**, *16*, 582–589. [CrossRef]
24. Hussain, H.; Ting, C.M.; Numan, F.; Ibrahim, M.N.; Izan, N.F.; Mohammad, M.M.; Sh-Hussain, H. Analysis of ECG biosignal recognition for client identifiction. In Proceedings of the 2017 IEEE International Conference on Signal and Image Processing Applications (ICSIPA), Kuching, Malaysia, 12–14 September 2017; pp. 15–20.
25. Nickel, C.; Brandt, H.; Busch, C. Classification of Acceleration Data for Biometric Gait Recognition on Mobile Devices. *Biosig* **2011**, *11*, 57–66.
26. Muheidat, F.; Tyrer, W.H.; Popescu, M. Walk Identification using a smart carpet and Mel-Frequency Cepstral Coefficient (MFCC) features. In Proceedings of the 2018 40th Annual International Conference of the IEEE Engineering in Medicine and Biology Society (EMBC), Honolulu, HI, USA, 18–21 July 2018; pp. 4249–4252.
27. Negi, S.S.; Kishor, N.; Negi, R.; Uhlen, K. Event signal characterization for disturbance interpretation in power grid. In Proceedings of the 2018 First International Colloquium on Smart Grid Metrology (SmaGriMet), Split, Croatia, 24–27 April 2018; pp. 1–5.
28. Xie, J.; Towsey, M.; Zhang, J.; Roe, P. Frog call classification: A survey. *Artif. Int. Rev.* **2018**, *49*, 375–391. [CrossRef]
29. Colonna, J.G.; Nakamura, E.F.; Rosso, O.A. Feature evaluation for unsupervised bioacoustic signal segmentation of anuran calls. *Expert Syst. Appl.* **2018**, *106*, 107–120. [CrossRef]
30. Luque, A.; Romero-Lemos, J.; Carrasco, A.; Barbancho, J. Non-sequential automatic classification of anuran sounds for the estimation of climate-change indicators. *Expert Syst. Appl.* **2018**, *95*, 248–260. [CrossRef]
31. Luque, A.; Romero-Lemos, J.; Carrasco, A.; Gonzalez-Abril, L. Temporally-aware algorithms for the classification of anuran sounds. *PeerJ* **2018**, *6*, e4732. [CrossRef]
32. Luque, A.; Romero-Lemos, J.; Carrasco, A.; Barbancho, J. Improving Classification Algorithms by Considering Score Series in Wireless Acoustic Sensor Networks. *Sensors* **2018**, *18*, 2465. [CrossRef] [PubMed]
33. Romero, J.; Luque, A.; Carrasco, A. Anuran sound classification using MPEG-7 frame descriptors. In Proceedings of the XVII Conferencia de la Asociación Española para la Inteligencia Artificial (CAEPIA), Salamanca, Spain, 14–16 September 2016; pp. 801–810.
34. Luque, A.; Gómez-Bellido, J.; Carrasco, A.; Personal, E.; Leon, C. Evaluation of the processing times in anuran sound classification. *Wireless Communications and Mobile Computing* **2017**. [CrossRef]
35. Luque, A.; Gómez-Bellido, J.; Carrasco, A.; Barbancho, J. Optimal Representation of Anuran Call Spectrum in Environmental Monitoring Systems Using Wireless Sensor Networks. *Sensors* **2018**, *18*, 1803. [CrossRef] [PubMed]
36. Hershey, S.; Chaudhuri, S.; Ellis, D.P.; Gemmeke, J.F.; Jansen, A.; Moore, R.C.; Plakal, M.; Platt, D.; Saurous, R.A.; Seybold, B.; et al. CNN architectures for large-scale audio classification. In Proceedings of the 2017 IEEE International Conference on Acoustics, Speech and Signal Processing (ICASSP), New Orleans, LA, USA, 5–9 March 2017; pp. 131–135.
37. Dai, W.; Dai, C.; Qu, S.; Li, J.; Das, S. Very deep convolutional neural networks for raw waveforms. In Proceedings of the 2017 IEEE International Conference on Acoustics, Speech and Signal Processing (ICASSP), New Orleans, LA, USA, 5–9 March 2017; pp. 421–425.
38. Strout, J.; Rogan, B.; Seyednezhad, S.M.; Smart, K.; Bush, M.; Ribeiro, E. Anuran call classification with deep learning. In Proceedings of the 2017 IEEE International Conference on Acoustics, Speech and Signal Processing (ICASSP), New Orleans, LA, USA, 5–9 March 2017; pp. 2662–2665.
39. Colonna, J.; Peet, T.; Ferreira, C.A.; Jorge, A.M.; Gomes, E.F.; Gama, J. Automatic classification of anuran sounds using convolutional neural networks. In Proceedings of the Ninth International Conference on Computer Science & Software Engineering, Porto, Portugal, 20–22 July 2016; pp. 73–78.
40. Podder, P.; Khan, T.Z.; Khan, M.H.; Rahman, M.M. Comparative performance analysis of hamming, hanning and blackman window. *Int. J. Comput. Appl.* **2014**, *96*, 1–7. [CrossRef]
41. O'shaughnessy, D. *Speech Communication: Human and Machine*, 2nd ed.; Wiley-IEEE Press: Hoboken, NJ, USA, 1999; ISBN 978-0-7803-3449-6.
42. Bhatia, R. *Fourier Series*; American Mathematical Society: Providence, RI, USA, 2005.

43. Broughton, S.A.; Bryan, K. *Discrete Fourier Analysis and Wavelets: Applications to Signal and Image Processing*; John Wiley & Sons: Hoboken, NJ, USA, 2018.
44. Rao, K.R.; Yip, P. *Discrete Cosine Transform: Algorithms, Advantages, Applications*; Academic Press: Cambridge, MA, USA, 2014.
45. Tan, L.; Jiang, J. *Digital Signal Processing: Fundamentals and Applications*; Academic Press: Cambridge, MA, USA, 2018.

© 2019 by the authors. Licensee MDPI, Basel, Switzerland. This article is an open access article distributed under the terms and conditions of the Creative Commons Attribution (CC BY) license (http://creativecommons.org/licenses/by/4.0/).

symmetry

MDPI

Article

Fuzzy Volterra Integro-Differential Equations Using General Linear Method

Zanariah Abdul Majid [1], Faranak Rabiei [2,*], Fatin Abd Hamid [1] and Fudziah Ismail [1]

1 Institute for Mathematical Research, Universiti Putra Malaysia, Serdang 43400, Malaysia; am_zana@upm.edu.my (Z.A.M.); fatinkd92@yahoo.com (F.A.H.); fudziah@upm.edu.my (F.I.)

2 School of Engineering, Monash University Malaysia, Jalan Lagoon Selatan, Bandar Sunway 47500, Malaysia

* Correspondence: faranak.rabiei@monash.edu or faranak.rabiei@gmail.com

Received: 26 December 2018; Accepted: 16 January 2019; Published: 15 March 2019

Abstract: In this paper, a fuzzy general linear method of order three for solving fuzzy Volterra integro-differential equations of second kind is proposed. The general linear method is operated using the both internal stages of Runge-Kutta method and multivalues of a multisteps method. The derivation of general linear method is based on the theory of B-series and rooted trees. Here, the fuzzy general linear method using the approach of generalized Hukuhara differentiability and combination of composite Simpson's rules together with Lagrange interpolation polynomial is constructed for numerical solution of fuzzy volterra integro-differential equations. To illustrate the performance of the method, the numerical results are compared with some existing numerical methods.

Keywords: fuzzy volterra integro-differential equations; fuzzy general linear method; fuzzy differential equations; generalized Hukuhara differentiability

1. Introduction

Fuzzy differential equations (FDEs) and fuzzy integral equations (FIEs) have been extensively studied in the past few years. They have appeared in many applications such as fuzzy matric spaces, population models, medicine, engineering problems, and others (see [1,2]). In the treatment of FDEs, one of the approaches was by using the definition of Hukuhara differentiability (see [3,4]). However, the Hukuhara differentiability experienced a disadvantage in its solutions. To overcome this, generalized Hukuhara differentiability was introduced by Bede and Gal in [5]. In the area of FIEs, the Rieman integral concept was proposed by Goetschel and Voxman in [6]. Another concept of integration is the Lebesgue concept by Kaleva in [7]. An early work in the numerical solutions of FDEs and FIEs is by Friedman et al. in [8]. Later, the area of interest in FIEs has been expanded into the fuzzy integro-differential equations (FIDEs). FIDEs take the form of both FDEs and FIEs. A particular class of FIDEs is known as fuzzy Volterra integro-differential equations (FVIDEs). The existence and uniqueness of FIDEs and FVIDEs solutions were investigated by Park and Jeong in [9], Hajighasemi et al. in [10], and Zeinali et al. in [11]. Mikaeilvand et al. in [12] presented the numerical examples of FVIDEs using the differential transform method. In [13], Allahviranloo et al. proposed a new technique to solve the FVIDEs using definition of generalized differentiability. Later, Allahviranloo et al. in [14] discussed the existence and uniqueness of second-order FVIDEs using the fuzzy kernel. Then Matinfar et al. in [15] solved the FVIDEs using the variational iteration method while Sahu and Saha Ray used Legendre wavelet method in [16].

In this work, we propose the numerical solutions of FVIDEs using the general linear method (GLM) introduced by Butcher in [17]. The GLM is a generalization of Runge-Kutta method (RK) and linear multistep method derived based on theory of B-series and definition of rooted trees. Recently, the GLM was studied for finding the numerical solutions of FDEs by Rabiei et al. in [18] and based on that, in this paper we develop the fuzzy third-order GLM together with suitable integration method to solve FVIDEs.

In Section 2, preliminaries on fuzzy numbers and theories are proposed. The concept of FVIDEs is discussed in Section 3. In Section 4, the general form of GLM is given followed by demonstration of the integration rules in Section 5. Then in Section 6, fuzzy version of the GLM combined with integration rules for FVIDE is developed while in Section 7, we derived the fuzzy RK method for FVIDEs using the same approaches used for GLM in Section 6. Section 8, some test problems are carried out to illustrate the efficiency of obtained method compared with a derived fuzzy RK method of order three in Section 7. Lastly, discussion and conclusion are presented in Section 9.

2. Preliminaries

In this section, some basic definitions on fuzzy numbers are given.

Definition 1 (see [19]). *Consider a fuzzy subset of the real line $u : \mathbb{R} \to [0,1]$. Then u is a fuzzy number if it satisfies the following properties:*

(i) *u is normal, that is $\exists\ x_0 \in \mathbb{R}$ with $u(x_0) = 1$;*
(ii) *u is fuzzy convex, that is $u(tx + (1-t)y) \geq min\ u(x), u(y)$, $\forall t \in [0,1], x, y \in \mathbb{R}$;*
(iii) *u is upper semicontinuous on $\mathbb{R}$, that is $\forall \varepsilon > 0\ \exists \delta > 0$ such that $u(x) - u(x_0) < \varepsilon, |x - x_0| < \delta$;*
(iv) *u is compactly supported, that is $cl\{x \in \mathbb{R};\ u(x) > 0\}$ is compact, where $cl(A)$ denotes the closure of the set A.*

Then $\mathbb{R}_F$ is called the space of fuzzy numbers.

Definition 2 (see [19]). *For $0 < r \leq 1$, we have*

$$[u]^r = \{x \in \mathbb{R};\ u(x) \geq r\},$$

and

$$[u]^0 = cl\{x \in \mathbb{R};\ u(x) > r\}\}.$$

Then the $[u]^r$ denotes the r-level set of the fuzzy number u. The 1-level will refer to the core while the 0-level refers to the support of the fuzzy number.

Proposition 1 (see [19]). *A fuzzy number u is a pair $u = (u^-, u^+)$ of functions $u^-, u^+ : [0,1] \to \mathbb{R}$, implying the end points of r-level set, following the conditions:*

(i) *$u_r^- \in \mathbb{R}$ is a bounded nondecreasing left-continuous function $\forall r \in [0,1]$ and right-continuous for $r = 0$;*
(ii) *$u_r^+ \in \mathbb{R}$ is a bounded nonincreasing left-continuous function $\forall r \in [0,1]$ and right-continuous for $r = 0$;*
(iii) *$u_r^- \leq u_r^+$ for $r = 1$, which implies $u_r^- \leq u_r^+$, $\forall r \in [0,1]$.*

Definition 3 (see [19]). *Let $u, v \in \mathbb{R}_F$, the distance $D(u,v)$ between two fuzzy intervals is defined by*

$$D_\infty(u,v) = \sup_{r \in [0,1]} max\{|u_r^- - v_r^-|, |u_r^+ - v_r^+|\}.$$

Then $D_\infty(u,v)$ is the Hausdorff distance between fuzzy numbers.

Proposition 2 (see [19]). *It is said that $D_\infty(u,v)$ is a metric space in $\mathbb{R}_F$ and the following properties hold:*

(i) *$D_\infty(u + w, v + w)) = D_\infty(u,v)$, $\forall u, v, w \in \mathbb{R}_F$;*
(ii) *$D_\infty(k \cdot u, k \cdot v) = |k| D_\infty(u,v)$, $\forall u, v \in \mathbb{R}_F, \forall k \in \mathbb{R}$;*
(iii) *$D_\infty(u + v, w + e) \leq D_\infty(u,w) + D_\infty(v,e)$, $\forall u, v, w, e \in \mathbb{R}_F$.*

Definition 4. *A function* $f : \mathbb{R} \to \mathbb{R}_{\mathcal{F}}$ *is said to be fuzzy continuous function if* f *exists for any fixed arbitrary* $g_0 \in \mathbb{R}$ *and* $\varepsilon > 0, \delta > 0$ *such that* $|g - g_0| < \delta \Longrightarrow D[f(g), f(g_0)] < \varepsilon$.

Definition 5. *Let* $x, y \in \mathbb{R}_{\mathcal{F}}$. *If there exists* $z \in \mathbb{R}_{\mathcal{F}}$ *such that* $x = y \oplus z$, *then* z *is called the Hukuhara difference (H-difference) of* x *and* y *and it is denoted by* $x \ominus y$. *(Please note that,* $x \ominus y \neq x + (-y)$*).*

Definition 6 (see [5]). *Let* $f : (a, b) \to \mathbb{R}_F$ *and* $x_0 \in (a, b)$. f *is known as strongly generalized differentiable at* x_0, *if there exists an element* $f'(x_0) \in \mathbb{R}_F$ *such that*

(i) *for all* $h > 0$ *sufficiently small,* $\exists f(x_0 + h) \ominus f(x_0)$, $f(x_0) \ominus f(x_0 - h)$ *and the limits in metric* D

$$\lim_{h \searrow 0} \frac{f(x_0 + h) \ominus f(x_0)}{h} = \lim_{h \searrow 0} \frac{f(x_0) \ominus f(x_0 - h)}{h} = f'(x_0), \tag{1}$$

is type-(i)*-differentiability on* (a, b),

(ii) *for all* $h > 0$ *sufficiently small,* $\exists f(x_0) \ominus f(x_0 + h)$, $f(x_0 - h) \ominus f(x_0)$ *and the limits in metric* D

$$\lim_{h \searrow 0} \frac{f(x_0) \ominus f(x_0 + h)}{(-h)} = \lim_{h \searrow 0} \frac{f(x_0 - h) \ominus f(x_0)}{(-h)} = f'(x_0). \tag{2}$$

is type-(ii)*-differentiability on* (a, b),

Theorem 1 (see [20]). *Let* $F : T \to \mathbb{R}_F$ *be a function and denote* $[F(t)]_r = [f_r(t), g_r(t)]$, *for each* $r \in [0, 1]$. *Then*

(i) *If* F *is differentiable in the first form (1), then* f_r *and* g_r *are differentiable functions and* $[F'(t)]_r = [f'_r(t), g'_r(t)]$,

(ii) *If* F *is differentiable in the second form (2), then* f_r *and* g_r *are differentiable functions and* $[F'(t)]_r = [g'_r(t), f'_r(t)]$.

Definition 7 (see [15]). *Let* $f : [a, b] \to \mathbb{R}_F$, *for each partition* $P = \{t_0, t_1, \ldots, t_n\}$ *of* $[a, b]$ *and for arbitrary* $\xi_i \in [t_i - 1, t_i], 1 \leq i \leq n$, *and suppose*

$$R_p = \sum_{i=1}^{n} (\xi_i)(t_i - t_{i-1}),$$
$$\triangle := max\{|t_i - t_{i-1}|, 1 \leq i \leq n\}.$$

The integration of $f(t)$ *over* $[a, b]$ *is*

$$\int_a^b f(t)dt = \lim_{\triangle \to 0} R_p,$$

given that in metric D, *the limit exists. The definite integral of fuzzy function* $f(t)$ *exists, if* $f(t)$ *is continuous function in metric* D, *and*

$$\left(\int_a^b f(t)dt\right)_r^- = \int_a^b f_r^-(t)dt,$$
$$\left(\int_a^b f(t)dt\right)_r^+ = \int_a^b f_r^+(t)dt.$$

3. Fuzzy Volterra Integro-Differential Equations

Consider the first order fuzzy initial value problems of second kind FVIDEs given by

$$y'(t) = f(t,y) + \int_0^x K(t,s)y(s)ds, \quad y(t_0) = y_0 \in \mathbb{R}_F, \tag{3}$$

and in short notation is given as

$$y'(t) = f\Big(t, y, \int_0^x K(t,s)y(s)ds\Big), \quad y(t_0) = y_0 \in \mathbb{R}_F, \tag{4}$$

where function $f : \mathbb{R} \times \mathbb{R}_F \to \mathbb{R}_F$, crisp function $K(t,s)$ are continuous and y_0 is a fuzzy number.

Using Theorem 1, and extending the characterization theorem in [19], FVIDE given in (4) for type (i)-differentiability is equivalent to the following system of ODEs:

$$\begin{cases} (y_r^-)'(t) = f_r^-(t, y_r^-, y_r^+) + \int_0^t K^-(t,s)y(s)ds = F(t, y_r^-, y_r^+, \int_0^t K^-(t,s)y(s)ds), \\ (y_r^+)'(t) = f_r^+(t, y_r^-, y_r^+) + \int_0^t K^+(t,s)y(s)ds = G(t, y_r^-, y_r^+, \int_0^t K^+(t,s)y(s)ds), \\ y_r^-(0) = (y_0)_r^-, \\ y_r^+(0) = (y_0)_r^+, \end{cases} \tag{5}$$

and for type (ii)-differentiability FVIDE (4) is equivalent to the system of ODEs as follows:

$$\begin{cases} (y_r^-)'(t) = f_r^+(t, y_r^-, y_r^+) + \int_0^t K^+(t,s)y(s)ds = G(t, y_r^-, y_r^+, \int_0^t K^+(t,s)y(s)ds), \\ (y_r^+)'(t) = f_r^-(t, y_r^-, y_r^+) + \int_0^t K^-(t,s)y(s)ds = F(t, y_r^-, y_r^+, \int_0^t K^-(t,s)y(s)ds), \\ y_r^-(0) = (y_0)_r^-, \\ y_r^+(0) = (y_0)_r^+, \end{cases} \tag{6}$$

where

$$K^-(t,s)y(s)ds = \begin{cases} K(t,s)y^-(s)ds, & K(t,s) \geq 0, \\ K(t,s)y^+(s)ds, & K(t,s) \leq 0, \end{cases}$$

$$K^+(t,s)y(s)ds = \begin{cases} K(t,s)y^+(s)ds, & K(t,s) \geq 0, \\ K(t,s)y^-(s)ds, & K(t,s) \leq 0. \end{cases}$$

4. General Linear Method

Consider the first order initial value problems

$$y'(x) = f(x, y(x)), \quad y(x_0) = y_0. \tag{7}$$

The general form of GLM (see [17]) is given as

$$Y_i = \sum_{j=1}^{s} a_{ij}hF_j + \sum_{j=1}^{r} u_{ij}y_j^{[n-1]}, \qquad i = 1, 2, \ldots, s, \tag{8}$$

$$y_i^{[n]} = \sum_{j=1}^{s} b_{ij}hF_j + \sum_{j=1}^{r} v_{ij}y_j^{[n-1]}, \qquad i = 1, 2, \ldots, r, \tag{9}$$

where n is the step number, $y_i^{[n]}$, $i = 1, 2, \ldots, r$ are the approximate solutions, Y_i, $i = 1, 2, \ldots, s$ is the stage values and F_i, $i = 1, 2, \ldots, s$ is the stage derivatives.

The algebraic coefficients a, u, b, and v of the proposed method here, are given from Rabiei et al. in [18] as shown in Table 1.

Table 1. Coefficients of third-order GLM.

			$u_{11} = 1$	$u_{12} = 0$
$a_{21} = \frac{13}{18}$			$u_{21} = \frac{7}{9}$	$u_{22} = \frac{2}{9}$
$a_{31} = \frac{-17}{9}$	$a_{32} = 2$		$u_{31} = \frac{17}{9}$	$u_{32} = \frac{-8}{9}$
$b_{11} = \frac{1}{6}$	$b_{12} = \frac{2}{3}$	$b_{13} = \frac{1}{6}$	$v_{11} = 1$	$v_{12} = 0$
$b_{21} = 0$	$b_{22} = 0$	$b_{23} = 0$	$v_{21} = 1$	$v_{22} = 0$

5. Simpson's Rule and Lagrange Interpolation Polynomial

In the FVIDE, the integral operator $z \approx \int_0^t K(t,s)y(s)ds$ need to be approximated first before applying the third-order fuzzy GLM. The range of integration is divided into two intervals as shown below.

$$\int_0^t K(t,s)y(s)ds = \int_0^{t_n} K(t,s)y(s)ds + \int_{t_n}^{t_{n+c}} K(t,s)y(s)ds, \tag{10}$$

where the grid points are calculated by $t_n = t_0 + nh$ with $h = \frac{T-t_0}{N}$ and $0 < n < N$. The value c is the value of coefficient for third-order GLM given in Table 1.

The composite Simpson's rule (Simpson's II method defined in [21]) is used to compute the integration in the interval $\int_0^{t_n} K(t,s)y(s)ds$. Meanwhile we compute the integration in the interval $\int_{t_n}^{t_{n+c}} K(t,s)y(s)ds$ using Lagrange's interpolation method. The Lagrange interpolating polynomial is determined by interpolating on set of points $\{t_{n-1}, t_n, t_{n+c}\}$. For points $\{t_{-1}, t_0, t_{\frac{1}{2}}\}$ the Lagrange interpolating polynomial is:

$$P(t) = \frac{(t-t_0)(t-t_{\frac{1}{2}})}{(t_{-1}-t_0)(t_{-1}-t_{\frac{1}{2}})}y_{-1} + \frac{(t-t_{-1})(t-t_{\frac{1}{2}})}{(t_0-t_{-1})(t_0-t_{\frac{1}{2}})}y_0 + \frac{(t-t_{-1})(t-t_0)}{(t_{\frac{1}{2}}-t_{-1})(t_{\frac{1}{2}}-t_0)}y_{1/2}. \tag{11}$$

Substituting $t_{-1} = -h$, $t_0 = 0$, and $t_{\frac{1}{2}} = \frac{1}{2}h$ into (11) gives

$$P(t) = \frac{2t\,(t-h/2)}{3h^2}y_{-1} - \frac{2\,(t+h)\,(t-h/2)}{h^2}y_0 + \frac{4\,(t+h)\,t}{3h^2}y_{1/2}. \tag{12}$$

Then integrate Equation (12) with limit from 0 to $\frac{h}{2}$, to produce

$$\int_{t_n}^{t_{n+\frac{1}{2}}} K(t,s)y(s)ds = h\Big\{ -\frac{1}{72}K(t,s)y(t_{n-1}) + \frac{7}{24}K(t,s)y(t_n) + \frac{2}{9}K(t,s)y(t_{n+\frac{1}{2}})\Big\}. \tag{13}$$

In the first step where $n = 0$, the value of $y(t_{-1}) = y(t_0 - h)$ is evaluated by using a fourth order RK method.

6. Fuzzy General Linear Method for Fuzzy Volterra Integro-Differential Equations

Rabiei et al. [18] proposed the fuzzy GLM for solving FDEs. The convergence of the method also was proven. Here, by using the third-order GLM derived in [18], we will apply the fuzzy GLM for solving FVIDEs. Consider the fuzzy Problem 4, we denote the initial value $y_0 \in \mathbb{R}_F$ with r-level sets

$$[y_0]_r = [y^-(t_0;r), y^+(t_0;r)], \quad r \in [0,1]. \tag{14}$$

The set of equally spaced grid points $t_0 < t_1 < t_2 \cdots < t_N = T$ is a set of interval T. The exact solutions are given as

$$[\tilde{Y}(t)]_r = [\tilde{Y}^-(t;r), \tilde{Y}^+(t;r)], \tag{15}$$

are approximated by

$$[y(t)]_r = [y^-(t;r), y^+(t;r)]. \tag{16}$$

The grid points are calculated by $t_n = t_0 + nh$ with $h = \frac{T-t_0}{N}$ and $0 < n < N$. Thus, we have the exact and approximate solutions at t_n set as

$$[\tilde{Y}(t_n)]_r = [\tilde{Y}^-(t_n;r), \tilde{Y}^+(t_n;r)], \tag{17}$$

$$[y(t_n)]_r = [y^-(t_n;r), y^+(t_n;r)]. \tag{18}$$

The third-order fuzzy GLM for solving FVIDEs based on type (i)-differentiability is given by the formulae:

$$y_i^-(t_{n+1};r) = \sum_{j=1}^{s=3} b_{ij} h F_j(t_n, y(t_n;r), z(t_n;r)) + \sum_{j=1}^{r=2} v_{ij} y_j^-(t_n;r), \quad i = 1, \dots, r, \tag{19}$$

$$y_i^+(t_{n+1};r) = \sum_{j=1}^{s=3} b_{ij} h G_j(t_n, y(t_n;r), z(t_n;r)) + \sum_{j=1}^{r=2} v_{ij} y_j^+(t_n;r), \quad i = 1, \dots, r, \tag{20}$$

where

$$\begin{aligned}
Y_1^-(y(t_n;r)) &= u_{11} y_1^-(t_n;r) + u_{12} y_2^-(t_n;r), \\
Y_1^+(y(t_n;r)) &= u_{11} y_1^+(t_n;r) + u_{12} y_2^+(t_n;r), \\
Y_2^-(y(t_n;r)) &= a_{21} h F_1(t_n, y(t_n;r), z(t_n;r)) + u_{21} y_1^-(t_n;r) + u_{22} y_2^-(t_n;r), \\
Y_2^+(y(t_n;r)) &= a_{21} h G_1(t_n, y(t_n;r), z(t_n;r)) + u_{21} y_1^+(t_n;r) + u_{22} y_2^+(t_n;r), \\
Y_3^-(y(t_n;r)) &= a_{31} h F_1(t_n, y(t_n;r), z(t_n;r)) + a_{32} h F_2(t_n, y(t_n;r), z(t_n;r)) \\
&\quad + u_{31} y_1^-(t_n;r) + u_{32} y_2^-(t_n;r), \\
Y_3^+(y(t_n;r)) &= a_{31} h G_1(t_n, y(t_n;r), z(t_n;r)) + a_{32} h G_2(t_n, y(t_n;r), z(t_n;r)) \\
&\quad + u_{31} y_1^+(t_n;r) + u_{32} y_2^+(t_n;r),
\end{aligned} \tag{21}$$

such that

$$\begin{aligned}
F_1(t_n, y(t_n;r)) &= min\Big\{f(t_n + c_1 h, u, v) | u \in [Y_1^-(y(t_n;r)), Y_1^+(y(t_n;r))], \\
&\qquad v \in [z_1^-(y(t_n;r)), z_1^+(y(t_n;r))]\Big\}, \\
G_1(t_n, y(t_n;r)) &= max\Big\{f(t_n + c_1 h, u, v) | u \in [Y_1^-(y(t_n;r)), Y_1^+(y(t_n;r))], \\
&\qquad v \in [z_1^-(y(t_n;r)), z_1^+(y(t_n;r))]\Big\}, \\
F_2(t_n, y(t_n;r)) &= min\Big\{f(t_n + c_2 h, u, v) | u \in [Y_2^-(y(t_n;r)), Y_2^+(y(t_n;r))], \\
&\qquad v \in [z_2^-(y(t_n;r)), z_2^+(y(t_n;r))]\Big\}, \\
G_2(t_n, y(t_n;r)) &= max\Big\{f(t_n + c_2 h, u, v) | u \in [Y_2^-(y(t_n;r)), Y_2^+(y(t_n;r))], \\
&\qquad v \in [z_2^-(y(t_n;r)), z_2^+(y(t_n;r))]\Big\}, \\
F_3(t_n, y(t_n;r)) &= min\Big\{f(t_n + c_3 h, u, v) | u \in [Y_3^-(y(t_n;r)), Y_3^+(y(t_n;r))], \\
&\qquad v \in [z_3^-(y(t_n;r)), z_3^+(y(t_n;r))]\Big\}, \\
G_3(t_n, y(t_n;r)) &= max\Big\{f(t_n + c_3 h, u, v) | u \in [Y_3^-(y(t_n;r)), Y_3^+(y(t_n;r))], \\
&\qquad v \in [z_3^-(y(t_n;r)), z_3^+(y(t_n;r))]\Big\},
\end{aligned} \tag{22}$$

and

$$
\begin{aligned}
z_1^-(y(t_n)) &= \int_0^{t_n} K^-(t,s)y(s;r)ds, \\
z_1^+(y(t_n)) &= \int_0^{t_n} K^+(t,s)y(s;r)ds, \\
z_2^-(y(t_n)) &= \int_0^{t_n} K^-(t,s)y(s;r)ds + \int_{t_n}^{t_{n+c_2}} K^-(t,s)y(s)ds, \\
z_2^+(y(t_n)) &= \int_0^{t_n} K^+(t,s)y(s;r)ds + \int_{t_n}^{t_{n+c_2}} K^+(t,s)y(s)ds, \\
z_3^-(y(t_n)) &= \int_0^{t_n} K^-(t,s)y(s)ds + \int_{t_n}^{t_{n+c_3}} K^-(t,s)y(s)ds, \\
z_3^+(y(t_n)) &= \int_0^{t_n} K^+(t,s)y(s)ds + \int_{t_n}^{t_{n+c_3}} K^+(t,s)y(s)ds.
\end{aligned} \tag{23}
$$

Meanwhile the fuzzy third-order GLM for solving FVIDEs based on type (ii)-differentiability is given by the formulae:

$$
y_i^-(t_{n+1};r) = \sum_{j=1}^{s=3} b_{ij}hG_j(t_n, y(t_n;r), z(t_n;r)) + \sum_{j=1}^{r=2} v_{ij}y_j^-(t_n;r), \quad i = 1,\ldots,r, \tag{24}
$$

$$
y_i^+(t_{n+1};r) = \sum_{j=1}^{s=3} b_{ij}hF_j(t_n, y(t_n;r), z(t_n;r)) + \sum_{j=1}^{r=2} v_{ij}y_j^+(t_n;r), \quad i = 1,\ldots,r, \tag{25}
$$

where

$$
\begin{aligned}
Y_1^-(y(t_n;r)) &= u_{11}y_1^-(t_n;r) + u_{12}y_2^-(t_n;r), \\
Y_1^+(y(t_n;r)) &= u_{11}y_1^+(t_n;r) + u_{12}y_2^+(t_n;r), \\
Y_2^-(y(t_n;r)) &= a_{21}hG_1(t_n, y(t_n;r), z(t_n;r)) + u_{21}y_1^-(t_n;r) + u_{22}y_2^-(t_n;r), \\
Y_2^+(y(t_n;r)) &= a_{21}hF_1(t_n, y(t_n;r), z(t_n;r)) + u_{21}y_1^+(t_n;r) + u_{22}y_2^+(t_n;r), \\
Y_3^-(y(t_n;r)) &= a_{31}hG_1(t_n, y(t_n;r), z(t_n;r)) + a_{32}hG_2(t_n, y(t_n;r), z(t_n;r)) \\
&\quad + u_{31}y_1^-(t_n;r) + u_{32}y_2^-(t_n;r), \\
Y_3^+(y(t_n;r)) &= a_{31}hF_1(t_n, y(t_n;r), z(t_n;r)) + a_{32}hF_2(t_n, y(t_n;r), z(t_n;r)) \\
&\quad + u_{31}y_1^+(t_n;r) + u_{32}y_2^+(t_n;r),
\end{aligned} \tag{26}
$$

where F_1, F_2, F_3, G_1,G_2 and G_3 are same as (22).

7. Fuzzy Runge-Kutta Method for Fuzzy Volterra Integro-Differential Equations

In this section, we will develop the fuzzy version of third-order RK method to solve the FVIDEs. The RK method is combined with suitable integration methods to deal with the integral part. It is appropriate to apply the composite Simpson's rule and Lagrange's method in Section 5 similarly. The coefficients (see [22]) for RK method is represented in Table 2. The general form of RK method for solving Equation (7) is given by

$$
y_{n+1} = y_n + h\sum_{i=1}^{s} B_i k_i, \quad 1 \le n \le N-1, \tag{27}
$$

where

$$k_1 = f(x_n, y_n),$$
$$k_i = f(x_n + C_i h, y_n + h \sum_{j=1}^{i-1} A_{ij} k_j), \quad i = 2, 3, \ldots, s. \tag{28}$$

Table 2. Coefficients of third-order RK method.

$C_1 = 0$			
$C_2 = \frac{1}{2}$	$A_{21} = \frac{1}{2}$		
$C_3 = 1$	$A_{31} = -1$	$A_{32} = 2$	
	$B_1 = \frac{1}{6}$	$B_2 = \frac{2}{3}$	$B_3 = \frac{1}{6}$

The formulae of a third-order RK method for FVIDEs based on type (i)-differentiability is given as follows:

$$y^-(t_{n+1}; r) = y^-(t_n; r) + hF(t_n, (y(t_n; r))), \tag{29}$$

$$y^+(t_{n+1}; r) = y^+(t_n; r) + hG(t_n, (y(t_n; r))), \tag{30}$$

while for type (ii)-differentiability is given as:

$$y^-(t_{n+1}; r) = y^-(t_n; r) + hG(t_n, (y(t_n; r))), \tag{31}$$

$$y^+(t_{n+1}; r) = y^+(t_n; r) + hF(t_n, (y(t_n; r))), \tag{32}$$

where

$$F(t_n, (y(t_n; r))) = \Big\{B_1 k_1^-(y(t_n; r)) + B_2 k_2^- y((t_n; r)) + B_3 k_3^-(y(t_n; r))\Big\},$$
$$G(t_n, (y(t_n; r))) = \Big\{B_1 k_1^+((y(t_n; r)) + B_2 k_2^+(y(t_n; r)) + B_3 k_3^+(y(t_n; r))\Big\}, \tag{33}$$

where

$$k_1^-(y(t_n; r)) = min\Big\{f(t, u, v) | u \in [y^-(t_n; r), y^+(t_n; r)], v \in [z_1^-(y(t_n; r)), z_1^+(y(t_n; r))]\Big\},$$
$$k_1^+(y(t_n; r)) = max\Big\{f(t, u, v) | u \in [y^-(t_n; r), y^+(t_n; r)], v \in [z_1^-(y(t_n; r)), z_1^+(y(t_n; r))]\Big\},$$
$$k_2^-(y(t_n; r)) = min\Big\{f(t + C_2 h, u, v) | u \in [w_1^-(t_n; r), w_1^+(t_n; r)], v \in [z_2^-(y(t_n; r)), z_2^+(y(t_n; r))]\Big\},$$
$$k_2^+(y(t_n; r)) = max\Big\{f(t + C_2 h, u, v) | u \in [w_1^-(t_n; r), w_1^+(t_n; r)], v \in [z_2^-(y(t_n; r)), z_2^+(y(t_n; r))]\Big\}, \tag{34}$$
$$k_3^-(y(t_n; r)) = min\Big\{f(t + C_3 h, u, v) | u \in [w_2^-(t_n; r), w_2^+(t_n; r)], v \in [z_3^-(y(t_n; r)), z_3^+(y(t_n; r))]\Big\},$$
$$k_3^+(y(t_n; r)) = max\Big\{f(t + C_3 h, u, v) | u \in [w_2^-(t_n; r), w_2^+(t_n; r)], v \in [z_3^-(y(t_n; r)), z_3^+(y(t_n; r))]\Big\},$$

such that

$$w_1^-(t_n; r) = y^-(t_n; r) + hA_{21} k_1^-,$$
$$w_1^+(t_n; r) = y^+(t_n; r) + hA_{21} k_1^+,$$
$$w_2^-(t_n; r) = y^-(t_n; r) + h(A_{31} k_1^- + A_{32} k_2^-), \tag{35}$$
$$w_2^+(t_n; r) = y^+(t_n; r) + h(A_{31} k_1^+ + A_{32} k_2^+),$$

and

$$
\begin{aligned}
z_1^-(y(t_n)) &= \int_0^{t_n} K^-(t,s)y(s;r)ds, \\
z_1^+(y(t_n)) &= \int_0^{t_n} K^+(t,s)y(s;r)ds, \\
z_2^-(y(t_n)) &= \int_0^{t_n} K^-(t,s)y(s;r)ds + \int_{t_n}^{t_{n+C_2}} K^-(t,s)y(s)ds, \\
z_2^+(y(t_n)) &= \int_0^{t_n} K^+(t,s)y(s;r)ds + \int_{t_n}^{t_{n+C_2}} K^+(t,s)y(s)ds, \\
z_3^-(y(t_n)) &= \int_0^{t_{n+1}} K^-(t,s)y(s)ds, \\
z_3^+(y(t_n)) &= \int_0^{t_{n+1}} K^+(t,s)y(s)ds.
\end{aligned}
\tag{36}
$$

8. Numerical Results

We tested the fuzzy GLM to illustrate the efficiency of the method. Comparison is made between fuzzy versions of GLM and the RK method, Variational iteration method and homotopy perturbation method. The efficiency of method is shown in terms of error which is estimated by $\underline{E}(t;r) = |\underline{y}(t;r) - \underline{Y}(t;r)|$ and $\overline{E}(t;r) = |\overline{y}(t;r) - \overline{Y}(t;r)|$. List of abbreviations used in the tabulated results are as follows:

r	r-level set of fuzzy numbers,
$\tilde{Y}^-$	Left bound of exact solution,
$\tilde{Y}^+$	Right bound of exact solution,
y^-	Left bound of approximate solution,
y^+	Right bound of approximate solution,
E^-	Left bound of error computed ($\lvert y^- - \tilde{Y}^-\rvert$),
E^+	Right bound of error computed ($\lvert y^+ - \tilde{Y}^+\rvert$),
GLM	Third-order general linear method from this paper,
RK	Third-order Runge-Kutta method from Section 7,
VIM	Variational iteration method from [15],
HAM	Homotopy perturbation method from [15].

8.1. Problem 1

Consider the following FVIDEs (see [15])

$$
\begin{aligned}
y'(t) &= C\frac{1}{12t}(36-5t^4) + \int_0^t (t^2+s^2)y(s;r)ds, \\
C &= [(r^5+2r)t^3, (6-3r^3)t^3], \quad y(0) = [0,0], \qquad 0 \le s \le t \le 1.
\end{aligned}
$$

The equivalent system of ODEs based on (i)-differentiability:

$$
\begin{aligned}
(y^-)'(t;r) &= \frac{1}{12}rt^2(r^4+2)(36-5t^4) + \int_0^t (t^2+s^2)y^-(s;r)ds, \quad y^-(0;r) = 0, \\
(y^+)'(t;r) &= \frac{1}{4}t^2(r^3-2)(5t^4-36) + \int_0^t (t^2+s^2)y^+(s;r)ds, \quad y^+(0;r) = 0.
\end{aligned}
$$

Exact solutions :

$$
\begin{aligned}
\tilde{Y}^-(t;r) &= (r^5+2r)t^3, \\
\tilde{Y}^+(t;r) &= (6-3r^3)t^3.
\end{aligned}
$$

For Problem 1, the graphical results of GLM together with exact solutions are shown in Figure 1. The comparison of numerical results of GLM with existing methods is given in Table 3 and Figure 2.

Table 3. Comparison between GLM and existing methods for solving Problem 1.

	GLM		RK		VIM	HPM
r	$E^-(t=0.5;r)$	$E^-(t=1;r)$	$E^-(t=0.5;r)$	$E^-(t=1;r)$	$E^-(t=0.5;r)$	$E^-(t=0.5;r)$
0.0	0	0	0	0	0	0
0.1	9.956810(−11)	1.209550(−9)	4.823713(−10)	7.462263(−9)	7.6675(−11)	2.5125(−6)
0.2	1.992855(−10)	2.420914(−9)	9.654662(−10)	1.493572(−8)	1.5347(−10)	5.0288(−6)
0.3	2.998988(−10)	3.643163(−9)	1.452902(−9)	2.247632(−8)	2.3095(−10)	7.5676(−6)
0.4	4.033500(−10)	4.899886(−9)	1.954086(−9)	3.022960(−8)	3.1061(−10)	1.0178(−5)
0.5	5.133730(−10)	6.236430(−9)	2.487103(−9)	3.847537(−8)	3.9534(−10)	1.2954(−5)
0.6	6.360880(−10)	7.727180(−9)	3.081621(−9)	4.767250(−8)	4.8984(−10)	1.6051(−5)
0.7	7.806100(−10)	9.482820(−8)	3.781770(−9)	5.850381(−8)	6.0113(−10)	1.9698(−5)
0.8	9.596280(−10)	1.165754(−8)	4.649055(−9)	7.192064(−8)	7.3899(−10)	2.4215(−5)
0.9	1.190025(−9)	1.445635(−8)	5.765231(−9)	8.918787(−8)	9.1641(−10)	3.0029(−5)
1.0	1.493445(−9)	1.814234(−8)	7.235207(−9)	1.119283(−7)	1.1501(−9)	3.7686(−5)
r	$E^+(t=0.5;r)$	$E^+(t=1;r)$	$E^+(t=0.5;r)$	$E^+(t=1;r)$	$E^+(t=0.5;r)$	$E^+(t=0.5;r)$
0.0	2.986892(−9)	3.628465(−8)	1.447042(−8)	2.238567(−7)	2.3001(−9)	7.5371(−5)
0.1	2.985401(−9)	3.626651(−8)	1.446318(−8)	2.237446(−7)	2.2990(−9)	7.5333(−5)
0.2	2.974945(−9)	3.613950(−8)	1.441254(−8)	2.229612(−7)	2.2909(−9)	7.5070(−5)
0.3	2.946571(−9)	3.579484(−8)	1.427506(−8)	2.208346(−7)	2.2691(−9)	7.4354(−5)
0.4	2.891310(−9)	3.512354(−8)	1.400736(−8)	2.166932(−7)	2.2265(−9)	7.2959(−5)
0.5	2.800214(−9)	3.401690(−8)	1.356601(−8)	2.098656(−7)	2.1564(−9)	7.0660(−5)
0.6	2.664307(−9)	3.236592(−8)	1.290761(−8)	1.996801(−7)	2.0517(−9)	6.7231(−5)
0.7	2.474641(−9)	3.006184(−8)	1.198874(−8)	1.854652(−7)	1.9057(−9)	6.2444(−5)
0.8	2.222250(−9)	2.699583(−8)	1.076599(−8)	1.665493(−7)	1.7113(−9)	5.6076(−5)
0.9	1.898170(−9)	2.305891(−8)	9.195954(−8)	1.422609(−7)	1.4617(−9)	4.7899(−5)
1.0	1.493445(−9)	1.814234(−8)	7.235207(−8)	1.119283(−7)	1.1501(−9)	3.7686(−5)

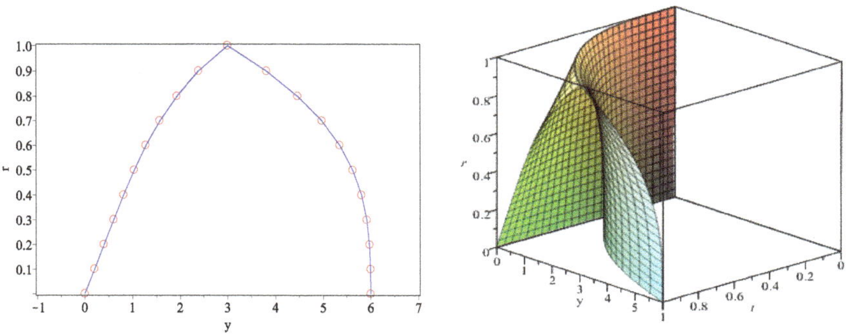

(**a**) (**b**)

Figure 1. (**a**) Approximate solution of GLM (circle) and exact solution (line) at $t = 1.0$; (**b**) 3D-plot of GLM for Problem 1.

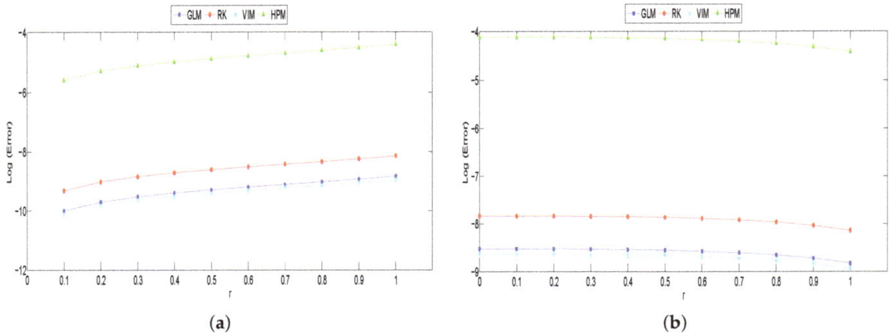

Figure 2. (**a**) Graph of Log (Error) versus r at $t = 0.5$ for left bound of solutions; (**b**) Graph of Log (Error) versus r at $t = 0.5$ for right bound of solutions for Problem 1.

8.2. Problem 2

Consider the following FVIDEs (see [12])

$$y'(t) = C + \int_0^t y(s;r)ds,$$
$$C = [r-1, 1-r], \quad y(0) = [0,0], \qquad 0 \le s \le t \le 1.$$

The equivalent system of ODEs based on (i)-differentiability:

$$(y^-)'(t;r) = (r-1) + \int_0^t y^-(s;r)ds, \quad y^-(0;r) = 0,$$
$$(y^+)'(t;r) = (1-r) + \int_0^t y^+(s;r)ds, \quad y^+(0;r) = 0.$$

Exact solutions :

$$\tilde{Y}^-(t;r) = (r-1)\sinh(t),$$
$$\tilde{Y}^+(t;r) = (1-r)\sinh(t).$$

For Problem 2, the graph of approximate solutions and 3D-plot of GLM are represented in Figure 3. Table 4 and Figure 4, show the numerical results using GLM compared with RK method.

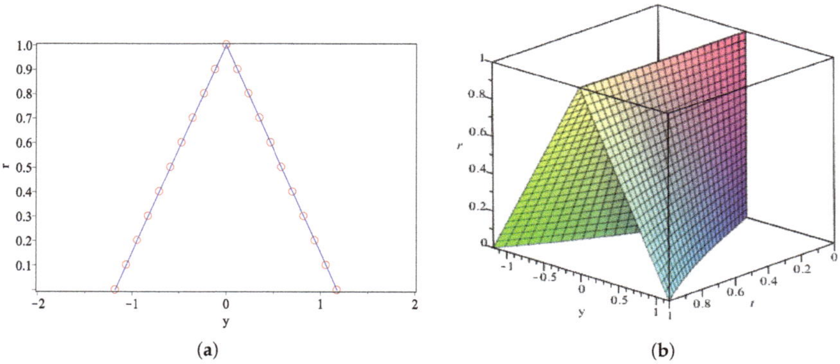

Figure 3. (**a**) Approximate solution of GLM (circle) and exact solution (line) at $t = 1.0$; (**b**) 3D-plot of GLM for Problem 2.

Table 4. Comparison between GLM and RK for solving Problem 2.

	GLM		RK	
r	$E^{-}(t=0.5;r)$	$E^{-}(t=1;r)$	$E^{-}(t=0.5;r)$	$E^{-}(t=1;r)$
0.0	2.069650(−10)	8.391900(−10)	1.431575(−9)	6.462170(−9)
0.1	1.862700(−10)	7.552900(−10)	1.288422(−9)	5.815940(−9)
0.2	1.655720(−10)	6.713560(−10)	1.145253(−9)	5.169731(−9)
0.3	1.655720(−10)	5.874450(−10)	1.145253(−9)	4.523523(−9)
0.4	1.241780(−10)	5.035170(−10)	8.589440(−10)	3.877302(−9)
0.5	1.034820(−10)	4.195950(−10)	7.157840(−10)	3.231079(−9)
0.6	8.278600(−11)	3.356810(−10)	5.726270(−10)	2.584868(−9)
0.7	6.208900(−11)	2.517620(−10)	4.294700(−10)	1.938651(−9)
0.8	4.139400(−11)	1.678420(−10)	2.863150(−10)	1.292426(−9)
0.9	2.069650(−11)	8.391900(−10)	1.431575(−10)	6.462170(−9)
1.0	0	0	0	0
r	$E^{+}(t=0.5;r)$	$E^{+}(t=1;r)$	$E^{+}(t=0.5;r)$	$E^{+}(t=1;r)$
0.0	2.069650(−10)	8.391900(−10)	1.431575(−9)	6.462170(−9)
0.1	1.862700(−10)	7.552900(−10)	1.288422(−9)	5.815940(−9)
0.2	1.655720(−10)	6.713560(−10)	1.145253(−9)	5.169731(−9)
0.3	1.448800(−10)	5.874450(−10)	1.002105(−9)	4.523523(−9)
0.4	1.241780(−10)	5.035170(−10)	8.589440(−10)	3.877302(−9)
0.5	1.034820(−10)	4.195950(−10)	7.157840(−10)	3.231079(−9)
0.6	8.278600(−11)	3.356810(−10)	5.726270(−10)	2.584868(−9)
0.7	6.208900(−11)	2.517620(−10)	4.294700(−10)	1.938651(−9)
0.8	4.139400(−11)	1.678420(−10)	2.863150(−10)	1.292426(−9)
0.9	2.069650(−11)	8.391900(−10)	1.431575(−10)	6.462170(−9)
1.0	0	0	0	0

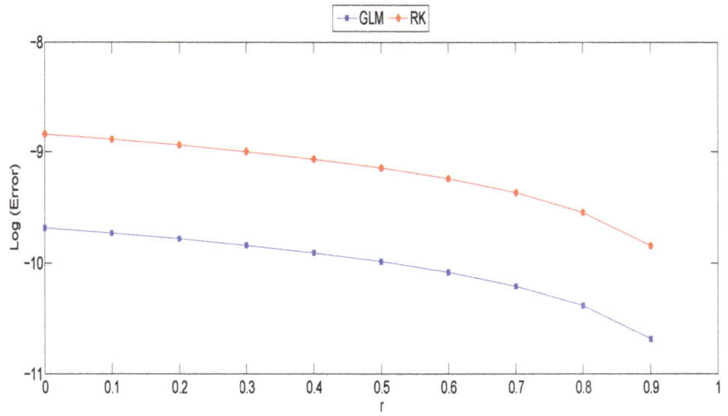

Figure 4. Graph of Log (Error) versus r at $t = 0.5$ for left bound of solutions and right bound of solutions for Problem 2.

8.3. Problem 3

Consider the following FVIDEs (see [23])

$$y'(t) = C + \int_0^t (-y(s))ds,$$
$$C = [2(r-2)\sin(t), 2(2-3r)\sin(t)], \quad y(0) = [3r-2, 2-r], \qquad 0 \le t \le 1.$$

The equivalent system of ODEs based on (i)-differentiability:

$$(y^-)'(t;r) = 2(r-2)\sin(t) + \int_0^1 (-1)y^+(s;r)ds, \quad y^-(0;r) = 3r-2,$$
$$(y^+)'(t;r) = 2(2-3r)\sin(t) + \int_0^1 (-1)y^-(s;r)ds, \quad y^+(0;r) = 2-r.$$

Exact solutions based on (i)-differentiability:

$$\tilde{Y}^-(t;r) = -rt\sin(t) + (2-r)\cos(t) + 2(r-1)(\exp(t)+\exp(-t)),$$
$$\tilde{Y}^+(t;r) = -rt\sin(t) + (3r-2)\cos(t) + 2(1-r)(\exp(t)+\exp(-t)).$$

The equivalent system of ODEs based on (ii)-differentiability:

$$(y^-)'(t;r) = 2(2-3r)\sin(t) + \int_0^1 (-1)y^-(s;r)ds, \quad y^-(0;r) = 3r-2,$$
$$(y^+)'(t;r) = 2(r-2)\sin(t) + \int_0^1 (-1)y^+(s;r)ds, \quad y^+(0;r) = 2-r.$$

Exact solutions based on (ii)-differentiability:

$$\tilde{Y}^-(t;r) = (3r-2)(\cos(t) - t\sin(t)),$$
$$\tilde{Y}^+(t;r) = (2-r)(\cos(t) - t\sin(t)).$$

For Problem 3, the graph of approximate solution compared with exact solution is given in Figure 5. Also the numerical results of GLM are compared with RK method using the both types of differentiabilities. The comparison of obtained results based on type (i)-differentiability are presented in Table 5 and Figure 6 whereas the results obtained based on type (ii)-differentiability are given in Table 6 and Figure 7. 3D-plots of GLM based on types (i) and (ii)-differentiability are shown in Figure 8.

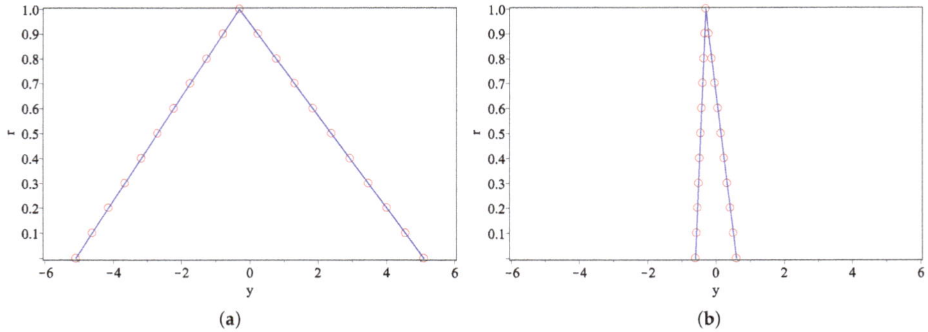

Figure 5. (**a**) Approximate solution of GLM (circle) and exact solution (line) at $t = 1.0$ using type (i)-differentiability; (**b**) Approximate solution of GLM (circle) and exact solution (line) at $t = 1.0$ using type (ii)-differentiability for Problem 3.

Table 5. Comparison between GLM and RK for solving Problem 3 based on (i)-differentiability.

	GLM		RK	
r	$E^-(t=0.5;r)$	$E^-(t=1;r)$	$E^-(t=0.5;r)$	$E^-(t=1;r)$
0.0	9.591260(−9)	1.703350(−8)	3.884529(−8)	8.745231(−8)
0.1	8.252610(−9)	1.499912(−8)	3.333684(−8)	7.665465(−8)
0.2	6.913930(−9)	1.296471(−8)	2.782840(−8)	6.585705(−8)
0.3	5.575350(−9)	1.093031(−8)	2.231994(−8)	5.505935(−8)
0.4	4.236740(−9)	8.895910(−9)	1.681141(−8)	4.426157(−8)
0.5	2.898090(−9)	6.861520(−9)	1.130299(−8)	3.346385(−8)
0.6	1.559467(−9)	4.827130(−9)	5.794547(−9)	2.266616(−8)
0.7	2.208310(−10)	2.792740(−9)	2.860890(−10)	1.186848(−8)
0.8	1.117786(−9)	7.583400(−9)	5.222359(−9)	1.070790(−8)
0.9	2.456414(−9)	1.276027(−9)	1.073082(−8)	9.726932(−8)
1.0	3.795042(−9)	3.310421(−9)	1.623928(−8)	2.052464(−8)
r	$E^+(t=0.5;r)$	$E^+(t=1;r)$	$E^+(t=0.5;r)$	$E^+(t=1;r)$
0.0	9.591260(−9)	1.703350(−8)	3.884529(−8)	8.745231(−8)
0.1	9.011600(−9)	1.566117(−8)	3.658467(−8)	8.075957(−8)
0.2	8.431990(−9)	1.428888(−8)	3.432415(−8)	7.406693(−8)
0.3	7.852400(−9)	1.291656(−8)	3.206350(−8)	6.737409(−8)
0.4	7.272740(−9)	1.154423(−8)	2.980293(−8)	6.068132(−8)
0.5	6.693130(−9)	1.017195(−8)	2.754225(−8)	5.398850(−8)
0.6	6.113500(−9)	8.799650(−9)	2.528165(−8)	4.729569(−8)
0.7	5.533890(−9)	7.427330(−9)	2.302113(−8)	4.060305(−8)
0.8	4.954250(−9)	6.054989(−9)	2.076049(−8)	3.391018(−8)
0.9	4.374664(−9)	4.682732(−9)	1.849986(−8)	2.721740(−8)
1.0	3.795042(−9)	3.310421(−9)	1.623928(−8)	2.052464(−8)

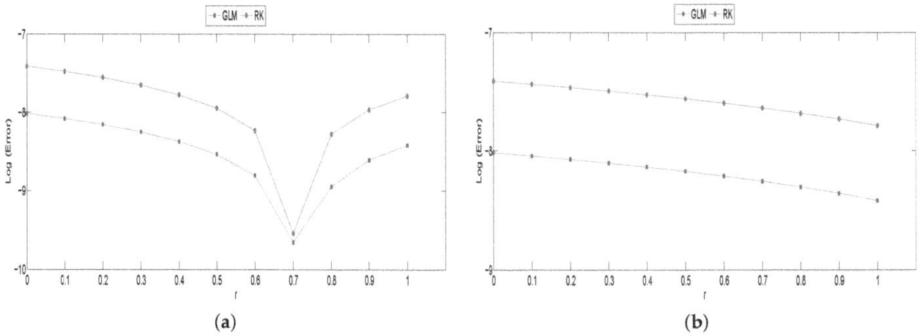

Figure 6. (**a**) Graph of Log (Error) versus r at $t = 0.5$ for left bound of solutions using type (i)-differentiability; (**b**) Graph of Log (Error) versus r at $t = 0.5$ for right bound of solutions using type (i)-differentiability for Problem 3.

Table 6. Comparison between GLM and RK for solving Problem 3 based on (ii)-differentiability.

	GLM		RK	
r	$E^-(t=0.5;r)$	$E^-(t=1;r)$	$E^-(t=0.5;r)$	$E^-(t=1;r)$
0.0	7.590100(−9)	6.620856(−9)	3.247856(−8)	4.104928(−8)
0.1	6.451570(−9)	5.627722(−9)	2.760677(−8)	3.489188(−8)
0.2	5.313055(−9)	4.634590(−9)	2.273503(−8)	2.873452(−8)
0.3	4.174528(−9)	3.641453(−9)	1.786324(−8)	2.257713(−8)
0.4	3.036036(−9)	2.648343(−9)	1.299143(−8)	1.641972(−8)
0.5	1.897518(−9)	1.655208(−9)	8.119643(−9)	1.026232(−8)
0.6	7.590100(−10)	6.620856(−10)	3.247856(−9)	4.104928(−9)
0.7	3.795042(−10)	3.310421(−10)	1.623928(−9)	2.052464(−9)
0.8	1.518015(−9)	1.324166(−9)	6.495721(−9)	8.209863(−8)
0.9	2.656530(−9)	2.317297(−9)	1.136750(−8)	1.436725(−8)
1.0	3.795042(−9)	3.310421(−9)	1.623928(−8)	2.052464(−8)
r	$E^+(t=0.5;r)$	$E^+(t=1;r)$	$E^+(t=0.5;r)$	$E^+(t=1;r)$
0.0	7.590100(−9)	6.620856(−9)	3.247856(−8)	4.104928(−8)
0.1	7.210590(−9)	6.289804(−9)	3.085463(−8)	3.899682(−8)
0.2	6.831070(−9)	5.958752(−9)	2.923064(−8)	3.694430(−8)
0.3	6.451570(−9)	5.627722(−9)	2.760677(−8)	3.489188(−8)
0.4	6.072060(−9)	5.296667(−9)	2.598287(−8)	3.283943(−8)
0.5	5.692550(−9)	4.965625(−9)	2.435890(−8)	3.078694(−8)
0.6	5.313055(−9)	4.634590(−9)	2.273503(−8)	2.873452(−8)
0.7	4.933553(−9)	4.303548(−9)	2.111106(−8)	2.668203(−8)
0.8	4.554036(−9)	3.972496(−9)	1.948713(−8)	2.462956(−8)
0.9	4.174528(−9)	3.641453(−9)	1.786324(−8)	2.257713(−8)
1.0	3.795042(−9)	3.310421(−9)	1.623928(−8)	2.052464(−8)

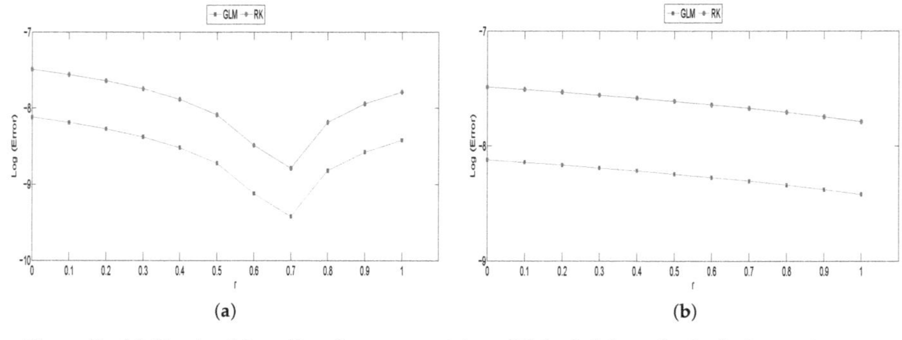

Figure 7. (**a**) Graph of Log (Error) versus r at $t = 0.5$ for left bound of solutions using type (ii)-differentiability; (**b**) Graph of Log (Error) versus r at $t = 0.5$ for right bound of solutions using type (ii)-differentiability for Problem 3.

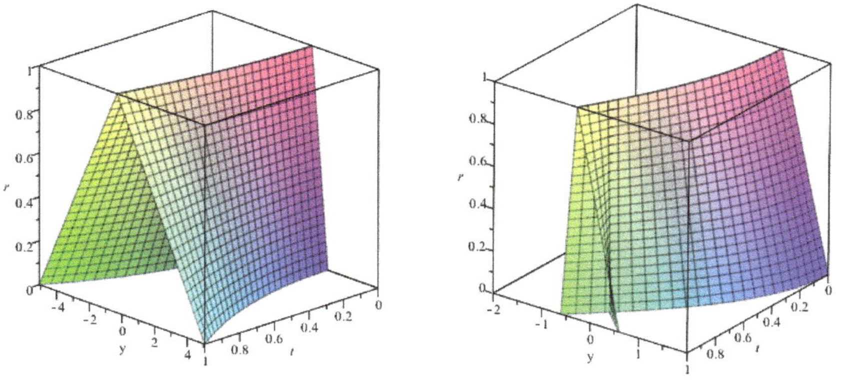

Figure 8. (**a**) 3D-plot of GLM using type (i)-differentiability for Problem 3; (**b**) 3D-plot of GLM using type (ii)-differentiability for Problem 3.

9. Discussion and Conclusions

Using the fuzzy GLM and fuzzy RK derived in this paper, the three FVIDEs given in Problems 1–3 are solved and the numerical results are shown. In addition, the obtained numerical results are compared with two other existing methods, variational iteration method and homotopy perturbation method, for Problem 1 at $t = 0.5$.

For Problem 1, in Table 3, it is observed that the left bound of errors at $t = 0.5$ obtained by the fuzzy GLM is competitive with the fuzzy RK for $r = 0.2, 0.9, 1.0$. At $t = 1$, the left bound errors from the fuzzy GLM is comparable with fuzzy RK when $r = 0.1, 0.7, 0.8, 0.9$. However, for the rest of errors fuzzy GLM achieved better accuracy compared to fuzzy RK. Moreover, the errors for the right bound of fuzzy GLM at both $t = 0.5$ and $t = 1$ are found to be one decimal place better than the fuzzy RK method. Meanwhile, the results acquired by GLM are almost the same with the results acquired by VIM. In comparison between GLM and HPM, GLM clearly outperformed the HPM.

For Problem 2, in Table 4, the fuzzy GLM is competitive with the fuzzy RK only when $r = 0.4$ and $r = 0.5$, though for the rest of r-levels the fuzzy GLM outperformed the fuzzy RK again by one decimal place better. For Problem 3, both types of differentiability are applied to solve this problem. In Table 5 by using type (i)-differentiability, there are some competitive results between the fuzzy GLM and fuzzy RK. However, in Table 6 by using type (ii)-differentiability, for almost all r-levels the fuzzy GLM gave more accurate results than the fuzzy RK.

Graphical illustrations of approximated solutions by fuzzy GLM in comparison with the exact solutions and 3D-plots of GLM for solving FVIDEs are presented in Figures 1, 3, 5 and 8. The graphs shown that the GLM performed the accurate results. Moreover, in Figure 5b the approximate solutions of GLM based on type (ii)-differentiability showed smaller bound compared to the solutions based on type (i)-differentiability in Figure 5a. Considering that there exists a negative function in Problem 3, therefore the (ii)-differentiability approach is preferred. Graphs of comparison in terms of errors at $t = 0.5$ between fuzzy GLM and other methods are showed as well. The fuzzy GLM is seen competitive with the VIM in Figure 2 meanwhile from Figures 4, 6, and 7, the fuzzy GLM is the more accurate method than HPM and RK3.

In conclusion, the fuzzy GLM combined with Simpson's II method and Lagrange interpolation polynomials is an efficient numerical method for solving FVIDEs.

Author Contributions: F.R. and F.A.H. conceived, designed, performed the experiments and wrote the paper. Z.A.M. analyzed the designed experiments and data, F.I. checked the results and edited the paper.

Funding: This research was funded by Research University Grant from Research Management Centre of Universiti Putra Malaysia.

Acknowledgments: This work was part of FRGS research grant project (Ref: FRGS/2/2013/SG04/UPM/02/1) and the authors wish to thank on Ministry of Higher Education Malaysia for supporting this research.

Conflicts of Interest: We declare that there does not exists any conflict of interest regarding this paper.

References

1. Khastan, A.; Ivaz, K. Numerical solution of fuzzy differential equations by Nyström method. *Chaos Solitons Fractals* **2009**, *41*, 859–868. [CrossRef]
2. Abu-Arqub, O.; El-Ajou, A.; Momani, S.; Shawagfeh, N. Analytical solutions of fuzzy initial value problems by HAM. *Appl. Math. Inf. Sci.* **2013**, *7*, 1903. [CrossRef]
3. Puri, M.L.; Ralescu, D.A. Differentials of fuzzy functions. *J. Math. Anal. Appl.* **1983**, *91*, 552–558. [CrossRef]
4. Buckley, J.J.; Feuring, T. Fuzzy differential equations. *Fuzzy Sets Syst.* **2000**, *110*, 43–54. [CrossRef]
5. Bede, B.; Gal, S.G. Generalizations of the differentiability of fuzzy-number-valued functions with applications to fuzzy differential equations. *Fuzzy Sets Syst.* **2005**, *151*, 581–599. [CrossRef]
6. Goetschel, R.; Voxman, W. Elementary fuzzy calculus. *Fuzzy Sets Syst.* **1986**, *18*, 31–43. [CrossRef]
7. Kaleva, O. Fuzzy differential equations. *Fuzzy Sets Syst.* **1987**, *24*, 301–317. [CrossRef]
8. Friedman, M.; Ma, M.; Kandel, A. Numerical solutions of fuzzy differential and integral equations. *Fuzzy Sets Syst.* **1999**, *106*, 35-48. [CrossRef]
9. Park, J.Y.; Jeong, J.U. On existence and uniqueness of solutions of fuzzy integrodifferential equations. *Indian J. Pure Appl. Math.* **2003**, *34*, 1503–1512.
10. Hajighasemi, S.; Allahviranloo, T.; Khezerloo, M.; Khorasany, M.; Salahshour, S. Existence and uniqueness of solutions of fuzzy Volterra integro-differential equations.In Proceedings of the International Conference on Information Processing and Management of Uncertainty in Knowledge-Based Systems, Cadiz, Spain, 11–15 June 2010; pp. 491–500.
11. Zeinali, M.; Shahmorad, S.; Mirnia, K. Fuzzy integro-differential equations: Discrete solution and error estimation. *Iranian J. Fuzzy Syst.* **2013**, *10*, 107–122.
12. Mikaeilvand, N.; Khakrangin, S.; Allahviranloo, T. Solving fuzzy Volterra integro-differential equation by fuzzy differential transform method. In Proceedings of the 7th Conference of the European Society for Fuzzy Logic and Technology, Aix-Les-Bains, France, 18–22 July 2011; Atlantis Press: Paris, France, 2011; pp. 891–896.
13. Allahviranloo, T.; Abbasbandy, S.; Sedaghatfar, O. A New Method for Solving Fuzzy Integro-Differential Equation Under Generalized Differentiability. *Neural Comput. Appl.* **2012**, *21*, 191–196. [CrossRef]
14. Allahviranloo, T.; Khezerloo, M.; Sedaghatfar, O.; Salahshour, S. Toward the existence and uniqueness of solutions of second-order fuzzy volterra integro-differential equations with fuzzy kernel. *Neural Comput. Appl.* **2013**, *22*, 133–141. [CrossRef]
15. Matinfar, M.; Ghanbari, M.; Nuraei, R. Numerical solution of linear fuzzy Volterra integro-differential equations by variational iteration method. *J. Intell. Fuzzy Syst.* **2013**, *24*, 575–586.
16. Sahu, P.K.; Saha Ray, S. Two-dimensional Legendre wavelet method for the numerical solutions of fuzzy integro-differential equations. *J. Intell. Fuzzy Syst.* **2015**, *28*, 1271–1279.
17. Butcher, J.C. General linear methods. *Acta Numer.* **2006**, *15*, 157–256. [CrossRef]
18. Rabiei, F.; Abd Hamid, F.; Rashidi, M.M.; Ismail, F. Numerical simulation of fuzzy differential equations using general linear method and B-series. *Adv. Mech. Eng.* **2017**, *9*, 1–16. [CrossRef]
19. Bede, B. *Mathematics of Fuzzy Sets and Fuzzy Logic*; Springer: Berlin, Germany, 2013; 295p.
20. Chalco-Cano, Y.; Román-Flores, H. On new solutions of fuzzy differential equations. *Chaos Solitons Fractals* **2008**, *38*, 112–119. [CrossRef]
21. Linz, P. *Analytical and Numerical Methods for Volterra Equations*; Siam: Philadelphia, PA, USA, 1985; 227p.

22. Butcher, J.C. *Numerical Methods or Ordinary Differential Equations Second Edition*; John Wiley and Sons: New York, NY, USA, 2008; 48p.
23. Ghanbari, M. A new approach for solving fuzzy linear Volterra integro-differential equations. *Iran. J. Fuzzy Syst.* **2016**, *13*, 69–87.

© 2019 by the authors. Licensee MDPI, Basel, Switzerland. This article is an open access article distributed under the terms and conditions of the Creative Commons Attribution (CC BY) license (http://creativecommons.org/licenses/by/4.0/).

symmetry

MDPI

Article

Upper Bound of the Third Hankel Determinant for a Subclass of q-Starlike Functions

Shahid Mahmood [1], Hari M. Srivastava [2,3], Nazar Khan [4], Qazi Zahoor Ahmad [4,*], Bilal Khan [4] and Irfan Ali [5]

1 Department of Mechanical Engineering, Sarhad University of Science and Information Technology, Ring Road, Peshawar 25000, Pakistan; shahidmahmood757@gmail.com

2 Department of Mathematics and Statistics, University of Victoria, Victoria, BC V8W 3R4, Canada; harimsri@math.uvic.ca

3 Department of Medical Research, China Medical University Hospital, China Medical University, Taichung 40402, Taiwan

4 Department of Mathematics, Abbottabad University of Science and Technology, Abbottabad 22010, Pakistan; nazarmaths@gmail.com (N.K.); bilalmaths789@gmail.com (B.K.)

5 Department of Mathematical Sciences, Balochistan University of Information Technology, Engineering and Management Sciences, Quetta 87300, Pakistan; irfan.ali@buitms.edu.pk

* Correspondence: zahoorqazi5@gmail.com; Tel.: +92-334-96-60-162

Received: 1 January 2019; Accepted: 23 February 2019; Published: 7 March 2019

Abstract: The main purpose of this article is to find the upper bound of the third Hankel determinant for a family of q-starlike functions which are associated with the Ruscheweyh-type q-derivative operator. The work is motivated by several special cases and consequences of our main results, which are pointed out herein.

Keywords: analytic functions; Hadamard product; starlike functions; q-derivative (or q-difference) operator; Hankel determinant; q-starlike functions

MSC: Primary 05A30, 30C45; Secondary 11B65, 47B38

1. Introduction

We denote by $\mathcal{A}(\mathbb{U})$ the class of functions which are analytic in the open unit disk

$$\mathbb{U} = \{z : z \in \mathbb{C} \quad \text{and} \quad |z| < 1\},$$

where $\mathbb{C}$ is the complex plane. Let $\mathcal{A}$ be the class of analytic functions having the following normalized form:

$$f(z) = z + \sum_{n=2}^{\infty} a_n z^n \qquad (\forall\, z \in \mathbb{U}) \tag{1}$$

in the open unit disk $\mathbb{U}$, centered at the origin and normalized by the conditions given by

$$f(0) = 0 \quad \text{and} \quad f'(0) = 1.$$

In addition, let $\mathcal{S} \subset \mathcal{A}$ be the class of functions which are univalent in $\mathbb{U}$. The class of starlike functions in $\mathbb{U}$ will be denoted by $\mathcal{S}^*$, which consists of normalized functions $f \in \mathcal{A}$ that satisfy the following inequality:

$$\Re\left(\frac{zf'(z)}{f(z)}\right) > 0, \qquad (\forall\, z \in \mathbb{U}). \tag{2}$$

If two functions f and g are analytic in $\mathbb{U}$, we say that the function f is subordinate to g and write in the form:

$$f \prec g \quad \text{or} \quad f(z) \prec g(z),$$

if there exists a Schwarz function w which is analytic in $\mathbb{U}$, with

$$w(0) = 0 \qquad \text{and} \qquad |w(z)| < 1,$$

such that

$$f(z) = g(w(z)).$$

In particular, if the function g is univalent in $\mathbb{U}$, then it follows that (cf., e.g., [1]; see also [2])

$$f(z) \prec g(z) \quad (z \in \mathbb{U}) \Rightarrow f(0) = g(0) \quad \text{and} \quad f(\mathbb{U}) \subset g(\mathbb{U}).$$

Moreover, for two analytic functions f and g given by

$$f(z) = z + \sum_{n=2}^{\infty} a_n z^n \qquad (\forall\, z \in \mathbb{U})$$

and

$$g(z) = z + \sum_{n=2}^{\infty} b_n z^n \qquad (\forall\, z \in \mathbb{U}),$$

the convolution (or the Hadamard product) of f and g is defined as follows:

$$f(z) * g(z) = z + \sum_{n=2}^{\infty} a_n b_n z^n.$$

We next denote by $\mathcal{P}$ the class of analytic functions p which are normalized by

$$p(z) = 1 + \sum_{n=1}^{\infty} p_n z^n, \tag{3}$$

such that

$$\Re(p(z)) > 0 \qquad (z \in \mathbb{U}).$$

We now recall some essential definitions and concept details of the basic or quantum (q-) calculus, which are used in this paper. We suppose throughout the paper that $0 < q < 1$ and that

$$\mathbb{N} = \{1, 2, 3, \cdots\} = \mathbb{N}_0 \setminus \{0\} \qquad (\mathbb{N}_0 = \{0, 1, 2, 3, \cdots\}).$$

Definition 1. *Let $q \in (0,1)$ and define the q-number $[\lambda]_q$ by*

$$[\lambda]_q = \begin{cases} \dfrac{1 - q^{\lambda}}{1 - q} & (\lambda \in \mathbb{C}) \\[2ex] \sum\limits_{k=0}^{n-1} q^k = 1 + q + q^2 + \cdots + q^{n-1} & (\lambda = n \in \mathbb{N}). \end{cases}$$

Definition 2. *Let $q \in (0,1)$ and define the q-factorial $[n]_q!$ by*

$$[n]_q! = \begin{cases} 1 & (n = 0) \\ \prod_{k=1}^{n-1} [k]_q & (n \in \mathbb{N}). \end{cases}$$

Definition 3. *Let $q \in (0,1)$ and define the generalized q-Pochhammer symbol $[\lambda]_{q,n}$ by*

$$[\lambda]_{q,n} = \begin{cases} 1 & (n = 0) \\ \prod_{k=0}^{n} [\lambda + k]_q & (n \in \mathbb{N}). \end{cases}$$

Definition 4. *For $\omega > 0$, let the q-gamma function $\Gamma_q(\omega)$ be defined by*

$$\Gamma_q(\omega + 1) = [\omega]_q \Gamma_q(\omega) \quad \text{and} \quad \Gamma_q(1) := 1.$$

Definition 5. (see [3,4]) *The q-derivative (or the q-difference) operator D_q of a function f in a given subset of $\mathbb{C}$ is defined by*

$$(D_q f)(z) = \begin{cases} \dfrac{f(z) - f(qz)}{(1-q)z} & (z \neq 0) \\ f'(0) & (z = 0), \end{cases} \tag{4}$$

provided that $f'(0)$ exists.

We note from Definition 5 that

$$\lim_{q \to 1-} (D_q f)(z) = \lim_{q \to 1-} \frac{f(qz) - f(z)}{(1-q)z} = f'(z),$$

for a differentiable function f in a given subset of $\mathbb{C}$. It is readily deduced from (1) and (4) that

$$(D_q f)(z) = 1 + \sum_{n=2}^{\infty} [n]_q a_n z^{n-1}. \tag{5}$$

The operator D_q plays a vital role in the investigation and study of numerous subclasses of the class of analytic functions of the form given in Definition 5. A q-extension of the class of starlike functions was first introduced in [5] by using the q-derivative operator (see Definition 6 below). A background of the usage of the q-calculus in the context of Geometric Funciton Theory was actually provided and the basic (or q-) hypergeometric functions were first used in Geometric Function Theory by Srivastava (see, for details, [6]). Some recent investigations associated with the q-derivative operator D_q in analytic function theory can be found in [7–13] and the references cited therein.

Definition 6. (see [5]) *A function $f \in \mathcal{A}(\mathbb{U})$ is said to belong to the class $\mathcal{S}_q^*$ if*

$$f(0) = f'(0) - 1 = 0 \tag{6}$$

and

$$\left| \frac{z}{f(z)} (D_q f)(z) - \frac{1}{1-q} \right| \leqq \frac{1}{1-q} \quad (\forall\, z \in \mathbb{U}). \tag{7}$$

The notation $\mathcal{S}_q^*$ was first used by Sahoo et al. (see [14]).

It is readily observed that, as $q \to 1-$, the closed disk given

$$\left| w - \frac{1}{1-q} \right| \leqq \frac{1}{1-q}$$

becomes the right-half plane and the class $\mathcal{S}_q^*$ reduces to $\mathcal{S}^*$. Equivalently, by using the principle of subordination between analytic functions, we can rewrite the conditions in (6) and (7) as follows (see [15]):

$$\frac{z}{f(z)} \left(D_q f \right)(z) \prec \widehat{p} \qquad \left(\widehat{p} = \frac{1+z}{1-qz} \right).$$

Definition 7. (see [16]) *For a function $f \in \mathcal{A}(\mathbb{U})$, the Ruscheweyh-type q-derivative operator is defined as follows:*

$$\mathcal{R}_q^{\delta} f(z) = \phi(q, \delta+1; z) * f(z) = z + \sum_{n=2}^{\infty} \psi_{n-1} a_n z^n \quad (z \in \mathbb{U};\ \delta > -1), \tag{8}$$

where

$$\phi(q, \delta+1; z) = z + \sum_{n=2}^{\infty} \psi_{n-1} z^n \tag{9}$$

and

$$\psi_{n-1} = \frac{\Gamma_q(\delta+n)}{[n-1]_q! \Gamma_q(\delta+1)} = \frac{[n+1]_{n-1,q}}{[n-1]_q!}. \tag{10}$$

From (8) it can be seen that

$$\mathcal{R}_q^0 f(z) = f(z) \quad \text{and} \quad \mathcal{R}_q^1 f(z) = z D_q f(z),$$

$$\mathcal{R}_q^m f(z) = \frac{z D_q^m f(z) \left(z^{m-1} f(z) \right)}{[m]_q!} \qquad (m \in \mathbb{N}),$$

$$\lim_{q \to 1-} \phi(q, \delta+1; z) = \frac{z}{(1-z)^{\delta+1}}$$

and

$$\lim_{q \to 1-} \mathcal{R}_q^{\delta} f(z) = f(z) * \frac{z}{(1-z)^{\delta+1}}.$$

This shows that, in case of $q \to 1-$, the Ruscheweyh-type q-derivative operator reduces to the Ruscheweyh derivative operator $D^{\delta} f(z)$ (see [17]). From (8) the following identity can easily be derived:

$$z D_q \mathcal{R}_q^{\delta} f(z) = \left(1 + \frac{[\delta]_q}{q^{\delta}} \right) \mathcal{R}_q^{\delta+1} f(z) - \frac{[\delta]_q}{q^{\delta}} \mathcal{R}_q^{\delta} f(z). \tag{11}$$

If $q \to 1-$, then

$$z \left(\mathcal{R}^{\delta} f(z) \right)' = (1+\delta) \mathcal{R}^{\delta+1} f(z) - \delta \mathcal{R}^{\delta} f(z).$$

Now, by using the Ruscheweyh-type q-derivative operator, we define the following class of q-starlike functions.

Definition 8. *For $f \in \mathcal{A}(\mathbb{U})$, we say that f belongs to the class $\mathcal{RS}_q^*(\delta)$ if the following inequality holds true:*

$$\left| \frac{z D_q \mathcal{R}_q^{\delta} f(z)}{f(z)} - \frac{1}{1-q} \right| \leqq \frac{1}{1-q} \qquad (z \in \mathbb{U};\ \delta > -1)$$

or, equivalently, we have (see [15])

$$\frac{zD_q\mathcal{R}_q^{\delta}f(z)}{f(z)} \prec \frac{1+z}{1-qz} \tag{12}$$

by using the principle of subordination.

Let $n \geqq 0$ and $j \geqq 1$. The jth Hankel determinant is defined as follows:

$$\mathcal{H}_j(n) = \begin{vmatrix} a_n & a_{n+1} & \cdot & \cdot & \cdot & a_{n+j-1} \\ a_{n+1} & \cdot & & & & \cdot \\ \cdot & \cdot & & & & \cdot \\ \cdot & \cdot & & & & \cdot \\ \cdot & \cdot & & & & \cdot \\ a_{n+j-1} & \cdot & \cdot & \cdot & \cdot & a_{n+2(j-1)} \end{vmatrix}$$

The above Hankel determinant has been studied by several authors. In particular, sharp upper bounds on $\mathcal{H}_2(2)$ were obtained by several authors (see, for example, [18–21]) for various classes of normalized analytic functions. It is well-known that the Fekete-Szegö functional $|a_3 - a_2^2| = \mathcal{H}_2(1)$. This functional is further generalized as $|a_3 - \mu a_2^2|$ for some real or complex μ. In fact, Fekete and Szegö gave sharp estimates of $|a_3 - \mu a_2^2|$ for real μ and $f \in \mathcal{S}$, the class of normalized univalent functions in $\mathbb{U}$. It is also known that the functional $|a_2a_4 - a_3^2|$ is equivalent to $\mathcal{H}_2(2)$. Babalola [22] studied the Hankel determinant $\mathcal{H}_3(1)$ for some subclasses of analytic functions. In the present investigation, our focus is on the Hankel determinant $\mathcal{H}_3(1)$ for the above-defined function class $\mathcal{RS}_q^*(\delta)$.

2. A Set of Lemmas

Each of the following lemmas will be needed in our present investigation.

Lemma 1. (see [23]) *Let*

$$p(z) = 1 + c_1z + c_2z^2 + \cdots$$

be in the class $\mathcal{P}$ of functions with positive real part in $\mathbb{U}$. Then, for any complex number v,

$$\left|c_2 - vc_1^2\right| \leqq \begin{cases} -4v+2 & (v \leqq 0) \\ 2 & (0 \leqq v \leqq 1) \\ 4v-2 & (v \geqq 1). \end{cases} \tag{13}$$

When $v < 0$ or $v > 1$, the equality holds true in (13) if and only if

$$p(z) = \frac{1+z}{1-z}$$

or one of its rotations. If $0 < v < 1$, then the equality holds true in (13) if and only if

$$p(z) = \frac{1+z^2}{1-z^2}$$

or one of its rotations. If $v = 0$, the equality holds true in (13) if and only if

$$p(z) = \left(\frac{1+\rho}{2}\right)\frac{1+z}{1-z} + \left(\frac{1-\rho}{2}\right)\frac{1-z}{1+z} \qquad (0 \leqq \rho \leqq 1)$$

or one of its rotations. If $v = 1$, then the equality in (13) holds true if $p(z)$ is a reciprocal of one of the functions such that the equality holds true in the case when $v = 0$.

Lemma 2. (see [24,25]) *Let*

$$p(z) = 1 + p_1 z + p_2 z^2 + \cdots$$

be in the class $\mathcal{P}$ of functions with positive real part in $\mathbb{U}$. Then

$$2p_2 = p_1^2 + x\left(4 - p_1^2\right)$$

for some x, $|x| \leqq 1$ and

$$4p_3 = p_1^3 + 2\left(4 - p_1^2\right)p_1 x - \left(4 - p_1^2\right)p_1 x^2 + 2\left(4 - p_1^2\right)\left(1 - |x|^2\right)z$$

for some z $(|z| \leqq 1)$.

Lemma 3. (see [26]) *Let*

$$p(z) = 1 + p_1 z + p_2 z^2 + \cdots$$

be in the class $\mathcal{P}$ of functions positive real part in $\mathbb{U}$. Then

$$|p_k| \leqq 2 \qquad (k \in \mathbb{N})$$

and the inequality is sharp.

3. Main Results

In this section, we will prove our main results. Throughout our discussion, we assume that

$$q \in (0,1) \quad \text{and} \quad \delta > -1.$$

Our first main result is stated as follows.

Theorem 1. *Let $f \in \mathcal{RS}_q^*(\delta)$ be of the form (1). Then*

$$\left|a_3 - \mu a_2^2\right| \leqq \begin{cases} \dfrac{(1+q+q^2)\,\psi_1^2 - \mu\,(1+q)^2\,\psi_2}{q^2\psi_2\psi_1^2} & \left(\mu < \dfrac{(q^2+1)\,\psi_1^2}{(1+q)^2\,\psi_2}\right) \\[2ex] \dfrac{1}{q\psi_2} & \left(\dfrac{(q^2+1)\,\psi_1^2}{(1+q)^2\,\psi_2} \leqq \mu \leqq \dfrac{\psi_1^2}{\psi_2}\right) \\[2ex] \dfrac{\mu\,(1+q)^2\,\psi_2 - (1+q+q^2)\,\psi_1^2}{q^2\psi_2\psi_1^2} & \left(\mu > \dfrac{\psi_1^2}{\psi_2}\right), \end{cases}$$

where ψ_{n-1} is given by (10).

It is also asserted that, for

$$\frac{(q^2+1)\,\psi_1^2}{(1+q)^2\,\psi_2} \leqq \mu \leqq \frac{(1+q+q^2)\,\psi_1^2}{(1+q)^2\,\psi_2},$$

$$|a_3 - \mu a_2^2| + \left(\mu - \frac{(q^2+1)\,\psi_1^2}{(1+q)^2\,\psi_2}\right)|a_2|^2 \leq \frac{1}{q\psi_2}$$

and that , for

$$\frac{(1+q+q^2)\,\psi_1^2}{(1+q)^2\,\psi_2} \leqq \mu \leqq \frac{\psi_1^2}{\psi_2},$$

$$\left|a_3-\mu a_2^2\right|+\left(\frac{\psi_1^2-\mu\psi_2}{\psi_2}\right)|a_2|^2 \leqq \frac{1}{q\psi_2}.$$

Proof. If $f \in \mathcal{RS}_q^*(\delta)$, then it follows from (12) that

$$\frac{zD_q\mathcal{R}_q^\delta f(z)}{f(z)} \prec \phi(z), \tag{14}$$

where

$$\phi(z)=\frac{1+z}{1-qz}.$$

We define a function $p(z)$ by

$$p(z)=\frac{1+w(z)}{1-w(z)}=1+p_1z+p_2z^2+p_3z^3+\cdots.$$

It is clear that $p \in \mathcal{P}$. From the above equation, we have

$$w(z)=\frac{p(z)-1}{p(z)+1}.$$

From (14), we find that

$$\frac{zD_q\mathcal{R}_q^\delta f(z)}{f(z)}=\phi(w(z)),$$

together with

$$\phi(w(z))=\frac{2p(z)}{(1-q)p(z)+1+q}.$$

Now

$$\begin{aligned}
&\frac{2p(z)}{(1-q)p(z)+1+q}\\
&\quad=1+\frac{1}{2}(1+q)p_1z+\left\{\frac{1}{2}(q+1)p_2-\frac{1}{4}(1-q^2)p_1^2\right\}z^2\\
&\quad+\left\{\frac{1}{2}(1+q)p_3-\frac{1}{2}(1-q^2)p_1p_2+\frac{1}{8}(1+q)(1-q)^2p_1^3\right\}z^3\\
&\quad+\left\{\frac{1}{2}(1+q)p_4=\frac{1}{4}\left(1-q^2\right)p_2^2-\frac{1}{2}\left(1-q^2\right)p_1p_3\right.\\
&\quad\left.+\frac{3}{8}(1+q)(q-1)^2p_1^2p_2+\frac{1}{16}(1+q)(1-q)^3p_1^4\right\}z^4+\cdots.
\end{aligned}$$

Similarly, we get

$$\begin{aligned}
\frac{zD_q\mathcal{R}_q^\delta f(z)}{\mathcal{R}_q^\delta f(z)} = {} & 1+qa_2\psi_1z+\left\{\left(q+q^2\right)\psi_2a_3-q\psi_1^2a_2^2\right\}z^2+\left\{\left(q+q^2+q^3\right)\psi_3a_4\right.\\
&\left.-\left(2q+q^2\right)\psi_1\psi_2a_2a_3+q\psi_1^3a_2^3\right\}z^3+\left\{\left(q+q^2+q^3+q^4\right)\psi_5a_5\right.\\
&-\left(2q+q^2+q^3\right)\psi_2\psi_3a_2a_4-\left(q+q^2\right)\psi_2^2a_3^2\\
&\left.+\left(3q+q^2\right)\psi_1^2\psi_2a_2^2a_3-q\psi_1^4a_2^4\right\}z^4+\cdots,
\end{aligned}$$

Therefore, we have

$$a_2 = \frac{(1+q)}{2q\psi_1} p_1, \tag{15}$$

$$a_3 = \frac{1}{2q\psi_2} p_2 + \frac{(q^2+1)}{4q^2\psi_2} p_1^2 \tag{16}$$

and

$$\begin{aligned} a_4 &= \frac{(1+q)}{2q(1+q+q^2)\psi_3} p_3 - \frac{(1+q)(q-2)(2q+1)}{4q^2(1+q+q^2)\psi_3} p_1 p_2 \\ &+ \frac{(1+q)(q^2+1)(q^2-q+1)}{8q^3(1+q+q^2)\psi_3} p_1^3. \end{aligned} \tag{17}$$

We thus obtain

$$\left|a_3 - \mu a_2^2\right| = \frac{1}{2q\psi_2} \left| p_2 - \left(\frac{\mu (1+q)^2 \psi_2 - (1+q^2)\psi_1^2}{2q\psi_1^2} \right) p_1^2 \right|. \tag{18}$$

Finally, by applying Lemma 1 and Equation (13) in conjunction with (18), we obtain the result asserted by Theorem 1. □

We now state and prove Theorem 2 below.

Theorem 2. *Let $f \in \mathcal{RS}_q^*(\delta)$ be of the form (1). Then*

$$\left|a_2 a_4 - a_3^2\right| \leqq \frac{1}{q^2\psi_2^2}.$$

Proof. From (15)–(17), we obtain

$$\begin{aligned} a_2 a_4 - a_3^2 = & \left(\frac{(1+q)^2}{4q^2(1+q+q^2)\psi_1\psi_3} \right) p_1 p_3 - \left(\frac{(1+q)^2(q-2)(2q+1)}{8q^3(1+q+q^2)\psi_1\psi_3} + \frac{(q^2+1)}{4q^3\psi_2^2} \right) p_1^2 p \\ & - \left(\frac{1}{4q^2\psi_2^2} \right) p_2^2 + \left(-\frac{(q^2+1)^2}{16q^4\psi_2^2} + \frac{(1+q)^2(q^2+1)(q^2-q+1)}{16q^3(1+q+q^2)\psi_1\psi_3} \right) p_1^4. \end{aligned}$$

By using Lemma 2, we have

$$\begin{aligned} a_2 a_4 - a_3^2 = & \left(\frac{(1+q)^2(q^2+1)(q^2-q+1)}{16q^3(1+q+q^2)\psi_1\psi_3} - \frac{(q^2+1)^2}{16q^4\psi_2^2} \right) p_1^4 \\ & + \left(\frac{(1+q)^2}{16q^2(1+q+q^2)\psi_1\psi_3} \right) p_1 \left\{ p_1^3 + 2p_1\left(4-p_1^2\right)x \right. \\ & \left. - p_1\left(4-p_1^2\right)x^2 + 2\left(4-p_1^2\right)\left(1-|x|^2\right)z \right\} + \left(\frac{(q^2+1)}{8q^3\psi_2^2} \right. \\ & \left. \cdot \frac{(1+q)^2(q-2)(2q+1)}{16q^3(1+q+q^2)\psi_1\psi_3} \right) p_1^2 \left\{ \left(p_1^2 + \left(4-p_1^2\right)x \right) \right\} \\ & - \left(\frac{1}{16q^2\psi_2^2} \right) \left\{ p_1^4 + \left(4-p_1^2\right)^2 x^2 + 2p_1^2\left(4-p_1^2\right)x \right\}. \end{aligned}$$

Now, taking the moduli and replacing $|x|$ by ρ and p_1 by p, we have

$$\begin{aligned} \left|a_2a_4 - a_3^2\right| &\leqq \frac{1}{\Lambda(q)}\left[\omega(q)\,p^4 + 2q(1+q)^2\psi_2^2 p\left(4-p^2\right)\right. \\ &\quad + \Omega(q)\left(4-p^2\right)p^2\rho + \left(q(q+1)^2\psi_2^2p^2 + q\left(4-p^2\right)\right. \\ &\quad \left.\left.\cdot\left(1+q+q^2\right)\psi_1\psi_3 - 2q(1+q)^2\psi_2^2 p\right)\left(4-p^2\right)\rho^2\right] \\ &= F(p,\rho), \end{aligned} \tag{19}$$

where

$$\Lambda(q) = 16q^3\left(1+q+q^2\right)\psi_1\psi_3\psi_2^2,$$

$$\begin{aligned} \omega(q) = \Big|&\left(3+3q-q^3+q^4\right)(1+q)^2\psi_2^2 - \left(1+3q+2q^2+2q^3+q^4\right) \\ &\cdot\left(1+q+q^2\right)\psi_1\psi_3\Big| \end{aligned}$$

and

$$\Omega(q) = \left|(1+q)^2\left(2q^2-5q-2\right)\psi_2^2 + 2q\left(q^2+2\right)\left(1+q+q^2\right)\psi_1\psi_3\right|.$$

Upon differentiating both sides (19) with respect to ρ, we have

$$\begin{aligned} \frac{\partial F(p,\rho)}{\partial\rho} &= \left(\frac{1}{\Lambda(q)}\right)\left[\Omega(q)\left(4-p^2\right)p^2 + 2\left(q(q+1)^2\psi_2^2p^2 + q\left(4-p^2\right)\right.\right. \\ &\quad \left.\left.\cdot\left(1+q+q^2\right)\psi_1\psi_3 - 2q(1+q)^2\psi_2^2p\right)\left(4-p^2\right)\rho\right]. \end{aligned}$$

It is clear that

$$\frac{\partial F(p,\rho)}{\partial\rho} > 0,$$

which show that $F(p,\rho)$ is an increasing function of ρ on the closed interval $[0,1]$. This implies that the maximum value occurs at $\rho = 1$. This implies that

$$\max\{F(p,\rho)\} = F(p,1) =: G(p).$$

We now observe that

$$\begin{aligned} G(p) &= \left(\frac{1}{\Lambda(q)}\right)\left[\left(\omega(q) - \Omega(q) - q(q+1)^2\psi_2^2 + \left(q+q^2+q^3\right)\psi_1\psi_3\right)p^4\right. \\ &\quad + \left(4\Omega(q) + 4q(q+1)^2\psi_2^2 - 8\left(q+q^2+q^3\right)\psi_1\psi_3\right)p^2 \\ &\quad + 16\left(q+q^2+q^3\right)\psi_1\psi_3 \\ &= G(p). \end{aligned} \tag{20}$$

By differentiating both sides of (20) with respect to p, we have

$$\begin{aligned} G'(p) = \left(\frac{1}{\Lambda(q)}\right)&\left[4\left(\omega(q) - \Omega(q) - q(q+1)^2\psi_2^2 + \left(q+q^2+q^3\right)\psi_1\psi_3\right)p^3\right. \\ &\left. +2\left(4\Omega(q) + 4q(q+1)^2\psi_2^2 - 8\left(q+q^2+q^3\right)\psi_1\psi_3\right)p\right]. \end{aligned}$$

Differentiating the above equation once again with respect to p, we get

$$G''(p)=\left(\frac{1}{\Lambda(q)}\right)\left[12\left(\omega(q)-\Omega(q)-q(q+1)^2\psi_2^2+\left(q+q^2+q^3\right)\psi_1\psi_3\right)p^2\right.$$
$$\left.+2\left(4\Omega(q)+4q(q+1)^2\psi_2^2-8\left(q+q^2+q^3\right)\psi_1\psi_3\right)\right].$$

For $p=0$, this shows that the maximum value of $(G(p))$ occurs at $p=0$. Hence, we obtain

$$\left|a_2a_4-a_3^2\right|\leqq\frac{1}{q^2\psi_2^2}.$$

The proof of Theorem 2 is thus completed. □

If, in Theorem 2, we let $q\longrightarrow 1-$ and put $\delta=1$, then we are led to the following known result.

Corollary 1. (see [18]) *Let $f\in\mathcal{S}^*$. Then*

$$\left|a_2a_4-a_3^2\right|\leqq 1,$$

and the inequality is sharp.

Theorem 3. *Let $f\in\mathcal{RS}_q^*(\delta)$. Then*

$$|a_2a_3-a_4|\leqq\frac{(1+q)\,\kappa(q)}{\psi_1\psi_2\psi_3\left(q^2+q^3+q^4\right)},$$

where

$$\kappa(q)=\left|\left(1+q+q^2\right)^2\psi_3-\left(q^4-3q+6q^2+q+1\right)\psi_1\psi_2\right|. \tag{21}$$

Proof. Using the values given in (15) and (16) we have

$$\begin{aligned}a_2a_3-a_4&=\left(\frac{(1+q)\left(q^2+1\right)}{8q^3\psi_1\psi_2}-\frac{(1+q)\left(q^2+1\right)\left(q^2-q+1\right)}{8\psi_3\left(q^2+q^3+q^4\right)}\right)p_1^3\\&+\left(\frac{(1+q)}{4q^2\psi_1\psi_2}-\frac{(q-2)(2q+1)(1+q)}{4\psi_3\left(q^2+q^3+q^4\right)}\right)p_1p_2\\&-\left(\frac{(1+q)}{2\left(q+q^2+q^3\right)\psi_3}\right)p_3.\end{aligned} \tag{22}$$

We now use Lemma 2 and assume that $p_1\leqq 2$. In addition, by Lemma 3, we let $p_1=p$ and assume without restriction that $p\in[0,2]$. Then, by taking the moduli and applying the trigonometric inequality on (22) with $\rho=|x|$, we obtain

$$\begin{aligned}|a_2a_3-a_4|\leqq&\left(\frac{(1+q)}{8\left(q^3+q^4+q^5\right)\psi_1\psi_2\psi_3}\right)\left[\kappa(q)\,p^3+\eta(q)\,p(4-p^2)\rho\right.\\&\left.+2q^2\psi_1\psi_2(4-p^2)+q^2\psi_1\psi_2\,(p-2)\,(4-p^2)\rho^2\right]\\=:&\ F(\rho),\end{aligned}$$

where

$$\eta(q)=\left|\left(q+q^2+q^3\right)\psi_3+\left(2q^3-q^2-2q\right)\psi_1\psi_2\right|$$

and $\kappa(q)$ is given by (21). Differentiating $F(\rho)$ with respect to ρ, we have

$$\begin{aligned} F'(\rho) &= \left(\frac{(1+q)}{8(q^3+q^4+q^5)\psi_1\psi_2\psi_3}\right)\left[\eta(q)\,p(4-p^2)+2q^2\psi_1\psi_2(p-2)(4-p^2)\rho\right] \\ &> 0. \end{aligned}$$

This implies that $F(\rho)$ is an increasing function of ρ on the closed interval $[0,1]$. Hence, we have

$$F(\rho) \leqq F(1) \qquad (\forall\, \rho \in [0,1]),$$

that is,

$$\begin{aligned} F(\rho) \leqq &\left(\frac{(1+q)}{8(q^3+q^4+q^5)\psi_1\psi_2\psi_3}\right)\left[\left(\kappa(q)-\eta(q)-q^2\psi_1\psi_2\right)p^3 \right. \\ &\left. +\left(4\eta(q)+4q^2\psi_1\psi_2\right)p\right] \\ &=: G(p). \end{aligned}$$

Since $p \in [0,2]$, $p=2$ is a point of maximum. We thus obtain

$$G(p) \leqq \frac{(1+q)\,\kappa(q)}{(q^3+q^4+q^5)\psi_1\psi_2\psi_3},$$

which corresponds to $\rho = 1$ and $p = 2$ and it is the desired upper bound. □

For $\delta = 1$ and $q \to 1-$, we obtain the following special case of Theorem 3.

Corollary 2. (see [22]) *Let $f \in S^*$. Then*

$$|a_2a_3 - a_4| \leqq 2.$$

Finally, we prove Theorem 4 below.

Theorem 4. *Let $f \in \mathcal{RS}_q^*(\delta)$. Then*

$$\mathcal{H}_3(1) \leqq \left[\frac{(1+q+q^2)}{q^4\psi_2^3} + \frac{\varkappa(q)\,\kappa(q)}{q^5(1+q+q^2)^2\psi_1\psi_2\psi_3^2} + \frac{\tau(q)}{q^5(1+q+q^2+q^3)(1+q+q^2)\psi_2\psi_4}\right],$$

where

$$\varkappa(q) = (1+q)^2\left(q^4-3q^3+6q^2+q+1\right), \tag{23}$$

$$\tau(q) = (1+q)\left(4q^7+2q^6+6q^5+7q^4+13q^3-q-1\right) \tag{24}$$

and $\kappa(q)$ is given by (21).

Proof. Since

$$\mathcal{H}_3(1) \leqq |a_3|\left|a_2a_4-a_3^2\right| + |a_4|\,|a_2a_3-a_4| + |a_5|\left|a_3-a_2^2\right|,$$

by using Lemma 3, we have

$$|a_4| \leqq \frac{(1+q)(1+q+6q^2-3q^3+q^4)}{q^3(1+q+q^2)\psi_3}$$

and

$$|a_5| \leqq \frac{\tau(q)}{q^4(1+q+q^2+q^3)(1+q+q^2)\psi_4},$$

where $\tau(q)$ is given by (24). Now, by applying Theorems 1–3, we have the required result asserted by Theorem 4. □

4. Conclusions

By making use of the basic or quantum (q-) calculus, we have introduced a Ruscheweyh-type q-derivative operator. This Ruscheweyh-type q-derivative operator is then applied to define a certain subclass of q-starlike functions in the open unit disk $\mathbb{U}$. We have successfully derived the upper bound of the third Hankel determinant for this family of q-starlike functions which are associated with the Ruscheweyh-type q-derivative operator. Our main results are stated and proved as Theorems 1–4. These general results are motivated essentially by their several special cases and consequences, some of which are pointed out in this presentation.

Author Contributions: All authors contributed equally to the present investigation.

Funding: This work is partially supported by Sarhad University of Science and I.T, Ring Road, Peshawar 2500, Pakistan.

Acknowledgments: The first author would like to acknowledge Salim ur Rehman, V.C. Sarhad University of Science & I. T, for providing excellent research and academic environment.

Conflicts of Interest: The authors declare no conflict of interest.

References

1. Miller, S.S.; Mocanu, P.T. Differential subordination and univalent functions. *Mich. Math. J.* **1981**, *28*, 157–171. [CrossRef]
2. Miller, S.S.; Mocanu, P.T. Mocanu. In *Differential Subordination: Theory and Applications*; Series on Monographs and Textbooks in Pure and Applied Mathematics, No. 225; Marcel Dekker Incorporated: New York, NY, USA; Basel, Switzerland, 2000.
3. Jackson, F.H. On q-definite integrals. *Quart. J. Pure Appl. Math.* **1910**, *41*, 193–203.
4. Jackson, F.H. q-difference equations. *Am. J. Math.* **1910**, *32*, 305–314. [CrossRef]
5. Ismail, M.E.H.; Merkes, E.; Styer, D. A generalization of starlike functions. *Complex Var. Theory Appl.* **1990**, *14*, 77–84. [CrossRef]
6. Srivastava, H.M. Univalent functions, fractional calculus, and associated generalized hypergeometric functions. In *Univalent Functions, Fractional Calculus, and Their Applications*; Srivastava, H.M., Owa, S., Eds.; Halsted Press (Ellis Horwood Limited, Chichester); John Wiley and Sons: New York, NY, USA; Chichester, UK; Brisbane, Australian; Toronto, ON, Canada, 1989; pp. 329–354.
7. Srivastava, H.M.; Ahmad, Q.Z.; Khan, N.; Khan, N.; Khan, B. Hankel and Toeplitz determinants for a subclass of q-starlike functions associated with a general conic domain. *Mathematics* **2019**, *7*, 181. [CrossRef]
8. Srivastava, H.M.; Tahir, M.; Khan, B.; Ahmad, Q.Z.; Khan, N. Some general classes of q-starlike functions associated with the Janowski functions. *Symmetry* **2019**, *11*, 292. [CrossRef]
9. Mahmood, S.; Jabeen, M.; Malik, S.N.; Srivastava, H.M.; Manzoor, R.; Riaz, S.M.J. Some coefficient inequalities of q-starlike functions associated with conic domain defined by q-derivative. *J. Funct. Spaces* **2018**, *2018*, 8492072. [CrossRef]
10. Srivastava, H.M.; Bansal, D. Close-to-convexity of a certain family of q-Mittag-Leffler functions. *J. Nonlinear Var. Anal.* **2017**, *1*, 61–69.
11. Aldweby, H.; Darus, H. Some subordination results on q-analogue of Ruscheweyh differential operator. *Abstr. Appl. Anal.* **2014**, *2014*, 958563. [CrossRef]
12. Ezeafulukwe U.A.; Darus, M. A note on q-calculus. *Fasc. Math.* **2015**, *55*, 53–63. [CrossRef]
13. Ezeafulukwe. U.A.; Darus, M. Certain properties of q-hypergeometric functions. *Int. J. Math. Math. Sci.* **2015**, *2015*, 489218. [CrossRef]

14. Sahoo, S.K.; Sharma, N.L. On a generalization of close-to-convex functions. *Ann. Pol. Math.* **2015**, *113*, 93–108. [CrossRef]
15. Uçar, H.E.Ö. Coefficient inequality for q-starlike functions. *Appl. Math. Comput.* **2016**, *276*, 122–126.
16. Kanas S.; Răducanu, D. Some class of analytic functions related to conic domains. *Math. Slovaca* **2014**, *64*, 1183–1196. [CrossRef]
17. Ruscheweyh, S. New criteria for univalent functions. *Proc. Am. Math. Soc.* **1975**, *49*, 109–115. [CrossRef]
18. Janteng, J.; Abdulhalim, S.; Darus, M. Hankel determinant for starlike and convex functions. *Int. J. Math. Anal.* **2007**, *1*, 619–625.
19. Mishra, A.K.; Gochhayat, P. Second Hankel determinant for a class of analytic functions defined by fractional derivative. *Int. J. Math. Math. Sci.* **2008**, *2008*, 1–10. [CrossRef]
20. Singh, G.; Singh, G. On the second Hankel determinant for a new subclass of analytic functions. *J. Math. Sci. Appl.* **2014**, *2*, 1–3.
21. Srivastava, H.M., Altinkaya, Ş. and Yalçın, S. Hankel determinant for a subclass of bi-univalent functions defined by using a symmetric q-derivative operator. *Filomat* **2018**, *32*, 503–516. [CrossRef]
22. Babalola, K.O. On $\mathcal{H}_3(1)$ Hankel determinant for some classes of univalent functions. *Inequal. Theory Appl.* **2007**, *6*, 1–7.
23. Ma, W.C.; Minda, D. A. unified treatment of some special classes of univalent functions. In *Proceedings of the Conference on Complex Analysis (Tianjin, 1992)*; Li, Z., Ren, F., Yang, L., Zhang, S., Eds.; International Press: Cambridge, UK, 1994; pp. 157–169.
24. Libera, R.J.; Zlotkiewicz, E.J. Early coefficient of the inverse of a regular convex function. *Proc. Am. Math. Soc.* **1982**, *85*, 225–230. [CrossRef]
25. Libera, R.J.; Zlotkiewicz, E.J. Coefficient bounds for the inverse of a function with derivative in $\mathcal{P}$. *Proc. Am. Math. Soc.* **1983**, *87*, 251–257. [CrossRef]
26. Duren, P.L. *Univalent Functions;* Grundlehren der Mathematischen Wissenschaften, Band 259; Springer: New York, NY, USA; Berlin/Heidelberg, Germany; Tokyo, Japan, 1983.

© 2019 by the authors. Licensee MDPI, Basel, Switzerland. This article is an open access article distributed under the terms and conditions of the Creative Commons Attribution (CC BY) license (http://creativecommons.org/licenses/by/4.0/).

symmetry

MDPI

Article

Fractional Telegraph Equation and Its Solution by Natural Transform Decomposition Method

Hassan Eltayeb [1,*], Yahya T. Abdalla [2], Imed Bachar [1] and Mohamed H. Khabir [2]

1 Department of Mathematics, College of Science, King Saud University, P.O. Box 2455, Riyadh 11451, Saudi Arabia; abachar@ksu.edu.sa

2 Department of Mathematics, College of Science, Sudan University of Science and Technology, P.O. Box 407, Khartoum 11111, Sudan; amynt2005@gmail.com (Y.T.A.); khabir11@gmail.com (M.H.K.)

* Correspondence: hgadain@ksu.edu.sa; Tel.: +966-536345057

Received: 11 January 2019; Accepted: 26 February 2019; Published: 6 March 2019

Abstract: In this work, the natural transform decomposition method (NTDM) is applied to solve the linear and nonlinear fractional telegraph equations. This method is a combined form of the natural transform and the Adomian decomposition methods. In addition, we prove the convergence of our method. Finally, three examples have been employed to illustrate the preciseness and effectiveness of the proposed method.

Keywords: natural transform; Adomian decomposition method; Caputo fractional derivative; generalized mittag-leffler function

1. Introduction

The fractional calculus (non-integer) plays an important role in applied mathematics and other fields such as science, physics and engineering. It describes the smallest details of natural phenomena, which is better than using a calculus integer. In [1] the fractional telegraph equation is obtained from the classical telegraph equation by replacing the second-order distance derivative with the fractional derivative $(0 < \alpha \leq 2)$ given to it. The telegraph equation describes the signal propagation of an electrical signal in transmission cable lines in general. Recently, many researchers and engineers have done excellent work to solve the fractional telegraph equation by different methods, such as the Laplace transform method [2], Laplace transform variational iteration method [3], double Laplace transform method [4], variational iteration method [5], Adomian decomposition method [6], Mixture of a new integral transform and homotopy perturbation method (HPM) [7], homotopy analysis method (HAM) [8], Chebyshev tau method [9], and the method of separating variables [10]. The natural transform Adomian decomposition method (NTDM) is a combination of the natural transform method and Adomian decomposition method. The main aim of this article is to use the (NTDM) to obtain the approximate solution of linear and nonlinear fractional telegraph equations. The natural transform Adomian decomposition method is a sturdy mathematical method for solving linear and nonlinear fractional telegraph equation and is an amelioration of the existing methods.

2. Preliminaries

Definition 1 ([11]). *The Adomian decomposition method is defined as*

$$A_n = \frac{1}{n!}\frac{d^n}{d\lambda^n}\left[F\left(\sum_{i=1}^{n}\psi_i\lambda^i\right)\right]_{\lambda=0}, \qquad n = 0, 1, 2, \ldots \tag{1}$$

where the function $F(\psi)$ is a nonlinear term and λ a formal parameter.

Definition 2 ([12]). *The natural transform of a function $f(t) > 0$ and $f(t) = 0$ for $t < 0$ is defined by*

$$\mathbb{N}^{+}[f(t)] = R(s,u) = \frac{1}{u}\int_0^{\infty} e^{\frac{-st}{u}} f(t)dt; \;\; s,u > 0 \tag{2}$$

where s and u are the transform variables.

Definition 3 ([12,13]). *The inverse natural transform of a function is defined by*

$$\mathbb{N}^{-}[R(s,u)] = f(t) = \tfrac{1}{2\pi i}\int_{c-i\infty}^{c+i\infty} e^{\frac{st}{u}} R(s,u)ds$$

Definition 4 ([14]). *The natural transform of $\frac{\partial^{\alpha} f(x,t)}{\partial t^{\alpha}}$ w.r.t (t) can be calculated as*

$$\mathbb{N}^{+}[\frac{\partial^{\alpha} f(x,t)}{\partial t^{\alpha}}] = \frac{s^{\alpha}}{u^{\alpha}}R(s,u) - \sum_{k=0}^{n-1}\frac{s^{\alpha-k-1}}{u^{\alpha-k}}\left[\frac{\partial^{\alpha} f(x,0)}{\partial t^{\alpha}}\right] \tag{3}$$

Definition 5 ([14]). *The natural transform of Mittag-Leffler function $E_{\alpha,\beta}$ is defined as follows*

$$\mathbb{N}^{+}[f(x,t)] = \int_0^{\infty} e^{-st} f(x,ut)dt = \sum_{k=0}^{\infty}\frac{u^{k+1}\Gamma(k+\beta)}{s^{k+1}\Gamma(\alpha k+\beta)} \tag{4}$$

Definition 6 ([15]). *A two parameter function of the Mittag-Leffler type is defined by the series expansion*

$$E_{\alpha,\beta}(z) = \sum_{k=0}^{\infty}\frac{z^k}{\Gamma(\alpha k+\beta)}, \qquad (\alpha > 0, \beta > 0) \tag{5}$$

3. Natural Transform Adomian Decomposition Method Linear and Nonlinear Telegraph Equations (NTADM)

In this section, we will study two problems as follows:

First Problem: linear fractional telegraph equations

In this part, we derive the main idea of the natural transform decomposition method to find the general solution for linear fractional telegraph equations.

We consider the following general multiterm fractional telegraph equation

$$\begin{gathered}\frac{\partial^{\alpha}\psi(x,t)}{\partial t^{\alpha}} = \frac{\partial^{2}\psi(x,t)}{\partial x^{2}} - \frac{\partial\psi(x,t)}{\partial t} - \psi(x,t) + h(x,t), \\ 0 < \alpha \leq 2 \text{ and } x,t \geq 0\end{gathered} \tag{6}$$

subject to

$$\psi(x,0) = f_1(x) \text{ and } \psi_t(x,0) = f_2(x) \tag{7}$$

where $h(x,t)$ is given function. The new technique of natural transform Adomian decomposition is based on the following steps. By applying the definition of natural transform to Equation (6), we get

$$\tfrac{s^{\alpha}}{u^{\alpha}}R(x,s,u) - \tfrac{s^{\alpha-1}}{u^{\alpha}}\psi(x,0) - \tfrac{s^{\alpha-2}}{u^{\alpha-1}}\psi_t(x,0) = \mathbb{N}^{+}\left[\tfrac{\partial^{2}\psi(x,t)}{\partial x^{2}} - \tfrac{\partial\psi(x,t)}{\partial t} - \psi(x,t) + h(x,t)\right], \tag{8}$$

substituting the initial conditions Equation (7) into Equation (8), we obtain

$$R(x,s,u) = \tfrac{1}{s}f_1(x) + \tfrac{u}{s^2}f_2(x) + \tfrac{u^{\alpha}}{s^{\alpha}}\mathbb{N}^{+}\left[\tfrac{\partial^{2}\psi(x,t)}{\partial x^{2}} - \tfrac{\partial\psi(x,t)}{\partial t} - \psi(x,t) + h(x,t)\right]. \tag{9}$$

Now, implementing the inverse natural transform for Equation (9) we obtain the general solution of Equation (6) as follows:

$$\psi(x,t) = \Phi(x,t) + \mathbb{N}^{-1}\left[\frac{u^\alpha}{s^\alpha}\mathbb{N}^+\left[\frac{\partial^2\psi(x,t)}{\partial x^2} - \frac{\partial\psi(x,t)}{\partial t} - \psi(x,t)\right]\right], \tag{10}$$

where

$$\Phi(x,t) = \mathbb{N}^{-1}\left[f_1(x) + tf_2(x) + \frac{u^\alpha}{s^\alpha}N^+\left[h(x,t)\right]\right], \tag{11}$$

the natural transform decomposition method defined the solution of $\psi(x,t)$ by the infinite series

$$\psi(x,t) = \sum_{n=0}^{\infty}\psi_n(x,t). \tag{12}$$

The solution of Equation (10) is given by

$$\sum_{n=0}^{\infty}\psi_n(x,t) = \Phi(x,t) + \mathbb{N}^{-1}\left[\frac{u^\alpha}{s^\alpha}\mathbb{N}^+\left[\sum_{n=0}^{\infty}\frac{\partial^2\psi_n(x,t)}{\partial x^2} - \sum_{n=0}^{\infty}\frac{\partial\psi_n(x,t)}{\partial t} - \sum_{n=0}^{\infty}\psi_n(x,t)\right]\right]. \tag{13}$$

Here we assume that the inverse natural transform of each term in the right side of Equation (9) exists. The initial term

$$\psi_0(x,t) = \Phi(x,t), \tag{14}$$

consequently, the first few components can be written as

$$\begin{aligned}
\psi_1(x,t) &= \mathbb{N}^{-1}\left[\tfrac{u^\alpha}{s^\alpha}\mathbb{N}^+\left[\tfrac{\partial^2\psi_0(x,t)}{\partial x^2} - \tfrac{\partial\psi_0(x,t)}{\partial t} - \psi_0(x,t)\right]\right]\\
\psi_2(x,t) &= \mathbb{N}^{-1}\left[\tfrac{u^\alpha}{s^\alpha}\mathbb{N}^+\left[\tfrac{\partial^2\psi_1(x,t)}{\partial x^2} - \tfrac{\partial\psi_1(x,t)}{\partial t} - \psi_1(x,t)\right]\right]\\
\psi_3(x,t) &= \mathbb{N}^{-1}\left[\tfrac{u^\alpha}{s^\alpha}\mathbb{N}^+\left[\tfrac{\partial^2\psi_2(x,t)}{\partial x^2} - \tfrac{\partial\psi_2(x,t)}{\partial t} - \psi_2(x,t)\right]\right]\\
&\ \ \vdots
\end{aligned} \tag{15}$$

then we have

$$\psi_{n+1}(x,t) = \mathbb{N}^{-1}\left[\frac{u^\alpha}{s^\alpha}\mathbb{N}^+\left[\frac{\partial^2\psi_n(x,t)}{\partial x^2} - \frac{\partial\psi_n(x,t)}{\partial t} - \psi_n(x,t)\right]\right], \ n \geq 0 \tag{16}$$

Second Problem Nonlinear fractional telegraph equation:

We consider the general form of nonlinear fractional telegraph equation:

$$\begin{gathered}
\frac{\partial^\alpha\psi(x,t)}{\partial t^\alpha} = \frac{\partial^2\psi(x,t)}{\partial x^2} - \frac{\partial\psi(x,t)}{\partial t} - N\psi(x,t) + h(x,t),\\
0 < \alpha \leq 2 \ \text{and} \ x,t \geq 0
\end{gathered} \tag{17}$$

with the initial conditions

$$\psi(x,0) = g_1(x) \text{ and } \psi_t(x,0) = g_2(x), \tag{18}$$

where N is a nonlinear, $h(x,t)$ is a source term. By applying the definition of natural transform for Equation (17), we have

$$\tfrac{s^\alpha}{u^\alpha}R(x,s,u) - \tfrac{s^{\alpha-1}}{u^\alpha}\psi(x,0) - \tfrac{s^{\alpha-2}}{u^{\alpha-1}}\psi_t(x,0) = \mathbb{N}^+\left[\tfrac{\partial^2\psi(x,t)}{\partial x^2} - \tfrac{\partial\psi(x,t)}{\partial t} - N\psi(x,t) + h(x,t)\right], \tag{19}$$

by substituting initial conditions Equation (18) into Equation (19), we obtain

$$R(x,s,u) = \frac{1}{s}g_1(x) + \frac{u}{s^2}g_2(x) + \frac{u^\alpha}{s^\alpha}\mathbb{N}^+\left[\frac{\partial^2\psi(x,t)}{\partial x^2} - \frac{\partial\psi(x,t)}{\partial t} - N\psi(x,t) + h(x,t)\right]. \quad (20)$$

Now, implementing the inverse natural transform for Equation (20), we obtain the general solution of Equation (17) in the form of,

$$\psi(x,t) = \Phi(x,t) + \mathbb{N}^{-1}\left[\frac{u^\alpha}{s^\alpha}\mathbb{N}^+\left[\frac{\partial^2\psi(x,t)}{\partial t^2} - \frac{\partial\psi(x,t)}{\partial t} - N\psi(x,t)\right]\right], \quad (21)$$

where

$$\Phi(x,t) = \mathbb{N}^{-1}\left[g_1(x) + tg_2(x) + \frac{u^\alpha}{s^\alpha}\mathbb{N}^+\left[h(x,t)\right]\right], \quad (22)$$

here we assume that the inverse natural transform of each term in the right side of Equation (22) exists. The natural transform decomposition method consists of calculating the solution in a series form

$$\psi(x,t) = \sum_{n=0}^{\infty}\psi_n(x,t), \quad (23)$$

the nonlinear term $N\psi(x,t)$ becomes

$$N\psi(x,t) = \sum_{n=0}^{\infty}A_n, \quad (24)$$

where A_n defined by Equation (1). By substituting Equations (23) and (24) into Equation (21) we get

$$\sum_{n=0}^{\infty}\psi_n(x,t) = \Phi(x,t) + \mathbb{N}^{-1}\left[\frac{u^\alpha}{s^\alpha}\mathbb{N}^+\left[\sum_{n=0}^{\infty}\frac{\partial^2\psi_n(x,t)}{\partial x^2} - \sum_{n=0}^{\infty}\frac{\partial\psi_n(x,t)}{\partial t} - \sum_{n=0}^{\infty}A_n\right]\right], \quad (25)$$

by using the recursive relation

$$\psi_0(x,t) = \Phi(x,t) \quad (26)$$

consequently, the first few components can be written as

$$\begin{aligned}
\psi_1(x,t) &= \mathbb{N}^{-1}\left[\tfrac{u^\alpha}{s^\alpha}\mathbb{N}^+\left[\tfrac{\partial^2\psi_0(x,t)}{\partial x^2} - \tfrac{\partial\psi_0(x,t)}{\partial t} - A_0\right]\right] \\
\psi_2(x,t) &= \mathbb{N}^{-1}\left[\tfrac{u^\alpha}{s^\alpha}\mathbb{N}^{+1}\left[\tfrac{\partial^2\psi_1(x,t)}{\partial x^2} - \tfrac{\partial\psi_1(x,t)}{\partial t} - A_1\right]\right] \\
\psi_3(x,t) &= \mathbb{N}^{-1}\left[\tfrac{u^\alpha}{s^\alpha}\mathbb{N}^+\left[\tfrac{\partial^2\psi_2(x,t)}{\partial x^2} - \tfrac{\partial\psi_2(x,t)}{\partial t} - A_2\right]\right], \\
&\vdots
\end{aligned} \quad (27)$$

then we have

$$\psi_{n+1}(x,t) = \mathbb{N}^{-1}\left[\frac{u^\alpha}{p^\alpha}\mathbb{N}^+\left[\frac{\partial^2\psi_n(x,t)}{\partial t^2} + \frac{\partial\psi_n(x,t)}{\partial t} + A_n\right]\right], \ n \geq 0 \quad (28)$$

the solution $\psi_n(x,t)$ can be written as convergent series

$$\psi(x,t) = \sum_{n=0}^{\infty}\psi_n(x,t). \quad (29)$$

4. Convergence Analysis

In this section, the sufficient condition that guarantees existence of a unique solution is introduced and we discuss the convergence of the solution.

In next theorem we follow [16]

Theorem 1. *(Uniqueness theorem):* Equation (28) has a unique solution whenever $0 < \varepsilon < 1$ where $\varepsilon = \frac{(L_1+L_2+L_3)t^{\alpha+1}}{(\alpha-1)!}$

Proof of Theorem 1. Let $E = (C[I], \|.\|)$ be the Banach space of all continuous functions on $I = [0, T]$ with the norm $\|.\|$, we define a mapping $F : E \to E$ where

$$\psi_{n+1}(x,t) = \Phi(x,t) + \mathbb{N}^{-1}\left[\frac{u^\alpha}{s^\alpha}\mathbb{N}^+\left[L\left[\psi_n(x,t)\right] + M\left[\psi_n(x,t)\right] + N\left[\psi_n(x,t)\right]\right]\right], \; n \geq 0$$

where $L[\psi(x,t)] \equiv \frac{\partial^2\psi(x,t)}{\partial x^2}$ *and* $M[\psi(x,t)] \equiv \frac{\partial\psi(x,t)}{\partial t}$. Now suppose $M[\psi(x,t)]$ and $L[\psi(x,t)]$ is also Lipschitzian with $\left|M\psi - M\widehat{\psi}\right| < L_1\left|\psi - \widehat{\psi}\right|$and $\left|L\psi - L\widehat{\psi}\right| < L_2\left|\psi - \widehat{\psi}\right|$ where L_1 and L_2 is Lipschitz constant respectively and $\psi, \widehat{\psi}$ is different values of the function.

$$\left\|F\psi - F\widehat{\psi}\right\| = \max_{t\in I}\left|\begin{array}{c} \mathbb{N}^{-1}\left[\frac{u^\alpha}{s^\alpha}\mathbb{N}^+\left[L\left[\psi(x,t)\right] + M\left[\psi(x,t)\right] + N\left[\psi(x,t)\right]\right]\right] \\ -\mathbb{N}^{-1}\left[\frac{u^\alpha}{s^\alpha}\mathbb{N}^+\left[L\left[\widehat{\psi}(x,t)\right] + M\left[\widehat{\psi}(x,t)\right] + N\left[\widehat{\psi}(x,t)\right]\right]\right] \end{array}\right|,$$

$$\leq \max_{t\in I}\left|\begin{array}{c} \mathbb{N}^{-1}\left[\frac{u^\alpha}{p^\alpha}\mathbb{N}^+\left[L\left[\psi(x,t)\right] - L\left[\widehat{\psi}(x,t)\right]\right]\right] \\ +\mathbb{N}^{-1}\left[\frac{u^\alpha}{s^\alpha}\mathbb{N}^+\left[M\left[\psi(x,t)\right] - M\left[\widehat{\psi}(x,t)\right]\right]\right] \\ +\mathbb{N}^{-1}\left[\frac{u^\alpha}{s^\alpha}\mathbb{N}^+\left[N\left[\psi(x,t)\right] - N\left[\widehat{\psi}(x,t)\right]\right]\right] \end{array}\right|,$$

$$\leq \max_{t\in I}\left[\begin{array}{c} L_1\mathbb{N}^{-1}\left[\frac{u^\alpha}{s^\alpha}\mathbb{N}^+\left|\psi(x,t) - \widehat{\psi}(x,t)\right|\right] \\ +L_2\mathbb{N}^{-1}\left[\frac{u^\alpha}{s^\alpha}\mathbb{N}^+\left|\psi(x,t) - \widehat{\psi}(x,t)\right|\right] \\ +L_3\mathbb{N}^{-1}\left[\frac{u^\alpha}{s^\alpha}\mathbb{N}^+\left|\psi(x,t) - \widehat{\psi}(x,t)\right|\right] \end{array}\right],$$

$$\leq \max_{t\in I}(L_1 + L_2 + L_3)\left[\mathbb{N}^{-1}\left[\frac{u^\alpha}{s^\alpha}\mathbb{N}^+\left|\psi(x,t) - \widehat{\psi}(x,t)\right|\right]\right],$$

$$\leq (L_1 + L_2 + L_3)\left[\mathbb{N}^{-1}\left[\frac{u^\alpha}{s^\alpha}\mathbb{N}^+\left\|\psi(x,t) - \widehat{\psi}(x,t)\right\|\right]\right],$$

$$= \frac{(L_1+L_2+L_3)t^{(\alpha-1)}}{(\alpha-1)!}\left\|\psi(x,t) - \widehat{\psi}(x,t)\right\|.$$

Under the condition $0 < \varepsilon < 1$, the mapping is contraction. Therefore, by Banach fixed point theorem for contraction, there exists a unique solution to Equation (29). This ends the proof of Theorem 1. □

Theorem 2. ***(Convergence Theorem):*** The solution of Equations (6) and (18) in general forum will be convergence.

Proof of Theorem 2. Let S_n be the nth partial sum, i.e., $S_n = \sum_{i=0}^{n}\psi_i(x,t)$. We shall prove that $\{S_n\}$ is a Cauchy sequence in Banach space E. By using a new formulation of Adomian polynomials we get

$$R(S_n) = \widehat{A}_n + \sum_{r=0}^{n-1} \widehat{A}_r$$

$$N(S_n) = \widehat{A}_n + \sum_{c=0}^{n-1} \widehat{A}_c$$

$$\|S_n - S_m\| = \max_{t\in I} |S_n - S_m| = \max_{t\in I} \left|\sum_{i=m+1}^{n} \widehat{\psi}_i(x,t)\right|, p = 1,2,3,...$$

$$\leq \max_{t\in I} \left| \begin{array}{c} \mathbb{N}^{-1}\left[\frac{u^\alpha}{s^\alpha}\mathbb{N}^+\left[\sum_{i=m+1}^{n} L\left[\psi_{n-1}(x,t)\right]\right]\right] \\ +\mathbb{N}^{-1}\left[\frac{u^\alpha}{s^\alpha}\mathbb{N}^+\left[\sum_{i=m+1}^{n} M\left[\psi_{n-1}(x,t)\right]\right]\right] \\ +\mathbb{N}^{-1}\left[\frac{u^\alpha}{s^\alpha}\mathbb{N}^+\left[\sum_{i=m+1}^{n} A_{n-1}(x,t)\right]\right] \end{array} \right|,$$

$$= \max_{t\in I} \left| \begin{array}{c} \mathbb{N}^{-1}\left[\frac{u^\alpha}{s^\alpha}\mathbb{N}^+\left[\sum_{i=m}^{n-1} L\left[\psi_{n}(x,t)\right]\right]\right] \\ +\mathbb{N}^{-1}\left[\frac{u^\alpha}{s^\alpha}\mathbb{N}^+\left[\sum_{i=m}^{n-1} M\left[\psi_{n}(x,t)\right]\right]\right] \\ +\mathbb{N}^{-1}\left[\frac{u^\alpha}{s^\alpha}\mathbb{N}^+\left[\sum_{i=m+1}^{n} A_{n}(x,t)\right]\right] \end{array} \right|,$$

$$\leq \max_{t\in I} \left| \begin{array}{c} \mathbb{N}^{-1}\left[\frac{u^\alpha}{s^\alpha}\mathbb{N}^+\left[\sum_{i=m}^{n-1} L(S_{n-1}) - L(S_{m-1})\right]\right] \\ +\mathbb{N}^{-1}\left[\frac{u^\alpha}{s^\alpha}\mathbb{N}^+\left[\sum_{i=m}^{n-1} M(S_{n-1}) - M(S_{m-1})\right]\right] \\ +\mathbb{N}^{-1}\left[\frac{u^\alpha}{s^\alpha}\mathbb{N}^+\left[\sum_{i=m+1}^{n} N(S_{n-1}) - N(S_{m-1})\right]\right] \end{array} \right|,$$

$$\leq \max_{t\in I} \left| \begin{array}{c} \mathbb{N}^{-1}\left[\frac{u^\alpha}{s^\alpha}\mathbb{N}^+\left[L(S_{n-1}) - L(S_{m-1})\right]\right] \\ +\mathbb{N}^{-1}\left[\frac{u^\alpha}{s^\alpha}\mathbb{N}^+\left[M(S_{n-1}) - R(S_{m-1})\right]\right] \\ +\mathbb{N}^{-1}\left[\frac{u^\alpha}{s^\alpha}\mathbb{N}^+\left[N(S_{n-1}) - N(S_{m-1})\right]\right] \end{array} \right|,$$

$$\begin{aligned} &\leq L_1 \max_{t\in I} \mathbb{N}^{-1}\left|\left[\frac{u^\alpha}{s^\alpha}\mathbb{N}^+\left[(S_{n-1}) - (S_{m-1})\right]\right]\right|, \\ &+L_2 \max_{t\in I} \mathbb{N}^{-1}\left|\left[\frac{u^\alpha}{s^\alpha}\mathbb{N}^+\left[(S_{n-1}) - (S_{m-1})\right]\right]\right|, \\ &+L_3 \max_{t\in I} \mathbb{N}^{-1}\left|\left[\frac{u^\alpha}{s^\alpha}\mathbb{N}^+\left[(S_{n-1}) - (S_{m-1})\right]\right]\right|. \end{aligned}$$

$$= \frac{(L_1 + L_2 + L_3)t^{(\alpha-1)}}{(\alpha-1)!} \|S_{n-1} + S_{m-1}\|$$

Let $n = m + 1$; then

$$\|S_{m+1} - S_m\| \leq \varepsilon \|S_m - S_{m-1}\| \leq \varepsilon^2 \|S_{m-1} - S_{m-2}\| \leq ... \leq \varepsilon^m \|S_1 - S_0\|.$$

where $\varepsilon = \frac{(L_1+L_2+L_3)t^{(\alpha-1)}}{(\alpha-1)!}$ similarly, we have, from the triangle inequality we have

$$\begin{aligned} \|S_n - S_m\| &\leq \|S_{m+1} - S_m\| + \|S_{m+2} - S_{m+1}\| + ... + \|S_n - S_{n-1}\|, \\ &\leq \left[\varepsilon^m + \varepsilon^{m+1} + ... + \varepsilon^{n-1}\right] \leq \|S_1 + S_0\|, \\ &\leq \varepsilon^m \left(\frac{1-\varepsilon^{n-m}}{\varepsilon}\right) \|\psi_1\|, \end{aligned}$$

since $0 < \varepsilon < 1$ we have $(1 - \varepsilon^{n-m}) < 1$: then,

$$\|S_n - S_m\| \leq \frac{\varepsilon^m}{1-\varepsilon} \max_{t\in I} \|\psi_1\|.$$

However, $|\psi_1| < \infty$ (since $\psi(x,t)$ is bounded) so, as $m \to \infty$ then $\|S_n - S_m\| \to 0$, hence $\{S_n\}$ is a Cauchy sequence in E so, the series $\sum_{n=0}^{\infty} \psi_n$ converges and the proof is complete. □

Theorem 3. ***(Error estimate:)*** The maximum absolute truncation error of the series solution Equation (28) to Equation (6) is estimated to be:

$$\max_{t\in I}\left|\psi(x,t)-\sum_{n=1}^{m}\psi_n(x,t)\right|\leq\frac{\varepsilon^m}{1-\varepsilon}\max_{t\in I}\|\psi_1\|,$$

Proof of Theorem 3. From Equation (28) and Theorem 2 we have

$$|S_n-S_m|\leq\frac{\varepsilon^m}{1-\varepsilon}\max_{t\in I}\|\psi_1\|,$$

as $n\to\infty$ then $S_n\to\psi(x,t)$ so we have

$$\|\psi(x,t)-S_m\|\leq\frac{\varepsilon^m}{1-\varepsilon}\max_{t\in I}\|\psi_1(x,t)\|,$$

finally, the maximum absolute truncation error in the interval I is

$$\max_{t\in I}\left|\psi(x,t)-\sum_{n=1}^{m}\psi_n(x,t)\right|\leq\max_{t\in I}\frac{\varepsilon^m}{1-\varepsilon}|\psi_1(x,t)|=\frac{\varepsilon^m}{1-\varepsilon}\|\psi_1(x,t)\|.$$

Thus, completing the proof of Theorem (3). □

5. Numerical Examples

In this section, we demonstrate the applicability of the previous method by the following examples.

Example 1. *Consider the following space-fractional homogenous telegraph equation:*

$$\frac{\partial^\alpha\psi(x,t)}{\partial t^\alpha}=\frac{\partial^2\psi(x,t)}{\partial x^2}-\frac{\partial\psi(x,t)}{\partial t}-\psi(x,t),\qquad x,t\geq0\ \text{and}\ 0<\alpha\leq2 \tag{30}$$

with the initial conditions

$$\psi(x,0)=e^{-x}\text{ and }\psi_t(x,0)=-e^{-x}. \tag{31}$$

Solution 1

Applying natural transform for Equation (30) w.r.t (t) on both sides, we get

$$\frac{s^\alpha}{u^\alpha}R(x,s,u)-\frac{s^{\alpha-1}}{u^\alpha}\psi(x,0)-\frac{s^{\alpha-2}}{u^{\alpha-1}}\psi_t(x,0)=\mathbb{N}^+\left[\frac{\partial^2\psi(x,t)}{\partial x^2}-\frac{\partial\psi(x,t)}{\partial t}-\psi(x,t)\right], \tag{32}$$

simplify and substitute the condition Equation (31), we get

$$R(x,s,u)=\frac{1}{s}e^x-\frac{u}{s^2}e^x+\frac{u^\alpha}{s^\alpha}\mathbb{N}^+\left[\frac{\partial^2\psi(x,t)}{\partial x^2}-\frac{\partial\psi(x,t)}{\partial t}-\psi(x,t)\right], \tag{33}$$

using the inverse natural transform for Equation (33), we have

$$\psi(x,t)=e^x-te^x+\mathbb{N}^{-1}\left[\frac{u^\alpha}{s^\alpha}\mathbb{N}^+\left[\frac{\partial^2\psi(x,t)}{\partial x^2}-\frac{\partial\psi(x,t)}{\partial t}-\psi(x,t)\right]\right], \tag{34}$$

the correction function for Equation (34), is given by

$$\sum_{n=0}^{\infty}\psi_{n+1}(x,t)=e^{-x}-te^{-x}+\mathbb{N}^{-1}\left[\frac{u^{\alpha}}{s^{\alpha}}\mathbb{N}^{+}\left[\sum_{n=0}^{\infty}\frac{\partial^{2}\psi_{n}(x,t)}{\partial x^{2}}-\sum_{n=0}^{\infty}\frac{\partial\psi_{n}(x,t)}{\partial t}-\sum_{n=0}^{\infty}\psi_{n}(x,t)\right]\right], \quad (35)$$

the initial term

$$\psi_{0}(x,t)=e^{x}-te^{x}, \quad (36)$$

then we have

$$\psi_{n+1}(x,t)=\mathbb{N}^{-1}\left[\frac{u^{\alpha}}{s^{\alpha}}\mathbb{N}^{+}\left[\sum_{n=0}^{\infty}\frac{\partial^{2}\psi_{n}(x,t)}{\partial x^{2}}-\sum_{n=0}^{\infty}\frac{\partial\psi_{n}(x,t)}{\partial t}-\sum_{n=0}^{\infty}\psi_{n}(x,t)\right]\right], n\geq 0 \quad (37)$$

the first 3rd terms is given by

$$\begin{aligned}\psi_{1}(x,t)&=\tfrac{t^{\alpha}}{\Gamma(\alpha+1)}e^{x},\\ \psi_{2}(x,t)&=-\tfrac{t^{2\alpha-1}}{\Gamma(2\alpha)}e^{x},\\ \psi_{3}(x,t)&=\tfrac{t^{3\alpha-2}}{\Gamma(3\alpha-1)}e^{x},\end{aligned} \quad (38)$$

then general form is successive approximation is given by

$$\psi_{n}(x,t)=e^{x}\left(1-t+\frac{t^{\alpha}}{\Gamma(\alpha+1)}-\frac{t^{2\alpha-1}}{\Gamma(2\alpha)}+\frac{t^{3\alpha-2}}{\Gamma(3\alpha-1)}-...\right), \quad (39)$$

$$\psi_{n}(x,t)=e^{x}\left[1+\sum_{k=0}^{\infty}(-1)^{k+1}\left[\frac{t^{k\alpha-k+1}}{\Gamma(k\alpha-k+2)}\right]\right], \quad (40)$$

when $\alpha=2$ we get

$$\psi(x,t)=e^{x-t}. \quad (41)$$

Example 2. *Consider the following space-fractional non-homogenous telegraph equation:*

$$\begin{aligned}\tfrac{\partial^{\alpha}\psi(x,t)}{\partial t^{\alpha}}&=\tfrac{\partial^{2}\psi(x,t)}{\partial x^{2}}-\tfrac{\partial\psi(x,t)}{\partial t}-\psi(x,t)+x^{2}+t-1,\\ &x,t\geq 0 \text{ and } 0<\alpha\leq 2\end{aligned} \quad (42)$$

with the initial conditions

$$\psi(x,0)=x^{2} \text{ and } \psi_{t}(x,0)=1. \quad (43)$$

Solution 2

Applying natural transform for both sides of Equation (42), we have

$$\tfrac{s^{\alpha}}{u^{\alpha}}R(x,s,u)-\tfrac{s^{\alpha-1}}{u^{\alpha}}\psi(x,0)-\tfrac{s^{\alpha-2}}{u^{\alpha-1}}\psi_{t}(x,0)=\mathbb{N}^{+}\left[\tfrac{\partial^{2}\psi(x,t)}{\partial x^{2}}-\tfrac{\partial\psi(x,t)}{\partial t}-\psi(x,t)\right]+\mathbb{N}^{+}\left[x^{2}+t-1\right], \quad (44)$$

by simplifying and substitute the conditions, we obtain

$$R(x,s,u)=\frac{1}{s}x^{2}+\frac{u}{s^{2}}+\frac{u^{\alpha}}{s^{\alpha+1}}x^{2}+\frac{u^{\alpha+1}}{s^{\alpha+2}}-\frac{u^{\alpha}}{s^{\alpha+1}}+\frac{u^{\alpha}}{s^{\alpha}}\mathbb{N}^{+}\left[\frac{\partial^{2}\psi(x,t)}{\partial t^{2}}-\frac{\partial\psi(x,t)}{\partial t}-\psi(x,t)\right]. \quad (45)$$

On using inverse natural transform Equation (45), we have

$$\psi(x,t)=x^{2}+t+\tfrac{t^{\alpha}}{\Gamma(\alpha+1)}x^{2}+\tfrac{t^{\alpha+1}}{\Gamma(\alpha+2)}-\tfrac{t^{\alpha}}{\Gamma(\alpha+1)}+\mathbb{N}^{-}\left[\tfrac{u^{\alpha}}{s^{\alpha}}\mathbb{N}^{+}\left[\tfrac{\partial^{2}\psi(x,t)}{\partial x^{2}}-\tfrac{\partial\psi(x,t)}{\partial t}-\psi(x,t)\right]\right], \quad (46)$$

therefore

$$\begin{aligned} \sum_{n=0}^{\infty} \psi_n(x,t) &= x^2 + t + \frac{t^{\alpha}}{\Gamma(\alpha+1)}x^2 + \frac{t^{\alpha+1}}{\Gamma(\alpha+2)} - \frac{t^{\alpha}}{\Gamma(\alpha+1)} \\ &+ \mathbb{N}^{-}\left[\frac{u^{\alpha}}{s^{\alpha}}\mathbb{N}^{+}\left[\sum_{n=0}^{\infty}\frac{\partial^2\psi_n(x,t)}{\partial x^2} - \sum_{n=0}^{\infty}\frac{\partial\psi_n(x,t)}{\partial t} - \sum_{n=0}^{\infty}\psi_n(x,t)\right]\right], \end{aligned} \tag{47}$$

the initial term

$$\psi_0(x,t) = x^2 + t + \frac{t^{\alpha}}{\Gamma(\alpha+1)}x^2 + \frac{t^{\alpha+1}}{\Gamma(\alpha+2)} - \frac{t^{\alpha}}{\Gamma(\alpha+1)}, \tag{48}$$

then we have

$$\sum_{n=0}^{\infty}\psi_{n+1}(x,t) = \mathbb{N}^{-1}\left[\frac{u^{\alpha}}{s^{\alpha}}\mathbb{N}^{+}\left[\sum_{n=0}^{\infty}\frac{\partial^2\psi_n(x,t)}{\partial x^2} - \sum_{n=0}^{\infty}\frac{\partial\psi_n(x,t)}{\partial t} - \sum_{n=0}^{\infty}\psi_n(x,t)\right]\right], \quad n \geq 0 \tag{49}$$

Now the components of the series solution are given by

$$\begin{aligned} \psi_1(x,t) &= \frac{t^{\alpha}}{\Gamma(\alpha+1)} - \frac{t^{\alpha}}{\Gamma(\alpha+1)}x^2 - \frac{t^{\alpha+1}}{\Gamma(\alpha+2)} + 2\frac{t^{2\alpha}}{\Gamma(2\alpha+1)} \\ &- \frac{t^{2\alpha+1}}{\Gamma(2\alpha+2)} + \frac{t^{2\alpha-1}}{\Gamma(2\alpha)} - \frac{t^{2\alpha}}{\Gamma(2\alpha+1)}x^2 - \frac{t^{2\alpha-1}}{\Gamma(2\alpha)}x^2, \\ \psi_2(x,t) &= -2\frac{t^{2\alpha}}{\Gamma(2\alpha+1)} + \frac{t^{2\alpha+1}}{\Gamma(2\alpha+2)} - \frac{t^{2\alpha-1}}{\Gamma(2\alpha)} + \frac{t^{2\alpha}}{\Gamma(2\alpha+1)}x^2 \\ &+ \frac{t^{2\alpha-1}}{\Gamma(2\alpha)}x^2 - 5\frac{t^{3\alpha-1}}{\Gamma(3\alpha+2)} - 3\frac{t^{3\alpha}}{\Gamma(3\alpha+1)} + 2\frac{t^{3\alpha-1}}{\Gamma(3\alpha)}x^2 \\ &- \frac{t^{3\alpha-2}}{\Gamma(3\alpha-1)} + \frac{t^{3\alpha+1}}{\Gamma(3\alpha+2)} + \frac{t^{3\alpha}}{\Gamma(3\alpha+1)}x^2 + \frac{t^{3\alpha-2}}{\Gamma(3\alpha-1)}x^2, \\ \psi_3(x,t) &= 5\frac{t^{3\alpha-1}}{\Gamma(3\alpha+2)} + 3\frac{t^{3\alpha}}{\Gamma(3\alpha+1)} - 2\frac{t^{3\alpha-1}}{\Gamma(3\alpha)}x^2 + \frac{t^{3\alpha-2}}{\Gamma(3\alpha-1)} - \frac{t^{3\alpha+1}}{\Gamma(3\alpha+2)} \\ &- \frac{t^{3\alpha}}{\Gamma(3\alpha+1)}x^2 - \frac{t^{3\alpha-2}}{\Gamma(3\alpha-1)}x^2 + \frac{t^{4\alpha}}{\Gamma(4\alpha+1)} - \frac{t^{4\alpha}}{\Gamma(4\alpha+1)}x^2 + 7\frac{t^{4\alpha-1}}{\Gamma(4\alpha)} - 3\frac{t^{4\alpha-1}}{\Gamma(4\alpha)}x^2 \\ &- 2\frac{t^{4\alpha+1}}{\Gamma(4\alpha)} + 5\frac{t^{4\alpha-1}}{\Gamma(4\alpha)} + 8\frac{t^{4\alpha-2}}{\Gamma(4\alpha-1)} - 3\frac{t^{4\alpha-2}}{\Gamma(4\alpha-1)}x^2 + \frac{t^{4\alpha-3}}{\Gamma(4\alpha-2)} - \frac{t^{4\alpha-3}}{\Gamma(4\alpha-2)}x^2, \end{aligned} \tag{50}$$

Substituting Equations (48) and (50) into Equation (47) gives the solution in a series form by

$$
\begin{aligned}
\sum_{n=0}^{\infty} \psi_n(x,t) &= \psi_0(x,t) + \psi_1(x,t) + \psi_2(x,t) + \ldots \\
\psi(x,t) &= x^2 + t + \frac{t^{\alpha}}{\Gamma(\alpha+1)}x^2 + \frac{t^{\alpha+1}}{\Gamma(\alpha+2)} - \frac{t^{\alpha}}{\Gamma(\alpha+1)} \\
&+ \frac{t^{\alpha}}{\Gamma(\alpha+1)} - \frac{t^{\alpha}}{\Gamma(\alpha+1)}x^2 - \frac{t^{\alpha+1}}{\Gamma(\alpha+2)} + 2\frac{t^{2\alpha}}{\Gamma(2\alpha+1)} \\
&- \frac{t^{2\alpha+1}}{\Gamma(2\alpha+2)} + \frac{t^{2\alpha-1}}{\Gamma(2\alpha)} - \frac{t^{2\alpha}}{\Gamma(2\alpha+1)}x^2 - \frac{t^{2\alpha-1}}{\Gamma(2\alpha)}x^2 \\
&- 2\frac{t^{2\alpha}}{\Gamma(2\alpha+1)} + \frac{t^{2\alpha+1}}{\Gamma(2\alpha+2)} - \frac{t^{2\alpha-1}}{\Gamma(2\alpha)} + \frac{t^{2\alpha}}{\Gamma(2\alpha+1)}x^2 \\
&+ \frac{t^{2\alpha-1}}{\Gamma(2\alpha)}x^2 - 5\frac{t^{3\alpha-1}}{\Gamma(3\alpha+2)} - 3\frac{t^{3\alpha}}{\Gamma(3\alpha+1)} + 2\frac{t^{3\alpha-1}}{\Gamma(3\alpha)}x^2 \\
&- \frac{t^{3\alpha-2}}{\Gamma(3\alpha-1)} + \frac{t^{3\alpha+1}}{\Gamma(3\alpha+2)} + \frac{t^{3\alpha}}{\Gamma(3\alpha+1)}x^2 + \frac{t^{3\alpha-2}}{\Gamma(3\alpha-1)}x^2 \\
&+ 5\frac{t^{3\alpha-1}}{\Gamma(3\alpha+2)} + 3\frac{t^{3\alpha}}{\Gamma(3\alpha+1)} - 2\frac{t^{3\alpha-1}}{\Gamma(3\alpha)}x^2 + \frac{t^{3\alpha-2}}{\Gamma(3\alpha-1)} - \frac{t^{3\alpha+1}}{\Gamma(3\alpha+2)} \\
&- \frac{t^{3\alpha}}{\Gamma(3\alpha+1)}x^2 - \frac{t^{3\alpha-2}}{\Gamma(3\alpha-1)}x^2 + \frac{t^{4\alpha}}{\Gamma(4\alpha+1)} - \frac{t^{4\alpha}}{\Gamma(4\alpha+1)}x^2 + 7\frac{t^{4\alpha-1}}{\Gamma(4\alpha)} - 3\frac{t^{4\alpha-1}}{\Gamma(4\alpha)}x^2 \\
&- 2\frac{t^{4\alpha+1}}{\Gamma(4\alpha)} + 5\frac{t^{4\alpha-1}}{\Gamma(4\alpha)} + 8\frac{t^{4\alpha-2}}{\Gamma(4\alpha-1)} - 3\frac{t^{4\alpha-2}}{\Gamma(4\alpha-1)}x^2 + \frac{t^{4\alpha-3}}{\Gamma(4\alpha-2)} - \frac{t^{4\alpha-3}}{\Gamma(4\alpha-2)}x^2
\end{aligned}
$$

at $\alpha = 2,$ we obtain the exact solution of standard telegraph equation

$$\psi(x,t) = t + x^2 \tag{51}$$

Example 3. *Consider the following space-fractional nonlinear telegraph equation:*

$$\begin{gathered}\frac{\partial^{\alpha}\psi(x,t)}{\partial t^{\alpha}} = \frac{\partial^2\psi(x,t)}{\partial x^2} + \frac{\partial\psi(x,t)}{\partial t} - \psi^2(x,t) + x\psi(x,t)\psi_x(x,t), \\ x,t \geq 0 \text{ and } 0 < \alpha \leq 2\end{gathered} \tag{52}$$

with the initial conditions

$$\psi(x,0) = x \text{ and } \psi_t(x,0) = x. \tag{53}$$

Solution 3

By taking natural transform for Equation (52), we have

$$\frac{s^{\alpha}}{u^{\alpha}}R(x,s,u) - \frac{s^{\alpha-1}}{u^{\alpha}}\psi(x,0) - \frac{s^{\alpha-2}}{u^{\alpha-1}}\psi_t(x,0) = \mathbb{N}^{+}\left[\frac{\partial\psi^2(x,t)}{\partial x^2} + \frac{\partial\psi(x,t)}{\partial t} - \psi^2(x,t) + x\psi(x,t)\psi_x(x,t)\right], \tag{54}$$

arrangement and substitute the initial condition, we get

$$R(x,s,u) = \frac{1}{s}x + \frac{u}{s^2}x + \left[\frac{u^{\alpha}}{s^{\alpha}}\mathbb{N}^{+}\left[\frac{\partial\psi^2(x,t)}{\partial x^2} + \frac{\partial\psi(x,t)}{\partial t} - \psi^2(x,t) + x\psi(x,t)\psi_x(x,t)\right]\right], \tag{55}$$

applying the inverse natural transform for Equation (55), we have

$$\psi(x,t) = x + tx + \mathbb{N}^{-}\left[\frac{u^{\alpha}}{s^{\alpha}}\mathbb{N}^{+}\left[\frac{\partial\psi^2(x,t)}{\partial x^2} + \frac{\partial\psi(x,t)}{\partial t} - \psi^2(x,t) + x\psi(x,t)\psi_x(x,t)\right]\right], \tag{56}$$

hence

$$\sum_{n=0}^{\infty}\psi_{n+1}(x,t) = (x+tx) + \mathbb{N}^{-}\left[\frac{u^{\alpha}}{s^{\alpha}}\mathbb{N}^{+}\left[\sum_{n=0}^{\infty}\frac{\partial^2\psi_n(x,t)}{\partial x^2} + \sum_{n=0}^{\infty}\frac{\partial\psi_n(x,t)}{\partial t} - \sum_{n=0}^{\infty}A_n(x,t) + x\sum_{n=0}^{\infty}B_n(x,t)\right]\right], \tag{57}$$

the initial term

$$\psi_0(x,t) = (x + tx). \tag{58}$$

Now the components of the series solution are given by

$$\psi_1(x,t) = \mathbb{N}^{-}\left[\frac{u^{\alpha}}{s^{\alpha}}\mathbb{N}^{+}\left[\frac{\partial^2\psi_0(x,t)}{\partial x^2} + \frac{\partial\psi_0(x,t)}{\partial t} - A_0(x,t) + xB_0(x,t)\right]\right], \tag{59}$$

$$\psi_1(x,t) = \left(\frac{t^{\alpha}}{\Gamma(\alpha+1)}x\right),$$

$$\psi_2(x,t) = \left(\frac{t^{\alpha+1}}{\Gamma(\alpha+2)}x\right), \tag{60}$$

$$\psi_3(x,t) = \left(\frac{t^{\alpha+2}}{\Gamma(\alpha+3)}x\right).$$

Since

$$\psi_n(x,t) = \psi_0(x,t) + \psi_1(x,t) + \psi_2(x,t) + \psi_3(x,t) + ... \tag{61}$$

$$\psi(x,t) = x + tx + \frac{t^{\alpha}}{\Gamma(\alpha+1)}x + \frac{t^{\alpha+1}}{\Gamma(\alpha+2)}x + \frac{t^{\alpha+2}}{\Gamma(\alpha+3)}x + ... \tag{62}$$

by substituting $\alpha = 2$ in Equation (62), we obtain the exact solution of standard telegraph equation in the following form:

$$\psi(x,t) = xe^t \tag{63}$$

6. Numerical Result

In this section, we shall illustrate the accuracy and efficiency of the (NTDM) by comparing the approximate and exact solution.

Figure 1 confirm the accuracy and efficiency of the natural transform and Adomian decomposition method and discuss the behavior of exact solution and approximate solutions Equation (30) obtained by (NTDM) for the special case $\alpha = 2$ for example (1). We see that Table 1 illustrated the absolute error by computing $\psi = |\psi - \psi_{10}|$ where ψ is the exact solution and ψ_{10} is approximate solution of Equation (30) obtained by truncating the respective solution series Equation (40) at ψ_{10}. Approximate solutions converge very swiftly to the exact solutions in only 10th order approximations i.e., approximate solutions are nearly identical to the exact solutions. The accuracy of the result can be amelioration by generating more terms of the approximate solutions.

Figure 2 shows the exact solution and the approximate solution Equation (30) obtained by natural transform and Adomian decomposition method when α decreasing then the ψ decreasing.

Table 2 discuss the solution of Example 1 by choosing different values of $t = \{0, 0.5, 1, 1.5, 2\}$ and the values of $\psi(x,t)$ decreasing when t increasing for different values of $\alpha = 1.99, 1.98$ and 1.97.

Figure 3 shows when setting $\alpha = 2$ in the nth approximations and canceling noise terms yields the exact solution $\psi = |\psi - \psi_{10}|$ as $n \to \infty$. The analytical solution for the exact solution and the approximate solution Equation (42) obtained by natural transform and Adomian decomposition method. In addition, the exact solution is presented graphically in Figure 3.

The exact and approximate solutions of Equation (52) are presented graphically in Figure 4, the approximate solution is given at $\alpha = 1.99, 1.98$ and 1.97. The value of the solution satisfies Equation (52) see in Table 3 for the values $\alpha = 1.99, 1.98$ and 1.97.

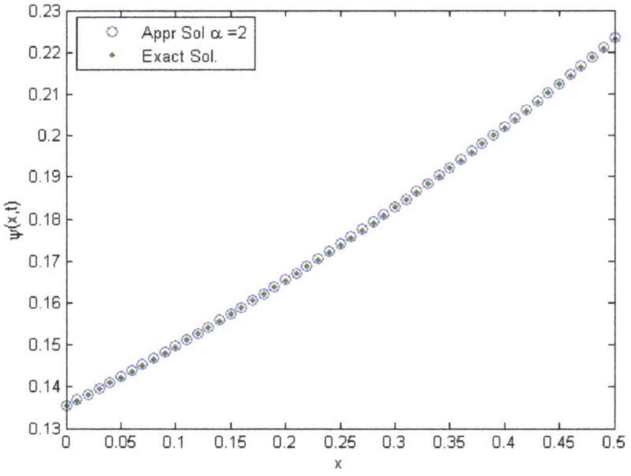

Figure 1. The Exact and Approximate Solutions of $\psi(x,t)$ for Example 1 for $\alpha = 2$.

Table 1. Exact and Approximate Solution of $\psi(x,t)$ for Example 1.

t	Exact Solution	Approximate Sol	$\psi = \lvert\psi - \psi_{10}\rvert$
0.0	1.648721270700128	1.648721270700128	0.0
0.5	1.0	1.000000000040401	4.040101586 $e-11$
1.0	0.606530659712633	0.606530742852590	8.313995659 $e-8$
1.5	0.367879441171442	0.367886690723836	7.249552393 $e-6$
2.0	0.223130160148429	0.223303762933655	1.569783692 $e-4$

Table 2. Approximate Solution of $\psi(x,t)$ for Example 1.

t	$\alpha = 1.99$	$\alpha = 1.98$	$\alpha = 1.97$
0.0	1.648721270700128	1.648721270700128	1.648721270700128
0.5	1.002243362235993	1.004498274095028	1.006764389796932
1.0	0.609569757949665	0.612585971492061	0.615579284214978
1.5	0.369222108754806	0.371788163947318	0.370522335669580
2.0	0.221291547575669	0.219269844467107	0.217240584657229

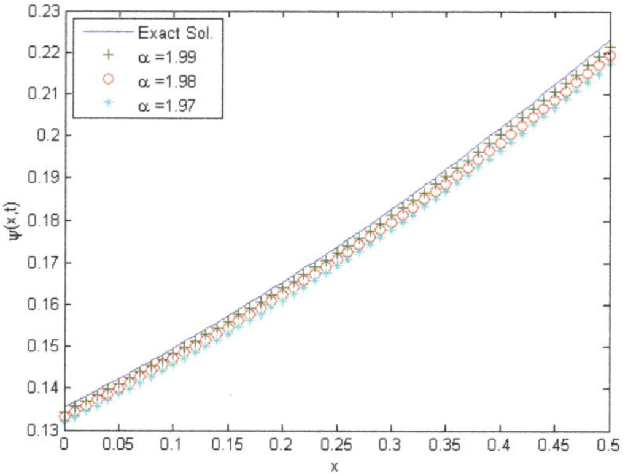

Figure 2. The Exact Solutions and Approximate Solutions of $\psi(x,t)$ for Example 1 for different value for α.

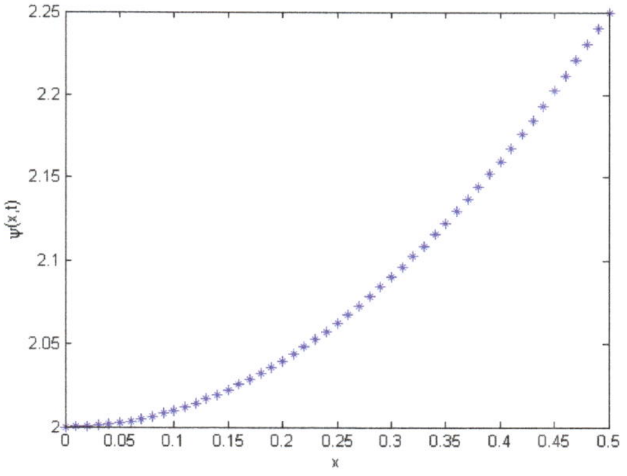

Figure 3. The exact solution of $\psi(x,t)$ for Example 2.

Table 3. Approximate solution of $\psi(x,t)$ for Example 3.

t	Exact Solution	$\alpha = 1.95$	$\alpha = 1.90$	$\alpha = 1.85$
0.0	0.5	0.5	0.5	1.5
5.0	0.824360635350064	0.830817752645242	0.837755999175080	0.845202109327201
1.0	1.359140914229523	1.378259288907402	1.398076433466764	1.418592017094073
1.5	2.240844535169032	2.276244404149126	2.312171661003479	2.348587393416824
2.0	3.694528049465325	3.748855797997422	3.803171995755493	3.857406067787722

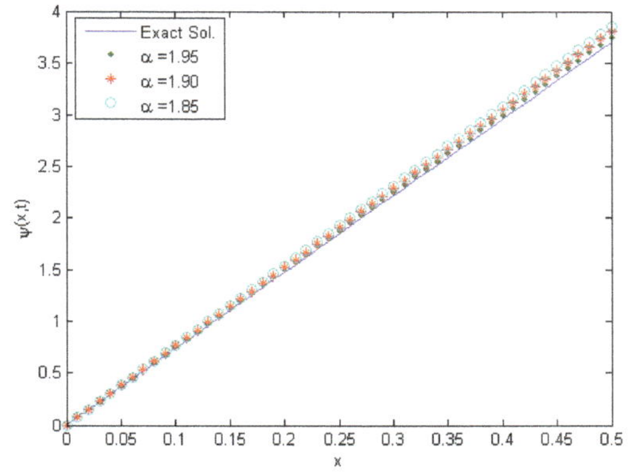

Figure 4. The approximate solutions of $\psi(x,t)$ for Example 3 for $\alpha = 1.95$, $\alpha = 1.90$, $\alpha = 1.85$ and exact solution.

7. Conclusions

We have successfully applied the natural transform and Adomian decomposition method to obtain the approximate solutions of the fractional telegraph equation. The (NTDM) give us small error and high convergence. As seen in Tables 1–3, errors are very small, and sometimes deflate as shown in Table 3. These techniques lead us to say that the method is accurate and efficient according to theoretical analysis and examples 3 and 4 the exact solution and approximate solution of $\psi(x,t)$ are equal at $\alpha = 2$ the absolute error equal zero.

Author Contributions: Investigation H.E., Y.T.A. Methodology, I.B. writing—review and editing, M.H.K. investigation.

Funding: The authors would like to extend their sincere appreciation to the Deanship of Scientific Research at King Saud University for its funding this Research group No. (RG-1440-030).

Conflicts of Interest: The authors declare no conflict of interest.

References

1. Hosseini, V.R.; Chen, W.; Avazzadeh, Z. Numerical solution of fractional telegraph equation by using radial basis functions. *Eng. Anal. Bound. Elem.* **2014**, *38*, 31–39. [CrossRef]
2. Kumar, S. A new analytical modelling for fractional telegraph equation via Laplace transform. *Appl. Math. Model.* **2014**, *38*, 3154–3163. [CrossRef]
3. Alawad, F.A.; Yousif, E.A.; Arbab, A.I. A new technique of Laplace variational iteration method for solving space-time fractional telegraph equations. *Int. J. Differ. Equ.* **2013**, *2013*, 256593. [CrossRef]
4. Dhunde, R.R.; Waghmare, G. Double Laplace transform method for solving space and time fractional telegraph equations. *Int. J. Math. Math. Sci.* **2016**, *2016*, 1414595. [CrossRef]
5. Biazar, J.; Shafiof, S. A simple algorithm for calculating Adomian polynomials. *Int. J. Contemp. Math. Sci.* **2007**, *2*, 975–982. [CrossRef]
6. Garg, M.; Sharma, A. Solution of space-time fractional telegraph equation by Adomian decomposition method. *J. Inequal. Spec. Funct.* **2011**, *2*, 1–7.
7. Kashuri, A.; Fundo, A.; Kreku, M. Mixture of a new integral transform and homotopy perturbation method for solving nonlinear partial differential equations. *Adv. Pure Math.* **2013**, *3*, 317. [CrossRef]
8. Xu, H.; Liao, S.-J.; You, X.-C. Analysis of nonlinear fractional partial differential equations with the homotopy analysis method. *Commun. Nonlinear Sci. Numer. Simul.* **2009**, *14*, 1152–1156. [CrossRef]
9. Saadatmandi, A.; Dehghan, M. Numerical solution of hyperbolic telegraph equation using the Chebyshev tau method. *Numer. Methods Part. Differ. Equ.* **2010**, *26*, 239–252. [CrossRef]
10. Chen, J.; Liu, F.; Anh, V. Analytical solution for the time-fractional telegraph equation by the method of separating variables. *J. Math. Anal. Appl.* **2008**, *338*, 1364–1377. [CrossRef]
11. Wazwaz, A.-M. *Partial Differential Equations And Solitary Waves Theory*; Springer Science & Business Media: Berlin/Heidelberg, Germany, 2010.
12. Belgacem, F.B.M.; Silambarasan, R. Theory of natural transform. *Math. Eng. Sci. Aerosp. J.* **2012**, *3*, 99–124.
13. Maitama, S. A hybrid natural transform homotopy perturbation method for solving fractional partial differential equations. *Int. J. Differ. Equ.* **2016**, *2016*, 9207869. [CrossRef]
14. Shah, K.; Khalil, H.; Khan, R.A. Analytical solutions of fractional order diffusion equations by natural transform method. *Iranian J. Sci. Technol. Trans. A Sci.* **2016**, 1–12. [CrossRef]
15. Podlubny, I. *Fractional Differential Equations: An Introduction to Fractional Derivatives, Fractional Differential Equations, to Methods of Their Solution and Some of Their Applications*; Elsevier: Amsterdam, The Netherlands, 1998; Volume 198.
16. El-Kalla, I. Convergence of the Adomian method applied to a class of nonlinear integral equations. *Appl. Math. Lett.* **2008**, *21*, 372–376. [CrossRef]

© 2019 by the authors. Licensee MDPI, Basel, Switzerland. This article is an open access article distributed under the terms and conditions of the Creative Commons Attribution (CC BY) license (http://creativecommons.org/licenses/by/4.0/).

symmetry

MDPI

Article

Some Difference Equations for Srivastava's λ-Generalized Hurwitz–Lerch Zeta Functions with Applications

Asifa Tassaddiq

College of Computer and Information Sciences Majmaah University, Al Majmaah 11952, Saudi Arabia; a.tassaddiq@mu.edu.sa

Received: 25 December 2018; Accepted: 19 February 2019; Published: 1 March 2019

Abstract: In this article, we establish some new difference equations for the family of λ-generalized Hurwitz–Lerch zeta functions. These difference equations proved worthwhile to study these newly defined functions in terms of simpler functions. Several authors investigated such functions and their analytic properties, but no work has been reported for an estimation of their values. We perform some numerical computations to evaluate these functions for different values of the involved parameters. It is shown that the direct evaluation of involved integrals is not possible for the large values of parameter s; nevertheless, using our new difference equations, we can evaluate these functions for the large values of s. It is worth mentioning that for the small values of this parameter, our results are 100% accurate with the directly computed results using their integral representation. Difference equations so obtained are also useful for the computation of some new integrals of products of λ-generalized Hurwitz–Lerch zeta functions and verified to be consistent with the existing results. A derivative property of Mellin transforms proved fundamental to present this investigation.

Keywords: analytic number theory; λ-generalized Hurwitz–Lerch zeta functions; derivative properties; recurrence relations; integral representations; Mellin transform

---

1. Introduction

In this paper, we practice the customary symbolizations:

$$N := \{1, 2, \cdots\};\ N_0 := N \cup \{0\};\ Z^- := \{-1, -2, \cdots\};\ Z_0^- := Z^- \cup \{0\}, \tag{1}$$

where Z^- is the set of integers. The involved symbols R, R^+, and C represent the set of real, positive real, and complex numbers, consistently.

The Hurwitz–Lerch zeta function has always been a topic of motivation for several researchers due to its impact in analytic number theory and other applied sciences. Recently, Srivastava presented a considerably new universal family of Hurwitz–Lerch zeta functions defined by [1] (p. 1487, Equation (1.14)):

$$\begin{aligned}
&\Phi^{(\rho_1,\ldots,\rho_p,\sigma_1,\ldots,\sigma_q)}_{\lambda_1,\ldots,\lambda_p,\mu_1,\ldots,\mu_q}(z, s, a; b, \lambda) \\
&= \frac{1}{\Gamma(s)} \int_0^\infty t^{s-1} \exp\left(-at - \frac{b}{t^\lambda}\right) {}_p\Psi^*{}_q\left[\begin{matrix} (\lambda_1, \rho_1), \ldots, (\lambda_p, \rho_p) \\ (\mu_1, \sigma_1), \ldots, (\mu_p, \sigma_p) \end{matrix}; ze^{-t}\right] dt; \\
&(min[R(a), R(s)] > 0; R(b) \geqq 0; \lambda \geqq 0)
\end{aligned} \tag{2}$$

so that, evidently, one can get the subsequent connection with the extended Hurwitz–Lerch zeta functions $\Phi^{(\rho_1,\ldots,\rho_p,\sigma_1,\ldots,\sigma_q)}_{\lambda_1,\ldots,\lambda_p,\mu_1,\ldots,\mu_q}(z,s,a)$ defined by the authors of [2] (p. 503, Equation (6.2)) (see also References [3,4]):

$$\Phi^{(\rho_1,\ldots,\rho_p,\sigma_1,\ldots,\sigma_q)}_{\lambda_1,\ldots,\lambda_p,\mu_1,\ldots,\mu_q}(z,s,a;0,\lambda) = \Phi^{(\rho_1,\ldots,\rho_p,\sigma_1,\ldots,\sigma_q)}_{\lambda_1,\ldots,\lambda_p,\mu_1,\ldots,\mu_q}(z,s,a) = e^b\Phi^{(\rho_1,\ldots,\rho_p,\sigma_1,\ldots,\sigma_q)}_{\lambda_1,\ldots,\lambda_p,\mu_1,\ldots,\mu_q}(z,s,a;b,0). \quad (3)$$

In the above Equation (2), ${}_p\Psi^*_q$ where ($p, q \in \mathbb{N}_0$) is the standard Fox–Wright function defined by the authors of [4] (p. 2219, Equation (1)) (see also References [3] (p. 516, Equation (1)) and [2] (p. 493, Equation (2.1)):

$${}_p\Psi^*_q\left[\begin{matrix}(\lambda_1,\rho_1),\ldots, & (\lambda_p,\rho_p) \\ (\mu_1,\sigma_1),\ldots, & (\mu_q,\sigma_q)\end{matrix};z\right] = \sum_{\chi=0}^{\infty}\frac{([\lambda_p])_{\rho_p\chi}}{([\mu_q])_{\sigma_q\chi}}\frac{z^\chi}{\chi!}. \quad (4)$$

Pochhammer symbols $([\lambda_p])_{\rho_p n} := [\lambda_1]_{\rho_p n}\cdots[\lambda_p]_{\rho_p n}$ symbolize the shifted factorial defined in terms of the basic Gamma function as follows:

$$(\lambda)_\rho = \frac{\Gamma(\lambda+\rho)}{\Gamma(\lambda)} = \begin{cases} 1 & (\rho = 0, \lambda \in \mathbb{C}\setminus\{0\}) \\ \lambda(\lambda+1)\ldots(\lambda+\chi-1) & (\rho = \chi \in \mathbb{N}; \lambda \in \mathbb{C}), \end{cases} \quad (5)$$

$$\Delta := \sum_{j=1}^{q}\sigma_j - \sum_{j=1}^{p}\rho_j \text{ and } \nabla := \left(\prod_{j=1}^{p}\rho_j^{-\rho_j}\right)\cdot\left(\prod_{j=1}^{q}\sigma_j^{\sigma_j}\right).$$

The series given by Equation (4) converges in the entire complex z-plane for $\Delta > -1$; and if $\Delta = 0$, the series (Equation (4)) converges only for $|z| < \nabla$. For more detailed discussion of such functions, we refer the interested reader to also see References [5–9].

The analysis of Srivastava's λ-generalized Hurwitz–Lerch zeta functions and its different forms have attracted noteworthy concern, and many papers have subsequently appeared on this subject. Jankov et al. [10] and Srivastava et al. [3] discussed some inequalities for different cases of λ-generalized Hurwitz–Lerch zeta functions. Srivastava et al. [11] introduced a nonlinear operator related with the λ-generalized Hurwitz–Lerch zeta functions to analyze the inclusion properties of definite subclass of special type of meromorphic functions. Srivastava and Gaboury [12] deliberated on new expansion formulas for such functions (see, for details, References [13,14]; see also the further thoroughly associated studies cited in each of these publications). Luo and Raina [4] discussed some new inequalities involving Srivastava's λ-generalized Hurwitz–Lerch zeta functions and obtained the following series representation [4] (p. 2221, Equation (6)):

$$\Phi^{(\rho_1,\ldots,\rho_p,\sigma_1,\ldots,\sigma_q)}_{\lambda_1,\ldots,\lambda_p,\mu_1,\ldots,\mu_q}(z,s,a;b,\lambda) = \frac{1}{\lambda\Gamma(s)}\sum_{\chi=0}^{\infty}\frac{([\lambda_p])_{\rho_p\chi}}{([\mu_q])_{\sigma_q\chi}}Z^{\frac{s}{\lambda}}_{\frac{1}{\lambda}}(a+\chi)^\lambda b\frac{z^\chi}{\chi!(\chi+a)^s},$$
$$\left(\lambda j \in R (j = 1,..,p) \text{ and } \mu_j \in R\setminus Z - 0\ (j = 1,\ldots,q); \rho j > 0(j,\ldots,p); \sigma j > 0(j = 1,\ldots,q); 1+\Delta \geq 0\right). \quad (6)$$

Srivastava beautifully described important results about the zeta and related functions in an expository article [15]. Choi et al. [16] further discussed these functions by introducing one more variable. Srivastava et al. [17] presented an innovative integral transform connected with the λ-extended Hurwitz–Lerch zeta function. More recently, Tassaddiq [18] obtained a new representation for this family of the λ-generalized Hurwitz–Lerch zeta functions in terms of complex delta functions such that the definition of these functions is formalized over the space of entire test functions denoted by Z. The author also listed and discussed all the possible special cases of Srivastava's λ-generalized Hurwitz–Lerch zeta functions [18] (p. 4) in the form of a table. For the purposes of our present investigation, this table is given on the next page. For any use of the special cases of the generalized Hurwitz–Lerch zeta functions, the reader is referred to this table. For more detailed study of zeta and related functions, we refer the interested reader to References [19–40] and further bibliography cited therein.

Table 1. Different special cases of λ-generalized Hurwitz–Lerch zeta functions [18].

$\boldsymbol{min[\Re(a),\Re(s)] > 0; \Re(b) \geqq 0; \lambda \geqq 0;}$ $\boldsymbol{\rho = \rho_1,\ldots,\rho_p; \sigma = \sigma_1,\ldots,\sigma_q; \lambda^* = \lambda_1,\ldots,\lambda_p; \mu = \mu_1,\ldots,\mu_q}$		$\boldsymbol{(p-1 = q = 0; \lambda_1 = \mu; \rho_1 = 1)}$					$\boldsymbol{(p-1 = q = 0; \lambda_1 = \mu; \rho_1 = 1)}$	
			$\lambda = 1$	$\mu = 1$	$\lambda = \mu = 1$	$b = 0$	$b = 0$	$\mu = 1; b = 0$
λ-Generalized Hurwitz–Lerch Zeta Functions	$\Phi^{(\rho;\sigma)}_{\lambda^*,\mu}(\pm z, s, a; b, \lambda)$ [1], (p. 1487), Equation (1.14)	$\Theta^{\lambda}_{\mu}(\mp z, s, a; b)$ [41], (p. 90), Equation (6) and [42]	$\Phi^*_{\mu}(\pm z, s, a, b)$	$\Phi_b(\pm z, s, a, \lambda)$	$\Phi_b(\pm z, s, a)$	$\Phi^{(\rho;\sigma)}_{\lambda^*,\mu}(\pm z, s, a)$ ([1], p. 1486, Equation (1.11)) & [2]	$\Phi^*_{\mu}(\pm z, s, a)$ ([43], p. 100, Equation (1.5))	$\Phi(\pm z, s, a)$ ([44], p. 27, Equation (1.11))
$\Phi^{(\rho;\sigma)}_{\lambda^*,\mu}(\pm e^{-x}, s, a; b, \lambda)$ λ-Generalized Extended Fermi–Dirac and Extended Bose–Einstein Functions	$\Theta^{(\rho;\sigma)}_{\lambda^*,\mu}(x, s, a; b, \lambda)$	$\Theta^{\lambda}_{\mu}(x, s, a; b)$	$\Theta^*_{\mu}(x, s, a, b)$	$\Theta_b(x, s, a; \lambda)$	$\Theta_b(x, s, a)$	$\Theta^{(\rho;\sigma)}_{\lambda^*,\mu}(x, s, a)$	$\Theta^*_{\mu}(x, s, a)$ ([43], p. 12, Equation (45))	$\Theta_a(x; s)$ ([45], p. 9, Equation (3.14))
	$\Psi^{(\rho;\sigma)}_{\lambda^*,\mu}(x, s, a; b, \lambda)$	$\Psi^{\lambda}_{\mu}(x, s, a; b)$	$\Psi^*_{\mu}(x, s, a, b)$	$\Psi_b(x, s, a, \lambda)$	$\Psi_b(x, s, a)$	$\Psi^{(\rho;\sigma)}_{\lambda^*,\mu}(x, s, a)$	$\Psi^*_{\mu}(x, s, a)$ ([43], p. 12, Equation (45))	$\Psi_a(x; s)$ [45], p. 115, Equation (4.4)
$\Phi^{(\rho;\sigma)}_{\lambda^*,\mu}(\pm z, s, 1; b, \lambda)$ λ-Generalized Polylogarithm Functions	$\mathrm{Li}^{(\rho;\sigma)}_{\lambda^*,\mu}(\pm z, s; b, \lambda)$	$\mathrm{Li}^{\lambda}_{\mu}(\mp z, s, a; b)$	$\mathrm{Li}^*_{\mu}(z, s, b)$	$\mathrm{Li}_b(z, s, \lambda)$	$\mathrm{Li}_b(z, s)$	$\mathrm{Li}^{(\rho;\sigma)}_{\lambda^*,\mu}(z, s)$	$\mathrm{Li}^*_{\mu}(z, s)$ ([43], p. 12, Equation (47))	$\mathrm{Li}_s(z)$ [44], (Chapter 1)
$\Phi^{(\rho;\sigma)}_{\lambda^*,\mu}(\pm e^{-x}, s+1, 1; b, \lambda)$ λ-Generalized Fermi–Dirac and Bose–Einstein Functions	$\mathrm{F}^{(\rho;\sigma)}_{\lambda^*,\mu}(x, s; b, \lambda)$	$\mathrm{F}^{\lambda}_{\mu}(x, s, a; b)$	$\mathrm{F}^*_{\mu}(x, s, b)$	$\mathrm{F}_b(x, s, \lambda)$	$\mathrm{F}_b(x, s)$	$\mathrm{F}^{(\rho;\sigma)}_{\lambda^*,\mu}(x, s)$	$\mathrm{F}^*_{\mu}(x, s)$ ([43], p. 12, Equation (47))	$\mathrm{F}_s(x)$ ([45], p. 109, Equation (1.12)
	$\mathrm{B}^{(\rho;\sigma)}_{\lambda^*,\mu}(x, s; b, \lambda)$	$\mathrm{B}^{\lambda}_{\mu}(x, s, a; b)$	$B^*_{\mu}(x, s, b)$	$B_b(x, s, \lambda)$	$B_b(x, s)$	$\mathrm{B}^{(\rho;\sigma)}_{\lambda^*,\mu}(x, s)$	$\mathrm{B}^*_{\mu}(x, s)$ ([43], p. 12, Equation (45))	$\mathrm{B}_s(x)$ ([45], p. 109, Equation (1.12))
$\Phi^{(\rho;\sigma)}_{\lambda^*,\mu}(\pm 1, s, a; b, \lambda)$ λ-Generalized Hurwitz zeta Functions	$\zeta^{(\rho;\sigma)}_{\lambda^*,\mu}(s, a; b, \lambda)$	$\zeta^{\lambda}_{\mu}(s, a; b)$	$\zeta^*_{\mu}(s, a, b)$	$\zeta_b(s, a, \lambda)$	$\zeta_b(s, a)$ [46], p. 308	$\zeta^{(\rho;\sigma)}_{\lambda^*,\mu}(s, a)$	$\zeta^*_{\mu}(s, a)$ [43]	$\zeta(s, a)$ [44], (Chapter 1)
$\Phi^{(\rho;\sigma)}_{\lambda^*,\mu}(\pm 1, s, 1; b, \lambda)$ λ-Generalized Riemann Zeta Functions	$\zeta^{(\rho;\sigma)}_{\lambda^*,\mu}(s)$	$\zeta^{\lambda}_{\mu}s; b)$	$\zeta^*_{\mu}(s, b)$	$\zeta_b(s, \lambda)$	$\zeta_b(s)$ [46], p. 308	$\zeta^{(\rho;\sigma)}_{\lambda^*,\mu}(s)$	$\zeta^*_{\mu}(s)$ [43]	$\zeta(s)$ [44], (Chapter 1)

In this research, our focus is to establish some new difference equations for the family of λ-generalized Hurwitz–Lerch zeta functions and its special cases by following the approach of Tassaddiq and Qadir [33]. From the above discussion and Table 1, we can notice that several authors presented and studied worthwhile generalizations of the Hurwitz–Lerch zeta functions. They obtained various analytic formulas, integral, and series representations. However, as we deeply study Riemann zeta functions, we know their values, their graphs, and several other important aspects. We could not develop this approach for these generalizations. Bayad and Chiki [43] obtained reduction and duality formulas of the generalized Hurwitz–Lerch zeta functions. Their results contain the earlier obtained results of Choi [47]. These reduction formulas were concerned with the reduction of one parameter that represent the generalized Hurwitz–Lerch zeta $\Phi^*_\mu(z,s,a)$ and Hurwitz zeta functions $\zeta^*_\mu(s,a)$ in terms of Hurwitz–Lerch zeta $\Phi(z,s,a)$ and Hurwitz zeta $\zeta(s,a)$ functions, respectively. The difference equations presented here have the advantage of reducing the generalized Hurwitz–Lerch zeta $\Phi^*_\mu(z,s,a)$ and the generalized Hurwitz zeta functions $\zeta^*_\mu(s,a)$ in terms of basic polylogarithm $\mathrm{Li}_s(z)$ and zeta functions $\zeta(s)$, respectively. That means we have reduced one more parameter and our results are simple enough to evaluate these functions for different values of the involved parameters. By following the approach developed in this paper, we can initiate a deeper analysis of these functions that will enhance their applications. The Riemann hypothesis is a well-known unsolved problem in analytic number theory [22]. It states that "all the non-trivial zeros of the zeta function exist on the real line $s=\frac{1}{2}$". These zeros seem to be complex conjugates and hence symmetric on this line. The integrals of the zeta function and its generalizations are vital in the study of Riemann hypothesis and for the investigation of zeta functions themselves. The study of distributions in statistical inference and reliability theory [1,48,49] also involves such integrals. Difference equations obtained in this investigation are worthwhile to evaluate integrals of products of the family of λ-generalized Hurwitz–Lerch zeta functions that are consistent with the existing results.

The plan of the paper as follows: We present some new difference equations involving the λ-generalized Hurwitz–Lerch zeta functions in Section 2 and obtain similar results for other related functions. We discuss some applications of these difference equations in Section 3 by evaluating some special cases of the function. Based upon the results of Section 2, we evaluate new integrals of products of these functions in Section 4. We conclude our results in the last Section 5 by highlighting some future directions of this work.

Throughout this investigation, conditions on the parameters will be considered standard as given in Equations (1)–(6) and Table 1 unless otherwise stated.

2. Results

New Difference Equation of the λ-Generalized Hurwitz–Lerch Zeta Functions

Theorem 1. *Prove that λ-Generalized Hurwitz–Lerch zeta functions satisfy the following relation:*

$$\Gamma(s)\Phi^{\lambda_1,\ldots,\lambda_p,\mu_1,\ldots,\mu_q}_{(\lambda_1+\rho_1,\ldots,\lambda_p\rho_p,\mu_1+\sigma_1,\ldots,\mu_q+\sigma_q)}(z,s,a+1;b,\lambda)= \frac{(\mu_1)_{\sigma_1}\cdots(\mu_q)_{\sigma_q}}{z.(\lambda_1)_{\rho_1}\cdots(\lambda_p)_{\rho_p}}\left[\begin{array}{c} b\lambda\Gamma(s-\lambda-1)\Phi^{(\rho_1,\ldots,\rho_p\sigma_1,\ldots,\sigma_q)}_{\lambda_1,\ldots,\lambda_p,\mu_1,\ldots,\mu_q}(z,s-\lambda-1,a,b)+\Gamma(s) \\ \left[\Phi^{(\rho_1,\ldots,\rho_p\sigma_1,\ldots,\sigma_q)}_{\lambda_1,\ldots,\lambda_p,\mu_1,\ldots,\mu_q}(z,s-1,a;b,\lambda)-a\Phi^{(\rho_1,\ldots,\rho_p\sigma_1,\ldots,\sigma_q)}_{\lambda_1,\ldots,\lambda_p,\mu_1,\ldots,\mu_q}(z,s,a;b,\lambda)\right]\end{array}\right]. \tag{7}$$

Proof: Consider the function:

$$f(t)=\exp\left(-at-\frac{b}{t^{\lambda}}\right)\,{}_p\Psi^*_q\left[\begin{array}{ll}(\lambda_1,\rho_1),\ldots, & (\lambda_p,\rho_p) \\ (\mu_1,\sigma_1),\ldots, & (\mu_q,\sigma_q)\end{array};ze^{-t}\right] \tag{8}$$

and differentiate Equation (8) to get:

$$
\begin{aligned}
&\frac{d}{dt}\left[\exp\left(-at-\frac{b}{t^{\lambda}}\right)p^{\Psi^*}q\left[\begin{matrix}(\lambda_1,\rho_1),\ldots, & (\lambda_p,\rho_p)\\ (\mu_1,\sigma_1),\ldots, & (\mu_q,\sigma_q)\end{matrix};ze^{-t}\right]\right]=\\
&-a.\exp\left(-at-\frac{b}{t^{\lambda}}\right)p^{\Psi^*}q\left[\begin{matrix}(\lambda_1,\rho_1),\ldots, & (\lambda_p,\rho_p)\\ (\mu_1,\sigma_1),\ldots, & (\mu_q,\sigma_q)\end{matrix};ze^{-t}\right]\\
&+b\lambda\frac{\exp\left(-at-\frac{b}{t^{\lambda}}\right)}{t^{\lambda+1}}p^{\Psi^*}q\left[\begin{matrix}(\lambda_1,\rho_1),\ldots, & (\lambda_p,\rho_p)\\ (\mu_1,\sigma_1),\ldots, & (\mu_q,\sigma_q)\end{matrix};ze^{-t}\right]\\
&-z\frac{(\lambda_1)_{\rho_1}\cdots\cdots(\lambda_p)_{\rho_p}}{(\mu_1)_{\sigma_1}\cdots\cdots(\mu_q)_q}\exp\left(-(a+1)t-\frac{b}{t^{\lambda}}\right)p^{\Psi^*}q\left[\begin{matrix}(\lambda_1+\rho_1,\rho_1),\ldots, & (\lambda_p+\rho_p,\rho_p)\\ (\mu_1+\sigma_1,\sigma_1),\ldots, & (\mu_q+\sigma_q,\sigma_q)\end{matrix};ze^{-t}\right]
\end{aligned}
\tag{9}
$$

so that:

$$
\begin{aligned}
f'(t)=&\exp\left(-at-\frac{b}{t^{\lambda}}\right)p^{\Psi^*}q\left[\begin{matrix}(\lambda_1,\rho_1),\ldots, & (\lambda_p,\rho_p)\\ (\mu_1,\sigma_1),\ldots, & (\mu_q,\sigma_q)\end{matrix}\ ze^{-t}\right]\left[-a+\frac{b\lambda}{t^{\lambda+1}}\right]\\
&-z\frac{(\lambda_1)_{\rho_1}\cdots\cdots(\lambda_p)_{\rho_p}}{(\mu_1)_{\sigma_1}\cdots\cdots(\mu_q)_q}\exp\left(-(a+1)t-\frac{b}{t^{\lambda}}\right)p^{\Psi^*}q\left[\begin{matrix}(\lambda_1+\rho_1,\rho_1),\ldots, & (\lambda_p+\rho_p,\rho_p)\\ (\mu_1+\sigma_1,\sigma_1),\ldots, & (\mu_q+\sigma_q,\sigma_q)\end{matrix};ze^{-t}\right],
\end{aligned}
\tag{10}
$$

where we have used the usual differentiation and the derivative property, which is obtained on the same lines as given by Reference [1] (p. 1492, Equation (3.1)):

$$
\begin{aligned}
&\frac{d}{dt}\left[p^{\Psi^*}q\left[\begin{matrix}(\lambda_1,\rho_1),\ldots, & (\lambda_p,\rho_p)\\ (\mu_1,\sigma_1),\ldots, & (\mu_q,\sigma_q)\end{matrix};ze^{-t}\right]\right]=\\
&-ze^{-t}\frac{(\lambda_1)_{\rho_1}\cdots\cdots(\lambda_p)_{\rho_p}}{(\mu_1)_{\sigma_1}\cdots\cdots(\mu_q)_q}p^{\Psi^*}q\left[\begin{matrix}(\lambda_1+\rho_1,\rho_1),\ldots, & (\lambda_p+\rho_p,\rho_p)\\ (\mu_1+\sigma_1,\sigma_1),\ldots, & (\mu_q+\sigma_q,\sigma_q)\end{matrix};ze^{-t}\right].
\end{aligned}
\tag{11}
$$

Taking Mellin transform on both sides of Equation (8) and using the defining integral representation as given in Equation (2), we can write:

$$
\Gamma(s)\Phi^{(\rho_1,\ldots,\rho_p,\sigma_1,\ldots,\sigma_q)}_{\lambda_1,\ldots,\lambda_p,\mu_1,\ldots,\mu_q}(z,s,a;b,\lambda)=M\left[\exp\left(-at-\frac{b}{t^{\lambda}}\right)p^{\Psi^*}q\left[\begin{matrix}(\lambda_1,\ \rho_1),\ldots, & (\lambda_p,\ \rho_p)\\ (\mu_1,\sigma_1),\ldots, & (\mu_p,\sigma_p)\end{matrix}\ ze^{-t}\right];s\right]. \tag{12}
$$

Using the derivative property of Mellin transform given by, see [50] (Chapter 10):

$$
M[\,u'(y);\ \tau]\ =\ -(\tau-1)M[u(y);\ \tau-1] \tag{13}
$$

we obtain the following equation:

$$
\begin{aligned}
&\Gamma(s)\left[a\Phi^{(\rho_1,\ldots,\rho_p,\sigma_1,\ldots,\sigma_q)}_{\lambda_1,\ldots,\lambda_p,\mu_1,\ldots,\mu_q}(z,s,a;b,\lambda)-\Phi^{(\rho_1,\ldots,\rho_p,\sigma_1,\ldots,\sigma_q)}_{\lambda_1,\ldots,\lambda_p,\mu_1,\ldots,\mu_q}(z,s-1,a;b,\lambda)\right]=\\
&b\lambda\Gamma(s-\lambda-1)\Phi^{(\rho_1,\ldots,\rho_p,\sigma_1,\ldots,\sigma_q)}_{\lambda_1,\ldots,\lambda_p,\mu_1,\ldots,\mu_q}(z,s-\lambda-1,a,b)\\
&-\frac{(\lambda_1)_{\rho_1}\cdots\cdots(\lambda_p)_{\rho_p}}{(\mu_1)_{\sigma_1}\cdots\cdots(\mu_q)_q}z\Gamma(s)\Phi^{\lambda_1,\ldots,\lambda_p,\mu_1,\ldots,\mu_q}_{(\lambda_1+\rho_1,\ldots,\lambda_p\rho_p,\mu_1+\sigma_1,\ldots,\mu_q+\sigma_q)}(z,s,a+1;b,\lambda)
\end{aligned}
\tag{14}
$$

which leads to:

$$
\begin{aligned}
&\frac{(\lambda_1)_{\rho_1}\cdots\cdots(\lambda_p)_{\rho_p}}{(\mu_1)_{\sigma_1}\cdots\cdots(\mu_q)_q}z\Gamma(s)\Phi^{\lambda_1,\ldots,\lambda_p,\mu_1,\ldots,\mu_q}_{(\lambda_1+\rho_1,\ldots,\lambda_p\rho_p,\mu_1+\sigma_1,\ldots,\mu_q+\sigma_q)}(z,s,a+1;b,\lambda)=\\
&b\lambda\Gamma(s-\lambda-1)\Phi^{(\rho_1,\ldots,\rho_p,\sigma_1,\ldots,\sigma_q)}_{\lambda_1,\ldots,\lambda_p,\mu_1,\ldots,\mu_q}(z,s-\lambda-1,a,b)\\
&+\Gamma(s)\left[\Phi^{(\rho_1,\ldots,\rho_p,\sigma_1,\ldots,\sigma_q)}_{\lambda_1,\ldots,\lambda_p,\mu_1,\ldots,\mu_q}(z,s-1,a;b,\lambda)-a\Phi^{(\rho_1,\ldots,\rho_p,\sigma_1,\ldots,\sigma_q)}_{\lambda_1,\ldots,\lambda_p,\mu_1,\ldots,\mu_q}(z,s,a;b,\lambda)\right].
\end{aligned}
\tag{15}
$$

After some simple modifications, one can arrive at the required result of Equation (7). □

Remark 1. *We can obtain similar results for other related functions as listed in Table 1 by considering different parameter values in the resulting corollaries.*

Corollary 1. *λ-Generalized Extended Fermi–Dirac functions have the following representation:*

$$\Gamma(s)\Theta^{(\rho_1,\ldots,\rho_p,\sigma_1,\ldots,\sigma_q)}_{\lambda_1+\rho_1,\ldots,\lambda_p+\rho_p,\mu_1+\sigma_1,\ldots,\mu_q\sigma_q}(x,s,a+1;b,\lambda) = \frac{e^x.(\mu_1)_{\sigma_1}\ldots.(\mu_q)_{\sigma_q}}{(\lambda_1)_{\rho_1}\ldots.(\lambda_p)_{\rho_p}}\left[\begin{array}{c}\Gamma(s)\left[a\Theta^{(\rho_1,\ldots,\rho_p,\sigma_1,\ldots,\sigma_q)}_{\lambda_1,\ldots,\lambda_p,\mu_1,\ldots,\mu_q}(x,s,a;b,\lambda)-\Theta^{(\rho_1,\ldots,\rho_p,\sigma_1,\ldots,\sigma_q)}_{\lambda_1,\ldots,\lambda_p,\mu_1,\ldots,\mu_q}(x,s-1,a;b,\lambda)\right]\\ -b\lambda\Gamma(s-\lambda-1)\Theta^{(\rho_1,\ldots,\rho_p,\sigma_1,\ldots,\sigma_q)}_{\lambda_1,\ldots,\lambda_p,\mu_1,\ldots,\mu_q}(x,s-\lambda-1,a,b,\lambda)\end{array}\right] \tag{16}$$

and λ-Generalized Extended Bose–Einstein functions have the following representation:

$$\begin{aligned}\Gamma(s)\Psi^{(\rho_1,\ldots,\rho_p,\sigma_1,\ldots,\sigma_q)}_{\lambda_1+\rho_1,\ldots,\lambda_p+\rho_p,\mu_1+\sigma_1,\ldots,\mu_q+\sigma_q}(x,s,a+1;b,\lambda)&\\ =\frac{e^x.(\mu_1)_{\sigma_1}\ldots.(\mu_q)_{\sigma_q}}{(\lambda_1)_{\rho_1}\ldots.(\lambda_p)_{\rho_p}}&\Big[b\lambda\Gamma(s-\lambda-1)\Psi^{(\rho_1,\ldots,\rho_p,\sigma_1,\ldots,\sigma_q)}_{\lambda_1,\ldots,\lambda_p,\mu_1,\ldots,\mu_q}(x,s-\lambda-1,a,b,\lambda)\\ +\Gamma(s)&\left[\Psi^{(\rho_1,\ldots,\rho_p,\sigma_1,\ldots,\sigma_q)}_{\lambda_1,\ldots,\lambda_p,\mu_1,\ldots,\mu_q}(x,s-1,a;b,\lambda)-a\Psi^{(\rho_1,\ldots,\rho_p,\sigma_1,\ldots,\sigma_q)}_{\lambda_1,\ldots,\lambda_p,\mu_1,\ldots,\mu_q}(x,s,a;b,\lambda)\right]\Big].\end{aligned} \tag{17}$$

Proof. The results follow directly from Equation (7) upon replacing $\mathbf{z}\longrightarrow\pm\mathbf{e}^{-\mathbf{x}}$ and using the parallel case given in row 2 and column 2 of Table 1. □

Corollary 2. *λ-Generalized Fermi–Dirac functions have the following representation:*

$$\Gamma(s)\Theta^{(\rho_1,\ldots,\rho_p,\sigma_1,\ldots,\sigma_q)}_{\lambda_1+\rho_1,\ldots,\lambda_p+\rho_p,\mu_1+\sigma_1,\ldots,\mu_q+\sigma_q}(x,s,2;b,\lambda) = \frac{e^x.(\mu_1)_{\sigma_1}\ldots.(\mu_q)_{\sigma_q}}{(\lambda_1)_{\rho_1}\ldots.(\lambda_p)_{\rho_p}}\left[\begin{array}{c}b\lambda\Gamma(s-\lambda-1)F^{(\rho_1,\ldots,\rho_p,\sigma_1,\ldots,\sigma_q)}_{\lambda_1,\ldots,\lambda_p,\mu_1,\ldots,\mu_q}(x,s-\lambda-1,b,\lambda)\\ +\Gamma(s)\left[F^{(\rho_1,\ldots,\rho_p,\sigma_1,\ldots,\sigma_q)}_{\lambda_1,\ldots,\lambda_p,\mu_1,\ldots,\mu_q}(x,s-1;b,\lambda)-F^{(\rho_1,\ldots,\rho_p,\sigma_1,\ldots,\sigma_q)}_{\lambda_1,\ldots,\lambda_p,\mu_1,\ldots,\mu_q}(x,s;b,\lambda)\right]\end{array}\right] \tag{18}$$

and λ-Generalized Bose–Einstein functions have the following representation:

$$\begin{aligned}\Gamma(s)\Psi^{(\rho_1,\ldots,\rho_p,\sigma_1,\ldots,\sigma_q)}_{\lambda_1+\rho_1,\ldots,\lambda_p+\rho_p,\mu_1+\sigma_1,\ldots,\mu_q+\sigma_q}(x,s,2;b,\lambda)&\\ =\frac{e^x.(\mu_1)_{\sigma_1}\ldots.(\mu_q)_{\sigma_q}}{(\lambda_1)_{\rho_1}\ldots.(\lambda_p)_{\rho_p}}&\Big[b\lambda\Gamma(s-\lambda-1)B^{(\rho_1,\ldots,\rho_p,\sigma_1,\ldots,\sigma_q)}_{\lambda_1,\ldots,\lambda_p,\mu_1,\ldots,\mu_q}(x,s-\lambda-1,b,\lambda)\\ +\Gamma(s)&\left[B^{(\rho_1,\ldots,\rho_p,\sigma_1,\ldots,\sigma_q)}_{\lambda_1,\ldots,\lambda_p,\mu_1,\ldots,\mu_q}(x,s-1;b,\lambda)-B^{(\rho_1,\ldots,\rho_p,\sigma_1,\ldots,\sigma_q)}_{\lambda_1,\ldots,\lambda_p,\mu_1,\ldots,\mu_q}(x,s;b,\lambda)\right]\Big].\end{aligned} \tag{19}$$

Proof. The results follow directly from Equation (7) upon replacing $z\longrightarrow\pm e^{-x}$; $a\longrightarrow 1$ and taking the item from Table 1 corresponding to these parameter values. □

Corollary 3. *λ-Generalized Polylogarithm functions have the following representation:*

$$\Gamma(s)Li^{(\rho_1,\ldots,\rho_p,\sigma_1,\ldots,\sigma_q)}_{\lambda_1+\rho_1,\ldots,\lambda_p+\rho_p,\mu_1+\sigma_1,\ldots,\mu_q+\sigma_q}(z,s,2;b,\lambda) = \frac{(\mu_1)_{\sigma_1}\ldots.(\mu_q)_{\sigma_q}}{z.(\lambda_1)_{\rho_1}\ldots.(\lambda_p)_{\rho_p}}\left[\begin{array}{c}b\lambda\Gamma(s-\lambda-1)Li^{(\rho_1,\ldots,\rho_p,\sigma_1,\ldots,\sigma_q)}_{\lambda_1,\ldots,\lambda_p,\mu_1,\ldots,\mu_q}(z,s-\lambda-1,b,\lambda)\\ +\Gamma(s)\left[Li^{(\rho_1,\ldots,\rho_p,\sigma_1,\ldots,\sigma_q)}_{\lambda_1,\ldots,\lambda_p,\mu_1,\ldots,\mu_q}(z,s-1;b,\lambda)-Li^{(\rho_1,\ldots,\rho_p,\sigma_1,\ldots,\sigma_q)}_{\lambda_1,\ldots,\lambda_p,\mu_1,\ldots,\mu_q}(z,s;b,\lambda)\right]\end{array}\right]. \tag{20}$$

Proof. The result follows directly from Equation (7) upon replacing $a \longrightarrow 1$ and considering the specific case of these parameter values from Table 1. □

Corollary 4. *λ-Generalized Hurwitz zeta functions have the following representation:*

$$\begin{aligned}\Gamma(s)\zeta^{(\rho_1,\ldots,\rho_p,\sigma_1,\ldots,\sigma_q)}_{\lambda_1+\rho_1,\ldots,\lambda_p+\rho_p,\mu_1+\sigma_1,\ldots,\mu_q+\sigma_q}(s,a+1;b,\lambda) \\ = \frac{(\mu_1)_{\sigma_1}\cdots(\mu_q)_{\sigma_q}}{(\lambda_1)_{\rho_1}\cdots(\lambda_p)_{\rho_p}}\Big[b\lambda\Gamma(s-\lambda-1)\zeta^{(\rho_1,\ldots,\rho_p,\sigma_1,\ldots,\sigma_q)}_{\lambda_1,\ldots,\lambda_p,\mu_1,\ldots,\mu_q}(s-\lambda-1,a;b,\lambda) \\ +\Gamma(s)\Big[\zeta^{(\rho_1,\ldots,\rho_p,\sigma_1,\ldots,\sigma_q)}_{\lambda_1,\ldots,\lambda_p,\mu_1,\ldots,\mu_q}(s-1,a;b,\lambda)-a\zeta^{(\rho_1,\ldots,\rho_p,\sigma_1,\ldots,\sigma_q)}_{\lambda_1,\ldots,\lambda_p,\mu_1,\ldots,\mu_q}(s,a;b,\lambda)\Big]\Big].\end{aligned} \tag{21}$$

Proof. The result follows directly from Equation (7) upon replacing $z \longrightarrow 1$ and in view of the defined item from Table 1 dependable on these parameter values. □

Corollary 5. *λ-Generalized Riemann zeta functions have the following representation:*

$$\begin{aligned}\zeta^{(\rho_1,\ldots,\rho_p,\sigma_1,\ldots,\sigma_q)}_{\lambda_1+\rho_1,\ldots,\lambda_p+\rho_p,\mu_1+\sigma_1,\ldots,\mu_q+\sigma_q}(s,2;b,\lambda) \\ = \frac{(\mu_1)_{\sigma_1}\cdots(\mu_q)_{\sigma_q}}{(\lambda_1)_{\rho_1}\cdots(\lambda_p)_{\rho_p}}\left[\begin{array}{c} b\lambda\Gamma(s-\lambda-1)\zeta^{(\rho_1,\ldots,\rho_p,\sigma_1,\ldots,\sigma_q)}_{\lambda_1,\ldots,\lambda_p,\mu_1,\ldots,\mu_q}(s-\lambda-1;b,\lambda) \\ +\Gamma(s)\Big[\zeta^{(\rho_1,\ldots,\rho_p,\sigma_1,\ldots,\sigma_q)}_{\lambda_1,\ldots,\lambda_p,\mu_1,\ldots,\mu_q}(s-1;b,\lambda)-\zeta^{(\rho_1,\ldots,\rho_p,\sigma_1,\ldots,\sigma_q)}_{\lambda_1,\ldots,\lambda_p,\mu_1,\ldots,\mu_q}(s;b,\lambda)\Big]\end{array}\right].\end{aligned} \tag{22}$$

Proof. The result follows directly from Equation (7) upon replacing $z \longrightarrow 1; a \longrightarrow 1$ and with reference to the definite element from Table 1 stable with these parameter values. □

Remark 2. *We can get similar representations for other special cases of these functions by considering different parameter variations in view of Table 1 column-wise.*

Note that by taking $b = 0$ in the above results, we can get the following formulae for unified extended Hurwitz–Lerch zeta functions $\Phi^{(\rho_1,\ldots,\rho_p,\sigma_1,\ldots,\sigma_q)}_{\lambda_1,\ldots,\lambda_p,\mu_1,\ldots,\mu_q}(z,s,a;0,\lambda)$:

$$\begin{aligned}\Phi^{(\rho_1,\ldots,\rho_p,\sigma_1,\ldots,\sigma_q)}_{\lambda_1+\rho_1,\ldots,\lambda_p+\rho_p,\mu_1+\sigma_1,\ldots,\mu_q+\sigma_q}(z,s,a+1) \\ = \frac{(\mu_1)_{\sigma_1}\cdots(\mu_q)_{\sigma_q}}{z.(\lambda_1)_{\rho_1}\cdots(\lambda_p)_{\rho_p}}\Big[\Phi^{(\rho_1,\ldots,\rho_p,\sigma_1,\ldots,\sigma_q)}_{\lambda_1,\ldots,\lambda_p,\mu_1,\ldots,\mu_q}(z,s-1,a)-a\Phi^{(\rho_1,\ldots,\rho_p,\sigma_1,\ldots,\sigma_q)}_{\lambda_1,\ldots,\lambda_p,\mu_1,\ldots,\mu_q}(z,s,a)\Big].\end{aligned} \tag{23}$$

Next, by selecting $p-1 = q = 0; \lambda_1 = \mu \neq 0;\ 1 = \rho_1,$ in the above results, we can get the following result for unified Hurwitz–Lerch zeta functions $\Phi^*_\mu(z,s,a)$:

$$\int_0^\infty \frac{t^{s-1}e^{-(a+1)t}}{(1-ze^{-t})^{\mu+1}}dt = \Gamma(s)\Phi^*_{\mu+1}(z,s,a+1) = \frac{\Gamma(s)}{\mu z}\left[\Phi^*_\mu(z,s-1,a)-a\Phi^*_\mu(z,s,a)\right]. \tag{24}$$

Next, we note that by taking $\mu = 1$, we get for Hurwitz–Lerch zeta functions:

$$\int_0^\infty \frac{t^{s-1}e^{-(a+1)t}}{(1-ze^{-t})^2}dt = \Gamma(s)\Phi^*_2(z,s,a+1) = \frac{\Gamma(s)[\Phi(z,s-1,a)-a\Phi(z,s,a)]}{z}. \tag{25}$$

If we consider the same parameter values as above but with $b \neq 0$, then we can find the following new results for the extended Riemann and Hurwitz zeta functions:

$$\begin{aligned}\mu z\Gamma(s)\Theta^{\lambda}_{\mu+1}(z,s,a+1,b) \\ &= b\lambda\Gamma(s-\lambda-1)\Theta^{\lambda}_{\mu}(z,s-\lambda-1,a,b) \\ &+ \Gamma(s)\left[\Theta^{\lambda}_{\mu}(z,s-1,a,b) - a\Theta^{\lambda}_{\mu}(z,s,a,b)\right]\end{aligned} \tag{26}$$

$$\begin{aligned}\int_0^\infty \frac{t^{s-1}e^{-(a+1)t-\frac{b}{t}}}{(1-ze^{-t})^2}dt &= \Gamma(s)\Phi^*_2(z,s,a+1;b,1) \\ &= \frac{b\Gamma(s-2)}{z}\Phi_b(z,s-2,a) + \frac{\Gamma(s)}{z}[\Phi_b(z,s-1,a) - a\Phi_b(z,s,a)]\end{aligned} \tag{27}$$

$$\int_0^\infty \frac{t^{s-1}e^{-2t-\frac{b}{t}}}{(1-e^{-t})^2}dt = \Gamma(s)\zeta^*_2(s,2;b,1) = b\Gamma(s-2)\zeta_b(s-2) + \Gamma(s)[\zeta_b(s-1) - \zeta_b(s)]. \tag{28}$$

3. Some applications of the difference equation

In this section, we consider some interesting special cases of difference equations. On one side, these are useful to know the values of generalized Hurwitz zeta functions $\Phi^*_{\mu}(z,s,a)$ in terms of zeta functions, and on the other, they lead to the computation of some elementary integrals that are nontrivial to obtain for small values of $\mu = 2,3,4,5$ and the large values of s.

Taking $\mu = 2$ and $a \longrightarrow a+1$ in Equation (24), we get:

$$\int_0^\infty \frac{t^{s-1}e^{-(a+2)t}}{(1-ze^{-t})^3}dt = \Gamma(s)\Phi^*_3(z,s,a+2) = \frac{\Gamma(s)}{1.2.z}[\Phi^*_2(z,s-1,a+1) - (a+1)\Phi^*_2(z,s,a+1)]. \tag{29}$$

Next, making use of Equation (25) on the right-hand side of the above Equation (29) leads to the following form of $\Phi^*_3(z,s,a+2)$ in terms of the Hurwitz–Lerch zeta function:

$$\begin{aligned}\int_0^\infty \frac{t^{s-1}e^{-(a+2)t}}{(1-ze^{-t})^3}dt &= \frac{\Gamma(s)}{1.2.z^2}[\Phi(z,s-2,a) - a\Phi(z,s-1,a) - (a+1)\Phi(z,s-1,a) + a(a+1)\Phi(z,s,a)] \\ &= \frac{\Gamma(s)}{1.2.z^2}[\Phi(z,s-2,a) - (2a+1)\Phi(z,s-1,a) + a(a+1)\Phi(z,s,a)].\end{aligned} \tag{30}$$

Now we consider some interesting special cases of the above Equation (30).

For $z = 1$, it leads to the following representation in terms of the Hurwitz zeta function:

$$\int_0^\infty \frac{t^{s-1}e^{-(a+2)t}}{(1-e^{-t})^3}dt = \frac{\Gamma(s)}{1.2}[\zeta(s-2,a) - (2a+1)\zeta(s-1,a) + a(a+1)\zeta(s,a)]; \tag{31}$$

$$\int_0^\infty \frac{t^{s-1}e^{-(a+2)t}}{(1-e^{-t})^3}dt = \frac{\Gamma(s)}{1.2}[\zeta(s-2,a) - (2a+1)\zeta(s-1,a) + a(a+1)\zeta(s,a)]. \tag{32}$$

For $a = 1$, we get the following representation in terms of the polylogarithm function:

$$\int_0^\infty \frac{t^{s-1}e^{-3t}}{(1-ze^{-t})^3}dt = \frac{\Gamma(s)}{1.2.z^3}[Li_{s-2}(z) - 3Li_{s-1}(z) + 2Li_s(z)]. \tag{33}$$

For $a = 1,\ z = 1$, it leads to the following relation in terms of the zeta function:

$$\int_0^\infty \frac{t^{s-1}e^{-3t}}{(1-e^{-t})^3}dt = \frac{\Gamma(s)}{1.2}[2\zeta(s) + \zeta(s-2) - 3\zeta(s-1)];\ (s \neq 1,2,3). \tag{34}$$

For $s = 4$ in Equation (34), we get the following integral:

$$\int_0^\infty \frac{t^3e^{-3t}}{(1-e^{-t})^3}dt = \frac{\Gamma(4)}{1.2}[2\zeta(4) + \zeta(2) - 3\zeta(3)]. \tag{35}$$

Similarly, by considering different values of s, we can produce the following Tables 2 and 3 of values. These computations show that Mathematica is unable to compute the involved integral on a commonly available computer for the large values of s, but it can be done using these new difference equations. For small values of s, our results are 100% accurate with the direct computed results

Table 2. Computation of $\int_0^\infty \frac{t^{s-1}e^{-3t}}{(1-e^{-t})^3}dt$.

s	Direct Evaluation by Mathematica	Using difference Equation (34)
4	0.610229	0.610229
30	4.29669×10^{16}	4.29669×10^{16}
40	1.67783×10^{27}	1.67783×10^{27}
45	8.998×10^{32}	8.998×10^{32}
46	1.3497×10^{34}	1.3497×10^{34}
48	3.24227×10^{36}	3.24227×10^{36}
48.5	1.29353×10^{37}	1.29353×10^{37}
48.9	3.92781×10^{37}	3.92781×10^{37}
49	5.18762×10^{37}	5.18762×10^{37}
52	2.40071×10^{41}	2.40071×10^{41}
56	2.426×10^{46}	2.426×10^{46}
160	Unable to compute	1.349×10^{206}
220	Unable to compute	1.121×10^{314}
400	Unable to compute	2.269×10^{675}

Putting $\mu = 3,\ a \longrightarrow a+2$ in Equation (24), we get:

$$\begin{aligned}&\int_0^\infty \frac{t^{s-1}e^{-(a+3)t}}{(1-ze^{-t})^{3+1}}dt = \Gamma(s)\Phi^*_{3+1}(z,s,a+3)\\ &= \frac{\Gamma(s)}{3.z}\left[\Phi^*_3(z,s-1,a+2) - (a+2)\Phi^*_3(z,s,a+2)\right].\end{aligned} \tag{36}$$

Next, combining the above two results (Equations (30) and (36)), we get the following representation in terms of the Hurwitz–Lerch zeta function

$$\begin{aligned}&\int_0^\infty \frac{t^{s-1}e^{-(a+3)t}}{(1-ze^{-t})^4}dt\\ &= \frac{\Gamma(s)}{1.2.3.z^3}\left[\begin{array}{c}\Phi(z,s-3,a) - 3(a+1)\Phi(z,s-2,a) + (3a^2+6a+2)\Phi(z,s-1,a)\\ -a(a+1)(a+2)\Phi(z,s,a)\end{array}\right].\end{aligned} \tag{37}$$

Some interesting special cases: For $z = 1$, it leads to the following representation in terms of the Hurwitz zeta function:

$$\begin{aligned}&\int_0^\infty \frac{t^{s-1}e^{-(a+3)t}}{(1-e^{-t})^4}dt = \Gamma(s)\zeta^*_4(s,a+3)\\ &= \frac{\Gamma(s)}{1.2.3}\left[\begin{array}{c}\zeta(s-3,a) - 3(a+1)\zeta(s-2,a) + (3a^2+6a+2)\zeta(s-1,a)\\ -a(a+1)(a+2)\zeta(s,a)\end{array}\right].\end{aligned} \tag{38}$$

For $a = 1$, we get the following representation in terms of the polylogarithm function:

$$\begin{aligned}&\int_0^\infty \frac{t^{s-1}e^{-4t}}{(1-ze^{-t})^4}dt = \Gamma(s)\Phi^*_4(z,s,4)\\ &= \frac{\Gamma(s)}{1.2.3.z^4}\left[Li_{s-3}(z) - 6Li_{s-2}(z) + 11Li_{s-1}(z) - 6Li_s(z)\right].\end{aligned} \tag{39}$$

For $a = 1,\ z = 1$, it leads to the following relation in terms of the zeta function:

$$\int_0^\infty \frac{t^{s-1}e^{-4t}}{(1-e^{-t})^4}dt = \Gamma(s)\zeta_4^*(s,4)$$
$$= \frac{\Gamma(s)}{1.2.3}\left[\ \zeta(s-3) - 6\zeta(s-2) + 11\zeta(s-1) - 6\zeta(s)\ \right];\ (s \neq 1,2,3,4). \tag{40}$$

For s = 5, we have:

$$\int_0^\infty \frac{t^4 e^{-4t}}{(1-e^{-t})^4}dt = \Gamma(5)\zeta_4^*(4,4)$$
$$= \frac{\Gamma(5)}{1.2.3}\left[\ \zeta(2) - 6\zeta(3) + 11\zeta(4) - 6\zeta(5)\ \right]. \tag{41}$$

Table 3. Computation of $\int_0^\infty \frac{t^{s-1}e^{-4t}}{(1-e^{-t})^4}dt$.

s	Direct Evaluation by Mathematica	By using difference Equation (40)
90	1.07719×10^{82}	1.07719×10^{82}
100	5.8077×10^{95}	5.8077×10^{95}
140	Unable to compute	1.78191×10^{156}
160	Unable to compute	1.37955×10^{186}

Similarly, putting $\mu = 4$, $a \longrightarrow a+3$ in Equation (24), we get:

$$\int_0^\infty \frac{t^{s-1}e^{-(a+4)t}}{(1-ze^{-t})^{4+1}}dt = \Gamma(s)\Phi_{4+1}^*(z,s,a+4)$$
$$= \frac{\Gamma(s)}{4.z}\left[\Phi_4^*(z,s-1,a+3) - (a+3)\Phi_4^*(z,s,a+3)\right]. \tag{42}$$

Next, combining the above two results (Equations (38) and (42)), we get:

$$\int_0^\infty \frac{t^{s-1}e^{-(a+4)t}}{(1-ze^{-t})^5}dt$$
$$= \frac{\Gamma(s)}{1.2.3.4.z^4}\left[\begin{array}{c} \Phi(z,s-3,a) + \left(6a^2+18a+11\right)\Phi(z,s-2,a) \\ -\left(4a^3+18a^2+22a+6\right)\Phi(z,s-2,a) + a(a+1)(a+2)(a+3)\Phi(z,s,a) \end{array}\right]. \tag{43}$$

Some interesting special cases:
For z = 1:

$$\int_0^\infty \frac{t^{s-1}e^{-(a+4)t}}{(1-e^{-t})^5}dt = \Gamma(s)\zeta_5^*(s,a+4)$$
$$= \frac{\Gamma(s)}{1.2.3.4}\left[\begin{array}{c} \zeta(s-4,a) - 2(2a+3)\zeta(s-3,a) + \left(6a^2+18a+11\right)\zeta(s-2,a) \\ -\left(4a^3+18a^2+22a+6\right)\zeta(s-2,a) + a(a+1)(a+2)(a+3)\zeta(s,a) \end{array}\right]. \tag{44}$$

For a = 1:

$$\int_0^\infty \frac{t^{s-1}e^{-5t}}{(1-ze^{-t})^5}dt = \Gamma(s)\Phi_5^*(z,s,5)$$
$$= \frac{\Gamma(s)}{1.2.3.4.z^5}\left[\begin{array}{c} Li_{s-4}(z) - 10Li_{s-3}(z) + 37Li_{s-2}(z) \\ -50Li_{s-2}(z) + 24Li_s(z) \end{array}\right]. \tag{45}$$

For $a = 1,\ z = 1$:

$$\int_0^\infty \frac{t^{s-1}e^{-5t}}{(1-e^{-t})^5}dt = \Gamma(s)\zeta_5^*(s,5)$$
$$= \frac{\Gamma(s)}{1.2.3.4}\left[\begin{array}{c} \zeta(s-4) - 10\zeta(s-3) + 37\zeta(s-2) \\ -50\zeta(s-2) + 24\zeta(s) \end{array}\right];(s \neq 1,2,3,4,5). \tag{46}$$

Now put $s = 6$:

$$\int_0^\infty \frac{t^5 e^{-5t}}{(1-e^{-t})^5}dt = \Gamma(6)\zeta_5^*(6,5)$$
$$= \frac{\Gamma(6)}{1.2.3.4}\left[\begin{array}{c} \zeta(2) - 10\zeta(3) + 37\zeta(4) \\ -50\zeta(5) + 24\zeta(6) \end{array}\right]. \tag{47}$$

Continuing in this way, by putting $\mu = 5$, $a \longrightarrow a+4$ in Equation (24), we get:

$$\begin{aligned} &\int_0^\infty \frac{t^{s-1}e^{-(a+5)t}}{(1-ze^{-t})^{5+1}}dt = \Gamma(s)\Phi^*_{5+1}(z,s,a+5) \\ &= \frac{\Gamma(s)}{5.z}\left[\Phi^*_5(z,s-1,a+4) - (a+4)\Phi^*_5(z,s,a+4)\right]. \end{aligned} \tag{48}$$

Next, combining the above two results of Equations (43) and (48), we can get $\Phi^*_6(z,s,a+4)$ in terms of Hurwitz–Lerch zeta functions. Similarly, for nonzero values of z, for example, $z = 0.3; a = 1$; $\mu = 3$ in Equation (33) we have:

$$\int_0^\infty \frac{\left(e^{-3\,t}t^4\right)}{\left(1 - \frac{3\,e^{-t}}{10}\right)^3}dt = 0.125061 \tag{49}$$

$$\frac{\Gamma(5)\left(Li_3(0.3) - 3\,Li_4(0.3) + 2\,Li_5(0.3)\right)}{2\,(0.3)^3} = 0.125061 \tag{50}$$

and so on and so forth.

4. Integrals of products of the family of λ-Generalized Hurwitz–Lerch zeta functions

By means of the basic Parseval's identity of Mellin transform [50] (Chapter 10) and difference equations obtained in Section 2, we can get the following integral formulae in view of Equation (2) and column 3 of Table 1. For example, for the generalized Hurwitz–Lerch zeta functions $\Theta^\lambda_\mu(z,s,a,b)$:

$$\frac{1}{2\pi i}\int_{c-i\infty}^{c+i\infty}\Gamma(s)\Gamma(w-s)\Theta^\lambda_\mu(z,s,a,b)\Theta^\lambda_\delta(z,w-s,a,b) = \int_0^\infty \frac{t^{w-1}e^{-2at-\frac{2b}{t^\lambda}}}{(1-ze^{-t})^{\mu+\delta}}dt \tag{51}$$

that leads to the following by simply replacing $\mu \longrightarrow \mu+\delta-1$ in Equation (26):

$$\Gamma(\omega)\Theta^\lambda_{\mu+\delta}(z,\omega,2a,2b) = \frac{1}{z(\mu-1+\delta)}\left[\begin{array}{c} 2b\lambda\Gamma(\omega-\lambda-1)\Theta^\lambda_{\mu-1+\delta}(z,\omega-\lambda-1,2a-1,2b) \\ +\Gamma(\omega)\Theta^\lambda_{\mu-1+\delta}(z,\omega-1,2a-1,2b) \\ -(2a-1)\Gamma(\omega)\Theta^\lambda_{\mu-1+\delta}(z,w,2a-1,2b) \end{array}\right]. \tag{52}$$

Therefore, we get the following new integral formulae for other related cases given in column 3 of Table 1 and Equations (51) and (52):

$$\begin{aligned} &\frac{1}{2\pi i}\int_{c-i\infty}^{c+i\infty}\Gamma(s)\Gamma(w-s)\Theta^\lambda_\mu(z,s,a,b)\Theta^\lambda_\delta(z,w-s,a,b) = \int_0^\infty \frac{t^{w-1}e^{-2at-\frac{2b}{t^\lambda}}}{(1+e^{-x}e^{-t})^{\mu+\delta}}dt \\ &= \frac{e^x}{\mu-1+\delta}\left[\begin{array}{c} \Gamma(\omega)\left[(2a-1)\Theta^\lambda_{\mu-1+\delta}(x,w,2a-1,2b) - \Theta^\lambda_{\mu-1+\delta}(x,\omega-1,2a-1,2b)\right] \\ -2b\lambda\Gamma(\omega-\lambda-1)\Theta^\lambda_{\mu-1+\delta}(x,\omega-\lambda-1,2a-1,\,2b) \end{array}\right] \end{aligned} \tag{53}$$

$$\begin{aligned} &\frac{1}{2\pi i}\int_{c-i\infty}^{c+i\infty}\Gamma(s)\Gamma(w-s)F^\lambda_\mu(x,s,b)F^\lambda_\delta(x,w-s,b) = \int_0^\infty \frac{t^{w-1}e^{-2t-\frac{2b}{t^\lambda}}}{(1+e^{-x}e^{-t})^{\mu+\delta}}dt \\ &= \frac{e^x}{\mu-1+\delta}\left[\begin{array}{c} \Gamma(\omega)\left[F^\lambda_{\mu-1+\delta}(x,w,2b) - F^\lambda_{\mu-1+\delta}(x,\omega-1,2b)\right] \\ -2b\lambda\Gamma(\omega-\lambda-1)F^\lambda_{\mu-1+\delta}(x,\omega-\lambda-1,\,2b) \end{array}\right] \end{aligned} \tag{54}$$

$$\begin{aligned} &\frac{1}{2\pi i}\int_{c-i\infty}^{c+i\infty}\Gamma(s)\Gamma(w-s)\Psi^\lambda_\mu(z,s,a,b)\Psi^\lambda_\delta(z,w-s,a,b) = \int_0^\infty \frac{t^{w-1}e^{-2at-\frac{2b}{t^\lambda}}}{(1-e^{-x}e^{-t})^{\mu+\delta}}dt \\ &= \frac{e^x}{\mu-1+\delta}\left[\begin{array}{c} 2b\lambda\Gamma(\omega-\lambda-1)\Psi^\lambda_{\mu-1+\delta}(x,\omega-\lambda-1,2a-1,\,2b) \\ +\Gamma(\omega)\left[\Psi^\lambda_{\mu-1+\delta}(x,\omega-1,2a-1,2b)\right] - (2a-1)\Psi^\lambda_{\mu-1+\delta}(x,w,2a-1,2b) \end{array}\right] \end{aligned} \tag{55}$$

$$\frac{1}{2\pi i}\int_{c-i\infty}^{c+i\infty}\Gamma(s)\Gamma(w-s)B_{\mu}^{\lambda}(x,s,b)B_{\delta}^{\lambda}(x,w-s,b)=\int_0^{\infty}\frac{t^{w-1}e^{-2t-\frac{2b}{t^{\lambda}}}}{(1-e^{-x}e^{-t})^{\mu+\delta}}dt$$
$$=\frac{e^x}{\mu-1+\delta}\begin{bmatrix}2b\lambda\Gamma(\omega-\lambda-1)B_{\mu-1+\delta}^{\lambda}(x,\omega-\lambda-1,\ 2b)\\+\Gamma(\omega)\left[B_{\mu-1+\delta}^{\lambda}(x,\omega-1,2b)-B_{\mu-1+\delta}^{\lambda}(x,w,2b)\right]\end{bmatrix}\tag{56}$$

$$\frac{1}{2\pi i}\int_{c-i\infty}^{c+i\infty}\Gamma(s)\Gamma(w-s)Li_{\mu}^{\lambda}(z,s,b)Li_{\delta}^{\lambda}(z,w-s,b)=\int_0^{\infty}\frac{t^{w-1}e^{-2t-\frac{2b}{t^{\lambda}}}}{(1-ze^{-t})^{\mu+\delta}}dt$$
$$=\frac{1}{(\mu-1+\delta)z}\begin{bmatrix}2b\lambda\Gamma(\omega-\lambda-1)Li_{\mu-1+\delta}^{\lambda}(x,\omega-\lambda-1,\ 2b)\\+\Gamma(\omega)\left[Li_{\mu-1+\delta}^{\lambda}(x,\omega-1,2b)-Li_{\mu-1+\delta}^{\lambda}(x,w,2b)\right]\end{bmatrix}\tag{57}$$

$$\frac{1}{2\pi i}\int_{c-i\infty}^{c+i\infty}\Gamma(s)\Gamma(w-s)\zeta_{\mu}^{\lambda}(s,a;b)\zeta_{\delta}^{\lambda}(w-s,a;b)=\int_0^{\infty}\frac{t^{w-1}e^{-2at-\frac{2b}{t^{\lambda}}}}{(1-e^{-t})^{\mu+\delta}}dt$$
$$=\frac{1}{(\mu-1+\delta)}\begin{bmatrix}2b\lambda\Gamma(\omega-\lambda-1)\zeta_{\mu-1+\delta}^{\lambda}(\omega-\lambda-1,2a-1,\ 2b)\\+\Gamma(\omega)\left[\zeta_{\mu-1+\delta}^{\lambda}(w-1,2a-1,2b)-(2a-1)\zeta_{\mu-1+\delta}^{\lambda}(w,2a-1,2b)\right]\end{bmatrix}\tag{58}$$

$$\frac{1}{2\pi i}\int_{c-i\infty}^{c+i\infty}\Gamma(s)\Gamma(w-s)\zeta_{\mu}^{\lambda}(s;b)\zeta_{\delta}^{\lambda}(w-s;b)=\int_0^{\infty}\frac{t^{w-1}e^{-2t-\frac{2b}{t^{\lambda}}}}{(1-e^{-t})^{\mu+\delta}}dt$$
$$=\frac{1}{(\mu-1+\delta)}\begin{bmatrix}2b\lambda\Gamma(\omega-\lambda-1)\zeta_{\mu-1+\delta}^{\lambda}(\omega-\lambda-1,2b)\\+\Gamma(\omega)\left[\zeta_{\mu-1+\delta}^{\lambda}(w-1,2b)-\zeta_{\mu-1+\delta}^{\lambda}(w,2b)\right]\end{bmatrix}.\tag{59}$$

Next, for $b=0$, we can get the following new formulae in view of column 8 of Table 1 and Equation (53):

$$\frac{1}{2\pi i}\int_{c-i\infty}^{c+i\infty}\Gamma(s)\Gamma(w-s)\Phi_{\mu}^{*}(z,s,a)\ \Phi_{\delta}^{*}(z,w-s,a)=\int_0^{\infty}\frac{t^{w-1}e^{-2at}}{(1-ze^{-t})^{\mu+\delta}}dt$$
$$=\frac{\Gamma(\omega)}{(\mu-1+\delta)z}\left[\Phi_{\mu-1+\delta}^{*}\ (z,\omega-1,2a-1)-(2a-1)\Phi_{\mu-1+\delta}^{*}(z,w,2a-1)\right]\tag{60}$$

$$\frac{1}{2\pi i}\int_{c-i\infty}^{c+i\infty}\Gamma(s)\Gamma(w-s)\Theta_{\mu}^{*}(x,s,a)\Theta_{\delta}^{*}(x,w-s,a)=\int_0^{\infty}\frac{t^{w-1}e^{-2at}}{(1+e^{-x}e^{-t})^{\mu+\delta}}dt$$
$$=\frac{e^x\Gamma(\omega)}{\mu-1+\delta}\left[(2a-1)\Theta_{\mu-1+\delta}^{*}(x,w,2a-1)-\Theta_{\mu-1+\delta}^{*}(x,\omega-1,2a-1)\right]\tag{61}$$

$$\frac{1}{2\pi i}\int_{c-i\infty}^{c+i\infty}\Gamma(s)\Gamma(w-s)F_{\mu}^{*}(x,s)F_{\delta}^{*}(x,w-s)=\int_0^{\infty}\frac{t^{w-1}e^{-2t}}{(1+e^{-x}e^{-t})^{\mu+\delta}}dt$$
$$=\frac{e^x\Gamma(\omega)}{\mu-1+\delta}\left[F_{\mu-1+\delta}^{*}(x,w)-F_{\mu-1+\delta}^{*}(x,\omega-1)\right]\tag{62}$$

$$\frac{1}{2\pi i}\int_{c-i\infty}^{c+i\infty}\Gamma(s)\Gamma(w-s)\Psi_{\mu}^{*}(x,s,a)\Psi_{\delta}^{*}(x,w-s,a)=\int_0^{\infty}\frac{t^{w-1}e^{-2at}}{(1-e^{-x}e^{-t})^{\mu+\delta}}dt$$
$$=\frac{e^x\Gamma(\omega)}{\mu-1+\delta}\left[\Psi_{\mu-1+\delta}^{*}(x,\omega-1,2a-1)-(2a-1)\Psi_{\mu-1+\delta}^{*}(x,w,2a-1)\right]\tag{63}$$

$$\frac{1}{2\pi i}\int_{c-i\infty}^{c+i\infty}\Gamma(s)\Gamma(w-s)B_{\mu}^{*}(x,s)B_{\delta}^{*}(x,w-s)=\int_0^{\infty}\frac{t^{w-1}e^{-2t}}{(1-e^{-x}e^{-t})^{\mu+\delta}}dt$$
$$=\frac{e^x\Gamma(\omega)}{\mu-1+\delta}\left[B_{\mu-1+\delta}^{*}(x,\omega-1)-B_{\mu-1+\delta}^{*}(x,w)\right]\tag{64}$$

$$\frac{1}{2\pi i}\int_{c-i\infty}^{c+i\infty}\Gamma(s)\Gamma(w-s)Li_{\mu}^{*}(z,s)\ Li_{\delta}^{*}(z,w-s)=\int_0^{\infty}\frac{t^{w-1}e^{-2t}}{(1-ze^{-t})^{\mu+\delta}}dt$$
$$=\frac{\Gamma(\omega)}{(\mu-1+\delta)z}\left[Li_{\mu-1+\delta}^{*}\ (z,\omega-1)-Li_{\mu-1+\delta}^{*}(z,w)\right]\tag{65}$$

$$\frac{1}{2\pi i}\int_{c-i\infty}^{c+i\infty}\Gamma(s)\Gamma(w-s)\zeta_{\mu}^{*}(s,a)\ \zeta_{\delta}^{*}(w-s,a)ds=\int_0^{\infty}\frac{t^{w-1}e^{-2at}}{(1-e^{-t})^{\mu+\delta}}dt$$
$$=\frac{\Gamma(\omega)}{(\mu-1+\delta)}\left[\zeta_{\mu-1+\delta}^{*}\ (\omega-1,a)-(2a-1)\zeta_{\mu-1+\delta}^{*}(w,a)\right]\tag{66}$$

$$\frac{1}{2\pi i}\int_{c-i\infty}^{c+i\infty}\Gamma(s)\Gamma(w-s)\zeta_{\mu}^{*}(s)\ \zeta_{\delta}^{*}(w-s)ds=\int_0^{\infty}\frac{t^{w-1}e^{-2t}}{(1-e^{-t})^{\mu+\delta}}dt$$
$$=\frac{\Gamma(\omega)}{(\mu-1+\delta)}\left[\zeta_{\mu-1+\delta}^{*}\ (\omega-1)-\zeta_{\mu-1+\delta}^{*}(w)\right].\tag{67}$$

Next, if we consider $\delta = \mu = \lambda = 1; b \neq 0$ in Equation (53), we can obtain the following new integral formulae in view of column 6 of Table 1:

$$\begin{aligned} &\frac{1}{2\pi i}\int_{c-i\infty}^{c+i\infty}\Gamma(s)\Gamma(w-s)\Phi_b(z,s,a)\ \Phi_b(z,w-s,a)ds = \int_0^\infty \frac{t^{w-1}e^{-2at-\frac{2b}{t}}}{(1-ze^{-t})^2}dt \\ &= \frac{2b\Gamma(\omega-2)\Phi_{2b}(z,\omega-2,2a-1)+\Gamma(\omega)[\Phi_{2b}(z,\omega-1,2a-1)-(2a-1)\Phi_{2b}(z,w,2a-1)]}{z} \end{aligned} \tag{68}$$

$$\begin{aligned} &\frac{1}{2\pi i}\int_{c-i\infty}^{c+i\infty}\Gamma(s)\Gamma(w-s)\mathrm{Li}_b(z,s)\ Li_b(z,w-s)ds = \int_0^\infty \frac{t^{w-1}e^{-2t-\frac{2b}{t}}}{(1-ze^{-t})^2}dt \\ &= \frac{2b\Gamma(\omega-2)\mathrm{Li}_{2b}(z,\omega-2)+\Gamma(\omega)[\mathrm{Li}_{2b}(z,\omega-1)-\mathrm{Li}_{2b}(z,w)]}{z} \end{aligned} \tag{69}$$

$$\begin{aligned} &\frac{1}{2\pi i}\int_{c-i\infty}^{c+i\infty}\Gamma(s)\Gamma(w-s)\zeta_b(s,a)\ \zeta_b(w-s,a)ds = \int_0^\infty \frac{t^{w-1}e^{-2at-\frac{2b}{t}}}{(1-e^{-t})^2}dt \\ &= 2b\Gamma(\omega-2)\zeta_{2b}(\omega-2,2a-1)+\Gamma(\omega)[\zeta_{2b}(\omega-1,2a-1)-(2a-1)\zeta_{2b}(w,2a-1)] \end{aligned} \tag{70}$$

$$\begin{aligned} &\frac{1}{2\pi i}\int_{c-i\infty}^{c+i\infty}\Gamma(s)\Gamma(w-s)\zeta_b(s)\ \zeta_b(w-s)ds = \int_0^\infty \frac{t^{w-1}e^{-2t-\frac{2b}{t}}}{(1-e^{-t})^2}dt \\ &= 2b\Gamma(\omega-2)\zeta_{2b}(\omega-2)+\Gamma(\omega)[\zeta_{2b}(\omega-1)-\zeta_{2b}(w)] \end{aligned}. \tag{71}$$

Similarly, by considering the different parameter values consistent with the results obtained in Section 2, one can obtain more integral formulae for the family of zeta and associated functions. One model is the following by means of Theorem 1:

$$\begin{aligned} &\frac{1}{2\pi i}\int_{c-i\infty}^{c+i\infty}\Gamma(s)\Gamma(w-s)\Phi_{\lambda_1,\ldots,\lambda_p,\mu_1,\ldots,\mu_q}^{(\rho_1,\ldots,\rho_p,\sigma_1,\ldots,\sigma_q)}(z,s,a,b;\lambda)\Phi_{\lambda_1,\ldots,\lambda_p,\mu_1,\ldots,\mu_q}^{(\rho_1,\ldots,\rho_p,\sigma_1,\ldots,\sigma_q)}(z,w-s,a,b;\lambda)ds \\ &= \frac{(\mu_1)_{\sigma_1}\cdots\cdots(\mu_q)_{\sigma_q}}{z(\lambda_1)_{\rho_1}\cdots\cdots(\lambda_p)_{\rho_p}}\left[\begin{array}{c} 2b\lambda\Phi_{\lambda_1,\ldots,\lambda_p,\mu_1,\ldots,\mu_q}^{(\rho_1,\ldots,\rho_p,\sigma_1,\ldots,\sigma_q)}\Gamma(w-\lambda-1)(z,w-\lambda-1,2a-1,2b) \\ +\Gamma(\omega)\left[\begin{array}{c}\Phi_{\lambda_1,\ldots,\lambda_p,\mu_1,\ldots,\mu_q}^{(\rho_1,\ldots,\rho_p,\sigma_1,\ldots,\sigma_q)}(z,w-1,2a-1;2b,\lambda)- \\ (2a-1)\Phi_{\lambda_1,\ldots,\lambda_p,\mu_1,\ldots,\mu_q}^{(\rho_1,\ldots,\rho_p,\sigma_1,\ldots,\sigma_q)}(z,w,2a-1;2b,\lambda)\end{array}\right]\end{array}\right] \end{aligned} \tag{72}$$

and for $b = 0$, it leads to:

$$\begin{aligned} &\frac{1}{2\pi i}\int_{c-i\infty}^{c+i\infty}\Gamma(s)\Gamma(w-s)\Phi_{\lambda_1,\ldots,\lambda_p,\mu_1,\ldots,\mu_q}^{(\rho_1,\ldots,\rho_p,\sigma_1,\ldots,\sigma_q)}(z,s,a)\Phi_{\lambda_1,\ldots,\lambda_p,\mu_1,\ldots,\mu_q}^{(\rho_1,\ldots,\rho_p,\sigma_1,\ldots,\sigma_q)}(z,w-s,a)ds \\ &= \frac{\Gamma(\omega)(\mu_1)_{\sigma_1}\cdots\cdots(\mu_q)_{\sigma_q}}{z(\lambda_1)_{\rho_1}\cdots\cdots(\lambda_p)_{\rho_p}}\left[\begin{array}{c}\Phi_{\lambda_1,\ldots,\lambda_p,\mu_1,\ldots,\mu_q}^{(\rho_1,\ldots,\rho_p,\sigma_1,\ldots,\sigma_q)}(z,w-1,2a-1;2b,\lambda)- \\ (2a-1)\Phi_{\lambda_1,\ldots,\lambda_p,\mu_1,\ldots,\mu_q}^{(\rho_1,\ldots,\rho_p,\sigma_1,\ldots,\sigma_q)}(z,w,2a-1;2b,\lambda)\end{array}\right]. \end{aligned} \tag{73}$$

5. Discussion and Future Directions

In this study, we obtained some recurrence relations for the newly defined family of the λ-generalized Hurwitz–Lerch zeta functions using the familiar Mellin transforms. These relations proved valuable to acquire new integral formulae involving the family of zeta functions. The outcomes were also confirmed with the previous obtained results as special cases. It is remarkable that the recurrence relations obtained in this research work are worthwhile to achieve simple relations such as Equations (34) and (40) that express special cases of λ-generalized Hurwitz–Lerch zeta functions in terms of Riemann zeta functions, so that we can evaluate the values of these functions. By following the method, we can obtain significant new results by considering the further specific values of the involved parameters. This is useful for the further analysis of these functions by plotting the graphs and deriving different series and asymptotic representations, etc. This work is in progress and would be a part of some future research.

λ-generalized Hurwitz–Lerch zeta functions analytically generalize the functions of the zeta family and offer consideration for some further presumable new members of this family that are not discussed in the literature. This aspect is most suitable for attaining new consequences from one key result. Our foremost results produce simultaneously important new results for a class of well-studied functions by applying the new difference equations. The Bose–Einstein and Fermi–Dirac functions are of fundamental importance in quantum statistics that contracts by means of two specific categories of spin symmetry, that is, fermions and bosons. Fitting together these functions here with the λ-generalized Hurwitz–Lerch zeta functions yields substantial new identities for them that provides clues regarding the forthcoming applications of these difference equations in quantum physics and associated fields. This practice to acquire the outcomes by making use of new difference equations explores the required simplicity that always inspires hope. We have discussed here the direct consequences of our results. It is remarked that the method established in this research is in fact noteworthy for the analysis and study of these higher transcendental functions.

Funding: This research was funded by Deanship of Scientific Research at Majmaah University, Project Number No. 1440-56.

Acknowledgments: The author would like to thank Deanship of Scientific Research at Majmaah University for supporting this work under Project Number No. 1440-56. The author is also thankful to the anonymous reviewers for their useful comments. They really improved the quality of this manuscript.

Conflicts of Interest: The authors declare no conflict of interest.

References

1. Srivastava, H.M. A new family of the λ-generalized Hurwitz-Lerch Zeta functions with applications. *Appl. Math. Inf. Sci.* **2014**, *8*, 1485–1500. [CrossRef]
2. Srivastava, H.M.; Saxena, R.K.; Pogany, T.K.; Saxena, R. Integral and computational representations of the extended Hurwitz-Lerch Zeta function. *Integral Transforms Spec. Funct.* **2011**, *22*, 487–506. [CrossRef]
3. Srivastava, H.M.; Jankov, D.; Pogany, T.K.; Saxena, R.K. Two-sided inequalities for the extended Hurwitz-Lerch Zeta function. *Comput. Math. Appl.* **2011**, *62*, 516–522. [CrossRef]
4. Luo, M.-J.; Raina, R.K. New Results for Srivastava's λ-Generalized Hurwitz-Lerch Zeta Function. *Filomat* **2017**, *31*, 2219–2229. [CrossRef]
5. Kilbas, A.A.; Saigo, M. *H-Transforms: Theory and Applications*; Chapman and Hall (CRC Press Company): Boca Raton, FL, USA; London, UK; New York, NY, USA; Washington, DC, USA, 2004.
6. Tassaddiq, A. A new representation of the k-gamma functions. *Mathematics* **2019**, *10*, 133. [CrossRef]
7. Mathai, A.M.; Saxena, R.K. *Generalized Hypergeometric Functions with Applications in Statistics and Physical Sciences*; Lecture Notes in Mathematics; Springer-Verlag: Berlin/Heidelberg, Germany; New York, NY, USA, 1973.
8. Mathai, A.M.; Saxena, R.K.; Haubold, H.J. *The H-Function: Theory and Applications*; Springer: New York, NY, USA; Dordrecht, The Netherlands; Heidelberg, Germany; London, UK, 2010.
9. Olver, F.W.J.; Lozier, D.W.; Boisvert, R.F.; Clark, C.W. (Eds.) *NIST Handbook of Mathematical Functions*; [With 1 CD-ROM (Windows, Macintosh and UNIX)]; U.S. Department of Commerce, National Institute of Standards and Technology: Washington, DC, USA, 2010.
10. Jankov, D.; Pogany, T.K.; Saxena, R.K. An extended general Hurwitz-Lerch Zeta function as a Mathieu (a, λ)-series. *Appl. Math. Lett.* **2011**, *24*, 1473–1476.
11. Srivastava, H.M.; Ghanim, F.; El-Ashwah, R.M. Inclusion properties of certain subclass of univalent meromorphic functions defined by a linear operator associated with the -generalized Hurwitz-Lerch.zeta function. *Asian-Eur. J. Math.* **2017**, *3*, 34–50.
12. Srivastava, H.M.; Gaboury, S. New expansion formulas for a family of the λ-generalized Hurwitz-Lerch zeta functions. *Int. J. Math. Math. Sci.* **2014**, *2014*, 131067. [CrossRef] [PubMed]
13. Srivastava, H.M.; Gaboury, S.; Ghanim, F. Certain subclasses of meromorphically univalent functions defined by a linear operator associated with the λ-generalized Hurwitz-Lerch zeta function. *Integral Transforms Spec. Funct.* **2015**, *26*, 258–272. [CrossRef]
14. Srivastava, H.M.; Gaboury, S.; Ghanim, F. Some further properties of a linear operator associated with the λ-generalized Hurwitz-Lerch zeta function related to the class of meromorphically univalent functions. *Appl. Math. Comput.* **2015**, *259*, 1019–1029.
15. Srivastava, H.M. Some properties and results involving the zeta and associated functions. *Funct. Anal. Approx. Comput.* **2015**, *7*, 89–133.

16. Choi, J.; Parmar, R.K. An Extension of the Generalized Hurwitz-Lerch Zeta Function of Two Variables. *Filomat* **2017**, *31*, 91–96. [CrossRef]
17. Srivastava, H.M.; Jolly, N.; Kumar, B.; Manish Jain, R. A new integral transform associated with the λ-extended Hurwitz–Lerch zeta function. *Revista de la Real Academia de Ciencias Exactas Físicas y Naturales Series A Matemáticas* **2018**. [CrossRef]
18. Tassaddiq, A. A New Representation for Srivastava's λ-Generalized Hurwitz-Lerch Zeta Functions. *Symmetry* **2018**, *10*, 733. [CrossRef]
19. Abramowitz, M.; Stegun, I.A. (Eds.) *HandbooK of Mathematical Functions with Formulas, Graphs, and Mathematical Tables*; Applied Mathematics Series; National Bureau of Standards: Washington, DC, USA, 1964; Volume 55, Reprinted by Dover Publications, New York, NY, USA, 1965.
20. Apostol, T.M. *Introduction to Analytic Number Theory*; Springer-Verlag: Berlin/Heidelberg, Germany; New York, NY, USA; 1976.
21. Chaudhry, M.A.; Qadir, A.; Tassaddiq, A. A new generalization of the Riemann zeta function. *Adv. Differ. Equ.* **2011**, *2011*. [CrossRef]
22. Titchmarsh, E.C. *The Theory of the Riemann Zeta Function*; Oxford University Press: Oxford, UK, 1951.
23. Raina, R.K.; Srivastava, H.M. Certain results associated with the generalized Riemann zeta functions. *Revista Tecnica de la Facultad de Ingenieria Universidad del Zulia* **1995**, *18*, 301–304.
24. Srivastava, H.M. Some formulas for the Bernoulli and Euler polynomials at rational arguments. *Math. Proc. Camb. Philos. Soc.* **2000**, *129*, 77–84. [CrossRef]
25. Srivastava, H.M. Generating relations and other results associated with some families of the extended Hurwitz-Lerch Zeta functions. *SpringerPlus* **2013**, *2*, 67.
26. Srivastava, H.M.; Choi, J. *Series Associated with the Zeta and Related Functions*; Kluwer Academic Publishers: Dordrecht, The Netherlands; Boston, MA, USA; London, UK, 2001.
27. Srivastava, H.M.; Choi, J. *Zeta and q-Zeta Functions and Associated Series and Integrals*; Elsevier Science Publishers: Amsterdam, The Netherlands; London, UK; New York, NY, USA, 2012.
28. Srivastava, H.M.; Gupta, K.C.; Goyal, S.P. *The H-Functions of One and Two Variables with Applications*; South Asian Publishers: New Delhi, India; Madras, India, 1982.
29. Srivastava, H.M.; Karlsson, P.W. *Multiple Gaussian Hypergeometric Series*; Halsted Press (Ellis Horwood Limited): Chichester, UK; John Wiley and Sons: New York, NY, USA; Chichester, UK; Brisbane, Australia; Toronto, ON, Canada, 1985.
30. Srivastava, H.M.; Manocha, H.L. *A Treatise on Generating Functions*; Halsted Press (Ellis Horwood Limited): Chichester, UK; John Wiley and Sons: New York, NY, USA; Chichester, UK; Brisbane, Australia; Toronto, ON, Canada, 1984.
31. Srivastava, H.M. Some generalizations and basic (or *q*-) extensions of the Bernoulli, Euler and Genocchipolynomials. *Appl. Math. Inform. Sci.* **2011**, *5*, 390–444.
32. Tassaddiq, A. Some Representations of the Extended Fermi-Dirac and Bos-Einstein Functions with Applications. Ph.D. Dissertation, National University of Sciences and Technology Islamabad, Islamabad, Pakistan, 2011.
33. Tassaddiq, A.; Qadir, A. Fourier transform representation of the extended Fermi-Dirac and Bose-Einstein functions with applications to the family of the zeta and related functions. *Integral Transforms Spec. Funct.* **2011**, *22*, 453–466. [CrossRef]
34. Tassaddiq, A. A New Representation of the Extended Fermi-Dirac and Bose-Einstein Functions. *Int. J. Math. Appl.* **2017**, *5*, 435–446.
35. Gaboury, S.; Bayad, A. Series representations at special values of generalized Hurwitz-Lerch zeta function. *Abstr. Appl. Anal.* **2013**, *2013*, 975615. [CrossRef]
36. Garg, M.; Jain, K.; Kalla, S.L. A further study of general Hurwitz-Lerch zeta function. *Algebras Groups Geom.* **2008**, *25*, 311–319.
37. Garg, M.; Jain, K.; Srivastava, H.M. Some relationships between the generalized Apostol-Bernoulli polynomials and Hurwitz-Lerch Zeta functions. *Integral Transforms Spec. Funct.* **2006**, *17*, 803–815. [CrossRef]
38. Gupta, P.L.; Gupta, R.C.; Ong, S.-H.; Srivastava, H.M. A class of Hurwitz-Lerch Zeta distributions and their applications in reliability. *Appl. Math. Comput.* **2008**, *196*, 521–531. [CrossRef]
39. Lin, S.-D.; Srivastava, H.M. Some families of the Hurwitz-Lerch Zeta functions and associated fractional derivative and other integral representations. *Appl. Math. Comput.* **2004**, *154*, 725–733. [CrossRef]

40. Lin, S.-D.; Srivastava, H.M.; Wang, P.-Y. Some expansion formulas for a class of generalized Hurwitz-Lerch Zeta functions. *Integral Transforms Spec. Funct.* **2006**, *17*, 817–827. [CrossRef]
41. Raina, R.K.; Chhajed, P.K. Certain results involving a class of functions associated with the Hurwitz zeta function. *Acta Math. Univ. Comen.* **2004**, *73*, 89–100.
42. Srivastava, H.M.; Luo, M.-J.; Raina, R.K. New results involving a class of generalized Hurwitz-Lerch zeta functions and their applications. *Turk. J. Anal. Number Theory* **2013**, *1*, 26–35. [CrossRef]
43. Bayad, A.; Chikhi, J. Reduction and duality of the generalized Hurwitz-Lerch zetas. *Fixed Point Theory Appl.* **2013**, *2013*, 82. [CrossRef]
44. Erdelyi, A.; Magnus, W.; Oberhettinger, F.; Tricomi, F.G. *Higher Transcendental Functions*; McGraw-Hill Book Company: New York, NY, USA; Toronto, ON, Canada; London, UK, 1953.
45. Srivastava, H.M.; Chaudhry, M.A.; Qadir, A.; Tassaddiq, A. Some extensions of the Fermi-Dirac and Bose-Einstein functions with applications to zeta and related functions. *Russ. J. Math. Phys.* **2011**, *18*, 107–121. [CrossRef]
46. Chaudhry, M.A.; Zubair, S.M. *On a Class of Incomplete Gamma Functions with Applications*; Chapman and Hall (CRC Press Company): Boca Raton, FL, USA; London, UK; New York, NY, USA; Washington, DC, USA, 2001.
47. Choi, J. Explicit formulas for Bernoulli polynomials of order n. *Indian J. Pure Appl. Math.* **1996**, *27*, 667–674.
48. Lippert, R.A. A probabilistic interpretation of the Hurwitz zeta function. *Adv. Math.* **1993**, *97*, 278–284.
49. Saxena, R.K.; Pogany, T.K.; Saxena, R.; Jankov, D. On generalized Hurwitz-Lerch Zeta distributions occurring in statistical inference. *Acta Univ. Sapientiae Math.* **2011**, *3*, 43–59.
50. Zayed, A.I. *Handbook of Functions and Generalized Function Transforms*; CRC Press: Boca Raton, FL, USA, 1996.

© 2019 by the author. Licensee MDPI, Basel, Switzerland. This article is an open access article distributed under the terms and conditions of the Creative Commons Attribution (CC BY) license (http://creativecommons.org/licenses/by/4.0/).

symmetry

MDPI

Article

Some General Classes of q-Starlike Functions Associated with the Janowski Functions

Hari M. Srivastava [1,2,*], Muhammad Tahir [3], Bilal Khan [3], Qazi Zahoor Ahmad [3] and Nazar Khan [3]

1 Department of Mathematics and Statistics, University of Victoria, Victoria, BC V8W 3R4, Canada
2 Department of Medical Research, China Medical University Hospital, China Medical University, Taichung 40402, Taiwan
3 Department of Mathematics, Abbottabad University of Science and Technology, Abbottabad 22010, Pakistan; tahirmuhammad778@gmail.com (M.T.); bilalmaths789@gmail.com (B.K.); zahoorqazi5@gmail.com (Q.Z.A.); nazarmaths@gmail.com (N.K.)
* Correspondence: harimsri@math.uvic.ca; Tel.: +1-250-477-6960

Received: 27 January 2019 ; Accepted: 19 February 2019; Published: 23 February 2019

Abstract: By making use of the concept of basic (or q-) calculus, various families of q-extensions of starlike functions of order α in the open unit disk $\mathbb{U}$ were introduced and studied from many different viewpoints and perspectives. In this paper, we first investigate the relationship between various known classes of q-starlike functions that are associated with the Janowski functions. We then introduce and study a new subclass of q-starlike functions that involves the Janowski functions. We also derive several properties of such families of q-starlike functions with negative coefficients including (for example) distortion theorems.

Keywords: analytic functions; univalent functions; starlike and q-starlike functions; q-derivative (or q-difference) operator; sufficient conditions; distortion theorems; Janowski functions

MSC: Primary 05A30, 30C45; Secondary 11B65, 47B38

---

1. Introduction

The basic (or q-) calculus is the ordinary classical calculus without the notion of limits, while q stands for the quantum. The application of the q-calculus was initiated by Jackson [1,2]. Later, geometrical interpretation of the q-analysis was recognized through studies on quantum groups. It also suggests a relation between integrable systems and q-analysis. Aral and Gupta [3–5] defined and studied the q-analogue of the Baskakov-Durrmeyer operator, which is based on the q-analogue of the beta function. Some other important q-generalizations and q-extensions of complex operators are the q-Picard and the q-Gauss-Weierstrass singular-integral operators, which are discussed in [6–8].

In Geometric Function Theory, several subclasses of the normalized analytic function class $\mathcal{A}$ have already been analyzed and investigated through various perspectives. The q-calculus provides valuable tools that have been extensively used in order to examine several subclasses of the normalized analytic function class $\mathcal{A}$ in the open unit disk $\mathbb{U}$. Ismail et al. [9] were the first to use the q-derivative operator D_q in order to study a certain q-analogue of the class $\mathcal{S}^*$ of starlike functions in $\mathbb{U}$ (see Definition 6 below). Mohammed and Darus [10] studied the approximation and geometric properties of these q-operators in some subclasses of analytic functions in a compact disk. These q-operators are defined by using the convolution of normalized analytic functions and q-hypergeometric functions, where several interesting results were obtained (see [11,12]). Certain basic properties of the q-close-to-convex functions were studied by Raghavendar and Swaminathan [13]. Aral et al. [14] successfully studied the applications of the q-calculus in operator theory. Kanas and Raducanu [15] used the fractional

q-calculus operators in investigations of certain classes of functions, which are analytic in the open unit disk $\mathbb{U}$ by using the idea of the canonical domain. The coefficient inequality problems for q-closed-to-convex functions with respect to Janowski starlike functions were studied recently (see, for example, [16]). In the year 2016, Wongsaijai and Sukantamala [17] published a paper, in which they generalized certain subclasses of starlike functions in a systematic way. In fact, they made a very significant usage of the q-calculus basically in the context of Geometric Function Theory. Moreover, the generalized basic (or q-) hypergeometric functions were first used in Geometric Function Theory in a book chapter by Srivastava (see, for details [18], (p. 347 et seq.); see also [19]).

Motivated by the works of Wongsaijai and Sukantamala [17] and other related works cited above in this paper, we shall consider three new subfamilies of q-starlike functions with respect to Janowski functions. Several properties and characteristics, for example, sufficient conditions, inclusion results, distortion theorems, and radius problems, shall be discussed in this investigation. We shall also point out some relevant connections of our results with the existing results.

We denote by $\mathcal{H}(\mathbb{U})$ the class of functions that are analytic in the open unit disk:

$$\mathbb{U} = \{z : z \in \mathbb{C} \quad \text{and} \quad |z| < 1\},$$

where $\mathbb{C}$ is the set of complex numbers. Let $\mathcal{A}$ be the subclass of functions $f \in \mathcal{H}(\mathbb{U})$, which are represented by the following Taylor-Maclaurin series expansion:

$$f(z) = z + \sum_{n=2}^{\infty} a_n z^n \qquad (z \in \mathbb{U}) \tag{1}$$

that is, which satisfy the normalization condition given by

$$f(0) = f'(0) - 1 = 0.$$

Furthermore, let $\mathcal{S}$ be the class of functions in $\mathcal{A}$, which are univalent in $\mathbb{U}$.

The familiar class of starlike functions in $\mathbb{U}$ will be denoted by $\mathcal{S}^*$, which consists of normalized functions $f \in \mathcal{S}$ that satisfy the following conditions:

$$f \in \mathcal{S} \quad \text{and} \quad \Re\left(\frac{zf'(z)}{f(z)}\right) > 0 \qquad (\forall\, z \in \mathbb{U}). \tag{2}$$

For two functions f and g, which are analytic in $\mathbb{U}$, we say that the function f is subordinate to g and write

$$f \prec g \quad \text{or} \quad f(z) \prec g(z),$$

if there exists a Schwarz function w, which is analytic in $\mathbb{U}$ with

$$w(0) = 0 \quad \text{and} \quad |w(z)| < 1$$

such that

$$f(z) = g(w(z)).$$

In particular, if the function g is univalent in $\mathbb{U}$, then we have the following equivalence (cf., e.g., [20]; see also [21]):

$$f(z) \prec g(z) \quad (z \in \mathbb{U}) \Longleftrightarrow f(0) = g(0) \quad \text{and} \quad f(\mathbb{U}) \subset g(\mathbb{U}).$$

We next denote by $\mathcal{P}$ the class of analytic functions p in $\mathbb{U}$, which are normalized by

$$p(z) = 1 + \sum_{n=1}^{\infty} p_n z^n, \tag{3}$$

such that

$$\Re\{p(z)\} > 0.$$

In the next section (Section 2), we first give some basic definitions and concept details. Thereafter we will demonstrate three (presumably new) subclasses of the class $\mathcal{S}_q^*$ of q-starlike functions associated with the Janowski functions.

2. A Set of Definitions

Throughout this paper, we suppose that $0 < q < 1$ and that

$$\mathbb{N} = \{1, 2, 3, \cdots\} = \mathbb{N}_0 \setminus \{0\} \qquad (\mathbb{N}_0 := \{0, 1, 2, \cdots\}).$$

Definition 1. (See [22]) *A given function h with $h(0) = 1$ is said to belong to the class $\mathcal{P}[A, B]$ if and only if*

$$h(z) \prec \frac{1 + Az}{1 + Bz} \qquad (-1 \leqq B < A \leqq 1).$$

The analytic function class $\mathcal{P}[A, B]$ was introduced by Janowski [22], who showed that $h(z) \in \mathcal{P}[A, B]$ if and only if there exists a function $p \in \mathcal{P}$ such that

$$h(z) = \frac{(A+1)\,p(z) - (A-1)}{(B+1)\,p(z) - (B-1)} \qquad (-1 \leqq B < A \leqq 1).$$

Definition 2. *A function $f \in \mathcal{S}$ is said to belong to the class $\mathcal{S}^*[A, B]$ if and only if there exists a function $p \in \mathcal{P}$ such that*

$$\frac{zf'(z)}{f(z)} = \frac{(A+1)\,p(z) - (A-1)}{(B+1)\,p(z) - (B-1)} \qquad (-1 \leqq B < A \leqq 1). \tag{4}$$

Definition 3. *Let $q \in (0, 1)$, and define the q-number $[\lambda]_q$ by*

$$[\lambda]_q = \begin{cases} \dfrac{1 - q^\lambda}{1 - q} & (\lambda \in \mathbb{C}) \\[2ex] \sum\limits_{k=0}^{n-1} q^k = 1 + q + q^2 + \cdots + q^{n-1} & (\lambda = n \in \mathbb{N}). \end{cases}$$

Definition 4. *Let $q \in (0, 1)$, and define the q-factorial $[n]_q!$ by*

$$[n]_q! = \begin{cases} 1 & (n = 0) \\[2ex] \prod\limits_{k=1}^{n} [k]_q & (n \in \mathbb{N}). \end{cases}$$

Definition 5. (See [1,2]) *The q-derivative (or the q-difference) operator $D_q f$ of a function f is defined, in a given subset of $\mathbb{C}$, by*

$$(D_q f)(z) = \begin{cases} \dfrac{f(z) - f(qz)}{(1 - q)z} & (z \neq 0) \\[2ex] f'(0) & (z = 0), \end{cases} \tag{5}$$

provided that $f'(0)$ exists.

We note from Definition 5 that

$$\lim_{q\to 1-}\left(D_qf\right)(z)=\lim_{q\to 1-}\frac{f(z)-f(qz)}{(1-q)z}=f'(z)$$

for a function f, which is differentiable in a given subset of $\mathbb{C}$. It is readily deduced from (1) and (5) that

$$\left(D_qf\right)(z)=1+\sum_{n=2}^{\infty}[n]_q\,a_nz^{n-1}. \tag{6}$$

Definition 6. (See [9]) *A function $f\in\mathcal{S}$ is said to belong to the class $\mathcal{S}_q^*$ of q-starlike functions in $\mathbb{U}$ if*

$$f\left(0\right)=f'\left(0\right)-1=0 \tag{7}$$

and

$$\left|\frac{z}{f\left(z\right)}\left(D_qf\right)\left(z\right)-\frac{1}{1-q}\right|\leqq\frac{1}{1-q}\qquad\left(z\in\mathbb{U}\right). \tag{8}$$

We readily observe that, as $q\to 1-$, the closed disk:

$$\left|w-\frac{1}{1-q}\right|\leqq\frac{1}{1-q}$$

becomes the right-half complex plane, and the class $\mathcal{S}_q^*$ of q-starlike functions in $\mathbb{U}$ reduces to the familiar class $\mathcal{S}^*$ of normalized starlike functions with respect to the origin $(z=0)$. Equivalently, by using the principle of subordination between analytic functions, we can rewrite the conditions in (7) and (8) as follows (see [16]):

$$\frac{z}{f\left(z\right)}\left(D_qf\right)\left(z\right)\prec\widehat{p}(z)\qquad\left(\widehat{p}(z):=\frac{1+z}{1-qz}\right). \tag{9}$$

We now introduce three (presumably new) subclasses of the class $\mathcal{S}_q^*$ of q-starlike functions associated with the Janowski functions in the following way.

Definition 7. *A function $f\in\mathcal{A}$ is said to belong to the class $\mathcal{S}_{(q,1)}^*\left[A,B\right]$ if and only if*

$$\Re\left(\frac{\left(B-1\right)\dfrac{zD_qf\left(z\right)}{f\left(z\right)}-\left(A-1\right)}{\left(B+1\right)\dfrac{zD_qf\left(z\right)}{f\left(z\right)}-\left(A+1\right)}\right)\geqq 0.$$

We call $\mathcal{S}_{(q,1)}^*\left[A,B\right]$ the class of q-starlike functions of Type 1 associated with the Janowski functions.

Definition 8. *A function $f\in\mathcal{A}$ is said to belong to the class $f\in\mathcal{S}_{(q,2)}^*\left[A,B\right]$ if and only if*

$$\left|\frac{\left(B-1\right)\dfrac{zD_qf\left(z\right)}{f\left(z\right)}-\left(A-1\right)}{\left(B+1\right)\dfrac{zD_qf\left(z\right)}{f\left(z\right)}-\left(A+1\right)}-\frac{1}{1-q}\right|<\frac{1}{1-q}.$$

We call $\mathcal{S}_{(q,2)}^*\left[A,B\right]$ the class of q-starlike functions of Type 2 associated with the Janowski functions.

Definition 9. *A function $f \in \mathcal{A}$ is said to belong to the class $f \in \mathcal{S}^*_{(q,3)}[A,B]$ if and only if*

$$\left| \frac{(B-1)\dfrac{zD_qf(z)}{f(z)} - (A-1)}{(B+1)\dfrac{zD_qf(z)}{f(z)} - (A+1)} - 1 \right| < 1.$$

We call $\mathcal{S}^*_{(q,3)}[A,B]$ the class of q-starlike functions of Type 3 associated with the Janowski functions.

Each of the following special cases of the above-defined q-starlike functions:

$$\mathcal{S}^*_{(q,1)}[A,B], \quad \mathcal{S}^*_{(q,2)}[A,B] \quad \text{and} \quad \mathcal{S}^*_{(q,3)}[A,B]$$

is worthy of note.

I. If we put

$$A = 1 - 2\alpha \quad (0 \leqq \alpha < 1) \quad \text{and} \quad B = -1$$

in Definition 7, we get the class $\mathcal{S}^*_{(q,1)}(\alpha)$, which was introduced and studied by Wongsaijai and Sukantamala (see [17], Definition 1).

II. If we put

$$A = 1 - 2\alpha \quad (0 \leqq \alpha < 1) \quad \text{and} \quad B = -1,$$

in Definition 8, we are led to the class $\mathcal{S}^*_{(q,2)}(\alpha)$, which was introduced and studied by Wongsaijai and Sukantamala (see [17], Definition 2).

III. If we put

$$A = 1 - 2\alpha \quad (0 \leqq \alpha < 1) \quad \text{and} \quad B = -1$$

in Definition 9, we have the class $\mathcal{S}^*_{(q,3)}(\alpha)$, which was introduced and studied by Wongsaijai and Sukantamala (see [17], Definition 3).

IV. If we put

$$A = 1 - 2\alpha \quad (0 \leqq \alpha < 1) \quad \text{and} \quad B = -1$$

in Definition 8, we obtain the class $\mathcal{S}^*_q(\alpha)$, which was introduced and studied by Agrawal and Sahoo [23].

V. If we put

$$A = 1 \quad \text{and} \quad B = -1$$

in Definition 8, we get the class $\mathcal{S}^*_q$ introduced and studied by Ismail et al. [9].

VI. In Definition 8, if we let $q \to 1-$ and put $A = \lambda$ and $B = 0$, then we will arrive at the function class, studied by Ponnusamy and Singh (see [24]).

Geometrically, for $f \in \mathcal{S}^*_{(q,k)}[A,B] \quad (k = 1,2,3)$, the quotient:

$$\frac{zD_qf(z)}{f(z)}$$

lies in the domains $\Omega_j \quad (j = 1,2,3)$ given by

$$\Omega_1 = \left\{ w : w \in \mathbb{C} \quad \text{and} \quad \Re(w) > \frac{A-1}{B-1} \right\},$$

$$\Omega_2 = \left\{ w : w \in \mathbb{C} \quad \text{and} \quad \left| w - \frac{2+q(A-1)}{(B-1)q+(B+3)} \right| < \frac{A+1}{(B-1)q+(B+3)} \right\}$$

and

$$\Omega_3 = \left\{ w : w \in \mathbb{C} \quad \text{and} \quad \left| w - \frac{2}{B+3} \right| < \frac{A+1}{B+3} \right\},$$

respectively.

In this paper, many properties and characteristics, for example sufficient conditions, inclusion results, distortion theorems, and radius problems, are discussed. We also indicate relevant connections of our results with a number of other related works on this subject.

3. Main Results and Their Demonstration

We first derive the inclusion results for the following generalized q-starlike functions:

$$\mathcal{S}^*_{(q,1)}[A,B], \quad \mathcal{S}^*_{(q,2)}[A,B] \qquad \text{and} \qquad \mathcal{S}^*_{(q,3)}[A,B],$$

which are associated with the Janowski functions.

Theorem 1. *If* $-1 \leqq B < A \leqq 1$, *then*

$$\mathcal{S}^*_{(q,3)}[A,B] \subset \mathcal{S}^*_{(q,2)}[A,B] \subset \mathcal{S}^*_{(q,1)}[A,B].$$

Proof. First of all, we suppose that $f \in \mathcal{S}^*_{(q,3)}[A,B]$. Then, by Definition 9, we have

$$\left| \frac{(B-1)\frac{zD_qf(z)}{f(z)} - (A-1)}{(B+1)\frac{zD_qf(z)}{f(z)} - (A+1)} - 1 \right| < 1,$$

so that

$$\left| \frac{(B-1)\frac{zD_qf(z)}{f(z)} - (A-1)}{(B+1)\frac{zD_qf(z)}{f(z)} - (A+1)} - 1 \right| + \frac{q}{1-q} < 1 + \frac{q}{1-q}. \tag{10}$$

By using the triangle inequality and Equation (10), we find that

$$\left| \frac{(B-1)\frac{zD_qf(z)}{f(z)} - (A-1)}{(B+1)\frac{zD_qf(z)}{f(z)} - (A+1)} - \frac{1}{1-q} \right| < \frac{1}{1-q}. \tag{11}$$

The last expression in (11) now implies that $f \in \mathcal{S}^*_{(q,2)}[A,B]$, that is, that

$$\mathcal{S}^*_{(q,3)}[A,B] \subset \mathcal{S}^*_{(q,2)}[A,B].$$

Next, we let $f \in \mathcal{S}^*_{(q,2)}[A,B]$, so that

$$f \in \mathcal{S}^*_{(q,2)}[A,B] \iff \left| \frac{(B-1)\frac{zD_qf(z)}{f(z)} - (A-1)}{(B+1)\frac{zD_qf(z)}{f(z)} - (A+1)} - \frac{1}{1-q} \right| < \frac{1}{1-q},$$

by Definition 8.

Since

$$\frac{1}{1-q} > \left| \frac{(B-1)\frac{zD_qf(z)}{f(z)} - (A-1)}{(B+1)\frac{zD_qf(z)}{f(z)} - (A+1)} - \frac{1}{1-q} \right|$$

$$= \left| \frac{1}{1-q} - \frac{(B-1)\frac{zD_qf(z)}{f(z)} - (A-1)}{(B+1)\frac{zD_qf(z)}{f(z)} - (A+1)} \right|,$$

we have

$$\Re\left(\frac{(B-1)\frac{zD_qf(z)}{f(z)} - (A-1)}{(B+1)\frac{zD_qf(z)}{f(z)} - (A+1)} \right) > 0 \qquad (z \in \mathbb{U}).$$

This last equation now shows that $f \in \mathcal{S}^*_{(q,1)}[A,B]$, that is, that

$$\mathcal{S}^*_{(q,2)}[A,B] \subset \mathcal{S}^*_{(q,1)}[A,B],$$

which completes the proof of Theorem 1. □

As a special case of Theorem 1, if we put

$$A = 1 - 2\alpha \quad (0 \leqq \alpha < 1) \qquad \text{and} \qquad B = -1,$$

we get the following known result due to Wongsaijai and Sukantamala (see [17]).

Corollary 1. (See [17]) *For* $0 \leqq \alpha < 1$,

$$\mathcal{S}^*_{q,3}(\alpha) \subset \mathcal{S}^*_{q,2}(\alpha) \subset \mathcal{S}^*_{q,1}(\alpha).$$

Next, we present a remarkable simple characterization of functions in the class $\mathcal{S}^*_{(q,2)}[A,B]$ of q-starlike functions of Type 2 associated with the Janowski functions.

Theorem 2. *Let* $f \in \mathcal{A}$. *Then* $f \in \mathcal{S}^*_{(q,2)}[A,B]$ *if and only if*

$$\left| \frac{f(qz)}{f(z)} - \frac{\sigma}{(B-1)q + (B+3)} \right| \leqq \frac{(A+1)(1-q)}{(B-1)q + (B+3)},$$

where

$$\sigma = (A-1)q^2 + (B-A+2)q + B + 1.$$

Proof. The proof of Theorem 2 can be easily obtained from the fact that

$$\frac{zD_qf(z)}{f(z)} = \left(\frac{1}{1-q}\right)\left(1 - \frac{f(qz)}{f(z)}\right)$$

and Definition 8 of the class $\mathcal{S}^*_{(q,2)}[A,B]$ of q-starlike functions of Type 2 associated with the Janowski functions. □

Upon setting

$$A = 1 - 2\alpha \qquad \text{and} \qquad B = -1$$

in Theorem 2, we get the following known result.

Corollary 2. (See [23]) *Let $f \in \mathcal{A}$. Then, $f \in \mathcal{S}^*_q(\alpha)$ if and only if*

$$\left| \frac{f(qz)}{f(z)} - \alpha q \right| \leqq 1 - \alpha.$$

Our next result is directly obtained by using Theorem 1 and a known result given in [23].

Theorem 3. *The classes*

$$\mathcal{S}^*_{(q,1)}[A,B], \quad \mathcal{S}^*_{(q,2)}[A,B] \qquad \text{and} \qquad \mathcal{S}^*_{(q,3)}[A,B]$$

of the generalized q-starlike functions of Type 1, Type 2, and Type 3, respectively, satisfy the following properties:

$$\bigcap_{0<q<1} \mathcal{S}^*_{(q,1)}[A,B] = \bigcap_{0<q<1} \mathcal{S}^*_{(q,2)}[A,B] = \mathcal{S}^*[A,B]$$

and

$$\bigcap_{0<q<1} \mathcal{S}^*_{(q,1)}[A,B] = \bigcap_{0<q<1} \mathcal{S}^*_{(q,3)}[A,B] \subset \mathcal{S}^*[A,B].$$

Finally, by means of a coefficient inequality, we give a sufficient condition for the class $\mathcal{S}^*_{(q,3)}[A,B]$ of generalized q-starlike functions of Type 3, which also provides a corresponding sufficient condition for the classes $\mathcal{S}^*_{(q,1)}[A,B]$ and $\mathcal{S}^*_{(q,2)}[A,B]$ of Type 1 and Type 2, respectively.

Theorem 4. *A function $f \in \mathcal{A}$ and of the form (1) is in the class $\mathcal{S}^*_{(q,3)}[A,B]$ if it satisfies the following coefficient inequality:*

$$\sum_{n=2}^{\infty} \Big(2q\,[n-1]_q + [n]_q\,(B+1) + (A+1) \Big) |a_n| < |B - A|. \tag{12}$$

4. Analytic Functions with Negative Coefficients

In this section, we introduce new subclasses of q-starlike functions associated with the Janowski functions, which involve negative coefficients. Let $\mathcal{T}$ be a subset of $\mathcal{A}$ consisting of functions with a negative coefficient, that is,

$$f(z) = z - \sum_{n=2}^{\infty} |a_n|\, z^n \qquad (a_n \geqq 0). \tag{13}$$

We also let

$$\mathcal{TS}^*_{(q,k)}[A,B] := \mathcal{S}^*_{(q,k)}[A,B] \cap \mathcal{T} \qquad (k = 1,2,3). \tag{14}$$

Theorem 5. *If $-1 \leqq B < A \leqq 1$, then*

$$\mathcal{TS}^*_{(q,1)}[A,B] = \mathcal{TS}^*_{(q,2)}[A,B] = \mathcal{TS}^*_{(q,3)}[A,B].$$

Proof. In view of Theorem 1, it is sufficient here to show that

$$\mathcal{TS}^{*}_{(q,1)}[A,B] \subset \mathcal{TS}^{*}_{(q,3)}[A,B].$$

Indeed, if we assume that $f \in \mathcal{TS}^{*}_{(q,1)}[A,B]$, then we have

$$\Re\left(\frac{(B-1)\frac{zD_qf(z)}{f(z)}-(A-1)}{(B+1)\frac{zD_qf(z)}{f(z)}-(A+1)}\right) \geqq 0,$$

so that

$$\Re\left(\frac{(B-1)\frac{zD_qf(z)}{f(z)}-(A-1)}{(B+1)\frac{zD_qf(z)}{f(z)}-(A+1)}-1\right) \geqq -1.$$

After a simple calculation, we thus find that

$$\frac{2\left[f(z)-zD_qf(z)\right]}{(B+1)zD_qf(z)-(A+1)f(z)} \geqq -1,$$

that is, that

$$-\frac{2\sum\limits_{n=2}^{\infty}\left([n]_q-1\right)a_nz^n}{(A-B)+\sum\limits_{n=2}^{\infty}\left([n]_q(B+1)-(A+1)\right)a_nz^n} \geqq -1,$$

which can be written as follows:

$$\frac{2\sum\limits_{n=2}^{\infty}\left([n]_q-1\right)a_nz^n}{(A-B)+\sum\limits_{n=2}^{\infty}\left([n]_q(B+1)-(A+1)\right)a_nz^n} < 1. \tag{15}$$

The last expression in (15) implies that

$$\frac{2\sum\limits_{n=2}^{\infty}\left([n]_q-1\right)a_nz^n}{|B-A|+\sum\limits_{n=2}^{\infty}\left([n]_q(B+1)+(A+1)\right)a_nz^n} \leqq 1,$$

which satisfies (12). By Theorem 4, the proof of Theorem 5 is completed. □

In its special case, when

$$A = 1-2\alpha \quad (0 \leqq \alpha < 1) \qquad \text{and} \qquad B = -1.$$

Theorem 5 reduces to the following known result.

Corollary 3. (See [17], Theorem 8) *If* $0 \leqq \alpha < 1$, *then*

$$\mathcal{TS}^{*}_{(q,1)}(\alpha) = \mathcal{TS}^{*}_{(q,2)}(\alpha) = \mathcal{TS}^{*}_{(q,3)}(\alpha).$$

The assertions of Theorem 5 imply that the Type 1, Type 2, and Type 3 generalized q-starlike functions associated with the Janowski functions are exactly the same. For convenience, therefore, we state the following distortion theorem by using the notation $\mathcal{TS}^{*}_{(q,k)}[A,B]$ in which it is tacitly assumed that $k = 1,2,3$.

Theorem 6. *If* $f \in \mathcal{TS}^{*}_{(q,k)}[A,B]$ $(k=1,2,3)$, *then*

$$r-\left(\frac{|B-A|}{\Lambda(2,A,B,q)}\right)r^{2} \leqq |f(z)| \leqq r+\left(\frac{|B-A|}{\Lambda(2,A,B,q)}\right)r^{2}$$

$$(|z|=r \ \ (0<r<1)),$$

where

$$\Lambda(n,A,B,q)=2\left([n]_{q}-1\right)+[n]_{q}(B+1)+(A+1) \qquad (n \in \mathbb{N}\setminus\{1\}). \tag{16}$$

Proof. We note that the following inequality follows from Theorem 4:

$$\Lambda(2,A,B,q)\sum_{n=2}^{\infty}|a_{n}| \leqq \sum_{n=2}^{\infty}\Lambda(n,A,B,q)|a_{n}| < |B-A|,$$

which yields

$$|f(z)| \leqq r+\sum_{n=2}^{\infty}|a_{n}|r^{n} \leqq r+r^{2}\sum_{n=2}^{\infty}|a_{n}| \leqq r+\left(\frac{|B-A|}{\Lambda(2,A,B,q)}\right)r^{2}.$$

Similarly, we have

$$|f(z)| \geqq r-\sum_{n=2}^{\infty}|a_{n}|r^{n} \geqq r-r^{2}\sum_{n=2}^{\infty}|a_{n}| \geqq r-\left(\frac{|B-A|}{\Lambda(2,A,B,q)}\right)r^{2}.$$

We have thus completed the proof of Theorem 6. □

In its special case, when

$$A=1-2\alpha \quad (0 \leqq \alpha < 1) \qquad \text{and} \qquad B=-1,$$

if we let $q \longrightarrow 1-$, Theorem 6 reduces to the following known result.

Corollary 4. (See [25]) *If* $f \in \mathcal{TS}^{*}(\alpha)$, *then*

$$r-\left(\frac{1-\alpha}{2-\alpha}\right)r^{2} \leqq |f(z)| \leqq r+\left(\frac{1-\alpha}{2-\alpha}\right)r^{2} \qquad (|z|=r \ \ (0<r<1)).$$

The following result (Theorem 7) can be proven by using arguments similar to those that were already presented in the proof of Theorem 6, so we choose to omit the details of our proof of Theorem 7.

Theorem 7. *If* $f \in \mathcal{TS}^{*}_{(q,k)}[A,B]$ $(k=1,2,3)$, *then*

$$1-\left(\frac{2|B-A|}{\Lambda(2,A,B,q)}\right)r \leqq |f'(z)| \leqq 1+\left(\frac{2|B-A|}{\Lambda(2,A,B,q)}\right)r$$

$$(|z|=r \ \ (0<r<1)),$$

where $\Lambda(n,A,B,q)$ *is given by* (16).

In its special case, when

$$A=1-2\alpha \quad (0 \leqq \alpha < 1) \qquad \text{and} \qquad B=-1,$$

if we let $q \longrightarrow 1-$, Theorem 6 reduces to the following known result.

Corollary 5. (See [25]) *If $f \in \mathcal{TS}^*(\alpha)$, then*

$$1 - \left(\frac{2(1-\alpha)}{2-\alpha}\right) r \leqq \left|f'(z)\right| \leqq 1 + \left(\frac{2(1-\alpha)}{2-\alpha}\right) r \qquad (|z| = r \ (0 < r < 1)).$$

Remark 1. *By using Theorem 4, it is easy to see that the function:*

$$f_0(z) = z - \frac{|B-A| - \epsilon}{2q[n-1]_q + [n]_q(B+1) + (A+1)} z^n \in \mathcal{TS}_{(q,k)}[A,B] \tag{17}$$

where

$$0 < \epsilon < \frac{n|B-A| - 2q[n-1]_q + [n]_q(B+1) + (A+1)}{n}$$

and

$$2q[n-1]_q + \left([n]_q(B+1) + (A+1)\right) < n(|B-A| - \epsilon),$$

but

$$f_0'(z) = 0$$

at

$$z_0 = \left[\frac{2q[n-1]_q + \left([n]_q(B+1) + (A+1)\right)}{n(|B-A| - \epsilon)}\right]^{\frac{1}{n-1}} \left(\cos\left(\frac{2k\pi}{n-1}\right) + i\sin\left(\frac{2k\pi}{n-1}\right)\right) \in \mathbb{U}.$$

That is, $f_0(z) \notin \mathcal{S}$ and also $f_0(z) \notin \mathcal{S}^*$. Therefore, it is interesting to study the radius of univalency and starlikeness of class $\mathcal{TS}_{(q,k)}[A,B]$.

Theorem 8. *Let $f \in \mathcal{TS}_{(q,k)}[A,B]$ $(k = 1,2,3)$. Then, f is univalent and starlike in $|z| < r_0$, where*

$$r_0 = \min_{2 \leq n \leq M_0} \left[\frac{2q[n-1]_q + [n]_q(B+1) + (A+1)}{n|B-A|}\right]^{\frac{1}{n-1}} \tag{18}$$

and M_0 satisfies the following inequality:

$$M_0 > \exp\left(1 + \left|\ln \frac{(1-q)|B-A|}{(B+3) + (A-1)(1-q)}\right|\right).$$

Proof. To prove Theorem 8, it is sufficient to show that

$$\left|f'(z) - 1\right| < 1 \quad (|z| \leqq r_0).$$

Now, we have

$$\left|f'(z) - 1\right| = \left|-\sum_{n=2}^{\infty} n a_n z^{n-1}\right| \leqq \sum_{n=2}^{\infty} n|a_n||z|^{n-1}.$$

Thus,

$$\left|f'(z) - 1\right| < 1$$

if

$$\sum_{n=2}^{\infty} n|a_n||z|^{n-1} < 1. \tag{19}$$

In light of Theorem 4, the inequality in (19) will be true if

$$n|z|^{n-1} \leqq \frac{2q[n-1]_q + \left([n]_q(B+1) + (A+1)\right)}{|B-A|}. \tag{20}$$

Solving the inequality in (20) for z, we have

$$|z| \leqq \left[\frac{2q\,[n-1]_q + \left([n]_q\,(B+1) + (A+1)\right)}{n\,|B-A|}\right]^{\frac{1}{n-1}}. \tag{21}$$

Next, we need to find $M_0 \in \mathbb{N}$ satisfying (18). Let $f : [2,\infty) \longrightarrow \mathbb{R}^+$ be the function defined by

$$f(x) = \left[\frac{2q\,[x-1]_q + [x]_q\,(B+1) + (A+1)}{x\,|B-A|}\right]^{\frac{1}{x-1}} \tag{22}$$

Differentiating on both sides of (22) logarithmically, we have

$$\begin{aligned} f'(x) = \frac{f(x)}{(x-1)^2}\Bigg[&\ln x - \frac{(x-1)(B+3)(q^x \ln q)}{(B+3)(1-q^x) + (A-1)(1-q)} \\ &+ \ln\frac{(1-q)\,|B-A|}{(B+3)(1-q^x) + (A-1)(1-q)} - \frac{x-1}{x}\Bigg]. \end{aligned} \tag{23}$$

It is easy to see that the second term of (23) is positive. Since

$$\sup_{x\geqq 2}\left|\ln\frac{(1-q)\,|B-A|}{(B+3)(1-q^x) + (A-1)(1-q)}\right| = \left|\frac{(1-q)\,|B-A|}{(B+3) + (A-1)(1-q)}\right|$$

and

$$\sup_{x\geqq 2}\frac{x-1}{x} = 1$$

then the third and the last term in (23) can be dominated by $\ln x$ when x is sufficiently large. This implies that f is an increasing function on $[M_0, \infty]$, where

$$M_0 > \exp\left(1 + \left|\ln\frac{(1-q)\,|B-A|}{(B+3) + (A-1)(1-q)}\right|\right).$$

Therefore, the radius of univalence can be defined by

$$\begin{aligned} r_0 &= \inf_{n\geqq 2}\left[\frac{2q\,[n-1]_q + [n]_q\,(B+1) + (A+1)}{n\,|B-A|}\right]^{\frac{1}{n-1}} \\ &= \min_{2\leqq n\leqq M_0}\left[\frac{2q\,[n-1]_q + [n]_q\,(B+1) + (A+1)}{n\,|B-A|}\right]^{\frac{1}{n-1}} \end{aligned} \tag{24}$$

In view of (24), the proof of our Theorem is now completed. □

If, in Theorem 8, we let

$$B = -1 \quad \text{and} \quad A = (1-2\alpha)$$

we are led to the following known result:

Corollary 6. [17] *Let $f \in \mathcal{TS}_q$. Then, f is univalent and starlike in $|z| < r_0$, where*

$$r_0 = \min_{2\leqq n\leqq M_0}\left[\frac{[n]_q - \alpha}{n\,(1-\alpha)}\right]^{\frac{1}{n-1}} \tag{25}$$

and M_0 satisfies the following inequality:

$$M_0 > \exp\left(1 + \left|\ln \frac{(1-q)(1-\alpha)}{q + (1-q)(1-\alpha)}\right|\right).$$

Now, below, we give an example that validates Theorem 8.

Example 1. *Consider the class $\mathcal{TS}_{(0.75,k)}[0,\lambda]$ with $\lambda = 0.99$. By Theorem 8, we obtain the radius of univalency of class $\mathcal{TS}_{(0.75,k)}[0,\lambda]$, given by*

$$\begin{aligned} r_0 &= \min_{2 \leqq n \leqq \exp(1+|\ln 0.08256880734|)} \left[\frac{3[n-1]_q - 0.01}{(0.99)\,n}\right]^{\frac{1}{n-1}} \\ &= \min_{2 \leqq n \leqq 33} \left[\frac{3[n-1]_q - 0.01}{(0.99)\,n}\right]^{\frac{1}{n-1}} = 0.9691405946. \end{aligned}$$

Now, we consider the sharpness example function $f_0(z)$ defined in (17) with $n = 2$ and $\epsilon = 0.001$, that is,

$$f_0(z) = z - \frac{0.989}{5.24} z^2.$$

Obviously, $f_0(z)$ is locally univalent on $\mathbb{U}_{0.9691405946}$ because $f'(z_0) = 0$ at $z_0 = 2.649140540$ outside the open disk $\mathbb{U}_{0.9691405946}$. By applying Theorem 8, function $f_0(z)$ is univalent on $\mathbb{U}_{0.9691405946}$.

The next Theorem (Theorem 9) can be derived by working in a similar way as in Theorem 8; here, we omit the proof.

Theorem 9. *Let $f \in \mathcal{TS}_{(q,k)}[A,B]$ $(k = 1,2,3)$. Then f is starlike of order α in $|z| < r_1$, where*

$$r_1 = \min_{2 \leqq n \leqq M_1} \left[\frac{\left(2q[n-1]_q + [n]_q(B+1) + (A+1)\right)(1-\alpha)}{(n-\alpha)|B-A|}\right]^{\frac{1}{n-1}} \tag{26}$$

and M_1 satisfies the following inequality:

$$M_1 > \exp\left(1 + \left|\ln \frac{(1-q)|B-A|}{((B+3) + (A-1)(1-q))(1-\alpha)}\right|\right).$$

5. Conclusions

In our present investigation, we first defined certain new subclasses of q-starlike functions, which are associated with the Janowski function. We then discussed many properties and characteristics of each of these subclasses of q-starlike functions including, for example, sufficient conditions, inclusion results, distortion theorems, and radius problems. For the motivation and validity of our results, we have also pointed out relevant connections with those that were given in earlier works.

Author Contributions: All authors contributed equally to the present investigation.

Funding: This research received no external funding.

Acknowledgments: The authors would like to express their sincere thanks to the referees for many valuable suggestions regarding a previous version of this paper.

Conflicts of Interest: The authors declare no conflict of interest.

References

1. Jackson, F.H. On q-definite integrals. *Quart. J. Pure Appl. Math.* **1910**, *41*, 193–203.
2. Jackson, F.H. q-difference equations. *Amer. J. Math.* **1910**, *32*, 305–314. [CrossRef]
3. Aral, A.; Gupta, V. On q-Baskakov type operators. *Demonstr. Math.* **2009**, *42*, 109–122.
4. Aral, A.; Gupta, V. On the Durrmeyer type modification of the q-Baskakov type operators. *Nonlinear Anal. Theory Methods Appl.* **2010**, *72*, 1171–1180. [CrossRef]
5. Aral, A.; Gupta, V. Generalized q-Baskakov operators. *Math. Slovaca* **2011**, *61*, 619–634. [CrossRef]
6. Anastassiou, G.A.; Gal, S.G. Geometric and approximation properties of some singular integrals in the unit disk. *J. Inequal. Appl.* **2006**, *2006*, 1–19. [CrossRef]
7. Anastassiou, G.A.; Gal, S.G. Geometric and approximation properties of generalized singular integrals in the unit disk. *J. Korean Math. Soc.* **2006**, *43*, 425–443. [CrossRef]
8. Aral, A. On the generalized Picard and Gauss Weierstrass singular integrals. *J. Comput. Anal. Appl.* **2006**, *8*, 249–261.
9. Ismail, M.E.H.; Merkes, E.; Styer, D. A generalization of starlike functions. *Complex Var. Theory Appl.* **1990**, *14*, 77–84. [CrossRef]
10. Mohammed, A.; Darus, M. A generalized operator involving the q-hypergeometric function. *Mat. Vesnik* **2013**, *65*, 454–465.
11. Dweby, H.A.; Darus, M. On harmonic meromorphic functions associated with basic hypergeometric functions. *Sci. World J.* **2013**, *2013*, 1–7. [CrossRef] [PubMed]
12. Dweby, H.A.; Darus, M. A subclass of harmonic univalent functions associated with q-analogue of Dziok-Srivastava operator. *ISRN Math. Anal.* **2013**, *2013*, 1–6.
13. Raghavendar, K.; Swaminathan, A. Close-to-convexity of basic hypergeometric functions using their Taylor coefficients. *J. Math. Appl.* **2012**, *35*, 111–125. [CrossRef]
14. Aral, A.; Gupta, V.; Agarwal, R.P. *Applications of q-Calculus in Operator Theory*; Springer: New York, NY, USA, 2013.
15. Kanas S.; Răducanu, D. Some class of analytic functions related to conic domains. *Math. Slovaca* **2014**, *64*, 1183–1196. [CrossRef]
16. Uçar, H.E.Ö. Coefficient inequality for q-starlike functions. *Appl. Math. Comput.* **2016**, *276*, 122–126.
17. Wongsaijai B.; Sukantamala, N. Certain properties of some families of generalized starlike functions with respect to q-calculus. *Abstr. Appl. Anal.* **2016**, *2016*, 1–8. [CrossRef]
18. Srivastava, H.M. Univalent functions, fractional calculus, and associated generalized hypergeometric functions. In *Univalent Functions, Fractional Calculus, and Their Applications*; Srivastava, H.M., Owa, S., Eds.; Halsted Press (Ellis Horwood Limited, Chichester); John Wiley and Sons: New York, NY, USA; Chichester, UK; Brisbane, Australia; Toronto, ON, Canada, 1989; pp. 329–354.
19. Srivastava, H.M.; Bansal, D. Close-to-convexity of a certain family of q-Mittag-Leffler functions. *J. Nonlinear Var. Anal.* **2017**, *1*, 61–69.
20. Miller, S.S.; Mocanu, P.T. Differential subordination and univalent functions. *Mich. Math. J.* **1918**, *28*, 157–171. [CrossRef]
21. Miller, S.S.; Mocanu, P.T. *Differential Subordination: Theory and Applications*; Series on Monographs and Textbooks in Pure and Applied Mathematics, No. 225; Marcel Dekker Incorporated: New York, NY, USA; Basel, Switzerland, 2000.
22. Janowski, W. Some extremal problems for certain families of analytic functions. *Annal. Polon. Math.* **1973**, *28*, 297–326. [CrossRef]
23. Agrawal, S.; Sahoo, S.K. A generalization of starlike functions of order α. *Hokkaido Math. J.* **2017**, *46*, 15–27. [CrossRef]
24. Ponnusamy, S.; Singh, V. Criteria for strongly starlike functions. *Complex Var. Theory Appl.* **1997**, *34*, 267–291. [CrossRef]
25. Silverman, H. Univalent functions with negative coefficients. *Proc. Am. Math. Soc.* **1975**, *51*, 109–116. [CrossRef]

© 2019 by the authors. Licensee MDPI, Basel, Switzerland. This article is an open access article distributed under the terms and conditions of the Creative Commons Attribution (CC BY) license (http://creativecommons.org/licenses/by/4.0/).

symmetry

MDPI

Article

Some Subclasses of Uniformly Univalent Functions with Respect to Symmetric Points

Shahid Mahmood [1], Hari M. Srivastava [2,3] and Sarfraz Nawaz Malik [4,*]

1. Department of Mechanical Engineering, Sarhad University of Science and I.T, Ring Road, Peshawar 25000, Pakistan; shahidmahmood757@gmail.com
2. Department of Mathematics and Statistics, University of Victoria, Victoria, BC V8W 3R4, Canada; harimsri@math.uvic.ca
3. Department of Medical Research, China Medical University Hospital, China Medical University, Taichung 40402, Taiwan
4. Department of Mathematics, COMSATS University Islamabad, Wah Campus 47040, Pakistan

* Correspondence: snmalik110@yahoo.com

Received: 11 January 2019; Accepted: 18 February 2019; Published: 22 February 2019

Abstract: This article presents the study of certain analytic functions defined by bounded radius rotations associated with conic domain. Many geometric properties like coefficient estimate, radii problems, arc length, integral representation, inclusion results and growth rate of coefficients of Taylor's series representation are investigated. By varying the parameters in results, several well-known results in literature are obtained as special cases.

Keywords: functions of bounded boundary and bounded radius rotations; subordination; functions with positive real part; uniformly starlike and convex functions

---

1. Introduction

Let $\mathcal{A}$ denote the family of complex valued functions f which are holomorphic (analytic) in $\mathfrak{E} = \{z \in \mathbb{C} : |z| < 1\}$ and are normalized through the conditions $f(0) = 0$ and $f'(0) = 1$. That is, for $f \in \mathcal{A}$, one may have its series form

$$f(z) = z + \sum_{k=2}^{\infty} a_k z^k, \quad z \in \mathfrak{E}. \tag{1}$$

The class $\mathcal{UCV}$ is comprised those univalent functions $f(z)$ by which every circular arc $\mathfrak{C} \subset \mathfrak{E}$, with center at $\mathfrak{E}$, is mapped onto the convex arc and such functions are known as uniformly convex functions. This class was first introduced by Goodman [1]. The interesting analytic condition of class $\mathcal{UCV}$ was given in [2] and is stated as follows:

$$\mathcal{UCV} = \left\{ f \in \mathcal{A} : \Re\left\{1 + \frac{zf''(z)}{f'(z)}\right\} > \left|\frac{zf''(z)}{f'(z)}\right|, \ z \in \mathfrak{E} \right\}.$$

Kanas et al. [3] further generalized the class $\mathcal{UCV}$ by introducing the class of k-uniformly convex functions, named as k-$\mathcal{UCV}$, $k \geq 0$ and the class k-$\mathcal{ST}$ of corresponding k-starlike functions. The class k-$\mathcal{UCV}$ is defined as follows:

$$k\text{-}\mathcal{UCV} = \left\{ f \in \mathcal{A} : \Re\left\{1 + \frac{zf''(z)}{f'(z)}\right\} > k\left|\frac{zf''(z)}{f'(z)}\right|, \ z \in \mathfrak{E} \right\}.$$

They, in addition, discussed these classes geometrically and established connections with the conic domains

$$\mathcal{G}_k = \left\{u + iv\,; u^2 > k^2\left((u-1)^2 + v^2\right)\right\}. \tag{2}$$

It is important to mention that the class k-$\mathcal{UCV}$ was studied much earlier with some extra conditions but without geometrical interpretation. The class k-$\mathcal{UCV}$ is defined geometrically in a way that the common region of $\mathfrak{E}$ and the disk $|\mathfrak{D}| \leq k$ is mapped onto a convex domain by these univalent functions. Thus, the notion of convexity got the generalized version of k-uniform convexity. If $k = 0$, Then, the center $\mathfrak{D}$ shifts to origin and thus k-$\mathcal{UCV}$ takes the form of $\mathcal{C}$, the family of convex univalent functions.

The domain $\mathcal{G}_k$ represents conic regions for certain values of parameter k, that is, it gives an elliptic region for $k > 1$, the hyperbolic region (right branch) for $0 < k < 1$ and the parabolic region when $k = 1$. For more details, see [3–6]. The domain $\mathcal{G}_{k,\beta}$, which is generalization of $\mathcal{G}_k$ is given as:

$$\mathcal{G}_{k,\beta} = (1-\beta)\,\mathcal{G}_k + \beta,$$

where

$$\beta = \begin{cases} [0,1)\,, & \text{if} \quad k \in [0,1]\,, \\ \left[0, 1 - \frac{\sqrt{k^2-1}}{k}\right), & \text{if } k > 1. \end{cases} \tag{3}$$

For details, see [7]. The function which gives the boundary curves of these conical regions is denoted by $\varphi_{k,\beta}(z)$ which is holomorphic in $\mathfrak{E}$ and maps $\mathfrak{E}$ onto $\mathcal{G}_{k,\beta}$ such that $\varphi_{k,\beta}(z) = 1$ and $\varphi'_{k,\beta}(0) > 1$ and is defined as:

$$\varphi_{k,\beta}(z) = \begin{cases} \frac{1+(1-2\beta)z}{1-z}, & k = 0, \\ 1 + \frac{2(1-\beta)}{\pi^2}\left(\log\frac{1+\sqrt{z}}{1-\sqrt{z}}\right)^2, & k = 1, \\ 1 + \frac{2(1-\beta)}{1-k^2}\sinh^2\left[\left(\frac{2}{\pi}\cos^{-1}k\right)\tan^{-1}\mathrm{h}\sqrt{z}\right], & 0 < k < 1, \\ 1 + \frac{(1-\beta)}{k^2-1}\sin\left[\frac{\pi}{2R(t)}\int_0^{\frac{u(z)}{\sqrt{t}}}\frac{1}{\sqrt{1-x^2}\sqrt{1-(tx)^2}}dx\right] + \frac{1}{k^2-1}, & k > 1. \end{cases} \tag{4}$$

For the detailed study of this function, we refer the readers to see [3,6].

Let k-$\mathcal{P}(\beta)$ denote the family of holomorphic functions $q(z)$ with $q(0) = 1$ and $q(z) \prec \varphi_{k,\beta}(z)$ for $z \in \mathfrak{E}$, where the notion "$\prec$" denotes the familiar subordinations. It is pertinent to have

$$k\text{-}\mathcal{P}(\beta) \subset \mathcal{P}\left(\frac{k+\beta}{1+k}\right) \subset \mathcal{P},$$

where $\mathcal{P}$ is the family of functions with a positive real part. In addition, for $q \in k$-$\mathcal{P}(0)$, we have

$$|\arg q(z)| \leq \frac{\lambda\pi}{2},$$

where

$$\lambda = \frac{2}{\pi}\tan^{-1}(1/k). \tag{5}$$

Therefore, one may write

$$q(z) = h^{\lambda}(z), \quad h(z) \in \mathcal{P}.$$

Definition 1. *Let the function $q(z)$ be holomorphic in $\mathfrak{E}$ with $q(0) = 1$. Then, $q \in k\text{-}\mathcal{P}_m(\beta)$, if for $m \geq 2$, $k \geq 0$, $z \in \mathfrak{E}$ and β is given by Equation (3), we have*

$$q(z) = \left(\frac{m}{4} + \frac{1}{2}\right) q_1(z) - \left(\frac{m}{4} - \frac{1}{2}\right) q_2(z),$$

where $q_1(z), q_2(z) \in k\text{-}\mathcal{P}(\beta)$ [8].

Taking $k = 0$ and $\beta = 0$, the class $\mathcal{P}_m$ introduced by Pinchuk [9] is obtained. In addition, $k\text{-}\mathcal{P}_2(\beta) = k\text{-}\mathcal{P}(\beta)$, $0\text{-}\mathcal{P}_m(\beta) = \mathcal{P}_m(\beta)$ and $0\text{-}\mathcal{P}_2(\beta) = \mathcal{P}(\beta)$, where $\mathcal{P}_m(\beta)$ and $\mathcal{P}(\beta)$ were introduced in [9].

It is noted that $k\text{-}\mathcal{P}_m(\beta)$ is a convex set. Noor [8] introduced the classes $k\text{-}\mathcal{UV}^m(\beta)$ and $k\text{-}\mathcal{UR}^m(\beta)$ of k-uniformly bounded boundary and radius rotation of order β corresponding to the class $k\text{-}\mathcal{P}_m(\beta)$.

Now, we consider the following new subclasses of holomorphic functions.

Definition 2. *A function $f \in \mathcal{A}$ is known to be in $k\text{-}\mathcal{UR}_s^m(\beta)$, $k \geq 0$, $m \geq 2$ and β is given by Equation (3), if*

$$\frac{2zf'(z)}{f(z) - f(-z)} \in k\text{-}\mathcal{P}_m(\beta), \quad (z \in \mathfrak{E}).$$

Definition 3. *A function $f \in \mathcal{A}$ is known to be in the class $k\text{-}\mathcal{B}_s^m(\alpha, \beta)$, $\alpha > 0$, $k \geq 0$, $m \geq 2$ and β is given by Equation (3), if there exists $g \in k\text{-}\mathcal{UR}_s^m(\beta)$ such that*

$$\Re\left\{\frac{zf'(z)}{f(z)}\left(\frac{2f(z)}{g(z) - g(-z)}\right)^{\alpha}\right\} > k\left|\frac{zf'(z)}{f(z)}\left(\frac{2f(z)}{g(z) - g(-z)}\right)^{\alpha} - 1\right|,$$

or equivalently

$$\frac{zf'(z)}{f(z)}\left(\frac{2f(z)}{g(z) - g(-z)}\right)^{\alpha} \in k\text{-}\mathcal{P}(0).$$

It is pertinent to note that, by assigning specific values to parameters α, β, m and k in $k\text{-}\mathcal{UR}_s^m(\beta)$ and $k\text{-}\mathcal{B}_s^m(\alpha, \beta)$, several well-known subclasses of holomorphic and univalent functions are obtained, from which some are listed below:

1. $0\text{-}\mathcal{UR}_s^m(\beta) = \mathcal{R}_s^m(\beta)$, introduced by Bhargava et al. [10].
2. For $m = 2$ and $\alpha = 0$, we obtain the class $k\text{-}\mathcal{ST}_s(\beta)$, and $k\text{-}\mathcal{UK}_s(\beta)$, for details, we refer to [8].
3. $0\text{-}\mathcal{UR}_s^2(0) = \mathcal{S}_s^*$, for details, see [11].

Throughout the article, we shall consider, unless otherwise stated, that $m \geq 2$, $\alpha > 0$, $k \geq 0$ and β is given by Equation (3).

2. Preliminary Lemmas

Lemma 1. *[12] Let $k \in [0, \infty)$ and $\varphi_{k,\beta}(z)$ be defined by Equation (4). If*

$$\varphi_{k,\beta}(z) = 1 + Q_1 z + Q_2 z^2 + \cdots,$$

Then,

$$Q_1 = \begin{cases} \frac{2\beta A^2}{1-k^2} & 0 \leq k < 1, \\ \frac{8\beta}{\pi^2} & k = 1, \\ \frac{\pi^2 \beta}{4\sqrt{t}(k^2-1)R^2(t)(1+t)} & k > 1, \end{cases} \tag{6}$$

and

$$Q_2 = \begin{cases} \frac{(A^2+2)}{3} Q_1 & 0 \leq k < 1, \\ \frac{2}{3} Q_1 & k = 1, \\ \frac{4R^2(t)(t^2+6t+1)-\pi^2}{24\sqrt{t}R^2(t)(1+t)} Q_1 & k > 1, \end{cases} \tag{7}$$

where

$$A = \frac{2\cos^{-1} k}{\pi},$$

and $t \in (0,1)$ is taken such that $k = \cosh\left(\frac{\pi R'(t)}{R(t)}\right)$, $R(t)$ is the Legendre's complete elliptic integral of the first kind.

To proceed our main results, the following Lemmas proved by Pommerenke [13] and Golusin [14] are needed.

Lemma 2. *Let the holomorphic function $p \in \mathcal{P}$. Then [13]*

$$\frac{1}{2\pi}\int_0^{2\pi} |p(z)|^2 \, d\theta \leq \frac{1+3r^2}{1-r^2}.$$

Lemma 3. *Let the function $s_1(z)$ be starlike in $\mathfrak{E}$. Then [14],*
(i) : there exists ξ with $|\xi| = r$ such that for all z, $|z| = r$

$$|z-\xi| \, |s_1(z)| \leq \frac{2r^2}{1-r^2}$$

(ii)

$$\frac{r}{(1+r)^2} \leq s_1(z) \leq \frac{r}{(1-r)^2}.$$

3. Main Results

Theorem 1. *Let $f \in k\text{-}\mathcal{UR}_s^m(\beta)$. Then, the odd function*

$$\phi(z) = \frac{f(z) - f(-z)}{2}$$

belongs to $k\text{-}\mathcal{UR}^m(\beta)$.

Proof. Let $f \in k\text{-}\mathcal{UR}_s^m(\beta)$ and consider

$$\phi(z) = \frac{f(z) - f(-z)}{2}.$$

Logarithmic differentiation of the above relation yields

$$\frac{\phi'(z)}{\phi(z)} = \frac{f'(z) + f'(-z)}{f(z) - f(-z)},$$

or, equivalently,

$$\frac{z\phi'(z)}{\phi(z)} = \frac{1}{2}\left[q(z) + q(-z)\right],$$

where

$$q(z)=\frac{2zf'(z)}{f(z)-f(-z)} \quad \text{and} \quad q(-z)=\frac{2(-z)f'(-z)}{f(-z)-f(z)}.$$

Because $f(z)\in k\text{-}\mathcal{UR}_s^m(\beta)$, then, there exist $p_1(z),\ p_2(z)\in k\text{-}\mathcal{P}(\beta)$ such that

$$q(z)=\frac{2zf'(z)}{f(z)-f(-z)}=\left(\frac{m}{4}+\frac{1}{2}\right)p_1(z)-\left(\frac{m}{4}-\frac{1}{2}\right)p_2(z).$$

Therefore, we have

$$\frac{z\phi'(z)}{\phi(z)}=\left(\frac{m}{4}+\frac{1}{2}\right)\frac{p_1(z)+p_1(-z)}{2}-\left(\frac{m}{4}-\frac{1}{2}\right)\frac{p_2(z)+p_2(-z)}{2}.$$

Since $k\text{-}\mathcal{P}(\beta)$ is a convex set, we have

$$\frac{p_i(z)+p_i(-z)}{2}\in k\text{-}\mathcal{P}(\beta),\ \ i=1,2.$$

Thus, we have that

$$\frac{z\phi'(z)}{\phi(z)}\in k\text{-}\mathcal{P}_m(\beta),\ \ (z\in\mathfrak{E}),$$

and hence $\phi(z)\in k\text{-}\mathcal{UR}^m(\beta)$. □

When we take $m=2$, the following result, proved by Noor [8], is obtained.

Corollary 1. *Let $f\in k\text{-}\mathcal{ST}_s(\beta)$. Then,*

$$\phi(z)=\frac{1}{2}\left[f(z)-f(-z)\right]$$

belongs to $k\text{-}\mathcal{ST}(\beta)$.

Corollary 2. *Let $f\in\mathcal{R}_s^m(\beta)$. Then,*

$$\phi(z)=\frac{1}{2}\left[f(z)-f(-z)\right]$$

belongs to $\mathcal{R}^m(\beta)$.

Theorem 2. *If $f\in k\text{-}\mathcal{UR}_s^m(\beta)$, then*

$$f'(z)=\frac{p(z)}{2}\exp\left\{\int_0^z\frac{1}{2\xi}\left(p(\xi)+p(-\xi)-2\right)d\xi\right\} \tag{8}$$

for some $p(z)\in k\text{-}\mathcal{P}_m(\beta)$.

Proof. Let $f\in k\text{-}\mathcal{UR}_s^m(\beta)$. Then, by definition, one may have

$$\frac{2zf'(z)}{f(z)-f(-z)}=p(z),\ p(z)\in k\text{-}\mathcal{P}_m(\beta). \tag{9}$$

Simple computation leads us to

$$\frac{f(z)-f(-z)}{z}=\exp\left\{\int_0^z\frac{1}{2\xi}\left(p(\xi)+p(-\xi)-2\right)d\xi\right\}. \tag{10}$$

Using (9) in (10), we can easily obtain (8). □

When we take $m = 2$, the above result takes the following form, proved by Noor [8].

Corollary 3. *If $f \in k\text{-}\mathcal{ST}_s(\beta)$, then*

$$f'(z) = \frac{p(z)}{2} \exp\left\{ \int_0^z \frac{1}{2\xi} \left(p(\xi) + p(-\xi) - 2\right) d\xi \right\}$$

for $p(z) \in k\text{-}\mathcal{P}(\beta)$.

When $m = 2$, $k = 0$ and $\beta = 0$. Then, we have the following result, proved in [11].

Theorem 3. *Let $f \in k\text{-}\mathcal{UR}_s^m(\beta)$ be of the form (1). Then,*

$$|a_2| \leq \frac{m}{8} |Q_1|, \tag{11}$$

where Q_1 is given by (6).

Proof. Let $f \in k\text{-}\mathcal{UR}_s^m(\beta)$ and let it be of the form (9). Then,

$$f''(z) = \frac{p'(z)}{2} \exp\left\{ \int_0^z \frac{p(\xi) + p(-\xi) - 2}{2\xi} d\xi \right\} + \frac{p(z)}{2} \left\{ \frac{f(z) - f(-z)}{z} \right\}'. \tag{12}$$

From (12), we have $f''(0) = \frac{p'(0)}{2}$. It is well known that $|p'(0)|$ in the class $k\text{-}\mathcal{P}_m(\beta)$ is $|p'(0)| \leq \frac{m}{2} |Q_1|$, where Q_1 is given by (6). Thus, we get (11). □

Corollary 4. *The following disk is contained in the range of every function from $k\text{-}\mathcal{UR}_s^m(\beta)$.*

$$|w| < \frac{8}{16 + m|Q_1|},$$

where Q_1 is given by (6).

Proof. According to the Koebe theorem, each omitted value w satisfies

$$|w| > \frac{1}{2 + |a_2|}. \tag{13}$$

Using (13) and Theorem 3, we get the required result. □

By using the similar technique as used in [11], we have the following result.

Theorem 4. *Let $f \in k\text{-}\mathcal{UR}_s^m(\beta)$. Then, for $z = re^{i\theta}$ and $0 \leq \theta_1 < \theta_2 \leq 2\pi$,*

$$\int_{\theta_1}^{\theta_2} \Re\left(\frac{zf'(z)}{f(z)}\right) d\theta > -(1 - \beta_1)\left(\frac{m}{2} - 1\right)\pi,$$

for

$$\beta_1 = \frac{\beta + k}{1 + k}. \tag{14}$$

Theorem 5. *Let $f(z) \in k\text{-}\mathcal{B}_s^m(\alpha, \beta)$. Then, for $z = re^{i\theta}$,*

$$\int_{\theta_1}^{\theta_2} \Re J(\alpha, f(z))\, d\theta > -\left(\alpha(1-\beta_1)\left(\frac{m}{2}-1\right)+\sigma\right)\pi, \tag{15}$$

where $0 \leq \theta_1 < \theta_2 \leq 2\pi$, β_1 is defined by (14) and

$$J(\alpha, f(z)) = \left(1 + \frac{zf''(z)}{f'(z)}\right) + (\alpha - 1)\frac{zf'(z)}{f(z)}. \tag{16}$$

Proof. Let

$$\frac{zf'(z)}{f(z)}\left(\frac{2f(z)}{g(z)-g(-z)}\right)^{\alpha} = h^{\sigma}(z),$$

where $h(z) \in \mathcal{P}$,

$$\frac{(zf'(z))'}{f'(z)} + (\alpha-1)\frac{zf'(z)}{f(z)} = \frac{\sigma z h'(z)}{h(z)} + \frac{\alpha z \phi'(z)}{\phi(z)},$$

$$\int_{\theta_1}^{\theta_2}\left[\frac{(zf'(z))'}{f'(z)} + (\alpha-1)\frac{zf'(z)}{f(z)}\right] d\theta = \sigma\int_{\theta_1}^{\theta_2}\frac{zh'(z)}{h(z)}d\theta + \alpha\int_{\theta_1}^{\theta_2}\frac{z\phi'(z)}{\phi(z)}d\theta,$$

where $\phi(z)$ is an odd function of the form

$$\phi(z) = \frac{1}{2}\left[g(z) - g(-z)\right].$$

Since $g(z) \in k\text{-}\mathcal{UR}_s^m(\beta)$ and by Theorem 1 $\phi(z) \in k\text{-}\mathcal{UR}^m(\beta) \subset \mathcal{R}^m(\beta_1)$, therefore, by using Theorem 4, we have

$$\int_{\theta_1}^{\theta_2} \Re\left(\frac{z\phi'(z)}{\phi(z)}\right) d\theta > -(1-\beta_1)\left(\frac{m}{2}-1\right)\pi. \tag{17}$$

In addition, we observe that, for $h(z) \in \mathcal{P}$,

$$\begin{aligned}\frac{\partial}{\partial\theta}\arg h\left(re^{i\theta}\right) &= \frac{\partial}{\partial\theta}\Re\left\{-i\ln h\left(re^{i\theta}\right)\right\} \\ &= \Re\left\{\frac{re^{i\theta}h'\left(re^{i\theta}\right)}{h\left(re^{i\theta}\right)}\right\}.\end{aligned}$$

Therefore,

$$\int_{\theta_1}^{\theta_2}\Re\left\{\frac{re^{i\theta}h'\left(re^{i\theta}\right)}{h\left(re^{i\theta}\right)}\right\}d\theta = \arg h\left(re^{i\theta_2}\right) - \arg h\left(re^{i\theta_1}\right),$$

which takes the form

$$\left|\int_{\theta_1}^{\theta_2}\Re\left\{\frac{re^{i\theta}h'\left(re^{i\theta}\right)}{h\left(re^{i\theta}\right)}\right\}d\theta\right| = \left|\arg h\left(re^{i\theta_2}\right) - \arg h\left(re^{i\theta_1}\right)\right|.$$

This implies that

$$\max_{h\in\mathcal{P}(\beta)}\left|\int_{\theta_1}^{\theta_2}\Re\left(\frac{re^{i\theta}h'\left(re^{i\theta}\right)}{h\left(re^{i\theta}\right)}\right)d\theta\right| = \max_{h\in\mathcal{P}(\beta)}\left|\arg h\left(re^{i\theta_2}\right) - \arg h\left(re^{i\theta_1}\right)\right|. \tag{18}$$

Since $h(z) \in \mathcal{P}$, thus

$$\left| h(z) - \frac{1+r^2}{1-r^2} \right| \leq \frac{2r}{1-r^2}.$$

Thus, the values $h(z)$ are contained in the circle of Apollonius with diameter end points $\frac{1-r}{1-r}$ and $\frac{1+r}{1-r}$ and radius $\frac{2r}{1-r^2}$. Thus, the maximum of $|\arg h(z)|$ is attained at points where tangent ray from origin to the circle can be drawn, that is, when

$$\arg h(z) = \pm \sin^{-1}\left(\frac{2r}{1-r^2}\right).$$

Now,

$$\max_{h \in \mathcal{P}(\beta)} \left| \int_{\theta_1}^{\theta_2} \Re\left(\frac{re^{i\theta} h'(re^{i\theta})}{h(re^{i\theta})}\right) d\theta \right| \leq 2\sin^{-1}\left(\frac{2r}{1-r^2}\right).$$

This implies that

$$\max_{h \in \mathcal{P}(\beta)} \left| \int_{\theta_1}^{\theta_2} \Re\left(\frac{re^{i\theta} h'(re^{i\theta})}{h(re^{i\theta})}\right) d\theta \right| \leq \pi - 2\cos^{-1}\left(\frac{2r}{1-r^2}\right). \tag{19}$$

Thus,

$$\int_{\theta_1}^{\theta_2} \Re J(\alpha, f(z))\, d\theta > -\left(\alpha(1-\beta_1)\left(\frac{m}{2}-1\right)+\sigma\right)\pi + 2\sigma\cos^{-1}\left(\frac{2r}{1-r^2}\right),$$

which gives

$$\int_{\theta_1}^{\theta_2} \Re J(\alpha, f(z))\, d\theta > -\left(\alpha(1-\beta_1)\left(\frac{m}{2}-1\right)+\sigma\right)\pi, \qquad (r \to 1).$$

This completes the proof. □

For talking $k = 0$, we obtain the integral representation for the class $\mathcal{T}_s^m(\beta)$.

Corollary 5. *Let $f \in \mathcal{T}_s^m(\beta)$. Then, for $z = re^{i\theta}$,*

$$\int_{\theta_1}^{\theta_2} \Re\left(1 + \frac{zf''(z)}{f'(z)}\right) d\theta > -\sigma\pi,$$

where $0 \leq \theta_1 < \theta_2 \leq 2\pi$.

Theorem 6. *Let $f \in k\text{-}\mathcal{B}_s^m(\alpha, \beta)$. Then, for $\frac{\alpha}{2-\sigma}(m+2)(1-\beta_1) > 1$,*

$$L_r f(z) \leq \begin{cases} C(\alpha, \sigma, m, \beta_1)\, \mathfrak{M}^{1-\alpha}(r) \left(\frac{1}{1-r}\right)^{\alpha\left(\frac{m}{2}+1\right)(1-\beta_1)-1+\sigma}, & 0 < \alpha \leq 1, \\ C(\alpha, \sigma, m, \beta_1)\, \mathfrak{m}^{1-\alpha}(r) \left(\frac{1}{1-r}\right)^{\alpha\left(\frac{m}{2}+1\right)(1-\beta_1)-1+\sigma}, & \alpha > 1, \end{cases}$$

where $\mathfrak{M}(r) = \max_{|z|=r} |f(z)|$, $\mathfrak{m}(r) = \min_{|z|=r} |f(z)|$ and $C(\alpha, \sigma, m, \beta_1)$ is a constant depending upon α, σ, m and β_1 only.

Proof. We know that

$$L_r f(z) = \int_0^{2\pi} |zf'(z)|\, d\theta, \ z = re^{i\theta},\ 0 < r < 1.$$

Since $f \in k\text{-}\mathcal{B}_s^m(\alpha, \beta)$, thus

$$\frac{zf'(z)}{f^{1-\alpha}(z)} \left(\frac{2}{g(z) - g(-z)} \right)^{\alpha} = p^{\sigma}(z), \ \ p(z) \in \mathcal{P}.$$

By Theorem 1, we have for $g \in k\text{-}\mathcal{UR}_s^m(\beta)$, the function

$$\phi(z) = \frac{1}{2}\left[g(z) - g(-z)\right] \in k\text{-}\mathcal{UR}^m(\beta),$$

which yields

$$zf'(z) = (f(z))^{1-\alpha} (\phi(z))^{\alpha} p^{\sigma}(z).$$

Therefore, we have

$$\begin{aligned} L_r f(z) &\leq \int_0^{2\pi} |f(z)|^{1-\alpha} |\phi(z)|^{\alpha} |p(z)|^{\sigma} d\theta \\ &\leq \mathfrak{M}^{1-\alpha}(r) \int_0^{2\pi} |\phi(z)|^{\alpha} |p(z)|^{\sigma} d\theta. \end{aligned}$$

Since $\phi(z) \in k\text{-}\mathcal{UR}^m(\beta) \subset \mathcal{R}^m(\beta_1)$, we have

$$\phi(z) = \frac{(s_1(z))^{\left(\frac{m}{4}+\frac{1}{2}\right)}}{(s_2(z))^{\left(\frac{m}{4}-\frac{1}{2}\right)}}, \ \ s_1, s_2 \in k\text{-}\mathcal{UR}^2(\beta).$$

Since $k\text{-}\mathcal{UR}^2(\beta) \subset \mathcal{S}^*(\beta_1)$, so we can write

$$s_i(z) = z\left(\frac{\phi_i(z)}{z} \right)^{1-\beta_1}, \ \ \text{for } i = 1,2 \text{ and } \phi_i(z) \in \mathcal{S}^*.$$

Thus, for odd functions $s_1(z), s_2(z) \in \mathcal{S}^*(\beta_1)$, we have

$$\begin{aligned} L_r(f(z)) &\leq \mathfrak{M}^{1-\alpha}(r) \int_0^{2\pi} |z|^{\beta_1} \left| \frac{(\phi_1(z))^{(1-\beta_1)\left(\frac{m}{4}+\frac{1}{2}\right)}}{(\phi_2(z))^{(1-\beta_1)\left(\frac{m}{4}-\frac{1}{2}\right)}} \right|^{\alpha} |p(z)|^{\sigma} d\theta \\ &\leq \mathfrak{M}^{1-\alpha}(r) \int_0^{2\pi} \frac{|(\phi_1(z))|^{\alpha\left(\frac{m}{4}+\frac{1}{2}\right)(1-\beta_1)}}{|(\phi_2(z))|^{\alpha\left(\frac{m}{4}-\frac{1}{2}\right)(1-\beta_1)}} |p(z)|^{\sigma} d\theta \\ &\leq \mathfrak{M}^{1-\alpha}(r) \int_0^{2\pi} \frac{2^{\alpha\left(\frac{m}{2}-1\right)(1-\beta_1)}}{r^{\alpha\left(\frac{m}{4}-\frac{1}{2}\right)(1-\beta_1)}} |(\phi_1(z))|^{\alpha\left(\frac{m}{4}+\frac{1}{2}\right)(1-\beta_1)} |p(z)|^{\sigma} d\theta \\ &= \frac{\mathfrak{M}^{1-\alpha}(r)\, 2^{\alpha\left(\frac{m}{2}-1\right)(1-\beta_1)}}{r^{\alpha\left(\frac{m}{4}-\frac{1}{2}\right)(1-\beta_1)}} \int_0^{2\pi} |(\phi_1(z))|^{\alpha\left(\frac{m}{4}+\frac{1}{2}\right)(1-\beta_1)} |p(z)|^{\sigma} d\theta. \end{aligned}$$

Now, by making use of Holder's inequality, with $m_1 = 2/2-\sigma$ and $m_2 = 2/\sigma$ such that $(1/m_1) + (1/m_2) = 1$, we have

$$L_r(f(z)) \leq \frac{\mathfrak{M}^{1-\alpha}(r)\ \pi\, 2^{\alpha\left(\frac{m}{2}-1\right)(1-\beta_1)+1}}{r^{\alpha\left(\frac{m}{4}-\frac{1}{2}\right)(1-\beta_1)}} \left(\frac{1}{2\pi}\int_0^{2\pi} |p(z)|^2\, d\theta\right)^{\frac{\sigma}{2}}$$

$$\times \left(\frac{1}{2\pi}\int_0^{2\pi} |\phi_1(z)|^{\frac{\alpha}{2-\sigma}\left(\frac{m}{2}+1\right)(1-\beta_1)}\, d\theta\right)^{\frac{2-\sigma}{2}}.$$

By using Lemma 2 and distortion results, we obtain

$$L_r(f(z)) \leq \frac{\mathfrak{M}^{1-\alpha}(r)\ \pi\, 2^{\alpha\left(\frac{m}{2}-1\right)(1-\beta_1)+1}}{r^{\alpha\left(\frac{m}{4}-\frac{1}{2}\right)(1-\beta_1)}} \left(\frac{1+3r^2}{1-r^2}\right)^{\frac{\sigma}{2}} \left(\frac{1}{2\pi}\int_0^{2\pi} \frac{r^{\frac{\alpha}{2-\sigma}\left(\frac{m}{2}+1\right)(1-\beta_1)}}{|1-re^{i\theta}|^{\frac{2\alpha}{2-\sigma}\left(\frac{m}{2}+1\right)(1-\beta_1)}}\, d\theta\right)^{\frac{2-\sigma}{2}}$$

$$= \frac{\mathfrak{M}^{1-\alpha}(r)\ \pi\, 2^{\alpha\left(\frac{m}{2}-1\right)(1-\beta_1)+1}}{r^{\alpha\left(\frac{m}{4}-\frac{1}{2}\right)(1-\beta_1)}} r^{\frac{\alpha}{2-\sigma}\left(\frac{m}{2}+1\right)(1-\beta_1)} \left(\frac{1}{2\pi}\int_0^{2\pi} \frac{1}{|1-re^{i\theta}|^{\frac{2\alpha}{2-\sigma}\left(\frac{m}{2}+1\right)(1-\beta_1)}}\, d\theta\right)^{\frac{2-\sigma}{2}} \left(\frac{1+3r^2}{1-r^2}\right)^{\frac{\sigma}{2}}.$$

This implies that

$$\begin{aligned}
L_r(f(z)) &\leq \mathfrak{M}^{1-\alpha}(r)\ \pi^{\frac{\sigma}{2}}\, 2^{\alpha\left(\frac{m}{2}-1\right)(1-\beta_1)+1+\sigma} \left(\frac{1}{(1-r)^{\frac{2\alpha}{2-\sigma}\left(\frac{m}{2}+1\right)(1-\beta_1)-1}}\right)^{\frac{2-\sigma}{2}} \left(\frac{1}{1-r}\right)^{\frac{\sigma}{2}} \\
&= \mathfrak{M}^{1-\alpha}(r)\ \pi^{\frac{\sigma}{2}}\, 2^{\alpha\left(\frac{m}{2}-1\right)(1-\beta_1)+1+\sigma} \left(\frac{1}{1-r}\right)^{\left(\frac{2-\sigma}{2}\right)\left(\frac{2\alpha}{2-\sigma}\left(\frac{m}{2}+1\right)(1-\beta_1)-1\right)} \left(\frac{1}{1-r}\right)^{\frac{\sigma}{2}} \\
&= C(\alpha,\sigma,m,\beta_1)\, \mathfrak{M}^{1-\alpha}(r) \left(\frac{1}{1-r}\right)^{\alpha\left(\frac{m}{2}+1\right)(1-\beta_1)-1+\sigma},
\end{aligned}$$

where

$$C(\alpha,\sigma,m,\beta_1) = \pi^{\frac{\sigma}{2}}\, 2^{\alpha\left(\frac{m}{2}-1\right)(1-\beta_1)+1+\sigma}$$

is a constant depending upon α, σ, m and β_1 only. Similarly, for $\alpha > 1$, we have

$$L_r(f(z)) \leq C(\alpha,\sigma,m,\beta_1)\ \mathfrak{m}^{\alpha-1}(r) \left(\frac{1}{1-r}\right)^{\alpha\left(\frac{m}{2}+1\right)(1-\beta_1)-1+\sigma}.$$

□

Theorem 7. *Let $f \in k\text{-}\mathcal{B}_s^m(\alpha,\beta)$. Then, for $n \geq 2$ and $\frac{\alpha}{2-\sigma}(m+2)(1-\beta_1) > 1$,*

$$|a_n| \leq \begin{cases} C_1(\alpha,\sigma,m,\beta_1)\ \mathfrak{M}^{1-\alpha}(n)\,(n)^{\alpha\left(\frac{m}{2}+1\right)(1-\beta_1)-2+\sigma}, & 0 < \alpha \leq 1, \\ C_1(\alpha,\sigma,m,\beta_1)\ \mathfrak{m}^{\alpha-1}(n)\,(n)^{\alpha\left(\frac{m}{2}+1\right)(1-\beta_1)-2+\sigma}, & \alpha > 1, \end{cases}$$

where β_1 is given by (14) and $\mathfrak{m}$, $\mathfrak{M}$ are the same as in Theorem 6 and $C_1(\alpha,\sigma,m,\beta_1)$ is a constant.

Proof. Since $z = re^{i\theta}$, Cauchy theorem gives

$$na_n = \frac{1}{2\pi r^n}\int_0^{2\pi} zf'(z)\, e^{-in\theta}\, d\theta,$$

which reduces to

$$\begin{aligned} |na_n| &= \left| \frac{1}{2\pi r^n} \int_0^{2\pi} zf'(z)\, e^{-in\theta} d\theta \right| \\ &\leq \frac{1}{2\pi r^n} \int_0^{2\pi} \left| zf'(z)\, e^{-in\theta} \right| d\theta. \end{aligned}$$

Therefore,

$$n\,|a_n| \leq \frac{1}{2\pi r^n} L_r f(z).$$

Now, using Theorem 6 for $0 < \alpha \leq 1$, we have

$$n\,|a_n| \leq \frac{1}{2\pi r^n} C(\alpha,\sigma,m,\beta_1)\; \mathfrak{M}^{1-\alpha}(r) \left(\frac{1}{1-r}\right)^{\alpha\left(\frac{m}{2}+1\right)(1-\beta_1)-1+\sigma}.$$

Putting $r = 1 - \frac{1}{n}$, we have

$$|a_n| \leq C_1(\alpha,\sigma,m,\beta_1)\; \mathfrak{M}^{1-\alpha}(r)\,(n)^{\alpha\left(\frac{m}{2}+1\right)(1-\beta_1)-2+\sigma}.$$

Similarly, we obtain the required result for $\alpha > 1$. □

Theorem 8. *Let $f \in k\text{-}\mathcal{B}_s^m(\alpha,\beta)$. Then, for $\frac{\alpha}{2-\sigma}(m+2)(1-\beta_1) > 1$,*

$$||a_{n+1}| - |a_n|| \leq \begin{cases} \mathfrak{M}^{1-\alpha}(r)\; C_2(\alpha,\sigma,m,\beta_1)\,(n)^{\alpha\left(\frac{m}{2}+1\right)(1-\beta_1)+\sigma-3}, & 0 < \alpha \leq 1, \\ \mathfrak{m}^{1-\alpha}(r)\; C_2(\alpha,\sigma,m,\beta_1)\,(n)^{\alpha\left(\frac{m}{2}+1\right)(1-\beta_1)+\sigma-3}, & \alpha > 1, \end{cases}$$

where $\mathfrak{m}(r) = \min_{|z|=r} |f(z)|$, $\mathfrak{M}(r) = \max_{|z|=r} |f(z)|$ and $C_2(\alpha,\sigma,m,\beta_1)$ is a constant depending upon α,σ,m and β_1 only.

Proof. We know that, for $\xi \in \mathfrak{E}$ and $n \geq 1$,

$$|(n+1)\,\xi a_{n+1} - na_n| \leq \frac{1}{2\pi r^{n+1}} \int_0^{2\pi} |z-\xi|\,|zf'(z)|\, d\theta, \quad z = re^{i\theta},\ 0<r<1,\ 0 \leq \theta \leq 2\pi.$$

As $f \in k\text{-}\mathcal{B}_s^m(\alpha,\beta)$, thus

$$\frac{zf'(z)}{f(z)} \left(\frac{2f(z)}{g(z)-g(-z)}\right)^{\alpha} = p^{\sigma}(z), \quad p \in \mathcal{P}.$$

From Theorem 4, we have

$$\phi(z) = \frac{1}{2}\left[g(z)-g(-z)\right] \in k-\mathcal{UR}^m(\beta) \qquad \text{for } g \in k-\mathcal{UR}_s^m(\beta).$$

This leads us to

$$zf'(z) = (f(z))^{1-\alpha}\,(\phi(z))^{\alpha}\, p^{\sigma}(z).$$

Thus, for $\xi \in \mathfrak{E}$ and $n \geq 1$, we have

$$\left|(n+1)\,\xi a_{n+1} - n a_n\right| \leq \frac{M^{1-\alpha}(r)}{2\pi r^{n+1}} \int\limits_0^{2\pi} |z-\xi|\, |\phi(z)|^{\alpha}\, |p(z)|^{\sigma}\, d\theta.$$

Since $\phi(z) \in k\text{-}\mathcal{UR}^m(\beta) \subset \mathcal{R}^m(\beta_1)$, therefore, for $\phi_1(z), \phi_2(z) \in \mathcal{S}^*$, we have

$$\begin{aligned} \left|\xi a_{n+1}(n+1) - n a_n\right| &\leq \frac{M^{1-\alpha}(r)}{2\pi r^{n+1}} \int\limits_0^{2\pi} |z|^{\alpha\beta_1} |z-\xi| \left| \frac{(\phi_1(z))^{(1-\beta_1)\left(\frac{m}{4}+\frac{1}{2}\right)}}{(\phi_2(z))^{(1-\beta_1)\left(\frac{m}{4}-\frac{1}{2}\right)}} \right|^{\alpha} |p(z)|^{\sigma}\, d\theta \\ &\leq \frac{M^{1-\alpha}(r)}{2\pi r^{n+1}} \int\limits_0^{2\pi} |z-\xi| \frac{|(\phi_1(z))|^{\alpha(1-\beta_1)\left(\frac{m}{4}+\frac{1}{2}\right)}}{|(\phi_2(z))|^{\alpha(1-\beta_1)\left(\frac{m}{4}-\frac{1}{2}\right)}} |p(z)|^{\sigma}\, d\theta. \end{aligned}$$

Using Lemma 3(i), we have

$$\begin{aligned} &\left|(n+1)\,\xi a_{n+1} - n a_n\right| \leq \frac{2^{\alpha\left(\frac{m}{2}-1\right)(1-\beta_1)} M^{1-\alpha}(r)}{2\pi r^{n+1+\alpha\left(\frac{m}{4}-\frac{1}{2}\right)(1-\beta_1)}} \\ &\times \int\limits_0^{2\pi} |z-\xi|\, |(\phi_1(z))|\, |(\phi_1(z))|^{\alpha(1-\beta_1)\left(\frac{m}{4}+\frac{1}{2}\right)-1} |p(z)|^{\sigma} \mathrm{d}\theta. \end{aligned}$$

Now, using Lemma 3(ii), we have

$$\left|(n+1)\,\xi a_{n+1} - n a_n\right| \leq \frac{2^{\alpha\left(\frac{m}{2}-1\right)(1-\beta_1)} M^{1-\alpha}(r)}{2\pi r^{n-1+\alpha\left(\frac{m}{4}-\frac{1}{2}\right)(1-\beta_1)}(1-r)} \int\limits_0^{2\pi} |(\phi_1(z))|^{\alpha(1-\beta_1)\left(\frac{m}{4}+\frac{1}{2}\right)-1} |p(z)|^{\sigma} \mathrm{d}\theta.$$

Now, using Cauchy–Schwarz inequality, we have

$$\begin{aligned} \left|(n+1)\,\xi a_{n+1} - n a_n\right| &\leq \frac{2^{\alpha\left(\frac{m}{2}-1\right)(1-\beta_1)} M^{1-\alpha}(r)}{r^{n-1+\alpha\left(\frac{m}{4}-\frac{1}{2}\right)(1-\beta_1)}(1-r)} \left(\frac{1}{2\pi}\int\limits_0^{2\pi} |p(z)|^2\, d\theta\right)^{\frac{\sigma}{2}} \\ &\times \left(\frac{1}{2\pi}\int\limits_0^{2\pi} |(\phi_1(z))|^{\frac{\alpha(1-\beta_1)\left(\frac{m}{2}+1\right)-2}{2-\sigma}}\, d\theta\right)^{\frac{2-\sigma}{2}}. \end{aligned}$$

By using Lemma 2 and distortion results, we obtain

$$\begin{aligned} &\left|(n+1)\,\xi a_{n+1} - n a_n\right| \leq \frac{2^{\alpha\left(\frac{m}{2}-1\right)(1-\beta_1)} M^{1-\alpha}(r)}{r^{n-1+\alpha\left(\frac{m}{4}-\frac{1}{2}\right)(1-\beta_1)}(1-r)} \left(\frac{1+3r^2}{1-r^2}\right)^{\frac{\sigma}{2}} \\ &\times \left(\frac{1}{2\pi}\int\limits_0^{2\pi} \frac{r^{\frac{1}{2-\sigma}\left\{\alpha\left(\frac{m}{2}+1\right)(1-\beta_1)-2\right\}}}{|1-re^{i\theta}|^{\frac{2}{2-\sigma}\left\{\left(\alpha\frac{m}{2}+1\right)(1-\beta_1)-2\right\}}}\, d\theta\right)^{\frac{2-\sigma}{2}} \\ &\leq \frac{C_2(\alpha,\sigma,m,\beta_1) M^{1-\alpha}(r) r^{\alpha(1-\beta_1)-n}}{(1-r)^{1+\frac{\sigma}{2}}} \left(\frac{1}{(1-re^{i\theta})^{\frac{2}{2-\sigma}\left\{\alpha\left(\frac{m}{2}+1\right)(1-\beta_1)-2\right\}-1}}\right)^{\frac{2-\sigma}{2}} \\ &\leq \frac{C_2(\alpha,\sigma,m,\beta_1) M^{1-\alpha}(r)}{r^{n-1}(1-r)^{1+\frac{\sigma}{2}}} \left(\frac{1}{(1-r)^{\alpha\left(\frac{m}{2}+1\right)(1-\beta_1)+\frac{\sigma}{2}-3}}\right) \\ &\leq \frac{C_2(\alpha,\sigma,m,\beta_1) M^{1-\alpha}(r)}{r^{n-1}(1-r)^{\alpha\left(\frac{m}{2}+1\right)(1-\beta_1)+\sigma-2}}, \end{aligned}$$

where $C_2(\alpha,\sigma,m,\beta_1)$ is a constant. Now, putting $|\xi| = \frac{n}{n+1}$, we obtain

$$n\left|\left|a_{n+1}\right| - \left|a_n\right|\right| \leq \frac{C_2(\alpha,\sigma,m,\beta_1)\, M^{1-\alpha}(r)}{r^{n-1}(1-r)^{\alpha\left(\frac{m}{2}+1\right)(1-\beta_1)+\sigma-2}}.$$

Now, taking $r = 1 - \frac{1}{n}$ $(n \to \infty)$,we have

$$C_2(\alpha, \sigma, m, \beta_1)\,\mathfrak{M}^{1-\alpha}(r)\,(n)^{\alpha\left(\frac{m}{2}+1\right)(1-\beta_1)+\sigma-3}, \quad 0 < \alpha \leq 1.$$

Similarly for $\alpha > 1$, we have

$$||a_{n+1}| - |a_n|| \leq C_2(\alpha, \sigma, m, \beta_1)\ m^{\alpha-1}(r)\,(n)^{\alpha\left(\frac{m}{2}+1\right)(1-\beta_1)+\sigma-3}.$$

Thus, the result follows. □

Theorem 9. *Let $f \in k\text{-}\mathcal{B}_s^m(\alpha, \beta)$ for $\alpha > 0$. Then, $f(z)$ is $\frac{1}{\alpha}$−convex for $|z| < r_m^*$,*

$$r_m^* = \frac{2\alpha}{(\alpha m + 2\sigma - \alpha\beta_1 m) + \sqrt{(\alpha m + 2\sigma - \alpha\beta_1 m)^2 - 4\alpha^2(1 - 2\beta_1)}}, \quad \alpha > 0.$$

Proof. Let

$$zf'(z) = (f(z))^{1-\alpha}(\phi(z))^{\alpha} h^{\sigma}(z),$$

where $\frac{g(z)-g(-z)}{2} = \phi(z) \in k\text{-}\mathcal{UR}^m(\beta) \subset \mathcal{R}^m(\beta_1)$ and $h(z) \in \mathcal{P}$. Differentiating logarithmically, we obtain

$$\frac{1}{\alpha}\left(\frac{(zf'(z))'}{f'(z)}\right) + \left(1 - \frac{1}{\alpha}\right)\frac{zf'(z)}{f(z)} = \frac{z\phi'(z)}{\phi(z)} + \frac{\sigma}{\alpha}\frac{zh'(z)}{h(z)}.$$

We can write

$$\begin{aligned}\Re\left\{\frac{1}{\alpha}\left(\frac{(zf'(z))'}{f'(z)}\right) + \left(1 - \frac{1}{\alpha}\right)\frac{zf'(z)}{f(z)}\right\} &= \Re\left(\frac{z\phi'(z)}{\phi(z)}\right) + \frac{\sigma}{\alpha}\Re\left(\frac{zh'(z)}{h(z)}\right) \\ &> \Re\left(\frac{z\phi'(z)}{\phi(z)}\right) - \frac{\sigma}{\alpha}\left|\frac{zh'(z)}{h(z)}\right|.\end{aligned}$$

Now, using the distortion results for the classes $\mathcal{R}^m(\beta_1)$ and $\mathcal{P}$, we have

$$\begin{aligned}\Re\left\{\frac{1}{\alpha}\left(\frac{(zf'(z))'}{f'(z)}\right) + \left(1 - \frac{1}{\alpha}\right)\frac{zf'(z)}{f(z)}\right\} &\geq \beta_1 + \frac{(1-\beta_1)(1 - mr + r^2)}{1 - r^2} - \frac{2\sigma r}{\alpha(1 - r^2)} \\ &= \frac{\alpha\beta_1(1 - r^2) + \alpha(1 - \beta_1)(1 - mr + r^2) - 2\sigma r}{\alpha(1 - r^2)} \\ &\geq \frac{\alpha(1 - 2\beta_1)r^2 - (\alpha m + 2\sigma - \alpha\beta_1 m)r + \alpha}{\alpha(1 - r^2)},\end{aligned}$$

taking

$$\alpha(1 - 2\beta_1)r^2 - (\alpha m + 2\sigma - \alpha\beta_1 m)r + \alpha = 0,$$

$$r_m^* = \frac{(\alpha m + 2\sigma - \alpha\beta_1 m) \pm \sqrt{(\alpha m + 2\sigma - \alpha\beta_1 m)^2 - 4\alpha^2(1 - 2\beta_1)}}{2\alpha(1 - 2\beta_1)}.$$

Since $0 \leq r < 1$,

$$\begin{aligned}r_m^* &= \frac{(\alpha m + 2\sigma - \alpha\beta_1 m) - \sqrt{(\alpha m + 2\sigma - \alpha\beta_1 m)^2 - 4\alpha^2(1 - 2\beta_1)}}{2\alpha(1 - 2\beta_1)} \\ &= \frac{2\alpha}{(\alpha m + 2\sigma - \alpha\beta_1 m) + \sqrt{(\alpha m + 2\sigma - \alpha\beta_1 m)^2 - 4\alpha^2(1 - 2\beta_1)}}, \quad \alpha > 0.\end{aligned}$$

This completes the proof. □

4. Conclusions

In this article, we have presented certain analytic functions defined by bounded radius rotations associated with conic domain. We have investigated many geometric properties like coefficient estimate, radii problems, arc length, integral representation, inclusion results and growth rate of coefficients of Taylor's series representation. By varying the parameters in results, several well-known results in literature have been shown as special cases.

Author Contributions: Conceptualization, S.M. and H.M.S.; Formal analysis, S.M. and H.M.S.; Funding acquisition, S.M.; Investigation, S.M. and S.N.M.; Methodology, S.M. and S.N.M.; Supervision, H.M.S.; Validation, H.M.S.; Writing–original draft, S.N.M.; Writing–review and editing, S.N.M.

Funding: This work is partially supported by Sarhad University of Science and I.T, Ring Road, Peshawar 25000, Pakistan.

Conflicts of Interest: The authors have no conflict of interest.

References

1. Goodman, A.W. On Uniformly Convex Functions. *Ann. Pol. Math.* **1991**, *56*, 87–92. [CrossRef]
2. Ma, W.; Minda, D. Uniformly convex functions. *Ann. Pol. Math.* **1992**, *57*, 165–175. [CrossRef]
3. Kanas, S.; Wisniowska, A. Conic regions and k-uniform convexity. *J. Comput. Appl. Math.* **1999**, *105*, 327–336. [CrossRef]
4. El-Ashwah, R.; Thomas, D.K. Some subclasses of close-to-convex functions. *J. Ramanujan Math. Soc.* **1987**, *2*, 85–100.
5. Ali, R.M.; Ravichandran, V. Uniformly Convex and Uniformly Starlike Functions. *Ramanujan Math. Newsl.* **2011**, *21*, 16–30.
6. Kanas, S.; Srivastava, H.M. Linear operators associated with *k*-uniformly convex functions. *Integral Transforms Spec. Funct.* **2000**, *9*, 121–132. [CrossRef]
7. Noor, K.I.; Malik, S.N. On a new class of analytic functions associated with conic domain. *Comput. Math. Appl.* **2011**, *62*, 367–375. [CrossRef]
8. Noor, K.I. On uniformly univalent functions with respect to symmetrical points. *J. Inequal. Appl.* **2014**, *2014*, 254. [CrossRef]
9. Pinchuk, B. Functions of bounded boundary rotations. *Isr. J. Math.* **1971**, *10*, 6–16. [CrossRef]
10. Bhargava, S.; Rao, S.N. On a class of functions unifying the classes of Paatero, Robertson and others. *Int. J. Math. Math. Sci.* **1988**, *11*, 251–258. [CrossRef]
11. Sakaguchi, K. On a certain univalent mapping. *J. Math. Soc. Jpn.* **1959**, *11*, 72–75. [CrossRef]
12. Sim, Y.J.; Kwon, O.S.; Cho, N.E.; Srivastava, H.M. Some classes of analytic functions associated with conic regions. *Taiwan. J. Math.* **2012**, *16*, 387–408. [CrossRef]
13. Pommerenke, C. On close-to-convex analytic functions. *Trans. Am. Math. Soc.* **1965**, *114*, 176–186. [CrossRef]
14. Golusin, G.M. On distortion theorem and coefficients of univalent functions. *Rec. Math.* **1946**, *19*, 183–203.

© 2019 by the authors. Licensee MDPI, Basel, Switzerland. This article is an open access article distributed under the terms and conditions of the Creative Commons Attribution (CC BY) license (http://creativecommons.org/licenses/by/4.0/).

symmetry MDPI

Article

Existence Theory for Nonlinear Third-Order Ordinary Differential Equations with Nonlocal Multi-Point and Multi-Strip Boundary Conditions

Ahmed Alsaedi [1], Mona Alsulami [1,2], Hari M. Srivastava [3,4,*], Bashir Ahmad [1] and Sotiris K. Ntouyas [1,5]

1 Nonlinear Analysis and Applied Mathematics (NAAM)-Research Group, Department of Mathematics, Faculty of Science, King Abdulaziz University, P.O. Box 80203, Jeddah 21589, Saudi Arabia; aalsaedi@hotmail.com (A.A.); bashirahmad_qau@yahoo.com (B.A.)
2 Department of Mathematics, Faculty of Science, University of Jeddah, P.O. Box 80327, Jeddah 21589, Saudi Arabia; mralsolami@uj.edu.sa
3 Department of Mathematics and Statistics, University of Victoria, Victoria, BC V8W 3R4, Canada
4 Department of Medical Research, China Medical University Hospital, China Medical University, Taichung 40402, Taiwan
5 Department of Mathematics, University of Ioannina, 451 10 Ioannina, Greece; sntouyas@uoi.gr
* Correspondence: harimsri@math.uvic.ca

Received: 19 January 2019; Accepted: 20 February 2019; Published: 22 February 2019

Abstract: We investigate the solvability and Ulam stability for a nonlocal nonlinear third-order integro-multi-point boundary value problem on an arbitrary domain. The nonlinearity in the third-order ordinary differential equation involves the unknown function together with its first- and second-order derivatives. Our main results rely on the modern tools of functional analysis and are well illustrated with the aid of examples. An analogue problem involving non-separated integro-multi-point boundary conditions is also discussed.

Keywords: nonlinear boundary value problem; nonlocal; multi-point; multi-strip; existence; Ulam stability

1. Introduction

Consider a third-order ordinary differential equation of the form:

$$u'''(t) = f(t, u(t), u'(t), u''(t)), \ a < t < T, \ a, T \in \mathbb{R}, \tag{1}$$

supplemented with the boundary conditions:

$$\begin{aligned} \int_a^T u(s)ds &= \sum_{j=1}^m \gamma_j u(\sigma_j) + \sum_{i=1}^p \xi_i \int_{\rho_i}^{\rho_{i+1}} u(s)ds, \\ \int_a^T u'(s)ds &= \sum_{j=1}^m \mu_j u'(\sigma_j) + \sum_{i=1}^p \eta_i \int_{\rho_i}^{\rho_{i+1}} u'(s)ds, \\ \int_a^T u''(s)ds &= \sum_{j=1}^m \nu_j u''(\sigma_j) + \sum_{i=1}^p \omega_i \int_{\rho_i}^{\rho_{i+1}} u''(s)ds, \end{aligned} \tag{2}$$

where $f : [a, T] \times \mathbb{R}^3 \to \mathbb{R}$ is a continuous function, $a < \sigma_1 < \sigma_2 < \cdots < \sigma_m < \rho_1 < \rho_2 < \cdots < \rho_{p+1} < T$, and $\gamma_j, \mu_j, \nu_j \in \mathbb{R}^+$ $(j = 1, 2, \ldots, m)$, $\xi_i, \eta_i, \omega_i \in \mathbb{R}^+$ $(i = 1, 2, \ldots, p)$.

As a second problem, we study Equation (1) with the following type non-separated boundary conditions:

$$\begin{aligned} \alpha_1 u(a) + \alpha_2 u(T) &= \sum_{j=1}^{m} \gamma_j u(\sigma_j) + \sum_{i=1}^{p} \xi_i \int_{\rho_i}^{\rho_{i+1}} u(s)ds, \\ \beta_1 u'(a) + \beta_2 u'(T) &= \sum_{j=1}^{m} \mu_j u'(\sigma_j) + \sum_{i=1}^{p} \eta_i \int_{\rho_i}^{\rho_{i+1}} u'(s)ds, \\ \delta_1 u''(a) + \delta_2 u''(T) &= \sum_{j=1}^{m} \nu_j u''(\sigma_j) + \sum_{i=1}^{p} \omega_i \int_{\rho_i}^{\rho_{i+1}} u''(s)ds, \end{aligned} \tag{3}$$

where $\alpha_j, \beta_j, \delta_j \in \mathbb{R}$ $(j = 1, 2)$, while the rest of parameters are the same as fixed in the problem in Equations (1) and (2).

The subject of boundary value problems has been an interesting and important area of investigation in view of its varied application in applied sciences. One can find the examples in blood flow problems, underground water flow, chemical engineering, thermoelasticity, etc. For a detailed account of applications, see [1].

Nonlinear third-order ordinary differential equations frequently appear in the study of applied problems. In [2], the authors studied the existence of solutions for third-order nonlinear boundary value problems arising in nano-boundary layer fluid flows over stretching surfaces. In the study of magnetohydrodynamic flow of a second grade nanofluid over a nonlinear stretching sheet, the system of transformed governing equations involves a nonlinear third-order ordinary equation and is solved for local behavior of velocity distributions [3]. The investigation of the model of magnetohydrodynamic flow of second grade nanofluid over a nonlinear stretching sheet is also based on a nonlinear third-order ordinary differential equation [4].

During the last few decades, boundary value problems involving nonlocal and integral boundary conditions attracted considerable attention. In contrast to the classical boundary data, nonlocal boundary conditions help to model physical, chemical or other changes occurring within the given domain. For the study of heat conduction phenomenon in presence of nonclassical boundary condition, see [5]. The details on theoretical development of nonlocal boundary value problems can be found in the articles [6–10] and the references cited therein. On the other hand, integral boundary conditions play a key role in formulating the real world problems involving arbitrary shaped structures, for example, blood vessels in fluid flow problems [11–13]. For the recent development of the boundary value problems involving integral and multi-strip conditions, we refer the reader to the works [14–19].

In heat conduction problems, the concept of nonuniformity can be relaxed by using the boundary conditions of the form (2), which can accommodate the nonuniformities in form of points or sub-segments on the heat sources. In fact, the integro-multipoint conditions (2) can be interpreted as the sum of the values of the unknown function (e.g., temperature) at the nonlocal positions (points and sub-segments) is proportional to the value of the unknown function over the given domain. Moreover, in scattering problems, the conditions (2) can be helpful in a situation when the scattering boundary consists of finitely many sub-strips (finitely many edge-scattering problems). For details and applications in engineering problems, see [20–23].

In the present work, we derive the existence results for the problem in Equations (1) and (2) by applying Leray–Schauder nonlinear alternative and Krasnoselskii fixed-point theorem, while the uniqueness result is obtained with the aid of celebrated Banach fixed point theorem. These results are presented in Section 3. The Ulam type stability for the problem in Equations (1) and (2) is discussed in Section 4. In Section 5, we describe the outline for developing the existence theory for the problem in Equations (1) and (3). Section 2 contains the auxiliary lemmas related to the linear variants of the given problems, which lay the foundation for establishing the desired results. It is imperative to mention that the results obtained in this paper are new and yield several new results as special cases for appropriate choices of the parameters involved in the problems at hand.

2. Preliminary Result

In this section, we solve linear variants of the problems in Equations (1) and (2), and Equations (1) and (3).

Lemma 1. *For $g \in C([a,T],\mathbb{R})$ and $\Lambda \neq 0$, the unique solution of the problem consisting of the equation*

$$u'''(t) = g(t),\ t \in [a,T],$$

and the boundary condition in Equation (2) is

$$\begin{aligned} u(t) &= \int_a^t \frac{(t-s)^2}{2} g(s)ds \\ &-\frac{1}{\Lambda}\int_a^T \Big[A_1A_2\frac{(T-s)^3}{3!} + G_1(t)\frac{(T-s)^2}{2} + G_2(t)(T-s)\Big] g(s)ds \\ &+\frac{1}{\Lambda}\sum_{j=1}^{m}\int_a^{\sigma_j}\Big[\gamma_jA_1A_2\frac{(\sigma_j-s)^2}{2} + \mu_jG_1(t)(\sigma_j-s) + \nu_jG_2(t)\Big]g(s)ds \\ &+\frac{1}{\Lambda}\sum_{i=1}^{p}\int_{\rho_i}^{\rho_{i+1}}\Big[\int_a^s\Big(\xi_iA_1A_2\frac{(s-\tau)^2}{2} + \eta_iG_1(t)(s-\tau) + \omega_iG_2(t)\Big)g(\tau)d\tau\Big]ds, \end{aligned} \tag{4}$$

where

$$G_1(t) = A_1\Big(A_4(t-a) - A_5\Big),\ G_2(t) = A_3\Big(A_5 - A_4(t-a)\Big) - A_2\Big(A_6 - A_4\frac{(t-a)^2}{2}\Big), \tag{5}$$

$$\begin{cases} \Lambda = A_1A_2A_4,\ A_1 = \Big(T - a - \sum\limits_{i=1}^{p}\omega_i(\rho_{i+1}-\rho_i) - \sum\limits_{j=1}^{m}\nu_j\Big) \neq 0, \\ A_2 = \Big(T - a - \sum\limits_{i=1}^{p}\eta_i(\rho_{i+1}-\rho_i) - \sum\limits_{j=1}^{m}\mu_j\Big) \neq 0, \\ A_3 = \dfrac{(T-a)^2}{2} - \sum\limits_{i=1}^{p}\eta_i\Big(\dfrac{(\rho_{i+1}-a)^2}{2} - \dfrac{(\rho_i-a)^2}{2}\Big) - \sum\limits_{j=1}^{m}\mu_j(\sigma_j-a), \\ A_4 = \Big(T - a - \sum\limits_{i=1}^{p}\xi_i(\rho_{i+1}-\rho_i) - \sum\limits_{j=1}^{m}\gamma_j\Big) \neq 0, \\ A_5 = \dfrac{(T-a)^2}{2} - \sum\limits_{i=1}^{p}\xi_i\Big(\dfrac{(\rho_{i+1}-a)^2}{2} - \dfrac{(\rho_i-a)^2}{2}\Big) - \sum\limits_{j=1}^{m}\gamma_j(\sigma_j-a), \\ A_6 = \dfrac{(T-a)^3}{3!} - \sum\limits_{i=1}^{p}\xi_i\Big(\dfrac{(\rho_{i+1}-a)^3}{3!} - \dfrac{(\rho_i-a)^3}{3!}\Big) - \sum\limits_{j=1}^{m}\gamma_j\dfrac{(\sigma_j-a)^2}{2}. \end{cases} \tag{6}$$

Proof. Integrating $u'''(t) = g(t)$ repeatedly from a to t, we get

$$u(t) = c_0 + c_1(t-a) + c_2\frac{(t-a)^2}{2} + \int_a^t \frac{(t-s)^2}{2}g(s)ds, \tag{7}$$

where c_0, c_1 and c_2 are arbitrary unknown real constants. Moreover, from Equation (7), we have

$$u'(t) = c_1 + c_2(t-a) + \int_a^t (t-s)g(s)ds, \tag{8}$$

$$u''(t) = c_2 + \int_a^t g(s)ds. \tag{9}$$

Using the third condition of Equation (2) in Equation (9), we get

$$c_2 = \frac{1}{A_1}\Big[-\int_a^T (T-s)g(s)ds + \sum_{i=1}^{p} \omega_i \int_{\rho_i}^{\rho_{i+1}} \int_a^s g(\tau)d\tau ds + \sum_{j=1}^{m} \nu_j \int_a^{\sigma_j} g(s)ds \Big]. \tag{10}$$

Making use of the second condition of Equation (2) in Equation (8) together with Equation (10) yields

$$\begin{aligned} c_1 &= \frac{1}{A_2}\Big[-\int_a^T \frac{(T-s)^2}{2} g(s)ds + \sum_{i=1}^{p} \eta_i \int_{\rho_i}^{\rho_{i+1}} \int_a^s (s-\tau)g(\tau)d\tau ds \\ &+ \sum_{j=1}^{m} \mu_j \int_a^{\sigma_j} (\sigma_j - s)g(s)ds \Big] + \frac{A_3}{A_1A_2}\Big[-\int_a^T (T-s)g(s)ds + \sum_{i=1}^{p} \omega_i \int_{\rho_i}^{\rho_{i+1}} \int_a^s g(\tau)d\tau ds \\ &+ \sum_{j=1}^{m} \nu_j \int_a^{\sigma_j} g(s)ds \Big]. \end{aligned} \tag{11}$$

Finally, using the first condition of Equation (2) in Equation (7) together with Equations (10) and (11), we obtain

$$\begin{aligned} c_0 &= \frac{1}{\Lambda}\Big\{ \Big(A_3A_5 - A_2A_6\Big)\Big[-\int_a^T (T-s)g(s)ds + \sum_{i=1}^{p} \omega_i \int_{\rho_i}^{\rho_{i+1}} \int_a^s g(\tau)d\tau ds \\ &+ \sum_{j=1}^{m} \nu_j \int_a^{\sigma_j} g(s)ds \Big] - A_1A_5\Big[-\int_a^T \frac{(T-s)^2}{2} g(s)ds \\ &+ \sum_{i=1}^{p} \eta_i \int_{\rho_i}^{\rho_{i+1}} \int_a^s (s-\tau)g(\tau)d\tau ds + \sum_{j=1}^{m} \mu_j \int_a^{\sigma_j} (\sigma_j - s)g(s)ds \Big] \\ &+ A_1A_2\Big[-\int_a^T \frac{(T-s)^3}{3!} g(s)ds + \sum_{i=1}^{p} \xi_i \int_{\rho_i}^{\rho_{i+1}} \int_a^s \frac{(s-\tau)^2}{2} g(\tau)d\tau ds \\ &+ \sum_{j=1}^{m} \gamma_j \int_a^{\sigma_j} \frac{(\sigma_j - s)^2}{2} g(s)ds \Big] \Big\}. \end{aligned} \tag{12}$$

In Equations (10)–(12), we have used the notations in Equation (6). Inserting the values of c_0, c_1 *and* c_2 in Equation (7) completes the solution to Equation (4). By direct computation, one can obtain the converse of the Lemma. □

Lemma 2. *For* $h \in C([a,T],\mathbb{R})$*, the problem consisting of the equation* $u'''(t) = h(t), t \in [a,T]$ *and non-separated boundary conditions in Equation (3) is equivalent to the integral equation*

$$\begin{aligned} u(t) &= \int_a^t \frac{(t-s)^2}{2} h(s)ds \\ &- \frac{1}{\Delta} \int_a^T \Big[\alpha_2\zeta_1\zeta_2 \frac{(T-s)^2}{2} + \beta_2 P_1(t)(T-s) + \delta_2 P_2(t) \Big] h(s)ds \\ &+ \frac{1}{\Delta} \sum_{j=1}^{m} \int_a^{\sigma_j} \Big[\gamma_j\zeta_1\zeta_2 \frac{(\sigma_j - s)^2}{2} + \mu_j P_1(t)(\sigma_j - s) + \nu_j P_2(t) \Big] h(s)ds \\ &+ \frac{1}{\Delta} \sum_{i=1}^{p} \int_{\rho_i}^{\rho_{i+1}} \Big[\int_a^s \Big(\xi_i\zeta_1\zeta_2 \frac{(s-\tau)^2}{2} + \eta_i P_1(t)(s-\tau) + \omega_i P_2(t) \Big) h(\tau)d\tau \Big] ds, \end{aligned} \tag{13}$$

where

$$P_1(t) = \zeta_1\Big(\zeta_4(t-a) - \zeta_5\Big),\ P_2(t) = \zeta_3\Big(\zeta_5 - \zeta_4(t-a)\Big) - \zeta_2\Big(\zeta_6 - \zeta_4\frac{(t-a)^2}{2}\Big), \tag{14}$$

$$\begin{cases} \Delta = \zeta_1\zeta_2\zeta_4,\ \zeta_1 = \Big(\delta_1 + \delta_2 - \sum_{i=1}^{p}\omega_i(\rho_{i+1}-\rho_i) - \sum_{j=1}^{m}\nu_j\Big) \neq 0, \\ \zeta_2 = \Big(\beta_1 + \beta_2 - \sum_{i=1}^{p}\eta_i(\rho_{i+1}-\rho_i) - \sum_{j=1}^{m}\mu_j\Big) \neq 0, \\ \zeta_3 = \beta_2(T-a) - \sum_{i=1}^{p}\eta_i\Big(\frac{(\rho_{i+1}-a)^2}{2} - \frac{(\rho_i-a)^2}{2}\Big) - \sum_{j=1}^{m}\mu_j(\sigma_j - a), \\ \zeta_4 = \Big(\alpha_1 + \alpha_2 - \sum_{i=1}^{p}\xi_i(\rho_{i+1}-\rho_i) - \sum_{j=1}^{m}\gamma_j\Big) \neq 0, \\ \zeta_5 = \alpha_2(T-a) - \sum_{i=1}^{p}\xi_i\Big(\frac{(\rho_{i+1}-a)^2}{2} - \frac{(\rho_i-a)^2}{2}\Big) - \sum_{j=1}^{m}\gamma_j(\sigma_j - a), \\ \zeta_6 = \alpha_2\frac{(T-a)^2}{2} - \sum_{i=1}^{p}\xi_i\Big(\frac{(\rho_{i+1}-a)^3}{3!} - \frac{(\rho_i-a)^3}{3!}\Big) - \sum_{j=1}^{m}\gamma_j\frac{(\sigma_j-a)^2}{2}. \end{cases} \tag{15}$$

Proof. We omit the proof as it runs parallel to that of Lemma 1. □

3. Main Results

Let us set $\widehat{f}(t) = f(t, u(t), u'(t), u''(t))$ and introduce a fixed point problem equivalent to the problem in Equations (1) and (2) via Lemma 1 as follows

$$u = \mathcal{L}u, \tag{16}$$

where the operator $\mathcal{L} : \mathcal{H} \to \mathcal{H}$ is defined by

$$\begin{aligned} (\mathcal{L}u)(t) &= \int_a^t \frac{(t-s)^2}{2}\widehat{f}(s)ds \\ &-\frac{1}{\Lambda}\int_a^T \Big[A_1A_2\frac{(T-s)^3}{3!} + G_1(t)\frac{(T-s)^2}{2} + G_2(t)(T-s)\Big]\widehat{f}(s)ds \\ &+\frac{1}{\Lambda}\sum_{j=1}^{m}\int_a^{\sigma_j}\Big[\gamma_jA_1A_2\frac{(\sigma_j-s)^2}{2} + \mu_jG_1(t)(\sigma_j-s) + \nu_jG_2(t)\Big]\widehat{f}(s)ds \\ &+\frac{1}{\Lambda}\sum_{i=1}^{p}\int_{\rho_i}^{\rho_{i+1}}\int_a^s\Big[\xi_iA_1A_2\frac{(s-\tau)^2}{2} + \eta_iG_1(t)(s-\tau) + \omega_iG_2(t)\Big]\widehat{f}(\tau)d\tau ds. \end{aligned} \tag{17}$$

Here, $\mathcal{H} = \{u|u,\ u',\ u'' \in C([a,T],\mathbb{R})\}$ is the Banach space equipped with the norm $\|u\|_{\mathcal{H}} = \max_{t\in[a,T]}\Big\{|u(t)| + |u'(t)| + |u''(t)|\Big\} = \|u\| + \|u'\| + \|u''\|$. From Equation (17), we have

$$\begin{aligned} (\mathcal{L}u)'(t) &= \int_a^t (t-s)\widehat{f}(s)ds - \frac{1}{A_1A_2}\int_a^T \Big[A_1\frac{(T-s)^2}{2} + G_3(t)(T-s)\Big]\widehat{f}(s)ds \\ &+\frac{1}{A_1A_2}\sum_{j=1}^{m}\int_a^{\sigma_j}\Big[\mu_jA_1(\sigma_j-s) + \nu_jG_3(t)\Big]\widehat{f}(s)ds \\ &+\frac{1}{A_1A_2}\sum_{i=1}^{p}\int_{\rho_i}^{\rho_{i+1}}\int_a^s\Big[\eta_iA_1(s-\tau) + \omega_iG_3(t)\Big]\widehat{f}(\tau)d\tau ds, \end{aligned} \tag{18}$$

$$(\mathcal{L}u)''(t) = \int_a^t \widehat{f}(s)ds + \frac{1}{A_1}\Big[- \int_a^T (T-s)\widehat{f}(s)ds + \sum_{j=1}^{m} \int_a^{\sigma_j} \nu_j \widehat{f}(s)ds + \sum_{i=1}^{p} \int_{\rho_i}^{\rho_{i+1}} \int_a^s \omega_i \widehat{f}(\tau)d\tau ds \Big], \tag{19}$$

where

$$G_3(t) = A_2(t-a) - A_3. \tag{20}$$

Observe that the existence of the fixed points for the operator in Equation (16) implies the existence of solutions for the problem in Equations (1) and (2).

For the sake of computational convenience in the forthcoming analysis, we set

$$Q = Q_1 + Q_2 + Q_3, \tag{21}$$

where

$$\begin{aligned} Q_1 &= \frac{(T-a)^3}{3!} + \frac{1}{|A_4|}\Big[\frac{(T-a)^4}{4!} + \sum_{i=1}^{p} \xi_i \Big(\frac{(\rho_{i+1}-a)^4}{4!} - \frac{(\rho_i - a)^4}{4!}\Big) + \sum_{j=1}^{m} \gamma_j \frac{(\sigma_j - a)^3}{3!}\Big] \\ &+ \frac{b_1}{|\Lambda|}\Big[\frac{(T-a)^3}{3!} + \sum_{i=1}^{p} \eta_i \Big(\frac{(\rho_{i+1}-a)^3}{3!} - \frac{(\rho_i - a)^3}{3!}\Big) + \sum_{j=1}^{m} \mu_j \frac{(\sigma_j - a)^2}{2}\Big] \\ &+ \frac{b_2}{|\Lambda|}\Big[\frac{(T-a)^2}{2} + \sum_{i=1}^{p} \omega_i \Big(\frac{(\rho_{i+1}-a)^2}{2} - \frac{(\rho_i - a)^2}{2}\Big) + \sum_{j=1}^{m} \nu_j (\sigma_j - a)\Big], \end{aligned} \tag{22}$$

$$\begin{aligned} Q_2 &= \frac{(T-a)^2}{2} + \frac{1}{|A_2|}\Big[\frac{(T-a)^3}{3!} + \sum_{i=1}^{p} \eta_i \Big(\frac{(\rho_{i+1}-a)^3}{3!} - \frac{(\rho_i - a)^3}{3!}\Big) + \sum_{j=1}^{m} \mu_j \frac{(\sigma_j - a)^2}{2}\Big] \\ &+ \frac{b_3}{|A_1 A_2|}\Big[\frac{(T-a)^2}{2} + \sum_{i=1}^{p} \omega_i \Big(\frac{(\rho_{i+1}-a)^2}{2} - \frac{(\rho_i - a)^2}{2}\Big) + \sum_{j=1}^{m} \nu_j (\sigma_j - a)\Big], \end{aligned} \tag{23}$$

and

$$Q_3 = (T-a) + \frac{1}{|A_1|}\Big[\frac{(T-a)^2}{2} + \sum_{i=1}^{p} \omega_i \Big(\frac{(\rho_{i+1}-a)^2}{2} - \frac{(\rho_i - a)^2}{2}\Big) + \sum_{j=1}^{m} \nu_j (\sigma_j - a)\Big], \tag{24}$$

where $\max_{t\in[a,T]} |G_1(t)| = b_1$, $\max_{t\in[a,T]} |G_2(t)| = b_2$ *and* $\max_{t\in[a,T]} |G_3(t)| = b_3$ ($G_1(t)$, $G_2(t)$ are given by Equation (5) while $G_3(t)$ is defined in Equation (20)).

3.1. Existence of Solutions

In this subsection, we discuss the existence of solutions for the problem in Equations (1) and (2). In our first result, we make use of Krasnoselskii's fixed point theorem [24].

Theorem 1. *Let $f : [a,T] \times \mathbb{R}^3 \to \mathbb{R}$ be a continuous function satisfying the conditions:*

(H_1) $\Big|f(t,u,u',u'') - f(t,v,v',v'')\Big| \le \ell\Big(|u-v| + |u'-v'| + |u''-v''|\Big)$, $\forall t \in [a,T]$, $\ell > 0$, $u, v, u', v', u'', v'' \in \mathbb{R}$;

(H_2) *there exist a function $\varepsilon \in C([a,T], \mathbb{R}^+)$ with $\|\varepsilon\| = \sup_{t\in[a,T]} |\varepsilon(t)|$ such that*

$$|\widehat{f}(t)| = |f(t,u,u',u'')| \le \varepsilon(t), \ \ \forall (t,u,u',u'') \in [a,T] \times \mathbb{R}^3;$$

(H_3) $\ell\Big(Q - \frac{(T-a)}{6}\Big[6 + 3(T-a) + (T-a)^2\Big]\Big) < 1$, *where Q is given by Equation (21).*

Then, there exists at least one solution for the problem in Equations (1) and (2) on $[a,T]$.

Proof. Consider a closed ball $B_r = \{(u,u',u'') \ : \ \|u\|_{\mathcal{H}} \leq r, \ u, \ u', \ u'' \in C([a,T],\mathbb{R})\}$ for fixed $r \geq Q\|\varepsilon\|$ and introduce the operators $\mathcal{L}_1$ and $\mathcal{L}_2$ on B_r as follows:

$$(\mathcal{L}_1 u)(t) = \int_a^t \frac{(t-s)^2}{2}\widehat{f}(s)ds,$$

$$\begin{aligned}(\mathcal{L}_2 u)(t) &= -\frac{1}{\Lambda}\int_a^T \Big[A_1A_2\frac{(T-s)^3}{3!} + G_1(t)\frac{(T-s)^2}{2} + G_2(t)(T-s)\Big]\widehat{f}(s)ds \\ &+\frac{1}{\Lambda}\sum_{j=1}^m \int_a^{\sigma_j}\Big[\gamma_j A_1A_2\frac{(\sigma_j-s)^2}{2} + \mu_j G_1(t)(\sigma_j-s) + \nu_j G_2(t)\Big]\widehat{f}(s)ds \\ &+\frac{1}{\Lambda}\sum_{i=1}^p \int_{\rho_i}^{\rho_{i+1}}\int_a^s \Big[\xi_i A_1A_2\frac{(s-\tau)^2}{2} + \eta_i G_1(t)(s-\tau) + \omega_i G_2(t)\Big]\widehat{f}(\tau)d\tau ds.\end{aligned}$$

Moreover, we have

$$(\mathcal{L}_1 u)'(t) = \int_a^t (t-s)\widehat{f}(s)ds, \ (\mathcal{L}_1 u)''(t) = \int_a^t \widehat{f}(s)ds,$$

$$\begin{aligned}(\mathcal{L}_2 u)'(t) &= -\frac{1}{A_1A_2}\int_a^T \Big[A_1\frac{(T-s)^2}{2} + G_3(t)(T-s)\Big]\widehat{f}(s)ds \\ &+\frac{1}{A_1A_2}\sum_{j=1}^m \int_a^{\sigma_j}\Big[\mu_j A_1(\sigma_j-s) + \nu_j G_3(t)\Big]\widehat{f}(s)ds \\ &+\frac{1}{A_1A_2}\sum_{i=1}^p \int_{\rho_i}^{\rho_{i+1}}\int_a^s \Big[\eta_i A_1(s-\tau) + \omega_i G_3(t)\Big]\widehat{f}(\tau)d\tau ds,\end{aligned}$$

$$(\mathcal{L}_2 u)''(t) = \frac{1}{A_1}\Big[-\int_a^T (T-s)\widehat{f}(s)ds + \sum_{j=1}^m \int_a^{\sigma_j}\nu_j\widehat{f}(s)ds + \sum_{i=1}^p \int_{\rho_i}^{\rho_{i+1}}\int_a^s \omega_i\widehat{f}(\tau)d\tau ds\Big].$$

Notice that $\mathcal{L} = \mathcal{L}_1 + \mathcal{L}_2$. For $u, v \in B_r$, and $t \in [a,T]$, we have

$$\begin{aligned}&\|\mathcal{L}_1 u + \mathcal{L}_2 v\| \\ = \ &\sup_{t\in[a,T]}\Big\{\Big|\int_a^t \frac{(t-s)^2}{2} f(s,u(s),u'(s),u''(s))ds \\ &-\frac{1}{\Lambda}\int_a^T \Big[A_1A_2\frac{(T-s)^3}{3!} + G_1(t)\frac{(T-s)^2}{2} + G_2(t)(T-s)\Big]f(s,v(s),v'(s),v''(s))ds \\ &+\frac{1}{\Lambda}\sum_{j=1}^m \int_a^{\sigma_j}\Big[\gamma_j A_1A_2\frac{(\sigma_j-s)^2}{2} + \mu_j G_1(t)(\sigma_j-s) + \nu_j G_2(t)\Big]f(s,v(s),v'(s),v''(s))ds \\ &+\frac{1}{\Lambda}\sum_{i=1}^p \int_{\rho_i}^{\rho_{i+1}}\int_a^s \Big[\xi_i A_1A_2\frac{(s-\tau)^2}{2} + \eta_i G_1(t)(s-\tau) + \omega_i G_2(t)\Big]f(\tau,v(\tau),v'(\tau),v''(\tau))d\tau ds\Big|\Big\} \\ \leq \ &\|\varepsilon\|\sup_{t\in[a,T]}\Big\{\frac{(t-a)^3}{3!} + \frac{1}{|A_4|}\Big[\frac{(T-a)^4}{4!} + \sum_{i=1}^p \xi_i\Big(\frac{(\rho_{i+1}-a)^4}{4!} - \frac{(\rho_i-a)^4}{4!}\Big) + \sum_{j=1}^m \gamma_j\frac{(\sigma_j-a)^3}{3!}\Big] \\ &+\frac{|G_1(t)|}{|\Lambda|}\Big[\frac{(T-a)^3}{3!} + \sum_{i=1}^p \eta_i\Big(\frac{(\rho_{i+1}-a)^3}{3!} - \frac{(\rho_i-a)^3}{3!}\Big) + \sum_{j=1}^m \mu_j\frac{(\sigma_j-a)^2}{2}\Big] \\ &+\frac{|G_2(t)|}{|\Lambda|}\Big[\frac{(T-a)^2}{2} + \sum_{i=1}^p \omega_i\Big(\frac{(\rho_{i+1}-a)^2}{2} - \frac{(\rho_i-a)^2}{2}\Big) + \sum_{j=1}^m \nu_j(\sigma_j-a)\Big]\Big\} \leq \|\varepsilon\|Q_1,\end{aligned}$$

where Q_1 is given by Equation (22). In a similar manner, it can be shown that

$$\|(\mathcal{L}_1 u)' + (\mathcal{L}_2 v)'\| \leq \|\varepsilon\| Q_2, \ \|(\mathcal{L}_1 u)'' + (\mathcal{L}_2 v)''\| \leq \|\varepsilon\| Q_3,$$

where Q_2 and Q_3 are, respectively, given by Equations (23) and (24). Consequently, we obtain

$$\|\mathcal{L}_1 u + \mathcal{L}_2 v\|_{\mathcal{H}} \leq \|\varepsilon\| Q \leq r,$$

where we have used (H_2) and Equation (21). From the above inequality, it follows that $\mathcal{L}_1 u + \mathcal{L}_2 v \in B_r$. Thus, the first condition of Krasnoselskii's fixed point theorem [24] is satisfied. Next, we show that $\mathcal{L}_2$ is a contraction. For $u, v \in \mathbb{R}$, it follows by the assumption (H_1) that

$$\begin{aligned}
& \|\mathcal{L}_2 u - \mathcal{L}_2 v\| \\
\leq\ & \sup_{t\in[a,T]} \Big\{ \frac{1}{|\Lambda|} \int_a^T \Big[|A_1 A_2| \frac{(T-s)^3}{3!} + |G_1(t)| \frac{(T-s)^2}{2} + |G_2(t)|(T-s) \Big] \\
& \times \Big| f(s, u(s), u'(s), u''(s)) - f(s, v(s), v'(s), v''(s)) \Big| ds + \frac{1}{|\Lambda|} \sum_{j=1}^{m} \int_a^{\sigma_j} \Big[\gamma_j |A_1 A_2| \frac{(\sigma_j - s)^2}{2} \\
& + \mu_j |G_1(t)| (\sigma_j - s) + \nu_j |G_2(t)| \Big] \Big| f(s, u(s), u'(s), u''(s)) - f(s, v(s), v'(s), v''(s)) \Big| ds \\
& + \frac{1}{|\Lambda|} \sum_{i=1}^{p} \int_{\rho_i}^{\rho_{i+1}} \int_a^s \Big[\xi_i |A_1 A_2| \frac{(s-\tau)^2}{2} + \eta_i |G_1(t)| (s-\tau) + \omega_i |G_2(t)| \Big] \\
& \times \Big| f(\tau, u(\tau), u'(\tau), u''(\tau)) - f(\tau, v(\tau), v'(\tau), v''(\tau)) \Big| d\tau ds \Big\} \\
\leq\ & \ell \Big(\|u - v\| + \|u' - v'\| + \|u'' - v''\| \Big) \Big\{ \frac{1}{|A_4|} \Big[\frac{(T-a)^4}{4!} + \sum_{i=1}^{p} \xi_i \Big(\frac{(\rho_{i+1} - a)^4}{4!} - \frac{(\rho_i - a)^4}{4!} \Big) \\
& + \sum_{j=1}^{m} \gamma_j \frac{(\sigma_j - a)^3}{3!} \Big] + \frac{b_1}{|\Lambda|} \Big[\frac{(T-a)^3}{3!} + \sum_{i=1}^{p} \eta_i \Big(\frac{(\rho_{i+1} - a)^3}{3!} - \frac{(\rho_i - a)^3}{3!} \Big) + \sum_{j=1}^{m} \mu_j \frac{(\sigma_j - a)^2}{2} \Big] \\
& + \frac{b_2}{|\Lambda|} \Big[\frac{(T-a)^2}{2} + \sum_{i=1}^{p} \omega_i \Big(\frac{(\rho_{i+1} - a)^2}{2} - \frac{(\rho_i - a)^2}{2} \Big) + \sum_{j=1}^{m} \nu_j (\sigma_j - a) \Big] \Big\} \\
\leq\ & \ell \Big(Q_1 - \frac{(T-a)^3}{3!} \Big) \parallel u - v \parallel_{\mathcal{H}} .
\end{aligned}$$

Similarly, we can obtain

$$\|(\mathcal{L}_2 u)' - (\mathcal{L}_2 v)'\| \leq \ell \Big(Q_2 - \frac{(T-a)^2}{2} \Big) \parallel u - v \parallel_{\mathcal{H}},$$

and

$$\|(\mathcal{L}_2 u)'' - (\mathcal{L}_2 v)''\| \leq \ell \Big(Q_3 - (T-a) \Big) \parallel u - v \parallel_{\mathcal{H}} .$$

Thus, we get

$$\|\mathcal{L}_2 u - \mathcal{L}_2 v\|_{\mathcal{H}} \leq \ell \Big(Q - \frac{(T-a)}{6} \Big[6 + 3(T-a) + (T-a)^2 \Big] \Big) \parallel u - v \parallel_{\mathcal{H}},$$

which, in view of the condition (H_3), implies that $\mathcal{L}_2$ is a contraction. Thus, the second hypothesis of Krasnoselskii's fixed point theorem [24] is satisfied. Finally, we verify the third and last hypothesis of

Krasnoselskii's fixed point theorem [24] that $\mathcal{L}_1$ is compact and continuous. Observe that continuity of f implies that the operator $\mathcal{L}_1$ is continuous. In addition, $\mathcal{L}_1$ is uniformly bounded on B_r as

$$\|\mathcal{L}_1 u\|_{\mathcal{H}} \leq \|\varepsilon\| \Big[\frac{(T-a)^3}{3!} + \frac{(T-a)^2}{2} + (T-a)\Big].$$

Let us fix $\sup_{(t,u,u',u'')\in[a,T]\times B_r} |f(t,u,u',u'')| = \bar{f}$, and take $a < t_1 < t_2 < T$. Then,

$$\begin{aligned} |(\mathcal{L}_1 u)(t_2) - (\mathcal{L}_1 u)(t_1)| &= \Big| \int_a^{t_1} \Big[\frac{(t_2-s)^2}{2} - \frac{(t_1-s)^2}{2}\Big]\widehat{f}(s)ds \\ &\quad + \int_{t_1}^{t_2} \frac{(t_2-s)^2}{2}\widehat{f}(s)ds\Big| \\ &\leq \bar{f}\Big(\frac{(t_2-t_1)^3}{3} + \frac{1}{3!}\Big|(t_2-a)^3 - (t_1-a)^3\Big|\Big) \to 0 \text{ as } t_2 \to t_1, \end{aligned}$$

independently of $u \in B_r$. In addition, we have

$$\begin{aligned} |(\mathcal{L}_1 u)'(t_2) - (\mathcal{L}_1 u)'(t_1)| &= \Big|\int_a^{t_1} [(t_2-s) - (t_1-s)]\widehat{f}(s)ds + \int_{t_1}^{t_2} (t_2-s)\widehat{f}(s)ds\Big| \\ &\leq \bar{f}\Big|(t_2-t_1)(t_1-a) + \frac{(t_2-t_1)^2}{2}\Big| \to 0 \text{ as } t_2 \to t_1, \end{aligned}$$

independently of $u \in B_r$ and

$$|(\mathcal{L}_1 u)''(t_2) - (\mathcal{L}_1 u)''(t_1)| \leq \bar{f}(t_2 - t_1) \to 0 \text{ as } t_2 \to t_1,$$

independently of $u \in B_r$. From the preceding arguments, we deduce that $\mathcal{L}_1$ is relatively compact on B_r. Hence, the operator $\mathcal{L}_1$ is compact on B_r by the Arzelá–Ascoli theorem. Since all the hypotheses of Krasnoselskii's fixed point theorem [24] are verified, its conclusion applies to the problem in Equations (1) and (2). □

Remark 1. *When the role of the operators $\mathcal{L}_1$ and $\mathcal{L}_2$ is mutually interchanged, the condition (H_3) of Theorem 1 takes the form: $\ell\frac{(T-a)}{6}\Big[6 + 3(T-a) + (T-a)^2\Big] < 1$.*

In the next result, we make use of Leray–Schauder nonlinear alternative for single valued maps [25].

Theorem 2. *Suppose that $f : [a,T] \times \mathbb{R}^3 \to \mathbb{R}$ is a continuous function and the following conditions hold:*

(H_4) *$|\widehat{f}(t)| = |f(t,u,u',u'')| \leq p(t)\Psi(|u|)$, $\forall (t,u,u',u'') \in [a,T] \times \mathbb{R}^3$, where $p \in C([a,T],\mathbb{R}^+)$, and $\Psi : \mathbb{R}^+ \to \mathbb{R}^+$ is a nondecreasing function;*

(H_5) *there exists a positive constant N satisfying the inequality:*

$$\frac{N}{\|p\|\Psi(N)Q} > 1,$$

where Q is defined by Equation (21). Then, the problem in Equations (1) and (2) has at least one solution on $[a,T]$.

Proof. We verify the hypotheses of Leray–Schauder nonlinear alternative [25] in several steps. We first show that the operator $\mathcal{L} : \mathcal{H} \to \mathcal{H}$ defined by Equation (17) maps bounded sets into bounded sets in

$\mathcal{H}$. Let us consider a set $B_{\bar{r}} = \{(u,u',u'') : \|u\|_{\mathcal{H}} \le \bar{r},\ u,\ u',\ u'' \in C([a,T],\mathbb{R}),\ \bar{r} > 0\}$ and note that it is bounded in $\mathcal{H}$. Then, in view of the condition (H_4), we get

$$\begin{aligned}
\|(\mathcal{L}u)\| &= \sup_{t\in[a,T]} \Big\{ \Big| \int_a^t \frac{(t-s)^2}{2}\widehat{f}(s)ds \\
&\quad -\frac{1}{\Lambda}\int_a^T \Big[A_1A_2\frac{(T-s)^3}{3!} + G_1(t)\frac{(T-s)^2}{2} + G_2(t)(T-s)\Big]\widehat{f}(s)ds \\
&\quad +\frac{1}{\Lambda}\sum_{j=1}^{m}\int_a^{\sigma_j}\Big[\gamma_j A_1A_2\frac{(\sigma_j-s)^2}{2} + \mu_j G_1(t)(\sigma_j - s) + \nu_j G_2(t)\Big]\widehat{f}(s)ds \\
&\quad +\frac{1}{\Lambda}\sum_{i=1}^{p}\int_{\rho_i}^{\rho_{i+1}}\int_a^s\Big[\xi_i A_1A_2\frac{(s-\tau)^2}{2} + \eta_i G_1(t)(s-\tau) + \omega_i G_2(t)\Big]\widehat{f}(\tau)d\tau ds\Big|\Big\} \\
&\le \|p\|\Psi(\|u\|_{\mathcal{H}})Q_1 \le \|p\|\Psi(\bar{r})Q_1,
\end{aligned}$$

where Q_1 is given by Equation (22). Similarly, one can establish that

$$\|(\mathcal{L}u)'\| \le \|p\|\Psi(\bar{r})Q_2,\ \|(\mathcal{L}u)''\| \le \|p\|\Psi(\bar{r})Q_3,$$

where Q_2 and Q_3 are given by Equations (23) and (24), respectively. In view of the foregoing arguments, we have

$$\|(\mathcal{L}u)\|_{\mathcal{H}} \le \|p\|\Psi(\bar{r})Q,$$

where Q is given by Equation (21). Next, it is verified that the operator $\mathcal{L}$ maps bounded sets into equicontinuous sets in $\mathcal{H}$. Notice that $\mathcal{L}$ is continuous in view of the continuity of $\widehat{f}(t)$. Let $t_1, t_2 \in [a,T]$ with $t_1 < t_2$ and $u \in B_{\bar{r}}$. Then, we have

$$\begin{aligned}
&|(\mathcal{L}u)(t_2) - (\mathcal{L}u)(t_1)| \\
\le\ & \Big|\int_a^{t_1}\Big[\frac{(t_2-s)^2}{2} - \frac{(t_1-s)^2}{2}\Big]\widehat{f}(s)ds + \int_{t_1}^{t_2}\frac{(t_2-s)^2}{2}\widehat{f}(s)ds\Big| \\
&+\frac{1}{|\Lambda|}\int_a^T (t_2-t_1)\Big[|A_1A_4|\frac{(T-s)^2}{2} + \Big(|A_3A_4| + \frac{|A_2A_4|}{2}(t_2+t_1)\Big)(T-s)\Big]\widehat{f}(s)ds \\
&+\frac{1}{|\Lambda|}\sum_{j=1}^{m}\int_a^{\sigma_j}(t_2-t_1)\Big[\mu_j|A_1A_4|(\sigma_j-s) + \nu_j\Big(|A_3A_4| + \frac{|A_2A_4|}{2}(t_2+t_1^2)\Big)\Big]|\widehat{f}(s)|ds \\
&+\frac{1}{|\Lambda|}\sum_{i=1}^{p}\int_{\rho_i}^{\rho_{i+1}}\int_a^s(t_2-t_1)\Big[\eta_i|A_1A_4|(s-\tau) + \omega_i\Big(|A_3A_4| + \frac{|A_2A_4|}{2}(t_2+t_1)\Big)\Big]|\widehat{f}(\tau)|d\tau ds \\
\le\ & \|p\|\Psi(\bar{r})\Big\{\frac{(t_2-t_1)^3}{3} + \frac{1}{3!}\Big|(t_2-a)^3 - (t_1-a)^3\Big| \\
&+\frac{(t_2-t_1)}{|A_2|}\Big[\frac{(T-a)^3}{3!} + \sum_{j=1}^{m}\mu_j\frac{(\sigma_j-a)^2}{2} + \sum_{i=1}^{p}\eta_i\Big(\frac{(\rho_{i+1}-a)^3}{3!} - \frac{(\rho_i-a)^3}{3!}\Big)\Big] \\
&+\frac{1}{|\Lambda|}\Big(|A_3A_4|(t_2-t_1) + \frac{|A_2A_4|}{2}(t_2^2-t_1^2)\Big)\Big[\frac{(T-a)^2}{2} + \sum_{j=1}^{m}\nu_j(\sigma_j-a) \\
&+\sum_{i=1}^{p}\omega_i\Big(\frac{(\rho_{i+1}-a)^2}{2} - \frac{(\rho_i-a)^2}{2}\Big)\Big]\Big\} \to 0 \text{ as } (t_2-t_1)\to 0,
\end{aligned}$$

independently of $u \in B_{\bar{r}}$. Moreover, we have

$$\begin{aligned}|(\mathcal{L}u)'(t_2)-(\mathcal{L}u)'(t_1)| &\leq \|p\|\Psi(\bar{r})\Big\{\Big|(t_2-t_1)(t_1-a)+\frac{(t_2-t_1)^2}{2}\Big| \\ &\quad +\frac{(t_2-t_1)}{|A_1|}\Big[\frac{(T-a)^2}{2}+\sum_{j=1}^{m}\nu_j(\sigma_j-a) \\ &\quad +\sum_{i=1}^{p}\omega_i\Big(\frac{(\rho_{i+1}-a)^2}{2}-\frac{(\rho_i-a)^2}{2}\Big)\Big]\Big\}\to 0 \text{ as } (t_2-t_1)\to 0,\end{aligned}$$

independently of $u \in B_{\bar{r}}$ and

$$\begin{aligned}|(\mathcal{L}u)''(t_2)-(\mathcal{L}u)''(t_1)| &\leq \Big|\int_{t_1}^{t_2}\widehat{f}(s)ds\Big| \\ &\leq \|p\|\Psi(\bar{r})(t_2-t_1)\to 0 \text{ as } (t_2-t_1)\to 0,\end{aligned}$$

independently of $u \in B_{\bar{r}}$. In view of the foregoing arguments, the Arzelá–Ascoli theorem applies and hence the operator $\mathcal{L}:\mathcal{H}\to\mathcal{H}$ is completely continuous. The conclusion of Leray–Schauder nonlinear alternative [25] is applicable once we establish the boundedness of all solutions to the equation $u=\lambda\mathcal{L}u$ for $\lambda\in[0,1]$. Let u be a solution of the problem in Equations (1) and (2). Then, as before, one can find that

$$|u(t)|=|\lambda(\mathcal{L}u)(t)|\leq\|p\|\Psi(\|u\|_{\mathcal{H}})Q,$$

which can alternatively be written in the following form after taking the norm for $t\in[a,T]$:

$$\frac{\|u\|_{\mathcal{H}}}{\|p\|\Psi(\|u\|_{\mathcal{H}})Q}\leq 1.$$

By the assumption (H_5), we can find a positive number N such that $\|u\|_{\mathcal{H}}\neq N$. Introduce a set $U=\{u\in C([a,T],\mathbb{R}):\|u\|_{\mathcal{H}}<N\}$ such that the operator $\mathcal{L}:\overline{U}\to C([a,T],\mathbb{R})$ is continuous and completely continuous. In view of the the choice of U, there does not exist any $u\in\partial U$ satisfying $u=\lambda\mathcal{L}(u)$ for some $\lambda\in(0,1)$. Thus, it follows from the nonlinear alternative of Leray–Schauder nonlinear alternative [25] that $\mathcal{L}$ has a fixed point $u\in\overline{U}$ which corresponds a solution of the problem in Equations (1) and (2). □

3.2. Uniqueness of Solutions

In this subsection, the uniqueness of solutions for the problem in Equations (1) and (2) is established by means of contraction mapping principle due to Banach.

Theorem 3. *Let $f:[a,T]\times\mathbb{R}^3\to\mathbb{R}$ be a continuous function satisfying the assumption (H_1) with $\ell<Q^{-1}$, where Q is given by Equation (21). Then, there exists a unique solution for the problem in Equations (1) and (2) on $[a,T]$.*

Proof. Let us define a set $B_w=\{u,u',u''\in C([a,T],\mathbb{R}):\|u\|_{\mathcal{H}}\leq w\}$, where $w\geq\frac{QM}{1-\ell Q}$, $\sup_{t\in[a,T]}|f(t,0,0,0)|=M$, and show that $\mathcal{L}B_w\subset B_w$, where the operator $\mathcal{L}$ is defined by

Equation (17). For any $u \in B_w, t \in [a,T]$, one can find with the aid of the condition (H_1) that $|\widehat{f}(t)| \le \|u\|_{\mathcal{H}} + M \le \ell w + M$. Then, for $u \in B_w$, we have

$$\begin{aligned}
\|(\mathcal{L}u)\| &= \sup_{t\in[a,T]} \Bigg| \int_a^t \frac{(t-s)^2}{2}\widehat{f}(s)ds \\
&\quad -\frac{1}{\Lambda}\int_a^T \Big[A_1A_2\frac{(T-s)^3}{3!} + G_1(t)\frac{(T-s)^2}{2} + G_2(t)(T-s)\Big]\widehat{f}(s)ds \\
&\quad +\frac{1}{\Lambda}\sum_{j=1}^{m}\int_a^{\sigma_j}\Big[\gamma_jA_1A_2\frac{(\sigma_j-s)^2}{2} + \mu_jG_1(t)(\sigma_j-s) + \nu_jG_2(t)\Big]\widehat{f}(s)ds \\
&\quad +\frac{1}{\Lambda}\sum_{i=1}^{p}\int_{\rho_i}^{\rho_{i+1}}\int_a^s\Big[\xi_iA_1A_2\frac{(s-\tau)^2}{2} + \eta_iG_1(t)(s-\tau) + \omega_iG_2(t)\Big]\widehat{f}(\tau)d\tau ds\Bigg| \\
&\le \sup_{t\in[a,T]}\Bigg\{\frac{(t-a)^3}{3!} + \frac{1}{|A_4|}\Big[\frac{(T-a)^4}{4!} + \sum_{i=1}^{p}\xi_i\Big(\frac{(\rho_{i+1}-a)^4}{4!} - \frac{(\rho_i-a)^4}{4!}\Big) + \sum_{j=1}^{m}\gamma_j\frac{(\sigma_j-a)^3}{3!}\Big] \\
&\quad +\frac{|G_1(t)|}{|\Lambda|}\Big[\frac{(T-a)^3}{3!} + \sum_{i=1}^{p}\eta_i\Big(\frac{(\rho_{i+1}-a)^3}{3!} - \frac{(\rho_i-a)^3}{3!}\Big) + \sum_{j=1}^{m}\mu_j\frac{(\sigma_j-a)^2}{2}\Big] \\
&\quad +\frac{|G_2(t)|}{|\Lambda|}\Big[\frac{(T-a)^2}{2} + \sum_{i=1}^{p}\omega_i\Big(\frac{(\rho_{i+1}-a)^2}{2} - \frac{(\rho_i-a)^2}{2}\Big) + \sum_{j=1}^{m}\nu_j(\sigma_j-a)\Big]\Bigg\}(\ell w + M) \\
&\le Q_1(\ell w + M),
\end{aligned}$$

where Q_1 is given by Equation (22). In addition,

$$\|(\mathcal{L}u)'\| \le (\ell w + M)Q_2 \text{ and } \|(\mathcal{L}u)''\| \le (\ell w + M)Q_3,$$

where Q_2 and Q_3 are, respectively, given by Equations (23) and (24). Consequently, we have

$$\|(\mathcal{L}u)\|_{\mathcal{H}} \le (\ell w + M)Q \le w,$$

where Q is given by Equation (21). This shows that $\mathcal{L}B_w \subset B_w$. Next, it is shown that the operator $\mathcal{L}$ is a contraction. For that, let $u, v \in \mathcal{H}$. Then, we have

$$\begin{aligned}
\|\mathcal{L}u - \mathcal{L}v\| &= \sup_{t\in[0,T]}\Big|\mathcal{L}u(t) - \mathcal{L}v(t)\Big| \\
&\le \sup_{t\in[a,T]}\Bigg\{\int_a^t \frac{(t-s)^2}{2}\Big|f(s,u(s),u'(s),u''(s)) - f(s,v(s),v'(s),v''(s))\Big|ds \\
&\quad +\frac{1}{|\Lambda|}\int_a^T\Big[|A_1A_2|\frac{(T-s)^3}{3!} + |G_1(t)|\frac{(T-s)^2}{2} + |G_2(t)|(T-s)\Big] \\
&\quad \times\Big|f(s,u(s),u'(s),u''(s)) - f(s,v(s),v'(s),v''(s))\Big|ds \\
&\quad +\frac{1}{|\Lambda|}\sum_{j=1}^{m}\int_a^{\sigma_j}\Big[\gamma_j|A_1A_2|\frac{(\sigma_j-s)^2}{2} + \mu_j|G_1(t)|(\sigma_j-s) + \nu_j|G_2(t)|\Big] \\
&\quad \times\Big|f(s,u(s),u'(s),u''(s)) - f(s,v(s),v'(s),v''(s))\Big|ds
\end{aligned}$$

$$
\begin{aligned}
&+\frac{1}{|\Lambda|}\sum_{i=1}^{p}\int_{\rho_i}^{\rho_{i+1}}\int_a^s\Big[\xi_i|A_1A_2|\frac{(s-\tau)^2}{2}+\eta_i|G_1(t)|(s-\tau)+\omega_i|G_2(t)|\Big]\\
&\times\Big|f(\tau,u(\tau),u'(\tau),u''(\tau))-f(\tau,v(\tau),v'(\tau),v''(\tau))\Big|d\tau ds\Big\}\\
\leq\ & \ell\Big(\|u-v\|+\|u'-v'\|+\|u''-v''\|\Big)\Big\{\frac{(T-a)^3}{3!}\\
&+\frac{1}{|A_4|}\Big[\frac{(T-a)^4}{4!}+\sum_{i=1}^{p}\xi_i\Big(\frac{(\rho_{i+1}-a)^4}{4!}-\frac{(\rho_i-a)^4}{4!}\Big)+\sum_{j=1}^{m}\gamma_j\frac{(\sigma_j-a)^3}{3!}\Big]\\
&+\frac{b_1}{|\Lambda|}\Big[\frac{(T-a)^3}{3!}+\sum_{i=1}^{p}\eta_i\Big(\frac{(\rho_{i+1}-a)^3}{3!}-\frac{(\rho_i-a)^3}{3!}\Big)+\sum_{j=1}^{m}\mu_j\frac{(\sigma_j-a)^2}{2}\Big]\\
&+\frac{b_2}{|\Lambda|}\Big[\frac{(T-a)^2}{2}+\sum_{i=1}^{p}\omega_i\Big(\frac{(\rho_{i+1}-a)^2}{2}-\frac{(\rho_i-a)^2}{2}\Big)+\sum_{j=1}^{m}\nu_j(\sigma_j-a)\Big]\Big\}\\
\leq\ & \ell Q_1 \parallel u-v\parallel_{\mathcal{H}}.
\end{aligned}
$$

In a similar manner, one can obtain

$$\|(\mathcal{L}u)'-(\mathcal{L}v)'\|\leq \ell Q_2\parallel u-v\parallel_{\mathcal{H}},\ \|(\mathcal{L}u)''-(\mathcal{L}v)''\|\leq \ell Q_3\parallel u-v\parallel_{\mathcal{H}}.$$

Consequently, we deduce that

$$\|\mathcal{L}u-\mathcal{L}v\|_{\mathcal{H}}\leq \ell Q\parallel u-v\parallel_{\mathcal{H}},$$

which, in view of the given condition ($\ell < Q^{-1}$), shows that the operator $\mathcal{L}$ is a contraction. Thus, by the conclusion of Banach contraction mapping principle, the operator $\mathcal{L}$ has a unique fixed point, which implies that the problem in Equations (1) and (2) has a unique solution on $[a, T]$. □

3.3. Examples

Here, we illustrate the results obtained in the last subsections with the aid of examples.

Example 1. *Consider the following integral multi-point and multi-strip boundary value problem:*

$$u'''(t)=\frac{1}{45\sqrt{t^2+3}}\tan^{-1}u(t)+\frac{1}{162}\frac{|u'|}{(|u'|+1)}+\frac{1}{270t}\frac{|u''|^2}{(|u''|^2+1)}+\cos(t-1),\ t\in[1,4], \tag{25}$$

$$\begin{cases}\displaystyle\int_1^4 u(s)ds=\sum_{i=1}^{4}\xi_i\int_{\rho_i}^{\rho_{i+1}}u(s)ds+\sum_{j=1}^{3}\gamma_j u(\sigma_j),\\ \displaystyle\int_1^4 u'(s)ds=\sum_{i=1}^{4}\eta_i\int_{\rho_i}^{\rho_{i+1}}u'(s)ds+\sum_{j=1}^{3}\mu_j u'(\sigma_j),\\ \displaystyle\int_1^4 u''(s)ds=\sum_{i=1}^{4}\omega_i\int_{\rho_i}^{\rho_{i+1}}u''(s)ds+\sum_{j=1}^{3}\nu_j u''(\sigma_j),\end{cases} \tag{26}$$

where $a = 1$, $T = 4$, $m = 3$, $p = 4$, $\gamma_1 = 1/2$, $\gamma_2 = 7/10$, $\gamma_3 = 9/10$, $\mu_1 = 1/4$, $\mu_2 = 5/12$, $\mu_3 = 7/12$, $\nu_1 = 2/5$, $\nu_2 = 13/20$, $\nu_3 = 9/10$, $\sigma_1 = 7/4, \sigma_2 = 15/8$, $\sigma_3 = 16/8$, $\rho_1 = 5/2$, $\rho_2 = 8/3$, $\rho_3 = 17/6$, $\rho_4 = 18/6$, $\rho_5 = 19/6$, $\xi_1 = 3/4$, $\xi_2 = 25/28$, $\xi_3 = 29/28$, $\xi_4 = 33/28$, $\eta_1 = 2/7$, $\eta_2 =$

$23/56$, $\eta_3 = 15/28$, $\eta_4 = 37/56$, $\omega_1 = 1/5$, $\omega_2 = 2/5$, $\omega_3 = 3/5$, $\omega_4 = 4/5$. *Clearly,* $|f(t,u,u',u'')| \leq \frac{\pi}{90\sqrt{t^2+3}} + \frac{1}{270t} + \frac{163}{162}$ *and*

$$\left|f(t,u,u',u'') - f(t,v,v',v'')\right| \leq \ell\Big(|u-v| + |u'-v'| + |u''-v''|\Big)$$

with $\ell = 1/90$. *Using the given data, it is found that* $A_1 \approx 0.716667 \neq 0, A_2 \approx 1.434524 \neq 0, A_3 \approx 2.768849, A_4 \approx 0.257143 \neq 0, A_5 \approx 1.414087, A_6 \approx 2.512768,$ *and* $|\Lambda| \approx 0.264363$ *(*Λ *and* A_i $(i = 1, \ldots, 6)$ *are defined by Equation (6)),* $Q_1 \approx 35.810002, Q_2 \approx 18.708093, Q_3 \approx 12.638560$ *and* $Q \approx 67.156655$ *(*Q_1, Q_2*,* Q_3 *and* Q *are given by Equations (22), (23), (24) and (21), respectively). Furthermore, we note that all the conditions of Theorem 1 are satisfied with*

$$\ell\Big(Q - \frac{(T-a)}{6}\Big[6 + 3(T-a) + (T-a)^2\Big]\Big) \approx 0.612852 < 1.$$

Hence, the problem in Equations (25) and (26) has a solution on $[1,4]$ *by Theorem 1.*

Since $\ell Q \approx 0.746185 < 1$*, therefore the conclusion of Theorem 3 also applies to Equation (26).*

Example 2. *Consider the third-order ordinary differential equation*

$$u'''(t) = \frac{1}{18\sqrt{t+24}}\Big[\frac{1}{21\pi}\sin(3\pi u) + \frac{3}{4}u'(t) + \frac{|u''|}{|u''|+1}\Big],\ t \in [1,4] \tag{27}$$

supplemented with the boundary conditions in Equation (26). Evidently,

$$|f(t,u,u',u'')| \leq \frac{1}{18\sqrt{t+24}}\Big(\frac{|u|}{7} + \frac{3}{4}|u'(t)| + 1\Big).$$

Let us set $\Psi(\|u\|) = \frac{\|u\|}{7} + \frac{3}{4}\|u'\| + 1,\ p(t) = \frac{1}{18\sqrt{t+24}},\ (\|p\| = \frac{1}{90})$*. The condition* (H_5) *implies that* $N > 2.235673$*. In consequence, it follows by the conclusion of Theorem 2 that the problem (27) and (26) has at least one solution on* $[1,4]$*.*

4. Ulam Stability

This section is concerned with the Ulam stability of the problem in Equations (1) and (2) by considering its equivalent integral equation:

$$\begin{aligned} v(t) &= \int_a^t \frac{(t-s)^2}{2}\widehat{f}(s)ds \\ &-\frac{1}{\Lambda}\int_a^T \Big[A_1A_2\frac{(T-s)^3}{3!} + G_1(t)\frac{(T-s)^2}{2} + G_2(t)(T-s)\Big]\widehat{f}(s)ds \\ &+\frac{1}{\Lambda}\sum_{j=1}^{m}\int_a^{\sigma_j}\Big[\gamma_jA_1A_2\frac{(\sigma_j-s)^2}{2} + \mu_jG_1(t)(\sigma_j-s) + \nu_jG_2(t)\Big]\widehat{f}(s)ds \\ &+\frac{1}{\Lambda}\sum_{i=1}^{p}\int_{\rho_i}^{\rho_{i+1}}\int_a^s\Big[\xi_iA_1A_2\frac{(s-\tau)^2}{2} \\ &+\eta_iG_1(t)(s-\tau) + \omega_iG_2(t)\Big]\widehat{f}(\tau)d\tau ds. \end{aligned} \tag{28}$$

Let us introduce a continuous nonlinear operator $\chi : \mathcal{H} \to \mathcal{H}$ given by

$$\chi v(t) = v'''(t) - \widehat{f}(t).$$

Definition 1. *For each $\epsilon > 0$ and for each solution $v \in \mathcal{H}$, we call the problem in Equations (1) and (2) Ulam–Hyers stable provided that*

$$\|\chi v\| \leq \epsilon, \tag{29}$$

and there exists a solution $v_1 \in \mathcal{H}$ of Equation (1) such that $\|v_1 - v\| \leq \varrho \epsilon_1$ for positive real numbers ϱ and $\epsilon_1(\epsilon)$.

Definition 2. *Let there exist a function $\kappa \in C(\mathbb{R}^+, \mathbb{R}^+)$ and a solution $v_1 \in \mathcal{H}$ of Equation (1) with $|v_1(t) - v(t)| \leq \kappa(\epsilon), t \in [a, T]$ for each solution $v \in \mathcal{H}$ of Equation (1). Then, the problem in Equations (1) and (2) is called generalized Ulam–Hyers stable.*

Definition 3. *The problem in Equations (1) and (2) is said to be Ulam–Hyers–Rassias stable with respect to $\varphi \in C([a, T], \mathbb{R}^+)$ if*

$$|\chi v(t)| \leq \epsilon \varphi(t),\ t \in [a, T], \tag{30}$$

and there exists a solution $v_1 \in \mathcal{H}$ of Equation (1) such that

$$|v_1(t) - v(t)| \leq \varrho \epsilon_1 \varphi(t),\ t \in [a, T],$$

where $\epsilon, \varrho, \epsilon_1$ are the same as defined in Definition 1.

Theorem 4. *If (H_1) and the condition $\ell < Q^{-1}$ (see Theorem 3) are satisfied, then the problem in Equations (1) and (2) is both Ulam–Hyers and generalized Ulam–Hyers stable.*

Proof. Recall that $v_1 \in \mathcal{H}$ is a unique solution of Equation (1) by Theorem 3.6. Let $v \in \mathcal{H}$ be an other solution of (1) which satisfies Equation (29). For every solution $v \in \mathcal{H}$ (given by Equation (28)) of Equation (1), it is easy to see that χ and $\mathcal{L} - I$ are equivalent operators. Therefore, it follows from Equations (16) and (29) and the fixed point property of the operator $\mathcal{L}$ given by Equation (17) that

$$\begin{aligned} |v_1(t) - v(t)| &= |\mathcal{L}v_1(t) - \mathcal{L}v(t) + \mathcal{L}v(t) - v(t)| \leq |\mathcal{L}v_1(t) - \mathcal{L}v(t)| + |\mathcal{L}v(t) - v(t)| \\ &\leq \ell Q \|v_1 - v\|_{\mathcal{H}} + \epsilon, \end{aligned}$$

which, on taking the norm for $t \in [a, T]$ and solving for $\|v_1 - v\|_{\mathcal{H}}$, yields

$$\|v_1 - v\|_{\mathcal{H}} \leq \frac{\epsilon}{1 - \ell Q},$$

where $\epsilon > 0$ and $\ell Q < 1$ (given condition).

Letting $\epsilon_1 = \frac{\epsilon}{1-\ell Q}$, and $\varrho = 1$, the Ulam–Hyers stability condition holds true. Furthermore, one can notice that the generalized Ulam–Hyers stability condition also holds valid if we set $\kappa(\epsilon) = \frac{\epsilon}{1-\ell Q}$. □

Theorem 5. *Let the assumptions of Theorem 4 be satisfied and that there exists a function $\varphi \in C([a, T], \mathbb{R}^+)$ satisfying the condition in Equation (30). Then, the problem in Equations (1) and (2) is Ulam–Hyers–Rassias stable with respect to φ.*

Proof. As argued in the proof of Theorem 4, we can get

$$\|v_1 - v\|_{\mathcal{H}} \leq \epsilon_1 \|\varphi\|,$$

with $\epsilon_1 = \frac{\epsilon}{1-\ell Q}$. □

5. Existence Results for the Problem in Equations (1) and (3)

We only outline the idea for obtaining the existence and uniqueness results for the problem in Equations (1) and (3). In relation to the problem in Equations (1) and (3), we introduce an operator $\mathcal{S} : \mathcal{H} \to \mathcal{H}$ by Lemma 2 as

$$\begin{aligned}(\mathcal{S}u)(t) &= \int_a^t \frac{(t-s)^2}{2}\widehat{f}(s)ds \\ &-\frac{1}{\Delta}\int_a^T \Big[\alpha_2\zeta_1\zeta_2\frac{(T-s)^2}{2} + \beta_2 P_1(t)(T-s) + \delta_2 P_2(t)\Big]\widehat{f}(s)ds \\ &+\frac{1}{\Delta}\sum_{j=1}^m \int_a^{\sigma_j} \Big[\gamma_j\zeta_1\zeta_2\frac{(\sigma_j-s)^2}{2} + \mu_j P_1(t)(\sigma_j-s) + \nu_j P_2(t)\Big]\widehat{f}(s)ds \\ &+\frac{1}{\Delta}\sum_{i=1}^p \int_{\rho_i}^{\rho_{i+1}}\int_a^s \Big[\xi_i\zeta_1\zeta_2\frac{(s-\tau)^2}{2} + \eta_i P_1(t)(s-\tau) + \omega_i P_2(t)\Big]\widehat{f}(\tau)d\tau ds,\end{aligned} \tag{31}$$

where

$$P_1(t) = \zeta_1\Big(\zeta_4(t-a) - \zeta_5\Big),\ P_2(t) = \zeta_3\Big(\zeta_5 - \zeta_4(t-a)\Big) - \zeta_2\Big(\zeta_6 - \zeta_4\frac{(t-a)^2}{2}\Big),$$

and $\zeta_i (i = 1,\ldots,6)$ are given by Equation (15).

Moreover, we set

$$\begin{aligned}\Theta_1 &= \frac{(T-a)^3}{3!} + \frac{1}{|\zeta_4|}\Big[|\alpha_2|\frac{(T-a)^3}{3!} + \sum_{i=1}^p \xi_i\Big(\frac{(\rho_{i+1}-a)^4}{4!} - \frac{(\rho_i-a)^4}{4!}\Big) + \sum_{j=1}^m \gamma_j\frac{(\sigma_j-a)^3}{3!}\Big] \\ &+\frac{p_1}{|\Delta|}\Big[|\beta_2|\frac{(T-a)^2}{2} + \sum_{i=1}^p \eta_i\Big(\frac{(\rho_{i+1}-a)^3}{3!} - \frac{(\rho_i-a)^3}{3!}\Big) + \sum_{j=1}^m \mu_j\frac{(\sigma_j-a)^2}{2}\Big] \\ &+\frac{p_2}{|\Delta|}\Big[|\delta_2|(T-a) + \sum_{i=1}^p \omega_i\Big(\frac{(\rho_{i+1}-a)^2}{2} - \frac{(\rho_i-a)^2}{2}\Big) + \sum_{j=1}^m \nu_j(\sigma_j-a)\Big], \\ \Theta_2 &= \frac{(T-a)^2}{2} + \frac{1}{|\zeta_2|}\Big[|\beta_2|\frac{(T-a)^2}{2} + \sum_{i=1}^p \eta_i\Big(\frac{(\rho_{i+1}-a)^3}{3!} - \frac{(\rho_i-a)^3}{3!}\Big) + \sum_{j=1}^m \mu_j\frac{(\sigma_j-a)^2}{2}\Big] \\ &+\frac{p_3}{|\zeta_1\zeta_2|}\Big[|\delta_2|(T-a) + \sum_{i=1}^p \omega_i\Big(\frac{(\rho_{i+1}-a)^2}{2} - \frac{(\rho_i-a)^2}{2}\Big) + \sum_{j=1}^m \nu_j(\sigma_j-a)\Big], \\ \Theta_3 &= (T-a) + \frac{1}{|\zeta_1|}\Big[|\delta_2|(T-a) + \sum_{i=1}^p \omega_i\Big(\frac{(\rho_{i+1}-a)^2}{2} - \frac{(\rho_i-a)^2}{2}\Big) + \sum_{j=1}^m \nu_j(\sigma_j-a)\Big],\end{aligned} \tag{32}$$

where $\max_{t\in[a,T]}|P_1(t)| = p_1, \max_{t\in[a,T]}|P_2(t)| = p_2$ and $\max_{t\in[a,T]}|\zeta_2(t-a)-\zeta_3| = p_3$ ($P_1(t)$ and $P_2(t)$ are given by Equation (14)). With the aid of the operator $\mathcal{S}$ defined by Equation (31) and the notations in Equation (32), we can obtain the existence results (analog to the ones derived in Section 3) for the problem in Equations (1) and (3). As an example, we formulate the uniqueness result for the problem in Equations (1) and (3) as follows.

Theorem 6. *Let $f : [a,T]\times\mathbb{R}^3 \to \mathbb{R}$ be a continuous function satisfying the Lipschitz condition (H_1) with the Lipschitz constant ℓ_1 (instead of ℓ in (H_1)) such that $\ell_1(\Theta_1+\Theta_2+\Theta_3) < 1$, where Θ_1, Θ_2 and Θ_3 are given by (32). Then, the problem in Equations (1) and (3) has a unique solution on $[a,T]$.*

Now, we present an example illustrating Theorem 6.

Example 3. *Consider the following problem:*

$$\begin{cases} u'''(t) = \dfrac{1}{210}\sin u + \dfrac{1}{4\sqrt{t+440}}u'(t) + \dfrac{1}{168}\dfrac{|u''|}{(|u''|+1)} + e^{-t}, \ t \in [1,4], \\ \alpha_1 u(a) + \alpha_2 u(T) = \sum_{i=1}^{4} \xi_i \int_{\rho_i}^{\rho_{i+1}} u(s)ds + \sum_{j=1}^{3} \gamma_j u(\sigma_j), \\ \beta_1 u'(a) + \beta_2 u'(T) = \sum_{i=1}^{4} \eta_i \int_{\rho_i}^{\rho_{i+1}} u'(s)ds + \sum_{j=1}^{3} \mu_j u'(\sigma_j), \\ \delta_1 u''(a) + \delta_2 u''(T) = \sum_{i=1}^{4} \omega_i \int_{\rho_i}^{\rho_{i+1}} u''(s)ds + \sum_{j=1}^{3} \nu_j u''(\sigma_j), \end{cases} \quad (33)$$

where $\alpha_1 = 1/4$, $\alpha_2 = 1/2$, $\beta_1 = 1/5$, $\beta_2 = 3/8$, $\delta_1 = 1/3$, $\delta_2 = 2/3$. *The other constants are the same as chosen in example 3.7. Clearly,* $|f(t,u,u',u'') - f(t,v,v',v'')| \leq \ell_1(|u-v| + |u'-v'| + |u''-v''|)$, *with* $\ell_1 = 1/84$. *Using the given data, we find that* $|\zeta_1| \approx 1.283333 \neq 0, |\zeta_2| \approx 0.990476 \neq 0, |\zeta_3| \approx 0.606151, |\zeta_4| \approx 1.992857 \neq 0$, $|\zeta_5| \approx 1.585913, |\zeta_6| \approx 0.262769$, *and* $|\Delta| \approx 2.533142$ (Δ *and* ζ_i $(i = 1,\ldots,6)$ *are given by Equation (15)),* $\Theta_1 \approx 23.050129, \Theta_2 \approx 15.505245, \Theta_3 \approx 6.434525$ (Θ_1, Θ_2 *and* Θ_3 *are given by Equation (32)) and* $\ell_1(\Theta_1 + \Theta_2 + \Theta_3) \approx 0.535594 < 1$. *Obviously, all the conditions of Theorem 6 hold and therefore Theorem 6 applies to the problem in Equation (33).*

6. Conclusions

We developed the existence theory and Ulam stability for a third-order nonlinear ordinary differential equation equipped with: (i) nonlocal integral multi-point and multi-strip; and (ii) non-separated integro-multi-point boundary conditions. The results obtained in this paper are new and quite general, and lead to several new ones for appropriate choices of the parameters involved in the problems at hand. For example, letting $\gamma_j = \rho_j = \nu_j = 0, \forall j$ and $\xi_i = \eta_i = \omega_i = 0, \forall i$ in Equation (2), the results for the problem in Equations (1) and (2), respectively, correspond to the ones for: (i) nonlocal integral multi-strip boundary conditions; and (ii) nonlocal integral multi-point boundary conditions. Likewise, by fixing $\alpha_k = \beta_k = \delta_k = 0, k = 1,2$ in the results of this paper, we obtain the ones for a third-order differential equation with purely nonlocal multi-point and multi-strip boundary conditions. Setting $\gamma_j = \rho_j = \nu_j = \xi_i = \eta_i = \omega_i = 0, \forall j,i$ and $\alpha_k = \beta_k = \delta_k = 1, k = 1,2$, the results obtained for the problem in Equations (1) and (3) reduce to the ones for anti-periodic boundary conditions. In the nutshell, the work presented in this paper significantly contributes to the existing literature on the topic.

Author Contributions: All authors contributed equally in this work.

Funding: This research received no external funding.

Acknowledgments: The authors thank the reviewers for their useful remarks on our work.

Conflicts of Interest: The authors declare no conflict of interest.

References

1. Zheng, L.; Zhang, X. *Modeling and Analysis of Modern Fluid Problems. Mathematics in Science and Engineering*; Elsevier/Academic Press: London, UK, 2017.
2. Akyildiz, F.T.; Bellout, H.; Vajravelu, K.; van Gorder, R.A. Van Gorder, Existence results for third order nonlinear boundary value problems arising in nano boundary layer fluid flows over stretching surfaces. *Nonlinear Anal. Real World Appl.* **2011**, *12*, 2919–2930. [CrossRef]
3. Hayat, T.; Aziz, A.; Muhammad, T.; Ahmad, B. On magnetohydrodynamic flow of second grade nanofluid over a nonlinear stretching sheet. *J. Magn. Magn. Mater.* **2016**, *408*, 99–106. [CrossRef]
4. Hayat, T.; Kiyani, M.Z.; Ahmad, I.; Ahmad, B. On analysis of magneto Maxwell nano-material by surface with variable thickness. *Int. J. Mech. Sci.* **2017**, *131–132*, 1016–1025. [CrossRef]

5. Ionkin, N.I. The solution of a certain boundary value problem of the theory of heat conduction with a nonclassical boundary condition. *Differ. Uravn.* **1977**, *13*, 294–304. (In Russian)
6. Il'in, V.A.; Moiseev, E.I. Nonlocal boundary-value problem of the first kind for a Sturm-Liouville operator in its differential and finite difference aspects. *Diff. Equa.* **1987**, *23* , 803–810.
7. Eloe, P.W.; Ahmad, B. Positive solutions of a nonlinear nth order boundary value problem with nonlocal conditions. *Appl. Math. Lett.* **2005**, *18* , 521–527. [CrossRef]
8. Webb, J.R.L.; Infante, G. Positive solutions of nonlocal boundary value problems: A unified approach. *J. London Math. Soc.* **2006**, *74* , 673–693. [CrossRef]
9. Clark, S.; Henderson, J. Uniqueness implies existence and uniqueness criterion for non local boundary value problems for third-order differential equations. *Proc. Am. Math. Soc.* **2006**, *134*, 3363–3372. [CrossRef]
10. Graef, J.R.; Webb, J.R.L. Third order boundary value problems with nonlocal boundary conditions. *Nonlinear Anal.* **2009**, *71*, 1542–1551. [CrossRef]
11. Taylor, C.; Hughes, T.; Zarins, C. Finite element modeling of blood flow in arteries. *Comput. Methods Appl. Mech. Eng.* **1998**, *158*, 155–196. [CrossRef]
12. Nicoud, F.; Schfonfeld, T. Integral boundary conditions for unsteady biomedical CFD applications. *Int. J. Numer. Meth. Fluids* **2002**, *40*, 457–465. [CrossRef]
13. Ahmad, B.; Alsaedi, A.; Alghamdi, B.S. Analytic approximation of solutions of the forced Duffing equation with integral boundary conditions. *Nonlinear Anal. Real World Appl.* **2008**, *9*, 1727–1740. [CrossRef]
14. Ahmad, B.; Ntouyas, S.K. A study of higher-order nonlinear ordinary differential equations with four-point nonlocal integral boundary conditions. *J. Appl. Math. Comput.* **2012**, *39*, 1–2, 97–108. [CrossRef]
15. Henderson, J. Smoothness of solutions with respect to multi-strip integral boundary conditions for nth order ordinary differential equations. *Nonlinear Anal. Model. Control* **2014**, *19*, 396–412. [CrossRef]
16. Ahmad, B.; Ntouyas, S.K.; Alsulami, H. Existence of solutions or nonlinear nth-order differential equations and inclusions with nonlocal and integral boundary conditions via fixed point theory. *Filomat* **2014**, *28*, 2149–2162. [CrossRef]
17. Karaca, I.Y.; Fen, F.T. Positive solutions of nth-order boundary value problems with integral boundary conditions. *Math. Model. Anal.* **2015**, *20*, 188–204. [CrossRef]
18. Ahmad, B.; Alsaedi, A.; Al-Malki, N. On higher-order nonlinear boundary value problems with nonlocal multipoint integral boundary conditions. *Lith. Math. J.* **2016**, *56*, 143–163. [CrossRef]
19. Boukrouche, M.; Tarzia, D.A. A family of singular ordinary differential equations of the third order with an integral boundary condition. *Bound. Value Probl.* **2018**, *32*. [CrossRef]
20. Ahmad, B.; Hayat, T. Diffraction of a plane wave by an elastic knife-edge adjacent to a rigid strip. *Canad. Appl. Math. Quart.* **2001**, *9*, 303–316.
21. Wang, G.S.; Blom, A.F. A strip model for fatigue crack growth predictions under general load conditions. *Eng. Fract. Mech.* **1991**, *40*, 507–533. [CrossRef]
22. Yusufoglu, E.; Turhan, I. A mixed boundary value problem in orthotropic strip containing a crack. *J. Franklin Inst.* **2012**, *349*, 2750–2769. [CrossRef]
23. Renterghem, T.V.; Botteldooren, D.; Verheyen, K. Road traffic noise shielding by vegetation belts of limited depth. *J. Sound Vib.* **2012**, *331*, 2404–2425. [CrossRef]
24. Smart, D.R. *Fixed Point Theorems*; Cambridge Tracts in Mathematics: London, UK, 2005.
25. Granas, A.; Dugundji, J. *Fixed Point Theory*; Springer: New York, NY, USA, 2005.

© 2019 by the authors. Licensee MDPI, Basel, Switzerland. This article is an open access article distributed under the terms and conditions of the Creative Commons Attribution (CC BY) license (http://creativecommons.org/licenses/by/4.0/).

symmetry MDPI

Article

Geometric Properties of Certain Analytic Functions Associated with the Dziok-Srivastava Operator

Cai-Mei Yan [1] and Jin-Lin Liu [2,*]

1 Information Engineering College, Yangzhou University, Yangzhou 225002, China; cmyan@yzu.edu.cn
2 Department of Mathematics, Yangzhou University, Yangzhou 225002, China
* Correspondence: jlliu@yzu.edu.cn

Received: 20 January 2019; Accepted: 15 February 2019; Published: 19 February 2019

Abstract: The objective of the present paper is to derive certain geometric properties of analytic functions associated with the Dziok–Srivastava operator.

Keywords: analytic function; subordination; Dziok–Srivastava operator

2010 Mathematics Subject Classification: 30C45

1. Introduction

Throughout this paper, we assume that:

$$n, p \in \mathbb{N},\ -1 \leq B < A \leq 1,\ \alpha > 0 \text{ and } \beta < 1. \tag{1}$$

Let $A_n(p)$ denote the class of functions of the form:

$$f(z) = z^p + \sum_{k=n+p}^{\infty} a_k z^k \tag{2}$$

which are analytic in the open unit disk $U = \{z : |z| < 1\}$. If $f(z) = z^p + \sum_{k=n+p}^{\infty} a_k z^k \in A_n(p)$ and $g(z) = z^p + \sum_{k=n+p}^{\infty} b_k z^k \in A_n(p)$, then the Hadamard product (or convolution) of f and g is defined by:

$$(f * g)(z) = z^p + \sum_{k=n+p}^{\infty} a_k b_k z^k$$

For:

$$\alpha_j \in \mathbb{C}\ (j = 1, 2, \cdots, l) \text{ and } \beta_j \in \mathbb{C} \setminus \{0, -1, -2, \cdots\}\ (j = 1, 2, \cdots, m)$$

the generalized hypergeometric function ${}_lF_m(\alpha_1, \cdots, \alpha_l; \beta_1, \cdots, \beta_m; z)$ is defined by:

$${}_lF_m(\alpha_1, \cdots, \alpha_l; \beta_1, \cdots, \beta_m; z) = \sum_{k=0}^{\infty} \frac{(\alpha_1)_k \cdots (\alpha_l)_k}{(\beta_1)_k \cdots (\beta_m)_k} \frac{z^k}{k!}$$

$$(l \leq m + 1; l, m \in \mathbb{N}_0 = \mathbb{N} \cup \{0\}; z \in U)$$

where $(x)_k$ is the Pochhammer symbol given by $(x)_k = x(x+1) \cdots (x+k-1)$ for $k \in \mathbb{N}$ and $(x)_0 = 1$. Corresponding to the function $z^p\, {}_lF_m(\alpha_1, \cdots, \alpha_l; \beta_1, \cdots, \beta_m; z)$, the well-known Dziok–Srivastava operator [1] $H(\alpha_1, \cdots, \alpha_l; \beta_1, \cdots, \beta_m) : A_n(p) \to A_n(p)$ is defined by:

$$H(\alpha_1, \cdots, \alpha_l; \beta_1, \cdots, \beta_m) f(z) = (z^p\, {}_lF_m(\alpha_1, \cdots, \alpha_l; \beta_1, \cdots, \beta_m; z) * f(z)$$

$$(l \leq m+1; l, m \in \mathbb{N}_0; z \in U)$$

If $f \in A_n(p)$ is given by (2), then we have:

$$H(\alpha_1, \cdots, \alpha_l; \beta_1, \cdots, \beta_m)f(z) = z^p + \sum_{k=n+p}^{\infty} \frac{(\alpha_1)_k \cdots (\alpha_l)_k}{(\beta_1)_k \cdots (\beta_m)_k} \frac{a_k}{k!} z^k$$

For convenience, we write:

$$H_m^l(\alpha_1) = H(\alpha_1, \cdots, \alpha_l; \beta_1, \cdots, \beta_m) \quad (l \leq m+1; l, m \in \mathbb{N}_0)$$

It is noteworthy to mention that the Dziok–Srivastava operator is a generalization of certain linear operators considered in earlier investigations.

Next, we consider the function $h(A, B; z) = (1 + Az)/(1 + Bz)$ for $z \in U$. It is known that the function $h(A, B; z)$ is the conformal map of U onto a disk, symmetrical with respect to the real axis, which is centered at the point $(1 - AB)/(1 - B^2)$ $(B \neq \pm 1)$ and with its radius equal to $(A - B)/(1 - B^2)$ $(B \neq \pm 1)$. Furthermore, the boundary circle of this disk intersects the real axis at the points $(1 - A)/(1 - B)$ and $(1 + A)/(1 + B)$ with $B \neq \pm 1$.

Let $P[A, B]$ denote the class of functions of the form $p(z) = 1 + p_1 z + \cdots$, which are analytic in U and satisfy the subordination $p(z) \prec h(A, B; z)$. It is clear that $p \in P[A, B]$ if and only if:

$$\left| p(z) - \frac{1 - AB}{1 - B^2} \right| < \frac{A - B}{1 - B^2} \quad (-1 < B < A \leq 1; z \in U)$$

and:

$$\mathrm{Re}\, p(z) > \frac{1 - A}{2} \quad (B = -1; z \in U)$$

For two functions f and g analytic in U, f is said to be subordinate to g, written by $f(z) \prec g(z) \quad (z \in U)$, if there exists a Schwarz function w in U such that:

$$|w(z)| \leq |z| \quad \text{and} \quad f(z) = g(w(z)) \quad (z \in U)$$

Furthermore, if the function g is univalent in U, then:

$$f(z) \prec g(z) \quad (z \in U) \Longleftrightarrow f(0) = g(0) \quad \text{and} \quad f(U) \subset g(U)$$

Many properties of analytic functions have been investigated by several authors(see [1–11]). In this paper, we derive certain geometric properties of analytic functions associated with the well-known Dziok–Srivastava operator.

2. Main Results

Theorem 1. *Let f belong to the class $A_n(p)$. Furthermore, let:*

$$\frac{H_m^l(\alpha_1)f(z)}{z^p} \in P[A, B]. \tag{3}$$

Then:

$$\mathrm{Re}\left\{ \frac{H_m^l(\alpha_1)f(z)}{z^p} + \alpha z \left(\frac{H_m^l(\alpha_1)f(z)}{z^p} \right)' \right\}$$

$$\leq \begin{cases} \frac{1+(A+B+n\alpha(A-B))r^n+ABr^{2n}}{(1+Br^n)^2} & \text{if } M_n(A,B,\alpha,r) \leq 0, \quad (4) \\ \frac{L_n^2-4\alpha^2 K_A K_B}{4\alpha(A-B)r^{n-1}(1-r^2)K_B} & \text{if } M_n(A,B,\alpha,r) \geq 0, \quad (5) \end{cases}$$

where:

$$\begin{cases} K_A = 1 - A^2 r^{2n} + nAr^{n-1}(1-r^2), \\ K_B = 1 - B^2 r^{2n} + nBr^{n-1}(1-r^2), \\ L_n = 2\alpha(1 - ABr^{2n}) + n\alpha(A+B)r^{n-1}(1-r^2) + (A-B)r^{n-1}(1-r^2), \\ M_n(A,B,\alpha,r) = 2\alpha K_B(1+Ar^n) - L_n(1+Br^n). \end{cases} \tag{6}$$

The result is sharp.

Proof. For $z = 0$, the equality in (4) holds true. Thus, we assume that $0 < |z| = r < 1$. From (3), we can write:

$$\frac{H_m^l(\alpha_1)f(z)}{z^p} = \frac{1 + Az^n\varphi(z)}{1 + Bz^n\varphi(z)} \quad (z \in U), \tag{7}$$

where $\varphi(z)$ is analytic and $|\varphi(z)| \le 1$ in U. From (7), we have:

$$\begin{aligned} &\frac{H_m^l(\alpha_1)f(z)}{z^p} + \alpha z\left(\frac{H_m^l(\alpha_1)f(z)}{z^p}\right)' \\ &= \frac{H_m^l(\alpha_1)f(z)}{z^p} + \frac{\alpha(A-B)z^n(n\varphi(z) + z\varphi'(z))}{(1+Bz^n\varphi(z))^2} \\ &= \frac{H_m^l(\alpha_1)f(z)}{z^p} + \frac{n\alpha}{A-B}(A - BH_m^l(\alpha_1)f(z)/z^p)(H_m^l(\alpha_1)f(z)/z^p - 1) \\ &+ \frac{\alpha(A-B)z^{n+1}\varphi'(z)}{(1+Bz^n\varphi(z))^2} \end{aligned} \tag{8}$$

By using the Carathéodory inequality:

$$|\varphi'(z)| \le \frac{1 - |\varphi(z)|^2}{1 - r^2},$$

we get:

$$\begin{aligned} \operatorname{Re}\left\{\frac{z^{n+1}\varphi'(z)}{(1+Bz^n\varphi(z))^2}\right\} &\le \frac{r^{n+1}(1 - |\varphi(z)|^2)}{(1-r^2)|1+Bz^n\varphi(z)|^2} \\ &= \frac{r^{2n}|A - BH_m^l(\alpha_1)f(z)/z^p|^2 - |H_m^l(\alpha_1)f(z)/z^p - 1|^2}{(A-B)^2 r^{n-1}(1-r^2)} \end{aligned} \tag{9}$$

Set $\frac{H_m^l(\alpha_1)f(z)}{z^p} = u + iv \quad (u, v \in \mathbb{R})$. Then, (8) and (9) give:

$$\begin{aligned} &\operatorname{Re}\left\{\frac{H_m^l(\alpha_1)f(z)}{z^p} + \alpha z\left(\frac{H_m^l(\alpha_1)f(z)}{z^p}\right)'\right\} \\ &\le \left(1 + n\alpha\frac{A+B}{A-B}\right)u - \frac{n\alpha A}{A-B} - \frac{n\alpha B}{A-B}(u^2 - v^2) \\ &+ \alpha\frac{r^{2n}((A-Bu)^2 + (Bv)^2) - ((u-1)^2 + v^2)}{(A-B)r^{n-1}(1-r^2)} \\ &= \left(1 + n\alpha\frac{A+B}{A-B}\right)u - \frac{n\alpha}{A-B}(A + Bu^2) \\ &+ \alpha\frac{r^{2n}(A-Bu)^2 - (u-1)^2}{(A-B)r^{n-1}(1-r^2)} + \frac{\alpha}{A-B}\left(nB - \frac{1 - B^2r^{2n}}{r^{n-1}(1-r^2)}\right)v^2 \end{aligned} \tag{10}$$

Note that:

$$\frac{1-B^2r^{2n}}{r^{n-1}(1-r^2)} \geq \frac{1-r^{2n}}{r^{n-1}(1-r^2)} = \frac{1}{r^{n-1}}(1+r^2+r^4+\cdots+r^{2(n-2)}+r^{2(n-1)})$$
$$= \frac{1}{2r^{n-1}}[(1+r^{2(n-1)})+(r^2+r^{2(n-2)})+\cdots+(r^{2(n-1)}+1)] \quad (11)$$
$$\geq n \geq nB$$

Using (10) and (11), we obtain:

$$\mathrm{Re}\{\frac{H_m^l(\alpha_1)f(z)}{z^p}+\alpha z(\frac{H_m^l(\alpha_1)f(z)}{z^p})'\} \leq \left(1+n\alpha\frac{A+B}{A-B}\right)u-\frac{n\alpha}{A-B}(A+Bu^2)$$
$$+\alpha\frac{r^{2n}(A-Bu)^2-(u-1)^2}{(A-B)r^{n-1}(1-r^2)} = \psi_n(u) \quad (12)$$

It is known that for $|\xi| \leq \sigma \ (\sigma < 1)$,

$$\left|\frac{1+A\xi}{1+B\xi}-\frac{1-AB\sigma^2}{1-B^2\sigma^2}\right| \leq \frac{(A-B)\sigma}{1-B^2\sigma^2} \quad (13)$$

and:

$$\frac{1-A\sigma}{1-B\sigma} \leq \mathrm{Re}\left\{\frac{1+A\xi}{1+B\xi}\right\} \leq \frac{1+A\sigma}{1+B\sigma} \quad (14)$$

Furthermore, (7) and (14) show that:

$$\frac{1-Ar^n}{1-Br^n} \leq \mathrm{Re}\{\frac{H_m^l(\alpha_1)f(z)}{z^p}\} \leq \frac{1+Ar^n}{1+Br^n}.$$

Now, we calculate the maximum value of $\psi_n(u)$ on the segment $\left[\frac{1-Ar^n}{1-Br^n}, \frac{1+Ar^n}{1+Br^n}\right]$. Obviously,

$$\psi_n'(u) = 1+n\alpha\frac{A+B}{A-B}-\frac{2n\alpha B}{A-B}u+2\alpha\frac{(1-ABr^{2n})-(1-B^2r^{2n})u}{(A-B)r^{n-1}(1-r^2)}$$

$$\psi_n''(u) = -\frac{2\alpha}{A-B}\left(nB+\frac{1-B^2r^{2n}}{r^{n-1}(1-r^2)}\right) < 0 \quad \text{(see (11))} \quad (15)$$

and $\psi_n'(u) = 0$ if and only if:

$$u = u_n = \frac{2\alpha(1-ABr^{2n})+n\alpha(A+B)r^{n-1}(1-r^2)+(A-B)r^{n-1}(1-r^2)}{2\alpha[1-B^2r^{2n}+nBr^{n-1}(1-r^2)]}$$
$$= \frac{L_n}{2\alpha K_B} \quad \text{(see (6))} \quad (16)$$

Since:

$$2\alpha K_B(1-Ar^n)-L_n(1-Br^n)$$
$$= 2\alpha[(1-Ar^n)(1-B^2r^{2n})-(1-Br^n)(1-ABr^{2n})]$$
$$-n\alpha r^{n-1}(1-r^2)[(A+B)(1-Br^n)-2B(1-Ar^n)]-(A-B)r^{n-1}(1-r^2)(1-Br^n)$$
$$= -2\alpha(A-B)r^n(1-Br^n)-n\alpha(A-B)r^{n-1}(1-r^2)(1+Br^n)$$
$$-(A-B)r^{n-1}(1-r^2)(1-Br^n) < 0$$

we see that:

$$u_n > \frac{1-Ar^n}{1-Br^n} \quad (17)$$

However, u_n is not always less than $\frac{1+Ar^n}{1+Br^n}$. The following two cases arise.

Case (I). $u_n \geq \frac{1+Ar^n}{1+Br^n}$, that is $M_n(A,B,\alpha,r) \leq 0$. In view of $\psi_n'(u_n) = 0$ and (15), the function $\psi_n(u)$ is increasing on the segment $\left[\frac{1-Ar^n}{1-Br^n}, \frac{1+Ar^n}{1+Br^n}\right]$. Thus, we deduce from (12) that, if $M_n(A,B,\alpha,r) \leq 0$, then:

$$\begin{aligned} \text{Re}\left\{\frac{H_m^l(\alpha_1)f(z)}{z^p} + \alpha z(\frac{H_m^l(\alpha_1)f(z)}{z^p})'\right\} &\leq \psi_n\left(\frac{1+Ar^n}{1+Br^n}\right) \\ &= \left(1+n\alpha\frac{A+B}{A-B}\right)\left(\frac{1+Ar^n}{1+Br^n}\right) \\ &\quad - \frac{n\alpha}{A-B}\left(A+B\left(\frac{1+Ar^n}{1+Br^n}\right)^2\right) \\ &= \frac{1+Ar^n}{1+Br^n} - \frac{n\alpha}{A-B}\left(1-\frac{1+Ar^n}{1+Br^n}\right)\left(A-B\frac{1+Ar^n}{1+Br^n}\right) \\ &= \frac{1+(A+B+n\alpha(A-B))r^n + ABr^{2n}}{(1+Br^n)^2} \end{aligned}$$

This gives (4).

Next, we consider the function f defined by:

$$\frac{H_m^l(\alpha_1)f(z)}{z^p} = \frac{1+Az^n}{1+Bz^n}$$

which satisfies the condition (3). It is easy to check that:

$$\frac{H_m^l(\alpha_1)f(r)}{r^p} + \alpha r(\frac{H_m^l(\alpha_1)f(r)}{r^p})' = \frac{1+(A+B+n\alpha(A-B))r^n + ABr^{2n}}{(1+Br^n)^2}$$

which implies that the inequality (4) is sharp.

Case (II). $u_n \leq \frac{1+Ar^n}{1+Br^n}$, that is $M_n(A,B,\alpha,r) \geq 0$. In this case, we easily have:

$$\text{Re}\left\{\frac{H_m^l(\alpha_1)f(z)}{z^p} + \alpha z(\frac{H_m^l(\alpha_1)f(z)}{z^p})'\right\} \leq \psi_n(u_n) \tag{18}$$

In view of (6), $\psi_n(u)$ in (12) can be written as:

$$\psi_n(u) = \frac{-\alpha K_B u^2 + L_n u - \alpha K_A}{(A-B)r^{n-1}(1-r^2)} \tag{19}$$

Therefore, if $M_n(A,B,\alpha,r) \geq 0$, then it follows from (16), (18), and (19) that:

$$\begin{aligned} \text{Re}\left\{\frac{H_m^l(\alpha_1)f(z)}{z^p} + \alpha z(\frac{H_m^l(\alpha_1)f(z)}{z^p})'\right\} &\leq \frac{-\alpha K_B u_n^2 + L_n u_n - \alpha K_A}{(A-B)r^{n-1}(1-r^2)} \\ &= \frac{L_n^2 - 4\alpha^2 K_A K_B}{4\alpha(A-B)r^{n-1}(1-r^2)K_B} \end{aligned}$$

To show the sharpness, we take:

$$\frac{H_m^l(\alpha_1)f(z)}{z^p} = \frac{1+Az^n\varphi(z)}{1+Bz^n\varphi(z)} \quad \text{and} \quad \varphi(z) = \frac{z-c_n}{1-c_n z}$$

where $c_n \in \mathbb{R}$ is determined by:

$$\frac{H_m^l(\alpha_1)f(r)}{r^p} = \frac{1 + Ar^n\varphi(r)}{1 + Br^n\varphi(r)} = u_n \in \left(\frac{1 - Ar^n}{1 - Br^n}, \frac{1 + Ar^n}{1 + Br^n}\right]$$

Clearly, $-1 < \varphi(r) \leq 1, -1 \leq c_n < 1, |\varphi(z)| \leq 1$ $(z \in U)$, and so, f satisfies the condition (3). Since:

$$\varphi'(r) = \frac{1 - c_n^2}{(1 - c_n r)^2} = \frac{1 - |\varphi(r)|^2}{1 - r^2}$$

from the above argument, we find that:

$$\frac{H_m^l(\alpha_1)f(r)}{r^p} + \alpha r(\frac{H_m^l(\alpha_1)f(r)}{z^p})' = \psi_n(u_n)$$

The proof of the theorem is now completed. □

Corollary 1. *Let $f \in A_1(p)$, and satisfy* $\text{Re}\{H_m^l(\alpha_1)f(z)/z^p\} > \beta$ $(\beta < 1; z \in U)$. *Then, for* $|z| = r < 1$,

$$\text{Re}\left\{\frac{H_m^l(\alpha_1)f(z)}{z^p} + \alpha z(\frac{H_m^l(\alpha_1)f(z)}{z^p})'\right\} \leq \beta + (1 - \beta)\frac{1 + 2\alpha r - r^2}{(1 - r)^2}$$

The result is sharp.

Proof. By considering $\frac{H_m^l(\alpha_1)f(z)/z^p - \beta}{1 - \beta}$ instead of $H_m^l(\alpha_1)f(z)/z^p$, we only need to prove the corollary for $\beta = 0$. Putting $n = A = 1$ and $B = -1$ in (6), we get:

$$K_1 = 2(1 - r^2), \quad K_{-1} = 0, \quad L_1 = 2\alpha(1 + r^2) + 2(1 - r^2)$$

and:

$$M_1(1, -1, \alpha, r) = -2(1 - r)[1 + \alpha - (1 - \alpha)r^2] \leq 0$$

Consequently, an application of (4) in Theorem 2.1 yields:

$$\text{Re}\left\{\frac{H_m^l(\alpha_1)f(z)}{z^p} + \alpha z(\frac{H_m^l(\alpha_1)f(z)}{z^p})'\right\} \leq \frac{1 + 2\alpha r - r^2}{(1 - r)^2}$$

The sharpness follows immediately from that of Theorem 1. □

Theorem 2. *Let α_j $(j = 1, 2, \cdots, l)$ and β_s $(s = 1, 2, \cdots, m)$ be positive real numbers. Furthermore, let* $f(z) = z^p + \sum_{k=n+p}^{\infty} a_k z^k \in A_n(p)$, *and satisfy:*

$$\frac{H_m^l(\alpha_1)f(z)}{z^p} + \alpha z\left(\frac{H_m^l(\alpha_1)f(z)}{z^p}\right)' \in P[A, B] \tag{20}$$

Then:

$$|a_k| \leq \frac{k!(A - B)(\beta_1)_k \cdots (\beta_m)_k}{(1 + \alpha(k - p))(\alpha_1)_k \cdots (\alpha_l)_k} \quad (k \geq n + p) \tag{21}$$

The result is sharp for each $k \geq n + p$.

Proof. It is well known that if:

$$g(z) = \sum_{k=1}^{\infty} b_k z^k \prec \varphi(z) \quad (z \in U)$$

where $g(z)$ is analytic in U and $\varphi(z) = z + \cdots$ is convex univalent in U, then $|b_k| \leq 1$ $(k = 1, 2, 3, \cdots)$.

From (20), we have:

$$\begin{aligned}\frac{1}{A-B}\left(\frac{H_m^l(\alpha_1)f(z)}{z^p} + \alpha z\left(\frac{H_m^l(\alpha_1)f(z)}{z^p}\right)' - 1\right) \\ = \frac{1}{A-B}\sum_{k=n+p}^{\infty}\frac{(1+\alpha(k-p))(\alpha_1)_k\cdots(\alpha_l)_k\cdot a_k}{k!(\beta_1)_k\cdots(\beta_m)_k}z^{k-p} \\ \prec \frac{z}{1+Bz} \quad (z \in U)\end{aligned} \tag{22}$$

In view of the function $\frac{z}{1+Bz}$ being convex univalent in U, it follows from (22) that:

$$\frac{(1+\alpha(k-p))(\alpha_1)_k\cdots(\alpha_l)_k}{k!(A-B)(\beta_1)_k\cdots(\beta_m)_k}|a_k| \leq 1 \quad (k \geq n+p)$$

which gives (21).

Next, we consider the function $f_{k-p}(z)$ defined by:

$$f_{k-p}(z) = z^p + (A-B)\sum_{q=1}^{\infty}\frac{(-B)^{q-1}(\beta_1)_{qk}\cdots(\beta_m)_{qk}(qk)!}{(1+\alpha q(k-p))(\alpha_1)_{qk}\cdots(\alpha_l)_{qk}}z^{q(k-p)+p} \quad (z \in U; k \geq n+p)$$

Since:

$$\frac{H_m^l(\alpha_1)f_{k-p}(z)}{z^p} + \alpha z\left(\frac{H_m^l(\alpha_1)f_{k-p}(z)}{z^p}\right)' = \frac{1+Az^{k-p}}{1+Bz^{k-p}} \prec \frac{1+Az}{1+Bz} \quad (z \in U)$$

and:

$$f_{k-p}(z) = z^p + \frac{k!(A-B)(\beta_1)_k\cdots(\beta_m)_k}{(1+\alpha(k-p))(\alpha_1)_k\cdots(\alpha_l)_k}z^k + \cdots$$

for each $k \geq n+p$, the proof of Theorem 2 is completed. □

Author Contributions: All authors contributed equally.

Funding: This work is supported by the National Natural Science Foundation of China (Grant No. 11571299).

Conflicts of Interest: The authors declare no conflict of interest.

References

1. Dziok, J.; Srivastava, H.M. Classes of analytic functions associated with the generalized hypergeometric function. *Appl. Math. Comput.* **1999**, *103*, 1–13. [CrossRef]
2. Chichra, P.N. New subclasses of the class of close-to-convex functions. *Proc. Am. Math. Soc.* **1977**, *62*, 37–43. [CrossRef]
3. Ali, R.M. On a subclass of starlike functions. *Rocky Mt. J. Math.* **1994**, *24*, 447–451. [CrossRef]
4. Dziok, J.; Srivastava, H.M. Certain subclasses of analytic functions associated with the generalized hypergeometric function. *Integral Transforms Spec. Funct.* **2003**, *14*, 7–18. [CrossRef]
5. Gao, C.-Y.; Zhou, S.-Q. Certain subclass of starlike functions. *Appl. Math. Comput.* **2007**, *187*, 176–182. [CrossRef]
6. H. Silverman, A class of bounded starlike functions, *Int. J. Math. Math. Sci.* **1994**, *17*, 249–252. [CrossRef]
7. Singh, R.; Singh, S. Convolution properties of a class of starlike functions. *Proc. Am. Math. Soc.* **1989**, *106*, 145–152. [CrossRef]
8. Srivastava, H.M. Some Fox-Wright generalized hypergeometric functions and associated families of convolution operators. *Appl. Anal. Discret. Math.* **2007**, *1*, 56–71.

9. Srivastava, H.M.; Frasin, B.A.; Pescar, V. Univalence of integral operators involving Mittag-Leffler functions. *Appl. Math. Inf. Sci.* **2017**, *11*, 635–641. [CrossRef]
10. Srivastava, H.M.; Yang, D.-G.; Xu, N.-E. Subordination for multivalent analytic functions associated with the Dziok–Srivastava operator. *Integral Transforms Spec. Funct.* **2009**, *20*, 581–606. [CrossRef]
11. Srivastava, H.M.; Prajapati, A.; Gochhayat, P. Third-order differential subordination and differential superordination results for analytic functions involving the Srivastava-Attiya operator. *Appl. Math. Inf. Sci.* **2018**, *12*, 469–481. [CrossRef]

© 2019 by the authors. Licensee MDPI, Basel, Switzerland. This article is an open access article distributed under the terms and conditions of the Creative Commons Attribution (CC BY) license (http://creativecommons.org/licenses/by/4.0/).

symmetry

MDPI

Article

Continuous Wavelet Transform of Schwartz Tempered Distributions in $S'(\mathbb{R}^n)$

Jagdish Narayan Pandey [1], Jay Singh Maurya [2] and Santosh Kumar Upadhyay [2,*] and Hari Mohan Srivastava [3,4,*]

1 School of Mathematics and Statistics, Carleton University, Ottawa, ON K1S 5B6, Canada; jimpandey1932@gmail.com
2 Department of Mathematical Sciences, Indian Institute of Technology (Banaras Hindu University), Varanasi-221005, India; jaysinghmaurya.rs.mat17@itbhu.ac.in
3 Department of Mathematics and Statistics, University of Victoria, Victoria, BC V8W 3R4, Canada
4 Department of Medical Research, China Medical University Hospital, China Medical University, Taichung 40402, Taiwan
* Correspondence: sk_upadhyay2001@yahoo.com (S.K.U.); harimsri@math.uvic.ca (H.M.S.); Tel.: +91-94501-12714 (S.K.U.); +1-250-472-5313 (H.M.S.)

Received: 11 January 2019; Accepted: 12 February 2019; Published: 15 February 2019

Abstract: In this paper, we define a continuous wavelet transform of a Schwartz tempered distribution $f \in S'(\mathbb{R}^n)$ with wavelet kernel $\psi \in S(\mathbb{R}^n)$ and derive the corresponding wavelet inversion formula interpreting convergence in the weak topology of $S'(\mathbb{R}^n)$. It turns out that the wavelet transform of a constant distribution is zero and our wavelet inversion formula is not true for constant distribution, but it is true for a non-constant distribution which is not equal to the sum of a non-constant distribution with a non-zero constant distribution.

Keywords: function spaces and their duals; distributions; tempered distributions; Schwartz testing function space; generalized functions; distribution space; wavelet transform of generalized functions; Fourier transform

---

1. Introduction

As studied in the earlier works (see, for example, [1–12], we define a Schwartz testing function space $S(\mathbb{R}^n)$ to consist of C^∞ functions ϕ defined on $\mathbb{R}^n$ and satisfying the following conditions:

$$\sup_{x\in\mathbb{R}^n}\left|x_n^{m_n}\cdots x_2^{m_2}x_1^{m_1}\frac{\partial^{k_n}}{\partial x_n}\frac{\partial^{k_{n-1}}}{\partial x_{n-1}}\cdots\frac{\partial^{k_2}}{\partial x_2}\frac{\partial^{k_1}}{\partial x_1}\phi(x_1,x_2,x_3,\cdots x_n)\right|<\infty \tag{1}$$

$|m|, |k| = 0, 1, 2, \cdots$.

The topology over $S(\mathbb{R}^n)$ is generated by the following sequence of semi-norms:

$$\{\gamma_{m,k}\}_{|m|,|k|=0}^{\infty},$$

where

$$\gamma_{m,k}(\phi) = \sup_{x\in\mathbb{R}^n}\left||x^m|\,\phi^{(k)}(x)\right|, \tag{2}$$

$$|m| = m_1 + m_2 + m_3 + \cdots + m_n,$$

$$|k| = k_1 + k_2 + k_3 + \cdots + k_n,$$

$$|x^m| = \left|x_1^{m_1}x_2^{m_2}x_3^{m_3}\cdots x_n^{m_n}\right|,$$

$$\phi^{(k)}(x) = \frac{\partial^{k_n}}{\partial x_n} \cdots \frac{\partial^{k_3}}{\partial x_3} \frac{\partial^{k_2}}{\partial x_2} \frac{\partial^{k_1}}{\partial x_1} \phi(x).$$

These collections of semi-norms in Equation (2) are separating which means that an element $\phi \in S(\mathbb{R}^n)$ is non-zero if and only if there exists at least one of the semi-norms $\gamma_{m,k}$ satisfying $\gamma_{m,k}(\phi) \neq 0$. A sequence $\{\phi_\nu\}_{\nu=1}^{\infty}$ in $S(\mathbb{R}^n)$ tends to ϕ in $S(\mathbb{R}^n)$ if and only if $\gamma_{m,k}(\phi_\nu - \phi) \to 0$ as ν goes to ∞ for each of the subscripts $|m|, |k| = 0, 1, 2, \cdots$, are as defined above. Now, one can verify that the function $e^{-(t_1^2+t_2^2+t_3^3+\cdots+t_n^2)} \in S(\mathbb{R}^n)$ and the sequence

$$\frac{\nu - 1}{\nu} e^{-(t_1^2+t_2^2+t_3^3+\cdots+t_n^2)} \to e^{-(t_1^2+t_2^2+t_3^3+\cdots+t_n^2)}$$

in $S(\mathbb{R}^n)$ as $\nu \to \infty$. The Dirac delta function $\delta(t)$ is defined here by

$$< \delta(t_1 - a_1, t_2 - a_2, t_3 - a_3, \cdots, t_n - a_n), \phi(t_1, t_2, t_3, \cdots, t_n) >= \phi(a_1, a_2, a_3, \cdots, a_n).$$

So, we have

$$< \delta(t_1, t_2, t_3, \cdots, t_n), \phi(t_1, t_2, t_3, \cdots, t_n) >= \phi(0, 0, 0, \cdots, 0) \qquad (\phi \in S(\mathbb{R}^n)).$$

It is easy to check that $\delta(t_1, t_2, \cdots, t_n)$ is a continuous linear functional on $S(\mathbb{R}^n)$. A regular distribution generated by a locally integrable function is an element of $S'(\mathbb{R}^n)$.

Our objective now is to find an element $\psi \in S(\mathbb{R}^n)$, which is a wavelet, so as to be able to define the wavelet transform of $f \in S'(\mathbb{R}^n)$ with respect to this kernel.

A function $\psi \in L^2(\mathbb{R}^n)$ is a window function if it satisfies the following conditions:

$$x_i\psi(x), x_i x_j \psi(x), \cdots, x_1 x_2 x_3 \cdots x_n \psi(x) \tag{3}$$

belonging to $L^2(\mathbb{R}^n)$. Here, $i, j, k, \cdots$ take on all assumed values $1, 2, 3, \cdots$ and all the lower suffixes in a term in Equation (3) are different. It has been proved by Pandey et al. [4,13] that a window function which is an element of $L^2(\mathbb{R}^n)$ belongs to $L^1(\mathbb{R}^n)$. It is easy to verify that every element of $S(\mathbb{R}^n)$ is a window function.

A window function ψ belonging to $L^2(\mathbb{R}^n)$ and satisfying the following condition:

$$\int_{-\infty}^{\infty} \psi(x_1, x_2, x_3, \cdots, x_i, \cdots, x_n) dx_i = 0 \qquad (\forall\, i = 1, 2, 3, \cdots, n) \tag{4}$$

also satisfies the admissibility condition given by

$$\int_{\mathbb{R}^n} \frac{|\hat{\psi}(\Lambda)|^2}{|\Lambda|} d\Lambda < \infty, \tag{5}$$

where

$$\hat{\psi}(\Lambda) = \hat{\psi}(\lambda_1, \lambda_2, \lambda_3, \cdots, \lambda_n),$$
$$|\Lambda| = |\lambda_1 \lambda_2 \cdots \lambda_n|$$

and $\hat{\psi}(\Lambda)$ is the Fourier transform of $\psi(x) \equiv \psi(x_1, x_2, x_3, \cdots, x_n)$ (see also a recent work [14]). Clearly, ψ in Equation (4) is a wavelet [13]. As an example, one can easily verify that the function given by

$$\psi(x) = x_1 x_2 \cdots x_n e^{-(x_1^2+x_2^2+x_3^3+\cdots+x_n^2)}$$

is a wavelet belonging to $S(\mathbb{R}^n)$. Let $s(\mathbb{R}^n)$ be a subspace of $S(\mathbb{R}^n)$ such that every element $\phi \in s(\mathbb{R}^n)$ satisfies Equation (4). Clearly, every element of $s(\mathbb{R}^n)$ is a wavelet [4].

Now, if $f \in S'(\mathbb{R}^n)$ and ψ is a wavelet belonging to $S(\mathbb{R}^n)$, the wavelet transform of f can be defined by

$$W_f(a,b) = \left\langle f(x), \frac{1}{\sqrt{|a|}}\psi(\frac{x-b}{a}) \right\rangle,$$

where $\langle \cdot, \cdot \rangle$ denotes the inner product and (a,b) is the argument of wavelet transform $W_f(a,b)$ of f with respect to wavelet ψ,

$$\psi\left(\frac{x-b}{a}\right) = \psi\left(\frac{x_1-b_1}{a_1}, \frac{x_2-b_2}{a_2}, \frac{x_3-b_3}{a_3}, \cdots, \frac{x_n-b_n}{a_n}\right)$$

$$(a_i \neq 0 \qquad (\forall\, i = 1,2,3,\cdots,n))$$

and

$$|a| = |a_1 a_2 a_3 \cdots a_n| .$$

Our objective next is to prove the following inversion formula:

$$\left\langle \frac{1}{C_\psi} \int_{\mathbb{R}^n}\int_{\mathbb{R}^n} W_f(a,b)\psi(\frac{t-b}{a})\frac{db\, da}{\sqrt{|a|}a^2}, \phi(t) \right\rangle \to \langle f, \phi \rangle, \quad \phi \in S(\mathbb{R}^n) \tag{6}$$

by interpreting the convergence in $S'(\mathbb{R}^n)$. Here, we have

$$C_\psi = (2\pi)^n \int_{\mathbb{R}^n} \frac{|\hat{\psi}(\wedge)|^2}{|\wedge|} d\wedge .$$

The derivation of the inversion formula given by the formula (6) is difficult. We, therefore, make an easy approach. The work on the multidimensional wavelet transform with positive scale $[a > 0]$ was done by Daubechies [15], Meyer [16], Pathak [17], and some others. Motivated by the earlier works [6,8,12], Pandey et al. [4] studied a generalization of these works and extended the multidimensional wavelet transform with real scale $[a \neq 0]$. In the year 1995, Holschneider [18] extended the multidimensional wavelet transform to Schwartz tempered distributions with positive scales $[a > 0]$. Recently, Weisz [19,20] studied the inversion formula for the continuous wavelet transform and found its convergence in L^p and Wiener amalgam spaces. Postnikov et al. [21] studied computational implementation of the inverse continuous wavelet transform without a requirement of the admissibility condition.

Our objective in this investigation is to extend the wavelet transform to Schwartz tempered distributions with real scale $[a \neq 0]$. The standard cut off of negative frequencies (which is required to apply continuous wavelet transform with $a > 0$) may result in a loss of information if the transformed functions were non-symmetric (in the Fourier space) mixture of real and imaginary frequency components. Our proposed and proven inversion formula is free from the mentioned defect. The main advantage of our work is a possible further practical utility of the proven result and the simplicity of calculation; in addition, our extension of the multidimensional wavelet inversion formula is the most general. In [4], it is proved that a window function $\psi(x) \in L^2(\mathbb{R}^n)$ is a wavelet if and only if the integral of ψ along each of the axes is zero; therefore, any $\psi(x) \in s(\mathbb{R}^n)$ is a wavelet. Hence, the wavelet transform of a constant distribution is zero.

We thus realize that two elements of $S'(\mathbb{R}^n)$ having equal wavelet transform will differ by a constant in general. Holschneider [18] uses the wavelet inversion formula for $f \in S'(\mathbb{R}^n)$, but he does not mention the wavelet inversion formula and its convergence in $S'(\mathbb{R}^n)$. Perhaps, he takes it for granted, as such an inversion formula is valid for elements of $L^2(\mathbb{R}^n)$ by interpreting convergence in $L^2(\mathbb{R}^n)$. Our objective in this paper is to fill up all these gaps. We will prove the inversion Formula (6) in Section 3.

2. Structure of Generalized Functions of Slow Growth

Elements of $S'(\mathbb{R}^n)$ are called tempered distributions or distributions of slow growth.

Definition 1. *A function $f(x)$ is said to be a function of slow growth in $\mathbb{R}^n$ if, for $m \geqq 0$, we have*

$$\int_{\mathbb{R}^n} |f(x)| \, (1 + |x|)^{-m} \, dx < \infty$$

and it determines a regular functional f in $S'(\mathbb{R}^n)$ by the formula given by

$$\langle f, \phi \rangle = \int_{\mathbb{R}^n} f(x)\phi(x)dx \; (\phi \in S(\mathbb{R}^n)). \tag{7}$$

It is easy to verify that the functional f defined by Equation (7) exists $\forall \phi \in S(\mathbb{R}^n)$ and that it is linear as well as continuous on $S(\mathbb{R}^n)$.

We now quote a theorem of Vladimirov proved in his book [8].

Theorem 1. *If $f \in S'(\mathbb{R}^n)$, then there exists a continuous function g of slow growth in $\mathbb{R}^n$ and an integer $m \geqq 0$ such that*

$$f(x) = D_1^m D_2^m D_3^m \cdots D_n^m g(x), \; \frac{\partial}{\partial x_i} \equiv D_i \tag{8}$$

or, equivalently,

$$f(x) = D^m g(x) \qquad (D := D_1 D_2 D_3 \cdots D_n). \tag{9}$$

The n-dimensional wavelet inversion formula for tempered distributions will now be proved very simply by using the structure Formula (9). This structure formula enables us to reduce the wavelet analysis problem relating to tempered distributions to the classical wavelet analysis problem of $L^2(\mathbb{R}^n)$ functions. Our wavelet inversion formula of $L^2(\mathbb{R}^n)$ functions will be used quite successfully in order to derive the wavelet inversion formula for the wavelet transform of tempered distributions.

3. Wavelet Transform of Tempered Distributions in $\mathbb{R}^n$ and Its Inversion

Henceforth, we assume that $a \neq 0$ implies each of the components $a_i \neq 0$ for all $i = 1, 2, 3, \cdots, n$ and that $a > 0$ means each of the component a_i of a is greater than zero. Moreover, $|a| > \epsilon$ will mean that $|a_i| > \epsilon$ for all $i = 1, 2, 3, \cdots, n$.

Let $\psi(x) = \psi(x_1, x_2, \cdots, x_n) \in S(\mathbb{R}^n)$. Then $\psi(x)$ is a window function and is a wavelet if and only if

$$\int_{-\infty}^{\infty} \psi(x_1, x_2, \cdots, x_i, \cdots, x_n) dx_i = 0 \qquad (\forall \, i = 1, 2, \cdots, n). \tag{10}$$

We define $\psi\left(\frac{x-b}{a}\right) \equiv \psi\left(\frac{x_1-b_1}{a_1}, \frac{x_2-b_2}{a_2}, \cdots, \frac{x_n-b_n}{a_n}\right)$, where a_i, b_i are real numbers and none of the a_i is zero. The wavelet transform $W_f(a,b)$ of f with respect to the kernel $\frac{1}{\sqrt{|a|}}\psi\left(\frac{x-b}{a}\right)$ is defined by

$$W_f(a,b) = \left\langle f(x), \frac{1}{\sqrt{|a|}} \psi\left(\frac{x-b}{a}\right) \right\rangle. \tag{11}$$

Here, we assume that

$$|a| = |a_1 a_2 a_3 \cdots a_n| \qquad (a_i \neq 0 \;\; (i = 1, 2, 3, \cdots, n)).$$

We now prove the following lemmas which will be used to prove the main inversion formula.

Lemma 1. (see [13]) *Let $\phi \in S(\mathbb{R}^n)$ and ψ be a wavelet belonging to $S(\mathbb{R}^n)$.*

$$\frac{1}{C_\psi} \int\limits_{a\in\mathbb{R}^n} \int\limits_{b\in\mathbb{R}^n} \int\limits_{t\in\mathbb{R}^n} (-D_t)^m \phi(t)\bar{\psi}\left(\frac{t-b}{a}\right) \psi\left(\frac{x_0-b}{a}\right) \frac{dt\, db\, da}{a^2|a|} = (-D_x)^m \phi(x)|_{x=x_0} \qquad (\forall\, x_0 \in \mathbb{R}^n).$$

This is called pointwise convergence of the wavelet inversion formula.

Lemma 2. *Let $\phi \in S(\mathbb{R}^n)$ and let ψ be a wavelet belonging to $S(\mathbb{R}^n)$. Then*

$$\frac{1}{C_\psi} \int\limits_{a\in\mathbb{R}^n} \int\limits_{b\in\mathbb{R}^n} \int\limits_{t\in\mathbb{R}^n} (-D_t)^m \phi(t)\bar{\psi}\left(\frac{t-b}{a}\right) \psi\left(\frac{x-b}{a}\right) \frac{dt\, db\, da}{a^2|a|}$$

converges to $(-D_x^m)\ \phi(x)$ uniformly for all $x \in \mathbb{R}^n$.

Proof. Let

$$F(\wedge) = \frac{1}{(2\pi)^{\frac{n}{2}}} \int\limits_{\mathbb{R}^n} (-D_t)^m \phi(t)\, e^{-i\, \wedge.t}\, dt$$

be the Fourier transform of $(-D_t)^m \phi(t)$. It follows that, in the sense of $L^2(\mathbb{R}^n)$ convergence [17],

$$\frac{1}{C_\psi} \int\limits_{a\in\mathbb{R}^n} \int\limits_{b\in\mathbb{R}^n} \int\limits_{c\in\mathbb{R}^n} (-D_t)^m \phi(t)\bar{\psi}\left(\frac{t-b}{a}\right) \psi\left(\frac{x-b}{a}\right) \frac{dt\, db\, da}{a^2|a|}$$

$$= \frac{1}{(2\pi)^{\frac{n}{2}}} \int\limits_{\mathbb{R}^n} F(\wedge) e^{i\wedge.x} d\wedge = (-D_x)^m \phi(x).$$

This convergence is also uniform by a Weierstrass M-test because

$$\left| \frac{1}{(2\pi)^{\frac{n}{2}}} \int\limits_{\mathbb{R}^n} F(\wedge)\, e^{i\wedge.x}\, d\wedge \right| \leqq \frac{1}{(2\pi)^{\frac{n}{2}}} \int\limits_{\mathbb{R}^n} |F(\wedge)| d\wedge < \infty$$

and

$$F(\wedge)\ \in\ S(\mathbb{R}^n).$$

□

Theorem 2. *Let $f \in S'(\mathbb{R}^n)$ and $W_f(a,b)$ be its wavelet transform defined by*

$$W_f(a,b) = \left\langle f(x),\ \frac{1}{\sqrt{|a|}}\psi\left(\frac{x-b}{a}\right) \right\rangle.$$

Then the inversion formula of the wavelet transform $W_f(a,b)$ is given by

$$\left\langle \frac{1}{C_\psi} \int\limits_{\mathbb{R}^n} \int\limits_{\mathbb{R}^n} W_f(a,b)\psi\left(\frac{t-b}{a}\right) \frac{db\, da}{\sqrt{|a|}a^2}, \phi(t) \right\rangle = \langle f, \phi \rangle \tag{12}$$

$$(\forall\, \phi \in S(\mathbb{R}^n)),$$

where the equality holds true almost everywhere.

Proof. Using the structure formula (9) for f, we find by distributional differentiation that

$$\begin{aligned} W_f(a,b) &= \left\langle D_x^m g(x), \frac{1}{\sqrt{|a|}} \psi\left(\frac{x-b}{a}\right)\right\rangle \\ &= \left\langle g(x), (-D_x)^m \frac{1}{\sqrt{|a|}} \psi\left(\frac{x-b}{a}\right)\right\rangle. \end{aligned}$$

Here, we have

$$(-D_x) = (-D_{x_1})(-D_{x_2})(-D_{x_3})\cdots(-D_{x_n})\, D_{x_i} \equiv \frac{\partial}{\partial x_i} \qquad (i = 1, 2, 3, \cdots, n).$$

We thus obtain

$$\begin{aligned} W_f(a,b) &= \left\langle g(x), (D_b)^m \frac{1}{\sqrt{|a|}} \psi\left(\frac{x-b}{a}\right)\right\rangle \\ D_b &= \left(\frac{\partial}{\partial b_1}\frac{\partial}{\partial b_2}\frac{\partial}{\partial b_3}\cdots\frac{\partial}{\partial b_n}\right). \end{aligned}$$

□

The expression on the left-hand side in (12) can be written as follows:

$$\begin{aligned} \Omega &:= \frac{1}{C_\psi} \int_{t\in\mathbb{R}^n}\int_{a\in\mathbb{R}^n}\int_{b\in\mathbb{R}^n}\int_{x\in\mathbb{R}^n} g(x) D_b^m \frac{1}{\sqrt{|a|}} \bar{\psi}\left(\frac{x-b}{a}\right) \psi\left(\frac{t-b}{a}\right) \bar{\phi}(t) dx\, db\, da\, dt \\ &= \frac{1}{C_\psi} \int_{t\in\mathbb{R}^n}\int_{a\in\mathbb{R}^n}\int_{x\in\mathbb{R}^n} g(x) \left[\int_{b\in\mathbb{R}^n} \left\{D_b^m \bar{\psi}\left(\frac{x-b}{a}\right)\right\} \psi\left(\frac{t-b}{a}\right) db\right] \frac{\bar{\phi}(t)\, dx\, da\, dt}{a^2|a|}. \end{aligned} \tag{13}$$

We now evaluate the integral in the big bracket by parts to find from Equation (13) that

$$\begin{aligned} \Omega &= \frac{1}{C_\psi} \int_{t\in\mathbb{R}^n}\int_{a\in\mathbb{R}^n}\int_{x\in\mathbb{R}^n} g(x) \left[\int_{b\in\mathbb{R}^n} \bar{\psi}\left(\frac{x-b}{a}\right) (-D_b)^m \psi\left(\frac{t-b}{a}\right) db\right] \frac{\bar{\phi}(t)\, dx\, da\, dt}{a^2|a|} \\ &= \frac{1}{C_\psi} \int_{t\in\mathbb{R}^n}\int_{a\in\mathbb{R}^n}\int_{x\in\mathbb{R}^n} g(x) \left[\int_{b\in\mathbb{R}^n} \bar{\psi}\left(\frac{x-b}{a}\right) (+D_t)^m \psi\left(\frac{t-b}{a}\right) db\right] \frac{\bar{\phi}(t)\, dx\, da\, dt}{a^2|a|}, \end{aligned}$$

which, upon inverting the order of integration with respect to a and t, yields

$$\begin{aligned} \Omega &= \frac{1}{C_\psi} \int_{a\in\mathbb{R}^n}\int_{t\in\mathbb{R}^n}\int_{b\in\mathbb{R}^n}\int_{x\in\mathbb{R}^n} g(x)\bar{\psi}\left(\frac{x-b}{a}\right) dx\, D_t^m\, \psi\left(\frac{t-b}{a}\right) db \bar{\phi}(t) \frac{dt\, da}{|a|^2|a|} \\ &= \frac{1}{C_\psi} \int_{a\in\mathbb{R}^n}\int_{b\in\mathbb{R}^n}\int_{x\in\mathbb{R}^n} g(x)\bar{\psi}\left(\frac{x-b}{a}\right) dx \int_{t\in\mathbb{R}^n} \psi\left(\frac{t-b}{a}\right) db\, (-D_t)^m\, \bar{\phi}(t) \frac{dt\, da}{|a|^2|a|}. \end{aligned} \tag{14}$$

In order to justify the inversion of the order of integration with respect to a and t, we first perform the integration in the region $[(a,t) : |a| > \epsilon, a, t \in \mathbb{R}^n]$, invert the order of integration and then let $\epsilon \to 0$. This existence of the triple integral in terms of b, a and t in Equation (14) is proved by using the Plancherel theorem with respect to the variable b. Thus, by using

$$C_\psi = (2\pi)^n \int_{\mathbb{R}^n} \frac{|\hat{\psi}(\wedge)|^2}{|\wedge|} d\wedge,$$

we notice that the variable a disappears from the denominator and every calculation goes on smoothly. Since the functions ϕ and ψ are elements of $S(\mathbb{R}^n)$, the Fubini's theorem can be applied in order to justify the above interchanges of the order of integration.

Now, Equation (14) can be written as follows:

$$\left\langle g(x), \frac{1}{C_\psi} \int\limits_{a\in\mathbb{R}^n} \int\limits_{b\in\mathbb{R}^n} \int\limits_{t\in\mathbb{R}^n} (-D_t)^m \phi(t)\bar{\psi}\left(\frac{t-b}{a}\right) dt\, \psi\left(\frac{x-b}{a}\right) \frac{db\, da}{|a|^2|a|} \right\rangle \tag{15}$$

$$= \langle g(x), (-D_x)^m \phi(x) \rangle, \tag{16}$$

by means of the wavelet inversion formula in $\mathbb{R}^n$ [4] and Lemma 2.

We note that the triple integral in Equation (15) converges uniformly to $(-D_x)^m \phi(x) \forall\, x \in \mathbb{R}^n$. Thus, Equation (15) becomes Equation (16):

$$\begin{aligned} \langle g(x), (-D_x)^m \phi(x) \rangle &= \langle (D_x)^m g(x),\ \phi(x) \rangle \\ &= \langle f(x),\ \phi(x) \rangle . \end{aligned}$$

4. Conclusions

In our present investigation, we have introduced and studied a continuous wavelet transform of a Schwartz tempered distribution $f \in S'(\mathbb{R}^n)$ with the wavelet kernel $\psi \in S(\mathbb{R}^n)$. We have successfully derived the corresponding wavelet inversion formula by interpreting convergence in the weak topology of $S'(\mathbb{R}^n)$.

We have found that the wavelet transform of a constant distribution is zero and also that our wavelet inversion formula is not true for constant distribution, but it is true for a non-constant distribution which is not equal to the sum of a non-constant distribution with a non-zero constant distribution. Our results and findings are stated and proved as Lemmas and Theorems.

Author Contributions: Conceptualization, J.N.P. and S.K.U.; Methodology, S.K.U. and J.S.M.; Software, J.S.M.; Validation, H.M.S.; Formal analysis, H.M.S.; Writing–original draft preparation, S.K.U. and J.S.M.; Writing–review and editing, S.K.U. and H.M.S.; Supervision, J.N.P. and H.M.S.

Funding: This research received no external funding.

Conflicts of Interest: The authors declare no conflict of interest.

References

1. Bremmermann, H. *Distributions, Complex Variables and Fourier Transforms*; Addison-Wesley Publishing Company Inc.: Boston, MA, USA, 1965.
2. Kantorovich, L.V.; Akilov, G.P. *Functional Analysis in Normed Spaces*; Macmillan: New York, NY, USA, 1963.
3. Pandey, J.N. *The Hilbert Transform of Schwartz Distributions and Applications*; John Wiley and Sons Inc.: Hoboken, NJ, USA, 1996.
4. Pandey, J.N.; Upadhyay, S.K. Continuous Wavelet transform and window functions. *Proc. Am. Math. Soc.* **2015**, *143*, 4750–4773. [CrossRef]
5. Schwartz, L. *Theorie des Distributions*; Hermann: Paris, France, 1966.
6. Shilov, G.E. *Generalized Functions and Partial Differential Equations*; Gordon and Breach Publishing Co.: Chelmsford, MA, USA, 2005.
7. Treves, F. *Topological Vector Spaces, Distributions and Kernels*; Academic Press: New York, NY, USA, 1997.
8. Vladimirov, V.S. *Generalized Functions in Mathematical Physics*; Yankovasky, G., Eds.; Central Books Ltd: London, UK, 1976, ISBN 9780714715452.
9. Walter, G.G.; Shen, X. *Wavelets and Other Orthogonal Systems*, 2nd ed.; Chapman and Hall (CRC Press): Boca Raton, FL, USA, 2009.
10. Yosida, K. *Functional Analysis*; Springer: Berlin, Germany, 1995.
11. Zemanian, A.H. *Distribution Theory and Transform Analysis*; McGraw-Hill Book Company: New York, NY, USA, 1965.
12. Zemanian, A.H. *Generalized Integral Transformation*; Inter Science Publishers, John Wiley and Sons Inc.: Hoboken, NJ, USA, 1968.

13. Pandey, J.N.; Jha, N.K.; Singh, O.P. The continuous wavelet transform in n-dimensions. *Internat. J. Wavelets Multiresolut. Inf. Process.* **2016**, *14*, 13. [CrossRef]
14. Srivastava, H.M.; Upadhyay, S.K.; Khatterwani, K. A family of pseudo-differential operators on the Schwartz space associated with the fractional Fourier transform, *Russian J. Math. Phys.* **24**, *2017*, 534–543. [CrossRef]
15. Daubechies, I. *Ten Lectures on Wavelets*; University of Lowell: Lowell, MA, USA, 1990.
16. Meyer, Y. *Wavelets and Operators*; Cambridge Studies in Advanced Mathematics (Series 37); Cambridge University Press: Cambridge, UK, 1992.
17. Pathak, R.S. *The Wavelet Transform*; Atlantic Press/World Scientific: Paris, France, 2009.
18. Holschneider, M. *Wavelets an Analysis Tool*; Oxford Science Publications, Clarendon (Oxford University) Press: Oxford, UK, 1995.
19. Weisz, F. Convergence of the inverse continuous wavelet transform in Wiener amalgam spaces. *Analysis* **2015**, *35*, 33–46. [CrossRef]
20. Weisz, F. Inversion Formulas for the Continuous Wavelet Transform. *Acta Math. Hungar.* **2013**, *138*, 237–258. [CrossRef]
21. Postnikov, E.B.; Lebedeva, E.A.; Lavrova, A.I. Computational implementation of the inverse continuous wavelet transform without a requirement of the admissibility condition. *Appl. Math. Comput.* **2016**, *282*, 128–136. [CrossRef]

© 2019 by the authors. Licensee MDPI, Basel, Switzerland. This article is an open access article distributed under the terms and conditions of the Creative Commons Attribution (CC BY) license (http://creativecommons.org/licenses/by/4.0/).

symmetry

MDPI

Article

A Dunkl–Type Generalization of Szász–Kantorovich Operators via Post–Quantum Calculus

Md. Nasiruzzaman [1], Aiman Mukheimer [2] and M. Mursaleen [3,*]

1 Department of Computer Science (SEST), Jamia Hamdard, New Delhi 110062, India; nasir3489@gmail.com
2 Department of Mathematics and General Sciences, Prince Sultan University, Riyadh 11586, Saudi Arabia; mukheimer@psu.edu.sa
3 Department of Mathematics, Aligarh Muslim University, Aligarh 202002, India
* Correspondence: mursaleenm@gmail.com

Received: 24 December 2018; Accepted: 11 February 2019; Published: 15 February 2019

Abstract: In this paper, we define the (p,q)-variant of Szász–Kantorovich operators via Dunkl-type generalization generated by an exponential function and study the Korovkin-type results. We also obtain the convergence of our operators in weighted space by the modulus of continuity, Lipschitz class, and Peetre's K-functionals. The extra parameter p provides more flexibility in approximation and plays an important role in symmetrizing these newly-defined operators.

Keywords: (p,q)-integers; Dunkl analogue; generating functions; generalization of exponential function; Szász operator; modulus of continuity

1. Introduction and Preliminaries

Bernstein [1] and q-Bernstein ([2,3]) operators have become very important tools in the study of approximation theory and several branches of applied sciences and engineering. For $[r]_{p,q} = \frac{p^r - q^r}{p-q}$, $r = 0, 1, 2, \cdots,\ 0 < q < p \leqq 1$, the (p,q)-Bernstein operators were introduced by Mursaleen et al. [4]:

$$B_r^{p,q}(g;y) = \frac{1}{p^{\frac{r(r-1)}{2}}} \sum_{m=0}^{r} \left[\begin{array}{c} r \\ m \end{array} \right]_{p,q} p^{\frac{m(m-1)}{2}} y^k \prod_{s=0}^{r-m-1} (p^s - q^s y) g\left(\frac{[m]_{p,q}}{p^{m-r}[r]_{p,q}} \right), \qquad y \in [0,1], \tag{1}$$

where $[r]_{p,q}$ denotes the (p,q)-integer.

The (p,q)-analogues of exponential functions are defined in two forms as follows:

$$e_r^{p,q}(y) = \sum_{r=0}^{\infty} p^{\frac{r(r-1)}{2}} \frac{y^r}{[r]_{p,q}!}, \quad E_r^{p,q}(y) = \sum_{r=0}^{\infty} q^{\frac{r(r-1)}{2}} \frac{y^r}{[r]_{p,q}!},$$

with the property that $e_r^{p,q}(y) E_r^{p,q}(-y) = 1$. In the case of $p = 1$, $e_r^{p,q}(y)$ and $E_r^{p,q}(y)$ reduce to q-analogues of exponential functions.

The Dunkl-type generalization of Szász operators [5] was introduced by Sucu [6] and the q-analogue by Ben Cheikh et al. [7]. Içöz [8] introduced the q-Dunkl analogue of Szász operators defined by:

$$D_\eta^q(g;y) = \frac{1}{e_\eta^q([r]_q y)} \sum_{m=0}^{\infty} \frac{([r]_q y)^m}{\gamma_\eta^q(m)} g\left(\frac{1 - q^{2\eta\theta_m + m}}{1 - q^r} \right) \tag{2}$$

where $\eta > -\frac{1}{2}$, $y \geqq 0$, $0 < q < 1$, $g \in C[0,\infty)$ and $C[0,\infty)$ is the set of all continuous functions defined on $[0,\infty)$.

The (p,q)- and q-Dunkl analogues have been studied by several authors (see [9–24]). For the most recent work on (p,q)-approximation, we refer to [25–27]. Recently. Alotaibi et al. [28] generalized the q-Dunkl analogue of Szász operators via (pq)-calculus as follows:

$$\mathcal{D}_{\eta}^{p,q}(g;y) = \frac{1}{e_{\eta}^{p,q}([r]_{p,q}y)} \sum_{m=0}^{\infty} \frac{([r]_{p,q}y)^m}{\gamma_{\eta}^{p,q}(m)} p^{\frac{m(m-1)}{2}} g\left(\frac{p^{2\eta\theta_m+m} - q^{2\eta\theta_m+m}}{p^{m-1}(p^r - q^r)}\right) \tag{3}$$

where for $q \in (0,1)$, $p \in (q,1]$, and $\eta > -\frac{1}{2}$, the (p,q)-Dunkl analogue of exponential functions is defined by:

$$e_{\eta}^{p,q} = \sum_{r=0}^{\infty} p^{\frac{r(r-1)}{2}} \frac{y^r}{\gamma_{\eta}^{p,q}(r)}, \quad y \in [0,\infty) \tag{4}$$

$$\gamma_{\eta}^{p,q}(r) = \frac{\prod_{i=0}^{[\frac{r+1}{2}]-1} p^{2\eta(-1)^{i+1}+1}\left((p^2)^i p^{2\eta+1} - (q^2)^i q^{2\eta+1}\right) \prod_{j=0}^{[\frac{r}{2}]-1} p^{2\eta(-1)^{j}+1}\left((p^2)^j p^2 - (q^2)^j q^2\right)}{(p-q)^r}, \tag{5}$$

$$\gamma_{\eta}^{p,q}(r+1) = \frac{p^{2\eta(-1)^{r+1}+1}(p^{2\eta\theta_{r+1}+r+1} - q^{2\eta\theta_{r+1}+r+1})}{(p-q)}\gamma_{\eta}^{p,q}(r), \tag{6}$$

$$\theta_r = \begin{cases} 0 & \text{for } r = 2\ell,\ \ell = 1,2,\cdots,n \\ 1 & \text{for } r = 2\ell+1,\ \ell = 1,2,\cdots,n. \end{cases} \tag{7}$$

and $[\frac{r}{2}]$ denotes the greatest integer function; also, we have:

$$(\alpha - \beta)_{p,q}^r = \begin{cases} \prod_{j=0}^{r-1}(p^j\alpha - q^j\beta) & \text{if } r = 1,2,\cdots,n \\ 1 & \text{if } r = 0. \end{cases}$$

Lemma 1. *For* $g(t) = 1,\ t,\ t^2$
1^*. $\mathcal{D}_{\eta}^{p,q}(1;y) = 1$;
2^*. $\mathcal{D}_{\eta}^{p,q}(t;y) = y$;
3^*. $y^2 + \frac{q^{2\eta}}{[r]_{p,q}}[1-2\eta]_{p,q}\frac{e_{\eta}^{p,q}(\frac{q}{p}[r]_{p,q}y)}{e_{\eta}^{p,q}([r]_{p,q}y)}y \leqq \mathcal{D}_{\eta}^{p,q}(t^2;y) \leqq y^2 + \frac{1}{[r]_{p,q}}[1+2\eta]_{p,q}y.$

2. New Operators and Estimations of Moments

In this section, we construct the (p,q)-variant of Szász–Kantorovich operators via Dunkl-type generalization as follows.

Definition 1. *For any* $y \in [0,\infty)$, $g \in C[0,\infty)$ $r \in \mathbb{N}$ *and* $0 < q < p \leqq 1$, *we define:*

$$\mathcal{K}_{\eta}^{p,q}(g;y) = \frac{[r]_{p,q}}{e_{\eta}^{p,q}([r]_{p,q}y)} \sum_{m=0}^{\infty} \frac{([r]_{p,q}y)^m}{\gamma_{\eta}^{p,q}(m)} p^{-(m+2\eta\theta_m)} p^{\frac{m(m-1)}{2}} \int_{q\mathcal{A}}^{q\mathcal{A}+\mathcal{B}} g\left(\frac{t}{qp^{m-1}}\right) d_{p,q}t. \tag{8}$$

We use the following relation:

$$[m+1+2\eta\theta_m]_{p,q} = q[m+2\eta\theta_m]_{p,q} + p^{m+2\eta\theta_m}, \tag{9}$$

$$\mathcal{A} = \frac{[m+2\eta\theta_m]_{p,q}}{[r]_{p,q}}, \quad \mathcal{B} = \frac{p^{m+2\eta\theta_m}}{[r]_{p,q}} \tag{10}$$

where the parameter $\eta \geqq 0$.

To show the uniform convergence of operators $\mathcal{K}_{\eta}^{p,q}(\,\cdot\,;\,\cdot\,)$, we take $q = q_r,\;\; p = p_r$ with $0 < q_r < 1$ and $q_r < p_r \leqq 1$ such that:

$$\lim_{r\to\infty} p_r \to 1, \quad \lim_{r\to\infty} q_r \to 1, \quad \lim_{r\to\infty} p_r^r \to u, \quad \lim_{r\to\infty} q_r^r \to v, \quad (0 < u, v \leqq 1). \tag{11}$$

For $p = 1$, these operators reduce to the operators defined in [29]. For $\eta = 0$, these are reduced to the (p,q)-variant of Kantorovich-type operators defined by [30].

Lemma 2. *Let $g(t) = g_i$ such that $g_i = t^{i-1}$ for $i = 1,2,3$. Then, we have:*
(1) $\mathcal{K}_{\eta}^{p,q}(g_1;y) = 1$
(2) $\mathcal{K}_{\eta}^{p,q}(g_2;y) \leqq \frac{2}{[2]_{p,q}}y + \frac{1}{[2]_{p,q}q[r]_{p,q}}$
(3) $\mathcal{K}_{\eta}^{p,q}(g_3;y) \leqq \frac{3}{[3]_{p,q}}y^2 + \frac{3}{[3]_{p,q}[r]_{p,q}}\left([1+2\eta]_{p,q} + \frac{1}{q[r]_{p,q}}\right)y + \frac{1}{[3]_{p,q}q^2[r]_{p,q}^2}$.

Proof. Using (9) and (10), we get:

$$\int_{q\mathcal{A}}^{q\mathcal{A}+\mathcal{B}} f\left(\frac{t}{qp^{k-1}}\right) d_{p,q}t = \begin{cases} \mathcal{B} & \text{for} \quad g(t) = g_1 \\[2ex] \frac{\mathcal{B}}{[2]_{p,q}p^{m-1}q}\,(2q\mathcal{A} + \mathcal{B}) & \text{for} \quad g(t) = g_2 \\[2ex] \frac{\mathcal{B}}{[3]_{p,q}p^{2(m-1)}q^2}\,(3q^2\mathcal{A}^2 + 3q\mathcal{A}\mathcal{B} + \mathcal{B}^2) & \text{for} \quad g(t) = g_3 \end{cases} \tag{12}$$

If we take $g(t) = g_1$, then from (12), we have:

$$\begin{aligned} \mathcal{K}_{\eta}^{p,q}(g_1;y) &= \frac{[r]_{p,q}}{e_{\eta}^{p,q}([r]_{p,q}y)} \sum_{m=0}^{\infty} \frac{([r]_{p,q}y)^m}{\gamma_{\eta}^{p,q}(m)} p^{-(m+2\eta\theta_m)} p^{\frac{m(m-1)}{2}} \int_{q\mathcal{A}}^{q\mathcal{A}+\mathcal{B}} d_{p,q}t \\ &= 1. \end{aligned}$$

For $g(t) = g_2$, (12) implies:

$$\begin{aligned} \mathcal{K}_{\eta}^{p,q}(g_2;y) &= \frac{1}{[2]_{p,q}q[r]_{p,q}} \frac{1}{e_{\eta}^{p,q}([r]_{p,q}y)} \sum_{m=0}^{\infty} \frac{([r]_{p,q}y)^m}{\gamma_{\eta}^{p,q}(m)} p^{-(m+2\eta\theta_m)} p^{\frac{m(m-1)}{2}} \\ &\times \; p^{1+2\eta\theta_m}\left(2q[m+2\eta\theta_m]_{p,q} + p^{m+2\eta\theta_m}\right) \\ &= \frac{2}{[2]_{p,q}[r]_{p,q}} \frac{1}{e_{\eta}^{p,q}([r]_{p,q}y)} \sum_{m=0}^{\infty} \frac{([r]_{p,q}y)^m}{\gamma_{\eta}^{p,q}(m)} p^{\frac{(m-1)(m-2)}{2}} [m+2\eta\theta_m]_{p,q} \\ &+ \frac{1}{[2]_{p,q}q[r]_{p,q}} \frac{1}{e_{\eta}^{p,q}([r]_{p,q}y)} \sum_{m=0}^{\infty} \frac{([r]_{p,q}y)^m}{\gamma_{\eta}^{p,q}(m)} p^{1+2\eta\theta_m} \\ &= \frac{2}{[2]_{p,q}} \frac{1}{e_{\eta}^{p,q}([r]_{p,q}y)} \sum_{m=0}^{\infty} \frac{([r]_{p,q}y)^m}{\gamma_{\eta}^{p,q}(m)} p^{\frac{m(m-1)}{2}} \left(\frac{p^{m+2\eta\theta_m} - q^{m+2\eta\theta_m}}{p^{m-1}(p^r - q^r)}\right) \\ &+ \frac{1}{[2]_{p,q}q[r]_{p,q}} \frac{1}{e_{\eta}^{p,q}([r]_{p,q}y)} \sum_{m=0}^{\infty} \frac{([r]_{p,q}y)^m}{\gamma_{\eta}^{p,q}(m)} p^{1+2\eta\theta_m}. \end{aligned}$$

Separating into even and odd terms, we get:

$$\begin{aligned} \mathcal{K}_{\eta}^{p,q}(g_2;y) &= \frac{2}{[2]_{p,q}}y + \frac{p}{[2]_{p,q}q[r]_{p,q}} \qquad \text{for} \quad r = 0,2,4,\cdots \\ \mathcal{K}_{\eta}^{p,q}(g_2;y) &= \frac{2}{[2]_{p,q}}y + \frac{p^{1+2\eta}}{[2]_{p,q}q[r]_{p,q}} \qquad \text{for} \quad r = 1,3,5,\cdots. \end{aligned}$$

Since $0 < q < p \leqq 1$, $\eta \geqq 0$, and $p^{1+2\eta} \leqq 1$, we have:

$$\mathcal{K}_\eta^{p,q}(g_2;y) \leqq \frac{2}{[2]_{p,q}}y + \frac{1}{q[2]_{p,q}[r]_{p,q}}.$$

Similarly for $g(t) = g_3$, we have:

$$\begin{aligned}
\mathcal{K}_\eta^{p,q}(g_3;y) &= \frac{3}{[3]_{p,q}[r]_{p,q}^3}\frac{1}{e_\eta^{p,q}([r]_{p,q}y)}\sum_{m=0}^{\infty}\frac{([r]_{p,q}y)^m}{\gamma_\eta^{p,q}(m)}p^{\frac{m(m-1)}{2}-2(m-1)}[m+2\eta\theta_m]_{p,q}^2 \\
&+ \frac{3}{[3]_{p,q}q[r]_{p,q}^3}\frac{1}{e_\eta^{p,q}([r]_{p,q}y)}\sum_{m=0}^{\infty}\frac{([r]_{p,q}y)^m}{\gamma_\eta^{p,q}(m)}p^{m+2\eta\theta_m}p^{\frac{m(m-1)}{2}-2(m-1)}[m+2\eta\theta_m]_{p,q} \\
&+ \frac{1}{[3]_{p,q}q^2[r]_{p,q}^2}\frac{1}{e_\eta^{p,q}([r]_{p,q}y)}\sum_{m=0}^{\infty}\frac{([r]_{p,q}y)^m}{\gamma_\eta^{p,q}(m)}p^{2(m+2\eta\theta_m)}p^{\frac{m(m-1)}{2}-2(m-1)} \\
&= \frac{3}{[3]_{p,q}[r]_{p,q}^2}\frac{1}{e_\eta^{p,q}([r]_{p,q}y)}\sum_{m=0}^{\infty}\frac{([r]_{p,q}y)^m}{\gamma_\eta^{p,q}(m)}p^{\frac{m(m-1)}{2}}\left(\frac{p^{m+2\eta\theta_m}-q^{m+2\eta\theta_m}}{p^{m-1}(p^r-q^r)}\right)^2 \\
&+ \frac{3}{[3]_{p,q}q[r]_{p,q}^2}\frac{1}{e_\eta^{p,q}([r]_{p,q}y)}\sum_{m=0}^{\infty}\frac{([r]_{p,q}y)^m}{\gamma_\eta^{p,q}(m)}p^{\frac{m(m-1)}{2}}\left(\frac{p^{m+2\eta\theta_m}-q^{m+2\eta\theta_m}}{p^{m-1}(p^r-q^r)}\right) \\
&+ \frac{1}{[3]_{p,q}q^2[r]_{p,q}^2}\frac{1}{e_\eta^{p,q}([r]_{p,q}y)}\sum_{m=0}^{\infty}\frac{([r]_{p,q}y)^m}{\gamma_\eta^{p,q}(m)}p^{\frac{m(m-1)}{2}}p^{2(1+2\eta\theta_m)}.
\end{aligned}$$

Hence, for $m = 0, 2, 4, \cdots$, we have:

$$\mathcal{K}_\eta^{p,q}(g_3;y) \leqq \frac{3}{[3]_{p,q}}y^2 + \frac{3}{[3]_{p,q}[r]_{p,q}}\left([1+2\eta]_{p,q} + \frac{p}{q[r]_{p,q}}\right)y + \frac{p^2}{q^2[3]_{p,q}[r]_{p,q}^2},$$

and for $m = 1, 3, 5, \cdots$,

$$\mathcal{K}_\eta^{p,q}(g_3;y) \leqq \frac{3}{[3]_{p,q}}y^2 + \frac{3}{[3]_{p,q}[r]_{p,q}}\left([1+2\eta]_{p,q} + \frac{p}{q[r]_{p,q}}\right)y + \frac{p^2}{[3]_{p,q}q^2[r]_{p,q}^2}.$$

Therefore,

$$\mathcal{K}_\eta^{p,q}(g_3;y) \leqq \frac{3}{[3]_{p,q}}y^2 + \frac{3}{[3]_{p,q}[r]_{p,q}}\left([1+2\eta]_{p,q} + \frac{1}{q[r]_{p,q}}\right)y + \frac{1}{[3]_{p,q}q^2[r]_{p,q}^2}.$$

This completes the proof of Lemma 2. □

Lemma 3. *Let $\chi_i = (t-y)^i$ for $i = 1,\ 2$. Then, we have:*

$$\mathcal{K}_\eta^{p,q}(\chi_i;y) \leqq \begin{cases} \left(\frac{2}{[2]_{p,q}} - 1\right)y + \frac{1}{[2]_{p,q}q[r]_{p,q}} & \text{for} \quad i = 1 \\ \left(\frac{3}{[3]_{p,q}} + 1 - \frac{4}{[2]_{p,q}}\right)y^2 \\ +\frac{1}{q[r]_{p,q}}\left(\frac{3}{[3]_{p,q}}\left(\frac{1}{[r]_{p,q}} + q[1+2\eta]_{p,q}\right) - \frac{2}{[2]_{p,q}}\right)y \\ +\frac{1}{[3]_{p,q}q^2[r]_{p,q}^2} & \text{for} \quad i = 2. \end{cases} \tag{13}$$

3. Main Results

In this section, we study the Korovkin-type approximation theorems for positive linear operators $\mathcal{K}_{\eta}^{p,q}(\,\cdot\,;\,\cdot\,)$ defined by (8). We denote the set of all bounded and continuous functions by $C_B[0,\infty)$ equipped with norm $\| g \|_{C_B} = \sup_{y\in[0,\infty)} | g(y) |$. We write:

$$\mathfrak{E} := \{g(y) : y \in [0,\infty), \frac{g(y)}{1+y^2} \text{ is convergent as } y \to \infty\}.$$

Let:

$$B_\sigma[0,\infty) = \{g : |g(y)| \leqq \mathcal{M}_g \sigma(y)\},$$
$$C_\sigma[0,\infty) = \{g : g \in B_\sigma[0,\infty) \cap C[0,\infty)\},$$
$$C_\sigma^k[0,\infty) = \left\{g : g \in C_\sigma[0,\infty) \text{ and } \lim_{y\to\infty} \frac{g(y)}{\sigma(y)} = k\right\},$$

where $\sigma(y)$ is the weight function given by $\sigma(y) = 1 + y^2$, k is a constant, and $\mathcal{M}_g$ depends on g. $C_\sigma[0,\infty)$ is equipped with the norm $||g||_\sigma = \sup_{y\in[0,\infty)} \frac{|g(y)|}{\sigma(y)}$.

Theorem 1. *Let q_r, p_r be the real numbers, with $q_r \in (0,1)$ and $p_r \in (q_r, 1]$ for every integer r, satisfying $(q_r) \to 1$ and $(p_r) \to 1$ as $r \to \infty$. Then, for every $g \in C[0,\infty) \cap \mathfrak{E}$,*

$$\lim_{r\to\infty} \mathcal{K}_{\eta}^{p_r,q_r}(g;y) = g(y)$$

uniformly on each compact subset of $[0,\infty)$.

Proof. For the proof of the uniform convergence of the operators $\mathcal{K}_{\eta}^{p_r,q_r}$ on each compact subset of $[0,\infty)$, we apply the well-known Korovkin theorem [31]. It is sufficient to show that $\lim_{r\to\infty} \mathcal{K}_{\eta}^{p_r,q_r}(g_i;y) = y^{i-1}$, where $g_i = t^{i-1}$ for $i = 1,2,3$.

Clearly, if $q_r \to 1$, $p_r \to 1$ as $r \to \infty$, then $\frac{1}{[r]_{p_r,q_r}} \to 0$, $\frac{r}{[r]_{p_r,q_r}} \to 1$. This yields that:

$$\lim_{r\to\infty} \mathcal{K}_{\eta}^{p_r,q_r}(g_1;y) = 1, \quad \lim_{r\to\infty} \mathcal{K}_{\eta}^{p_r,q_r}(g_2;y) = y, \quad \lim_{r\to\infty} \mathcal{K}_{\eta}^{p_r,q_r}(g_3;y) = y^2.$$

□

Theorem 2. *Let q_r, p_r be the real numbers, with $q_r \in (0,1)$ and $p_r \in (q_r, 1]$ for every integer r, satisfying $(q_r) \to 1$ and $(p_r) \to 1$ as $r \to \infty$. Then, for every $g \in C_\sigma^k[0,\infty)$, we have:*

$$\lim_{r\to\infty} \left|\left|\mathcal{K}_{\eta}^{p_r,q_r}(g;y) - g\right|\right|_\sigma = 0. \tag{14}$$

Proof. Suppose $g(t) \in C_\sigma^k[0,\infty)$ and $g(t) = g_\tau$, where $g_\tau = t^{\tau-1}$ for $\tau = 1,2,3$. Then, from the well-known Korovkin theorem, we have $\mathcal{K}_{\eta}^{p_r,q_r}(g_\tau;y) \to y^{\tau-1}$ $(r \to \infty)$ uniformly for each $\tau = 1,2,3$. Hence, from Lemma 2, we have:

$$\lim_{r\to\infty} \left|\left|\mathcal{K}_{\eta}^{p_r,q_r}(g_1;y) - 1\right|\right|_\sigma = 0. \tag{15}$$

For $\tau = 2$,

$$\begin{aligned}
&\left|\left|\mathcal{K}_{\eta}^{p_r,q_r}(g_2;y) - y\right|\right|_{\sigma} \\
&= \sup_{y \geqq 0} \frac{\left|\mathcal{K}_{\eta}^{p_r,q_r}(g_2;y) - y\right|}{1+y^2} \\
&\leqq \left(\frac{2}{[2]_{p_r,q_r}} - 1\right) \sup_{y \geqq 0} \frac{y}{1+y} + \frac{1}{q_r[2]_{p_r,q_r}[r]_{p_r,q_r}} \sup_{y \geqq 0} \frac{1}{1+y}.
\end{aligned}$$

Then:

$$\lim_{r\to\infty} \left|\left|\mathcal{K}_{\eta}^{p_r,q_r}(g_2;y) - y\right|\right|_{\sigma} = 0. \tag{16}$$

Similarly, if we take $\tau = 3$,

$$\begin{aligned}
&\left|\left|\mathcal{K}_{\eta}^{p_r,q_r}(g_3;y) - y^2\right|\right|_{\sigma} \\
&= \sup_{y \geqq 0} \frac{\left|\mathcal{K}_{\eta}^{p_r,q_r}(g_3;y) - y^2\right|}{1+y^2} \\
&\leqq \left(\frac{3}{[3]_{p_r,q_r}} - 1\right) \sup_{y \geqq 0} \frac{y^2}{1+y^2} \\
&+ \frac{3}{[3]_{p_r,q_r}[r]_{p_r,q_r}} \left([1+2\eta]_{p_r,q_r} + \frac{1}{q_r[r]_{p_r,q_r}}\right) \sup_{y \geqq 0} \frac{y}{1+y^2} \\
&+ \frac{1}{[3]_{p_r,q_r} q_r^2 [r]_{p_r,q_r}^2} \sup_{y \geqq 0} \frac{1}{1+y^2},
\end{aligned}$$

$$\lim_{r\to\infty} \left|\left|\mathcal{K}_{\eta}^{p_r,q_r}(g_3;y) - y^2\right|\right|_{\sigma} = 0. \tag{17}$$

This completes the proof. □

The modulus of continuity $\omega_b(g;\delta)$ of the function $g \in \tilde{C}[0,\infty)$ is defined by:

$$\omega_b(g;\delta) = \sup_{|t-y| \leq \delta;\, y,t \in [0,b]} \sup \mid g(t) - g(y) \mid \tag{18}$$

where $\tilde{C}[0,\infty)$ denotes the space of uniformly-continuous functions on $[0,\infty)$. It is obvious that $\lim_{\delta\to 0+} \omega_b(g;\delta) = 0$ and for $g \in C[0,\infty)$:

$$\mid g(t) - g(y) \mid \leq \left(\frac{\mid t-y \mid}{\delta} + 1\right) \omega_b(g;\delta). \tag{19}$$

Theorem 3. *Let q_r, p_r be the real numbers, with $q_r \in (0,1)$ and $p_r \in (q_r, 1]$ for every integer r, satisfying $(q_r) \to 1$ and $(p_r) \to 1$ as $r \to \infty$. Then, for every $g \in C_\sigma[0,\infty)$:*

$$\left|\mathcal{K}_{\eta}^{p_r,q_r}(g;y) - g(y)\right| \leqq 2\left(\omega_{b+1}(g;\delta_\eta(y)) + \mathcal{M}_g(1+b^2)\,(\delta_\eta(y))^2\right),$$

where $\delta_\eta(y) = \sqrt{\mathcal{K}_{\eta}^{p_r,q_r}(\chi_2;y)}$, $\mathcal{M}_g$ is a constant depending only on g and $\mathcal{K}_{\eta}^{p_r,q_r}(\chi_2;y)$ is defined by Lemma 3; and $[0,b+1] \subset [0,\infty), b > 0$.

Proof. Let $y \in [0,b]$ and $t > b+1$, with $t > 0$. Then, for $\delta > 0$, we have:

$$| g(t) - g(y) | \leqq \omega_{b+1}(g; | t-y |) \leqq \left(1 + \frac{| t-y |}{\delta}\right) \omega_{b+1}(g;\delta). \tag{20}$$

By applying the Cauchy–Schwarz inequality and the linearity of $\mathcal{K}_{\eta}^{p_r,q_r}$:

$$\mathcal{K}_{r,\eta}^{p_r,q_r} |g(t) - g(y); y| \leqq \left(\left(1 + \frac{1}{\delta}\mathcal{K}_{\eta}^{p_r,q_r}((t-y)^2; y)\right)^{\frac{1}{2}}\right)\omega_{b+1}(g;\delta). \tag{21}$$

For $t - y > 1$, we have:

$$\begin{aligned}
&|g(t) - g(y)| \\
&\quad \leqq \mathcal{M}_g\left(2 + y^2 + t^2\right) \\
&\quad \leqq \mathcal{M}_g\left(2 + 3y^2 + 2(t-y)^2\right) \leqq 2\mathcal{M}_g(1+b^2)(t-y)^2
\end{aligned}$$

$$\mathcal{K}_{\eta}^{p_r,q_r}(|g(t) - g(y)|;) \leqq 2\mathcal{M}_g(1+b^2)\mathcal{K}_{\eta}^{p_r,q_r}((t-y)^2; y). \tag{22}$$

From (21) and (22), we easily see that:

$$\begin{aligned}
&\left|\mathcal{K}_{\eta}^{p_r,q_r}(g;y) - g(y)\right| \\
&\qquad \leqq \mathcal{K}_{\eta}^{p_r,q_r}|g(t) - g(y); y| \\
&\qquad \leqq \left(\left(1 + \frac{1}{\delta}\mathcal{K}_{\eta}^{p_r,q_r}((t-y)^2; y)\right)^{\frac{1}{2}}\right)\omega_{b+1}(g;\delta) \\
&\qquad + 2\mathcal{M}_g(1+b^2)\mathcal{K}_{\eta}^{p_r,q_r}((t-y)^2; y) \\
&\qquad = \left(1 + \frac{1}{\delta}\mathcal{K}_{\eta}^{p_r,q_r}(\chi_2; y)\right)^{\frac{1}{2}}\omega_{b+1}(g;\delta) \\
&\qquad + 2\mathcal{M}_g(1+b^2)\mathcal{K}_{\eta}^{p_r,q_r}(\chi_2; y)
\end{aligned}$$

If we choose $\delta = \delta_{\eta}(y) = \sqrt{\mathcal{K}_{\eta}^{p_r,q_r}(\chi_2; y)}$, then we get our result. □

For any $g \in C[0,\infty]$, $\mathcal{L} > 0$, $0 < \nu \leq 1$ and $\gamma_1, \gamma_2 \in [0,\infty)$, we recall that:

$$Lip_{\mathcal{L}}(\nu) = \{g : | g(\gamma_1) - g(\gamma_2) | \leq \mathcal{L} | \gamma_1 - \gamma_2 |^{\nu}\}. \tag{23}$$

Theorem 4. *Let q_r, p_r be the real numbers, with $q_r \in (0,1)$ and $p_r \in (q_r, 1]$ for every integer r, satisfying $(q_r) \to 1$ and $(p_r) \to 1$ as $r \to \infty$. Then, for each $g \in Lip_{\mathcal{L}}(\nu)$, we have:*

$$\left|\mathcal{K}_{\eta}^{p_r,q_r}(g;y) - g(y)\right| \leqq \mathcal{L}\left(\delta_{\eta}(y)\right)^{\nu},$$

where $\delta_{\eta}(y)$ is defied by Theorem 3.

Proof. Using Theorem 4, (23), and the well-known Hölder's inequality, we get:

$$\begin{aligned}\left|\mathcal{K}_{\eta}^{p_r,q_r}(g;y)-g(y)\right| &\leqq \left|\mathcal{K}_{r,\eta}^{p_r,q_r}(g(t)-g(y);y)\right| \\ &\leqq \mathcal{K}_{\eta}^{p_r,q_r}\left(|g(t)-g(y)|\,;y\right) \\ &\leqq |\, \mathcal{L}\mathcal{K}_{\eta}^{p_r,q_r}\left(|t-y|^{\nu}\,;y\right) \\ &\leqq \mathcal{L}\left(\mathcal{K}_{\eta}^{p_r,q_r}(g_1;y)\right)^{\frac{2-\nu}{2}}\left(\mathcal{K}_{\eta}^{p_r,q_r}(|t-y|^2\,;y)\right)^{\frac{\nu}{2}} \\ &= \mathcal{L}\left(\mathcal{K}_{\eta}^{p_r,q_r}(\chi_2;y)\right)^{\frac{\nu}{2}}.\end{aligned}$$

This completes the proof of the theorem. □

We denote:

$$C_B^2[0,\infty)=\{\psi:\psi\in C_B[0,\infty) \quad \text{and} \quad \psi',\psi''\in C_B[0,\infty)\}, \tag{24}$$

$$||\psi||_{C_B^2(\mathbb{R}^+)}=||\psi||_{C_B[0,\infty)}+||\psi'||_{C_B[0,\infty)}+||\psi''||_{C_B[0,\infty)}, \tag{25}$$

$$||\psi||_{C_B[0,\infty)}=\sup_{y\in[0,\infty)}|\psi(y)|. \tag{26}$$

Theorem 5. *Let q_r, p_r be the real numbers, with $q_r\in(0,1)$ and $p_r\in(q_r,1]$ for every integer r, satisfying $(q_r)\to 1$ and $(p_r)\to 1$ as $r\to\infty$. Then:*

$$\left|\mathcal{K}_{\eta}^{p_r,q_r}(\psi;y)-\psi(y)\right|\leqq \Upsilon_{\eta}(y)||\psi||_{C_B^2[0,\infty)}, \tag{27}$$

where $\Upsilon_{\eta}(y)=\delta_n(y)\left(1+\frac{\delta_{\eta}(y)}{2}\right)$ and $\delta_{\eta}(y)$ is defined by Theorem 3.

Proof. From the Taylor series expansion for any $\psi\in C_B^2[0,\infty)$, we have:

$$\psi(t)=\psi(y)+\psi'(y)(t-y)+\psi''(\varphi)\,\frac{(t-y)^2}{2} \quad \text{for} \quad \varphi\in(y,t),$$

$$|\psi(t)-\psi(y)|\leqq \mathfrak{P}\,|\,t-y\,|+\frac{1}{2}\mathfrak{Q}(t-y)^2,$$

where:

$$\mathfrak{P}=\sup_{y[0,\infty)}|\psi'(y)|=||\psi'||_{C_B[0,\infty)}\leqq||\psi||_{C_B^2[0,\infty)},$$

$$\mathfrak{Q}=\sup_{y[0,\infty)}|\psi''(y)|=||\psi''||_{C_B[0,\infty)}\leqq||\psi||_{C_B^2[0,\infty)}.$$

Therefore,

$$|\psi(t)-\psi(y)|\leqq\left(|\,t-y\,|+\frac{1}{2}(t-y)^2\right)||\psi||_{C_B^2[0,\infty)}.$$

By applying the linearity of $\mathcal{K}_{\eta}^{p_r,q_r}$, we get:

$$\begin{aligned}
&\left|\mathcal{K}_{\eta}^{p_r,q_r}(\psi;y)-\psi(y)\right| \\
&\leqq \left(\mathcal{K}_{\eta}^{p_r,q_r}(\mid t-y\mid;y)+\frac{1}{2}\mathcal{K}_{\eta}^{p_r,q_r}((t-y)^2;y)\right)||\psi||_{C_B^2[0,\infty)} \\
&\leqq \left(\left(\mathcal{K}_{\eta}^{p_r,q_r}(\chi_2;y)\right)^{\frac{1}{2}}+\frac{1}{2}\mathcal{K}_{\eta}^{p_r,q_r}(\chi_2;y)\right)||\psi||_{C_B^2[0,\infty)} \\
&= \left(\delta_\eta(y)+\frac{(\delta_\eta(y))^2}{2}\right)||\psi||_{C_B^2[0,\infty)}.
\end{aligned}$$

This completes the proof of the theorem. □

Peetre's K-functional $K_2(g;\delta)$ for $\delta > 0$ (see [32]) is defined by:

$$K_2(g;\delta)=\inf_{y\in[0,\infty)}\left\{\left(\delta\left|\left|\psi''+||g-\psi||_{C_B[0,\infty)}\right|\right|_{C_B[0,\infty)}\right)\right\} \tag{28}$$

for all $\psi \in C_B^2[0,\infty)$.

For a given positive constant $\mathfrak{L} > 0$:

$$K_2(g;\delta)\leqq \mathfrak{L}\omega_2(g;\delta^{\frac{1}{2}}),$$

where the second-order modulus of continuity denoted by $\omega_2(g;\delta)$ is defined as:

$$\omega_2(g;\delta)=\sup_{0<h<\delta},\ \sup_{y\in[0,\infty)}|g(y)+g(y+2h)-2g(y+h)|. \tag{29}$$

Theorem 6. *Let q_r, p_r be the real numbers, with $q_r \in (0,1)$ and $p_r \in (q_r,1]$ for every integer r, satisfying $(q_r)\to 1$ and $(p_r)\to 1$ as $r\to\infty$. Then, for all $g\in C_B[0,\infty)$, we have:*

$$\begin{aligned}
&\left|\mathcal{K}_{\eta}^{p_r,q_r}(g;y)-g(y)\right| \\
&\leqq 2\mathcal{A}\left\{\omega_2\left(g;\sqrt{\frac{\Upsilon_\eta(y)}{2}}\right)+\min\left(1;\frac{\Upsilon_\eta(y)}{2}\right)||g||_{C_B[0,\infty)}\right\},
\end{aligned}$$

where $\mathcal{A}$ is a positive constant and $\Upsilon_\eta(y)$ is given in Theorem 5.

Proof. We take $\psi \in C_B^2[0,\infty)$ and apply Theorem (5). Thus:

$$\begin{aligned}
\left|\mathcal{K}_{\eta}^{p_r,q_r}(g;y)-g(y)\right| &\leqq \left|\mathcal{K}_{\eta}^{p_r,q_r}(g-\psi;y)\right|+\left|\mathcal{K}_{\eta}^{p_r,q_r}(\psi;y)-\psi(y)\right|+|g(y)-\psi(y)| \\
&\leqq 2||g-\psi||_{C_B[0,\infty)}+\Upsilon_\eta(y)||\psi||_{C_B^2[0,\infty)} \\
&= 2\left(||g-\psi||_{C_B[0,\infty)}+\frac{\Upsilon_\eta(y)}{2}||\psi||_{C_B^2[0,\infty)}\right).
\end{aligned}$$

By taking the infimum over all $\psi \in C_B^2[0,\infty)$ and using (28), we get:

$$\left|\mathcal{K}_{\eta}^{p_r,q_r}(g;y)-g(y)\right|\leqq 2K_2\left(g;\frac{\Upsilon_\eta(y)}{2}\right).$$

Now, from [33] for all $g \in C_B[0,\infty)$, we have the relation:

$$K_2(g;\delta) \leqq \mathcal{A}\{\min(1;\delta) + \omega_2(g;\sqrt{\delta})||g||_{C_B[0,\infty)}\},$$

where $\mathcal{A} > 0$ is an absolute constant. If we choose $\delta = \frac{\Upsilon_\eta(y)}{2}$, then we get the desired result. □

4. Conclusions

In this paper, we have studied the approximation results via Dunkl generalization of the Szász–Kantorovich operators in (p,q)-calculus. These types of modifications enable us to generalize error estimation rather than the classical and q-calculus on the interval $[0,\infty)$ obtained in [29]. We have also proven the Korovkin-type results and obtained the convergence of our operators in weighted space by the modulus of continuity, Lipschitz class, and Peetre's K-functionals. We have a more generalized version of the operators [29,30], and if we take $\eta = 0$ in (8), then the operators $\mathcal{K}_\eta^{p,q}$ reduce to the operators defined by [30].

Author Contributions: The authors contributed equally and significantly in writing this paper. All authors read and approved the final manuscript.

Funding: The second author would like to thank Prince Sultan University for funding this work through research group Nonlinear Analysis Methods in Applied Mathematics (NAMAM), Group Number RG-DES-2017-01-17.

Conflicts of Interest: The authors declare that they have no competing interests.

References

1. Bernstein, S.N. Démonstration du théoréme de Weierstrass fondée sur le calcul des probabilités. *Commun. Soc. Math. Kharkow* **1912**, *2*, 1–2.
2. Lupaş, A. A q-analogue of the Bernstein operator. *Univ. Cluj-Napoca Semin. Numer. Stat. Calculus* **1987**, *9*, 85–92.
3. Phillips, G.M. Bernstein polynomials based on the q- integers. The heritage of P.L. Chebyshev, A Festschrift in honor of the 70th-birthday of Professor T. J. Rivlin. *Ann. Numer. Math.* **1997**, *4*, 511–518.
4. Mursaleen, M.; Ansari, K.J.; Khan, A. On $(p;q)$-analogue of Bernstein operators. *Appl. Math. Comput.* **2015**, *266*, 874–882. [CrossRef]
5. Szász, O. Generalization of S. Bernstein's polynomials to the infinite interval. *J. Res. Natl. Bur. Stand.* **1950**, *45*, 239–245. [CrossRef]
6. Sucu, S. Dunkl analogue of Szász operators. *Appl. Math. Comput.* **2014**, *244*, 42–48. [CrossRef]
7. Cheikh, B.; Gaied, Y.; Zaghouani, M. A q-Dunkl-classical q-Hermite type polynomials. *Georgian Math. J.* **2014**, *21*, 125–137.
8. İçöz, G.; Çekim, B. Dunkl generalization of Szász operators via q-calculus. *J. Inequal. Appl.* **2015**, *2015*, 284. [CrossRef]
9. Acar, T.; Aral, A.; Mohiuddine, S.A. On Kantorovich modification of $(p;q)$-Baskakov operators. *J. Inequal. Appl.* **2016**, *2016*, 98. [CrossRef]
10. Acar, T.; Aral, A.; Mohiuddine, S.A. On Kantorovich modification of $(p;q)$-Bernstein operators. *Iran. J. Sci. Technol. Trans. A Sci.* **2018**, *42*, 1459–1464. [CrossRef]
11. Acar, T.; Mursaleen, M.; Mohiuddine, S.A. Stancu type $(p;q)$-Szász-Mirakyan-Baskakov operators. *Commun. Fac. Sci. Univ. Ank. Sér. A1 Math. Stat.* **2018**, *67*, 116–128.
12. Araci, S.; Duran, U.; Acikgöz, M.; Srivastava, H.M. A certain $(p;q)$-derivative operator and associated divided differences. *J. Inequal. Appl.* **2016**, *2016*, 301. [CrossRef]
13. Kadak, U. On weighted statistical convergence based on $(p;q)$-integers and related approyimation theorems for functions of two variables. *J. Math. Anal. Appl.* **2016**, *443*, 752–764. [CrossRef]
14. Kadak, U. Weighted statistical convergence based on generalized difference operator involving $(p;q)$-Gamma function and its applications to approyimation theorems. *J. Math. Anal. Appl.* **2017**, *448*, 1633–1650. [CrossRef]
15. Kadak, U.; Mishra, V.N.; Pandey, S. Chlodowsky type generalization of $(p;q)$-Szász operators involving Brenke type polynomials. *Rev. R. Acad. Cienc. Eyactas Fís. Nat. Ser. A Mat.* **2018**, *112*, 1443–1462. [CrossRef]

16. Mursaleen, M.; Nasiruzzaman, M.; Khan, A.; Ansari, K.J. Some approyimation results on Bleimann-Butzer-Hahn operators defined by $(p;q)$-integers. *Filomat* **2016**, *30*, 639–648. [CrossRef]
17. Mursaleen, M.; Nasiruzzaman, M.; Ashirbayev, N.; Abzhapbarov, A. Higher order generalization of Bernstein type operators defined by $(p;q)$-integers. *J. Comput. Anal. Appl.* **2018**, *25*, 817–829.
18. Milovanovic, G.V.; Mursaleen, M.; Nasiruzzaman, M. Modified Stancu type Dunkl generalization of Szász–Kantorovich operators. *Rev. R. Acad. Cienc. Eyactas Fís. Nat. Ser. A Mat.* **2018**, *112*, 135–151. [CrossRef]
19. Mishra, V.N.; Mursaleen, M.; Pandey, S.; Alotaibi, A. Approyimation properties of Chlodowsky variant of $(p;q)$-Bernstein-Stancu-Schurer operators. *J. Inequal. Appl.* **2017**, *2017*, 176. [CrossRef] [PubMed]
20. Mishra, V.N.; Pandey, S. On $(p;q)$-Baskakov-Durrmeyer-Stancu operators. *Adv. Appl. Clifford Algebras* **2017**, *27*, 1633–1646. [CrossRef]
21. Mursaleen, M.; Nasiruzzaman, M.; Alotaibi, A. On modified Dunkl generalization of Szász operators via q-calculus. *J. Inequal. Appl.* **2017**, *2017*, 38. [CrossRef] [PubMed]
22. Mursaleen, M.; Khan, A.; Srivastava, H.M.; Nisar, K.S. Operators constructed by means of q-Lagrange polynomials and A-statistical approyimation. *Appl. Math. Comput.* **2013**, *219*, 6911–6818.
23. Rao, N.; Wafi, A.; Acu, A.M. q -Szász-Durrmeyer type operators based on Dunkl analogue. *Complex Anal. Oper. Theory* **2018**. [CrossRef]
24. Srivastava, H.M.; Mursaleen, M.; Alotaibi, A.M.; Nasiruzzaman, M.; Al-Abied, A. Some approyimation results involving the q-Szasz-Mirakjan-Kantorovich type operators via Dunkl's generalization. *Math. Methods Appl. Sci.* **2017**, *40*, 5437–5452. [CrossRef]
25. Ansari, K.J.; Ahmad, I.; Mursaleen, M.; Hussain, I. On some statistical approyimation by $(p;q)$-Bleimann, Butzer and Hahn operators. *Symmetry* **2018**, *10*, 731. [CrossRef]
26. Jebreen, H.B.; Mursaleen, M.; Naaz, A. Approyimation by quaternion $(p;q)$-Bernstein polynomials and Voronovskaja type result on compact disk. *Adv. Differ. Equ.* **2018**, *2018*, 448. [CrossRef]
27. Mursaleen, M.; Naaz, A.; Khan, A. Improved approyimation and error estimations by King type $(p;q)$-Szasz-Mirakjan Kantorovich operators. *Appl. Math. Comput.* **2019**, *348*, 175–185.
28. Alotaibi, A.; Nasiruzzaman, M.; Mursaleen, M. A Dunkl type generalization of Szász operators via post-quantum calculus. *J. Inequal. Appl.* **2018**, *2018*, 287. [CrossRef]
29. Içõz, G.; Çekim, B. Stancu type generalization of Dunkl analogue of Szász-Kamtrovich operators. *Math. Methods Appl. Sci.* **2016**, *39*, 1803–1810. [CrossRef]
30. Mursaleen, M.; Alotaibi, A.; Ansari, K.J. On a Kantorovich Variant of $(p;q)$-Szász-Mirakjan operators. *J. Funct. Spaces* **2016**. [CrossRef]
31. Korovkin, P.P. On convergence of linear operators in the space of continuous functions. *Dokl. Akad. Nauk SSSR* **1953**, *90*, 961–964.
32. Peetre, J. *A Theory of Interpolation of Normed Spaces*; Noteas de Mathematica; Instituto de Mathemática Pura e Applicada, Conselho Nacional de Pesquidas: Rio de Janeiro, Brazil, 1968; Volume 39.
33. De Sole, A.; Kac, V.G. On integral representations of q- gamma and q- beta functions. *Atti Accad. Naz. Lincei Cl. Sci. Fis. Mat. Appl.* **2005**, *16*, 11–29.

© 2019 by the authors. Licensee MDPI, Basel, Switzerland. This article is an open access article distributed under the terms and conditions of the Creative Commons Attribution (CC BY) license (http://creativecommons.org/licenses/by/4.0/).

symmetry

MDPI

Article

Starlike Functions Related to the Bell Numbers

Nak Eun Cho [1,*], Sushil Kumar [2], Virendra Kumar [3], V. Ravichandran [4] and H. M. Srivastava [5,6,*]

1. Department of Applied Mathematics, Pukyong National University, Busan 48513, Korea
2. Bharati Vidyapeeth's College of Engineering, Delhi 110063, India; sushilkumar16n@gmail.com
3. Department of Mathematics, Ramanujan College, University of Delhi, Kalkaji, New Delhi 110019, India; vktmaths@yahoo.in
4. Department of Mathematics, National Institute of Technology, Tiruchirappalli, Tamil Nadu 620015, India; vravi68@gmail.com or ravic@nitt.edu
5. Department of Mathematics and Statistics, University of Victoria, Victoria, BC V8W 3R4, Canada
6. Department of Medical Research, China Medical University Hospital, China Medical University, Taichung 40402, Taiwan

* Correspondence: necho@pknu.ac.kr (N.E.C.); harimsri@math.uvic.ca (H.M.S.)

Received:19 January 2019; Accepted: 12 February 2019; Published: 13 February 2019

Abstract: The present paper aims to establish the first order differential subordination relations between functions with a positive real part and starlike functions related to the Bell numbers. In addition, several sharp radii estimates for functions in the class of starlike functions associated with the Bell numbers are determined.

Keywords: differential subordination; starlike functions; Bell numbers; radius estimate

MSC: 30C45; 30C55; 30C80

---

1. Introduction

Let $\mathcal{A}$ be a class of analytic functions f in the open unit disk $\mathbb{D} := \{z \in \mathbb{C} : |z| < 1\}$ and normalized by the conditions $f(0) = 0$ and $f'(0) = 1$. Suppose $\mathcal{S}$ is a subclass of $\mathcal{A}$ consisting of univalent functions. An analytic function f is subordinate to g, written as $f \prec g$, if there exists an analytic function $w : \mathbb{D} \to \mathbb{D}$ with $|w(z)| \leq |z|$ such that $f(z) = g(w(z))$ $(z \in \mathbb{D})$. Moreover, if g is univalent in $\mathbb{D}$, then the equivalent conditions for subordination can be written as $f(0) = g(0)$ and $f(\mathbb{D}) \subseteq g(\mathbb{D})$. By imposing some geometric and analytic conditions over the functions in the class $\mathcal{S}$, many authors considered several subclasses of $\mathcal{S}$. Various subclasses of starlike and convex functions were studied in the literature, and they can be unified by considering an analytic univalent function φ with a positive real part in $\mathbb{D}$, symmetric about the real axis and starlike with respect to $\varphi(0) = 1$, and $\varphi'(0) > 0$. Ma and Minda [1] studied the class

$$\mathcal{S}^*(\varphi) := \left\{ f \in \mathcal{A} : \frac{zf'(z)}{f(z)} \prec \varphi(z) \right\}.$$

The class $\mathcal{S}^*(\varphi)$ for various choice of the domain $\varphi(\mathbb{D})$ was considered in recent years. The class $\mathcal{S}^*[A, B] := \mathcal{S}^*((1 + Az)/(1 + Bz))(-1 \leq B < A \leq 1)$ was introduced by Janowski [2]. For $0 \leq \alpha \leq 1$, the class $\mathcal{S}^*(\alpha) := \mathcal{S}^*[1 - 2\alpha, -1]$ is the class of starlike functions of order α. Uralegaddi et al. [3] defined the class

$$\mathcal{M}(\beta) := \left\{ f \in \mathcal{A} : \operatorname{Re}\left(\frac{zf'(z)}{f(z)}\right) < \beta \ \ (\beta > 1) \right\} = \mathcal{S}^*\left(\frac{1 + (1 - 2\beta)z}{1 - z}\right).$$

Several authors considered various special cases of the class of Janowski starlike functions by considering some specific functions, namely $\varphi_q(z) := z + \sqrt{1 + z^2}$, $\varphi_0(z) := 1 +$

$(z/k)((k+z)/(k-z))$ $(k=\sqrt{2}+1)$, $\varphi_s(z) := 1+\sin z$, and $G_\alpha(z) := 1+z/(1-\alpha z^2)$. Some of those classes are: $\mathcal{S}_L^* := \mathcal{S}^*(\sqrt{1+z})$ [4], $\mathcal{S}_q^* := \mathcal{S}^*(\varphi_q(z))$ [5], $\mathcal{S}_e^* = \mathcal{S}^*(e^z)$ [6], $\mathcal{S}_R^* = \mathcal{S}^*(\varphi_0)$ [7], $\mathcal{S}_s^* = \mathcal{S}^*(\varphi_s)$ [8]) , $\mathcal{BS}^*(\alpha) := \mathcal{S}^*(G_\alpha(z)), 0 \le \alpha < 1$ [9,10]. For a brief survey on these classes, readers may refer to [11,12].

It should be noted that the special cases of φ, mentioned above, are univalent in the unit disk. In 2011, Dziok et al. [13,14] considered φ to be a non-univalent function associated with the Fibonacci numbers, defined by

$$\tilde{p}(z) := \varphi(z) = \frac{1+\tau^2 z^2}{1-\tau z-\tau^2 z^2}, \quad \tau := \left(1-\sqrt{5}\right)/2$$

which maps the unit disk $\mathbb{D}$ on to a shell-like domain in the right-half plane. Further, they defined the class $\mathcal{S}_F^* := \{f \in \mathcal{A} : zf'(z)/f(z) \prec \tilde{p}(z)\}$. The functions $f \in \mathcal{S}_F^*$ are starlike of order $\sqrt{5}/10$.

Motivated by the above defined classes, we consider a function associated with the Bell Numbers. For a fixed non-negative integer n, the Bell numbers B_n count the possible disjoint partitions of a set with n elements into non-empty subsets or, equivalently, the number of equivalence relations on it. The Bell numbers B_n satisfy a recurrence relation involving binomial coefficients $B_{n+1} = \sum_{k=0}^{n}\binom{n}{k}B_k$. Clearly $B_0 = B_1 = 1, B_2 = 2, B_3 = 5, B_4 = 15, B_5 = 52$, and $B_6 = 203$. For more details, see [15–21]. Kumar et al. [22] considered the function

$$Q(z) := e^{e^z-1} = \sum_{n=0}^{\infty} B_n \frac{z^n}{n!} = 1+z+z^2+\frac{5}{6}z^3+\frac{5}{8}z^4+\cdots \quad (z \in \mathbb{D})$$

which is starlike with respect to 1 and it's coefficients generate the Bell numbers. Kumar et al. [22] defined the class $\mathcal{S}_B^*$ by $\mathcal{S}_B^* := \mathcal{S}^*(Q)$. From [1], note that the function $f \in \mathcal{S}_B^*$ if and only if there exists an analytic function q, satisfying $q(z) \prec Q(z)$ $(z \in \mathbb{D})$, such that

$$f(z) = I(q(z)) = z\exp\left(\int_0^z \frac{q(t)-1}{t}dt\right).$$

The above representation shows that the functions in the class $\mathcal{S}_B^*$ can be seen as an integral transform $I(q(z))$ of the function q with $f(0) = 0$ and $f'(0) = 1$. The reader may refer to the paper [23] and the references cited therein for integral transform related works. The authors in [22] determined sharp coefficient bounds on the six initial coefficients, Hankel determinant, and on the first three consecutive higher order Schwarzian derivatives for functions in the class $\mathcal{S}_B^*$.

Let $\mathcal{P}$ be the class of analytic functions $p : \mathbb{D} \to \mathbb{C}$ with $p(0) = 1$ and $\operatorname{Re} p(z) > 0$ $(z \in \mathbb{D})$. In 1989, Nunokawa et al. [24] showed that if $1+zp'(z) \prec 1+z$, then $p(z) \prec 1+z$. In 2007, Ali et al. [25] computed the condition on β, in each case, for which

$$1+\frac{\beta zp'(z)}{p^j(z)} \prec \frac{1+Dz}{1+Ez} \quad (j=0,1,2) \text{ implies } p(z) \prec \frac{1+Az}{1+Bz},$$

$A, B, C, D, E, F \in [-1,1]$. Further, Ali et al. [26] determined some sufficient conditions for normalized analytic functions to lemniscate starlike functions. Recently, Kumar and Ravichandran [27] obtained sufficient conditions for first order differential subordinations so that the corresponding analytic function belongs to the class $\mathcal{P}$. In 2016, Tuneski [28] gave a criteria for analytic functions to be Janowski starlike. For more details, see [11,29–33].

Motivated by above works, in Section 2, using the theory of differential subordination developed by Miller and Mocanu, a sharp bound on parameter β is determined in each case so that $p(z) \prec Q(z)$, whenever $1+\beta zp'(z)/p^j(z) (j = 0, 1, 2)$ is subordinate to the function $\varphi_0(z)$ or $\sqrt{1+z}$ or $G_\alpha(z)$ or $(1+Az)/(1+Bz)$ or $\varphi_s(z)$ or $\varphi_q(z)$. Further, various sufficient conditions are obtained for $f \in \mathcal{A}$ to be in the class $\mathcal{S}_B^*$ as an application of these subordination results. In Section 3, $\mathcal{S}_B^*$-radius for the class of Janowski starlike functions and some other well-known classes of analytic functions are investigated.

2. Differential Subordinations

Theorem 1 provides estimate on β so that $p(z) \prec Q(z)$ holds, whenever $1+\beta zp'(z) \prec \varphi_0(z)$ or $\varphi_s(z)$ or $\sqrt{1+z}$ or $G_\alpha(z)$ or $(1+Az)/(1+Bz)$ or $\varphi_s(z)$ or $\varphi_q(z)$ or e^z.

To prove our main results, we need the following lemma due to Miller and Mocanu:

Lemma 1. ([32] Theorem 3.4h, p. 132) *Let q be analytic in $\mathbb{D}$ and let ψ and ν be analytic in a domain U containing $q(\mathbb{D})$ with $\psi(w) \neq 0$ when $w \in q(\mathbb{D})$. Set*

$$\mathcal{Q}(z) := zq'(z)\psi(q(z)) \text{ and } h(z) := \nu(q(z)) + \mathcal{Q}(z).$$

Suppose that

(i) either h is convex, or $\mathcal{Q}$ is starlike univalent in $\mathbb{D}$ and

(ii) $\operatorname{Re}\left(\dfrac{zh'(z)}{\mathcal{Q}(z)}\right) > 0$ for $z \in \mathbb{D}$.

If p is analytic in $\mathbb{D}$, with $p(0) = q(0)$, $p(\mathbb{D}) \subseteq U$ and

$$\nu(p(z)) + zp'(z)\psi(p(z)) \prec \nu(q(z)) + zq'(z)\psi(q(z)),$$

then $p \prec q$, and q is most dominant.

Theorem 1. *Let $l(e) = (1 - e^{(1-e)/e})^{-1}$, $0 < \alpha < 1$, $0 < B < A < 1$, and p be an analytic function defined in $\mathbb{D}$ with $p(0) = 1$.*

Set

$$\Upsilon_\beta(z, p(z)) = 1 + \beta zp'(z).$$

Then, the following are sufficient for $p(z) \prec Q(z)$.

(a) $\Upsilon_\beta(z, p(z)) \prec \varphi_0(z)$ for $\beta \geq l(e)(1 - \sqrt{2} + \log 2) \approx 0.59533$.
(b) $\Upsilon_\beta(z, p(z)) \prec \sqrt{1+z}$ for $\beta \geq l(e)(2(1 - log2)) \approx 1.30984$.
(c) $\Upsilon_\beta(z, p(z)) \prec G_\alpha(z)$ for $\beta \geq l(e)\frac{1}{2\sqrt{\alpha}} \log \frac{1+\sqrt{\alpha}}{1-\sqrt{\alpha}}$.
(d) $\Upsilon_\beta(z, p(z)) \prec \frac{1+Az}{1+Bz}$ for $\beta \geq l(e)\frac{A-B}{B} \log (1-B)^{-1}$.
(e) $\Upsilon_\beta(z, p(z)) \prec \varphi_s(z)$ for $\beta \geq l(e)\sum_{n=0}^{\infty} \frac{(-1)^n}{(2n+1)!(2n+1)} \approx 2.01905$.
(f) $\Upsilon_\beta(z, p(z)) \prec \varphi_q(z)$ for $\beta \geq l(e)(2 - \sqrt{2} - \log 2 + \log(1+\sqrt{2})) \approx 1.65198$.
(g) $\Upsilon_\beta(z, p(z)) \prec e^z$ for $\beta \geq l(e)\sum_{n=0}^{\infty} \frac{(-1)^{n-1}}{n!n} \approx 0.785166$.

The lower bound on β in each case is sharp.

Proof. Let the functions ν and ψ be defined by $\nu(w) = 1$ and $\psi(w) = \beta$.

(a) Define the function $q_\beta : \overline{\mathbb{D}} \to \mathbb{C}$ by

$$q_\beta(z) = 1 - \frac{1}{\beta k}\left(z + 2k \log\left(1 - \frac{z}{k}\right)\right)$$

is a solution of the differential equation $\beta zq'(z) = \varphi_0(z) - 1$ and is analytic in $\mathbb{D}$. Now consider the function

$$\mathcal{Q}(z) = zq'_\beta(z)\psi(q_\beta(z)) = \varphi_0(z) - 1 = \frac{k + z - 2k^2}{k - z}.$$

It can be easily seen that $\mathcal{Q}$ is starlike in $\mathbb{D}$ and the function h is defined by

$$h(z) := \nu(q(z)) + \mathcal{Q}(z) = 1 + \mathcal{Q}(z)$$

satisfies the following inequality

$$\operatorname{Re}\left(\frac{zh'(z)}{\mathcal{Q}(z)}\right) = \operatorname{Re}\left(\frac{z\mathcal{Q}'(z)}{\mathcal{Q}(z)}\right) > 0 \ (z \in \mathbb{D}).$$

Therefore, from Lemma 1, we conclude that

$$1 + \beta z p'(z) \prec 1 + \beta z q'_\beta(z) \text{ implies } p \prec q_\beta. \tag{1}$$

Now the subordination $p \prec Q$ holds if subordination $q_\beta \prec Q$. Thus, the subordination $q_\beta \prec Q$ holds if the inequalities

$$Q(-1) \le q_\beta(-1) \le q_\beta(1) \le Q(1)$$

hold and these yield a necessary condition for subordination $p \prec Q$ to hold. In view of the graph of the respective function, the necessary condition is also sufficient condition. The inequalities $q_\beta(-1) \ge Q(-1)$ and $q_\beta(1) \le Q(1)$ yield $\beta \ge \beta_1$ and $\beta \ge \beta_2$, where

$$\beta_1 = \frac{1 - \sqrt{2} + \log 2}{1 - e^{(1-e)/e}} \text{ and } \beta_2 = \frac{1 - \sqrt{2} - 2\log(2 - \sqrt{2})}{e^{(e-1)/e} - 1}.$$

Now the subordination $q_\beta \prec Q$ holds if $\beta \ge \max\{\beta_1, \beta_2\} = \beta_1$.

(b) The function

$$q_\beta(z) = \frac{\beta + 2(\sqrt{1+z} - \log(1 + \sqrt{1+z}) + \log 2 - 1)}{\beta}$$

is an analytic solution of the first order differential equation $\beta z q'(z) = \sqrt{1+z} - 1$ in $\mathbb{D}$. The function $\mathcal{Q}$ defined by $\mathcal{Q}(z) = z q'_\beta(z)\psi(q_\beta(z)) = \sqrt{1+z} - 1$ is starlike in $\mathbb{D}$ and the function $h(z) := \nu(q(z)) + \mathcal{Q}(z)$ satisfies $\operatorname{Re}(zh'(z)/\mathcal{Q}(z)) = \operatorname{Re}(z\mathcal{Q}'(z)/\mathcal{Q}(z)) > 0, z \in \mathbb{D}$. Therefore, in view of the subordination relation 1, the required subordination $p \prec Q$ holds if subordination $q_\beta \prec Q$ holds. Thus, the subordination $q_\beta \prec Q$ holds if the inequalities

$$Q(-1) \le q_\beta(-1) \le q_\beta(1) \le Q(1)$$

hold which in-turn yield a necessary condition for subordination $p \prec Q$. The inequalities $q_\beta(-1) \ge Q(-1)$ and $q_\beta(1) \le Q(1)$ yield $\beta \ge \beta_1 = 2(1 - \log 2)/1 - e^{(1-e)/e}$ and $\beta \ge \beta_2 = 2(\sqrt{2} - 1 + \log 2 - \log(1 + \sqrt{2}))/(e^{(1-e)/e} - 1)$, respectively. Therefore, the subordination $q_\beta \prec Q$ holds if $\beta \ge \max\{\beta_1, \beta_2\} = \beta_1$.

(c) The analytic function

$$q_\beta(z) = \frac{2\sqrt{\alpha}\beta + \log \frac{1+\sqrt{\alpha}z}{1-\sqrt{\alpha}z}}{2\sqrt{\alpha}\beta}$$

is a solution of the differential equation $\beta z q'_\beta(z) = G_\alpha(z) - 1$ in $\mathbb{D}$. Now computation shows that

$$\mathcal{Q}(z) = z q'_\beta(z)\psi(q_\beta(z)) = \frac{z}{1 - \alpha z^2}$$

is starlike in $\mathbb{D}$. Note that the function $h(z) := \nu(q(z)) + \mathcal{Q}(z) = 1 + \mathcal{Q}(z)$ satisfies $\operatorname{Re}(zh'(z)/\mathcal{Q}(z)) = \operatorname{Re}(z\mathcal{Q}'(z)/\mathcal{Q}(z)) > 0$ in $\mathbb{D}$. Therefore, in view of the subordination relation 1, the required subordination $p \prec Q$ holds if subordination $q_\beta \prec Q$. Similar to as in part (a), the desired subordination $p \prec Q$ holds if $\beta \ge \max\{\beta_1, \beta_2\} = \beta_1$, where $\beta_1 = l(e)g(\alpha)$ and $\beta_2 = -l(e)g(\alpha)$ such that

$$g(\alpha) = \frac{1}{2\sqrt{\alpha}} \log \frac{1 + \sqrt{\alpha}}{1 - \sqrt{\alpha}}.$$

(d) Consider the analytic function

$$q_\beta(z) = \frac{B\beta + (A-B)\log(1+Bz)}{B\beta}$$

which is a solution of differential equation

$$\beta z q'(z) = \frac{(A-B)z}{1+Bz}.$$

Since the function $(A-B)z/(1+Bz)$ is starlike in $\mathbb{D}$, it follows that $\mathcal{Q}(z) = zq'_\beta(z)\psi(q_\beta(z))$ is starlike in $\mathbb{D}$. The function $h : \mathbb{D} \to \mathbb{C}$ defined by $h(z) := \nu(q_\beta(z)) + \mathcal{Q}(z) = 1 + \mathcal{Q}(z)$ satisfies $\operatorname{Re}(zh'(z)/\mathcal{Q}(z)) > 0$ $(z \in \mathbb{D})$. Thus, as in previous case, the subordination $p \prec \mathcal{Q}$ holds if $\beta \geq \max\{\beta_1, \beta_2\} = \beta_1$, where

$$\beta_1 = \frac{(A-B)\log(1-B)^{-1}}{B(1-e^{(1-e)/e})} \quad \text{and} \quad \beta_2 = \frac{(A-B)\log(1+B)}{B(e^{(1-e)/e}-1)}.$$

(e) The differential equation

$$\frac{dq}{dz} = \frac{\sin z}{\beta z}$$

has an analytic solution

$$q_\beta(z) = 1 + \frac{1}{\beta}\sum_{n=0}^{\infty} \frac{(-1)^n z^{2n+1}}{(2n+1)!(2n+1)}$$

in $\mathbb{D}$. Now the function $\mathcal{Q}(z) = zq'_\beta(z)\psi(q_\beta(z)) = \sin z$ is starlike in $\mathbb{D}$ and the function $h(z) := \nu(q(z)) + \mathcal{Q}(z) = 1 + \mathcal{Q}(z)$, satisfies $\operatorname{Re}(zh'(z)/\mathcal{Q}(z)) = \operatorname{Re}(z\mathcal{Q}'(z)/\mathcal{Q}(z)) > 0$ holds. As in part (a), the desired subordination $p(z) \prec \mathcal{Q}(z)$ holds if $\beta \geq \max\{\beta_1, \beta_2\} = \beta_1$, where

$$\beta_1 = \frac{1}{(1-e^{(1-e)/e})}\sum_{n=0}^{\infty} \frac{(-1)^n}{(2n+1)!(2n+1)} \approx 2.01905$$

and

$$\beta_2 = \frac{1}{(e^{(e-1)}-1)}\sum_{n=0}^{\infty} \frac{(-1)^n}{(2n+1)!(2n+1)} \approx 0.206779.$$

(f) The differential equation

$$\frac{dq}{dz} = \frac{z+\sqrt{1+z^2}-1}{\beta z}$$

has an analytic solution

$$q_\beta(z) = \frac{\beta + (z+\sqrt{1+z^2} - \log(1+\sqrt{1+z^2}) - 1 + \log 2)}{\beta}.$$

Computation shows that the function

$$\mathcal{Q}(z) = zq'_\beta(z)\psi(q_\beta(z)) = z + \sqrt{1+z^2} - 1$$

is starlike in $\mathbb{D}$. As before, the function $h(z) := \nu(q(z)) + \mathcal{Q}(z)$ satisfies $\operatorname{Re}(zh'(z)/\mathcal{Q}(z)) > 0$, $z \in \mathbb{D}$. Therefore, the desired subordination $p \prec Q$ holds if $\beta \geq \max\{\beta_1, \beta_2\} = \beta_1$, where

$$\beta_1 = \frac{2-\sqrt{2}-\log 2 + \log(1+\sqrt{2})}{1-e^{(1-e)/e}} \approx 1.65198$$

and

$$\beta_2 = \frac{\sqrt{2} + \log 2 - \log(1+\sqrt{2})}{e^{(1-e)/e} - 1} \approx 0.267979.$$

(g) The differential equation

$$\frac{dq}{dz} = \frac{e^z - 1}{\beta z}$$

has an analytic solution

$$q_\beta(z) = 1 + \frac{1}{\beta} \sum_{n=0}^{\infty} \frac{z^n}{n!n}.$$

Note that the function $\mathcal{Q}(z) = zq'_\beta(z)\psi(q_\beta(z)) = e^z$ is starlike in the unit disk $\mathbb{D}$ and the function $h(z) := \nu(q(z)) + \mathcal{Q}(z) = 1 + \mathcal{Q}(z)$ satisfies $\mathrm{Re}\,(zh'(z)/\mathcal{Q}(z)) = \mathrm{Re}\,(z\mathcal{Q}'(z)/\mathcal{Q}(z)) > 0$. Now the subordination $p \prec Q$ holds if $\beta \geq \max\{\beta_1, \beta_2\} = \beta_1$, where

$$\beta_1 = \frac{1}{(1 - e^{(1-e)/e})} \sum_{n=0}^{\infty} \frac{(-1)^{n-1}}{n!n} \approx 0.785166 \quad \text{and} \quad \beta_2 = \frac{1}{(e^{(e-1)} - 1)} \sum_{n=0}^{\infty} \frac{1}{n!n} \approx 0.288069.$$

This ends the proof. □

Theorem 1 also provides the following various sufficient conditions for the normalized analytic functions f to be in the class $\mathcal{S}^*_B$.

Let function $f \in \mathcal{A}$ and set

$$\Upsilon_\beta\left(z, \frac{zf'(z)}{f(z)}\right) = 1 + \beta \frac{zf'(z)}{f(z)} \left(1 - \frac{zf'(z)}{f(z)} + \frac{zf''(z)}{f'(z)}\right).$$

If either of the following subordination holds

(a) $\Upsilon_\beta\left(z, \frac{zf'(z)}{f(z)}\right) \prec \varphi_0(z)\ (\beta \geq 0.59533)$,
(b) $\Upsilon_\beta\left(z, \frac{zf'(z)}{f(z)}\right) \prec \sqrt{1+z}\ (\beta \geq 1.30984)$,
(c) $\Upsilon_\beta\left(z, \frac{f'(z)}{f(z)}\right) \prec G_\alpha(z)\ (\beta \geq \frac{1}{(1-e^{(1-e)/e})} \frac{1}{2\sqrt{\alpha}} \log \frac{1+\sqrt{\alpha}}{1-\sqrt{\alpha}})$,
(d) $\Upsilon_\beta\left(z, \frac{f'(z)}{f(z)}\right) \prec \frac{1+Az}{1+Bz}\ (\beta \geq \frac{1}{(1-e^{(1-e)/e})} \frac{A-B}{B} \log(1-B)^{-1})$,
(e) $\Upsilon_\beta\left(z, \frac{zf'(z)}{f(z)}\right) \prec \varphi_s(z)\ (\beta \geq 2.01905)$,
(f) $\Upsilon_\beta\left(z, \frac{zf'(z)}{f(z)}\right) \prec \varphi_q(z)\ (\beta \geq 1.65198)$,
(g) $\Upsilon_\beta\left(z, \frac{zf'(z)}{f(z)}\right) \prec e^z\ (\beta \geq 0.785166)$,

then $f \in \mathcal{S}^*_B$.

The next result gives sharp lower bound on β such that subordination $p \prec Q$ holds, whenever $1 + \beta zp'(z)/p(z) \prec \varphi_0(z)$ or $\varphi_s(z)$ or $\sqrt{1+z}$ or $G_\alpha(z)$ or $(1+Az)/(1+Bz)$ or $\varphi_s(z)$ or $\varphi_q(z)$ or e^z.

Theorem 2. *Let $0 < \alpha < 1, 0 < B < A < 1$, and p be an analytic function defined in $\mathbb{D}$ with $p(0) = 1$. Set*

$$\Omega_\beta(z, p(z)) = 1 + \beta \frac{zp'(z)}{p(z)}.$$

Then, the following conditions are sufficient for subordination $p \prec Q$.

(a) $\Omega_\beta(z, p(z)) \prec \varphi_0(z)$ *for* $\beta \geq \frac{e(2(1+\sqrt{2})\log\sqrt{2}-1}{(e-1)(1+\sqrt{2})} \approx 0.441266$.
(b) $\Omega_\beta(z, p(z)) \prec \sqrt{1+z}$ *for* $\beta \geq \frac{2e(1-\log 2)}{e-1} \approx 0.970868$.
(c) $\Omega_\beta(z, p(z)) \prec G_\alpha(z)$ *for* $\beta \geq \frac{e}{2(e-1)\sqrt{\alpha}} \log \frac{1+\sqrt{\alpha}}{1-\sqrt{\alpha}}$.
(d) $\Omega_\beta(z, p(z)) \prec \frac{1+Az}{1+Bz}$ *for* $\beta \geq \frac{e}{B(e-1)}(A-B)\log(1-B)^{-1}$.

(e) $\Omega_\beta(z, p(z)) \prec \varphi_s(z)$ *for* $\beta \geq \frac{e}{e-1}\sum_{n=0}^{\infty}\frac{(-1)^n}{(2n+1)!(2n+1)} \approx 1.49655$.
(f) $\Omega_\beta(z, p(z)) \prec \varphi_q(z)$ *for* $\beta \geq \frac{e}{e-1}(2-\sqrt{2}+\log(1+\sqrt{2})-\log 2) \approx 1.22447$.
(g) $\Omega_\beta(z, p(z)) \prec e^z$ *for* $\beta \geq \frac{1}{e-1}\sum_{n=0}^{\infty}\frac{1}{n!n} \approx 0.766987$.

The lower bound on β *in each case is sharp.*

Proof. Let us define $\nu(w) = 1$ and $\psi(w) = \beta/w$ for all $w \in \mathbb{C}$.

(a) The function

$$q_\beta(z) = \exp\left(-\frac{1}{\beta k}\left(z + 2k\log\left(1-\frac{z}{k}\right)\right)\right)$$

satisfies the differential equation $\beta z q'(z)/q(z) = \varphi_0(z) - 1$. Clearly, the function $\mathcal{Q} : \overline{\mathbb{D}} \to$ defined by $\mathcal{Q}(z) = zq'_\beta(z)\psi(q_\beta(z)) = (z - 2k^2 + k)/(k-z)$ is starlike in $\mathbb{D}$. Further, the function $h(z) := \nu(q_\beta(z)) + \mathcal{Q}(z)$ satisfies $\mathrm{Re}(zh'(z)/\mathcal{Q}(z)) > 0$ $(z \in \mathbb{D})$. Thus, using Lemma 1, it follows that

$$1 + \beta\frac{zp'(z)}{p(z)} \prec 1 + \beta\frac{zq'_\beta(z)}{q_\beta(z)} \quad \text{implies } p \prec q_\beta. \tag{2}$$

Now using Theorem 1 (a), the subordination $p \prec Q$ holds if $\beta \geq \max\{\beta_1, \beta_2\} = \beta_1$, where

$$\beta_1 = \frac{(-1+2(1+\sqrt{2})\log\sqrt{2})e}{(e-1)(1+\sqrt{2})}$$

and

$$\beta_2 = -\frac{(1+2(1+\sqrt{2})\log(2-\sqrt{2}))}{(e-1)(1+\sqrt{2})}.$$

(b) The function

$$q_\beta(z) = \exp\left(\frac{2}{\beta}\left(\sqrt{1+z} - \log(1+\sqrt{1+z}) + \log 2 - 1\right)\right)$$

is a solution of the differential equation

$$\beta\frac{zq'(z)}{q(z)} = \sqrt{1+z} - 1.$$

Moreover, the function $\mathcal{Q}(z) = zq'_\beta(z)\psi(q_\beta(z)) = \sqrt{1+z} - 1$ is starlike in $\mathbb{D}$ and a computation shows that the function $h(z) := \nu(q(z)) + \mathcal{Q}(z)$ satisfies $\mathrm{Re}\,(zh'(z)/\mathcal{Q}(z)) > 0$ $(z \in \mathbb{D})$. Now the desired subordination $p \prec Q$ holds if $\beta \geq \max\{\beta_1, \beta_2\} = \beta_1$, where $\beta_1 = 2e(1-\log 2)/(e-1)$ and $\beta_2 = 2(-1+\sqrt{2}+\log 2 - \log(1+\sqrt{2}))/(e-1)$.

(c) Consider the function q_β defined by

$$q_\beta(z) = \exp\left(\frac{1}{2\sqrt{\alpha}\beta}\log\frac{1+\sqrt{\alpha}z}{1-\sqrt{\alpha}z}\right).$$

It can be verified that the function q_β is a solution of the differential equation

$$\beta\frac{zq'(z)}{q(z)} = \frac{1}{1-\alpha z^2}.$$

Now the function $\mathcal{Q}(z) = zq'_\beta(z)\psi(q_\beta(z)) = 1/(1-\alpha z^2)$ is starlike in $\mathbb{D}$ and the function $h(z) := \nu(q(z)) + \mathcal{Q}(z)$ satisfies $\mathrm{Re}\,(zh'(z)/\mathcal{Q}(z)) > 0$ $(z \in \mathbb{D})$. Now, as in previous cases, $p \prec Q$ holds only if $\beta \geq \max\{\beta_1, \beta_2\} = \beta_1$, where

$$\beta_1 = \frac{e}{2(e-1)\sqrt{\alpha}} \log \frac{1+\sqrt{\alpha}}{1-\sqrt{\alpha}} \text{ and } \beta_2 = \frac{1}{2(e-1)\sqrt{\alpha}} \log \frac{1+\sqrt{\alpha}}{1-\sqrt{\alpha}}.$$

(d) Let the function $q_\beta(z) = \exp\left((A-B)\log(1+Bz)/\beta B\right)$ be an analytic solution of the differential equation

$$1+\beta\frac{zq'(z)}{q(z)} = \frac{1+Az}{1+Bz}.$$

Now the desired subordination $p \prec Q$ holds if $\beta \geq \max\{\beta_1, \beta_2\} = \beta_1$, where $\beta_1 = e(A-B)\log(1-B)^{-1}/B(e-1)$ and $\beta_2 = e(A-B)\log(1+B)/B(e-1)$.

(e) The differential equation $\beta zq'(z)/q(z) = \sin z$ has an analytic solution given by

$$q_\beta(z) = \exp\left(\frac{1}{\beta}\sum_{n=0}^{\infty}\frac{(-1)^n z^{2n+1}}{(2n+1)!(2n+1)}\right).$$

As in part Theorem 2 (a), the subordination $p \prec Q$ holds if $\beta \geq \max\{\beta_1, \beta_2\} = \beta_1$ where

$$\beta_1 = \frac{e}{e-1}\sum_{n=0}^{\infty}\frac{(-1)^n}{(2n+1)!(2n+1)} \approx 1.49655$$

and

$$\beta_2 = \frac{1}{e-1}\sum_{n=0}^{\infty}\frac{(-1)^n}{(2n+1)!(2n+1)} \approx 0.55055.$$

(f) The solution of the differential equation

$$\frac{dq}{dz} = \frac{z+\sqrt{1+z^2}-1}{\beta z}$$

is given by

$$q_\beta(z) = \exp\left(\frac{z+\sqrt{1+z^2}-\log(1+\sqrt{1+z^2})-1+\log 2}{\beta}\right).$$

As in proof of Theorem 2 (a), the desired result holds if $\beta \geq \max\{\beta_1, \beta_2\} = \beta_1$, where $\beta_1 = e(2-\sqrt{2}+\log(1+\sqrt{2})-\log 2)/(e-1)$ and $\beta_2 = (\sqrt{2}-\log(1+\sqrt{2})+\log 2)/(e-1)$.

(g) The differential equation $\beta zq'(z)/q(z) = e^z - 1$ has a solution

$$q_\beta(z) = \exp\left(\frac{1}{\beta}\sum_{n=1}^{\infty}\frac{z^n}{n!n}\right)$$

analytic in $\mathbb{D}$. Thus, as previous, the subordination $p \prec Q$ holds if $\beta \geq \max\{\beta_1, \beta_2\} = \beta_2$, where

$$\beta_1 = \frac{e}{e-1}\sum_{n=0}^{\infty}\frac{(-1)^{n-1}}{n!n} \approx 0.581976 \text{ and } \beta_2 = \frac{1}{e-1}\sum_{n=0}^{\infty}\frac{1}{n!n} \approx 0.766987.$$

This ends the proof. □

Next, Theorem 2 also provides the following various sufficient conditions for the normalized analytic functions f to be in the class $\mathcal{S}^*_B$. Let the function $f \in \mathcal{A}$ and set

$$\Omega_\beta\left(z, \frac{zf'(z)}{f(z)}\right) = 1+\beta\left(1-\frac{zf'(z)}{f(z)}+\frac{zf''(z)}{f'(z)}\right).$$

If either of the following subordination conditions are fulfilled:

(a) $\Omega_\beta\left(z, \frac{zf'(z)}{f(z)}\right) \prec \varphi_0(z)$ ($\beta \geq 0.441266$),
(b) $\Omega_\beta\left(z, \frac{zf'(z)}{f(z)}\right) \prec \sqrt{1+z}$ ($\beta \geq 0.970868$),
(c) $\Omega_\beta\left(z, \frac{zf'(z)}{f(z)}\right) \prec G_\alpha(z)$ ($\beta \geq \frac{e}{2(e-1)\sqrt{\alpha}} \log \frac{1+\sqrt{\alpha}}{1-\sqrt{\alpha}}$),
(d) $\Omega_\beta\left(z, \frac{zf'(z)}{f(z)}\right) \prec \frac{1+Az}{1+Bz}$ ($\beta \geq \frac{e}{B(e-1)}(A-B)\log(1-B)^{-1}$),
(e) $\Omega_\beta\left(z, \frac{zf'(z)}{f(z)}\right) \prec \varphi_s(z)$ ($\beta \geq 1.49655$),
(f) $\Omega_\beta\left(z, \frac{zf'(z)}{f(z)}\right) \prec \varphi_q(z)$ ($\beta \geq 1.22447$),
(g) $\Omega_\beta\left(z, \frac{zf'(z)}{f(z)}\right) \prec e^z$ ($\beta \geq 0.766987$),

then $f \in \mathcal{S}^*_B$.

In the following theorem, the sharp lower bound on β is obtained so that the subordination $p \prec Q$ holds, whenever $1 + \beta zp'(z)/p^2(z) \prec \varphi_0(z)$ or $\varphi_s(z)$ or $\sqrt{1+z}$ or $G_\alpha(z)$ or $(1+Az)/(1+Bz)$ or $\varphi_s(z)$ or $\varphi_q(z)$ or e^z. These results can be proved by defining the functions $\nu, \psi : \mathbb{D} \to$ defined by $\nu(w) = 1$ and $\psi(w) = \beta/w^2$ and proceeding in a similar fashion as in the proofs of Theorems 1 and 2.

Theorem 3. *Let $0 < \alpha < 1$, $0 < B < A < 1$, and p be an analytic function defined in $\mathbb{D}$ with $p(0) = 1$. Set*

$$\Xi_\beta(z, p(z)) = 1 + \beta \frac{zp'(z)}{p^2(z)}.$$

Then, the following conditions are sufficient for $p \prec Q$.

(a) $\Xi_\beta(z, p(z)) \prec \varphi_0(z)$ *for* $\beta \geq \frac{1+2(\sqrt{2}+1)\log(2-\sqrt{2})}{(1+\sqrt{2})(e^{(1-e)}-1)} \approx 0.798642$.
(b) $\Xi_\beta(z, p(z)) \prec \sqrt{1+z}$ *for* $\beta \geq \frac{2(-1+\sqrt{2}+\log 2-\log(1+\sqrt{2}))}{1-e^{1-e}} \approx 0.550768$.
(c) $\Xi_\beta(z, p(z)) \prec G_\alpha(z)$ *for* $\beta \geq \frac{e^{e-1}}{e^{e-1}-1}\frac{1}{2\sqrt{\alpha}} \log \frac{1+\sqrt{\alpha}}{1-\sqrt{\alpha}}$.
(d) $\Xi_\beta(z, p(z)) \prec \frac{1+Az}{1+Bz}$ *for* $\beta \geq \frac{e^{(1-e)/e}}{1-e^{(1-e)/e}} \frac{(A-B)\log(1-B)^{-1}}{B}$.
(e) $\Xi_\beta(z, p(z)) \prec \varphi_s(z)$ *for* $\beta \geq \frac{e^{e-1}}{e^{e-1}-1}\sum_{n=0}^{\infty} \frac{(-1)^n}{(2n+1)!(2n+1)} \approx 1.15278$.
(f) $\Xi_\beta(z, p(z)) \prec \varphi_q(z)$ *for* $\beta \geq \frac{e^{e-1}}{e^{e-1}-1}(\sqrt{2} - \log(1+\sqrt{2}) + \log 2) \approx 1.49397$.
(g) $\Xi_\beta(z, p(z)) \prec e^z$ *for* $\beta \geq \frac{e^{e-1}}{e^{e-1}-1}\sum_{n=0}^{\infty} \frac{1}{n!n} \approx 1.60597$.

The lower bound on β in each case is sharp.

Let $f \in \mathcal{A}$ and set

$$\Xi_\beta\left(z, \frac{zf'(z)}{f(z)}\right) = 1 + \beta \left(\frac{zf'(z)}{f(z)}\right)^{-1} \left(1 - \frac{zf'(z)}{f(z)} + \frac{zf''(z)}{f'(z)}\right).$$

If either of the following subordination holds

(a) $\Xi_\beta\left(z, \frac{zf'(z)}{f(z)}\right) \prec \varphi_0(z)$ ($\beta \geq 0.798642$),
(b) $\Xi_\beta\left(z, \frac{zf'(z)}{f(z)}\right) \prec \sqrt{1+z}$ ($\beta \geq 0.550768$),
(c) $\Xi_\beta\left(z, \frac{zf'(z)}{f(z)}\right) \prec G_\alpha(z)$ ($\beta \geq \frac{e^{e-1}}{e^{e-1}-1}\frac{1}{2\sqrt{\alpha}} \log \frac{1+\sqrt{\alpha}}{1-\sqrt{\alpha}}$),
(d) $\Xi_\beta\left(z, \frac{zf'(z)}{f(z)}\right) \prec \frac{1+Az}{1+Bz}$ ($\beta \geq \frac{e^{(1-e)/e}}{1-e^{(1-e)/e}} \frac{(A-B)\log(1-B)^{-1}}{B}$),
(e) $\Xi_\beta\left(z, \frac{zf'(z)}{f(z)}\right) \prec \varphi_s(z)$ ($\beta \geq 1.15278$),
(f) $\Xi_\beta\left(z, \frac{zf'(z)}{f(z)}\right) \prec \varphi_q(z)$ ($\beta \geq 1.49397$),
(g) $\Xi_\beta\left(z, \frac{zf'(z)}{f(z)}\right) \prec e^z$ ($\beta \geq 1.60597$),

then $f \in \mathcal{S}^*_B$.

3. Radius Estimates

Let θ_1 and θ_2 be two sub-families of $\mathcal{A}$. The θ_1 radius of θ_2 is the largest number $\rho \in (0,1)$ such that $r^{-1}f(rz) \in \theta_1$, $0 < r \leq \rho$ for all $f \in \theta_2$. Grunsky [34] obtained the radius of starlikeness for functions in the class $\mathcal{S}$. Sokół [35] computed the radius of α-convexity and α-starlikeness for a class $\mathcal{S}^*_L$. In 2016, authors [7] determined the $\mathcal{S}^*_R$-radius for various subclasses of starlike functions. For more results on radius problems, see [36–41].

The main technique involved in tackling the $\mathcal{S}^*_B$-radius estimates for classes of functions f is the determination of the disk that contains the values of $zf'(z)/f(z)$. The associated technical lemma is achieved as:

Lemma 2. *Let $Q(z) := e^{e^z-1}$, $z \in \mathbb{D}$. Define the function $r : [e^{1/e-1}, e^{e-1}] \to \mathbb{R}^+$ by*

$$r(a) := \begin{cases} \frac{ea-e^{1/e}}{e}, & e^{\frac{1}{e}-1} \leq a \leq \frac{e^{1/e}+e^e}{2e}; \\ \frac{e^e-ea}{e}, & \frac{e^{1/e}+e^e}{2e} \leq a \leq e^{e-1}. \end{cases}$$

Then, the following holds:

$$\{w \in \mathbb{C} : |w-a| < r(a)\} \subset \Omega_B \subset \left\{w \in \mathbb{C} : |w-1| < \frac{e^e-e}{e}\right\}.$$

Proof. To prove the assertion, we let $z = e^{it}, t \in (-\pi, \pi]$. Therefore,

$$Q(e^{it}) = e^{e^{e^{it}}-1} = u(t) + iv(t)$$

with

$$u(t) := \cos\left(\sin(\sin t)e^{\cos t}\right)\exp\left(e^{\cos t}\cos(\sin t) - 1\right)$$

and

$$v(t) := \sin\left(\sin(\sin t)e^{\cos t}\right)\exp\left(e^{\cos(t)}\cos(\sin t) - 1\right).$$

Now, consider the square of the distance of an arbitrary point $(u(t), v(t))$ on the boundary of $\partial Q(\mathbb{D})$ from $(a, 0)$ and is given by

$$h(t) = d^2(t) = a^2 - 2ae^{e^{\cos t}\cos(\sin t)-1}\cos\left(\sin(\sin t)e^{\cos t}\right) + e^{2e^{\cos t}\cos(\sin t)-2}.$$

Now we need to prove $|w-a| < r(a)$ is the largest disk contained in $Q(\mathbb{D})$. For this, we need to show that $\min_{-\pi \leq t \leq \pi} d(t) = r(a)$. Since h is an even function, i.e., $h(t) = h(-t)$, we need to only consider the case when $t \in [0, \pi]$. Now $h'(t) = 0$ has three roots viz. 0, π and $t_0(a) \in (0, \pi)$. Among these roots, the root $t_0(a)$ depends on a and graphics reveals that h is increasing in the interval $[0, t_0(a)]$ and decreasing in $[t_0(a), \pi]$, and therefore, h attains its minimum either at 0 or π. Further computations give $h(\pi) = \left(ea - e^{1/e}\right)^2/e^2$ and $h(0) = (e^e - ea)^2/e^2$. Hence, we have

$$\min_{-\pi \leq t \leq \pi} h(t) = \min\{h(0), h(\pi)\} = \begin{cases} h(\pi), & e^{\frac{1}{e}-1} \leq a \leq \frac{e^{1/e}+e^e}{2e}; \\ h(0), & \frac{e^{1/e}+e^e}{2e} \leq a \leq e^{e-1}. \end{cases}$$

Therefore, we can write

$$\min_{-\pi \leq t \leq \pi} d(t) = \begin{cases} \frac{ea-e^{1/e}}{e}, & e^{\frac{1}{e}-1} \leq a \leq \frac{e^{1/e}+e^e}{2e}; \\ \frac{e^e-ea}{e}, & \frac{e^{1/e}+e^e}{2e} \leq a \leq e^{e-1}. \end{cases}$$

To find the circle of minimum radius with center at $(1,0)$ containing the domain $Q(\mathbb{D})$, we need to find the maximum distance from $(1,0)$ to an arbitrary point on the boundary of the domain $Q(\mathbb{D})$. The square of this distance function is given by

$$\phi(t) = -2e^{e^{\cos t}\cos(\sin t)-1}\cos\left(\sin(\sin t)e^{\cos t}\right) + e^{2e^{\cos t}\cos(\sin t)-2} + 1.$$

The equation $\phi'(t) = 0$ has two roots in $[0,\pi]$, namely 0 and π. It is easy to see that $\phi(0) = (e-e^e)^2/e^2$ and $\phi(\pi/2) = \left(e-e^{1/e}\right)^2/e^2$. Therefore,

$$\max\{\phi(0),\phi(\pi)\} = \phi(0) = \frac{(e-e^e)^2}{e^2}.$$

Hence, the radius of the smallest disk containing $Q(\mathbb{D})$ is $(e-e^e)/e$. This ends the proof. □

We now recall some classes and results related to them which are to be used for further development of this section. For $-1 \le B < A \le 1$, let

$$\mathcal{P}_n[A,B] := \left\{p(z) = 1 + \sum_{k=n}^{\infty} c_n z^n : \ p(z) \prec \frac{1+Az}{1+Bz}\right\}.$$

Let us denote $\mathcal{P}_n(\alpha) := \mathcal{P}_n[1-2\alpha,-1]$ and $\mathcal{P}_1(0) =: \mathcal{P}$. For $f \in \mathcal{A}$, if we set $p(z) = zf'(z)/f(z)$ and $p(z) = 1 + zf''(z)/f'(z)$, then the class $\mathcal{P}[A,B]$ is denoted by $\mathcal{S}^*[A,B]$ and $\mathcal{K}[A,B]$, respectively. These classes were introduced and studied by [2]. Further, let $\mathcal{S}^*(\alpha) := \mathcal{S}^*[1-2\alpha,-1]$.

The following results will be needed:

Lemma 3. [42] *If $p \in \mathcal{P}_n[A,B]$, then, for $|z| = r$,*

$$\left|p(z) - \frac{1-ABr^{2n}}{1-B^2r^{2n}}\right| \le \frac{(A-B)r^n}{1-B^2r^{2n}}.$$

In particular, if $p \in \mathcal{P}_n(\alpha)$, then, for $|z| = r$,

$$\left|p(z) - \frac{(1+(1-2\alpha))r^{2n}}{1-r^{2n}}\right| \le \frac{2(1-\alpha)r^n}{1-r^{2n}}.$$

Lemma 4. [43] *If $p \in \mathcal{P}_n(\alpha)$, then, for $|z| = r$,*

$$\left|\frac{zp'(z)}{p(z)}\right| \le \frac{2(1-\alpha)nr^n}{(1-r^n)(1+(1-2\alpha)r^n)}.$$

The main objective of this section is to determine the $\mathcal{S}^*_B$-radii constants for functions belonging to certain well-known subclasses of $\mathcal{A}$. Let $\mathcal{G}$ denote the class of functions $f \in \mathcal{S}$ for which $f(z)/z \in \mathcal{P}$. The following theorem gives the sharp $\mathcal{S}^*_B$-radius for the class $\mathcal{G}$.

Theorem 4. *Let $f \in \mathcal{G}$. Then, the sharp $\mathcal{S}^*_B$-radius is*

$$R_{\mathcal{S}^*_B}(\mathcal{G}) := \frac{e-e^{1/e}}{\sqrt{2e^2 - 2e^{1+\frac{1}{e}} + e^{2/e} + e}} \approx 0.222654.$$

Proof. Since $f \in \mathcal{G}$, therefore, $f(z)/z \in \mathcal{P}$. Then, from Lemma 2, we must have

$$\left|\frac{zf'(z)}{f(z)} - 1\right| \le \frac{2r}{1-r^2}$$

Therefore, $f \in \mathcal{S}^*_B$ if $2r/(1-r^2) \le (e - e^{1/e})/e$, or equivalently if

$$(e - e^{1/e})r^2 + 2er + e^{1/e} - e \le 0$$

which holds for all

$$r \le \frac{e - e^{1/e}}{\sqrt{2e^2 - 2e^{1+\frac{1}{e}} + e^{2/e} + e}} =: R_{\mathcal{S}^*_B}(\mathcal{G}) \approx 0.222654.$$

For verification of sharpness, consider the function $f(z) = z(1+z)/(1-z)$. Then, $f(z)/z \in \mathcal{P}$ and at $z = R_{\mathcal{S}^*_B}(\mathcal{G})$, we have

$$\frac{R_{\mathcal{S}^*_B}(\mathcal{G}) f'(R_{\mathcal{S}^*_B}(\mathcal{G}))}{f(R_{\mathcal{S}^*_B}(\mathcal{G}))} - 1 = \frac{R_{\mathcal{S}^*_B}(\mathcal{G})}{1 - R_{\mathcal{S}^*_B}(\mathcal{G})} = 1 - e^{\frac{1}{e}-1}.$$

Hence the result is sharp. □

In the following theorem, we shall investigate sharp $\mathcal{S}^*_B$-radius for the class $\mathcal{S}^*[A, B]$.

Theorem 5. *Let $f \in \mathcal{S}^*[A, B]$. Then,*

1. *for $0 \le B < A \le 1$, the sharp $\mathcal{S}^*_B$-radius for the class $\mathcal{S}^*[A, B]$ is*

$$R_{\mathcal{S}^*_B}(\mathcal{S}^*[A, B]) = \min\left\{1; \frac{\sqrt{e - e^{1/e}}}{\sqrt{eAB - e^{1/e}B^2}}; \frac{e^{1/e} - e}{e^{1/e}B - eA}\right\}.$$

2. *for $-1 \le B < 0 \le A \le 1$, the sharp $\mathcal{S}^*_B$-radius for the class $\mathcal{S}^*[A, B]$ is*

$$R_{\mathcal{S}^*_B}(\mathcal{S}^*[A, B]) = \min\left\{1; \sqrt{\frac{-2e + e^{1/e} + e^e}{-2eAB + e^{1/e}B^2 + e^e B^2}}; \frac{e^{1/e} - e}{e^{1/e}B - eA}\right\}.$$

Proof. Let $f \in \mathcal{S}^*[A, B]$. Then using Lemma 4, we see that f maps the disk $|z| \le r$ onto the disk

$$\left|\frac{zf'(z)}{f(z)} - \frac{1 - ABr^2}{1 - B^2r^2}\right| \le \frac{(A - B)r}{1 - B^2r^2}.$$

The center of the above disk is at $(c, 0)$ and the radius is R, where

$$c := \frac{1 - ABr^2}{1 - B^2r^2} \quad \text{and} \quad R := \frac{(A - B)r}{1 - B^2r^2}.$$

(1) We see that $c \le (e^{1/e} + e^e)/(2e)$ holds for all $0 \le B < A \le 1$ and $0 < r < 1$. Further, the condition $1 - e^{1/e} \le c$ is equivalent to

$$-eABr^2 + e^{1/e}B^2r^2 - e^{1/e} + e \ge 0$$

which holds for all

$$r \le \sqrt{\frac{e - e^{1/e}}{eAB - e^{1/e}B^2}} =: r_1.$$

Further computation shows that the condition $R \le (e^e a - e^{1/e})/e$ is equivalent to $eAr - e^{1/e}Br + e^{1/e} - e \le 0$ which holds for all

$$r \le \frac{e^{1/e} - e}{e^{1/e}B - eA} =: r_2.$$

Now from Lemma 2, $f \in \mathcal{S}^*_B$ for all $|z| \le R_{\mathcal{S}^*_B}(\mathcal{S}^*[A, B]) = \min\{1; r_1; r_2\}$.

(2) Let $-1 \leq B < 0 \leq A \leq 1$. Then we see that $e^{1/e-1} \leq c$ holds for all $0 < r < 1$. Further, $c \leq (e^e + e^{1/e})/2e$ is equivalent to

$$-2eABr^2 + e^{1/e}B^2r^2 + e^eB^2r^2 - e^{1/e} - e^e + 2e \leq 0$$

which holds for

$$r \leq \sqrt{\frac{-2e + e^{1/e} + e^e}{-2eAB + e^{1/e}B^2 + e^eB^2}} =: r_3.$$

Now, as in the previous case $R < (ec - e^{1/e})/e$ holds if $r \leq r_2$. Therefore, $\mathcal{S}^*_B$-radius for the class $\mathcal{S}^*[A, B]$ is $R_{\mathcal{S}^*_B}(\mathcal{S}^*[A, B]) = \min\{1; r_2; r_3\}$.

The equality holds in case of the function f_0 defined by

$$f_0(z) = \begin{cases} z(1 + Bz)^{\frac{A}{B}-1}, & B \neq 0; \\ ze^{Az}, & B = 0. \end{cases}$$

This ends the proof. □

Remark 1. *Let $f \in \mathcal{S}^*$. Then, since $\mathcal{S}^* = \mathcal{S}^*[0, -1]$, it follows from the above theorem, that the $\mathcal{S}^*_B$-radius for starlike functions is $r_4 := (e - e^{1/e})/(e + e^{1/e}) \approx 0.30594$. To see the sharpness, consider the Koebe function $k(z) = z/(1 - z)^2$. Then, at $z = r_4$, we have*

$$\frac{r_4 f'(r_4)}{f(r_4)} = \frac{1 + r_4}{1 - r_4} = e^{1-\frac{1}{e}}.$$

Because the function k is univalent too, it follows that the $\mathcal{S}^*_B$-radius for the class $\mathcal{S}$ and $\mathcal{S}^*$ is r_4. Therefore, the radius r_4 can not be increased. Thus, we have the following:

Corollary 1. *The sharp $\mathcal{S}^*_B$-radius for the classes $\mathcal{S}$ and $\mathcal{S}^*$ is $(e - e^{1/e})/(e + e^{1/e}) \approx 0.30594$.*

Let the class $\mathcal{F}_1$ be defined by

$$\mathcal{F}_1 := \left\{ f \in \mathcal{A} : \operatorname{Re}\frac{f(z)}{g(z)} > 0 \text{ and } \operatorname{Re}\frac{g(z)}{z} > 0,\ g \in \mathcal{A} \right\}.$$

The following theorem gives the sharp $\mathcal{S}^*_B$-radius for the class $\mathcal{F}_1$.

Theorem 6. *Let $f \in \mathcal{F}_1$. Then, the sharp $\mathcal{S}^*_B$-radius is*

$$R_{\mathcal{S}^*_B}(\mathcal{F}_1) = \frac{e - e^{1/e}}{\sqrt{5e^2 - 2e^{1+\frac{1}{e}} + e^{2/e}} + 2e} \approx 0.11557.$$

Proof. Since $f \in \mathcal{F}_1$, there is $g \in \mathcal{A}$ such that $\operatorname{Re}(g(z)/z) > 0$. Define the functions $p, h : \mathbb{D} \to \mathbb{C}$ by

$$p(z) = \frac{g(z)}{z} \text{ and } h(z) = \frac{f(z)}{g(z)}.$$

Then, through some assumptions, we have $p, h \in \mathcal{P}$. Now using Lemma 4, we get

$$\begin{aligned} \left|\frac{zf'(z)}{f(z)} - 1\right| &\leq \left|\frac{zh'(z)}{h(z)}\right| + \left|\frac{zp'(z)}{p(z)}\right| \\ &\leq \frac{4r}{1 - r^2} \leq \frac{e - e^{1/e}}{e}, \end{aligned}$$

this holds if and only if $(e-e^{1/e})r^2+4er+e^{1/e}-e\leq 0$, that is if

$$r\leq \frac{e-e^{1/e}}{\sqrt{5e^2-2e^{1+\frac{1}{e}}+e^{2/e}+2e}}=:R_{\mathcal{S}^*_B}(\mathcal{F}_1)\approx 0.11557.$$

Consider the functions f_2 and g_2 defined by

$$f_2(z)=z\left(\frac{1+z}{1-z}\right)^2 \quad \text{and} \quad g_2(z)=z\left(\frac{1+z}{1-z}\right).$$

Further, we have $\mathrm{Re}(f_2(z)/g_2(z))>0$ and $\mathrm{Re}(g_2(z)/z)>0$, and therefore $f\in\mathcal{F}_1$. Now a computation shows that, for $z=R_{\mathcal{S}^*_B}(\mathcal{F}_1)$,

$$\frac{R_{\mathcal{S}^*_B}(\mathcal{F}_1)f_2'(R_{\mathcal{S}^*_B}(\mathcal{F}_1))}{f_2(R_{\mathcal{S}^*_B}(\mathcal{F}_1))}-1=\frac{4R_{\mathcal{S}^*_B}(\mathcal{F}_1)}{1-R_{\mathcal{S}^*_B}(\mathcal{F}_1)^2}=1-e^{\frac{1}{e}-1}.$$

Hence the result is sharp. □

Let us define the class $\mathcal{F}_2$ by

$$\mathcal{F}_2:=\left\{f\in\mathcal{A}:\mathrm{Re}\,\frac{f(z)}{g(z)}>0 \text{ and } \mathrm{Re}\,\frac{g(z)}{z}>1/2,\ g\in\mathcal{A}\right\}.$$

The following theorem gives the sharp $\mathcal{S}^*_B$-radius for the class $\mathcal{F}_2$.

Theorem 7. *Let $f\in\mathcal{F}_2$. Then, the sharp $\mathcal{S}^*_B$-radius is*

$$\mathcal{S}^*_B(\mathcal{F}_2)=\frac{2\left(e-e^{1/e}\right)}{\sqrt{17e^2-12e^{1+\frac{1}{e}}+4e^{2/e}+3e}}\approx 0.145776.$$

Proof. Since $f\in\mathcal{F}_2$ and $g\in\mathcal{A}$ satisfies $\mathrm{Re}(g(z)/z)>1/2$. Now define the functions $p,h:\mathbb{D}\to\mathbb{C}$ by $p(z)=g(z)/z$ and $h(z)=f(z)/g(z)$. Then, it is clear that $p\in\mathcal{P}(1/2)$ and $h\in\mathcal{P}$. Further, since $f(z)=zp(z)h(z)$, it follows from Lemma 4, get

$$\left|\frac{zf'(z)}{f(z)}-1\right|\leq\frac{3r+r^2}{1-r^2}\leq\frac{e-e^{1/e}}{e}$$

provided $-e^{1/e}r^2+2er^2+3er+e^{1/e}-e\leq 0$. This holds for

$$r\leq\frac{2\left(e-e^{1/e}\right)}{\sqrt{17e^2-12e^{1+\frac{1}{e}}+4e^{2/e}+3e}}=:\mathcal{S}^*_B(\mathcal{F}_2)\approx 0.145776.$$

Thus, $f\in\mathcal{S}^*_B$ for $r\leq\mathcal{S}^*_B(\mathcal{F}_2)$.
For the sharpness of the result, consider the functions

$$f_3(z)=\frac{z(1+z)}{(1-z)^2} \quad \text{and} \quad g_3(z)=\frac{z}{1-z}.$$

Then, we see that $\mathrm{Re}(f_3(z)/g_3(z))>0$ and $\mathrm{Re}(g_3(z)/z)>1/2$, and therefore, $f\in\mathcal{F}_2$. Now from the definition of f_0, we see that at $z=\mathcal{S}^*_B(\mathcal{F}_2)$,

$$\frac{\mathcal{S}_B^*(\mathcal{F}_2)f_3'(\mathcal{S}_B^*(\mathcal{F}_2))}{f_3(\mathcal{S}_B^*(\mathcal{F}_2))}-1=\frac{3\mathcal{S}_B^*(\mathcal{F}_2)+\mathcal{S}_B^*(\mathcal{F}_2)^2}{1-\mathcal{S}_B^*(\mathcal{F}_2)^2}=1-e^{\frac{1}{e}-1}.$$

This confirms the sharpness of the result. □

Define the class $\mathcal{F}_3$ by

$$\mathcal{F}_3:=\left\{f\in\mathcal{A}:\left|\frac{f(z)}{g(z)}-1\right|<1 \text{ and } \operatorname{Re}\frac{g(z)}{z}>0,\ g\in\mathcal{A}\right\}.$$

The next result gives the sharp $\mathcal{S}_B^*$-radius for the class $\mathcal{F}_3$.

Theorem 8. *Let $f\in\mathcal{F}_3$. Then, the sharp $\mathcal{S}_B^*$-radius is*

$$\mathcal{S}_B^*(\mathcal{F}_3)=\frac{2\left(e-e^{1/e}\right)}{\sqrt{17e^2-12e^{1+\frac{1}{e}}+4e^{2/e}+3e}}\approx 0.145776.$$

Proof. Since $f\in\mathcal{F}_3$, it follows that $p\in\mathcal{P}$ and $h\in\mathcal{P}(1/2)$, where the functions $p,h:\mathbb{D}\to\mathbb{C}$ are defined by $p(z)=g(z)/z$ and $h(z)=g(z)/f(z)$. Now since $f(z)=zp(z)/h(z)$ from Lemma 4, we have

$$\left|\frac{zf'(z)}{f(z)}-1\right|\leq\frac{3r+r^2}{1-r^2}\leq\frac{e-e^{1/e}}{e}$$

which holds for all $r\leq\mathcal{S}_B^*(\mathcal{F}_3)$.

Consider the functions f_4 and g_4 defined by

$$f_4(z)=\frac{z(1+z)^2}{(1-z)} \text{ and } g_4(z)=\frac{z(1+z)}{1-z}.$$

The results are sharp, since at $z=\mathcal{S}_B^*(\mathcal{F}_3)$, we have

$$\frac{\mathcal{S}_B^*(\mathcal{F}_3)f_4'(\mathcal{S}_B^*(\mathcal{F}_3))}{f_4(\mathcal{S}_B^*(\mathcal{F}_3))}=2-e^{\frac{1}{e}-1}.$$

This completes the proof. □

Author Contributions: All authors contributed equally.

Funding: This research was funded by the Basic Science Research Program through the National Research Foundation of Korea (NRF) funded by the Ministry of Education, Science and Technology (No. 2016R1D1A1A09916450).

Conflicts of Interest: The authors declare no conflict of interest

References

1. Ma, W.C.; Minda, D. A unified treatment of some special classes of univalent functions. In Proceedings of the Conference on Complex Analysis, Tianjin, China, 19–23 June 1992.
2. Janowski, W. Extremal problems for a family of functions with positive real part and for some related families. *Ann. Polon. Math.* **1970**, *23*, 159–177. [CrossRef]
3. Uralegaddi, B.A.; Ganigi, M.D.; Sarangi, S.M. Univalent functions with positive coefficients. *Tamkang J. Math.* **1994**, *25*, 225–230.
4. Sokół, J.; Stankiewicz, J. Radius of convexity of some subclasses of strongly starlike functions. *Zeszyty Nauk. Politech. Rzeszowskiej Mat.* **1996**, *19*, 101–105.
5. Raina, R.K.; Sokół, J. Some properties related to a certain class of starlike functions. *C. R. Math. Acad. Sci.* **2015**, *353*, 973–978. [CrossRef]

6. Mendiratta, R.; Nagpal, S.; Ravichandran, V. On a subclass of strongly starlike functions associated with exponential function. *Bull. Malays. Math. Sci. Soc.* **2015**, *38*, 365–386. [CrossRef]
7. Kumar, S.; Ravichandran, V. A subclass of starlike functions associated with a rational function. *Southeast Asian Bull. Math.* **2016**, *40*, 199–212.
8. Cho, N.E.; Kumar, V.; Kumar, S.S.; Ravichandran, V. Radius problems for starlike functions associated with the Sine function. *Bull. Iran. Math. Soc.* **2018**. [CrossRef]
9. Kargar, R.; Ebadian, A.; Sokół, J. Radius problems for some subclasses of analytic functions. *Complex Anal. Oper. Theory* **2017**, *11*, 1639–1649. [CrossRef]
10. Kargar, R.; Ebadian, A.; Sokół, J. On Booth lemniscate and starlike functions. *Anal. Math. Phys.* **2017**. [CrossRef]
11. Cho, N.E.; Kumar, S.; Kumar, V.; Ravichandran, V. Differential subordination and radius estimates for starlike functions associated with the Booth lemniscate. *Turk. J. Math.* **2018**, *42*, 1380–1399.
12. Srivastava, H.M.; Owa, S. (Eds.) *Current Topics in Analytic Function Theory*; World Scientific Publishing Co., Inc.: River Edge, NJ, USA, 1992.
13. Dziok, J.; Raina, R.K.; Sokół, J. Certain results for a class of convex functions related to a shell-like curve connected with Fibonacci numbers. *Comput. Math. Appl.* **2011**, *61*, 2605–2613. [CrossRef]
14. Dziok, J.; Raina, R.K.; Sokół, J. On a class of starlike functions related to a shell-like curve connected with Fibonacci numbers. *Math. Comput. Model.* **2013**, *57*, 1203–1211. [CrossRef]
15. Bell, E.T. The iterated exponential integers. *Ann. Math.* **1938**, *39*, 539–557. [CrossRef]
16. Bell, E.T. Exponential polynomials. *Ann. Math.* **1934**, *35*, 258–277. [CrossRef]
17. Berndt, B.C. Ramanujan reaches his hand from his grave to snatch your theorems from you. *Asia Pac. Math. Newsl.* **2011**, *1*, 8–13.
18. Canfield, E.R. Engel's inequality for Bell numbers. *J. Combin. Theory Ser. A* **1995**, *72*, 184–187. [CrossRef]
19. Qi, F. An explicit formula for the Bell numbers in terms of the Lah and Stirling numbers. *Mediterr. J. Math.* **2016**, *13*, 2795–2800. [CrossRef]
20. Qi, F. Some inequalities for the Bell numbers. *Proc. Indian Acad. Sci. Math. Sci.* **2017**, *127*, 551–564. [CrossRef]
21. Srivastava, H.M.; Manocha, H.L. *A Treatise on Generating Functions*; Ellis Horwood Series: Mathematics and its Applications; Ellis Horwood Ltd.: Chichester, UK, 1984.
22. Kumar, V.; Cho, N.E.; Ravichandran, V.; Srivastava, H.M. Sharp coefficient bounds for starlike functions associated with the Bell numbers. *Math. Slovaca* **2019**, accepted.
23. Ali, R.M.; Ravichandran, V. Integral operators on Ma-Minda type starlike and convex functions. *Math. Comput. Model.* **2011**, *53*, 581–586. [CrossRef]
24. Nunokawa, M.; Obradović, M.; Owa, S. One criterion for univalency. *Proc. Am. Math. Soc.* **1989**, *106*, 1035–1037. [CrossRef]
25. Ali, R.M.; Ravichandran, V.; Seenivasagan, N. Sufficient conditions for Janowski starlikeness. *Int. J. Math. Math. Sci.* **2007**, *2007*, 62925. [CrossRef]
26. Ali, R.M.; Cho, N.E.; Ravichandran, V.; Kumar, S.S. Differential subordination for functions associated with the lemniscate of Bernoulli. *Taiwan. J. Math.* **2012**, *16*, 1017–1026. [CrossRef]
27. Kumar, S.; Ravichandran, V. Subordinations for Functions with Positive Real Part. *Complex Anal. Oper. Theory* **2018**, *12*, 1179–1191. [CrossRef]
28. Tuneski, N.; Bulboacă, T.; Jolevska-Tuneska, B. Sharp results on linear combination of simple expressions of analytic functions. *Hacet. J. Math. Stat.* **2016**, *45*, 121–128. [CrossRef]
29. Ahuja, O.P.; Kumar, S.; Ravichandran, V. Applications of first order differential subordination for functions with positive real part. *Stud. Univ. Babeş-Bolyai Math* **2018**, *63*, 303–311. [CrossRef]
30. Bohra, N.; Kumar, S.; Ravichandran, V. Some Special Differential Subordinations. *Hacet. J. Math. Stat.* **2018**, accepted. [CrossRef]
31. Dziok, J.; Raina, R.K.; Sokół, J. Applications of differential subordinations for norm estimates of an integral operator. *Proc. R. Soc. Edinb. Sect. A* **2018**, *148*, 281–291. [CrossRef]
32. Miller, S.S.; Mocanu, P.T. *Differential Subordinations: Theory and Applications*; Marcel Dekker: New York, NY, USA, 2000; Volume 225.
33. Marjono; Thomas, D.K. Subordination on δ-convex functions in a sector. *Honam Math. J.* **2001**, *23*, 41–50.
34. Grunsky, H. Neue abschätzungen zur konformen abbildung ein-und mehrfachzusammenhângender bereiche. *Schr. Deutsche Math.-Ver* **1934**, *43*, 140–143.

35. Sokół, J. Radius problems in the class $\mathcal{S}_L^*$. *Appl. Math. Comput.* **2009**, *214*, 569–573.
36. Ali, R.M.; Jain, N.K.; Ravichandran, V. On the radius constants for classes of analytic functions. *Bull. Malays. Math. Sci. Soc.* **2013**, *36*, 23–38.
37. Cho, N.E.; Kumar, S.; Kumar, V.; Ravichandran, V. Convolution and radius problems of analytic functions associated with the Tilted Carathedory functions. *Math. Commun.* **2019**, accepted.
38. Kowalczyk, J.; Les, E.; Sokół, J. Radius problems in a certain subclass of close-to-convex functions. *Houst. J. Math.* **2014**, *40*, 1061–1072.
39. Kumar, S.; Ravichandran, V. Functions defined by coefficient inequalities. *Malays. J. Math. Sci.* **2017**, *11*, 365–375.
40. Livingston, A.E. On the radius of univalence of certain analytic functions. *Proc. Am. Math. Soc.* **1966**, *17*, 352–357. [CrossRef]
41. MacGregor, T.H. The radius of univalence of certain analytic functions. *Proc. Am. Math. Soc.* **1963**, *14*, 514–520. [CrossRef]
42. Ravichandran, V.; Rønning, F.R.; Shanmugam, T.N. Radius of convexity and radius of starlikeness for some classes of analytic functions. *Complex Var. Theory Appl.* **1997**, *33*, 265–280. [CrossRef]
43. Shah, G.M. On the univalence of some analytic functions. *Pac. J. Math.* **1972**, *43*, 239–250. [CrossRef]

© 2019 by the authors. Licensee MDPI, Basel, Switzerland. This article is an open access article distributed under the terms and conditions of the Creative Commons Attribution (CC BY) license (http://creativecommons.org/licenses/by/4.0/).

symmetry

MDPI

Article

On Meromorphic Functions Defined by a New Operator Containing the Mittag–Leffler Function

Suhila Elhaddad and Maslina Darus *

Faculty of Science and Technology, Universiti Kebangsaan Malaysia, Bangi 43600, Selangor, Malaysia; suhila.e@yahoo.com
* Correspondence: maslina@ukm.edu.my

Received: 26 December 2018; Accepted: 30 January 2019; Published: 12 February 2019

Abstract: This study defines a new linear differential operator via the Hadamard product between a q-hypergeometric function and Mittag–Leffler function. The application of the linear differential operator generates a new subclass of meromorphic function. Additionally, the study explores various properties and features, such as convex properties, distortion, growth, coefficient inequality and radii of starlikeness. Finally, the work discusses closure theorems and extreme points.

Keywords: differential operator; q-hypergeometric functions; meromorphic function; Mittag–Leffler function; Hadamard product

2010 MSC: 30C45

---

1. Introduction

Let Σ denote the class of functions of the form

$$f(z) = z^{-1} + \sum_{j=1}^{\infty} a_j z^j, \tag{1}$$

which are analytic in the punctured open unit disk $\mathbb{U}^* = \{z : z \in \mathbb{C}, 0 < |z| < 1\} = \mathbb{U}/\{0\}$.

Let $\Sigma^*(\rho)$ and $\Sigma_k(\rho)$ denote the subclasses of Σ that are meromorphically starlike functions of order ρ and meromorphically convex functions of order ρ respectively. Analytically, a function f of the form (1) is in the class $\Sigma^*(\rho)$ if it satisfies

$$\mathcal{R}e\left\{-\frac{zf'(z)}{f(z)}\right\} > \rho \quad (z \in \mathbb{U}^*),$$

and $f \in \Sigma_k(\rho)$ if satisfies

$$\mathcal{R}e\left\{-\left(1 + \frac{zf''(z)}{f'(z)}\right)\right\} > \rho \quad (z \in \mathbb{U}^*).$$

The Hadamard product for two functions $f \in \Sigma$, defined by (1) and

$$g(z) = z^{-1} + \sum_{j=1}^{\infty} b_j z^j,$$

is given by

$$f(z) * g(z) = z^{-1} + \sum_{j=1}^{\infty} a_j b_j z^j. \tag{2}$$

For the two functions $f(z)$ and $g(z)$ analytic in $\mathbb{U}$, we say that $f(z)$ is subordinate to $g(z)$, written $f \prec g$ or $f(z) \prec g(z)$ $(z \in \mathbb{U})$, if there exists a Schwarz function $w(z)$ in $\mathbb{U}$ with $w(0) = 0$ and $|w(z)| < 1$ $(z \in \mathbb{U})$, such that $f(z) = g(w(z)), (z \in \mathbb{U})$.

For complex parameters $a_i, b_k, q\,(i = 1, ..., l, k = 1, ..., r, b_k \in \mathbb{C} \setminus \{0, -1, -2, ...\})$ the basic hypergeometric function (or q-hypergeometric function) ${}_l\Psi_r(z)$ is defined by:

$$ {}_l\Psi_r(a_1, ..., a_l; b_1, ..., b_r; q, z) = \sum_{j=0}^{\infty} \frac{(a_1, q)_j ... (a_l, q)_j}{(q, q)_j (b_1, q)_j ... (b_r, q)_j} \times \left[(-1)^j q^{\binom{j}{2}}\right]^{1+r-l} z^j, \tag{3} $$

with $\binom{j}{2} = j(j-1)/2$, where $q \neq 0$ when $l > r+1$ $(l, r \in \mathbb{N}_0 = \mathbb{N} \cup \{0\}, \mathbb{N} = \{1, 2, ...\})$, and $(a, q)_j$ is the q-analogue of the Pochhammer symbol $(a)_j$ defined by:

$$ (a, q)_j = \begin{cases} (1-a)(1-aq)(1-aq^2)...(1-aq^{j-1}), & j = 1, 2, 3, ..., \\ 1, & j = 0. \end{cases} $$

The hypergeometric series defined by (3) was initially introduced by Heine in 1846 and referred to as the Heines series. More details on q-theory are available in [1–3] for readers to refer to.

It is clear that

$$ \lim_{q \to 1^-} \left[{}_l\Psi_r(q^{a_1}, ..., q^{a_l}; q^{b_1}, ..., q^{b_r}; q, (q-1)^{1+r-l} z) \right] = {}_l F_r(a_1, ..., a_l; b_1, ..., b_r; z), $$

where ${}_lF_r(a_1, ..., a_l; b_1, ..., b_r; z)$ represents the generalised hypergeometric function (as shown in [4]).

Riemann, Gauss, Euler and others have conducted extensive studies of hypergeometric functions some hundreds years ago. The focus on this area is based mostly on the structural beauty and distinctive applications that this theory has, which include dynamic systems, mathematical physics, numeric analysis and combinatorics. Based on this, hypergeometric functions are utilised in various disciplines and this includes geometric function theory. One example that can be associated with the hypergeometric functions is the well-known Dziok–Srivastava operator [5,6] defined via the Hadamard product.

Now for $z \in \mathbb{U}$, $|q| < 1$, and $l = r + 1$, the q-hypergeometric function defined in (3) takes the following form:

$$ {}_l\Psi_r(a_1, ..., a_l; b_1, ..., b_r; q, z) = \sum_{j=0}^{\infty} \frac{(a_1, q)_j ... (a_l, q)_j}{(q, q)_j (b_1, q)_j ... (b_r, q)_j} z^j, \tag{4} $$

which converges absolutely in the open unit disk $\mathbb{U}$.

In reference to the function ${}_l\Psi_r(a_1, ..., a_l; b_1, ..., b_r; q, z)$ for meromorphic functions $f \in \Sigma$ that consist of functions in the form of (1), (see Aldweby and Darus [7], Murugusundaramoorthy and Janani [8]), as illustrated below, have recently introduced the q-analogue of the Liu–Srivastava operator

$$ \begin{aligned} {}_l\Upsilon_r(a_1, ..., a_l; b_1, ..., b_r; q, z) f(z) &= z^{-1} {}_l\Psi_r(a_1, ..., a_l; b_1, ..., b_r; q, z) * f(z) \\ &= z^{-1} + \sum_{j=1}^{\infty} \frac{\prod_{i=1}^{l} (a_i, q)_{j+1}}{(q, q)_{j+1} \prod_{k=1}^{r} (b_k, q)_{j+1}} a_j z^j. \end{aligned} \tag{5} $$

For convenience, we write

$$ {}_l\Upsilon_r(a_1, ..., a_l; b_1, ..., b_r; q, z) f(z) = {}_l\Upsilon_r(a_i, b_k; q, z) f(z). $$

Before going further, we state the well-known Mittag–Leffler function $E_{\alpha}(z)$, put forward by Mittag–Leffler [9,10], as well as Wiman's generalisation [11] $E_{\alpha,\beta}(z)$ given respectively as follows:

$$E_{\alpha}(z)=\sum_{j=0}^{\infty}\frac{z^{j}}{\Gamma(\alpha j+1)}, \tag{6}$$

and

$$E_{\alpha,\beta}(z)=\sum_{j=0}^{\infty}\frac{z^{j}}{\Gamma(\alpha j+\beta)}, \tag{7}$$

where $\alpha,\beta\in\mathbb{C}$, $Re(\alpha)>0$ and $Re(\beta)>0$.

There has been a growing focus on Mittag–Leffler-type functions in recent years based on the growth of possibilities for their application for probability, applied problems, statistical and distribution theory, among others. Further information about how the Mittag–Leffler functions are being utilised can be found in [12–18]. In most of our work related to Mittag–Leffler functions, we study the geometric properties, such as the convexity, close-to-convexity and starlikeness. Recent studies on the $E_{\alpha,\beta}(z)$ Mittag–Leffler function can be seen in [19]. Additionally, Ref. [20] presents findings related to partial sums for $E_{\alpha,\beta}(z)$.

The function given by (7) is not within the class Σ. Based on this, the function is then normalised as follows:

$$\begin{aligned}\Omega_{\alpha,\beta}(z) &= z^{-1}\Gamma(\beta)E_{\alpha,\beta}(z)\\ &= z^{-1}+\sum_{j=1}^{\infty}\frac{\Gamma(\beta)}{\Gamma(\alpha(j+1)+\beta)}z^{j}.\end{aligned} \tag{8}$$

Having use of the function $\Omega_{\alpha,\beta}(z)$ given by (8), a new operator $\mathfrak{D}_{\beta}^{\alpha,m}[a_l,b_r,\lambda]:\Sigma\rightarrow\Sigma$ is defined, in terms of Hadamard product, as follows:

$$\mathfrak{D}_{\beta}^{\alpha,0}[a_l,b_r,\lambda]f(z)={}_{l}\Upsilon_{r}(a_i,b_k;q,z)f(z)*\Omega_{\alpha,\beta}(z),$$

$$\mathfrak{D}_{\beta}^{\alpha,1}[a_l,b_r,\lambda]f(z)=(1-\lambda)({}_{l}\Upsilon_{r}(a_i,b_k;q,z)f(z)*\Omega_{\alpha,\beta}(z))+\lambda z({}_{l}\Upsilon_{r}(a_i,b_k;q,z)f(z)*\Omega_{\alpha,\beta}(z))',$$

$$\vdots$$

$$\mathfrak{D}_{\beta}^{\alpha,m}[a_l,b_r,\lambda]f(z)=\mathfrak{D}_{\beta}^{\alpha,1}(\mathfrak{D}_{\beta}^{\alpha,m-1}[a_l,b_r,\lambda]f(z)). \tag{9}$$

If $f\in\Sigma$, then from (9) we deduce that

$$\mathfrak{D}_{\beta}^{\alpha,m}[a_l,b_r,\lambda]f(z)=z^{-1}+\sum_{j=1}^{\infty}[1+(j-1)\lambda]^{m}\,\nabla_{(j+1,\alpha,\beta)}(a_l,b_r)a_jz^{j}, \tag{10}$$

where

$$\nabla_{(j+1,\alpha,\beta)}(a_l,b_r)=\frac{\prod_{i=1}^{l}(a_i,q)_{j+1}}{(q,q)_{j+1}\prod_{k=1}^{r}(b_k,q)_{j+1}}\left(\frac{\Gamma(\beta)}{\Gamma(\alpha(j+1)+\beta)}\right). \tag{11}$$

Remark 1. *It can be seen that, when specialising the parameters* $\lambda,l,r,m,\alpha,\beta,q,a_1,..,a_l$ *and* $b_1,...,b_r$, *it is observed that the defined operator* $\mathfrak{D}_{\beta}^{\alpha,m}[a_l,b_r,\lambda]f(z)$ *leads to various operators. Examples are presented for further illustration.*

- *For* $\lambda=1,l=1,r=0,\beta=1,\alpha=0,a_1=q$ *and* $q\rightarrow1$ *we get the operator* $I^{m}f(z)$ *studied by El-Ashwah and Aouf [21].*
- *For* $m=0,\alpha=0,\beta=1,a_i=q^{a_i},b_k=q^{b_k},a_i>0,b_k>0,\ (i=1,...,l;k=1,..,r,l=r+1)$ *and* $q\rightarrow1$ *we get the operator* $\mathcal{H}_{l,r}[a_i,b_k]f(z)$ *which was investigated by Liu and Srivastava [22].*

- *For* $m = 0, l = 2, r = 1, \beta = 1, \alpha = 0, a_2 = q$ *and* $q \to 1$ *we get the operator* $\mathcal{N}[a_1, b_1]f(z)$ *studied by Liu and Srivastava [23].*
- *For* $m = 0, l = 1, r = 0, \beta = 1, \alpha = 0, a_1 = \lambda + 1$ *and* $q \to 1$ *we get the operator* $\mathcal{D}^{\lambda} f(z) = (1/z(1-z)^{\lambda+1}) * f(z)$ $(\lambda > -1)$ *was introduced by Ganigi and Uralegaddi [24], and then it was generalised by Yang [25].*

A range of meromorphic function subclasses have been explored by, for example, Challab et al. [26], Elrifai et al. [27], Lashin [28], Liu and Srivastava [22] and others. These works have inspired our introduction of the new subclass $\mathcal{T}^{m}_{\alpha,\beta}(a_l, b_r, \lambda; D, H, d)$ of Σ, which involves the operator $\mathfrak{D}^{\alpha,m}_{\beta}[a_l, b_r, \lambda]f(z)$, and is shown as follows:

Definition 1. *For* $-1 \leq H < D \leq 1$, *the function* $f \in \Sigma$ *is in the class* $\mathcal{T}^{m}_{\alpha,\beta}(a_l, b_r, \lambda; D, H, d)$ *if it satisfies the inequality*

$$1 - \frac{1}{d}\left\{\frac{z(\mathfrak{D}^{\alpha,m}_{\beta}[a_l, b_r, \lambda]f(z))'}{\mathfrak{D}^{\alpha,m}_{\beta}[a_l, b_r, \lambda]f(z)} + 1\right\} \prec \frac{1+Dz}{1+Hz}, \tag{12}$$

or, equivalently, to:

$$\left|\frac{\dfrac{z(\mathfrak{D}^{\alpha,m}_{\beta}[a_l, b_r, \lambda]f(z))'}{\mathfrak{D}^{\alpha,m}_{\beta}[a_l, b_r, \lambda]f(z)} + 1}{H\dfrac{z(\mathfrak{D}^{\alpha,m}_{\beta}[a_l, b_r, \lambda]f(z))'}{\mathfrak{D}^{\alpha,m}_{\beta}[a_l, b_r, \lambda]f(z)} + [(D-H)d + H]}\right| < 1 \tag{13}$$

Let Σ^* denote the subclass of Σ consisting of functions of the form:

$$f(z) = z^{-1} + \sum_{j=1}^{\infty} |a_j| z^j. \tag{14}$$

Now, we define the class $\mathcal{T}^{m,*}_{\alpha,\beta}(a_l, b_r, \lambda; D, H, d)$ by

$$\mathcal{T}^{m,*}_{\alpha,\beta}(a_l, b_r, \lambda; D, H, d) = \mathcal{T}^{m}_{\alpha,\beta}(a_l, b_r, \lambda; D, H, d) \cap \Sigma^*.$$

2. Main Result

This section presents work to acquire sufficient conditions in which (14) gives the function f within the class $\mathcal{T}^{m,*}_{\alpha,\beta}(a_l, b_r, \lambda; D, H, d)$, as well as demonstrates that this condition is required for functions which belong to this class. In addition, linear combinations, growth and distortion bounds, closure theorems and extreme points are presented for the class $\mathcal{T}^{m,*}_{\alpha,\beta}(a_l, b_r, \lambda; D, H, d)$.

In our first theorem, we begin with the necessary and sufficient conditions for functions f in $\mathcal{T}^{m,*}_{\alpha,\beta}(a_l, b_r, \lambda; D, H, d)$.

Theorem 1. *Let the function* $f(z)$ *be of the form (14). Then the function* $f(z) \in \mathcal{T}^{m,*}_{\alpha,\beta}(a_l, b_r, \lambda; D, H, d)$ *if and only if*

$$\sum_{j=1}^{\infty} [(j+1)(1-H) - |d|(D-H)]\,[1 + (j-1)\lambda]^m\, \nabla_{(j+1,\alpha,\beta)}(a_l, b_r)|a_j| \leq |d|(D-H). \tag{15}$$

Proof. Suppose that the inequality (15) holds true, we obtain

$$\left|\frac{\dfrac{z(\mathfrak{D}_{\beta}^{\alpha,m}[a_l,b_r,\lambda]f(z))'}{\mathfrak{D}_{\beta}^{\alpha,m}[a_l,b_r,\lambda]f(z)}+1}{H\dfrac{z(\mathfrak{D}_{\beta}^{\alpha,m}[a_l,b_r,\lambda]f(z))'}{\mathfrak{D}_{\beta}^{\alpha,m}[a_l,b_r,\lambda]f(z)}+[(D-H)d+H]}\right|$$
$$=\left|\frac{z(\mathfrak{D}_{\beta}^{\alpha,m}[a_l,b_r,\lambda]f(z))'+\mathfrak{D}_{\beta}^{\alpha,m}[a_l,b_r,\lambda]f(z)}{Hz(\mathfrak{D}_{\beta}^{\alpha,m}[a_l,b_r,\lambda]f(z))'+[d(D-H)+H]\mathfrak{D}_{\beta}^{\alpha,m}[a_l,b_r,\lambda]f(z)}\right|$$
$$=\left|\frac{\sum_{j=1}^{\infty}(j+1)\,[1+(j-1)\lambda]^m\,\nabla_{(j+1,\alpha,\beta)}(a_l,b_r)|a_j|z^{j+1}}{d(D-H)+\sum_{j=1}^{\infty}[H(j+1)+d(D-H)]\,[1+(j-1)\lambda]^m\,\nabla_{(j+1,\alpha,\beta)}(a_l,b_r)|a_j|z^{j+1}}\right|$$
$$<1 \quad (z\in\mathbb{U}^*).$$

Then, by the maximum modulus theorem, we have $f(z)\in\mathcal{T}_{\alpha,\beta}^{m,*}(a_l,b_r,\lambda;D,H,d)$.

Conversely, assume that $f(z)$ is in the class $\mathcal{T}_{\alpha,\beta}^{m,*}(a_l,b_r,\lambda;D,H,d)$ with $f(z)$ of the form (14), then we find from (13) that

$$\left|\frac{z(\mathfrak{D}_{\beta}^{\alpha,m}[a_l,b_r,\lambda]f(z))'+\mathfrak{D}_{\beta}^{\alpha,m}[a_l,b_r,\lambda]f(z)}{Hz(\mathfrak{D}_{\beta}^{\alpha,m}[a_l,b_r,\lambda]f(z))'+[d(D-H)+H]\mathfrak{D}_{\beta}^{\alpha,m}[a_l,b_r,\lambda]f(z)}\right|$$
$$=\left|\frac{\sum_{j=1}^{\infty}(j+1)\,[1+(j-1)\lambda]^m\,\nabla_{(j+1,\alpha,\beta)}(a_l,b_r)|a_j|z^{j+1}}{d(D-H)+\sum_{j=1}^{\infty}[H(j+1)+d(D-H)]\,[1+(j-1)\lambda]^m\,\nabla_{(j+1,\alpha,\beta)}(a_l,b_r)|a_j|z^{j+1}}\right| \tag{16}$$
$$<1,$$

since the above inequality is genuine for all $z\in\mathbb{U}^*$, choose values of z on the real axis. After clearing the denominator in (16) and letting $z\to 1^-$ through real values, we get

$$\sum_{j=1}^{\infty}[(j+1)(1-H)-|d|(D-H)]\,[1+(j-1)\lambda]^m\,\nabla_{(j+1,\alpha,\beta)}(a_l,b_r)|a_j|\le |d|(D-H).$$

Thus, we obtain the desired inequality (15) of Theorem 1. □

Corollary 1. *If the function f of the form (14) is in the class $\mathcal{T}_{\alpha,\beta}^{m,*}(a_l,b_r,\lambda;D,H,d)$ then*

$$|a_j|\le\frac{|d|(D-H)}{[(j+1)(1-H)-|d|(D-H)]\,[1+(j-1)\lambda]^m\,\nabla_{(j+1,\alpha,\beta)}(a_l,b_r)} \quad (j\ge 1),$$

the result is sharp for the function

$$f(z)=z^{-1}+\frac{|d|(D-H)}{[(j+1)(1-H)-|d|(D-H)]\,[1+(j-1)\lambda]^m\,\nabla_{(j+1,\alpha,\beta)}(a_l,b_r)}z^j \quad (j\ge 1). \tag{17}$$

Growth and distortion bounds for functions belonging to the class $\mathcal{T}_{\alpha,\beta}^{m,*}(a_l,b_r,\lambda;D,H,d)$ will be given in the following result:

Theorem 2. *If a function f given by (14) is in the class* $\mathcal{T}^{m,*}_{\alpha,\beta}(a_l, b_r, \lambda; D, H, d)$ *then for* $|z| = r$, *we have:*

$$\frac{1}{r} - \frac{|d|(D-H)}{[2(1-H) - |d|(D-H)]\nabla_{(2,\alpha,\beta)}(a_l,b_r)} r \leq |f(z)| \leq \frac{1}{r} + \frac{|d|(D-H)}{[2(1-H) - |d|(D-H)]\nabla_{(2,\alpha,\beta)}(a_l,b_r)} r,$$

and

$$\frac{1}{r^2} - \frac{|d|(D-H)}{[2(1-B) - |d|(D-H)]\nabla_{(2,\alpha,\beta)}(a_l,b_r)} \leq |f^{'}(z)| \leq \frac{1}{r^2} + \frac{|d|(D-H)}{[2(1-H) - |d|(D-H)]\nabla_{(2,\alpha,\beta)}(a_l,b_r)}.$$

Proof. By Theorem 1,

$$[2(1-H) - |d|(D-H)]\nabla_{(2,\alpha,\beta)}(a_l,b_r) \sum_{j=1}^{\infty} |a_j|$$
$$\leq \sum_{j=1}^{\infty} [(j+1)(1-H) - |d|(D-H)]\,[1+(j-1)\lambda]^m\, \nabla_{(j+1,\alpha,\beta)}(a_l,b_r)|a_j| \leq |d|(D-H),$$

which yields:

$$\sum_{j=1}^{\infty} |a_j| \leq \frac{|d|(D-H)}{[2(1-H) - |d|(D-H)]\nabla_{(2,\alpha,\beta)}(a_l,b_r)}.$$

Therefore,

$$|f(z)| \leq \frac{1}{|z|} + |z| \sum_{j=1}^{\infty} |a_j| \leq \frac{1}{|z|} + \frac{|d|(D-H)}{[2(1-H) - |d|(D-H)]\nabla_{(2,\alpha,\beta)}(a_l,b_r)} |z|,$$

and

$$|f(z)| \geq \frac{1}{|z|} - |z| \sum_{j=1}^{\infty} |a_j| \geq \frac{1}{|z|} - \frac{|d|(D-H)}{[2(1-H) - |d|(D-H)]\nabla_{(2,\alpha,\beta)}(a_l.b_r)} |z|.$$

Now, by differentiating both sides of (14) with respect to z, we get:

$$|f'(z)| \leq \frac{1}{|z|^2} + \sum_{j=1}^{\infty} |a_j| \leq \frac{1}{|z|^2} + \frac{|d|(D-H)}{[2(1-H) - |d|(D-H)]\nabla_{(2,\alpha,\beta)}(a_l,b_r)},$$

and

$$|f'(z)| \geq \frac{1}{|z|^2} - \sum_{j=1}^{\infty} |a_j| \geq \frac{1}{|z|^2} - \frac{|d|(D-H)}{[2(1-H) - |d|(D-H)]\nabla_{(2,\alpha,\beta)}(a_l,b_r)}.$$

□

Next, we determine the radii of meromorphic starlikeness and convexity of order ρ for functions in the class $\mathcal{T}^{m,*}_{\alpha,\beta}(a_l, b_r, \lambda; D, H, d)$.

Theorem 3. *Let the function f given by (14) be in the class* $\mathcal{T}^{m,*}_{\alpha,\beta}(a_l, b_r, \lambda; D, H, d)$. *Thus, we have:*
(i) f is meromorphically starlike of order ρ *in the disc* $|z| < r_1$, *that is*

$$\mathcal{R}e\left\{-\frac{zf^{'}(z)}{f(z)}\right\} > \rho \quad (|z| < r_1, 0 \leq \rho < 1),$$

where

$$r_1 = \inf_{j\geq 1}\left\{\frac{(1-\rho)\,[(j+1)(1-H) - |d|(D-H)]\,[1+(j-1)\lambda]^m\, \nabla_{(j+1,\alpha,\beta)}(a_l,b_r)}{|d|(D-H)(j+\rho)}\right\}^{\frac{1}{j+1}}. \quad (18)$$

(ii) f is meromorphically convex of order ρ in the disc $|z| < r_2$, that is

$$\mathcal{R}e\left\{-\left(1+\frac{zf''(z)}{f'(z)}\right)\right\} > \rho \quad (|z| < r_2, 0 \leq \rho < 1),$$

where

$$r_2 = \inf_{j\geq 1}\left\{\frac{(1-\rho)\left[(j+1)(1-H)-|d|(D-H)\right]\left[1+(j-1)\lambda\right]^m \nabla_{(j+1,\alpha,\beta)}(a_l,b_r)}{j|d|(D-H)(j+\rho)}\right\}^{\frac{1}{j+1}}. \tag{19}$$

Proof. (i) From the definition (14), we can get:

$$\left|\frac{\frac{zf'(z)}{f(z)}+1}{\frac{zf'(z)}{f(z)}-1+2\rho}\right| \leq \frac{\sum_{j=1}^{\infty}(j+1)|a_j||z|^{j+1}}{2(1-\rho)-\sum_{j=1}^{\infty}(j-1+2\rho)|a_j||z|^{j+1}}.$$

Then, we have:

$$\left|\frac{\frac{zf'(z)}{f(z)}+1}{\frac{zf'(z)}{f(z)}-1+2\rho}\right| \leq 1 \quad (0 \leq \rho < 1),$$

if

$$\sum_{j=1}^{\infty}\left(\frac{j+\rho}{1-\rho}\right)|a_j||z|^{j+1} \leq 1. \tag{20}$$

Thus, by Theorem 1, the inequality (20) will be true if

$$\left(\frac{j+\rho}{1-\rho}\right)|z|^{j+1} \leq \frac{\left[(j+1)(1-H)-|d|(D-H)\right]\left[1+(j-1)\lambda\right]^m \nabla_{(j+1,\alpha,\beta)}(a_l,b_r)}{|d|(D-H)},$$

then

$$|z| \leq \left\{\frac{(1-\rho)\left[(j+1)(1-H)-|d|(D-H)\right]\left[1+(j-1)\lambda\right]^m \nabla_{(j+1,\alpha,\beta)}(a_l,b_r)}{|d|(D-H)(j+\rho)}\right\}^{\frac{1}{j+1}}.$$

The last inequality leads us immediately to the disc $|z| < r_1$, where r_1 is given by (18).
(ii) In order to prove the second affirmation of Theorem 3, we find from (14) that:

$$\left|\frac{2+\frac{zf''(z)}{f'(z)}}{\frac{zf''(z)}{f'(z)}+2\rho}\right| \leq \frac{\sum_{j=1}^{\infty}j(j+1)|a_j||z|^{j+1}}{2(1-\rho)-\sum_{j=1}^{\infty}j(j-1+2\rho)|a_j||z|^{j+1}}.$$

Thus, we have the desired inequality:

$$\left|\frac{2+\frac{zf''(z)}{f'(z)}}{\frac{zf''(z)}{f'(z)}+2\rho}\right| \leq 1 \quad (0 \leq \rho < 1),$$

if

$$\sum_{j=1}^{\infty}j\left(\frac{j+\rho}{1-\rho}\right)|a_j||z|^{j+1} \leq 1. \tag{21}$$

Thus, by Theorem 1, the inequality (21) will be true if

$$j\left(\frac{j+\rho}{1-\rho}\right)|z|^{j+1} \leq \frac{[(j+1)(1-H)-|d|(D-H)]\,[1+(j-1)\lambda]^m\,\nabla_{(j+1,\alpha,\beta)}(a_l,b_r)}{|d|(D-H)},$$

then

$$|z| \leq \left\{\frac{(1-\rho)\,[(j+1)(1-H)-|d|(D-H)]\,[1+(j-1)\lambda]^m\,\nabla_{(j+1,\alpha,\beta)}(a_l,b_r)}{j|d|(D-H)(j+\rho)}\right\}^{\frac{1}{j+1}}.$$

The last inequality readily yields the disc $|z|<r_2$, where r_2 is given by (19). □

The closure theorems and extreme points of the class $\mathcal{T}^{m,*}_{\alpha,\beta}(a_l,b_r,\lambda;D,H,d)$ will now be determined.

Theorem 4. *The class $\mathcal{T}^{m,*}_{\alpha,\beta}(a_l,b_r,\lambda;D,H,d)$ is closed under convex linear combinations.*

Proof. Assume that the functions

$$f_i(z)=z^{-1}+\sum_{j=1}^{\infty}|a_{j,i}|z^j \quad (i=1,2),$$

are in $\mathcal{T}^{m,*}_{\alpha,\beta}(a_l,b_r,\lambda;D,H,d)$. It suffices to show that the function h defined by

$$h(z)=(1-c)f_1(z)+cf_2(z) \quad (0\leq c\leq 1),$$

is in the class $\mathcal{T}^{m,*}_{\alpha,\beta}(a_l,b_r,\lambda;D,H,d)$, since

$$h(z)=z^{-1}+\sum_{j=1}^{\infty}\left[(1-c)|a_{j,1}|+c|a_{j,2}|\right]z^j \quad (0\leq c\leq 1).$$

In view of Theorem 1, we have:

$$\begin{aligned}
&\sum_{j=1}^{\infty}[(j+1)(1-H)-|d|(D-H)]\,[1+(j-1)\lambda]^m\,\nabla_{(j+1,\alpha,\beta)}\cdot\left\{(1-c)|a_{j,1}|+c|a_{j,2}|\right\}\\
&\quad=(1-c)\sum_{j=1}^{\infty}[(j+1)(1-H)-|d|(D-H)]\,[1+(j-1)\lambda]^m\,\nabla_{(j+1,\alpha,\beta)}|a_{j,1}|\\
&\qquad+c\sum_{j=1}^{\infty}[(j+1)(1-H)-|d|(D-H)]\,[1+(j-1)\lambda]^m\,\nabla_{(j+1,\alpha,\beta)}|a_{j,2}|\\
&\qquad\qquad\leq(1-c)|d|(D-H)+c|d|(D-H)=|d|(D-H),
\end{aligned}$$

which shows that $h(z)\in\mathcal{T}^{m,*}_{\alpha,\beta}(a_l,b_r,\lambda;D,H,d)$. □

Theorem 5. *Let $f_o(z)=\dfrac{1}{z}$ and*

$$f_j(z)=\frac{1}{z}+\frac{|d|(D-H)}{[(j+1)(1-H)-|d|(D-H)]\,[1+(j-1)\lambda]^m\,\nabla_{(j+1,\alpha,\beta)}}z^j \quad (j\geq 1).$$

Then $f\in\mathcal{T}^{m,*}_{\alpha,\beta}(a_l,b_r,\lambda;D,H,d)$ if and only if it can be expressed in the form

$$f(z)=\sum_{j=0}^{\infty}\nu_j f_j(z), \tag{22}$$

where

$$\nu_j \geq 0 \qquad \textit{and} \qquad \sum_{j=0}^{\infty} \nu_j = 1.$$

Proof. Let the function $f(z)$ be expressed in the form given by (22), then

$$f(z) = z^{-1} + \sum_{j=1}^{\infty} \nu_j \frac{|d|(D-H)}{[(j+1)(1-H) - |d|(D-H)]\,[1+(j-1)\lambda]^m\, \nabla_{(j+1,\alpha,\beta)}} z^j$$

and for this function, we have:

$$\sum_{j=1}^{\infty} [(j+1)(1-H) - |d|(D-H)]\,[1+(j-1)\lambda]^m\, \nabla_{(j+1,\alpha,\beta)}(a_l, b_r)$$
$$\times \nu_j \frac{|d|(D-H)}{[(j+1)(1-H) - |d|(D-H)]\,[1+(j-1)\lambda]^m\, \nabla_{(j+1,\alpha,\beta)}}$$
$$= \sum_{j=1}^{\infty} \nu_j |d|(D-H) = |d|(D-H)(1-\nu_0) \leq |d|(D-H)$$

The condition (15) is satisfied. Thus, $f \in \mathcal{T}_{\alpha,\beta}^{m,*}(a_l, b_r, \lambda; D, H, d)$
Conversely, we suppose that $f \in \mathcal{T}_{\alpha,\beta}^{m,*}(a_l, b_r, \lambda; D, H, d)$, since

$$|a_j| \leq \frac{|d|(D-H)}{[(j+1)(1-H) - |d|(D-H)]\,[1+(j-1)\lambda]^m\, \nabla_{(j+1,\alpha,\beta)}} \qquad (j \geq 1),$$

we set

$$\nu_j = \frac{[(j+1)(1-H) - |d|(D-H)]\,[1+(j-1)\lambda]^m\, \nabla_{(j+1,\alpha,\beta)}}{|d|(D-H)} |a_j|, \quad (j \geq 1),$$

and

$$\nu_0 = 1 - \sum_{j=1}^{\infty} \nu_j,$$

so it follows that

$$f(z) = \sum_{j=0}^{\infty} \nu_j f_j(z).$$

This completes the assertion of Theorem 5. □

3. Conclusions

Studying the theory of analytical functions has been an area of concern for many researchers. A more specific field is the study of inequalities in complex analysis. Literature review indicates lots of studies based on the classes of analytical functions. The interplay of geometry and analysis represents a very important aspect in complex function theory study. This rapid growth is directly linked to the relation that exists between analytical structure and geometric behaviour. Motivated by this approach, in the current study, we have introduced a new meromorphic function subclass which is related to both the Mittag–Leffler function and q-hypergeometric function, and we have obtained sufficient and necessary conditions in relation to this subclass. Linear combinations, distortion theory and other properties are also explored. For further research we could study the certain classes related to functions with respect to symmetric points associated with hypergeometric and Mittag–Leffler functions.

Author Contributions: Funding acquisition, M.D.; Investigation, S.E.; Methodology, S.E.; Supervision, M.D.; Writing—review & editing, M.D.

Funding: This research was funded by Universiti Kebangsaan Malaysia, grant number GUP-2017-064.

Conflicts of Interest: The authors declare no conflict of interest.

References

1. Exton, H. *q-Hypergeometric Functions and Applications, Ellis Horwood Series: Mathematics and Its Applications*; Ellis Horwood: Chichester, UK, 1983.
2. Gasper, G.; Rahman, M. *Basic Hypergeometric Series*; Cambridge University Press: Cambridge, UK, 1990.
3. Srivastava, H.M. Some generalizations and basic (or q-) extensions of the Bernoulli, Euler and Genocchi polynomials. *Appl. Math. Inf. Sci.* **2011**, *5*, 390–444.
4. Srivastava, H.M.; Karlsson, P.W. *Multiple Gaussian Hypergeometric Series*; Ellis Horwood: Chichester, UK, 1985.
5. Dziok, J.; Srivastava, H.M. Classes of analytic functions associated with the generalized hypergeometric function. *Appl. Math. Comput.* **1999**, *103*, 1–13. [CrossRef]
6. Dziok, J.; Srivastava, H.M. Certain subclasses of analytic functions associated with the generalized hypergeometric function. *Integral Transforms Spec. Funct.* **2003**, *14*, 7–18. [CrossRef]
7. Aldweby, H.; Darus, M. Integral operator defined by q-analogue of Liu–Srivastava operator. *Stud. Univ. Babes-Bolyai Math.* **2013**, *58*, 529–537.
8. Murugusundaramoorthy, G.; Janani, T. Meromorphic parabolic starlike functions associated with q-hypergeometric series. *ISRN Math. Anal.* **2014**, *2014*, 923607. [CrossRef]
9. Mittag–Leffler, G.M. Sur la nouvelle fonction $E_\alpha(x)$. *CR Acad. Sci. Paris* **1903**, *137*, 554–558.
10. Mittag–Leffler, G.M. Sur la representation analytique d'une branche uniforme d'une fonction monogene. *Acta Math.* **1905**, *29*, 101–181. [CrossRef]
11. Wiman, A. Über den fundamentalsatz in der teorie der funktionen $E_\alpha(x)$. *Acta Math.* **1905**, *29*, 191–201. [CrossRef]
12. Attiya, A.A. Some applications of Mittag–Leffler function in the unit disk. *Filomat* **2016**, *30*, 2075–2081. [CrossRef]
13. Gupta, I.S.; Debnath, L. Some properties of the Mittag–Leffler functions. *Integral Transforms Spec. Funct.* **2007**, *18*, 329–336. [CrossRef]
14. Răducanu, D. Third-Order differential subordinations for analytic functions associated with generalized Mittag–Leffler functions. *Mediterr. J. Math.* **2017**, *14*, 167. [CrossRef]
15. Rehman, H.; Darus, M.; Salah, J. Coefficient properties involving the generalized k-Mittag–Leffler functions. *Transylv. J. Math. Mech.* **2017**, *9*, 155–164.
16. Salah, J.; Darus, M. A note on generalized Mittag–Leffler function and application. *Far East J. Math. Sci.* **2011**, *48*, 33–46.
17. Srivastava, H.M.; Frasin, B.A.; Pescar, V. Univalence of integral operators involving Mittag–Leffler functions. *Appl. Math. Inf. Sci.* **2017**, *11*, 635–641. [CrossRef]
18. Srivastava, H.M.; Tomovski, Ž. Fractional calculus with an integral operator containing a generalized Mittag–Leffler function in the kernel. *Appl. Math. Comput.* **2009**, *211*, 198–210. [CrossRef]
19. Bansal, D.; Prajapat, J.K. Certain geometric properties of the Mittag–Leffler functions. *Complex Var. Elliptic Equ.* **2016**, *61*, 338–350. [CrossRef]
20. Răducanu, D. On partial sums of normalized Mittag–Leffler functions. *An. Şt. Univ. Ovidius Constanţa* **2017**, *25*, 123–133. [CrossRef]
21. El-Ashwah, R.M.; Aouf, M.K. Hadamard product of certain meromorphic starlike and convex functions. *Comput. Math. Appl.* **2009**, *57*, 1102–1106. [CrossRef]
22. Liu, J.-L.; Srivastava, H.M. Classes of meromorphically multivalent functions associated with the generalized hypergeometric function. *Math. Comput. Model.* **2004**, *39*, 21–34. [CrossRef]
23. Liu, J.-L.; Srivastava, H.M. A linear operator and associated families of meromorphically multivalent functions. *J. Math. Anal. Appl.* **2001**, *259*, 566–581. [CrossRef]
24. Ganigi, M.R.; Uralegaddi, B.A. New criteria for meromorphic univalent functions. *Bulletin Mathèmatique de la Sociètè des Sciences Mathèmatiques de la Rèpublique Socialiste de Roumanie Nouvelle Sèerie* **1989**, *33*, 9–13.
25. Yang, D. On a class of meromorphic starlike multivalent functions. *Bull. Inst. Math. Acad. Sin.* **1996**, *24*, 151–157.
26. Challab, K.; Darus, M.; Ghanim, F. On a certain subclass of meromorphic functions defined by a new linear differential operator. *J. Math. Fund. Sci.* **2017**, *49*, 269–282. [CrossRef]

27. Elrifai, E.A.; Darwish, H.E.; Ahmed, A.R. On certain subclasses of meromorphic functions associated with certain differential operators. *Appl. Math. Lett.* **2012**, *25*, 952–958. [CrossRef]
28. Lashin, A.Y. On certain subclasses of meromorphic functions associated with certain integral operators. *Comput. Math. Appl.* **2010**, *59*, 524–531 [CrossRef]

© 2019 by the authors. Licensee MDPI, Basel, Switzerland. This article is an open access article distributed under the terms and conditions of the Creative Commons Attribution (CC BY) license (http://creativecommons.org/licenses/by/4.0/).

symmetry

MDPI

Article

Hermite-Type Collocation Methods to Solve Volterra Integral Equations with Highly Oscillatory Bessel Kernels

Chunhua Fang [1,*], Guo He [2,*] and Shuhuang Xiang [3]

1 College of Mathematics, Hunan Institute of Science and Technology, Yueyang 414006, China
2 College of Economics, Jinan University, Guangzhou 510632, China
3 School of Mathematics and Statistics, Central South University, Changsha 410083, China; xiangshu@csu.edu.cn
* Correspondence: fangchunhuamath@163.com (C.F.); heguo261@126.com (G.H.)

Received: 20 December 2018; Accepted: 29 January 2019; Published: 1 February 2019

Abstract: In this paper, we present two kinds of Hermite-type collocation methods for linear Volterra integral equations of the second kind with highly oscillatory Bessel kernels. One method is direct Hermite collocation method, which used direct two-points Hermite interpolation in the whole interval. The other one is piecewise Hermite collocation method, which used a two-points Hermite interpolation in each subinterval. These two methods can calculate the approximate value of function value and derivative value simultaneously. Both methods are constructed easily and implemented well by the fast computation of highly oscillatory integrals involving Bessel functions. Under some conditions, the asymptotic convergence order with respect to oscillatory factor of these two methods are established, which are higher than the existing results. Some numerical experiments are included to show efficiency of these two methods.

Keywords: Volterra integral equations; highly oscillatory Bessel kernel; Hermite interpolation; direct Hermite collocation method; piecewise Hermite collocation method

---

1. Introduction

Volterra integral equations arise from many mathematical problems in engineering and physics [1–3]. For example, the numerical solution of a scalar retarded potential integral equation posted on an infinite flat surface,

$$\int_{\mathbb{R}^2} \frac{u(x', t - |x' - x|)}{|x' - x|} dx' = a(x,t) \qquad on\ \mathbb{R}^2 \times (0,T),$$

where u and a satisfy the causality condition $u \equiv 0$, $a \equiv 0$ for all $t \leq 0$. The continuous Fourier transform (CFT) of a function $g \in L^2(\mathbb{R}^2)$ is $\tilde{g} \in L^2(\mathbb{R}^2)$ defined by $\tilde{g}(\vec{\omega}) = \int_{\mathbb{R}^2} g(x) e^{-ix\vec{\omega}} dx$. When $a(\cdot,t), u(\cdot,t) \in L^2(\mathbb{R}^2)$ for $t \in (0,T)$, by taking CFT, Davies and Duncan [2] reformulated it as the following Volterra integral equation of the first kind with highly oscillatory Bessel kernel,

$$2\pi \int_0^t \tilde{u}(\vec{\omega}, t - R) J_0(\omega R) dR = \tilde{a}(\vec{\omega}, t), \quad for\ \ \vec{\omega} \in \mathbb{R}^2, t \in (0,T), \tag{1}$$

where $J_m(x)$ is the first-kind Bessel function of order m, which is the solution of the Bessel equation $\frac{d^2y}{dx^2} + \frac{1}{x}\frac{dy}{dx} + (1 - \frac{m^2}{x^2})y = 0$. In 2005, for the study of the problem of the electromagnetic scattering from a large cavity, G. Bao and W. W. Sun [1] reformulated (1) as a Volterra integral equation with Cauchy singular and highly oscillatory Hankel kernel.

The Bessel kernel of the above Equation (1) has a parameter ω. Obviously, when $\omega \gg 1$, the Bessel kernel function becomes highly oscillatory. When resort to numerical solutions of Equation (1), the computation of integrals involved Bessel kernel functions is inevitable. However, the classical quadrature rules, such as Newton-Cotes rule, Clenshaw-Curtis rule or Gauss rule, are failed to calculate this kind of integral. Hence, adopting suitable quadrature rules for the corresponding highly oscillatory integrals plays an important role in obtaining the numerical solution.

The function $J_0(\omega(x-t))$ satisfies the condition of Theorem 2.1.8 ([4], p. 64). Upon differentiation with respect to x, the first-kind Volterra integral Equation (1) can be rewritten as the second-kind Volterra integral equations. In this paper, we treat the following Volterra integral equation of the second kind with highly oscillatory Bessel kernel

$$u(x) - \int_0^x J_m(\omega(x-t))u(t)dt = f(x), \quad x \in [0,1], t \in I := [0,x], \tag{2}$$

where $u(x)$ is an unknown function, $f(x)$ is a given smooth function, J_m is the Bessel function of the first kind of order $m \geq 0$ and the frequency ω is a parameter. When $\omega \gg 1$, the Bessel kernel function is highly oscillatory, and this makes solving Equation (2) a challenging problem.

In recent years, there has been tremendous interest in developing methods for solving highly oscillatory Volterra integral equation, such as discontinuous Galerkin method [5], Filon-type method [6,7], collocation method [4,8,9], collocation boundary value method [10,11], collocation method on uniform mesh [12], collocation method on graded mesh [13].

Xiang and Brunner [14] presented three methods: direct Filon method, piecewise constant collocation method and piecewise linear collocation method for the equation,

$$u(x) - \int_0^x J_m(\omega(x-t))\frac{u(t)}{(x-t)^\alpha}dt = f(x), \quad x \in [0,1], t \in I := [0,x], 0 \leq \alpha < 1, f(x) \in C^1[0,1].$$

Based on the asymptotic analysis of the solution, they gave corresponding convergence rates in terms of the frequency for these methods. For the case of the $\alpha = 0, f \in C^2[0,1]$, Fang et al. [15] showed that the optimal convergence with respect to the ω are $O(\omega^{-2}), O(\omega^{-3/2}), O(\omega^{-2})$ respectively. These three methods, same as other existing methods, are constructed by original integral equation or its equivalent equation. Since only the function value in start point is used, which leads to low error precision. In this paper, we present two kinds of Hermite-type collocation methods by combining original integral equation and its differential equation. The new methods will use the values of function and derivative function in start point, which gets higher error precision than that of the above three methods.

The rest of the paper is organized as follows. In Section 2, we present two efficient methods for Equation (2): direct Hermite collocation method and piecewise Hermite collocation method. We show the error bound with respect to the frequency ω In Section 3. In Section 4, several numerical examples are included to verify the results of theoretical analysis. It is observed from numerical experiments that these methods have higher accuracy as compared with the Direct Filon method in [14].

2. Hermite-Type Collocation Methods

2.1. Direct Hermite Collocation Method (Algorithm 1)

Differentiate both sides of Equation (2),

$$u'(x) - J_m(0)u(x) - \int_0^x (J_m(\omega(x-t))'u(t)dt = f'(x). \tag{3}$$

Since

$$(J_m(\omega(x-t))' = \begin{cases} \frac{\omega}{2}(-J_{m+1}(\omega(x-t)) + J_{m-1}(\omega(x-t))), & m > 0, \\ -\omega J_1(\omega(x-t)), & m = 0, \end{cases} \tag{4}$$

it follows that for the case $m = 0$,

$$u'(x) - J_0(0)u(x) + \omega \int_0^x (J_1(\omega(x-t))u(t)dt = f'(x), \tag{5}$$

and for $m > 0$,

$$u'(x) - J_m(0)u(x) + \frac{\omega}{2} \int_0^x (J_{m+1}(\omega(x-t)) - J_{m-1}(\omega(x-t)))u(t)dt = f'(x). \tag{6}$$

Let us denote the Hermite interpolant polynomial between $u(0)$ and $u(x_j)$ by

$$u_h(x) = H_{0j}u(0) + H_{1j}u(x_j) + H_{2j}u'(0) + H_{3j}u'(x_j),$$

where the polynomials

$$H_{0j} = \left(1 + \frac{2x}{x_j}\right)\left(\frac{x-x_j}{x_j}\right)^2, \quad H_{1j} = \left(1 + 2\frac{x-x_j}{-x_j}\right)\left(\frac{x}{x_j}\right)^2,$$
$$H_{2j} = x\left(\frac{x-x_j}{-x_j}\right)^2, \qquad H_{3j} = (x-x_j)\left(\frac{x}{x_j}\right)^2,$$

mean the fundamental polynomials with respect to the nodes 0 and x_j. Then the collocation systems follow

$$u_j^d - \int_0^{x_j} J_m(\omega(x_j - t))(H_{0j}u_0 + H_{1j}u_j^d + H_{2j}u_0' + H_{3j}u_j'^d)dt = f_j, \tag{7}$$

$$u_j'^d - J_m(0)u_j^d + \frac{\omega}{2}\int_0^{x_j}(J_{m+1}(\omega(x_j-t)) - J_{m-1}(\omega(x_j-t))(H_{0j}u_0 + H_{1j}u_j^d + H_{2j}u_0' + H_{3j}u_j'^d)dt = f_j', \tag{8}$$

where u_j^d denotes an approximation of $u(x_j)$, $u_j'^d$ denotes an approximation of $u'(x_j)$. That is

$$(1 - I(1,j,m))u_j^d - I(3,j,m)u_j'^d = f_j + I(0,j,m)u_0 + I(2,j,m)u_0', \tag{9}$$

$$\begin{aligned}&\left(-J_m(0) + \frac{\omega}{2}(I(1,j,m+1) - I(1,j,m-1))\right)u_j^d + \left(1 + \frac{\omega}{2}(I(3,j,m+1) - I(3,j,m-1))\right)u_j'^d \\ &= f_j' - \frac{\omega}{2}(I(0,j,m+1) - I(0,j,m-1))u_0 - \frac{\omega}{2}(I(2,j,m+1) - I(2,j,m-1))u_0'.\end{aligned} \tag{10}$$

Solving these systems, we get direct Hermite appromximate schemes for $m = 0$,

$$u_j^d = \frac{(f_j + I(0,j,0)u_0 + I(2,j,0)u_0')(1+\omega I(3,j,1)) + (f_j' - \omega I(0,j,1)u_0 - \omega I(2,j,1)u_0')I(3,j,0)}{(1 - I(1,j,0))(1+\omega I(3,j,1)) + I(3,j,0)(-1+\omega I(1,j,1))}, \tag{11}$$

$$u_j'^d = \frac{(f_j + I(0,j,0)u_0 + I(2,j,0)u_0')(1-\omega I(1,j,1)) + (f_j' - \omega I(0,j,1)u_0 - \omega I(2,j,1)u_0')(1 - I(1,j,0))}{-I(3,j,0)(1-\omega I(1,j,1)) + (1+\omega I(3,j,0))(1 - I(1,j,0))}, \tag{12}$$

for $m > 0$,

$$u_j^d = \frac{b_1 a_{22} - b_2 a_{12}}{a_{11}a_{22} - a_{21}a_{12}}, \qquad u_j'^d = \frac{a_{11}b_2 - a_{21}b_1}{a_{11}a_{22} - a_{21}a_{12}}, \tag{13}$$

where

$$a_{11} = 1 - I(1,j,m), \quad a_{12} = -I(3,j,m), \quad a_{21} = -J_m(0) + \frac{\omega}{2}(I(1,j,m+1) - I(1,j,m-1)),$$
$$a_{22} = 1 + \frac{\omega}{2}(I(3,j,m+1) - I(3,j,m-1)), \quad b_1 = f_j + I(0,j,m)u_0 + I(2,j,m)u_0',$$
$$b_2 = f_j' - \frac{\omega}{2}(I(0,j,m+1) - I(0,j,m-1))u_0 - \frac{\omega}{2}(I(2,j,m+1) - I(2,j,m-1))u_0'.$$

$I(k,j,m)$ denotes the moment

$$I(k,j,m) = \int_0^{x_j} H_{kj} J_m(\omega(x_j - t))dt \qquad k = 0,1,2,3.$$

The specific calculation formula follows

$$I(0,j,m) = \frac{3}{x_j^2} L(2,m,\omega,x_j) - \frac{2}{x_j^3} L(3,m,\omega,x_j), \tag{14}$$

$$I(1,j,m) = L(0,m,\omega,x_j) - \frac{3}{x_j^2} L(2,m,\omega,x_j) + \frac{2}{x_j^3} L(3,m,\omega,x_j), \tag{15}$$

$$I(2,j,m) = \frac{1}{x_j} L(2,m,\omega,x_j) - \frac{1}{x_j^2} L(3,m,\omega,x_j), \tag{16}$$

$$I(3,j,m) = -L(1,m,\omega,x_j) + \frac{2}{x_j} L(2,m,\omega,x_j) - \frac{1}{x_j^2} L(3,m,\omega,x_j). \tag{17}$$

The moments $L[\mu,m,\omega,a] = \int_0^a t^\mu J_m(\omega t)dt$ can be efficiently calculated by

$$L[\mu,m,\omega,a] = \frac{2^\mu \Gamma\left(\frac{m+\mu+1}{2}\right)}{a^2 \omega^{\mu+1} \Gamma\left(\frac{m-\mu+1}{2}\right)} + \frac{(m+\mu-1)J_m(\omega a)s^{(2)}_{\mu-1,m-1}(\omega a) - J_{m-1}(\omega a)s^{(2)}_{\mu,m}(\omega a)}{a\omega^\mu}, \tag{18}$$

where $\Gamma(x) = \int_0^\infty e^{-t}t^{x-1}dt$ denotes the Gamma function and $s^{(2)}_{\mu,\nu}(z)$ denotes the Lommel function of the second kind [16,17]. Once ω is large, the Lommel function can be efficiently approximated by truncating

$$s^{(2)}_{\mu,\nu}(z) = z^{\mu-1}[1 - \frac{(\mu-1)^2-\nu^2}{z^2} + \ldots + (-1)^p \frac{[(\mu-1)^2-\nu^2]\ldots[(\mu-2p+1)^2-\nu^2]}{z^{2p}}] + O(z^{\mu-2p-2}) \tag{19}$$

Algorithm 1: direct Hermite collocation method.

1. `Compute` $L[i,m,\omega,x_j], i = 0,1,2,3$ by (18);
2. `Compute` $I(k,j,m), k = 0,1,2,3$ by (14)–(17) ;
3. `Compute` u_j^d and $u_j'^d$ by (13).

2.2. *Piecewise Hermite Collocation Method*

To obtain higher-order approximations, a direct improvement of the direct Hermite collocation method is the composite Hermite collocation method, which is so-called piecewise Hermite collocation method (Algorithm 2), that is split the interval into several bins and apply the formula over each bin independently of the other.

Without loss of generality, suppose that $I_\triangle = \{x_j = j * h : j = 0,1,\cdots,N\}$ is a uniform nodal point and $\hat{u}(x)$ is an approximation of $u(x)$ such that $\hat{u}(x)|[x_{j-1},x_j]$is the Hermite interpolant polynomial between $u(x_{j-1})$ and $u(x_j)$ for $j = 1,\ldots,N$.

Let us define

$$\hat{u}(x) = \hat{H}_{0j}u(x_{j-1}) + \hat{H}_{1j}u(x_j) + \hat{H}_{2j}u'(x_{j-1}) + \hat{H}_{3j}u'(x_j),$$

where the polynomials

$$\begin{aligned}
\hat{H}_{0j} &= \left(1+2\tfrac{x-x_{j-1}}{x_j-x_{j-1}}\right)\left(\tfrac{x-x_j}{x_j-x_{j-1}}\right)^2 = \left(1+2\tfrac{x-x_{j-1}}{h}\right)\left(\tfrac{x-x_j}{h}\right)^2,\\
\hat{H}_{1j} &= \left(1+2\tfrac{x_j-x}{x_j-x_{j-1}}\right)\left(\tfrac{x-x_{j-1}}{x_j-x_{j-1}}\right)^2 = \left(1+2\tfrac{x_j-x}{h}\right)\left(\tfrac{x-x_{j-1}}{h}\right)^2,\\
\hat{H}_{2j} &= (x-x_{j-1})\left(\tfrac{x-x_j}{x_j-x_{j-1}}\right)^2 = (x-x_{j-1})\left(\tfrac{x-x_j}{h}\right)^2, \quad \hat{H}_{3j} = (x-x_j)\left(\tfrac{x-x_{j-1}}{x_j-x_{j-1}}\right)^2 = (x-x_j)\left(\tfrac{x-x_{j-1}}{h}\right)^2
\end{aligned}$$

denote the fundamental polynomials with respect to the nodes x_{j-1} and x_j. Then the collocation systems follow

$$u_j - \sum_{i=1}^{j-1}\int_{x_{i-1}}^{x_i} J_m(\omega(x_j-t))\hat{u}_i(t)dt - \int_{x_{j-1}}^{x_j} J_m(\omega(x_j-t))\hat{u}_j(t)dt = f_j, \tag{20}$$

$$\begin{aligned}
u_j' - J_m(0)u_j &+ \frac{\omega}{2}\sum_{i=1}^{j-1}\int_{x_{i-1}}^{x_i} (J_{m+1}(\omega(x_j-t)) - J_{m-1}(\omega(x_j-t)) * \hat{u}_i(t)dt\\
&+ \frac{\omega}{2}\int_{x_{j-1}}^{x_j} (J_{m+1}(\omega(x_j-t)) - J_{m-1}(\omega(x_j-t)) * \hat{u}_j(t)dt = f_j'.
\end{aligned} \tag{21}$$

This leads to the piecewise Hermite collocation method

$$\begin{bmatrix} b_{11} & b_{12} \\ b_{21} & b_{22} \end{bmatrix} \begin{bmatrix} u_j \\ u_j' \end{bmatrix} = \begin{bmatrix} r_1 \\ r_2 \end{bmatrix}, \tag{22}$$

where

$$\begin{aligned}
b_{11} &= 1 - A_{jj1m}, \quad b_{12} = -A_{jj3m}, \quad b_{21} = -J_m(0) + \frac{\omega}{2}(A_{jj1(m+1)} - A_{jj1(m-1)}),\\
b_{22} &= 1 + \frac{\omega}{2}(A_{jj3(m+1)} - A_{jj3(m-1)}),\\
r_1 &= f_j + \sum_{i=1}^{j-1}(A_{ij0m}u_{i-1} + A_{ij1m}u_i + A_{ij2m}u_{i-1}' + A_{ij3m}u_i') + A_{jj0m}u_{j-1} + A_{jj2m}u_{j-1}',\\
r_2 &= f_j' - \frac{\omega}{2}\sum_{i=1}^{j-1}(A_{ij0(m+1)} - A_{ij0(m-1)})u_{i-1} + (A_{ij1(m+1)} - A_{ij1(m-1)})u_i\\
&+ (A_{ij2(m+1)} - A_{ij2(m-1)})u_{i-1}' + (A_{ij3(m+1)} - A_{ij3(m-1)})u_i'\\
&- \frac{\omega}{2}\left((A_{jj0(m+1)} - A_{jj0(m-1)})u_{j-1} + (A_{jj2(m+1)} - A_{jj2(m-1)})u_{j-1}'\right),
\end{aligned} \tag{23}$$

A_{ijkm} denotes the moment

$$A_{ijkm} = \int_{x_{i-1}}^{x_i} \hat{H}_{ki}J_m(\omega(x_j-t))dt \qquad k = 0,1,2,3.$$

The specific calculation formula is following that

$$
\begin{aligned}
A_{ij0m} =& (L(0,m,\omega,(j-i+1)h) - L(0,m,\omega,(j-i)h))\,(2j-2i+3)(j-i)^2\\
&- (L(1,m,\omega,(j-i+1)h) - L(1,m,\omega,(j-i)h))\,(j-i+1)(j-i)6/h\\
&+ (L(2,m,\omega,(j-i+1)h) - L(2,m,\omega,(j-i)h))\,3(2j-2i+1)/h^2\\
&- (L(3,m,\omega,(j-i+1)h) - L(3,m,\omega,(j-i)h))\,2/h^3,\\
A_{ij1m} =& (L(0,m,\omega,(j-i+1)h) - L(0,m,\omega,(j-i)h))\,(j-i+1)^2(-2j+2i+1)\\
&+ (L(1,m,\omega,(j-i+1)h) - L(1,m,\omega,(j-i)h))\,(j-i+1)(j-i)6/h\\
&- (L(2,m,\omega,(j-i+1)h) - L(2,m,\omega,(j-i)h))\,3(2j-2i+1)/h^2\\
&+ (L(3,m,\omega,(j-i+1)h) - L(3,m,\omega,(j-i)h))\,2/h^3,\\
A_{ij2m} =& (L(0,m,\omega,(j-i+1)h) - L(0,m,\omega,(j-i)h))\,(j-i+1)(j-i)^2h\\
&- (L(1,m,\omega,(j-i+1)h) - L(1,m,\omega,(j-i)h))\,(3j-3i+2)(j-i)\\
&+ (L(2,m,\omega,(j-i+1)h) - L(2,m,\omega,(j-i)h))\,(3j-3i+1)/h\\
&- (L(3,m,\omega,(j-i+1)h) - L(3,m,\omega,(j-i)h))\,/h^2,\\
A_{ij3m} =& (L(0,m,\omega,(j-i+1)h) - L(0,m,\omega,(j-i)h))\,(j-i+1)^2(j-i)h\\
&- (L(1,m,\omega,(j-i+1)h) - L(1,m,\omega,(j-i)h))\,(j-i+1)(3j-3i+1)\\
&+ (L(2,m,\omega,(j-i+1)h) - L(2,m,\omega,(j-i)h))\,(3j-3i+2)/h\\
&- (L(3,m,\omega,(j-i+1)h) - L(3,m,\omega,(j-i)h))\,/h^2.
\end{aligned}
\tag{24}
$$

Algorithm 2: piecewise Hermite collocation method.

1. Compute $L[i,m,\omega,x_j], i = 0,1,2,3$ by (18);
2. Compute $A_{ijkm}, k = 0,1,2,3$ by (24);
3. Compute u_j and u_j' by (22).

3. Error Analyses

Firstly, we introduce some useful lemmas, which will be used to prove theorems for the later analyses.

Lemma 1 ([15], Lemma 1). *For any integers $\mu, v \geq 0$ and $x \in (0,1]$, the following integral*

$$\omega \int_0^x J_\mu(\omega t) J_v(\omega(x-t))dt \tag{25}$$

is uniformly bounded with respect to $\omega > 0$.

Lemma 2 ([15], Lemma 2). *Suppose $g_\omega(t) \in C[0,1]$ and $g_\omega(t) = O(1)$ as $\omega \to \infty$. Then for any $v > 0$ and $x \in (0,1]$, it is true that the integral*

$$\int_0^x \frac{g_\omega(t) J_v(\omega t)}{t} dt \tag{26}$$

is uniformly bounded with respect to $\omega > 0$.

Lemma 3 ([18], Lemma 2.1). *For any $\omega \gg 1, m \geq 0$ and $h_\omega(t)$ satisfies*

- *$\int_0^1 |h'_\omega(s)|ds$ is integrable;*
- *$\int_0^1 |h'_\omega(s)|ds$ and $h_\omega(t)$ are bounded in $\omega \in (0,\infty]$ for fixed t, respectively,*

it is true that

$$\left|\int_0^1 h_\omega(t)t^k J_m(\omega t)dt\right| \le \begin{cases} K_1\omega^{-1-k}, & -1 < k < \frac{1}{2}, \\ K_2\omega^{-3/2}, & k \ge \frac{1}{2}, \end{cases} \tag{27}$$

where the constants K_1 and K_2 are independent of ω.

Let $\mathscr{A} : C(I) \to C(I)$ denote the linear Volterra integral operator defined by

$$(\mathscr{A}u)(t) := \int_0^x J_m(\omega(x-t))u(t)dt, \ \ x \in [0,1], t \in I := [0,x],$$

and $\mathscr{I}$ denote identity operator. Then Equation (2) can be reformulated more compactly as

$$(\mathscr{I} - \mathscr{A})\, u = f. \tag{28}$$

To get the expression of (1)–(3) order derivatives of the solution of (2), we first discuss the relation between the integral operator $\mathscr{A}$ and the differential operator D.

Theorem 1. *Assume $f \in C^3[0,1]$. The Volterra operator $\mathscr{A}^n := C(I) \to C(I)$ defined by $(\mathscr{A}^n u)(x) := \int_0^x K_n(t,x)u(t)dt, \ n \ge 1$, where $K_n(t,x)$ are the iterated kernels. Then the solution of (2) satisfies*

$$u = \sum_{j=0}^{\infty} \mathscr{A}^j f, \tag{29}$$

$$Du = \sum_{j=0}^{\infty} \left(f(0)\mathscr{A}^{j-1}r + \mathscr{A}^j Df\right), \tag{30}$$

$$D^2u = \sum_{j=0}^{\infty} \left(f(0)r(0)\mathscr{A}^{j-2}r + f(0)\mathscr{A}^{j-1}Dr + f'(0)\mathscr{A}^{j-1}r + \mathscr{A}^j D^2 f\right), \tag{31}$$

$$\begin{aligned} D^3u = \sum_{j=0}^{\infty} \Big(& f(0)r(0)\left(r(0)\mathscr{A}^{j-3}r + \mathscr{A}^{j-2}Dr\right) + f(0)\left(r'(0)\mathscr{A}^{j-2}r + \mathscr{A}^{j-1}D^2r\right) \\ & + f'(0)\left(r(0)\mathscr{A}^{j-2}r + \mathscr{A}^{j-1}Dr\right) + \left(f''(0)\mathscr{A}^{j-1}r + \mathscr{A}^j D^3 f\right)\Big). \end{aligned} \tag{32}$$

where, $r(x) = J_m(\omega x)$ and $\mathscr{A}^j = 0$ if $j < 0$. Moreover, we have both of $\|Du\|_\infty$, $\|D^2u - f(0)Dr - D^2 f\|_\infty$ and $\|D^3u - f(0)r(0)Dr - f(0)D^2r - f'(0)Dr - D^3 f\|_\infty$ are uniformly bounded with respect to ω.

Proof.

$$\mathscr{A}^j f = \int_0^x J_m(\omega(x-s_1)) \int_0^{s_1} J_m(\omega(s_1-s_2)) \cdots \int_0^{s_{j-1}} J_m(\omega(s_{j-1}-s)) f(s) ds ds_{j-1} \dots ds_1. \tag{33}$$

Let $s_1' = x - s_1, s_2' = s_1 - s_2, \dots, s_{j-1}' = s_{j-2} - s_{j-1}, s' = s_{j-1} - s$, it follows that

$$\mathscr{A}^j f = \int_0^x J_m(\omega s_1') \int_0^{x-s_1'} J_m(\omega s_2') \cdots \int_0^{x-\sum_{k=1}^{j-1} s_k'} J_m(\omega s') f\left(x - \sum_{k=1}^{j-1} s_k' - s'\right) ds' ds_{j-1}' \dots ds_1'. \tag{34}$$

Then

$$\begin{aligned} D\mathscr{A}^j f = & f(0) \int_0^x J_m(\omega s_1') \int_0^{x-s_1'} J_m(\omega s_2') \cdots \int_0^{x-\sum_{k=1}^{j-2} s_k'} J_m(\omega s_{j-1}') J_m\left(\omega(x - \sum_{k=1}^{j-1} s_k')\right) ds_{j-1}' \dots ds_1' \\ & + \int_0^x J_m(\omega s_1') \int_0^{x-s_1'} J_m(\omega s_2') \cdots \int_0^{x-\sum_{k=1}^{j-1} s_k'} J_m(\omega s') f'\left(x - \sum_{k=1}^{j-1} s_k' - s'\right) ds' ds_{j-1}' \dots ds_1' \\ = & f(0)\mathscr{A}^{j-1}r + \mathscr{A}^j Df. \end{aligned} \tag{35}$$

Since

$$u = \sum_{j=0}^{\infty} \mathscr{A}^j f, \tag{36}$$

this series is uniformly absolutely convergent, therefore we can differentiate it term by term

$$Du = \sum_{j=0}^{\infty} \left(f(0)\mathscr{A}^{j-1}r + \mathscr{A}^j Df \right),$$

$$D^2u = \sum_{j=0}^{\infty} \left(f(0)r(0)\mathscr{A}^{j-2}r + f(0)\mathscr{A}^{j-1}Dr + f'(0)\mathscr{A}^{j-1}r + \mathscr{A}^j D^2 f \right),$$

$$D^3u = \sum_{j=0}^{\infty} \Big(f(0)r(0)(r(0)\mathscr{A}^{j-3}r + \mathscr{A}^{j-2}Dr) + f(0)(r'(0)\mathscr{A}^{j-2}r + \mathscr{A}^{j-1}D^2r)$$
$$+ f'(0)(r(0)\mathscr{A}^{j-2}r + \mathscr{A}^{j-1}Dr) + (f''(0)\mathscr{A}^{j-1}r + \mathscr{A}^j D^3 f) \Big),$$

where $\mathscr{A}^j = 0$ if $j < 0$. □

If we define

$$\|\mathscr{A}^j\| := sup \frac{\|\mathscr{A}^j \phi\|_\infty}{\|\phi\|_\infty} = \max_{x \in I} \int_0^x |K_j(x,s)| ds$$

and recall that $\|\mathscr{A}^j \phi\|_\infty \le \|\mathscr{A}^j\| \|\phi\|_\infty$, we find

Remark 1.

$$\|\mathscr{A}^j\| \le \max\{|J_m(\omega(x-s))| : (x,s) \in I \times (0,x)\}/j! \le 1/j!,$$

$$\|Du\| \le \sum_{j=0}^{\infty} \left(f(0)\|\mathscr{A}^{j-1}\|\|r\| + \|\mathscr{A}^j\|\|Df\| \right),$$

$$\left\| D^2u - f(0)\sum_{j=0}^{\infty} \mathscr{A}^{j-1}Dr \right\| \le \sum_{j=0}^{\infty} \left(f(0)r(0)\|\mathscr{A}^{j-2}\|\|r\| + f'(0)\|\mathscr{A}^{j-1}\|\|r\| + \|\mathscr{A}^j\|\|D^2 f\| \right),$$

$$\|D^3u - \sum_{j=0}^{\infty}(f(0)r(0)\mathscr{A}^{j-2}Dr - f(0)\mathscr{A}^{j-1}D^2r - f'(0)\mathscr{A}^{j-1}Dr) \le \sum_{j=0}^{\infty}(|f(0)r^2(0)|\|\mathscr{A}^{j-3}\|\|r\|$$
$$+|f(0)r'(0)|\|\mathscr{A}^{j-2}\|\|r\| + f'(0)r(0)|\|\mathscr{A}^{j-2}\|\|r\| + |f''(0)|\|\mathscr{A}^{j-1}\|\|r\| + \|\mathscr{A}^j\|\|D^3 f\|,$$

then, we have $\|Du\|_\infty$, $\left\|D^2u - f(0)Dr - D^2 f\right\|_\infty$ and $\|D^3u - f(0)r(0)Dr - f(0)D^2r - f'(0)Dr - D^3 f\|_\infty$ are uniform bounded with respect to ω.

Theorem 2. *Assuming $f \in C^3[0,1]$, the pointwise error of the direct Hermite collocation method for (2) satisfies*

$$|u(x_i) - u_i^d| = \begin{cases} O(\omega^{-3}), f(0) = 0 \\ O(\omega^{-2}), f(0) \neq 0 \end{cases} \quad \omega \to \infty, \quad i = 1,2,3,\dots,N. \tag{37}$$

$$|u'(x_i) - u_i'^d| = \begin{cases} O(\omega^{-2}), f(0) = 0 \\ O(\omega^{-1}), f(0) \neq 0 \end{cases} \quad \omega \to \infty, \quad i = 1,2,3,\dots,N. \tag{38}$$

Proof. We only prove a situation $m > 0$. For the case $m = 0$, the proof method is similar.

By the definition of the direct Hermite collocation method, for any $x_i \in I_N$, it follows that

$$\begin{cases} E(x_j) - \int_0^{x_j} J_m(\omega(x_j - t))E(t)dt = 0, \\ E'(x_j) - J_m(0)E(t) + \frac{\omega}{2} \int_0^{x_j} (J_{m+1}(\omega(x_j - t)) - J_{m-1}(\omega(x_j - t)))E(t)dt = 0, \end{cases} \tag{39}$$

where $E(x) = u(x) - u_h(x)$ be the error function. Interpolating E(x) at $x = 0$ and $x = x_j$, we have

$$E(x) = H_{1j}E(x_j) + H_{3j}E'(x_j) + R(x), \tag{40}$$

where $R(x)$ denotes the remainder of the Hermite interpolation. As we know $E(x)$ satisfies that $E(0) = E'(0) = 0$. Substituting (40) into (39), we are led to

$$\begin{cases} E(x_j) - \int_0^{x_j} J_m(\omega(x_j - t))(H_{1j}E(x_j) + H_{3j}E'(x_j) + R(t))dt = 0, \\ E'(x_j) - J_m(0)E(t) \\ +\frac{\omega}{2} \int_0^{x_j} (J_{m+1}(\omega(x_j - t)) - J_{m-1}(\omega(x_j - t)))(H_{1j}E(x_j) + H_{3j}E'(x_j) + R(t))dt = 0. \end{cases} \tag{41}$$

That is

$$\begin{cases} (1 - \int_0^{x_j} J_m(\omega(x_j - t))H_{1j}dt)E(x_j) - \int_0^{x_j} J_m(\omega(x_j - t))H_{3j}dtE'(x_j) = \int_0^{x_j} J_m(\omega(x_j - t)R(t)dt \\ \left(-J_m(0) + \frac{\omega}{2} \int_0^{x_j} (J_{m+1}(\omega(x_j - t)) - J_{m-1}(\omega(x_j - t)))H_{1j}dt\right) E(x_j) \\ \quad + \left(1 + \frac{\omega}{2} \int_0^{x_j} (J_{m+1}(\omega(x_j - t)) - J_{m-1}(\omega(x_j - t)))H_{3j}dt\right) E'(x_j) \\ \quad = -\frac{\omega}{2} \int_0^{x_j} (J_{m+1}(\omega(x_j - t)) - J_{m-1}(\omega(x_j - t)))R(t)dt. \end{cases} \tag{42}$$

Therefore, the error $E(x_j)$ can be computed by

$$E(x_j) = \frac{Q_1}{Q_3}, \quad E'(x_j) = \frac{Q_2}{Q_3}, \tag{43}$$

where

$$\begin{aligned} Q_1 = & \int_0^{x_j} J_m(\omega(x_j - t))R(t)dt * \left(1 + \frac{\omega}{2} \int_0^{x_j} (J_{m+1}(\omega(x_j - t)) - J_{m-1}(\omega(x_j - t)))H_{3j}dt\right) \\ & - \int_0^{x_j} J_m(\omega(x_j - t))H_{3j}dt * \frac{\omega}{2} \int_0^{x_j} (J_{m+1}(\omega(x_j - t)) - J_{m-1}(\omega(x_j - t)))R(t)dt \\ Q_2 = & \int_0^{x_j} J_m(\omega(x_j - t))R(t)dt * \left(J_m(0) - \frac{\omega}{2} \int_0^{x_j} (J_{m+1}(\omega(x_j - t)) - J_{m-1}(\omega(x_j - t)))H_{1j}dt\right) \\ & - \left(1 - \int_0^{x_j} J_m(\omega(x_j - t))H_{1j}dt\right) * \frac{\omega}{2} \int_0^{x_j} (J_{m+1}(\omega(x_j - t)) - J_{m-1}(\omega(x_j - t)))R(t)dt \\ Q_3 = & \left(1 - \int_0^{x_j} J_m(\omega(x_j - t))H_{1j}dt\right) * \left(1 + \frac{\omega}{2} \int_0^{x_j} (J_{m+1}(\omega(x_j - t)) - J_{m-1}(\omega(x_j - t)))H_{3j}dt\right) \\ & - \int_0^{x_j} J_m(\omega(x_j - t))H_{3j}dt * \left(J_m(0) - \frac{\omega}{2} \int_0^{x_j} (J_{m+1}(\omega(x_j - t)) - J_{m-1}(\omega(x_j - t)))H_{1j}dt\right). \end{aligned}$$

Defining $R(x_j - t) = S(t)$, then $S(0) = S'(0) = S(x_j) = S'(x_j)$. From Lemma 1 to Lemma 3, we can easily get $Q_3 = O(1)$ with respect to ω. What shall we do is prove that

$$\int_0^{x_j} J_m(\omega s)S(s)ds = \begin{cases} O(\omega^{-3}), & f(0) = 0 \\ O(\omega^{-2}), & f(0) \neq 0 \end{cases}.$$

Using integration by parts twice, we get

$$\begin{aligned}\int_0^{x_j} J_m(\omega s)S(s)ds &= \int_0^{x_j} S(s)d\frac{s^{m+1}J_{m+1}(\omega s)}{\omega s^{m+1}} \\ &= \frac{1}{\omega^2}\int_0^{x_j}\left(S''(s) - (2m+3)\frac{S'(s)}{s} + (m+1)(m+3)\frac{S(s)}{s^2}\right)J_{m+2}(\omega s)ds.\end{aligned}$$

Denote

$$J = \int_0^{x_j}\left(S''(s) - (2m+3)\frac{S'(s)}{s} + (m+1)(m+3)\frac{S(s)}{s^2}\right)J_{m+2}(\omega s)ds.$$

So, we only need to prove that $J = \begin{cases} O(\omega^{-1}) & f(0) = 0 \\ O(1) & f(0) \neq 0 \end{cases}$.

In the following, we show that the convergence degree of J with respect to ω.

Letting

$$F(s) = S''(s) - (2m+3)\frac{S'(s)}{s} + (m+1)(m+3)\frac{S(s)}{s^2},$$

then we have

$$F(0) = S''(0) - (2m+3)S''(0) + (m+1)(m+3)\frac{S''(0)}{2},$$

$$\begin{aligned}J &= \int_0^{x_j} F(s)J_{m+2}(\omega s)ds \\ &= \int_0^{x_j}(F(s) - F(0))J_{m+2}(\omega s)ds + \int_0^{x_j} F(0)J_{m+2}(\omega s)ds \\ &= \frac{1}{\omega}\left(J_{m+3}(\omega s)(F(s) - F(0))|_{s=0}^{s=x_j} - \int_0^{x_j}(F'(s) - (m+3)\frac{F(s) - F(0)}{s})J_{m+3}(\omega s)ds\right) \\ &\quad + F(0)\int_0^{x_j} J_{m+2}(\omega s)ds.\end{aligned}$$

Observing that

$$\begin{aligned}F'(s) &= S'''(s) - (2m+3)\frac{S''(s)s - S'(s)}{s^2} + (m+1)(m+3)\left(\frac{S'(s)}{s^2} - \frac{2S(s)}{s^3}\right), \\ \frac{F(s) - F(0)}{s} &= \frac{S''(s)s - S'(0)}{s} - (2m+3)\left(\frac{S'(s)}{s} - S''(0)\right) + (m+1)(m+3)\left(\frac{S'(s)}{s^2} - \frac{S''(0)}{2}\right).\end{aligned}$$

Notice that

$$\begin{aligned}S'''(s) &= u'''(s) - u_h'''(s) \\ &= u'''(s) + c_1 \cdot u_0' + c_2 \cdot u_j',\end{aligned}$$

where c_1 and c_2 are some constants independent of ω. For $u_h(x)$ is cubic polynomial, we can easily show that $u_h'''(s) = O(1)$ with respect to ω. According to Theorem 1 it follows that $\|D^3u - f(0)r(0)Dr - f(0)D^2r - f'(0)Dr - D^3f\|_\infty$. Together with Lemma 3 we can easily get

$$\int_0^{x_j} (u'''(s) - f(0)r(0)Dr - f(0)D^2r - f'(0)Dr - D^3f)J_m(\omega(x_j - s)ds = \begin{cases} O(1), & f(0) = 0, \\ O(\omega), & f(0) \neq 0. \end{cases}$$

That is

$$\int_0^{x_j} S'''(s)J_m(\omega(x_j - s))ds = \begin{cases} O(1), & f(0) = 0, \\ O(\omega), & f(0) \neq 0. \end{cases}$$

Then

$$J = \begin{cases} O(\omega^{-1}), & f(0) = 0, \\ O(1), & f(0) \neq 0. \end{cases}$$

Therefore, we can get

$$|u(x_i) - u_i^d| = \begin{cases} O(\omega^{-3}), f(0) = 0 \\ O(\omega^{-2}), f(0) \neq 0 \end{cases} \quad \omega \to \infty, \quad i = 1, 2, 3, \ldots, N. \tag{44}$$

$$|u'(x_i) - u_i'^d| = \begin{cases} O(\omega^{-2}), f(0) = 0 \\ O(\omega^{-1}), f(0) \neq 0 \end{cases} \quad \omega \to \infty, \quad i = 1, 2, 3, \ldots, N. \tag{45}$$

□

Theorem 3. *Assuming $f \in C^3(I)$, the error of the piecewise Hermite collocation method for (2) satisfies*

$$|u(x_i) - u_i| = \begin{cases} O(\omega^{-3}h), f(0) = 0 \\ O(\omega^{-2}h), f(0) \neq 0 \end{cases} \quad \omega \to \infty, \quad i = 1, 2, 3, \ldots, N. \tag{46}$$

$$|u'(x_i) - u_i'| = \begin{cases} O(\omega^{-2}h), f(0) = 0 \\ O(\omega^{-1}h), f(0) \neq 0 \end{cases} \quad \omega \to \infty, \quad i = 1, 2, 3, \ldots, N. \tag{47}$$

Proof. For the piecewise Hermite collocation method, $u(x_j)$ satisfies

$$u(x_j) - \sum_{k=1}^{j-1} \int_{x_{k-1}}^{x_k} J_m(\omega(x_j - t))u(t)dt - \int_{x_{j-1}}^{x_j} J_m(\omega(x_j - t))u(t)dt = f(x_j). \tag{48}$$

Combining the above equation with

$$u_j - \sum_{k=1}^{j-1} \int_{x_{k-1}}^{x_k} J_m(\omega(x_j - t))\hat{u}_k(t)dt - \int_{x_{j-1}}^{x_j} J_m(\omega(x_j - t))\hat{u}_j(t)dt = f(x_j), \tag{49}$$

we get

$$\varepsilon_j = \frac{\sum_{k=1}^{j-1} \varepsilon_k \int_{x_{k-1}}^{x_k} J_m(\omega(x_j - t))dt + \sum_{k=1}^{j} \int_{x_{k-1}}^{x_k} J_m(\omega(x_j - t))r_k(t)dt}{1 - \int_0^{x_1} J_m(\omega t)dt}, \tag{50}$$

where $\varepsilon_j = u(x_j) - u_j, j = 1, 2, \cdots, N$ and $r_k(t) = (u(t) - \hat{u}_k(t))|_t \in [x_{k-1}, x_k]$. An argument similar to the one used in Theorem 2 shows that

$$\frac{\sum_{k=1}^{j} \int_{x_{k-1}}^{x_k} J_m(\omega(x_j - t))r_k(t)dt}{1 - \int_0^{x_1} J_m(\omega t)dt} = \begin{cases} O(\omega^{-3}h), f(0) = 0 \\ O(\omega^{-2}h), f(0) \neq 0 \end{cases}, \tag{51}$$

the desired result is then found by employing the generalized discrete Gronwall inequality ([4], p. 95). Similarly, one can derive the convergence order of $|u'(x_i) - u_i'|$. □

4. Numerical Examples

From Section 4, we can see that direct Hermite collocation method and piecewise Hermite collocation method are very efficient for solving the second-kind Volterra integral equation with highly oscillatory Bessel kernel. They possess the property that the higher oscillation, the higher accuracy. In this section, based on the Formulas (11), (13) and (22), we present some preliminary numerical experiments to verify the result of theoretical analysis. The experiments are performed on a 1.86 GHz PC with 2 GB of RAM. We are using the R2016a version of the MATLAB system. The following Direct Filon method (DF) is presented in paper [14].

Example 1. *Consider the following equation*

$$u(x) - \int_0^x J_m(\omega(x-t))u(t)dt = f(x) \quad \text{with } x \in I = [0,1], \tag{52}$$

where $f(x) = e^x - \int_0^x J_m(\omega(x-t))e^t dt$. *The analytic solution is* $u(x) = e^x$.

In Table 1, we compare the relative error of $u(x)$ from the DF method, piecewise linear collocation method, direct Hermite collocation, and piecewise Hermite collocation method. In Table 2, for fixed ω, we compare the relative error of $u(x)$ from the piecewise linear collocation method and piecewise Hermite collocation method when the steps are different. In Figures 1–3, we can see the convergence rate with respect to ω of these methods.

Table 1. Relative errors of $u(x)$ in N–point approximations to the Example 1 by the DF method, the piecewise linear method (PL), the direct Hermite method(DH) and the piecewise Hermite collocation method(PH). The step is 0.1 for piecewise method and the test point is 0.8.

$\omega\backslash$*Method*	*DF*	*PL*	*DH*	*PH*
10	6.85×10^{-3}	6.86×10^{-5}	8.80×10^{-5}	1.14×10^{-8}
100	8.89×10^{-5}	1.06×10^{-5}	1.56×10^{-7}	2.08×10^{-9}
1000	9.38×10^{-7}	1.31×10^{-7}	1.56×10^{-10}	3.08×10^{-12}
10,000	9.36×10^{-9}	1.46×10^{-9}	1.57×10^{-13}	3.39×10^{-15}

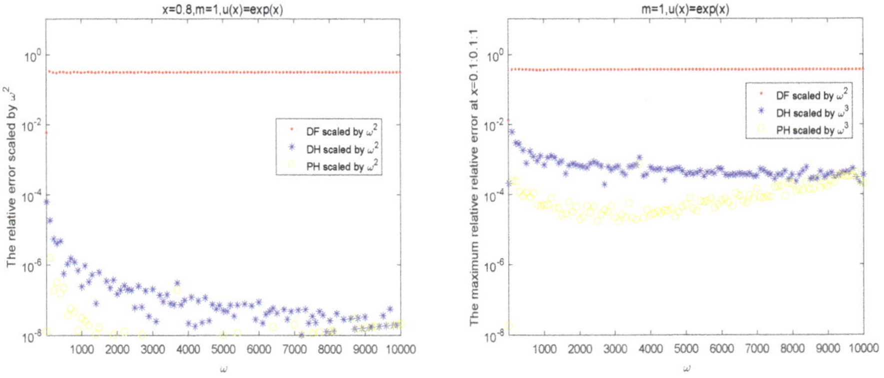

Figure 1. The relative errors of $u(x)$ for DF method, direct Hermite collocation method (DH) and piecewise Hermite collocation method (PH) at point $x = 0.8$ (**left**), the maximum relative errors at collocation points x = 0.1:0.1:1 (**right**).

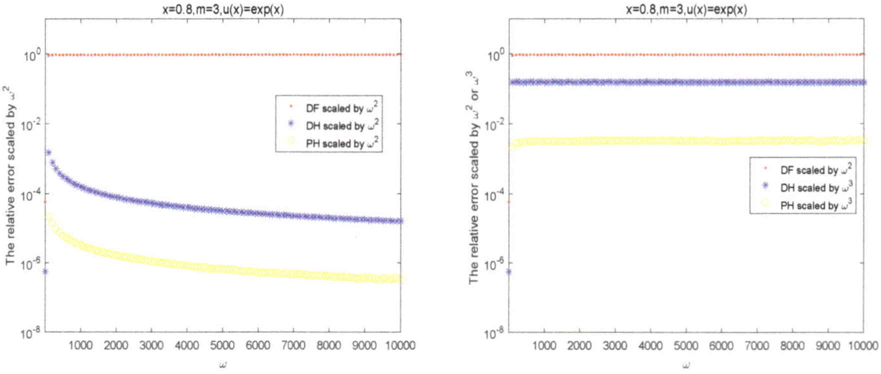

Figure 2. The relative errors of $u(x)$ at point $x = 0.8$ for DF method, direct Hermite collocation method (DH), piecewise Hermite collocation method (PH).

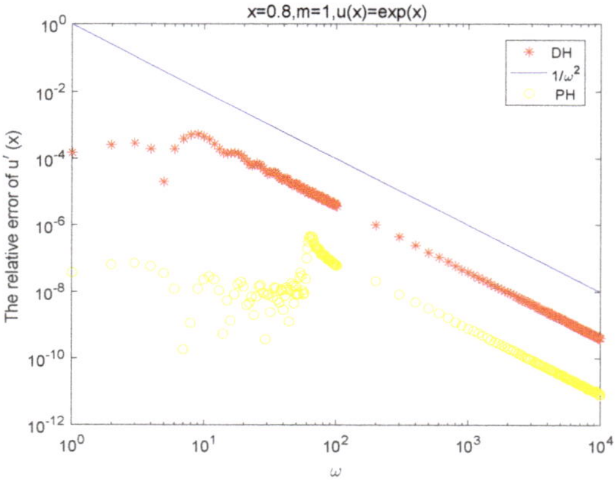

Figure 3. The relative error of $u'(x)$.

Table 2. Relative errors of $u(x)$ in N–point approximations to the Example 1 by the PL method and the piecewise Hermite collocation method(PH). where $\omega = 1000$ and the test point is 0.8.

$Method \backslash h$	0.2	0.1	0.05	0.01
PL	2.71×10^{-7}	1.31×10^{-7}	5.69×10^{-8}	1.10×10^{-8}
PH	1.21×10^{-11}	3.08×10^{-12}	7.44×10^{-13}	1.03×10^{-14}

Example 2. *Consider the following equation,*

$$u(x) - \int_0^x J_3(\omega(x-t))u(t)dt = f(x) \quad with\ x \in I = [0,1], \tag{53}$$

where $f(x) = \frac{1}{1+x^2} - \int_0^x J_3(\omega(x-t))\frac{1}{1+t^2}dt$. *The analytic solution is* $u(x) = \frac{1}{1+x^2}$.

We can see the numerical solutions from the Tables 3 and 4 and Figures 4 and 5.

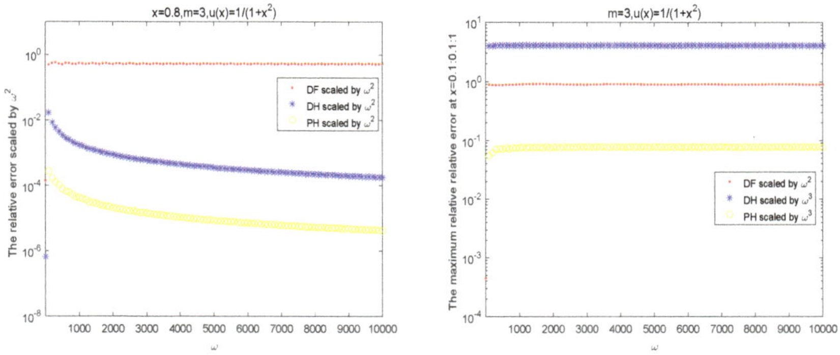

Figure 4. The relative errors of $u(x)$ for DF method, direct Hermite collocation method (DH) and piecewise Hermite collocation method (PH) at point $x = 0.8$ (**left**), the maximum relative errors at collocation points $x = 0.1{:}0.1{:}1$ (**right**).

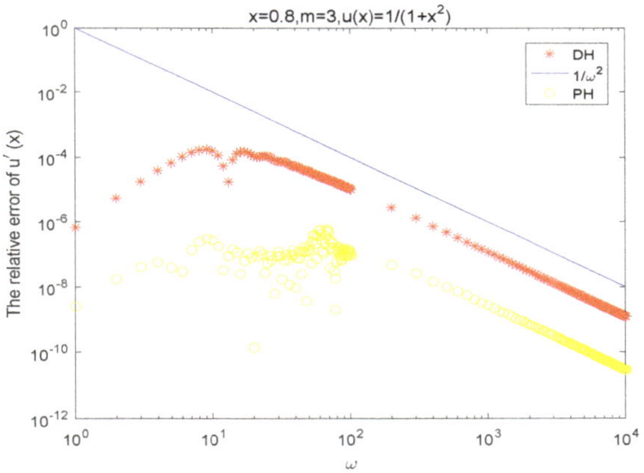

Figure 5. The relative error of $u'(x)$.

Table 3. Relative errors of $u(x)$ in N–point approximations to the Example 2 by the DF method, the PL method, the DH method, and the piecewise Hermite collocation method (PH). The step is 0.1 for piecewise method and the test point is 0.8.

$\omega \backslash Method$	DF	PL	DH	PH
10	1.13×10^{-2}	6.83×10^{-5}	4.34×10^{-4}	2.22×10^{-7}
100	5.10×10^{-5}	7.23×10^{-6}	1.68×10^{-6}	2.53×10^{-8}
1000	5.12×10^{-7}	6.77×10^{-8}	1.75×10^{-9}	4.11×10^{-11}
10,000	5.32×10^{-9}	9.68×10^{-10}	1.75×10^{-12}	4.21×10^{-14}

Table 4. Relative errors of $u(x)$ in N–point approximations to the Example 2 by the PL method and the piecewise Hermite collocation method (PH). where $\omega = 10{,}000$ and the test point is 0.8.

$Method \backslash h$	0.2	0.1	0.05	0.01
PL	1.64×10^{-9}	9.68×10^{-10}	4.91×10^{-10}	8.78×10^{-11}
PH	1.85×10^{-13}	4.21×10^{-14}	1.00×10^{-14}	0

Example 3. *Consider the following equation,*

$$u(x) - \int_0^x J_m(\omega(x-t))u(t)dt = f(x) \quad with\ x \in I = [0,1], \tag{54}$$

where $f(x) = \sin(x) - \int_0^x J_2(\omega(x-t))\sin(t)dt$. *The analytic solution is* $u(x) = \sin(x)$.

Results of these calculations are given in Table 5 and Figures 6 and 7.

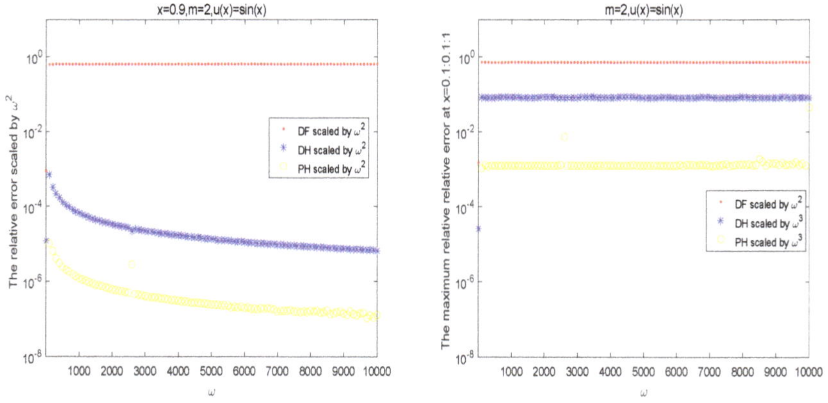

Figure 6. The relative errors of $u(x)$ for DF method, direct Hermite collocation method (DH) and piecewise Hermite collocation method (PH) at point $x = 0.9$ (**left**), the maximum relative errors at collocation points x = 0.1:0.1:1 (**right**).

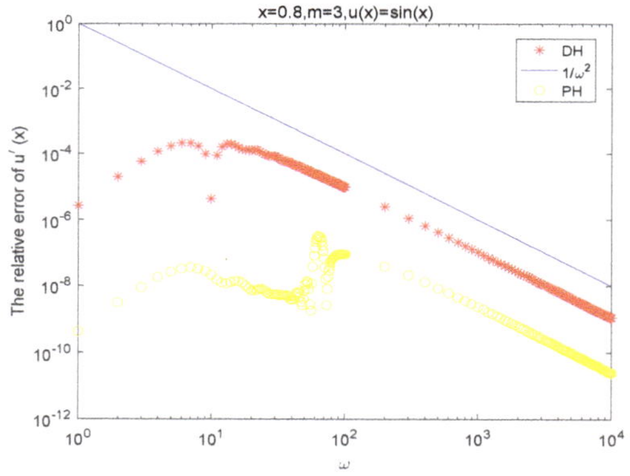

Figure 7. The relative error of $u'(x)$.

Table 5. Relative errors of $u(x)$ in N–point approximations to the Example 3 by the DF method and the PL method and the DH method, and the piecewise Hermite collocation method (PH). The step is 0.1 for piecewise method and the test point is 0.9.

$\omega\backslash Method$	DF	PL	DH	PH
10	5.02×10^{-3}	7.35×10^{-5}	7.88×10^{-5}	1.23×10^{-8}
100	6.31×10^{-5}	6.83×10^{-6}	7.01×10^{-8}	1.03×10^{-9}
1000	6.38×10^{-7}	8.92×10^{-8}	6.62×10^{-11}	1.20×10^{-12}
10,000	6.35×10^{-9}	9.88×10^{-10}	6.56×10^{-14}	1.28×10^{-15}

Example 4. *Consider the following equation,*

$$u(x) - \int_0^x J_3(\omega(x-t))u(t)dt = f(x) \quad with\ x \in I = [0,1], \tag{55}$$

where $f(x) = (x-0.5)^{3.1} - \int_0^x J_3(\omega(x-t))(t-0.5)^{3.1}dt$. *The analytic solution is* $u(x) = (x-0.5)^{3.1}$.

We can see the numerical solutions from the Figure 8.

From above examples, as can be seen, there is a good agreement between the present result and the exact solution. The Hermite-type collocation methods are better than DF method and PL collocation method. For Hermite-type collocation methods, the higher oscillation, the higher accuracy. For fixed frequency, the error is decrease with the increase of nodes.

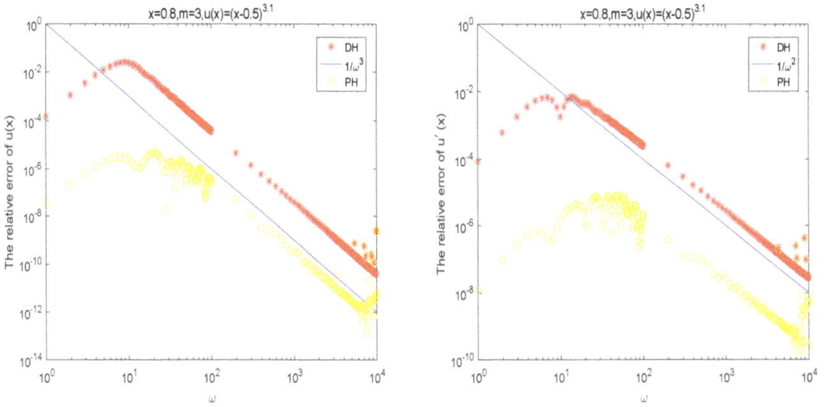

Figure 8. The relative errors of $u(x)$ and $u'(x)$ for direct Hermite collocation method (DH) and piecewise Hermite collocation method (PH) .

5. Conclusions

Collocation methods are efficient in solving Volterra integral equation with highly oscillatory kernel. In this paper, we present two collocation methods: DH collocation method and piecewise Hermite collocation method. The first conclusion to be drawn from the numerical evidence presented earlier is that Hermite-type collocation methods are higher efficient than existent collocation methods. Both methods can calculate the approximate value of function value and derivative value simultaneously. Finally, while we have considered only the case of Bessel kernel in this paper, the Hermite-type collocation methods can be extended to Fourier kernel.

In the future work, we will study better methods to solve the Volterra integral equations with different kernel and Fredholm integral equations.

Author Contributions: C.F., G.H. and S.X. conceived and designed the experiments; C.F. performed the experiments; G.H. analyzed the data; C.F. contributed reagents/materials/analysis tools; C.F. and G.H. wrote the paper.

Funding: This work is supported partly by National Natural Science Foundation of China (No. 11701171, 11771454), the Scientific Research Fund of Hunan Provincial Education Department (No. 17B113), the Hunan Provincial Natural Science Foundation of China (No. 2016JJ4037), the Aid program for Science and Technology Innovative Research Team in Higher Educational Institutions of Hunan Province, the Fundamental Research Funds for the Central Universities (No. 21618333), and the Opening Project of Guangdong Province Key Laboratory of Computational Science at the Sun Yat-sen University (No. 2018010) .

Acknowledgments: The authors are grateful to the anonymous referees for their useful comments and constructive suggestions for improvement of this paper.

Conflicts of Interest: The authors declare no conflict of interest.

References

1. Bao, G.; Sun, W.W. A fast algorithm for the electromagnetic scattering from a large cavity. *SIAM J. Sci. Comput.* **2005**, *27*, 553–574. [CrossRef]
2. Davis, P.J.; Duncan, D.B. Stability and convergence of collocation schemes for retarded potential integral equations. *SIAM J. Numer. Anal.* **2004**, *42*, 1167–1188. [CrossRef]
3. Langdon, S.; Chandler-Wilde, S.N. A wavenumber independent boundary element method for an acoustic scattering problem. *SIAM J. Numer. Anal.* **2006**, *43*, 2450–2477. [CrossRef]
4. Brunner, H. *Collocation Methods for Volterra Integral and Related Functional Equations*; Cambridge University Press: Cambridge, UK, 2004.
5. Brunner, H.; Davis, P.J.; Duncan D.B. Discontinuous Galerkin approximations for Volterra integral equations of the first kind. *IMA J. Numer. Anal.* **2009**, *29*, 856–881. [CrossRef]
6. Ma, J.; Fang, C.; Xiang, S. Modified asymptotic orders of the direct Filon method for a class of Volterra integral equations. *J. Comput. Appl. Math.* **2015**, *281*, 120–125. [CrossRef]
7. Wang, H.; Xiang, S. Asymptotic expansion and Filon-type methods for a Volterra integral equation with a highly oscillatory kernel. *IMA J. Numer. Anal.* **2011**, *31*, 469–490. [CrossRef]
8. Brunner, H. On Volterra integral operators with highly oscillatory kernels. *Discret. Contin. Dyn. Syst.* **2014**, *34*, 915–929. [CrossRef]
9. Ma, J.; Liu, H. On the Convolution Quadrature Rule for Integral Transforms with Oscillatory Bessel Kernels. *Symmetry* **2018**, *10*, 239. [CrossRef]
10. Chen, H; Zhang, C. Boundary value methods for Volterra integral and integro-differential equations. *Appl. Math. Comp.* **2011**, *218*, 2619–2630. [CrossRef]
11. Ma, J.; Xiang, S. A Collocation Boundary Value Method for Linear Volterra Integral Equations. *J. Sci. Comput.* **2017**, *71*, 1–20. [CrossRef]
12. Xiang, S.; Wu, Q. Numerical solutions to Volterra integral equations of the second kind with oscillatory trigonometric kernels. *Appl. Math. Comp.* **2013**, *223*, 34–44. [CrossRef]
13. Wu, Q. On graded meshes for weakly singular Volterra integral equations with oscillatory trigonometric kernels. *J. Comput. Appl. Math.* **2014**, *263*, 370–376. [CrossRef]
14. Xiang, S.; Brunner, H. Efficient methods for Volterra integral equations with highly oscillatory Bessel kernels. *BIT Numer. Math.* **2013**, *53*, 241–263. [CrossRef]
15. Fang, C.; Ma, J.; Xiang M. On Filon methods for a class of Volterra integral equations with highly oscillatory Bessel kernels. *Appl. Math. Comp.* **2015**, *268*, 783–792. [CrossRef]
16. Watson, G.N. *A Treatise on the Theory of Bessel Functions*; Cambridge University Press: Cambridge, UK, 1952.
17. Xiang, S.; Wang, H. Fast integration of highly oscillatory integrals with exotic oscillators. *Math. Comp.* **2010**, *79*, 829–844. [CrossRef]
18. Ma, J.; Xiang, S.; Kang, H. on the convergence rates of Filon methods for the solution of a Volterra integral equation with a highly oscillatory Bessel kernel. *Appl. Math. Lett.* **2013**, *26*, 699–705. [CrossRef]

Sample Availability: Samples of the compounds ...... are available from the authors.

© 2019 by the authors. Licensee MDPI, Basel, Switzerland. This article is an open access article distributed under the terms and conditions of the Creative Commons Attribution (CC BY) license (http://creativecommons.org/licenses/by/4.0/).

symmetry

MDPI

Article

Certain Results of q-Sheffer–Appell Polynomials

Ghazala Yasmin [1], Abdulghani Muhyi [1] and Serkan Araci [2,*]

1 Department of Applied Mathematics, Aligarh Muslim University, Aligarh 202002, India; ghazala30@gmail.com (G.Y.); muhyi2007@gmail.com (A.M.)

2 Department of Economics, Faculty of Economics, Administrative and Social Sciences, Hasan Kalyoncu University, TR-27410 Gaziantep, Turkey

* Correspondence: serkan.araci@hku.edu.tr or mtsrkn@hotmail.com

Received: 12 December 2018; Accepted: 28 January 2019; Published: 1 February 2019

Abstract: In this paper, the class of q-Sheffer–Appell polynomials is introduced. The generating function, series definition, determinant definition and some other identities of this class are established. Certain members of q-Sheffer–Appell polynomials are investigated and some properties of these members are derived. In addition, the class of 2D q-Sheffer–Appell polynomials is introduced. Further, the graphs of some members of q-Sheffer–Appell polynomials and 2D q-Sheffer–Appell polynomials are plotted for different values of indices by using Matlab.

Keywords: q-Sheffer–Appell polynomials; generating relations; determinant definition; recurrence relation; q-Hermite–Bernoulli polynomials; q-Hermite–Euler polynomials; q-Hermite–Genocchi polynomials

2010 Mathematics Subject Classification: 05A30; 11B83; 11B68

---

1. Introduction and Preliminaries

The subject of q-calculus leads to a new method for computations and classifications of q-special functions. It was launched in the 1920s. However, it has gained importance and considerable popularity during the last three decades [1–9]. In the last decades, q-calculus has been developed into an interdisciplinary subject and served as a bridge between physics and mathematics. The recent interest in the subject is due to the fact that q-series has popped in such various areas as quantum groups, statistical mechanics, transcendental number theory, etc. The definitions and notations of q-calculus reviewed here are taken from [10] (see also [11,12]).

The q-analog of the Pochhammer symbol $(\delta)_\kappa$, also called a q-shifted factorial, are defined by

$$(\delta;\mathrm{q})_0 = 1,\ (\delta;\mathrm{q})_\kappa = \prod_{r=0}^{\kappa-1}(1-\delta\mathrm{q}^r), \kappa \in \mathbb{N},\ \delta \in \mathbb{C}. \tag{1}$$

The q-analogs of a complex number δ and of the factorial function are given as follows:

$$[\delta]_\mathrm{q} = \frac{1-\mathrm{q}^\delta}{1-\mathrm{q}},\ \mathrm{q} \in \mathbb{C} - \{1\},\ \delta \in \mathbb{C}, \tag{2}$$

$$[\kappa]_\mathrm{q} = \sum_{\nu=1}^{\kappa} \mathrm{q}^{\nu-1},\ [0]_\mathrm{q} = 0,\ [\kappa]_\mathrm{q}! = \prod_{\nu=1}^{\kappa}[\nu]_\mathrm{q} = [1]_\mathrm{q}[2]_\mathrm{q}[3]_\mathrm{q}...[\kappa]_\mathrm{q},\ [0]_\mathrm{q}! = 1,\ \kappa \in \mathbb{N}, \mathrm{q} \in \mathbb{C}\backslash\{0,1\}. \tag{3}$$

The q-binomial coefficients $\begin{bmatrix}\kappa\\ \nu\end{bmatrix}_\mathrm{q}$ are defined by

$$\begin{bmatrix}\kappa\\ \nu\end{bmatrix}_\mathrm{q} = \frac{(\mathrm{q};\mathrm{q})_\kappa}{(\mathrm{q};\mathrm{q})_\nu(\mathrm{q};\mathrm{q})_{\kappa-\nu}} = \frac{[\kappa]_\mathrm{q}!}{[\nu]_\mathrm{q}!\ [\kappa-\nu]_\mathrm{q}!},\quad \nu = 0,1,2,...,\kappa. \tag{4}$$

The q-analog of the classical derivative $D\,u$ of a function u at a point $0 \neq \tau \in \mathbb{C}$ is given as

$$D_q u(\tau) = \frac{u(\tau) - u(q\tau)}{\tau - q\tau}, \quad 0 < |q| < 1, \quad \tau \neq 0. \tag{5}$$

In addition, we note that

$$(i) \quad \lim_{q \to 0} D_q u(\tau) = \frac{du(\tau)}{d\tau}, \text{where } \frac{d}{d\tau} \text{ denotes the classical ordinary derivative,} \tag{6}$$

$$(ii) \quad D_q(a_1 u(\tau) + a_2\, v(\tau)) = a_1 D_q u(\tau) + a_2 D_q v(\tau), \tag{7}$$

$$(iii) \quad D_q(uv)(\tau) = u(q\tau) D_q v(\tau) + v(\tau) D_q u(\tau) = u(\tau) D_q v(\tau) + D_q u(\tau) v(q\tau), \tag{8}$$

$$(vi) \quad D_q\left(\frac{u(\tau)}{v(\tau)}\right) = \frac{v(\tau) D_q u(\tau) - u(\tau) D_q v(\tau)}{v(\tau) v(q\tau)} = \frac{v(q\tau) D_q u(\tau) - u(q\tau) D_q v(\tau)}{v(\tau) v(q\tau)}. \tag{9}$$

The q-exponential functions $e_q(\tau)$ and $E_q(\tau)$ are defined as:

$$e_q(\tau) = \sum_{\kappa=0}^{\infty} \frac{\tau^\kappa}{[\kappa]_q!} := \frac{1}{((1-q)\tau; q)_\infty}, \quad 0 < |q| < 1, |\tau| < |1-q|^{-1}, \tag{10}$$

$$E_q(\tau) = \sum_{\kappa=0}^{\infty} q^{\frac{1}{2}\kappa(\kappa-1)} \frac{\tau^\kappa}{[\kappa]_q!} := (-(1-q); q)_\infty, \quad 0 < |q| < 1, \; \tau \in \mathbb{C}. \tag{11}$$

which satisfy the following properties:

$$D_q e_q(\tau) = e_q(\tau), \; D_q E_q(\tau) = E_q(q\tau), \tag{12}$$

$$e_q(\tau) E_q(-\tau) = E_q(\tau) e_q(-\tau) = 1. \tag{13}$$

The class of Appell polynomials was introduced and characterized completely by Appell [13]. Further, Throne [14], Sheffer [15] and Varma [16] studied this class of polynomials from different point of views. Sharma and Chak [17] introduced a q-analog for the class of Appell polynomials and called this sequence of polynomials as q-Harmonic. Later, Al-Salam [1] established the class of q-Appell polynomials $\{\mathcal{A}_{\kappa,q}(z)\}_{\kappa=0}^{\infty}$ and investigated some of its properties. These polynomials appear in several problems of theoretical physics, applied mathematics, approximation theory and many other branches of mathematics. The polynomials $\mathcal{A}_{\kappa,q}(z)$ (of degree κ) are called q-Appell polynomials provided that they satisfy the following q-differential equation

$$D_{q,z}\{\mathcal{A}_{\kappa,q}(z)\} = [\kappa]_q \mathcal{A}_{\kappa-1,q}(z), \; \kappa = 0,1,2,3,...; \; q \in \mathbb{C}, 0 < |q| < 1. \tag{14}$$

The generating function for the q-Appell polynomials $\mathcal{A}_{\kappa,q}(z)$ is given as:

$$\mathcal{A}_q(\tau)\, e_q(z\tau) = \sum_{\kappa=0}^{\infty} \mathcal{A}_{\kappa,q}(z) \frac{\tau^\kappa}{[\kappa]_q!}, \tag{15}$$

where

$$\mathcal{A}_q(\tau) = \sum_{\kappa=0}^{\infty} \mathcal{A}_{\kappa,q} \frac{\tau^\kappa}{[\kappa]_q!}, \quad \mathcal{A}_q(\tau) \neq 0; \; \mathcal{A}_{0,q} = 1, \tag{16}$$

is an analytic function at $\tau = 0$ and $\mathcal{A}_{\kappa,q} := \mathcal{A}_{\kappa,q}(0)$ denotes the q-Appell numbers.

We note that the function $\mathcal{A}_q(\tau)$ is called the determining function for the set $\mathcal{A}_{\kappa,q}(z)$. Based on suitable selection for the function $\mathcal{A}_q(\tau)$, different members belonging to the family of q-Appell polynomial $\mathcal{A}_{\kappa,q}(z)$ can be obtained. These members along with their notations, names and generating functions are listed in Table 1.

Table 1. Certain members of q-Appell family.

S. No.	$A_q(\tau)$	Generating Functions	Polynomials
I.	$A_q(\tau)=\frac{\tau}{(e_q(\tau)-1)}$	$\frac{\tau}{(e_q(\tau)-1)}e_q(z\tau)=\sum_{\kappa=0}^{\infty}\mathfrak{B}_{\kappa,q}(z)\frac{\tau^\kappa}{[\kappa]_q!}$	The q-Bernoulli polynomials [2,18,19]
II.	$A_q(\tau)=\frac{[2]_q}{(e_q(\tau)+1)}$	$\frac{[2]_q}{(e_q(\tau)+1)}e_q(z\tau)=\sum_{\kappa=0}^{\infty}\mathcal{E}_{\kappa,q}(z)\frac{\tau^\kappa}{[\kappa]_q!}$	The q-Euler polynomials [3,19,20]
III.	$A_q(\tau)=\frac{[2]_q\tau}{(e_q(\tau)+1)}$	$\frac{[2]_q\tau}{(e_q(\tau)+1)}e_q(z\tau)=\sum_{\kappa=0}^{\infty}\mathcal{G}_{\kappa,q}(z)\frac{\tau^\kappa}{[\kappa]_q!}$,	The q-Genocchi polynomials [7,19,21]

In 1978, Roman and Rota [22] used the umbral calculus to define the sequence of Sheffer polynomials whose their characteristics proved that this new proposed family of polynomials is equivalent to the family of polynomials of type zero, which was previously introduced by Sheffer [23]. Later, Roman [24] proposed a similar umbral approach under the area of nonclassical umbral calculus which is called q-umbral calculus. Recently, Kim et al. [5] introduced the q-Sheffer polynomials (qSP) $s_{\kappa,q}(z)$ for $(v(\tau),u(\tau))$ by means of the following generation function:

$$\frac{1}{v(u^{-1}(\tau))}\,e_q(zu^{-1}(\tau))=\sum_{\kappa=0}^{\infty}s_{\kappa,q}(z)\,\frac{\tau^\kappa}{[\kappa]_q!},\quad \text{for all } z\in\mathbb{C}, \tag{17}$$

where $u^{-1}(\tau)$ is the compositional inverse of $u(\tau)$.

In addition, the q-Sheffer polynomials may be alternatively defined as:

$$\phi_q(\tau)\,e_q(zH(\tau))=\sum_{\kappa=0}^{\infty}s_{\kappa,q}(z)\,\frac{\tau^\kappa}{[\kappa]_q!}, \tag{18}$$

where

$$\phi_q(\tau)=\sum_{\kappa=0}^{\infty}\phi_{\kappa,q}\,\frac{\tau^\kappa}{[\kappa]_q!}\quad\text{and}\quad H(\tau)=\sum_{\kappa=0}^{\infty}H_{\kappa,q}\,\frac{\tau^\kappa}{[\kappa]_q!}. \tag{19}$$

In view of Equations (17) and (18), we have

$$\phi_q(\tau)=\frac{1}{v(u^{-1}(\tau))}\quad\text{and}\quad H(\tau)=u^{-1}(\tau). \tag{20}$$

The q-Sheffer polynomials for the pair $(\phi(\tau),\tau)_q$ is called the q-Appell polynomials $\mathcal{A}_{\kappa,q}(z)$ and for the pair $(1,H(\tau))_q$ becomes the q-associated Sheffer polynomials $s_{\kappa,q}(z)$.

Recently, Duran et al. [25] introduced the q-Hermite polynomials (qHP) $\mathcal{H}_{\kappa,q}(z)$ by means of the following generating function:

$$e_q([2]_q z\tau)e_q(-\tau^2)=\sum_{\kappa=0}^{\infty}\mathcal{H}_{\kappa,q}(z)\frac{\tau^\kappa}{[\kappa]_q!}. \tag{21}$$

In [25], (p, q)-number is defined by $[x]_{p,q}=\frac{p^x-q^x}{p-q}$. It is worth noting that $[x]_{p,q}=c[x]_q$ for some constant c in p. Thus, there is no need to deal with the family of (p, q)-Sheffer–Appell polynomials.

In the present article, a new family of q-Sheffer–Appell polynomials (qSAP) is introduced by means of generating functions, series and determinant definitions. Further, some results are obtained for some members of this family. In the next section, the q-Sheffer–Appell polynomials are introduced by means of the generating functions and series definition. In addition, the determinant definition and many interesting properties of these q-hybrid special polynomials are derived. In Section 3, we consider some members of q-Sheffer–Appell polynomials and obtain the determinant definitions and some other properties of these members. In Section 4, the class of 2D q-Sheffer–Appell polynomials (2DqSAP) is also introduced. In Section 5, the graphs of some members of q-Sheffer–Appell polynomials and 2D q-Sheffer–Appell polynomials are plotted for different values of indices by using Matlab.

2. q-Sheffer–Appell Polynomials

In this section, the generating function, series definition and determinant definition for the q-Sheffer–Appell polynomials ${}_{\mathrm{s}}\mathcal{A}_{\kappa,\mathrm{q}}(z)$ are introduced.

To establish the generating function for the qSAP by making use of replacement technique, the following result is proved:

Theorem 1. *The following generating function for the* q*-Sheffer–Appell polynomials* ${}_{\mathrm{s}}\mathcal{A}_{\kappa,\mathrm{q}}(z)$ *holds true:*

$$\mathcal{A}_{\mathrm{q}}(\tau)\phi_{\mathrm{q}}(\tau)\, e_{\mathrm{q}}(zH(\tau)) = \sum_{\kappa=0}^{\infty} {}_{\mathrm{s}}\mathcal{A}_{\kappa,\mathrm{q}}(z)\, \frac{\tau^{\kappa}}{[\kappa]_{\mathrm{q}}!}. \tag{22}$$

Proof. By expanding the q-exponential function $e_{\mathrm{q}}(z\tau)$ in the left hand side of Equation (15) and then replacing the powers of z, i.e., $z^0, z, z^2, ..., z^{\kappa}$, by the corresponding polynomials $\mathrm{s}_{0,\mathrm{q}}(z), \mathrm{s}_{1,\mathrm{q}}(z), \mathrm{s}_{2,\mathrm{q}}(z), ..., \mathrm{s}_{\kappa,\mathrm{q}}(z)$ in the left hand side and z by $\mathrm{s}_{1,\mathrm{q}}(z)$ in the right hand side of the resultant equation, we have

$$\mathcal{A}_{\mathrm{q}}(\tau)\left(1 + \mathrm{s}_{1,\mathrm{q}}(z)\frac{\tau}{[1]_{\mathrm{q}}!} + \mathrm{s}_{2,\mathrm{q}}(z)\frac{\tau^2}{[2]_{\mathrm{q}}!} + ... + \mathrm{s}_{\kappa,\mathrm{q}}(z)\frac{\tau^{\kappa}}{[\kappa]_{\mathrm{q}}!} + ...\right) = \sum_{\kappa=0}^{\infty} \mathcal{A}_{\kappa,\mathrm{q}}(\mathrm{s}_{1,\mathrm{q}}(z))\, \frac{\tau^{\kappa}}{[\kappa]_{\mathrm{q}}!}. \tag{23}$$

Further, summing up the series in left hand side and then using Equation (18) in the resultant equation, we get

$$\mathcal{A}_{\mathrm{q}}(\tau)\phi_{\mathrm{q}}(\tau)\, e_{\mathrm{q}}(zH(\tau)) = \sum_{\kappa=0}^{\infty} \mathcal{A}_{\kappa,\mathrm{q}}(\mathrm{s}_{1,\mathrm{q}}(z))\, \frac{\tau^{\kappa}}{[\kappa]_{\mathrm{q}}!}. \tag{24}$$

Finally, indicating resultant qSAP by ${}_{\mathrm{s}}\mathcal{A}_{\kappa,\mathrm{q}}(z)$, that is

$$\mathcal{A}_{\kappa,\mathrm{q}}(\mathrm{s}_{1,\mathrm{q}}(z)) = {}_{\mathrm{s}}\mathcal{A}_{\kappa,\mathrm{q}}(z), \tag{25}$$

the assertion in Equation (22) is proved. □

Next, we introduce the series definition for the qSAP ${}_{\mathrm{s}}\mathcal{A}_{\kappa,\mathrm{q}}(z)$ by proving the following result:

Theorem 2. *The* q*-Sheffer–Appell polynomials* ${}_{\mathrm{s}}\mathcal{A}_{\kappa,\mathrm{q}}(z)$ *are defined by the following series definition:*

$${}_{\mathrm{s}}\mathcal{A}_{\kappa,\mathrm{q}}(z) = \sum_{\nu=0}^{\kappa} \begin{bmatrix} \kappa \\ \nu \end{bmatrix}_{\mathrm{q}} \mathcal{A}_{\nu,\mathrm{q}}\; \mathrm{s}_{\kappa-\nu,\mathrm{q}}(z). \tag{26}$$

Proof. In view of Equations (16) and (18), Equation (22) can be written as:

$$\sum_{\nu=0}^{\infty} \mathcal{A}_{\nu,\mathrm{q}}\, \frac{\tau^{\nu}}{[\nu]_{\mathrm{q}}!} \sum_{\kappa=0}^{\infty} \mathrm{s}_{\kappa,\mathrm{q}}(z)\, \frac{\tau^{\kappa}}{[\kappa]_{\mathrm{q}}!} = \sum_{\kappa=0}^{\infty} {}_{\mathrm{s}}\mathcal{A}_{\kappa,\mathrm{q}}(z)\, \frac{\tau^{\kappa}}{[\kappa]_{\mathrm{q}}!}, \tag{27}$$

which on using the Cauchy product rule [26] gives

$$\sum_{\kappa=0}^{\infty}\sum_{\nu=0}^{\kappa} \begin{bmatrix} \kappa \\ \nu \end{bmatrix}_{\mathrm{q}} \mathcal{A}_{\nu,\mathrm{q}}\; \mathrm{s}_{\kappa-\nu,\mathrm{q}}(z)\, \frac{\tau^{\kappa}}{[\kappa]_{\mathrm{q}}!} = \sum_{\kappa=0}^{\infty} {}_{\mathrm{s}}\mathcal{A}_{\kappa,\mathrm{q}}(z)\, \frac{\tau^{\kappa}}{[\kappa]_{\mathrm{q}}!}. \tag{28}$$

Now, comparing the coefficients of identical powers of τ in above equation, we arrive at our assertion in Equation (26). □

Theorem 3. *The* q*-Sheffer–Appell polynomials* ${}_{\mathrm{s}}\mathcal{A}_{\kappa,\mathrm{q}}(z)$ *satisfy the following linear homogeneous recurrence relation:*

$${}_{\mathrm{s}}\mathcal{A}_{\kappa,\mathrm{q}}(z) = \frac{1}{[\kappa]_{\mathrm{q}}} \sum_{\nu=0}^{\kappa} \begin{bmatrix} \kappa \\ \nu \end{bmatrix}_{\mathrm{q}} (\alpha_{\nu} + z\beta_{\nu})\, {}_{\mathrm{s}}\mathcal{A}_{\kappa-\nu,\mathrm{q}}(z), \tag{29}$$

where

$$\tau\,\frac{\mathcal{A}_q(q\tau)(D_{q,\tau}\phi_q(\tau))+\phi_q(\tau)(D_{q,\tau}\mathcal{A}_q(\tau))}{\mathcal{A}_q(\tau)\phi_q(\tau)}=\sum_{\kappa=0}^{\infty}\alpha_\kappa\frac{\tau^\kappa}{[\kappa]_q!},$$
$$\tau\,\frac{\mathcal{A}_q(q\tau)\phi_q(q\tau)(D_{q,\tau}H(\tau))}{\mathcal{A}_q(\tau)\phi_q(\tau)}=\sum_{\kappa=0}^{\infty}\beta_\kappa\frac{\tau^\kappa}{[\kappa]_q!}. \tag{30}$$

Proof. Consider the generating function

$$\mathsf{F}_q(z,\tau)=\mathcal{A}_q(\tau)\phi_q(\tau)\,e_q(zH(\tau))=\sum_{\kappa=0}^{\infty}{}_{\mathsf{s}}\mathcal{A}_{\kappa,q}(z)\,\frac{\tau^\kappa}{[\kappa]_q!}. \tag{31}$$

Taking the q-derivative of Equation (31) partially with respect to τ, we get

$$\begin{aligned}D_{q,\tau}(\mathsf{F}_q(z,\tau))=&\{\mathcal{A}_q(q\tau)(D_{q,\tau}\phi_q(\tau))+\phi_q(\tau)(D_{q,\tau}\mathcal{A}_q(\tau))\}e_q(zH(\tau))\\&+z\,\mathcal{A}_q(q\tau)\phi_q(q\tau)(D_{q,\tau}H(\tau))e_q(zH(\tau))\end{aligned} \tag{32}$$

Now, factorizing $\mathsf{F}_q(z,\tau)$ from its left hand side and after that multiplying both sides by τ, it follows that

$$\begin{aligned}&\tau D_{q,\tau}(\mathsf{F}_q(z,\tau))\\&=\mathsf{F}_q(z,\tau)\left\{\tau\frac{\mathcal{A}_q(q\tau)\phi_q(\tau)(D_{q,\tau}\mathcal{A}_q(\tau))(D_{q,\tau}\phi_q(\tau))}{\mathcal{A}_q(\tau)\phi_q(\tau)}+z\tau\frac{\mathcal{A}_q(q\tau)\phi_q(q\tau)(D_{q,\tau}H(\tau))}{\mathcal{A}_q(\tau)\phi_q(\tau)}\right\}.\end{aligned} \tag{33}$$

In view of the assumption in Equations (30) and (31), Equation (33) can be expressed as

$$\sum_{\kappa=0}^{\infty}[\kappa]_q\,{}_{\mathsf{s}}\mathcal{A}_{\kappa,q}(z)\frac{\tau^\kappa}{[\kappa]_q!}=\sum_{\kappa=0}^{\infty}{}_{\mathsf{s}}\mathcal{A}_{\kappa,q}(z)\frac{\tau^\kappa}{[\kappa]_q!}\left\{\sum_{\kappa=0}^{\infty}\alpha_\kappa\frac{\tau^\kappa}{[\kappa]_q!}+z\sum_{\kappa=0}^{\infty}\beta_\kappa\frac{\tau^\kappa}{[\kappa]_q!}\right\}, \tag{34}$$

which on using the Cauchy product rule, gives

$$\sum_{\kappa=0}^{\infty}[\kappa]_q\,{}_{\mathsf{s}}\mathcal{A}_{\kappa,q}(z)\frac{\tau^\kappa}{[\kappa]_q!}=\sum_{\kappa=0}^{\infty}\sum_{\nu=0}^{\kappa}\begin{bmatrix}\kappa\\ \nu\end{bmatrix}_q(\alpha_\nu+z\beta_\nu)\,{}_{\mathsf{s}}\mathcal{A}_{\kappa-\nu,q}(z)\frac{\tau^\kappa}{[\kappa]_q!}. \tag{35}$$

Finally, equating the coefficients of identical powers of τ in above equation and after that dividing both sides of the resultant equation by $[\kappa]_q$, we get the assertion in Equation (29). □

Due to the importance of determinant form for the computational and applied purposes, we derive the determinant definition for the qSAP ${}_{\mathsf{s}}\mathcal{A}_{\kappa,q}(z)$.

Theorem 4. *The* q*-Sheffer–Appell polynomials* ${}_{\mathsf{s}}\mathcal{A}_{\kappa,q}(z)$ *of degree* κ *are defined by*

$$
{}_{\mathbb{s}}\mathcal{A}_{0,q}(z)=\frac{1}{\mathcal{B}_{0,q}}, \tag{36}
$$

$$
{}_{\mathbb{s}}\mathcal{A}_{\kappa,q}(z)=\frac{(-1)^{\kappa}}{(\mathcal{B}_{0,q})^{\kappa+1}}\begin{vmatrix} 1 & \mathbb{s}_{1,q}(z) & \mathbb{s}_{2,q}(z) & \dots & \mathbb{s}_{\kappa-1,q}(z) & \mathbb{s}_{\kappa,q}(z) \\ \mathcal{B}_{0,q} & \mathcal{B}_{1,q} & \mathcal{B}_{2,q} & \dots & \mathcal{B}_{\kappa-1,q} & \mathcal{B}_{\kappa,q} \\ 0 & \mathcal{B}_{0,q} & \begin{bmatrix}2\\1\end{bmatrix}_q\mathcal{B}_{1,q} & \dots & \begin{bmatrix}\kappa-1\\1\end{bmatrix}_q\mathcal{B}_{\kappa-2,q} & \begin{bmatrix}\kappa\\1\end{bmatrix}_q\mathcal{B}_{\kappa-1,q} \\ 0 & 0 & \mathcal{B}_{0,q} & \dots & \begin{bmatrix}\kappa-1\\2\end{bmatrix}_q\mathcal{B}_{\kappa-3,q} & \begin{bmatrix}\kappa\\2\end{bmatrix}_q\mathcal{B}_{\kappa-2,q} \\ . & . & . & \dots & . & . \\ . & . & . & \dots & . & . \\ 0 & 0 & 0 & \dots & \mathcal{B}_{0,q} & \begin{bmatrix}\kappa\\\kappa-1\end{bmatrix}_q\mathcal{B}_{1,q} \end{vmatrix}, \tag{37}
$$

$$
\mathcal{B}_{\kappa,q}=-\frac{1}{\mathcal{A}_{0,q}}\left(\sum_{\nu=1}^{\kappa}\begin{bmatrix}\kappa\\\nu\end{bmatrix}_q\mathcal{A}_{\nu,q}\mathcal{B}_{\kappa-\nu,q}\right), \qquad \kappa=1,2,3,\dots,
$$

where $\mathcal{B}_{0,q}\neq 0$, $\mathcal{B}_{0,q}=\frac{1}{\mathcal{A}_{0,q}}$ *and* $\mathbb{s}_{\kappa,q}(z)(\kappa=0,1,2,\dots,)$ *are the* q*-Sheffer polynomials of degree* κ.

Proof. Consider ${}_{\mathbb{s}}\mathcal{A}_{\kappa,q}(z)$ to be a sequence of the qSAP defined by Equation (22) and $\mathcal{A}_{\kappa,q}$, $\mathcal{B}_{\kappa,q}$ be two numerical sequences (the coefficients of q-Taylor's series expansions of functions) such that

$$
\mathcal{A}_q(\tau)=\mathcal{A}_{0,q}+\mathcal{A}_{1,q}\frac{\tau}{[1]_q!}+\mathcal{A}_{2,q}\frac{\tau^2}{[2]_q!}+\dots+\mathcal{A}_{\kappa,q}\frac{\tau^{\kappa}}{[\kappa]_q!}+\dots,\ \kappa=0,1,2,3,\dots;\ \mathcal{A}_{0,q}\neq 0, \tag{38}
$$

$$
\hat{\mathcal{A}}_q(\tau)=\mathcal{B}_{0,q}+\mathcal{B}_{1,q}\frac{\tau}{[1]_q!}+\mathcal{B}_{2,q}\frac{\tau^2}{[2]_q!}+\dots+\mathcal{B}_{\kappa,q}\frac{\tau^{\kappa}}{[\kappa]_q!}+\dots,\ \kappa=0,1,2,3,\dots;\ \mathcal{B}_{0,q}\neq 0, \tag{39}
$$

satisfying

$$
\mathcal{A}_q(\tau)\hat{\mathcal{A}}_q(\tau)=1. \tag{40}
$$

On using Cauchy product rule for the two series production $\mathcal{A}_q(\tau)\hat{\mathcal{A}}_q(\tau)$, we get

$$
\begin{aligned}
\mathcal{A}_q(\tau)\hat{\mathcal{A}}_q(\tau) &= \sum_{\kappa=0}^{\infty}\mathcal{A}_{\kappa,q}\frac{\tau^{\kappa}}{[\kappa]_q!}\sum_{\kappa=0}^{\infty}\mathcal{B}_{\kappa,q}\frac{\tau^{\kappa}}{[\kappa]_q!} \\
&= \sum_{\kappa=0}^{\infty}\sum_{\nu=0}^{\kappa}\begin{bmatrix}\kappa\\\nu\end{bmatrix}_q\mathcal{A}_{\nu,q}\mathcal{B}_{\kappa-\nu,q}\frac{\tau^{\kappa}}{[\kappa]_q!}.
\end{aligned}
$$

Consequently,

$$
\sum_{\nu=0}^{\kappa}\begin{bmatrix}\kappa\\\nu\end{bmatrix}_q\mathcal{A}_{\nu,q}\mathcal{B}_{\kappa-\nu,q}=\begin{cases}1, & if\ \kappa=0,\\ 0, & if\ \kappa>0.\end{cases} \tag{41}
$$

That is,

$$
\begin{cases}\mathcal{B}_{0,q}=\frac{1}{\mathcal{A}_{0,q}}, \\ \mathcal{B}_{\kappa,q}=-\frac{1}{\mathcal{A}_{0,q}}\left\{\sum_{\nu=1}^{\kappa}\begin{bmatrix}\kappa\\\nu\end{bmatrix}_q\mathcal{A}_{\nu,q}\mathcal{B}_{\kappa-\nu,q}\right\}, \quad \kappa=0,1,2,\dots\end{cases} \tag{42}
$$

Next, multiplying both sides of Equation (22) by $\hat{\mathcal{A}}_q(t)$, we get

$$
\mathcal{A}_q(\tau)\hat{\mathcal{A}}_q(\tau)\phi_q(\tau)\,e_q(zH(\tau))=\hat{\mathcal{A}}_q(\tau)\sum_{\kappa=0}^{\infty}{}_{\mathbb{s}}\mathcal{A}_{\kappa,q}(z)\,\frac{\tau^{\kappa}}{[\kappa]_q!}. \tag{43}
$$

Further, in view of Equations (18), (39) and (40), the above equation can be expressed as

$$\sum_{\kappa=0}^{\infty} \mathrm{s}_{\kappa,q}(z)\, \frac{\tau^{\kappa}}{[\kappa]_q!} = \sum_{\kappa=0}^{\infty} \mathcal{B}_{\kappa,q} \frac{\tau^{\kappa}}{[\kappa]_q!} \sum_{\kappa=0}^{\infty} {}_{\mathrm{s}}\mathcal{A}_{\kappa,q}(z)\, \frac{\tau^{\kappa}}{[\kappa]_q!}. \tag{44}$$

Now, on using Cauchy product rule for the two series in the right hand side of Equation (44), we obtain the following infinite system for the unknowns ${}_{\mathrm{s}}\mathcal{A}_{\kappa,q}(z)$:

$$\begin{cases} \mathcal{B}_{0,q}\ {}_{\mathrm{s}}\mathcal{A}_{0,q}(z) = 1, \\ \mathcal{B}_{1,q}\ {}_{\mathrm{s}}\mathcal{A}_{0,q}(z) + \mathcal{B}_{0,q}\ {}_{\mathrm{s}}\mathcal{A}_{1,q}(z) = \mathrm{s}_{1,q}(z) \\ \mathcal{B}_{2,q}\ {}_{\mathrm{s}}\mathcal{A}_{0,q}(z) + \begin{bmatrix}2\\1\end{bmatrix}_q \mathcal{B}_{1,q}\ {}_{\mathrm{s}}\mathcal{A}_{1,q}(z) + \mathcal{B}_{0,q}\ {}_{\mathrm{s}}\mathcal{A}_{2,q}(z) = \mathrm{s}_{2,q}(z), \\ \vdots \\ \mathcal{B}_{\kappa-1,q}\ {}_{\mathrm{s}}\mathcal{A}_{0,q}(z) + \begin{bmatrix}\kappa-1\\1\end{bmatrix}_q \mathcal{B}_{\kappa-2,q}\ {}_{\mathrm{s}}\mathcal{A}_{1,q}(z) + \ldots + \mathcal{B}_{0,q}\ {}_{\mathrm{s}}\mathcal{A}_{\kappa-1,q}(z) = \mathrm{s}_{\kappa-1,q}(z), \\ \mathcal{B}_{\kappa,q}\ {}_{\mathrm{s}}\mathcal{A}_{0,q}(z) + \begin{bmatrix}\kappa\\1\end{bmatrix}_q \mathcal{B}_{\kappa-1,q}\ {}_{\mathrm{s}}\mathcal{A}_{1,q}(z) + \ldots + \mathcal{B}_{0,q}\ {}_{\mathrm{s}}\mathcal{A}_{\kappa,q}(z) = \mathrm{s}_{\kappa,q}(z), \\ \vdots \end{cases} \tag{45}$$

Obviously, the first equation of the system in Equation (45) leads to our first assertion in Equation (36). The coefficient matrix of the system in Equation (45) is lower triangular, thus this assist us to obtain the unknowns ${}_{\mathrm{s}}\mathcal{A}_{\kappa,q}(z)$ by applying Cramer rule to the first $\kappa+1$ equations of the system in Equation (45). According to this, we can obtain

$${}_{\mathrm{s}}\mathcal{A}_{\kappa,q}(z) = \frac{\begin{vmatrix} \mathcal{B}_{0,q} & 0 & 0 & \ldots & 0 & 1 \\ \mathcal{B}_{1,q} & \mathcal{B}_{0,q} & 0 & \ldots & 0 & \mathrm{s}_{1,q}(z) \\ \mathcal{B}_{2,q} & \begin{bmatrix}2\\1\end{bmatrix}_q \mathcal{B}_{1,q} & \mathcal{B}_{0,q} & \ldots & 0 & \mathrm{s}_{2,q}(z) \\ . & . & . & \ldots & . & . \\ . & . & . & \ldots & . & . \\ \mathcal{B}_{\kappa-1,q} & \begin{bmatrix}\kappa-1\\1\end{bmatrix}_q \mathcal{B}_{\kappa-2,q} & \begin{bmatrix}\kappa-1\\2\end{bmatrix}_q \mathcal{B}_{\kappa-3,q} & \ldots & \mathcal{B}_{0,q} & \mathrm{s}_{\kappa-1,q}(z) \\ \mathcal{B}_{\kappa,q} & \begin{bmatrix}\kappa\\1\end{bmatrix}_q \mathcal{B}_{\kappa-1,q} & \begin{bmatrix}\kappa\\2\end{bmatrix}_q \mathcal{B}_{\kappa-2,q} & \ldots & \begin{bmatrix}\kappa\\\kappa-1\end{bmatrix}_q \mathcal{B}_{1,q} & \mathrm{s}_{\kappa,q}(z) \end{vmatrix}}{\begin{vmatrix} \mathcal{B}_{0,q} & 0 & 0 & \ldots & 0 & 1 \\ \mathcal{B}_{1,q} & \mathcal{B}_{0,q} & 0 & \ldots & 0 & 0 \\ \mathcal{B}_{2,q} & \begin{bmatrix}2\\1\end{bmatrix}_q \mathcal{B}_{1,q} & \mathcal{B}_{0,q} & \ldots & 0 & 0 \\ . & . & . & \ldots & . & . \\ . & . & . & \ldots & . & . \\ \mathcal{B}_{\kappa-1,q} & \begin{bmatrix}\kappa-1\\1\end{bmatrix}_q \mathcal{B}_{\kappa-2,q} & \begin{bmatrix}\kappa-1\\2\end{bmatrix}_q \mathcal{B}_{\kappa-3,q} & \ldots & \mathcal{B}_{0,q} & 0 \\ \mathcal{B}_{\kappa,q} & \begin{bmatrix}\kappa\\1\end{bmatrix}_q \mathcal{B}_{\kappa-1,q} & \begin{bmatrix}\kappa\\2\end{bmatrix}_q \mathcal{B}_{\kappa-2,q} & \ldots & \begin{bmatrix}\kappa\\\kappa-1\end{bmatrix}_q \mathcal{B}_{1,q} & \mathcal{B}_{0,q} \end{vmatrix}} \tag{46}$$

where $\kappa = 1, 2, 3, \ldots$, which on expanding the determinant in the denominator and taking the transpose of the determinant in the numerator, yields to

$$
{}_{\mathtt{s}}\mathcal{A}_{\kappa,q}(z) = \frac{1}{(\mathcal{B}_{0,q})^{\kappa+1}} \begin{vmatrix} \mathcal{B}_{0,q} & \mathcal{B}_{1,q} & \mathcal{B}_{2,q} & \cdots & \mathcal{B}_{\kappa-1,q} & \mathcal{B}_{\kappa,q} \\ 0 & \mathcal{B}_{0,q} & \begin{bmatrix} 2 \\ 1 \end{bmatrix}_q \mathcal{B}_{1,q} & \cdots & \begin{bmatrix} \kappa-1 \\ 1 \end{bmatrix}_q \mathcal{B}_{\kappa-2,q} & \begin{bmatrix} \kappa \\ 1 \end{bmatrix}_q \mathcal{B}_{\kappa-1,q} \\ 0 & 0 & \mathcal{B}_{0,q} & \cdots & \begin{bmatrix} \kappa-1 \\ 2 \end{bmatrix}_q \mathcal{B}_{\kappa-3,q} & \begin{bmatrix} \kappa \\ 2 \end{bmatrix}_q \mathcal{B}_{\kappa-2,q} \\ . & . & . & \cdots & . & . \\ . & . & . & \cdots & . & . \\ 0 & 0 & 0 & \cdots & \mathcal{B}_{0,q} & \begin{bmatrix} \kappa \\ \kappa-1 \end{bmatrix}_q \mathcal{B}_{1,q} \\ 1 & \mathtt{s}_{1,q}(z) & \mathtt{s}_{2,q}(z) & \cdots & \mathtt{s}_{\kappa-1,q}(z) & \mathtt{s}_{\kappa,q}(z) \end{vmatrix}. \tag{47}
$$

Finally, after κ circular row exchanges, i.e., after moving the jth row to the $(j+1)$th position for $j = 1, 2, 3, ..., \kappa - 1$, we arrive at our assertion in Equation (37). □

Theorem 5. *The following identity for the qSAP ${}_{\mathtt{s}}\mathcal{A}_{\kappa,q}(z)$ holds true:*

$$
{}_{\mathtt{s}}\mathcal{A}_{\kappa,q}(z) = \frac{1}{\mathcal{B}_{0,q}} \left(\mathtt{s}_{\kappa,q}(z) - \sum_{\nu=0}^{\kappa-1} \begin{bmatrix} \kappa \\ \nu \end{bmatrix}_q \mathcal{B}_{\kappa-\nu,q} \; {}_{\mathtt{s}}\mathcal{A}_{\nu,q}(z) \right), \quad \kappa = 1, 2, ... \tag{48}
$$

Proof. Expanding the determinant in Equation (37) with respect to the $(\kappa+1)$th row and using a similar approach used in ([27], Theorem 3.1), the assertion in Equation (48) is proved. □

3. Examples

Several members belonging to the q-Sheffer–Appell family ${}_{\mathtt{s}}\mathcal{A}_{\kappa,q}(z)$ can be derived by making suitable selections for the functions $\mathcal{A}_q(\tau)$, $\phi_q(\tau)$ and $H(\tau)$. The q-Hermite polynomials (qHP) $\mathcal{H}_{\kappa,q}(z)$ [25] are one of the important members of q-Sheffer family. In addition, the q-Bernoulli polynomials $\mathfrak{B}_{\kappa,q}(z)$, q-Euler polynomials $\mathcal{E}_{\kappa,q}(z)$ and q-Genocchi polynomials $\mathcal{G}_{\kappa,q}(z)$ are considerable members of the q-Appell family. In this section, we introduce the q-Hermite–Bernoulli polynomials ${}_{\mathcal{H}}\mathfrak{B}_{\kappa,q}(z)$, q-Hermite–Euler polynomials ${}_{\mathcal{H}}\mathcal{E}_{\kappa,q}(z)$ and q-Hermite–Genocchi polynomials ${}_{\mathcal{H}}\mathcal{G}_{\kappa,q}(z)$ by means of the generating functions, series definitions and also explore other properties of these members.

3.1. q-Hermite–Bernoulli Polynomials

Since, for $\mathcal{A}_q(\tau) = \frac{\tau}{e_q(\tau)-1}$, the qAP $\mathcal{A}_{\kappa,q}(z)$ reduce to the qBP $\mathfrak{B}_{\kappa,q}(z)$ (Table 1(I)) and for $\phi_q(\tau) = e_q(-\tau^2), H(\tau) = [2]_q\tau$ the qSP $\mathtt{s}_{\kappa,q}(z)$ reduce to qHP $\mathcal{H}_{\kappa,q}(z)$, for the same choices of $A_q(\tau), \phi_q(\tau)$ and $H(\tau)$, the qSAP ${}_{\mathtt{s}}\mathcal{A}_{\kappa,q}(z)$ reduce to qHBP ${}_{\mathcal{H}}\mathfrak{B}_{\kappa,q}(z)$. In view of Equation (22), the generating function for the qHBP ${}_{\mathcal{H}}\mathfrak{B}_{\kappa,q}(z)$ is given as:

$$
\frac{\tau}{e_q(\tau)-1} \, e_q([2]_q z\tau) e_q(-\tau^2) = \sum_{\kappa=0}^{\infty} {}_{\mathcal{H}}\mathfrak{B}_{\kappa,q}(z) \, \frac{\tau^\kappa}{[\kappa]_q!}. \tag{49}
$$

In view of Equation (26), the qHBP ${}_{\mathcal{H}}\mathfrak{B}_{\kappa,q}(z)$ of degree κ are defined by the series:

$$
{}_{\mathcal{H}}\mathfrak{B}_{\kappa,q}(z) = \sum_{\nu=0}^{\kappa} \begin{bmatrix} \kappa \\ \nu \end{bmatrix}_q \mathfrak{B}_{\nu,q} \mathcal{H}_{\kappa-\nu,q}(z). \tag{50}
$$

In view of Equation (48), the following identity for the qHBP ${}_{\mathcal{H}}\mathfrak{B}_{\kappa,q}(z)$ holds true:

$$
{}_{\mathcal{H}}\mathfrak{B}_{\kappa,q}(z) = \frac{1}{\mathcal{B}_{0,q}} \left(\mathcal{H}_{\kappa,q}(z) - \sum_{\nu=0}^{\kappa-1} \begin{bmatrix} \kappa \\ \nu \end{bmatrix}_q \mathcal{B}_{\kappa-\nu,q} \; {}_{\mathcal{H}}\mathfrak{B}_{\nu,q}(z) \right), \quad \kappa = 1, 2, ... \tag{51}
$$

Further, by taking $s_{\kappa,q}(z) = \mathcal{H}_{\kappa,q}(z)$, $\mathcal{B}_{0,q} = 1$ and $\mathcal{B}_{j,q} = \frac{1}{[j+1]_q} (j = 1,2,3,...)$ in Equations (36) and (37), we obtain the determinant definition of the qHBP ${}_{\mathcal{H}}\mathfrak{B}_{\kappa,q}(z)$ given as:

Definition 1. *The* q*-Hermite–Bernoulli polynomials* ${}_{\mathcal{H}}\mathfrak{B}_{\kappa,q}(z)$ *of degree* κ *are defined by*

$$ {}_{\mathcal{H}}\mathfrak{B}_{0,q}(z) = 1, \tag{52}$$

$$ {}_{\mathcal{H}}\mathfrak{B}_{\kappa,q}(z) = (-1)^{\kappa} \begin{vmatrix} 1 & \mathcal{H}_{1,q}(z) & \mathcal{H}_{2,q}(z) & \dots & \mathcal{H}_{\kappa-1,q}(z) & \mathcal{H}_{\kappa,q}(z) \\ 1 & \frac{1}{[2]_q} & \frac{1}{[3]_q} & \dots & \frac{1}{[\kappa]_q} & \frac{1}{[\kappa+1]_q} \\ 0 & 1 & \begin{bmatrix} 2 \\ 1 \end{bmatrix}_q \frac{1}{[2]_q} & \dots & \begin{bmatrix} \kappa-1 \\ 1 \end{bmatrix}_q \frac{1}{[\kappa-1]_q} & \begin{bmatrix} \kappa \\ 1 \end{bmatrix}_q \frac{1}{[\kappa]_q} \\ 0 & 0 & 1 & \dots & \begin{bmatrix} \kappa-1 \\ 2 \end{bmatrix}_q \frac{1}{[\kappa-2]_q} & \begin{bmatrix} \kappa \\ 2 \end{bmatrix}_q \frac{1}{[\kappa-1]_q} \\ . & . & . & \dots & . & . \\ . & . & . & \dots & . & . \\ 0 & 0 & 0 & \dots & 1 & \begin{bmatrix} \kappa \\ \kappa-1 \end{bmatrix}_q \frac{1}{[2]_q} \end{vmatrix}, \quad \kappa = 1,2,3,..., \tag{53}$$

where $\mathcal{H}_{\kappa,q}(z) (\kappa = 0,1,2,3,...)$ *are the* q*-Hermite polynomials of degree* κ.

Theorem 6. *The* q*-Hermite–Bernoulli polynomials* ${}_{\mathcal{H}}\mathfrak{B}_{\kappa,q}(z)$ *satisfy the following* q*-recurrence relations:*

$$ D_{q,z} {}_{\mathcal{H}}\mathfrak{B}_{\kappa,q}(z) = [2]_q [\kappa]_q {}_{\mathcal{H}}\mathfrak{B}_{\kappa-1,q}(z), \tag{54}$$

$$ D_{q,z}^{(k)} {}_{\mathcal{H}}\mathfrak{B}_{\kappa,q}(z) = \frac{[2]_q^k [\kappa]_q!}{[\kappa-k]_q!} {}_{\mathcal{H}}\mathfrak{B}_{\kappa-k,q}(z). \tag{55}$$

Proof. Applying the q-derivative with respect to z to both sides of Equation (49), we get

$$ \begin{aligned} \sum_{\kappa=0}^{\infty} D_{q,z} {}_{\mathcal{H}}\mathfrak{B}_{\kappa,q}(z) \frac{\tau^{\kappa}}{[\kappa]_q!} &= [2]_q \tau \frac{\tau}{e_q(t) - 1} e_q([2]_q z\tau) e_q(-\tau^2) \\ &= [2]_q \sum_{\kappa=0}^{\infty} [\kappa]_q {}_{\mathcal{H}}\mathfrak{B}_{\kappa-1,q}(z) \frac{\tau^{\kappa}}{[\kappa]_q!}. \end{aligned} \tag{56}$$

Now, equating the coefficient of like powers of τ in both sides of the above equation, we get the assertion in Equation (54). Similarly, on applying the q-derivative with respect to z to both sides of Equation (49) k times, we get the assertion in Equation (55). □

3.2. q-Hermite–Euler Polynomials

Since, for $\mathcal{A}_q(\tau) = \frac{[2]_q}{e_q(\tau)+1}$, the qAP $\mathcal{A}_{\kappa,q}(z)$ reduce to the qEP $\mathcal{E}_{\kappa,q}(z)$ (Table 1(II)) and for $\phi_q(\tau) = e_q(-\tau^2)$, $H(t) = [2]_q\tau$ the qSP $s_{\kappa,q}(z)$ reduce to qHP $\mathcal{H}_{\kappa,q}(z)$, for the same choices of $A_q(\tau)$, $\phi_q(\tau)$ and $H(\tau)$, the qSAP ${}_{s}\mathcal{A}_{\kappa,q}(z)$ reduce to qHEP ${}_{\mathcal{H}}\mathcal{E}_{\kappa,q}(z)$. In view of Equation (22), the generating function for the qHEP ${}_{\mathcal{H}}\mathcal{E}_{\kappa,q}(z)$ is given as:

$$ \frac{[2]_q}{e_q(\tau)+1} e_q([2]_q z\tau) e_q(-\tau^2) = \sum_{\kappa=0}^{\infty} {}_{\mathcal{H}}\mathcal{E}_{\kappa,q}(z) \frac{\tau^{\kappa}}{[\kappa]_q!}. \tag{57}$$

In view of Equation (26), the qHEP ${}_{\mathcal{H}}\mathcal{E}_{\kappa,q}(z)$ of degree κ are defined by the series:

$$ {}_{\mathcal{H}}\mathcal{E}_{\kappa,q}(z) = \sum_{\nu=0}^{\kappa} \begin{bmatrix} \kappa \\ \nu \end{bmatrix}_q \mathcal{E}_{\nu,q} \mathcal{H}_{\kappa-\nu,q}(z). \tag{58}$$

In view of Equation (48), the following identity for the qHEP ${}_{\mathcal{H}}\mathcal{E}_{\kappa,q}(z)$ holds true:

$$ {}_{\mathcal{H}}\mathcal{E}_{\kappa,q}(z) = \frac{1}{\mathcal{B}_{0,q}}\left(\mathcal{H}_{\kappa,q}(z) - \sum_{\nu=0}^{\kappa-1}\begin{bmatrix}\kappa\\ \nu\end{bmatrix}_q \mathcal{B}_{\kappa-\nu,q}\, {}_{\mathcal{H}}\mathcal{E}_{\nu,q}(z)\right), \quad \kappa = 1,2,\ldots. \tag{59} $$

Further, by taking $s_{\kappa,q}(z) = \mathcal{H}_{\kappa,q}(z)$, $\mathcal{B}_{0,q} = 1$ and $\mathcal{B}_{j,q} = \frac{1}{2}(j = 1,2,3,\ldots)$ in Equations (36) and (37), we obtain the determinant definition of the qHEP ${}_{\mathcal{H}}\mathcal{E}_{\kappa,q}(z)$ given as:

Definition 2. *The q-Hermite–Euler polynomials ${}_{\mathcal{H}}\mathcal{E}_{\kappa,q}(z)$ of degree κ are defined by*

$$ {}_{\mathcal{H}}\mathcal{E}_{0,q}(z) = 1, \tag{60} $$

$$ {}_{\mathcal{H}}\mathcal{E}_{\kappa,q}(z) = (-1)^{\kappa}\begin{vmatrix} 1 & \mathcal{H}_{1,q}(z) & \mathcal{H}_{2,q}(z) & \cdots & \mathcal{H}_{\kappa-1,q}(z) & \mathcal{H}_{\kappa,q}(z) \\ 1 & \frac{1}{2} & \frac{1}{2} & \cdots & \frac{1}{2} & \frac{1}{2} \\ 0 & 1 & \begin{bmatrix}2\\1\end{bmatrix}_q\frac{1}{2} & \cdots & \begin{bmatrix}\kappa-1\\1\end{bmatrix}_q\frac{1}{2} & \begin{bmatrix}\kappa\\1\end{bmatrix}_q\frac{1}{2} \\ 0 & 0 & 1 & \cdots & \begin{bmatrix}\kappa-1\\2\end{bmatrix}_q\frac{1}{2} & \begin{bmatrix}\kappa\\2\end{bmatrix}_q\frac{1}{2} \\ . & . & . & \cdots & . & . \\ . & . & . & \cdots & . & . \\ 0 & 0 & 0 & \cdots & 1 & \begin{bmatrix}\kappa\\ \kappa-1\end{bmatrix}_q\frac{1}{2} \end{vmatrix}, \quad \kappa = 1,2,3,\ldots, \tag{61} $$

where $\mathcal{H}_{\kappa,q}(z)(\kappa = 0,1,2,3,\ldots)$ are the q-Hermite polynomials of degree κ.

Theorem 7. *The q-Hermite–Euler polynomials ${}_{\mathcal{H}}\mathcal{E}_{\kappa,q}(z)$ satisfy the following q-recurrence relations:*

$$ D_{q,z}\,{}_{\mathcal{H}}\mathcal{E}_{\kappa,q}(z) = [2]_q[\kappa]_q\,{}_{\mathcal{H}}\mathcal{E}_{\kappa-1,q}(z), \tag{62} $$

$$ D_{q,z}^{(k)}\,{}_{\mathcal{H}}\mathcal{E}_{\kappa,q}(z) = \frac{[2]_q^k[\kappa]_q!}{[\kappa-k]_q!}\,{}_{\mathcal{H}}\mathcal{E}_{\kappa-k,q}(z). \tag{63} $$

Proof. Using a similar approach used in the proof of Theorem 6, we are led to the assertions in Equations (62) and (63). □

3.3. q-Hermite–Genocchi Polynomials

Since, for $\mathcal{A}_q(\tau) = \frac{[2]_q\tau}{e_q(\tau)+1}$, the qAP $\mathcal{A}_{\kappa,q}(z)$ reduce to the qGP $\mathcal{G}_{\kappa,q}(z)$ (Table 1(III)) and for $\phi_q(\tau) = e_q(-\tau^2), H(t) = [2]_q\tau$ the qSP $s_{\kappa,q}(z)$ reduce to qHP $\mathcal{H}_{\kappa,q}(z)$, for the same choices of $A_q(\tau), \phi_q(\tau)$ and $H(\tau)$, the qSAP ${}_s\mathcal{A}_{\kappa,q}(z)$ reduce to qHGP ${}_{\mathcal{H}}\mathcal{G}_{\kappa,q}(z)$ which in view of Equation (22) can be defined by means of following generating functions:

$$ \frac{[2]_q\,\tau}{e_q(\tau)+1}\,e_q([2]_qz\tau)e_q(-\tau^2) = \sum_{\kappa=0}^{\infty}{}_{\mathcal{H}}\mathcal{G}_{\kappa,q}(z)\,\frac{\tau^{\kappa}}{[\kappa]_q!}. \tag{64} $$

In view of Equation (26), the qHGP ${}_{\mathcal{H}}\mathcal{G}_{\kappa,q}(z)$ of degree κ are defined by the series:

$$ {}_{\mathcal{H}}\mathcal{G}_{\kappa,q}(z) = \sum_{\nu=0}^{\kappa}\begin{bmatrix}\kappa\\ \nu\end{bmatrix}_q \mathcal{G}_{\nu,q}\mathcal{H}_{\kappa-\nu,q}(z). \tag{65} $$

In view of Equation (48), the following identity for the qHGP ${}_{\mathcal{H}}\mathcal{G}_{\kappa,q}(z)$ holds true:

$$ {}_{\mathcal{H}}\mathcal{G}_{\kappa,q}(z) = \frac{1}{\mathcal{B}_{0,q}}\left(\mathcal{H}_{\kappa,q}(z) - \sum_{\nu=0}^{\kappa-1}\begin{bmatrix}\kappa\\ \nu\end{bmatrix}_q \mathcal{B}_{\kappa-\nu,q}\, {}_{\mathcal{H}}\mathcal{G}_{\nu,q}(z)\right), \quad \kappa = 1,2,\ldots. \tag{66} $$

Further, by taking $s_{\kappa,q}(z) = \mathcal{H}_{\kappa,q}(z)$, $\mathcal{B}_{0,q} = 1$ and $\mathcal{B}_{j,q} = \frac{1}{2[j+1]_q} (j = 1,2,3,...)$ in Equations (36) and (37), we obtain the determinant definition of the qHGP ${}_{\mathcal{H}}\mathcal{G}_{\kappa,q}(z)$ given as:

Definition 3. *The* q*-Hermite–Genocchi polynomials* ${}_{\mathcal{H}}\mathcal{G}_{\kappa,q}(z)$ *of degree* κ *are defined by*

$$ {}_{\mathcal{H}}\mathcal{G}_{0,q}(z) = 1, \tag{67}$$

$$ {}_{\mathcal{H}}\mathcal{G}_{\kappa,q}(z) = (-1)^{\kappa} \begin{vmatrix} 1 & \mathcal{H}_{1,q}(z) & \mathcal{H}_{2,q}(z) & \dots & \mathcal{H}_{\kappa-1,q}(z) & \mathcal{H}_{1,q}(z) \\ 1 & \frac{1}{2[2]_q} & \frac{1}{2[3]_q} & \dots & \frac{1}{2[\kappa]_q} & \frac{1}{2[\kappa+1]_q} \\ 0 & 1 & \begin{bmatrix} 2 \\ 1 \end{bmatrix}_q \frac{1}{2[2]_q} & \dots & \begin{bmatrix} \kappa-1 \\ 1 \end{bmatrix}_q \frac{1}{2[\kappa-1]_q} & \begin{bmatrix} \kappa \\ 1 \end{bmatrix}_q \frac{1}{2[\kappa]_q} \\ 0 & 0 & 1 & \dots & \begin{bmatrix} \kappa-1 \\ 2 \end{bmatrix}_q \frac{1}{2[\kappa-2]_q} & \begin{bmatrix} \kappa \\ 2 \end{bmatrix}_q \frac{1}{2[\kappa-1]_q} \\ . & . & . & \dots & . & . \\ . & . & . & \dots & . & . \\ 0 & 0 & 0 & \dots & 1 & \begin{bmatrix} \kappa \\ \kappa-1 \end{bmatrix}_q \frac{1}{2[2]_q} \end{vmatrix}, \quad \kappa = 1,2,3,..., \tag{68}$$

where $\mathcal{H}_{\kappa,q}(z) (\kappa = 0,1,2,3,...)$ *are the* q*-Hermite polynomials of degree* κ.

Theorem 8. *The* q*-Hermite–Genocchi polynomials* ${}_{\mathcal{H}}\mathcal{G}_{\kappa,q}(z)$ *satisfy the following* q*-recurrence relations:*

$$ D_{q,z}\,{}_{\mathcal{H}}\mathcal{G}_{\kappa,q}(z) = [2]_q [\kappa]_q\,{}_{\mathcal{H}}\mathcal{G}_{\kappa-1,q}(z), \tag{69}$$

$$ D_{q,z}^{(k)}\,{}_{\mathcal{H}}\mathcal{G}_{\kappa,q}(z) = \frac{[2]_q^k [\kappa]_q!}{[\kappa-k]_q!}\,{}_{\mathcal{H}}\mathcal{G}_{\kappa-k,q}(z). \tag{70}$$

Proof. Using a similar approach used in the proof of Theorem 6, we are led to the assertions in Equations (69) and (70). □

In the next section, we introduce a new class of the 2D q-Sheffer–Appell polynomials by means of generating function and series representation.

4. 2D q-Sheffer–Appell Polynomials

Recently, Keleshteri and Mahmudov [27] introduced the 2D q-Appell polynomials (2DqAP) $\{\mathcal{A}_{\kappa,q}(z_1,z_2)\}_{\kappa=0}^{\infty}$, which are defined by means of the generating functions:

$$ \mathcal{A}_q(\tau)\, e_q(z_1\tau) E_q(z_2\tau) = \sum_{\kappa=0}^{\infty} \mathcal{A}_{\kappa,q}(z_1,z_2) \frac{\tau^{\kappa}}{[\kappa]_q!}, \quad 0 < q < 1, \tag{71}$$

where

$$ \mathcal{A}_q(\tau) = \sum_{\kappa=0}^{\infty} \mathcal{A}_{\kappa,q} \frac{\tau^{\kappa}}{[\kappa]_q!}, \quad \mathcal{A}_q(\tau) \neq 0; \ \mathcal{A}_{0,q} = 1 \tag{72}$$

and $\mathcal{A}_{\kappa,q} := \mathcal{A}_{\kappa,q}(0,0)$ denotes the 2D q-Appell numbers.

Some members of the 2D q-Appell polynomials are listed in Table 2.

The approach used in the previous section is further exploited to introduce the 2D q-Sheffer–Appell polynomials (2DqSAP) and the focus is on deriving its generating functions and series definitions.

Table 2. Some members of 2D q-Appell polynomials.

S. No.	$A_q(\tau)$	Generating Functions	Polynomials
I.	$A_q(\tau)=\frac{\tau}{(e_q(\tau)-1)}$	$\frac{\tau}{(e_q(\tau)-1)}e_q(z_1\tau)E_q(z_2\tau)=\sum_{\kappa=0}^{\infty}\mathfrak{B}_{\kappa,q}(z_1,z_2)\frac{\tau^\kappa}{[\kappa]_q!}$	The 2D q-Bernoulli polynomials [21,28]
II.	$A_q(\tau)=\frac{[2]_q}{(e_q(\tau)+1)}$	$\frac{[2]_q}{(e_q(\tau)+1)}e_q(z_1\tau)E_q(z_2\tau)=\sum_{\kappa=0}^{\infty}\mathcal{E}_{\kappa,q}(z_1,z_2)\frac{\tau^\kappa}{[\kappa]_q!}$	The 2D q-Euler polynomials [21,28]
III.	$A_q(\tau)=\frac{[2]_q\tau}{(e_q(\tau)+1)}$	$\frac{[2]_q\tau}{(e_q(\tau)+1)}e_q(z_1\tau)E_q(z_2\tau)=\sum_{\kappa=0}^{\infty}\mathcal{G}_{\kappa,q}(z_1,z_2)\frac{\tau^\kappa}{[\kappa]_q!}$,	The 2D q-Genocchi polynomials [21,28]

To establish the generating function for the 2DqSAP, the following result is proved:

Theorem 9. *The following generating function for the 2D q-Sheffer–Appell polynomials ${}_{s}\mathcal{A}_{\kappa,q}(z_1,z_2)$ holds true:*

$$\mathcal{A}_q(\tau)\phi_q(\tau)\,e_q(z_1H(\tau))E_q(z_2\tau)=\sum_{\kappa=0}^{\infty}{}_{s}\mathcal{A}_{\kappa,q}(z_1,z_2)\,\frac{\tau^\kappa}{[\kappa]_q!}. \tag{73}$$

Proof. By expanding the first q-exponential function $e_q(z_1\tau)$ in the left hand side of Equation (71) and then replacing the powers of z_1 i.e., $z_1^0, z_1, z_1^2, ..., z_1^\kappa$ by the corresponding polynomials $s_{0,q}(z_1), s_{1,q}(z_1), s_{2,q}(z_1), ..., s_{\kappa,q}(z_1)$ in the left hand side and z_1 by $s_{1,q}(z_1)$ in the right hand side of the resultant equation, we have

$$\mathcal{A}_q(\tau)\left(1+s_{1,q}(z_1)\frac{\tau}{[1]_q!}+s_{2,q}(z_1)\frac{\tau^2}{[2]_q!}+...+s_{\kappa,q}(z_1)\frac{\tau^\kappa}{[\kappa]_q!}+...\right)E_q(z_2\tau)=\sum_{\kappa=0}^{\infty}\mathcal{A}_{\kappa,q}(s_{1,q}(z_1),z_2)\,\frac{\tau^\kappa}{[\kappa]_q!}. \tag{74}$$

Further, summing up the series in left hand side and then using Equation (18) in the resultant equation, we get

$$\mathcal{A}_q(\tau)\phi_q(\tau)\,e_q(z_1H(\tau))E_q(z_2\tau)=\sum_{\kappa=0}^{\infty}\mathcal{A}_{\kappa,q}(s_{1,q}(z_1),z_2)\,\frac{\tau^\kappa}{[\kappa]_q!}. \tag{75}$$

Finally, denoting the resultant qSAP in the right hand side of the above equation by ${}_{s}\mathcal{A}_{\kappa,q}(z_1,z_2)$, that is

$$\mathcal{A}_{\kappa,q}(s_{1,q}(z_1),z_2)={}_{s}\mathcal{A}_{\kappa,q}(z_1,z_2), \tag{76}$$

the assertion in Equation (22) is proved. □

Theorem 10. *The 2D q-Sheffer–Appell polynomials ${}_{s}\mathcal{A}_{\kappa,q}(z_1,z_2)$ are defined by the following series definitions:*

$${}_{s}\mathcal{A}_{\kappa,q}(z_1,z_2)=\sum_{\nu=0}^{\kappa}\begin{bmatrix}\kappa\\ \nu\end{bmatrix}_q q^{\frac{\nu(\nu-1)}{2}}z_2^{\nu}\,{}_{s}\mathcal{A}_{\kappa,q}(z_1). \tag{77}$$

Proof. Using Equations (11) and (1) in Equation (73), we get

$$\sum_{\kappa=0}^{\infty}{}_{s}\mathcal{A}_{\kappa,q}(z_1)\frac{\tau^\kappa}{[\kappa]_q!}\sum_{\nu=0}^{\infty}q^{\frac{\nu(\nu-1)}{2}}z_2^{\nu}\frac{\tau^\nu}{[\nu]_q!}=\sum_{\kappa=0}^{\infty}{}_{s}\mathcal{A}_{\kappa,q}(z_1,z_2)\,\frac{\tau^\kappa}{[\kappa]_q!}. \tag{78}$$

Now, using the Cauchy product rule in the left hand side of the above equation and then equating the coefficients of like powers of τ in both sides of the resultant equation, we get the assertion in Equation (77). □

Since for $\phi_q(\tau)=e_q(-\tau^2), H(\tau)=[2]_q\tau$ the qSP $s_{\kappa,q}(z)$ reduce to qHP $\mathcal{H}_{\kappa,q}(z)$, by making same choices for the functions $\phi_q(\tau)$ and $H(\tau)$ in Equations (73) and (77), we get

$$\mathcal{A}_q(\tau)e_q([2]_q z_1\tau)e_q(-\tau^2)E_q(z_2\tau) = \sum_{\kappa=0}^{\infty} {}_{\mathcal{H}}\mathcal{A}_{\kappa,q}(z_1,z_2)\,\frac{\tau^\kappa}{[\kappa]_q!}, \tag{79}$$

$$ {}_{\mathcal{H}}\mathcal{A}_{\kappa,q}(z_1,z_2) = \sum_{\nu=0}^{\kappa} \begin{bmatrix}\kappa\\ \nu\end{bmatrix}_q q^{\frac{\nu(\nu-1)}{2}} z_2^{\nu}\,{}_{\mathcal{H}}\mathcal{A}_{\kappa,q}(z_1). \tag{80}$$

Certain members belonging to the 2D q-Appell family are given in Table 2. By making suitable choices for the functions $\mathcal{A}_q(t)$ in Equations (79) and (80), the generating functions and series definitions for the corresponding member belonging to the 2D q-Hermite–Appell family can be obtained. The resultant 2D q-Hermite–Appell polynomials (2DqHAP) along with their generating functions and series definitions are given in Table 3.

Table 3. Certain members belonging to the 2DqHAP ${}_{\mathcal{H}}\mathcal{A}_{\kappa,q}(z_1,z_2)$.

S. No.	$A_q(\tau)$	Generating Functions	Series Definition	Polynomials
I.	$\frac{\tau}{(e_q(\tau)-1)}$	$\frac{\tau}{(e_q(\tau)-1)}e_q([2]_q z_1\tau)e_q(-\tau^2)E_q(z_2\tau) = \sum_{\kappa=0}^{\infty} {}_{\mathcal{H}}\mathfrak{B}_{\kappa,q}(z_1,z_2)\frac{\tau^\kappa}{[\kappa]_q!}$	${}_{\mathcal{H}}\mathfrak{B}_{\kappa,q}(z_1,z_2) = \sum_{\nu=0}^{\kappa}\begin{bmatrix}\kappa\\ \nu\end{bmatrix}_q q^{\frac{\nu(\nu-1)}{2}} z_2^{\nu}\,{}_{\mathcal{H}}\mathfrak{B}_{\kappa-\nu,q}(z_1)$	The 2D q-Hermite–Bernoulli polynomials
II.	$\frac{[2]_q}{(e_q(\tau)+1)}$	$\frac{[2]_q}{(e_q(\tau)+1)}e_q([2]_q z_1\tau)e_q(-\tau^2)E_q(z_2\tau) = \sum_{\kappa=0}^{\infty} {}_{\mathcal{H}}\mathcal{E}_{\kappa,q}(z_1,z_2)\frac{\tau^\kappa}{[\kappa]_q!}$	${}_{\mathcal{H}}\mathcal{E}_{\kappa,q}(z_1,z_2) = \sum_{\nu=0}^{\kappa}\begin{bmatrix}\kappa\\ \nu\end{bmatrix}_q q^{\frac{\nu(\nu-1)}{2}} z_2^{\nu}\,{}_{\mathcal{H}}\mathcal{E}_{\kappa-\nu,q}(z_1)$	The 2D q-Hermite–Euler polynomials
III.	$\frac{[2]_q\tau}{(e_q(\tau)+1)}$	$\frac{[2]_q\tau}{(e_q(\tau)+1)}e_q([2]_q z_1\tau)e_q(-\tau^2)E_q(z_2\tau) = \sum_{\kappa=0}^{\infty} {}_{\mathcal{H}}\mathcal{G}_{\kappa,q}(z_1,z_2)\frac{\tau^\kappa}{[\kappa]_q!}$,	${}_{\mathcal{H}}\mathcal{G}_{\kappa,q}(z_1,z_2) = \sum_{\nu=0}^{\kappa}\begin{bmatrix}\kappa\\ \nu\end{bmatrix}_q q^{\frac{\nu(\nu-1)}{2}} z_2^{\nu}\,{}_{\mathcal{H}}\mathcal{G}_{\kappa-\nu,q}(z_1)$	The 2D q-Hermite–Genocchi polynomials

5. Graphical Representation

In this section, the shapes of some members of the q-Sheffer–Appell polynomials and 2D q-Sheffer–Appell polynomials are displayed with the help of Matlab.

To draw the graphs of qHBP ${}_{\mathcal{H}}\mathfrak{B}_{\kappa,q}(z)$, qHEP ${}_{\mathcal{H}}\mathcal{E}_{\kappa,q}(z)$ and qHGP ${}_{\mathcal{H}}\mathcal{G}_{\kappa,q}(z)$, we considered the first four values of q-Hermite polynomials $\mathcal{H}_{\kappa,q}(z)$ [25]; the expressions of these polynomials are listed in Table 4.

Table 4. Expressions of the first four $\mathcal{H}_{\kappa,q}(z)$.

κ	0	1	2	3
$\mathcal{H}_{\kappa,q}(z)$	1	$[2]_q z$	$[2]_q^2 z^2 - [2]_q$	$[2]_q^3 z^3 - [3]_q[2]_q^2 z$

Next, setting $\kappa = 3$ in the determinant definitions in Equations (53), (61) and (68), we have

$${}_{\mathcal{H}}\mathfrak{B}_{3,q}(z) = (-1)^3 \begin{vmatrix} 1 & \mathcal{H}_{1,q}(z) & \mathcal{H}_{2,q}(z) & \mathcal{H}_{3,q}(z) \\ 1 & \frac{1}{[2]_q} & \frac{1}{[3]_q} & \frac{1}{[4]_q} \\ 0 & 1 & \begin{bmatrix}2\\1\end{bmatrix}_q \frac{1}{[2]_q} & \begin{bmatrix}3\\1\end{bmatrix}_q \frac{1}{[3]_q} \\ 0 & 0 & 1 & \begin{bmatrix}3\\2\end{bmatrix}_q \frac{1}{[2]_q} \end{vmatrix}, \tag{81}$$

$${}_{\mathcal{H}}\mathcal{E}_{3,q}(z) = (-1)^3 \begin{vmatrix} 1 & \mathcal{H}_{1,q}(z) & \mathcal{H}_{2,q}(z) & \mathcal{H}_{3,q}(z) \\ 1 & \frac{1}{2} & \frac{1}{2} & \frac{1}{2} \\ 0 & 1 & \begin{bmatrix}2\\1\end{bmatrix}_q \frac{1}{2} & \begin{bmatrix}3\\1\end{bmatrix}_q \frac{1}{2} \\ 0 & 0 & 1 & \begin{bmatrix}3\\2\end{bmatrix}_q \frac{1}{2} \end{vmatrix} \tag{82}$$

and

$$
{}_{\mathcal{H}}\mathcal{G}_{3,q}(z) = (-1)^3 \begin{vmatrix} 1 & \mathcal{H}_{1,q}(z) & \mathcal{H}_{2,q}(z) & \mathcal{H}_{3,q}(z) \\ 1 & \frac{1}{2[2]_q} & \frac{1}{2[3]_q} & \frac{1}{2[4]_q} \\ 0 & 1 & \begin{bmatrix}2\\1\end{bmatrix}_q \frac{1}{2[2]_q} & \begin{bmatrix}3\\1\end{bmatrix}_q \frac{1}{2[3]_q} \\ 0 & 0 & 1 & \begin{bmatrix}3\\2\end{bmatrix}_q \frac{1}{2[2]_q} \end{vmatrix}. \tag{83}
$$

Now, taking $q = \frac{1}{3}$ and using the expressions of the $\mathcal{H}_{\kappa,q}(z)$ in Table 4, Equations (81)–(83) become

$$
{}_{\mathcal{H}}\mathfrak{B}_{3,\frac{1}{3}}(z) = \frac{64}{27}z^3 - \frac{52}{27}z^2 - \frac{103}{9}z + \frac{1049}{720}, \tag{84}
$$

$$
{}_{\mathcal{H}}\mathcal{E}_{3,\frac{1}{3}}(z) = \frac{64}{27}z^3 - \frac{104}{81}z^2 - \frac{26}{9}z + \frac{17}{18}, \tag{85}
$$

$$
{}_{\mathcal{H}}\mathcal{G}_{3,\frac{1}{3}}(z) = \frac{64}{27}z^3 + \frac{11}{27}z^2 - \frac{931}{324}z - \frac{2129}{5760}. \tag{86}
$$

Similarly, we can obtain the values of ${}_{\mathcal{H}}\mathfrak{B}_{\kappa,q}(z)$, ${}_{\mathcal{H}}\mathcal{E}_{\kappa,q}(z)$ and ${}_{\mathcal{H}}\mathcal{G}_{\kappa,q}(z)$ for $\kappa = 1, 2$ and $q = \frac{1}{3}$ as:
For $\kappa = 2$, we get

$$
{}_{\mathcal{H}}\mathfrak{B}_{2,\frac{1}{3}}(z) = \frac{16}{9}z^2 - \frac{4}{3}z - \frac{199}{156}, \tag{87}
$$

$$
{}_{\mathcal{H}}\mathcal{E}_{2,\frac{1}{3}}(z) = \frac{16}{9}z^2 - \frac{8}{9}z - \frac{3}{2}, \tag{88}
$$

$$
{}_{\mathcal{H}}\mathcal{G}_{2,\frac{1}{3}}(z) = \frac{16}{9}z^2 - \frac{2}{3}z - \frac{931}{624}. \tag{89}
$$

For $\kappa = 1$, we get

$$
{}_{\mathcal{H}}\mathfrak{B}_{1,\frac{1}{3}}(z) = -\frac{3}{4} + \frac{4}{3}z, \tag{90}
$$

$$
{}_{\mathcal{H}}\mathcal{E}_{1,\frac{1}{3}}(z) = -\frac{1}{2} + \frac{4}{3}z, \tag{91}
$$

$$
{}_{\mathcal{H}}\mathcal{G}_{1,\frac{1}{3}}(z) = -\frac{3}{8} + \frac{4}{3}z. \tag{92}
$$

Further, setting $\kappa = 3, q = \frac{1}{3}$ in the series definitions of ${}_{\mathcal{H}}\mathfrak{B}_{\kappa,q}(z_1, z_2)$, ${}_{\mathcal{H}}\mathcal{E}_{\kappa,q}(z_1, z_2)$ and ${}_{\mathcal{H}}\mathcal{G}_{\kappa,q}(z_1, z_2)$ given in Table 3 and using the expressions of ${}_{\mathcal{H}}\mathfrak{B}_{\kappa,q}(z)$, ${}_{\mathcal{H}}\mathcal{E}_{\kappa,q}(z)$ and ${}_{\mathcal{H}}\mathcal{G}_{\kappa,q}(z)$ for $\kappa = 1, 2, 3$ from Equations (84)–(92), we have

$$
\begin{aligned}
{}_{\mathcal{H}}\mathfrak{B}_{3,\frac{1}{3}}(z_1, z_2) = {} & \frac{64}{27}z_1^3 - \frac{52}{27}z_1^2 - \frac{103}{9}z_1 + \frac{1049}{720} + \frac{304}{27}z_1^2 z_2 - \frac{76}{9}z_1 z_2 - \frac{3781}{468}z_2 \\
& - \frac{19}{36}z_2^2 + \frac{76}{81}z_1 z_2^2 + \frac{1}{729}z_2^3,
\end{aligned} \tag{93}
$$

$$
\begin{aligned}
{}_{\mathcal{H}}\mathcal{E}_{3,\frac{1}{3}}(z_1, z_2) = {} & \frac{64}{27}z_1^3 - \frac{104}{81}z_1^2 - \frac{26}{9}z_1 + \frac{17}{18} + \frac{304}{27}z_1^2 z_2 - \frac{152}{27}z_1 z_2 - \frac{19}{2}z_2 - \frac{19}{54}z_2^2 \\
& + \frac{76}{81}z_1 z_2^2 + \frac{1}{729}z_2^3,
\end{aligned} \tag{94}
$$

$$
\begin{aligned}
{}_{\mathcal{H}}\mathcal{G}_{3,\frac{1}{3}}(z_1, z_2) = {} & \frac{64}{27}z_1^3 + \frac{11}{27}z_1^2 - \frac{931}{324}z_1 - \frac{2129}{5760} + \frac{304}{27}z_1^2 z_2 - \frac{38}{9}z_1 z_2 - \frac{17689}{1872}z_2 \\
& - \frac{19}{72}z_2^2 + \frac{76}{81}z_1 z_2^2 + \frac{1}{729}z_2^3.
\end{aligned} \tag{95}
$$

Now, with the help of Matlab and using Equations (52), (60), (67), (84)–(95), we get the following Figures 1–6.

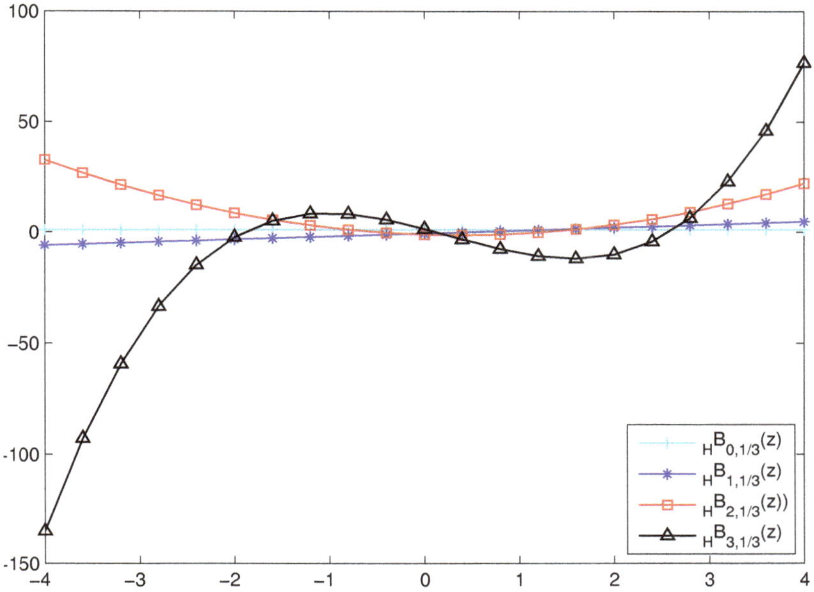

Figure 1. Graph of $_{\mathcal{H}}\mathfrak{B}_{\kappa,\mathrm{q}}(z)$.

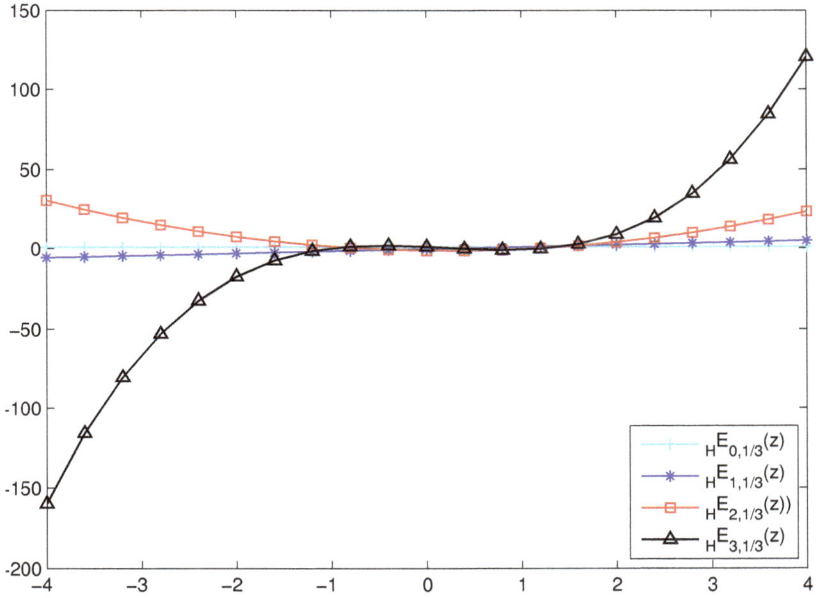

Figure 2. Graph of $_{\mathcal{H}}\mathcal{E}_{\kappa,\mathrm{q}}(z)$.

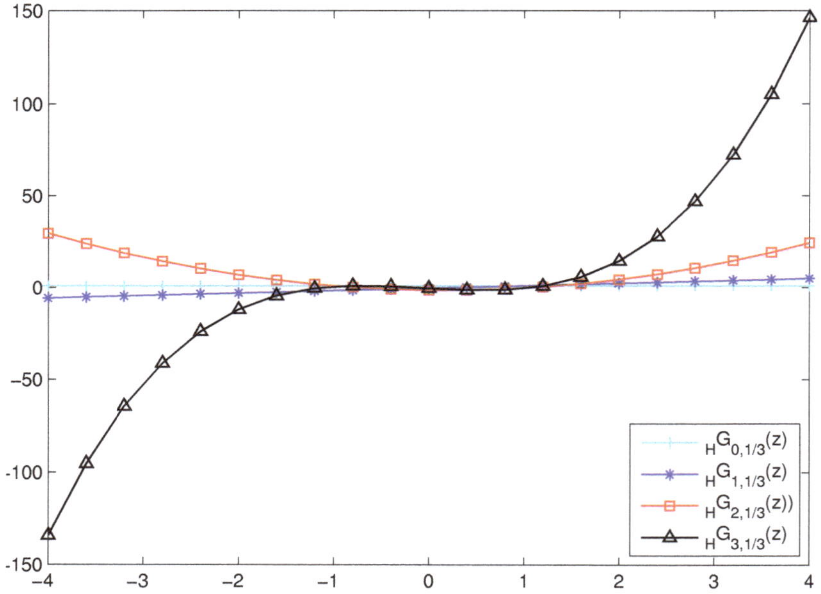

Figure 3. Graph of ${}_{\mathcal{H}}\mathcal{G}_{\kappa,\mathfrak{q}}(z)$.

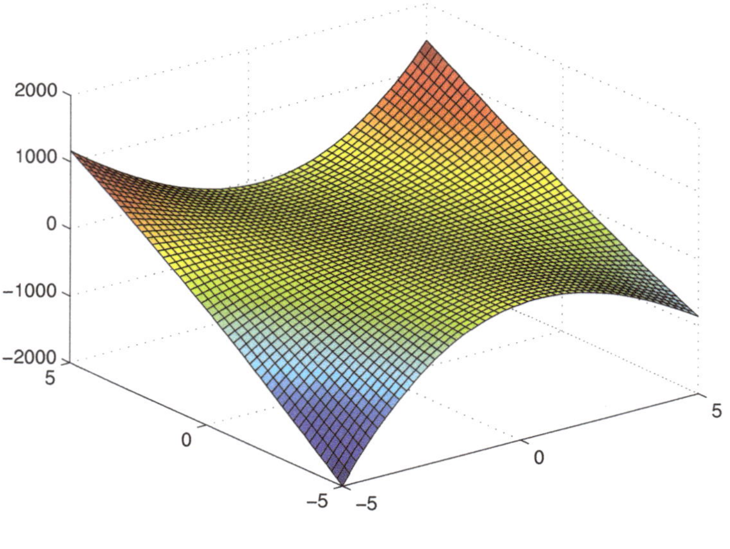

Figure 4. Surface plot of ${}_{\mathcal{H}}\mathfrak{B}_{3,\frac{1}{3}}(z_1, z_2)$.

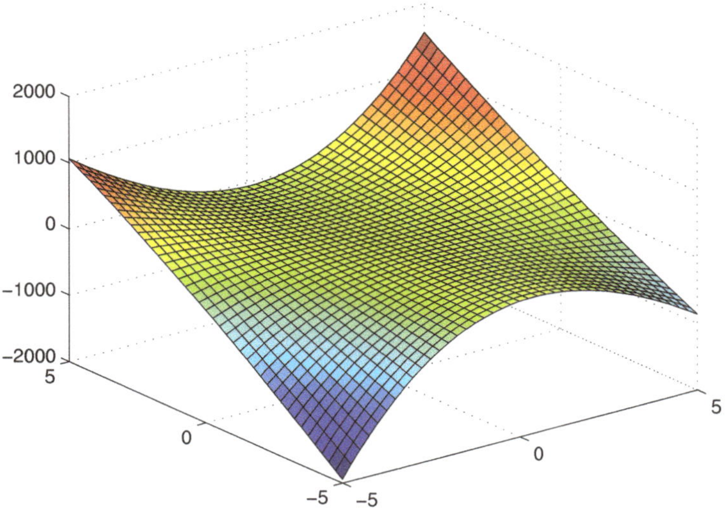

Figure 5. Surface plot of ${}_{\mathcal{H}}\mathcal{E}_{3,\frac{1}{3}}(z_1, z_2)$.

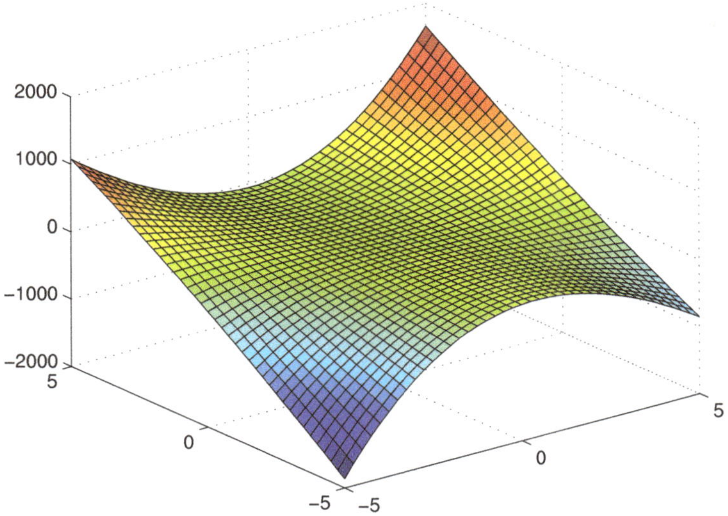

Figure 6. Surface plot of ${}_{\mathcal{H}}\mathcal{G}_{3,\frac{1}{3}}(z_1, z_2)$.

6. Further Remarks

It is worth noting that the results derived in the previous sections can be exploited to establish further new relations.

Let us consider the following relation

$$[2]_q^{-\kappa}\, D_{q,z}^{\kappa} e_q(-[2]_q z\tau) = (-\tau)^{\kappa} e_q(-[2]_q z\tau), \tag{96}$$

which, on replacing κ by 2κ and multiplying both sides of the resultant equation by $\frac{1}{[\kappa]_q!}$, gives

$$\frac{1}{[\kappa]_q!}[2]_q^{-2\kappa}\, D_{q,z}^{2\kappa} e_q(-[2]_q z\tau) = \frac{1}{[\kappa]_q!}(-\tau)^{2\kappa} e_q(-[2]_q z\tau). \tag{97}$$

Now, taking summation on both sides of the above equation and then multiplying both sides of the resultant equation by $\frac{\tau}{e_q(\tau)-1}$ and using Equation (49), we get

$$\sum_{\kappa=0}^{\infty} {}_{\mathcal{H}}\mathfrak{B}_{\kappa,q}(x)\,\frac{\tau^{\kappa}}{[\kappa]_q!} = \frac{\tau}{e_q(\tau)-1}\sum_{\kappa=0}^{\infty}\frac{[2]_q^{-2\kappa}}{[\kappa]_q!}\, D_{q,z}^{2\kappa} e_q([2]_q x\tau), \tag{98}$$

where $x=-z$.

Similarly, we can obtain the following results:

$$\sum_{\kappa=0}^{\infty} {}_{\mathcal{H}}\mathcal{E}_{\kappa,q}(x)\,\frac{\tau^{\kappa}}{[\kappa]_q!} = \frac{[2]_q}{e_q(\tau)+1}\sum_{\kappa=0}^{\infty}\frac{[2]_q^{-2\kappa}}{[\kappa]_q!}\, D_{q,z}^{2\kappa} e_q([2]_q x\tau), \tag{99}$$

$$\sum_{\kappa=0}^{\infty} {}_{\mathcal{H}}\mathcal{G}_{\kappa,q}(x)\,\frac{\tau^{\kappa}}{[\kappa]_q!} = \frac{[2]_q\,\tau}{e_q(\tau)+1}\sum_{\kappa=0}^{\infty}\frac{[2]_q^{-2\kappa}}{[\kappa]_q!}\, D_{q,z}^{2\kappa} e_q([2]_q x\tau), \tag{100}$$

where $x=-z$.

7. Conclusions

We would like to underline that the q-series and q-polynomials have many applications in different fields of mathematics, physics and engineering. In the present article, we demonstrate how a new replacement technique has been adopted to introduce mixed type q-special polynomials and different method to establish their q-recurrence relation.

To extend this new and significant approach, the hybrid class of the q-Sheffer–Appell polynomials and 2D q-Sheffer–Appell polynomials are introduced by means of series expansion and generating functions. The determinant form related to q-Sheffer–Appell polynomials are derived, which are important for the computational and applied purposes. This process can be used to establish further a wide variety of formulas and new relations for several other q-special polynomials.

The q-difference equation for the two iterated q-Appell and mixed type q-Appell polynomials are established in [29,30]. This aspect may be considered in future investigation.

Author Contributions: All authors contributed equally.

Funding: Serkan Araci was supported by the Research Fund of Hasan Kalyoncu University in 2019.

Acknowledgments: The authors are thankful to the reviewer(s) for several useful comments and suggestions towards the improvement of this paper.

Conflicts of Interest: The authors declare no conflict of interest.

References

1. Al-Salam, W.A. q-Appell polynomials. *Ann. Mat. Pura Appl.* **1967**, *4*, 31–45. [CrossRef]
2. Al-Salam, W.A. q-Bernoulli numbers and polynomials. *Math. Nachr.* **1959**, *17*, 239–260. [CrossRef]
3. Araci, S.; Acikgoz, M.; Diagana, T.; Srivastava, H.M. A novel approach for obtaining new identities for the λ extension of q-Euler polynomials arising from the q-umbral calculus. *J. Nonlinear Sci. Appl.* **2017**, *10*, 1316–1325. [CrossRef]

4. Cheon, G.-S.; Jung, J.-H. The q-Sheffer sequences of a new type and associated orthogonal polynomials. *Linear Algebra Appl.* **2016**, *491*, 171–186. [CrossRef]
5. Kim, D.S.; Kim, T.K. q-Bernoulli polynomials and q-umbral calculus. *Sci. China Math.* **2014**, *57*, 1867–1874. [CrossRef]
6. Kim, D.S.; Kim, T.K. Some identities of q-Euler polynomials arising from q-umbral calculus. *J. Inequal. Appl.* **2014**, *2014*, 12. [CrossRef]
7. Kim, T. A note on the q-Genocchi numbers and polynomials. *J. Inequal. Appl.* **2007**, *2007*, 071452. [CrossRef]
8. Kim, D.S.; Kim, T.; Komatsu, T.; Seo, J.-J. An umbral calculus approach to poly-Cauchy polynomials with a q parameter. *J. Comput. Anal. Appl.* **2015**, *18*, 762–792.
9. Kim, D.S.; Kim, T.; Lee, H.Y. p-adic q-integral on Z_p associated with Frobenius-type Eulerian polynomials and umbral calculus. *Adv. Stud. Contemp. Math.* **2013**, *23*, 243–251.
10. Andrews, G.E.; Askey, R.; Roy, R. *71th Special Functions of Encyclopedia of Mathematics and Its Applications*; Cambridge University Press: Cambridge, UK, 1999.
11. Aral, A.; Gupta, V.; Agarwal, R.P. *Applications of q-Calculus in Operator Theory*; Springer: New York, NY, USA, 2013.
12. Ernst, T. *A Comprehensive Treatment of q-Calculus*; Springer: Basel, Switzerland; Berlin/Heidelberg, Germany; New York, NY, USA; Dordrecht, The Netherlands; London, UK, 2012.
13. Appell, P. Sur une classe de polynômes. *Ann. Sci. Éc. Norm. Super.* **1880**, *9*, 119–144. [CrossRef]
14. Thorne, C.J. A property of Appell sets. *Am. Math. Mon.* **1945**, *52*, 191–193. [CrossRef]
15. Sheffer, I.M. Note on Appell polynomials. *Bull. Am. Math. Soc.* **1945**, *51*, 739–744. [CrossRef]
16. Varma, R.S. On Appell polynomials. *Proc. Am. Math. Soc.* **1951**, *2*, 593–596. [CrossRef]
17. Sharma, A.; Chak, A.M. The basic analogue of a class of polynomials. *Ann. Probab. Riv. Mat. Univ. Parma* **1954**, *5*, 325–337.
18. Ernst, T. q-Bernoulli and q-Euler polynomials, an umbral approach. *Int. J. Differ. Equ.* **2006**, *1*, 31–80.
19. Acikgoz, M.; Araci, S.; Duran, U. New extensions of some known special polynomials under the theory of multiple q-calculus. *Turkish J. Anal. Number Theory* **2015**, *3*, 128–139. [CrossRef]
20. Kim, T. q-Generalized Euler numbers and polynomials. *Russ. J. Math. Phys.* **2006**, *13*, 293–298. [CrossRef]
21. Mahmudov, N.I. On a class of q-Bernoulli and q-Euler polynomials. *Adv. Differ. Equ.* **2013**, *4*, 108. [CrossRef]
22. Roman, S.; Rota, G. The umbral calculus. *Adv. Math.* **1978**, *27*, 95–188. [CrossRef]
23. Sheffer, I.M. Some properties of polynomial sets of type zero. *Duke Math. J.* **1939**, *5*, 590–622. [CrossRef]
24. Roman, S. More on the umbral calculus, with emphasis on the q-umbral calculus. *J. Math. Anal. Appl.* **1985**, *107*, 222–254. [CrossRef]
25. Duran, U.; Acikgoz, M.; Esi, A.; Araci, S. A Note on the (p,q)-Hermite Polynomials. *Appl. Math. Inf. Sci.* **2018**, *12*, 227–231. [CrossRef]
26. Hardy, G.H. *Divergent Series*; American Mathematical Society: Providence, RI, USA, 2000; Volume 334.
27. Keleshteri, M.E.; Mahmudov, N.I. A study on q-Appell polynomials from determinant point of view. *Appl. Math. Comput.* **2015**, *260*, 351–369.
28. Carlitz, L. q-Bernoulli numbers and polynomials. *Duke Math. J.* **1948**, *15*, 987–1000. [CrossRef]
29. Riyasat, M.; Khan, S.; Nahid, T. q-difference equations for the composite 2D q-Appell polynomials and their applications. *Cogent Math.* **2017**, *4*, 1376972. [CrossRef]
30. Srivastava, H.M.; Khan, S.; Riyasat, M. q-Difference equations for the 2-Iterated q-Appell and mixed type q-Appell Polynomials. *Arab. J. Math.* **2018**, *5*, 1–15. [CrossRef]

© 2019 by the authors. Licensee MDPI, Basel, Switzerland. This article is an open access article distributed under the terms and conditions of the Creative Commons Attribution (CC BY) license (http://creativecommons.org/licenses/by/4.0/).

symmetry

MDPI

Article

The Order of Strongly Starlikeness of the Generalized α-Convex Functions

Yuan Yuan [1], Rekha Srivastava [2,*] and Jin-Lin Liu [3,*]

1 Department of Mathematics, Maanshan Teacher's College, Maanshan 243000, China; shurong123@163.com
2 Department of Mathematics and Statistics, University of Victoria, British Columbia, VIC V8W 3R4, Canada
3 Department of Mathematics, Yangzhou University, Yangzhou 225002, China
* Correspondence: rekhas@math.uvic.ca (R.S.); jlliu@yzu.edu.cn (J.-L.L.)

Received: 28 December 2018; Accepted: 10 January 2019; Published: 11 January 2019

Abstract: We consider the order of the strongly-starlikeness of the generalized α-convex functions. Some sufficient conditions for functions to be p-valently strongly-starlike are given.

Keywords: analytic; α-convex function; starlike function; strongly-starlike function; subordination

MSC: 30C45; 30C80

---

1. Introduction

Let $\mathbb{N}$, $\mathbb{R}$ and $\mathbb{C}$ denote the sets of positive integers, real numbers and complex numbers, respectively.

Definition 1. *A function f is called p-valent in a domain $\mathbb{D} \subset \mathbb{C}$ if the equation $f(z) = w$ has at most p roots in $\mathbb{D}$ for every complex number w, and there is a complex number w_0 such that $f(z) = w_0$ has exactly p roots in $\mathbb{D}$.*

Let $\mathcal{A}(p)$ denote the class of analytic functions in $\mathbb{U} = \{z : z \in \mathbb{C} \quad \text{and} \quad |z| < 1\}$ of the form:

$$f(z) = z^p + \sum_{n=1}^{\infty} a_{p+n} z^{p+n} \quad (p \in \mathbb{N}). \tag{1}$$

For $p = 1$, we denote $\mathcal{A} := \mathcal{A}(1)$.

Definition 2. *A function $f \in \mathcal{A}(p)$ is said to be p-valently starlike in $\mathbb{U}$ if it satisfies:*

$$\text{Re}\left\{\frac{zf'(z)}{f(z)}\right\} > 0 \quad (z \in \mathbb{U}). \tag{2}$$

We denote by $\mathcal{S}_p^*$ the subclass of $\mathcal{A}(p)$ consisting of all p-valently starlike functions in $\mathbb{U}$.

Definition 3. *If $f \in \mathcal{A}(p)$ satisfies:*

$$\left|\arg\left\{\frac{zf'(z)}{f(z)}\right\}\right| < \frac{\beta\pi}{2} \quad (z \in \mathbb{U}) \tag{3}$$

for some $\beta \in (0, 1]$, then the function f is called p-valently strongly-starlike of order β in $\mathbb{U}$. We denote this class by $\mathcal{SS}_p^*(\beta)$.

For $p = 1$, the class $\mathcal{SS}_1^*(\beta)$ was introduced by Brannan and Kirwan [1]. It is clear that:

$$\mathcal{SS}_p^*(\beta) \subset \mathcal{S}_p^* \quad \text{and} \quad \mathcal{SS}_p^*(1) = \mathcal{S}_p^*.$$

The strongly-starlike functions and related functions have been extensively studied by several authors (see, e.g., [1–16]).

We say that for functions f and g analytic in $\mathbb{U}$, g is subordinate to f, written $g \prec f$, if there exists a Schwarz function w such that $g(z) = f(w(z))$ for $z \in \mathbb{U}$. In particular, if f is univalent in $\mathbb{U}$, then:

$$g(z) \prec f(z) \quad (z \in \mathbb{U}) \Longleftrightarrow g(0) = f(0) \quad \text{and} \quad g(\mathbb{U}) \subset f(\mathbb{U}).$$

In [17], Mocanu first introduced the class:

$$\mathcal{M}(\alpha) = \left\{ f \in \mathcal{A} : \operatorname{Re}\left\{ \alpha\left(1 + \frac{zf''(z)}{f'(z)}\right) + (1-\alpha)\frac{zf'(z)}{f(z)} \right\} > 0, \quad \alpha \in \mathbb{R}, \quad z \in \mathbb{U} \right\} \tag{4}$$

of α-convex functions, which give a continuous passage from convex to starlike functions. He proved that every α-convex function is starlike. Recently, Nunokawa, Sokól and Trabka-Wieclaw [8] considered the generalized α-convex function class:

$$\mathcal{M}(\alpha,\beta) = \left\{ f \in \mathcal{A} : \left| \arg\left\{ \alpha\left(1 + \frac{zf''(z)}{f'(z)}\right) + (1-\alpha)\frac{zf'(z)}{f(z)} \right\} \right| < \frac{\beta\pi}{2}, \quad \alpha \in \mathbb{R}, \quad \beta \in (0,1], \ z \in \mathbb{U} \right\}.$$

In this paper, we shall further study the properties of the generalized α-convex functions. Several sufficient conditions for functions to be p-valently strongly starlike are obtained.

The following lemmas will be required in our investigation.

Lemma 1 (See [18]). *Let g be analytic and univalent in $\mathbb{U}$. Furthermore, let θ and φ be analytic in a domain $\mathbb{D} \supseteq g(\mathbb{U})$ with $\varphi(w) \neq 0$ for $w \in g(\mathbb{U})$. Put:*

$$Q(z) = zg'(z)\varphi(g(z)) \quad \text{and} \quad h(z) = \theta(g(z)) + Q(z)$$

and suppose that

(i) Q is univalent starlike in $\mathbb{U}$ and

(ii) $\operatorname{Re}\left\{\frac{zh'(z)}{Q(z)}\right\} = \operatorname{Re}\left\{\frac{\theta'(g(z))}{\varphi(g(z))} + \frac{zQ'(z)}{Q(z)}\right\} > 0 \quad (z \in \mathbb{U})$.

If q is analytic in $\mathbb{U}$ with $q(0) = g(0)$, $q(\mathbb{U}) \subset \mathbb{D}$ and:

$$\theta(q(z)) + zq'(z)\varphi(q(z)) \prec \theta(g(z)) + zg'(z)\varphi(g(z)) = h(z) \quad (z \in \mathbb{U}), \tag{5}$$

then $q(z) \prec g(z)$ $(z \in \mathbb{U})$. The function g is the best dominant of (5).

Lemma 2 (See [19]). *Let $p(z)$ be an analytic function in $\mathbb{U}$ of the form:*

$$p(z) = 1 + \sum_{n=m}^{\infty} c_n z^n, \quad c_m \neq 0, \quad m \geq 1,$$

with $p(z) \neq 0$ in $\mathbb{U}$. If there exists a point z_0, $|z_0| < 1$, such that:

$$|\arg\{p(z)\}| < \frac{\pi}{2}$$

for $|z| < |z_0|$ and:

$$|\arg\{p(z_0)\}| = \frac{\pi}{2},$$

then:

$$\frac{z_0 p'(z_0)}{p(z_0)} = il,$$

where:

$$l \geq \frac{m}{2}\left(a+\frac{1}{a}\right) \quad \text{when} \quad \arg\{p(z_0)\} = \frac{\pi}{2}$$

and:

$$l \leq -\frac{m}{2}\left(a+\frac{1}{a}\right) \quad \text{when} \quad \arg\{p(z_0)\} = -\frac{\pi}{2},$$

where $p(z_0) = \pm ia$, $a > 0$.

2. Main Results

Theorem 1. *Let* $\lambda_0, \lambda, \beta, a \in \mathbb{R}$ *satisfy* $\lambda \geq 0$, $\lambda_0 a \geq 0$, $0 < \beta \leq 1$ *and* $|a| \leq \frac{1}{\beta}$. *If* q *is analytic in* $\mathbb{U}$ *with* $q(0) = 1$ *and satisfies:*

$$\lambda_0(q(z))^a + \lambda q(z) + \frac{zq'(z)}{q(z)} \prec h(z) \quad (z \in \mathbb{U}), \tag{6}$$

where:

$$h(z) = \lambda_0\left(\frac{1+z}{1-z}\right)^{a\beta} + \lambda\left(\frac{1+z}{1-z}\right)^{\beta} + \frac{2\beta z}{1-z^2} \tag{7}$$

is (close-to-convex) univalent in $\mathbb{U}$, *then:*

$$|\arg\{q(z)\}| < \frac{\beta\pi}{2} \quad (z \in \mathbb{U}). \tag{8}$$

The bound β *in (8) is sharp for the function* q *defined by:*

$$q(z) = \left(\frac{1+z}{1-z}\right)^{\beta}. \tag{9}$$

Proof. We choose:

$$g(z) = \left(\frac{1+z}{1-z}\right)^{\beta}, \quad \theta(w) = \lambda_0 w^a + \lambda w \quad \text{and} \quad \varphi(w) = \frac{1}{w}$$

in Lemma 1. Then, the function g is analytic and convex univalent in $\mathbb{U}$ and:

$$|\arg\{g(z)\}| < \frac{\beta\pi}{2} \quad (z \in \mathbb{U}). \tag{10}$$

It is clear that φ and θ are analytic in a domain $\mathbb{D}$, which contains $g(\mathbb{U})$ and $q(\mathbb{U})$ with $\varphi(w) \neq 0$ for $w \in g(\mathbb{U})$. The function Q given by:

$$Q(z) = zg'(z)\varphi(g(z)) = \frac{2\beta z}{1-z^2}$$

is univalent starlike. Further, we have:

$$\begin{aligned}\theta(g(z)) + Q(z) &= \lambda_0\left(\frac{1+z}{1-z}\right)^{a\beta} + \lambda\left(\frac{1+z}{1-z}\right)^{\beta} + \frac{2\beta z}{1-z^2} \\ &= h(z),\end{aligned}$$

and so:

$$\frac{zh'(z)}{Q(z)} = \frac{\theta'(g(z))}{\varphi(g(z))} + \frac{zQ'(z)}{Q(z)}$$
$$= \lambda_0 a(g(z))^a + \lambda g(z) + \frac{zQ'(z)}{Q(z)}. \quad (11)$$

Furthermore, for $|a| \leq \frac{1}{\beta}$, we find that:

$$|\arg\{(g(z))^a\}| = \frac{|a|\beta\pi}{2} \leq \frac{\pi}{2} \quad (z \in \mathbb{U}). \quad (12)$$

Therefore, it follows from (10)–(12) that:

$$\mathrm{Re}\left\{\frac{zh'(z)}{Q(z)}\right\} > 0 \quad (z \in \mathbb{U}).$$

The other conditions of Lemma 1 are also satisfied. Hence, we conclude that:

$$q(z) \prec g(z) = \left(\frac{1+z}{1-z}\right)^{\beta} \quad (z \in \mathbb{U})$$

and the function g is the best dominant of (6).

Furthermore, for the function q defined by (9), we have:

$$\lambda_0(q(z))^a + \lambda q(z) + \frac{zq'(z)}{q(z)} = h(z)$$

and it follows that the bound β in (8) is sharp. The proof of Theorem 1 is completed. □

Theorem 2. *Let $\alpha > 0, 0 < \beta < 1$ and $\delta > 0$. If $f \in \mathcal{A}(p)$ satisfies $f(z)f'(z) \neq 0$ $(0 < |z| < 1)$ and:*

$$\left|\arg\left\{\alpha\left(1 + \frac{zf''(z)}{f'(z)} - \frac{zf'(z)}{f(z)}\right) + \left(\frac{zf'(z)}{pf(z)}\right)^{\delta}\right\}\right| < \frac{\beta\pi}{2} \quad (z \in \mathbb{U}), \quad (13)$$

then:

$$\left|\arg\left\{\frac{zf'(z)}{f(z)}\right\}\right| < \frac{\beta\pi}{2\delta} \quad (z \in \mathbb{U}). \quad (14)$$

In particular, if $\delta \geq 1$, then f is p-valently strongly starlike of order $\frac{\beta}{\delta}$. The bound $\frac{\beta\pi}{2}$ in (13) is the largest number such that (14) holds true.

Proof. One can see that the condition (13) is a generalization of the condition (4). For $f \in \mathcal{A}(p)$ satisfying $f(z)f'(z) \neq 0$ $(0 < |z| < 1)$, we define the function $p(z)$ by:

$$p(z) = \left(\frac{zf'(z)}{pf(z)}\right)^{\delta} \quad (z \in \mathbb{U}). \quad (15)$$

Then, $p(z)$ is analytic in $\mathbb{U}$ with $p(0) = 1$. The condition (13) becomes:

$$\left|\arg\left\{p(z) + \frac{\alpha}{\delta}\frac{zp'(z)}{p(z)}\right\}\right| < \frac{\beta\pi}{2} \quad (z \in \mathbb{U}). \quad (16)$$

Putting:

$$\lambda_0 = 0 \quad \text{and} \quad \lambda = \frac{\delta}{\alpha}$$

in Theorem 1 and using (16), we find that if:

$$\frac{\alpha}{\delta}\left\{\alpha\left(1+\frac{zf''(z)}{f'(z)}-\frac{zf'(z)}{f(z)}\right)+\left(\frac{zf'(z)}{pf(z)}\right)^{\delta}\right\}$$
$$=\frac{\delta}{\alpha}p(z)+\frac{zp'(z)}{p(z)}\prec h(z), \tag{17}$$

where:

$$h(z)=\frac{\delta}{\alpha}\left(\frac{1+z}{1-z}\right)^{\beta}+\frac{2\beta z}{1-z^2} \tag{18}$$

is (close-to-convex) univalent in $\mathbb{U}$, then (14) is true.

Letting $0<\theta<\pi$ and $x=\cot\frac{\theta}{2}$, we deduce that:

$$\begin{aligned}\arg\{h(e^{i\theta})\} &= \arg\left\{\frac{\delta}{\alpha}\left(\frac{1+e^{i\theta}}{1-e^{i\theta}}\right)^{\beta}+\frac{2\beta e^{i\theta}}{1-e^{2i\theta}}\right\}\\ &= \arg\left\{\frac{\delta}{\alpha}x^{\beta}e^{\frac{\beta\pi i}{2}}+\frac{\beta i}{2}\left(x+\frac{1}{x}\right)\right\}\\ &= \arctan\left\{\frac{\delta x^{\beta}\sin\left(\frac{\beta\pi}{2}\right)+\frac{\alpha\beta}{2}\left(x+\frac{1}{x}\right)}{\delta x^{\beta}\cos\left(\frac{\beta\pi}{2}\right)}\right\}\geq\frac{\beta\pi}{2}.\end{aligned} \tag{19}$$

Hence, in view of $h(e^{-i\theta})=\overline{h(e^{i\theta})}$, we deduce from (19) that $h(\mathbb{U})$ contains the sector $|\arg w|<\frac{\beta\pi}{2}$. Consequently, if $f\in\mathcal{A}(p)$ satisfies (13), then the subordination (17) holds true.

For the function f defined by:

$$f(z)=\exp\left(p\int_0^z\frac{1}{t}\left(\frac{1+t}{1-t}\right)^{\frac{\beta}{\delta}}dt\right)\in\mathcal{A}(p),$$

we find after some computations that f satisfies (14) and:

$$\frac{\alpha}{\delta}\left\{\alpha\left(1+\frac{zf''(z)}{f'(z)}-\frac{zf'(z)}{f(z)}\right)+\left(\frac{zf'(z)}{pf(z)}\right)^{\delta}\right\}=h(z),$$

which shows that the bound $\frac{\beta\pi}{2}$ in (13) is the largest number such that (14) holds true. The proof of Theorem 2 is completed. □

Theorem 3. *Let $\delta>0$ and $\alpha\in\mathbb{C}$. Assume that $-\frac{\pi}{2}<\varphi=\arg\{\alpha\}\leq 0$. If $f\in\mathcal{A}(p)$ satisfies $f(z)f'(z)\neq 0$ $(0<|z|<1)$ and:*

$$-\frac{\pi}{2}+\arctan\left\{\frac{|\alpha|\sin\varphi}{2\delta+|\alpha|\cos\varphi}\right\}<\arg\left\{\alpha\left(1+\frac{zf''(z)}{f'(z)}-\frac{zf'(z)}{f(z)}\right)+\left(\frac{zf'(z)}{pf(z)}\right)^{\delta}\right\}<\frac{\pi}{2}+\varphi \tag{20}$$

for $z\in\mathbb{U}$, then:

$$\left|\arg\left\{\frac{zf'(z)}{f(z)}\right\}\right|<\frac{\pi}{2\delta}\quad(z\in\mathbb{U}).$$

In particular, if $\delta\geq 1$, then f is p-valently strongly starlike of order $\frac{1}{\delta}$.

Proof. Define the function $p(z)$ by (15). Then, the condition (20) becomes:

$$-\frac{\pi}{2}+\arctan\left\{\frac{|\alpha|\sin\varphi}{2\delta+|\alpha|\cos\varphi}\right\}<\arg\left\{p(z)+\frac{\alpha}{\delta}\frac{zp'(z)}{p(z)}\right\}<\frac{\pi}{2}+\varphi \quad (z\in\mathbb{U}). \tag{21}$$

We want to prove that:

$$|\arg\{p(z)\}|<\frac{\pi}{2} \quad (z\in\mathbb{U}). \tag{22}$$

If there exists a point z_0 ($|z_0|<1$) such that:

$$|\arg\{p(z)\}|<\frac{\pi}{2} \quad (|z|<|z_0|)$$

and:

$$|\arg\{p(z_0)\}|=\frac{\pi}{2},$$

then from Lemma 2, we have:

$$\frac{z_0p'(z_0)}{p(z_0)}=il,$$

where $p(z_0)=\pm ai$, $a>0$ and:

$$l\geq\frac{m}{2}\left(a+\frac{1}{a}\right) \quad \text{when} \quad \arg\{p(z_0)\}=\frac{\pi}{2}$$

and:

$$l\leq-\frac{m}{2}\left(a+\frac{1}{a}\right) \quad \text{when} \quad \arg\{p(z_0)\}=-\frac{\pi}{2}.$$

For the case $\arg\{p(z_0)\}=-\frac{\pi}{2}$, we have $l<0$ and:

$$\begin{aligned}\arg\left\{p(z_0)+\frac{\alpha}{\delta}\frac{z_0p'(z_0)}{p(z_0)}\right\}&=\arg\left\{-ai+\frac{l\alpha}{\delta}i\right\}=-\frac{\pi}{2}+\arg\left\{a-\frac{l\alpha}{\delta}\right\}\\&=-\frac{\pi}{2}+\arctan\left\{\frac{\operatorname{Im}\left(a-\frac{l\alpha}{\delta}\right)}{\operatorname{Re}\left(a-\frac{l\alpha}{\delta}\right)}\right\}=-\frac{\pi}{2}+\arctan\left\{\frac{-l|\alpha|\sin\varphi}{a\delta-l|\alpha|\cos\varphi}\right\}\\&\leq-\frac{\pi}{2}+\arctan\left\{\frac{\frac{1}{2}\left(a+\frac{1}{a}\right)|\alpha|\sin\varphi}{a\delta+\frac{1}{2}\left(a+\frac{1}{a}\right)|\alpha|\cos\varphi}\right\}\\&\leq-\frac{\pi}{2}+Q(\alpha,\varphi),\end{aligned} \tag{23}$$

where:

$$Q(\alpha,\varphi)=\max_{a>0}\left\{\arctan\left(\frac{|\alpha|(a^2+1)\sin\varphi}{(2\delta+|\alpha|\cos\varphi)a^2+|\alpha|\cos\varphi}\right)\right\}.$$

The function:

$$g(a)=\frac{|\alpha|(a^2+1)\sin\varphi}{(2\delta+|\alpha|\cos\varphi)a^2+|\alpha|\cos\varphi}, \quad a>0$$

has a positive derivative:

$$g'(a)=\frac{-4a\delta|\alpha|\sin\varphi}{((2\delta+|\alpha|\cos\varphi)a^2+|\alpha|\cos\varphi)^2}\geq 0 \quad \text{for} \quad -\frac{\pi}{2}<\varphi\leq 0,$$

hence:

$$Q(\alpha,\varphi)=\lim_{a\to\infty}\arctan\{g(a)\}=\arctan\frac{|\alpha|\sin\varphi}{2\delta+|\alpha|\cos\varphi}.$$

Therefore, (23) becomes:

$$\arg\left\{p(z_0)+\frac{\alpha}{\delta}\frac{zp'(z_0)}{p(z_0)}\right\}\leq -\frac{\pi}{2}+\arctan\left\{\frac{|\alpha|\sin\varphi}{2\delta+|\alpha|\cos\varphi}\right\},$$

which contradicts (21). Thus,

$$|\arg\{p(z)\}|<\frac{\pi}{2}\quad (z\in\mathbb{U}).$$

For the case $\arg\{p(z_0)\}=\frac{\pi}{2}$, applying the same method as the above, we have $l>0$ and:

$$\arg\left\{p(z_0)+\frac{\alpha}{\delta}\frac{z_0p'(z_0)}{p(z_0)}\right\}=\arg\left\{ai+\frac{l\alpha}{\delta}i\right\}$$
$$=\frac{\pi}{2}+\arg\left\{a+\frac{l\alpha}{\delta}\right\}\geq\frac{\pi}{2}+\varphi.$$

This contradicts (21). Now, the proof of Theorem 3 is completed. □

Remark 1. *For $\varphi=0$ and $\delta=1$, Theorem 3 becomes the known result in [17] that every α-convex function is starlike.*

Applying the same method as the above, we can prove the following theorem.

Theorem 4. *Let $\delta>0$ and $\alpha\in\mathbb{C}$. Assume that $0\leq\varphi=\arg\{\alpha\}<\frac{\pi}{2}$. If $f\in\mathcal{A}(p)$ satisfies $f(z)f'(z)\neq 0$ $(0<|z|<1)$ and:*

$$-\frac{\pi}{2}+\varphi<\arg\left\{\alpha\left(1+\frac{zf''(z)}{f'(z)}-\frac{zf'(z)}{f(z)}\right)+\left(\frac{zf'(z)}{pf(z)}\right)^{\delta}\right\}<\frac{\pi}{2}+\arctan\left\{\frac{|\alpha|\sin\varphi}{2\delta+|\alpha|\cos\varphi}\right\}\tag{24}$$

for $z\in\mathbb{U}$, then:

$$\left|\arg\left\{\frac{zf'(z)}{f(z)}\right\}\right|<\frac{\pi}{2\delta}\quad (z\in\mathbb{U}).$$

In particular, if $\delta\geq 1$, then f is p-valently strongly starlike of order $\frac{1}{\delta}$.

Theorem 5. *Theorem 5. Let $0<\alpha<1$, $0<\beta<1$ and $\beta<\delta\leq 1$. If $f\in\mathcal{A}(p)$ satisfies $f(z)f'(z)\neq 0$ $(0<|z|<1)$ and:*

$$\left|\arg\left\{\alpha\left(1+\frac{zf''(z)}{f'(z)}\right)+(1-\alpha)\left(\frac{zf'(z)}{pf(z)}\right)^{\delta}\right\}\right|<\frac{\beta\pi}{2}\quad (z\in\mathbb{U}),\tag{25}$$

then:

$$\left|\arg\left\{\frac{zf'(z)}{f(z)}\right\}\right|<\frac{\beta\pi}{2\delta}\quad (z\in\mathbb{U}),\tag{26}$$

or f is p-valently strongly starlike of order $\frac{\beta}{\delta}$. The bound $\frac{\beta\pi}{2}$ in (25) is the largest number such that (26) holds true.

Proof. It is obvious that the condition (25) is a generalization of the condition (4). Defining the function $p(z)$ by (15), the condition (25) becomes:

$$\left|\arg\left\{p\alpha(p(z))^{\frac{1}{\delta}}+(1-\alpha)p(z)+\frac{\alpha}{\delta}\frac{zp'(z)}{p(z)}\right\}\right|<\frac{\beta\pi}{2}\quad (z\in\mathbb{U}).\tag{27}$$

Setting:

$$a = \frac{1}{\delta}, \quad \lambda_0 = p\delta \quad \text{and} \quad \lambda = \frac{\delta(1-\alpha)}{\alpha}$$

in Theorem 1 and using (27), we see that if:

$$\begin{aligned} &\frac{\alpha}{\delta}\left\{\alpha\left(1+\frac{zf''(z)}{f'(z)}\right)+(1-\alpha)\left(\frac{zf'(z)}{pf(z)}\right)^{\delta}\right\} \\ &= p\delta(p(z))^{\frac{1}{\delta}} + \frac{\delta(1-\alpha)}{\alpha}p(z) + \frac{zp'(z)}{p(z)} \prec h(z), \end{aligned} \tag{28}$$

where:

$$h(z) = p\delta\left(\frac{1+z}{1-z}\right)^{\frac{\beta}{\delta}} + \frac{\delta(1-\alpha)}{\alpha}\left(\frac{1+z}{1-z}\right)^{\beta} + \frac{2\beta z}{1-z^2} \tag{29}$$

is (close-to-convex) univalent in $\mathbb{U}$, then (26) holds true.

Letting $0 < \theta < \pi$ and $x = \cot\frac{\theta}{2}$, we have:

$$h(e^{i\theta}) = p\delta x^{\frac{\beta}{\delta}} e^{\frac{\beta\pi i}{2\delta}} + \frac{\delta(1-\alpha)}{\alpha}x^{\beta}e^{\frac{\beta\pi i}{2}} + \frac{\beta i}{2}\left(x+\frac{1}{x}\right)$$

and:

$$\arg\left\{h(e^{i\theta})\right\} = \arctan\left\{\frac{p\delta x^{\frac{\beta}{\delta}}\sin\left(\frac{\beta\pi}{2\delta}\right) + \frac{\delta(1-\alpha)}{\alpha}x^{\beta}\sin\left(\frac{\beta\pi}{2}\right) + \frac{\beta}{2}\left(x+\frac{1}{x}\right)}{p\delta x^{\frac{\beta}{\delta}}\cos\left(\frac{\beta\pi}{2\delta}\right) + \frac{\delta(1-\alpha)}{\alpha}x^{\beta}\cos\left(\frac{\beta\pi}{2}\right)}\right\}.$$

For $x > 0$, $0 < \alpha < 1$ and $0 < \frac{\beta}{\delta} < 1$, we deduce that:

$$\begin{aligned} \arg\left\{h(e^{i\theta})\right\} &\geq \arctan\left\{\frac{p\delta x^{\frac{\beta}{\delta}}\sin\left(\frac{\beta\pi}{2\delta}\right) + \frac{\delta(1-\alpha)}{\alpha}x^{\beta}\sin\left(\frac{\beta\pi}{2}\right)}{p\delta x^{\frac{\beta}{\delta}}\cos\left(\frac{\beta\pi}{2\delta}\right) + \frac{\delta(1-\alpha)}{\alpha}x^{\beta}\cos\left(\frac{\beta\pi}{2}\right)}\right\} \\ &= \arctan\left\{\tan\left(\frac{\beta\pi}{2}\right)\frac{p\delta x^{\frac{\beta}{\delta}}\cos\left(\frac{\beta\pi}{2\delta}\right)\tan\left(\frac{\beta\pi}{2\delta}\right)\cot\left(\frac{\beta\pi}{2}\right) + \frac{\delta(1-\alpha)}{\alpha}x^{\beta}\cos\left(\frac{\beta\pi}{2}\right)}{p\delta x^{\frac{\beta}{\delta}}\cos\left(\frac{\beta\pi}{2\delta}\right) + \frac{\delta(1-\alpha)}{\alpha}x^{\beta}\cos\left(\frac{\beta\pi}{2}\right)}\right\} \\ &\geq \frac{\beta\pi}{2}, \end{aligned} \tag{30}$$

since:

$$\tan\left(\frac{\beta\pi}{2\delta}\right)\cot\left(\frac{\beta\pi}{2}\right) \geq \tan\left(\frac{\beta\pi}{2}\right)\cot\left(\frac{\beta\pi}{2}\right) = 1 \quad \text{for} \quad \frac{1}{\delta} \geq 1.$$

In view of the proof of Theorem 2, we find from (30) that $h(\mathbb{U})$ contains the sector $|\arg w| < \frac{\beta\pi}{2}$. Hence, if $f \in \mathcal{A}(p)$ satisfies (25), then the subordination (28) holds true.

The sharpness part of the proof is similar to that in the proof of Theorem 2, and so, we omit it. The proof of Theorem 5 is completed. □

Author Contributions: All authors contributed equally.

Funding: This work is supported by the National Natural Science Foundation of China (Grant No. 11571299).

Acknowledgments: The authors would like to express sincere thanks to the referees for careful reading and suggestions, which helped us to improve the paper.

Conflicts of Interest: The authors declare no conflict of interest.

References

1. Brannan, D.A.; Kirwan, W.E. On some classes of bounded univalent functions. *J. Lond. Math. Soc.* **1969**, *1*, 431–443. [CrossRef]
2. Ali, M.F.; Vasudevarao, A. Logarithmic coefficients of some close-to-convex functions. *Bull. Aust. Math. Soc.* **2017**, *95*, 228–237. [CrossRef]
3. Ali, M.F.; Vasudevarao, A. Coefficient inequalities and Yamashita's conjecture for some classes of analytic functions. *J. Aust. Math. Soc.* **2016**, *100*, 1–20. [CrossRef]
4. Baricz, Á.; Szász, R. Close-to-convexity of some special functions and their derivatives. *Bull. Malaysian Math. Sci. Soc.* **2016**, *39*, 427–437. [CrossRef]
5. Gangadharan, A.; Ravichandran, V. Radii of convexity and strong starlikeness for some classes of analytic functions. *J. Math. Anal. Appl.* **1997**, *211*, 303–313. [CrossRef]
6. Liu, J.-L. Notes on Jung-Kim-Srivastava integral operator. *J. Math. Anal. Appl.* **2004**, *294*, 96–103. [CrossRef]
7. Nunokawa, M.; Owa, S.; Saitoh, H.; Ikeda, A.; Koike, N. Some results for strongly starlike functions. *J. Math. Anal. Appl.* **1997**, *212*, 98–106. [CrossRef]
8. Nunokawa, M.; Sokól, J.; Trabka-Wieclaw, K. On the order of strongly starlikeness in some classes of starlike functions. *Acta Math. Hung.* **2015**, *145*, 142–149. [CrossRef]
9. Nunokawa, M.; Thomas, D.K. On convex and starlike functions in a sector. *J. Aust. Math. Soc. Ser. A* **1996**, 60, 363–368. [CrossRef]
10. Obradorić, M.; Owa, S. Some sufficient conditions for strongly starlikeness. *Int. J. Math. Math. Sci.* **2000**, *24*, 643–647. [CrossRef]
11. Owa, S.; Srivastava, H.M.; Hayami, T.; Kuroki, K. A new general idea for starlike and convex functions. *Tamkang J. Math.* **2016**, *47*, 445–454.
12. Ponnusamy, S.; Singh, V. Criteria for strongly starlike functions. *Complex Var. Theory Appl.* **1997**, *34*, 267–291. [CrossRef]
13. Prajapat, J.K.; Raina, R.K.; Srivastava, H.M. Some inclusion properties for certain subclasses of strongly starlike and strongly convex functions involving a family of fractional integral operator. *Integr. Transforms Spec. Funct.* **2007**, *18*, 639–651. [CrossRef]
14. Shiraishi, H.; Owa, S.; Srivastava, H.M. Sufficient conditions for strongly Carathéodory functions. *Comput. Math. Appl.* **2011**, *62*, 2978–2987. [CrossRef]
15. Srivastava, H.M.; Yang, D.G.; Xu, N.E. Subordinations for multivalent analytic functions associated with the Dziok-Srivastava operator. *Integral Transforms Spec. Funct.* **2009**, *20*, 581–606. [CrossRef]
16. Yang, D.-G.; Liu, J.-L. Some subclasses of meromorphic and multivalent functions. *Ann. Polon. Math.* **2014**, *111*, 73–88. [CrossRef]
17. Mocanu, P.T. Une propriété de convexité géneralisée dans la theórie de la représentation conforme. *Mathematica* **1969**, *11*, 127–133.
18. Miller, S.S.; Mocanu, P.T. On some classes of first-order differential subordinations. *Michigan Math. J.* **1985**, *32*, 185–195. [CrossRef]
19. Nunokawa, M. On properties of non-Carathéodory function. *Proc. Jpn. Acad. Ser. A* **1992**, *68*, 152–153. [CrossRef]

© 2019 by the authors. Licensee MDPI, Basel, Switzerland. This article is an open access article distributed under the terms and conditions of the Creative Commons Attribution (CC BY) license (http://creativecommons.org/licenses/by/4.0/).

symmetry

MDPI

Article

On a Generalization of the Initial-Boundary Problem for the Vibrating String Equation

Djumaklich Amanov [1], Gafurjan Ibragimov [2] and Adem Kılıçman [2,3,*]

1 Institute of Mathematics of Academy of Sciences of Uzbekistan, Tashkent 100125, Uzbekistan; damanov@yandex.ru
2 Department of Mathematics and Institute for Mathematical Research, Universiti Putra Malaysia, Serdang 43400, Selangor, Malaysia; ibragimov@upm.edu.my
3 Department of Electrical and Electronic Engineering, Istanbul Gelisim University, Avcilar, Istanbul 34310, Turkey
* Correspondence: akilic@upm.edu.my; Tel.: +603-89466813

Received: 26 November 2018; Accepted: 7 January 2019; Published: 10 January 2019

Abstract: In the present paper, we study a generalization of the initial-boundary problem for the inhomogeneous vibrating string equation. The initial conditions include the higher order derivatives of the unknown function. The problem is studied under homogeneous boundary conditions of the first kind. The uniqueness and existence of a regular solution of the problem are proved. To prove the main result we use the spectral decomposition method.

Keywords: vibrating string equation; initial conditions; spectral decomposition; regular solution; the uniqueness of the solution; the existence of a solution

1. Introduction

The differential equations are used to model the real world application problems in science and engineering that involve several parameters as well as the change of variables with respect to others. Most of these problems will require the solution of initial and boundary conditions, that is, the solution to the differential equations are forced to satisfy certain conditions and data. However, to model most of the real world problems is very complicated task and in many forms it is also difficult to find the exact solution. Boundary value problems for the Laplace, Poisson and Helmholtz equations with boundary conditions containing the higher order derivatives were studied in works by Bavrin [1], Karachik [2–5], Sokolovskii [6]. In the papers [7–13], boundary problems, including higher derivatives on the boundary, were studied for the Poisson, Helmholtz, and biharmonic equations. It should be noted that unlike our work, in the mentioned papers [1–6,14], the higher order derivative is given on the entire boundary. For an inhomogeneous heat equation, an initial-boundary problem containing a higher order derivative in the presence of an initial condition was studied in [15]. Now we reconsider the following equation

$$\frac{\partial^2 u}{\partial t^2} - \frac{\partial^2 u}{\partial x^2} = f(x,t) \tag{1}$$

in the domain $\Omega = \{(x,y) \mid 0 < x < p,\ 0 < t < T\}$ where $f(x,t)$ is a given function. Then we try to find a solution of the Equation (1) in the domain Ω which satisfies the following conditions

$$u(0,t) = 0, \quad 0 \le t \le T, \tag{2}$$

$$u(p,t) = 0, \quad 0 \le t \le T, \tag{3}$$

$$\frac{\partial^k u(x,0)}{\partial t^k} = \varphi_k(x), \quad 0 \le x \le p, \tag{4}$$

$$\frac{\partial^{k+1} u(x,0)}{\partial t^{k+1}} = \psi_k(x), \quad 0 \le x \le p, \tag{5}$$

where $k \ge 1$ is a fixed integer number. For the case $k = 0$ and $f(x,t) = 0$, the problem (1)–(5) was studied in [16]. Further Tikhonov in [17] studied homogeneous heat equation with the boundary condition $\sum_{k=0}^{n} a_k \frac{\partial^k u(0,t)}{\partial x^k} = 0$ and the initial condition $u(x,0) = 0$ in the domain $(0 < x < \infty,\ t > 0)$. Similarly, Bitsadze in [14] studied the Laplace equation in an n-dimensional domain D under the condition

$$\frac{d^m u}{d\nu^m} = f(x), \ x \in \partial D$$

and proved its Fredholm property. There is also more related literature on the boundary conditions problem, see for example ([18–25]). In the present paper, we study a generalized initial-boundary problem (2)–(5) for the inhomogeneous vibrating string Equation (1). The initial conditions include the higher order derivatives of the unknown function. The problem is studied under homogeneous boundary conditions of the first kind. We prove the uniqueness and existence of a regular solution of the problem. To solve the problem (1)–(5), we apply the spectral decomposition method.

2. The Uniqueness of Solution

Theorem 1. *The solution of the problem (1)–(5) is unique if it exists.*

Proof. Let $f(x,t) = 0$ in $\overline{\Omega}$, $\varphi_k(x) = 0$, $\psi_k(x) = 0$ in $[0,p]$. We show that the homogeneous problem (1)–(5) has only the trivial solution. It is known [26], the functions

$$X_n(x) = \sqrt{\frac{2}{p}} \sin(\lambda_n x), \ \lambda_n = \frac{n\pi}{p}, \ n = 1,2,\ldots \tag{6}$$

form in $L_2(0,p)$ a complete orthonormal system. Following [27], we consider the functions

$$\alpha_n(t) = \int_0^p u(x,t) X_n(x) dx, \ 0 \le t \le T, \tag{7}$$

where $u(x,t)$ is the solution of the homogeneous equation corresponding to the Equation (1). Differentiating (7) twice with respect to t, we obtain from the corresponding homogeneous Equation (1)

$$\alpha_n''(t) + \lambda_n^2 \alpha_n(t) = 0. \tag{8}$$

The solution of (8) has the form

$$\alpha_n(t) = a_n \cos(\lambda_n t) + b_n \sin(\lambda_n t).$$

To find the unknown coefficients a_n and b_n, we use the homogeneous conditions (4) and (5), which lead to the following equations:

$$\alpha_n^{(k)}(t) = 0, \ \alpha_n^{(k+1)}(t) = 0. \tag{9}$$

It is not difficult to verify that

$$\begin{aligned} \alpha_n^{(k)}(t) &= \lambda_n^k \left[a_n \cos\left(\frac{\pi k}{2} + \lambda_n t\right) + b_n \sin\left(\frac{\pi k}{2} + \lambda_n t\right) \right], \\ \alpha_n^{(k+1)}(t) &= \lambda_n^{k+1} \left[a_n \cos\left(\frac{\pi(k+1)}{2} + \lambda_n t\right) + b_n \sin\left(\frac{\pi(k+1)}{2} + \lambda_n t\right) \right]. \end{aligned}$$

Using (9), we obtain the following system of equations to determine the unknown coefficients a_n and b_n:

$$a_n \cos\frac{\pi k}{2} + b_n \sin\frac{\pi k}{2} = 0,$$
$$a_n \cos\frac{\pi(k+1)}{2} + b_n \sin\frac{\pi(k+1)}{2} = 0,$$

whose determinant of coefficients is 1. Hence, that $\alpha_n(t) = 0$. By completeness of functions $X_n(x)$, the Equation (7) implies that $u(x,t) = 0$ in $\overline{\Omega}$. □

3. The Existence of Solution

We search the solution of (1) in the form

$$u(x,t) = \sum_{n=1}^{\infty} u_n(t) X_n(x). \tag{10}$$

Expand the functions $f(x,t)$, $\varphi_k(x)$, and $\psi_k(x)$ in Fourier series by functions $X_n(x)$:

$$f(x,t) = \sum_{n=1}^{\infty} f_n(t) X_n(x), \tag{11}$$

$$\varphi_k(x) = \sum_{n=1}^{\infty} \varphi_{kn} X_n(x), \tag{12}$$

$$\psi_k(x) = \sum_{n=1}^{\infty} \psi_{kn} X_n(x), \tag{13}$$

where

$$f_n(t) = \int_0^p f(x,t) X_n(x) dx, \tag{14}$$

$$\varphi_{kn} = \int_0^p \varphi_k(x) X_n(x) dx, \tag{15}$$

$$\psi_{kn} = \int_0^p \psi_k(x) X_n(x) dx. \tag{16}$$

Substituting (10) and (11) into (1), we obtain

$$u_n''(t) + \lambda_n^2 u_n(t) = f_n(t).$$

It can be shown that the solution of this equation satisfying the conditions

$$u_n^{(k)}(0) = \varphi_{kn}, \; u_n^{(k+1)}(0) = \psi_{kn},$$

is

$$
\begin{aligned}
u_n(t) &= \frac{\varphi_{kn}}{\lambda_n^k}\cos\left(\frac{\pi k}{2}-\lambda_n t\right)-\frac{\psi_{kn}}{\lambda_n^{k+1}}\sin\left(\frac{\pi k}{2}-\lambda_n t\right)\\
&+\sum_{s=0}^{\left[\frac{k+1}{2}\right]-1}\frac{(-1)^s}{\lambda_n^{k+1-2s}}f_n^{(k-1-2s)}(0)\sin\left(\frac{\pi k}{2}-\lambda_n t\right)\\
&-\sum_{s=0}^{\left[\frac{k}{2}\right]-1}\frac{(-1)^s}{\lambda_n^{k-2s}}f_n^{(k-2-2s)}(0)\cos\left(\frac{\pi k}{2}-\lambda_n t\right)\\
&+\frac{1}{\lambda_n}\int_0^t f_n(\tau)\sin(\lambda_n(t-\tau))d\tau.
\end{aligned}
\tag{17}
$$

Hereinafter $\sum_{s=0}^{m}(\dots)=0$ for $m<0$. Substituting (17) into (10), we get

$$
\begin{aligned}
u(x,t) &= \sum_{n=1}^{\infty}X_n(x)\Bigg\{\frac{\varphi_{kn}}{\lambda_n^k}\cos\left(\frac{\pi k}{2}-\lambda_n t\right)-\frac{\psi_{kn}}{\lambda_n^{k+1}}\sin\left(\frac{\pi k}{2}-\lambda_n t\right)\\
&+\sum_{s=0}^{\left[\frac{k+1}{2}\right]-1}\frac{(-1)^s}{\lambda_n^{k+1-2s}}f_n^{(k-1-2s)}(0)\sin\left(\frac{\pi k}{2}-\lambda_n t\right)\\
&-\sum_{s=0}^{\left[\frac{k}{2}\right]-1}\frac{(-1)^s}{\lambda_n^{k-2s}}f_n^{(k-2-2s)}(0)\cos\left(\frac{\pi k}{2}-\lambda_n t\right)\\
&+\frac{1}{\lambda_n}\int_0^t f_n(\tau)\sin(\lambda_n(t-\tau))d\tau\Bigg\}.
\end{aligned}
\tag{18}
$$

Using (18) we find the following derivatives of $u(x,t)$.

$$
\begin{aligned}
\frac{\partial^2 u}{\partial t^2} &= \sum_{n=1}^{\infty}X_n(x)\Bigg\{-\frac{\varphi_{kn}}{\lambda_n^{k-2}}\cos\left(\frac{\pi k}{2}-\lambda_n t\right)+\frac{\psi_{kn}}{\lambda_n^{k-1}}\sin\left(\frac{\pi k}{2}-\lambda_n t\right)\\
&-\sum_{s=0}^{\left[\frac{k+1}{2}\right]-1}\frac{(-1)^s}{\lambda_n^{k-1-2s}}f_n^{(k-1-2s)}(0)\sin\left(\frac{\pi k}{2}-\lambda_n t\right)\\
&+\sum_{s=0}^{\left[\frac{k}{2}\right]-1}\frac{(-1)^s}{\lambda_n^{k-2-2s}}f_n^{(k-2-2s)}(0)\cos\left(\frac{\pi k}{2}-\lambda_n t\right)+f_n(0)\cos(\lambda_n t)\\
&+\frac{1}{\lambda_n}f_n'(0)\sin(\lambda_n t)+\frac{1}{\lambda_n}\int_0^t f_n''(\tau)\sin(\lambda_n(t-\tau))d\tau\Bigg\}.
\end{aligned}
\tag{19}
$$

$$
\begin{aligned}
\frac{\partial^2 u}{\partial x^2} &= \sum_{n=1}^{\infty}X_n(x)\Bigg\{-\frac{\varphi_{kn}}{\lambda_n^{k-2}}\cos\left(\frac{\pi k}{2}-\lambda_n t\right)+\frac{\psi_{kn}}{\lambda_n^{k-1}}\sin\left(\frac{\pi k}{2}-\lambda_n t\right)\\
&-\sum_{s=0}^{\left[\frac{k+1}{2}\right]-1}\frac{(-1)^s}{\lambda_n^{k-1-2s}}f_n^{(k-1-2s)}(0)\sin\left(\frac{\pi k}{2}-\lambda_n t\right)\\
&+\sum_{s=0}^{\left[\frac{k}{2}\right]-1}\frac{(-1)^s}{\lambda_n^{k-2-2s}}f_n^{(k-2-2s)}(0)\cos\left(\frac{\pi k}{2}-\lambda_n t\right)-f_n(t)\\
&+f_n(0)\cos(\lambda_n t)+\frac{1}{\lambda_n}f_n'(0)\sin(\lambda_n t)+\frac{1}{\lambda_n}\int_0^t f_n''(\tau)\sin(\lambda_n(t-\tau))d\tau\Bigg\}.
\end{aligned}
\tag{20}
$$

$$
\begin{aligned}
\frac{\partial^k u}{\partial t^k} &= \sum_{n=1}^{\infty}X_n(x)\Bigg\{\varphi_{kn}\cos(\lambda_n t)+\frac{1}{\lambda_n}\psi_{kn}\sin(\lambda_n t)\\
&+\frac{(-1)^k}{\lambda_n}\int_0^t f_n^{(k)}(\tau)\sin\left[\pi k+\lambda_n(t-\tau)\right]d\tau\Bigg\},
\end{aligned}
\tag{21}
$$

$$
\begin{aligned}
\frac{\partial^{k+1} u}{\partial t^{k+1}} &= \sum_{n=1}^{\infty}X_n(x)\Bigg\{-\lambda_n\varphi_{kn}\sin(\lambda_n t)+\psi_{kn}\cos(\lambda_n t)+\frac{(-1)^k}{\lambda_n}f_n^{(k)}(0)\sin(\lambda_n t)\\
&+\frac{(-1)^{k+1}}{\lambda_n}\int_0^t f_n^{(k+1)}(0)\sin\left[\pi(k+1)+\lambda_n(t-\tau)\right]d\tau\Bigg\}.
\end{aligned}
\tag{22}
$$

Next we need to prove the absolute and uniformly convergence of the series (18)–(22). Below we prove several lemmas that are used in the proof of the existence theorem.

Lemma 1. *Let $f(x,t) \in C^1(\overline{\Omega})$, $f(0,t) = f(p,t) = 0$, $\dfrac{\partial f}{\partial x} \in Lip_\alpha[0,p]$ uniformly with respect to t and $0 < \alpha < 1$. Then the series (11) converges absolutely and uniformly in $\overline{\Omega}$.*

Proof. Integrating in parts (14) we find

$$f_n(t) = \frac{\sqrt{2p}}{n\pi} \int_0^p \frac{\partial f(x,t)}{\partial x} \cos(\lambda_n x) dx.$$

Then [28]

$$|f_n(t)| \leq \frac{c}{n^{1+\alpha}}, \; c = \frac{Kp^{\frac{3}{2}+\alpha}}{\pi\sqrt{2}},$$

where K is the Hölder constant. Since the series $\sum\limits_{n=1}^{\infty} \frac{1}{n^{1+\alpha}}$ converges, therefore the series (11) converges absolutely and uniformly in $\overline{\Omega}$. □

Lemma 2. *Let $\varphi_k(x) \in W_2^1(0,p)$, $\varphi_k(0) = \varphi_k(p) = 0$. Then the series (12) converges absolutely and uniformly in $[0,p]$.*

Proof. Integrating by parts (15) we obtain

$$\varphi_{kn} = \frac{1}{\lambda_n} \varphi_{kn}^{(1)}, \; \varphi_{kn}^{(1)} = \int_0^p \varphi_k'(x) \sqrt{\frac{2}{p}} \cos(\lambda_n x) dx.$$

Using the Bessel inequality [29], $\sum\limits_{n=1}^{\infty} \left|\varphi_{kn}^{(1)}\right|^2 \leq \|\varphi_k'\|_{L_2(0,p)}^2$ and the inequality $\sum\limits_{n=1}^{\infty} |\varphi_{kn}| = \frac{p}{\pi} \sum\limits_{n=1}^{\infty} \frac{1}{n} \left|\varphi_{kn}^{(1)}\right|$ and using the Hölder inequality for the sum [29] yields

$$\sum_{n=1}^{\infty} \frac{1}{n} \left|\varphi_{kn}^{(1)}\right| \leq \left(\sum_{n=1}^{\infty} \frac{1}{n^2}\right)^{\frac{1}{2}} \left(\sum_{n=1}^{\infty} \left|\varphi_{kn}^{(1)}\right|^2\right)^{\frac{1}{2}} \leq \frac{\pi}{\sqrt{6}} \|\varphi_k'\|_{L_2(0,p)}.$$

Here the equality $\sum\limits_{n=1}^{\infty} \frac{1}{n^2} = \frac{\pi^2}{6}$ was used. This implies the absolutely and uniformly convergence of the series (12) on $[0,p]$. □

Lemma 3. *Let $\psi_k(x) \in W_2^1(0,p)$, $\psi_k(0) = \psi_k(p) = 0$. Then the series (13) converges absolutely and uniformly on $[0,p]$.*

The proof is similar to the proof of Lemma 3.

Lemma 4. *Let $\varphi_k(x) \in W_2^1(0,p)$, $\varphi_k(0) = \varphi_k(p) = 0$. Then the series*

$$\sum_{n=1}^{\infty} \lambda_n \varphi_{kn} X_n(x) \sin(\lambda_n t) \tag{23}$$

converges absolutely and uniformly in $\overline{\Omega}$.

Proof. Integrating by parts (15) we obtain

$$\varphi_{kn} = \frac{1}{\lambda_n^2}\varphi_{kn}^{(2)}, \ \varphi_{kn}^{(2)} = \int\limits_0^p \varphi_k''(x)\sqrt{\frac{2}{p}}\sin(\lambda_n x)dx.$$

Using the Parseval equality [29],

$$\sum_{n=1}^{\infty}\left|\varphi_{kn}^{(2)}\right|^2 = \|\varphi_k''\|_{L_2(0,p)}^2,$$

we obtain

$$\sum_{n=1}^{\infty}\lambda_n|\varphi_{kn}| = \sum_{n=1}^{\infty}\frac{1}{\lambda_n}\left|\varphi_{kn}^{(2)}\right| \leq \left(\sum_{n=1}^{\infty}\frac{1}{\lambda_n^2}\right)^{\frac{1}{2}}\left(\sum_{n=1}^{\infty}\left|\varphi_{kn}^{(2)}\right|^2\right)^{\frac{1}{2}} = \frac{p}{\sqrt{6}}\|\varphi_k''\|_{L_2(0,p)}.$$

Hence, the series (23) converges absolutely and uniformly in $\overline{\Omega}$. □

Lemma 5. *If* $\frac{\partial^{k+1}f(x,t)}{\partial t^{k+1}} \in C\left(\overline{\Omega}\right)$, *then the series*

$$\sum_{n=1}^{\infty}\frac{X_n(x)}{\lambda_n}\int\limits_0^t f_n^{(m)}(\tau)\sin\left[\frac{\pi(k+m)}{2}+\lambda_n(t-m)\right]d\tau, \ m = 1,2,\ldots,k+1$$

converges absolutely and uniformly in $\overline{\Omega}$.

Proof. Applying the Hölder inequality for integrals [29], we get

$$\frac{1}{\lambda_n}\left|\int\limits_0^t f_n^{(m)}(\tau)\sin\left[\frac{\pi(k+m)}{2}+\lambda_n(t-m)\right]d\tau\right| \leq \frac{\sqrt{T}}{\lambda_n}\left\|f_n^{(m)}\right\|_{L_2(0,T)}.$$

Now applying the Hölder inequality for sums and the Bessel inequality, we find

$$\sum_{n=1}^{\infty}\frac{1}{\lambda_n}\left\|f_n^{(m)}\right\|_{L_2(0,T)} \leq \frac{p}{\sqrt{6}}\left\|\frac{\partial^m f}{\partial t^m}\right\|_{L_2(0,T)}.$$

□

Next, consider the following series

$$\sum_{n=1}^{\infty}X_n(x)\sum_{s=0}^{\left[\frac{k+1}{2}\right]-1}\frac{(-1)^s}{\lambda_n^{k+1-2s}}f_n^{(k+1-2s)}(0)\sin\left(\frac{\pi k}{2}-\lambda_n t\right), \tag{24}$$

$$\sum_{n=1}^{\infty}X_n(x)\sum_{s=0}^{\left[\frac{k}{2}\right]-1}\frac{(-1)^s}{\lambda_n^{k-2s}}f_n^{(k-2-2s)}(0)\cos\left(\frac{\pi k}{2}-\lambda_n t\right), \tag{25}$$

$$\sum_{n=1}^{\infty}X_n(x)\sum_{s=0}^{\left[\frac{k}{2}\right]-1}\frac{(-1)^s}{\lambda_n^{k-2-2s}}f_n^{(k-2-2s)}(0)\cos\left(\frac{\pi k}{2}-\lambda_n t\right), \tag{26}$$

$$\sum_{n=1}^{\infty}X_n(x)\sum_{s=0}^{\left[\frac{k+1}{2}\right]-1}\frac{(-1)^s}{\lambda_n^{k-1-2s}}f_n^{(k-1-2s)}(0)\sin\left(\frac{\pi k}{2}-\lambda_n t\right). \tag{27}$$

Lemma 6. *If* $\frac{\partial^{k-1} f(x,t)}{\partial t^{k-1}} \in C\left(\overline{\Omega}\right)$, $k \geq 1$, *then the series (24) converges absolutely and uniformly in* $\overline{\Omega}$.

Proof. Consider the series

$$\sum_{n=1}^{\infty} \sum_{s=0}^{\left[\frac{k+1}{2}\right]-1} \frac{\left|f_n^{(k-1-2s)}(0)\right|}{\lambda_n^{k+1-2s}}. \tag{28}$$

For $s = 0$ we have $\sum_{n=1}^{\infty} \frac{\left|f_n^{(k-1)}(0)\right|}{\lambda_n^{k+1}}$. The convergence of this series is obvious. Let $s = \left[\frac{k+1}{2}\right] - 1$. Then

$$(1)\ k-1-2s = \begin{cases} 1, & \text{if } k \text{ is even}, \\ 0, & \text{if } k \text{ is odd}, \end{cases} \qquad (2)\ k+1-2s = \begin{cases} 3, & \text{if } k \text{ is even}, \\ 2, & \text{if } k \text{ is odd}. \end{cases}$$

Therefore, the series (28) converges and so the series in (24) converges absolutely and uniformly in $\overline{\Omega}$. □

Lemma 7. *If* $\frac{\partial^{k-2} f(x,t)}{\partial t^{k-2}} \in C\left(\overline{\Omega}\right)$, $k \geq 2$, *then the series (25) converges absolutely and uniformly in* $\overline{\Omega}$.

Proof. Consider the series

$$\sum_{n=1}^{\infty} \sum_{s=0}^{\left[\frac{k}{2}\right]-1} \frac{\left|f_n^{(k-2-2s)}(0)\right|}{\lambda_n^{k-2s}}. \tag{29}$$

If $k = 2$, then $s = 0$, and $\sum_{n=1}^{\infty} \frac{\left|f_n^{(k-2)}(0)\right|}{\lambda_n^{k}}$ converges. It is easy to check that if $s = 0, 1, \ldots, \left(\left[\frac{k}{2}\right] - 1\right)$, then $k - 2 - 2s \geq 0$. Thus, the series in (29) converges. Therefore, the series (25) converges absolutely and uniformly in $\overline{\Omega}$. □

Lemma 8. *Let* $\frac{\partial^{k-2} f(x,t)}{\partial t^{k-2}} \in C\left(\overline{\Omega}\right)$, $k \geq 2$. *If either k is odd or k is even and*

$$\frac{\partial f(x,t)}{\partial x} \in C\left(\overline{\Omega}\right), \ f(0,t) = f(p,t) = 0, \tag{30}$$

then the series (26) converges absolutely and uniformly in $\overline{\Omega}$.

Proof. The proof is completed by showing that the series

$$\sum_{n=1}^{\infty} \sum_{s=0}^{\left[\frac{k}{2}\right]-1} \frac{\left|f_n^{(k-2-2s)}(0)\right|}{\lambda_n^{k-2-2s}} \tag{31}$$

is the convergent. Indeed, if we let $k \geq 2$ and $s = \left[\frac{k}{2}\right] - 1$, then

$$k-2-2s = \begin{cases} 0, & \text{if } k \text{ is even}, \\ 1, & \text{if } k \text{ is odd}. \end{cases}$$

Therefore, the series (31) converges for odd k. Then the series (26) converges absolutely and uniformly in $\overline{\Omega}$. If $k = 2$, then $s = 0$ and the series (31) takes the form

$$\sum_{n=1}^{\infty} |f_n(0)|, \ f_n(0) = \int_0^p f(x,0) X_n(x) dx. \tag{32}$$

In general, for any even k, the term of

$$\sum_{s=0}^{[\frac{k}{2}]-1} \frac{\left|f_n^{k-2-2s}(0)\right|}{\lambda_n^{k-2-2s}}$$

corresponding to $s = \left[\frac{k}{2}\right] - 1$ is $|f_n(0)|$. For $n = 1, 2, ...$, these terms in (31) form the series (32). Show that the series (32) converges. Indeed, integrating the last integral, we obtain by virtue of (30) that

$$f_n(0) = \frac{1}{\lambda_n} f_n^{(1,0)}(0), \ f_n^{(1,0)} = \int_0^p \frac{\partial f(x,0)}{\partial x} \sqrt{\frac{2}{p}} \cos(\lambda_n x) dx. \tag{33}$$

By using the Bessel inequality

$$\sum_{n=1}^{\infty} \left|f_n^{(1,0)}(0)\right|^2 \leq \left\|\frac{\partial f(x,0)}{\partial x}\right\|^2_{L_2(0,p)},$$

and therefore taking into account (33), we can see that

$$\begin{aligned} \sum_{n=1}^{\infty} |f_n(0)| &= \sum_{n=1}^{\infty} \frac{1}{\lambda_n} \left|f_n^{(1,0)}(0)\right| \\ &\leq \left(\sum_{n=1}^{\infty} \frac{1}{\lambda_n^2}\right)^{\frac{1}{2}} \left(\sum_{n=1}^{\infty} \left|f_n^{(1,0)}(0)\right|^2\right)^{\frac{1}{2}} \leq \frac{p}{\sqrt{6}} \left\|\frac{\partial f(x,0)}{\partial x}\right\|_{L_2(0,p)}. \end{aligned}$$

Thus, the series (26) converges absolutely and uniformly in $\overline{\Omega}$ for any even k. □

Lemma 9. *Let* $\frac{\partial^{k-1} f(x,t)}{\partial t^{k-1}} \in C\left(\overline{\Omega}\right)$, $k \geq 1$. *If either* k *is even or* k *is odd and conditions (30) are satisfied, then the series (27) converges absolutely and uniformly in* $\overline{\Omega}$.

Proof. In order to prove this Lemma it is sufficient to prove that the following series

$$\sum_{n=1}^{\infty} \sum_{s=0}^{[\frac{k+1}{2}]-1} \frac{\left|f_n^{(k-1-2s)}(0)\right|}{\lambda_n^{k-1-2s}}$$

convergent. Indeed, for $s = \left[\frac{k+1}{2}\right] - 1$, we have

$$k - 1 - 2s = \begin{cases} 0, & \text{if } k \text{ is even}, \\ 1, & \text{if } k \text{ is odd}. \end{cases}$$

The rest of the proof runs as the proof of Lemma 8. □

Theorem 2. *Let*

(1) $f(x,t) \in C^1\left(\overline{\Omega}\right)$, $f(0,t) = f(p,t) = 0$,
and $\dfrac{\partial f}{\partial x} \in Lip_\alpha[0,p]$ *uniformly with respect to* t, $0 < \alpha < 1$;
(2) $\dfrac{\partial^k f(x,t)}{\partial t^k} \in C\left(\overline{\Omega}\right)$, $\dfrac{\partial^{k+1} f(x,t)}{\partial t^{k+1}} \in L_2(\Omega)$;
(3) $\varphi_k(x) \in W_2^2(0,p)$, $\varphi_k(0) = \varphi_k(p) = 0$;
(4) $\psi_k(x) \in W_2^1(0,p)$, $\psi_k(0) = \psi_k(p) = 0$.

Then the series (18)–(22) converge absolutely and uniformly in $\overline{\Omega}$, *the solution (18) satisfies the Equation (1), conditions (2)–(5), and* $u(x,t) \in C_{x,t}^{2,k+1}\left(\overline{\Omega}\right)$.

Proof. The fact that the series (18)–(22) converge absolutely and uniformly follows from Lemmas 1–9. Properties of the function $X_n(x)$ imply that (18) satisfies the conditions (2) and (3). Passing to the limit as $t \to 0$ in equalities (21) and (22), we can see that (18) satisfies the conditions (4) and (5). Comparing series (19) and (20), we can see that (18) satisfies Equation (1). The fact that the series (20) and (22) converge uniformly and absolutely in $\overline{\Omega}$ imply that $u(x,t) \in C^{2,k+1}_{x,t}(\overline{\Omega})$ and that (18) satisfies the Equation (1). □

4. Conclusions

We have studied and generalized the initial-boundary problem for the inhomogeneous vibrating string equation. The problems studied in the present paper are the first work for hyperbolic equation which contain higher order derivatives of unknown function in initial conditions. This problem generalizes the classic initial-boundary value problems for hyperbolic equation. We have proved the uniqueness and existence of a regular solution of the problem. To prove the main result we have used the spectral decomposition method. In addition, we have explicitly presented the solution in the form of series. We also state that the extension to the multi variables form is an open question.

Author Contributions: The authors contributed equally and all authors read the manuscript and approved the final submission.

Funding: The authors are very grateful for the comments of the reviewers which helped to improve the present manuscript. Further the authors also acknowledge that this research was partially supported by Geran Putra Berimpak, UPM/700-2/1/GPB/2017/9590200.

Acknowledgments: The authors would like to thank the referees for the valuable comments that helped to improve the manuscript.

Conflicts of Interest: The authors declare no conflict of interest.

References

1. Bavrin, I.I. Operators for harmonic functions and applications. *Differ. Equat.* **1985**, *21*, 9–15.
2. Karachik, V.V.; Turmetov, B.K. A Problem for the Harmonic Equation. *Izv. Akad. Nauk UzSSR Ser. Fiz.-Mat. Nauk* **1990**, *1*, 17–21.
3. Karachik, V.V. On the solvability of the boundary value problem for the Helmholtz equation with normal derivatives of high order on the boundary. *Differ. Equat.* **1992**, *28*, 907–909.
4. Karachik, V.V. On a problem for the Poisson equation with normal derivatives of high order on the boundary. *Differ. Equat.* **1996**, *32*, 1501–1503.
5. Karachik, V.V. The generalized Neumann problem for harmonic functions in a half-space. *Differ. Equat.* **1999**, *35*, 1–6.
6. Sokolovskiy, V.B. On a generalization of the Neumann problem. *Differ. Equat.* **1999**, *34*, 714–716.
7. Karachik, V.V. Solution of the Dirichlet Problem with Polynomial Data for the Polyharmonic Equation in a Ball. *Differ. Equat.* **2015**, *51*, 1033–1042. [CrossRef]
8. Turmetov, B. Solvability of fractional analogues of the Neumann problem for a nonhomogeneous biharmonic equation. *Electron. J. Differ. Equat.* **2015**, *2015*, 1–21.
9. Amanov, D. On a generalization of the Dirichlet problem for the Poisson equation. *Bound. Value Probl.* **2016**, *2016*, 160. [CrossRef]
10. Turmetov, B.; Koshanova, M.; Usmanov, K. Solvability of boundary value problems for Poisson equation with Hadamard type boundary operator. *Electron. J. Differ. Equat.* **2016**, *2016*, 1–12.
11. Karachik, V.; Turmetov, B. Solvability of some Neumann-Type boundary value problem for biharmonic equations. *Electron. J. Differ. Equat.* **2017**, *2017*, 1–17.
12. Karachik, V.V. A Neumann-Type Problem for the Biharmonic Equation. *Sib. Adv. Math.* **2017**, *27*, 103–118. [CrossRef]
13. Karachik, V.V. Solving a problem of Robin-Type for Biharmonic Equation. *Russ. Math.* **2018**, *62*, 34–48. [CrossRef]
14. Bitsadze, A.V. On the Neumann Problem for Harmonic Functions (Russian). *Dokl. Akad. Nauk SSSR.* **1990**, *311*, 11–13; translation in *Soviet Math. Dokl.* **1990**, *41*, 193–195

15. Amanov, D. On a generalization of the first initial-boundary value problem for the heat conduction equation. *Contemp. Anal. Appl. Math.* **2014**, *2*, 88–97. [CrossRef]
16. Petrovskiy, I.G. *Lectures on Partial Differential Equations*; GITTL: Moscow-Leningrad, Russia, 1961.
17. Tikhonov, A.N. On boundary conditions containing derivatives of order greater than the order of the equation. *Mat. Sb.* **1950**, *26*, 35–56.
18. Agarwal, R.P. *Boundary Value Problems for High Ordinary Differential Equations*; World Scientific: Singapore, 1986.
19. Agarwal, R.P.; Regan, D.O. Boundary value problems for super linear second order Ordinary Delay Differential Equations. *J. Differ. Equat.* **1996**, *130*, 333–335. [CrossRef]
20. Agarwal, R.P.; Wong, F.H.; Lian, W.C. positive solutions of nonlinear singular Boundary value problems. *Appl. Math. Lett.* **1999**, *12*, 115–120. [CrossRef]
21. Kılıçman, A.; Eltayeb, H. A note on defining singular integral as distribution and partial differential equations with convolution terms. *Math. Comput. Model.* **2009**, *49*, 327–336. [CrossRef]
22. Kılıçman, A.; Ömer, A. On higher-order boundary value problems by using differential transformation method with convolution terms. *J. Frankl. Inst.* **2014**, *351*, 631–642. [CrossRef]
23. Kılıçman, A.; Eltayeb, H. Note on Partial Differential Equations with Non-Constant Coefficients and Convolution Method. *Appl. Math. Inf. Sci.* **2012**, *6*, 59–63.
24. Eltayeb, E.; Kılıçman, A. A Note on Solutions of Wave, Laplace's and Heat Equations with Convolution Terms by Using Double Laplace Transform. *Appl. Math. Lett.* **2008**, *21*, 1324–1329. [CrossRef]
25. Hussin, C.H.C.; Kılıçman, A. On the Solutions of Non-linear Higher-Order Boundary Value Problems by Using Differential Transformation Method and Adomian Decomposition Method. *Math. Probl. Eng.* **2011**, *2011*, 724927.
26. Il'in, V.A.; Poznyak, E.G. *Fundamentals of Mathematical Analysis*; Nauka: Moscow, Russia, 1973; Volume 2.
27. Moiseyev, Y.I. On the solution by a spectral method of a nonlocal boundary-value problem. *Differ. Equat.* **1999**, *35*, 1094–1100.
28. Bari, N.K. *Trigonometric Series*; Nauka: Moscow, Russia, 1961.
29. Lyusternik, L.A.; Sobolev, V.I. *Elements of Functional Analysis*; Nauka: Moscow, Russia, 1965.

© 2019 by the authors. Licensee MDPI, Basel, Switzerland. This article is an open access article distributed under the terms and conditions of the Creative Commons Attribution (CC BY) license (http://creativecommons.org/licenses/by/4.0/).

symmetry

MDPI

Article

Geometric Properties of Normalized Mittag–Leffler Functions

Saddaf Noreen [1], Mohsan Raza [1], Jin-Lin Liu [2,*] and Muhammad Arif [3]

1 Department of Mathematics, Government College University, Faisalabad 38000, Pakistan; saddafnoreen@yahoo.com (S.N.); mohsan976@yahoo.com (M.R.)
2 Department of Mathematics, Yangzhou University, Yangzhou 225002, China
3 Department of Mathematics, Abdul Wali Khan University Mardan, Mardan 23200, Pakistan; marifmaths@yahoo.com
* Correspondence: jlliu@yzu.edu.cn

Received: 7 December 2018; Accepted: 24 December 2018; Published: 3 January 2019

Abstract: The aim of this paper is to investigate certain properties such as convexity of order μ, close-to-convexity of order $(1+\mu)/2$ and starlikeness of normalized Mittag–Leffler function. We use some inequalities to prove our results. We also discuss the close-to-convexity of Mittag–Leffler functions with respect to certain starlike functions. Furthermore, we find the conditions for the above-mentioned function to belong to the Hardy space $\mathcal{H}^p$. Some of our results improve the results in the literature.

Keywords: analytic functions; Mittag–Leffler functions; starlike functions; convex functions; Hardy space

MSC: 30C45; 33E12.

---

1. introduction

The one parameter Mittag–Leffler function $\mathbb{E}_{\alpha}(z)$ defined by

$$\mathbb{E}_{\alpha}(z) = \sum_{m=0}^{\infty} \frac{z^m}{\Gamma(\alpha m + 1)} \tag{1}$$

was introduced by Mittag–Leffler [1]. This function of complex variable is entire. The series defined by Equation (1) converges in $\mathbb{C}$ when $\mathrm{Re}(\alpha) > 0$. Consider that the function $\mathbb{E}_{\alpha,\kappa}(z)$ which generalizes the function $\mathbb{E}_{\alpha}(z)$ is defined by

$$\mathbb{E}_{\alpha,\kappa}(z) = \sum_{m=0}^{\infty} \frac{z^m}{\Gamma(\alpha m + \kappa)}, \alpha, \kappa \in \mathbb{C},\ z \in \mathbb{C}. \tag{2}$$

It was introduced by Wiman [2] and was named as Mittag–Leffler type function. The series in Equation (2) converges in $\mathbb{C}$ when $Re(\alpha) > 0$ and $Re(\kappa) > 0$. Furthermore, the functions defined in (1) and (2) are entire of order $1/Re(\alpha)$ and of type 1, for more details, see [3]. The function $\mathbb{E}_{\alpha,\kappa}(z)$ and its analysis with its generalizations is increasingly becoming a rich research area in mathematics and its related fields. A number of researchers studied and analyzed the function given in (2) (see Wiman [2,4,5]). One can find this function in the study of kinetic equation of fractional order, Lévy flights, random walks, super-diffusive transport as well as in investigations of complex systems.

In a similar manner, the advanced studies of these functions reflect and highlight many vital properties of these functions. The function $\mathbb{E}_{\alpha,\kappa}(z)$ generalizes many functions such as

$$\begin{aligned}\mathbb{E}_{1,1}(z) &= e^z,\ \mathbb{E}_{1,2}(z) = \frac{e^z-1}{z},\\ \mathbb{E}_{2,1}(z) &= \cosh\left(\sqrt{z}\right),\ \mathbb{E}_{2,2}(z) = \frac{\sinh\left(\sqrt{z}\right)}{\sqrt{z}}.\end{aligned}$$

The interested readers are suggested to go through [6–9].

Let $\mathcal{A}$ be the family of all functions g having the form

$$g(z) = z + \sum_{m=2}^{\infty} a_m z^m, \tag{3}$$

and are analytic in $\mathcal{D} = \{z : |z| < 1\}$ and $\mathcal{S}$ denote the family of univalent functions from $\mathcal{A}$. The families of functions which are convex, starlike and close-to-convex of order μ, respectively, are defined as:

$$\begin{aligned}\mathcal{C}(\mu) &= \left\{g : g\in\mathcal{A} \text{ and } Re\left(1+\frac{zg''(z)}{g'(z)}\right) > \mu,\quad z\in\mathcal{D};\ 0\le\mu<1\right\},\\ \mathcal{S}^*(\mu) &= \left\{g : g\in\mathcal{A} \text{ and } Re\left(\frac{zg'(z)}{g(z)}\right) > \mu,\quad z\in\mathcal{D};\ 0\le\mu<1\right\},\end{aligned}$$

and

$$\mathcal{K}(\mu) = \left\{g : g\in\mathcal{A} \text{ and } Re\left(\frac{g'(z)}{h'(z)}\right) > \mu,\quad z\in\mathcal{D};\ 0\le\mu<1;\ h\in\mathcal{C}\right\}.$$

It is clear that $\mathcal{C}^*(0) = \mathcal{C}$, $\mathcal{S}^*(0) = \mathcal{S}^*$ and $\mathcal{K}(0) = \mathcal{K}$. Consider the class $\mathcal{H}$ of all analytic functions in $\mathcal{D}$ and $\mu < 1$, Baricz [10] introduced the classes

$$\mathcal{P}_\eta(\mu) = \left\{p : p\in\mathcal{H},\ p(0)=1,\ Re\left\{e^{i\eta}(p(z)-\mu)\right\} > 0,\quad z\in\mathcal{D},\ \eta\in\mathbb{R}\right\}$$

and

$$\mathcal{R}_\eta(\mu) = \left\{f : f\in\mathcal{A} \text{ and } Re\left\{e^{i\eta}(f'(z)-\mu)\right\} > 0,\quad z\in\mathcal{D},\ \eta\in\mathbb{R}\right\}.$$

For $\eta = 0$, we have the classes of analytic functions $\mathcal{P}_0(\alpha)$ and $\mathcal{R}_0(\alpha)$ respectively. Also for $\eta = 0$ and $\alpha = 0$, we have the classes $\mathcal{P}$ and $\mathcal{R}$.

For the functions $g\in\mathcal{A}$ given by (1) and $h\in\mathcal{A}$ given by

$$h(z) = z + \sum_{m=2}^{\infty} b_m z^m,$$

then the convolution (Hadamard product) of g and h is defined as:

$$(g*h)(z) = z + \sum_{m=2}^{\infty} a_m b_m z^m,\quad z\in\mathcal{D}.$$

It is clear that the function $\mathbb{E}_{\alpha,\kappa}(z)$ is not in class $\mathcal{A}$. Recently, Bansal and Prajapat [11] considered the normalization of the function $\mathbb{E}_{\alpha,\kappa}(z)$ given as

$$E_{\alpha,\kappa}(z) = \Gamma(\kappa)\, z\mathbb{E}_{\alpha,\kappa}(z) = z + \sum_{m=1}^{\infty}\frac{\Gamma(\kappa)\, z^{m+1}}{\Gamma(\alpha m+\kappa)},\quad \alpha,\kappa\in\mathbb{C},\ Re(\alpha)>0, \kappa\neq 0,-1,\cdots.$$

In this article, we investigate some geometric properties of function $E_{\alpha,\kappa}(z)$ with real parameters α and κ.

We need the following results in our investigations.

Lemma 1 ([12]). *If $g \in \mathcal{A}$ and*

$$\left|zg''(z)\right| < \frac{1-\mu}{4}, \quad z \in \mathcal{D};\ 0 \leq \mu < 1,$$

then

$$Re\left\{g'(z)\right\} > \frac{1+\mu}{2}, \quad z \in \mathcal{D};\ 0 \leq \mu < 1.$$

Lemma 2 ([13]). *Let $\kappa \in \mathbb{C}$ such that $Re(\kappa) > 0$, $c \in \mathbb{C}$ and $|c| \leq 1$, $c \neq -1$. If $h \in \mathcal{A}$ satisfies*

$$\left|c\,|z|^{2\beta} + \left(1 - |z|^{2\beta}\,\frac{zh''(z)}{\beta h'(z)}\right)\right| \leq 1, \quad z \in \mathcal{D},$$

then

$$C_\beta(z) = \left\{\beta \int_0^z t^{\beta-1} h'(t)dt\right\}^{1/\beta}, \quad z \in \mathcal{D}$$

is analytic and univalent in $\mathcal{D}$.

Lemma 3 ([14]). *Let $g(z) = z + a_2z^2 + ... + a_mz^m + ...$, be analytic in $\mathcal{D}$ and in addition $1 \geq 2a_2 \geq ... \geq ma_m \geq ... \geq 0$ or $1 \leq 2a_2 \leq ... \leq ma_m \leq ... \leq 2$, then $g(z)$ is in class $\mathcal{K}$ with respect to the function $z \to -\log(1-z)$. Also if the function $g(z) = z + 3a_3 + ... + a_{2m-1}z^{2m-1} + ...$, which is odd and analytic in $\mathcal{D}$ and satisfies in addition $1 \geq 3a_3 \geq ... \geq (2m+1)\,a_{2m+1} \geq ... \geq 0$ or $1 \leq 3a_3 \leq ... \leq (2m+1)\,a_{2m+1} \leq ... \leq 2$, then $g(z) \in \mathcal{S}$ in $\mathcal{D}$.*

Lemma 4 ([[15]). *If $g(z) = \sum_{m=1}^{\infty} a_mz^{m-1}$, such that $a_1 = 1$ and $a_m \geq 0$, $\forall m \geq 2$, is analytic in $\mathcal{D}$ and if $\{a_m\}_{m=1}^{\infty}$ is a sequence which is decreasing, i.e., $a_{m+2} + a_m - 2a_{m+1} \geq 0$ and $a_m - a_{m+1} \geq 0$, $\forall m \geq 1$, then*

$$Re\left\{\sum_{m=1}^{\infty} a_mz^{m-1}\right\} > \frac{1}{2}, \ \forall z \in \mathcal{D}.$$

Lemma 5 ([15]). *If $a_m \geq 0$, $\{ma_m\}$ and $\{ma_m - (m+1)a_{m+1}\}$ both are non-increasing, then the function g defined by (3) is in $\mathcal{S}^*$.*

2. Starlikeness, Convexity, Close-to-Convexity

Theorem 1. *Let $\alpha \geq \frac{3}{2}$ and $\kappa \geq \frac{3}{2}$. Then,*

$$Re\left\{\frac{E_{\alpha,\kappa}(z)}{z}\right\} > \frac{1}{2}, \ \textit{for } z \in \mathcal{D}.$$

Proof. For the proof of this result, we have to show that

$$\{a_m\}_{m=1}^{\infty} = \left\{\frac{\Gamma(\kappa)}{\Gamma(\alpha(m-1)+\kappa)}\right\}_{m=1}^{\infty}$$

is a decreasing sequence. Consider

$$\begin{aligned} a_m - a_{m+1} &= \frac{\Gamma(\kappa)}{\Gamma(\alpha(m-1)+\kappa)} - \frac{\Gamma(\kappa)}{\Gamma(\alpha m+\kappa)} \\ &= \Gamma(\kappa)\left\{\frac{\Gamma(\alpha m+\kappa) - \Gamma(\alpha(m-1)+\kappa)}{\Gamma(\alpha(m-1)+\kappa)\Gamma(\alpha m+\kappa)}\right\} > 0, \end{aligned}$$

where $\forall\ m \geq 1$, $\alpha \geq \frac{3}{2}$ and $\kappa \geq \frac{3}{2}$. Now, to show that $\{a_m\}_{m=1}^{\infty}$ is decreasing, we prove that $a_m + a_{m+2} \geq 2a_{m+1}$.

Take

$$a_m - 2a_{m+1} + a_{m+2} = \frac{\Gamma(\kappa)}{\Gamma(\alpha(m+1)+\kappa)} + \frac{\Gamma(\kappa)}{\Gamma(\alpha(m-1)+\kappa)} - \frac{2\Gamma(\kappa)}{\Gamma(\alpha m+\kappa)}$$

$$= \Gamma(\kappa)\left\{\frac{\begin{array}{c}\Gamma(\alpha m+\kappa)\Gamma(\alpha(m+1)+\kappa) - 2\Gamma(\alpha(m-1)+\kappa)\Gamma(\alpha(m+1)+\kappa)\\ +\Gamma(\alpha(m-1)+\kappa)\Gamma(\alpha m+\kappa)\end{array}}{\Gamma(\alpha(m-1)+\kappa)\Gamma(\alpha m+\kappa)\Gamma(\alpha(m+1)+\kappa)}\right\}$$

$$= \Gamma(\kappa)\left[\frac{\begin{array}{c}\Gamma(\alpha(m+1)+\kappa)\{\Gamma(\alpha m+\kappa) - 2\Gamma(\alpha(m-1)+\kappa)\}\\ +\Gamma(\alpha(m-1)+\kappa)\Gamma(\alpha m+\kappa)\end{array}}{\Gamma(\alpha(m-1)+\kappa)\Gamma(\alpha m+\kappa)\Gamma(\alpha(m+1)+\kappa)}\right].$$

The above expression is non negative $\forall\ m \geq 1$, $\alpha \geq \frac{3}{2}$ and $\kappa \geq \frac{3}{2}$, which shows that $\{a_m\}_{m=1}^{\infty}$ is decreasing and convex sequence. Now, from the Lemma 4, we have

$$Re\left(\sum_{m=1}^{\infty} b_m z^{m-1}\right) > \frac{1}{2}, \; z \in \mathcal{D},$$

which is equivalent to

$$Re\left(\frac{E_{\alpha,\kappa}(z)}{z}\right) > \frac{1}{2}, \; z \in \mathcal{D}.$$

□

Theorem 2. *Let $\alpha \geq 2.67$ and $\kappa \geq 1$. Then, $E_{\alpha,\kappa}(z)$ is starlike in the open unit disc $\mathcal{D}$.*

Proof. To show that $E_{\alpha,\kappa}(z)$ is starlike in $\mathcal{D}$, we prove that $\{ma_m\}$ and $\{ma_m - (m+1)a_{m+1}\}$ both are non-increasing in view of Lemma 5. Since $a_m \geq 0$ for the normalized Mittag–Leffler function under the given conditions, consider

$$\begin{aligned} ma_m - (m+1)a_{m+1} &= \frac{m\Gamma(\kappa)}{\Gamma(\alpha(m-1)+\kappa)} - \frac{(m+1)\Gamma(\kappa)}{\Gamma(\alpha m+\kappa)} \\ &= \Gamma(\kappa)\left\{\frac{m\Gamma(\alpha m+\kappa) - (m+1)\Gamma(\alpha(m-1)+\kappa)}{\Gamma(\alpha(m-1)+\kappa)\Gamma(\alpha m+\kappa)}\right\} > 0 \end{aligned}$$

for $m \geq 1$, $\alpha \geq 2.67$ and $\kappa \geq 1$. Now,

$$ma_m - 2(m+1)a_{m+1} + (m+2) = \frac{m\Gamma(\kappa)}{\Gamma(\alpha(m-1)+\kappa)} - \frac{2(m+1)\Gamma(\kappa)}{\Gamma(\alpha m+\kappa)} + \frac{(m+2)\Gamma(\kappa)}{\Gamma(\alpha(m+1)+\kappa)}$$

$$= \Gamma(\kappa)\left\{\frac{\begin{array}{c}-2(m+1)\Gamma(\alpha(m-1)+\kappa)\Gamma(\alpha(m+1)+\kappa)+\\ m\Gamma(\alpha m+\kappa)\Gamma(\alpha(m+1)+\kappa)+(m+2)\Gamma(\alpha(m-1)+\kappa)\Gamma(\alpha m+\kappa)\end{array}}{\Gamma(\alpha(m-1)+\kappa)\Gamma(\alpha m+\kappa)\Gamma(\alpha(m+1)+\kappa)}\right\}$$

$$= \Gamma(\kappa)\left[\frac{\begin{array}{c}\Gamma(\alpha(m+1)+\kappa)\{m\Gamma(\alpha m+\kappa)-2(m+1)\Gamma(\alpha(m-1)+\kappa)\}\\ +(m+2)\Gamma(\alpha(m-1)+\kappa)\Gamma(\alpha m+\kappa)\end{array}}{\Gamma(\alpha(m-1)+\kappa)\Gamma(\alpha m+\kappa)\Gamma(\alpha(m+1)+\kappa)}\right].$$

The above relation is non-negative $\forall\ m \geq 1$, $\alpha \geq 2.67$ and $\kappa \geq 1$. Thus, from Lemma 5, $E_{\alpha,\kappa}(z)$ is starlike in $\mathcal{D}$. □

Theorem 3. *Let $\alpha \geq 3.323$ and $\kappa \geq 1$. Then,*

$$Re\left\{E'_{\alpha,\kappa}(z)\right\} > \frac{1}{2}, \quad (z \in \mathcal{D}).$$

Proof. Consider

$$\begin{aligned} E_{\alpha,\kappa}(z) &= z + \sum_{m=2}^{\infty} \frac{\Gamma(\kappa) z^m}{\Gamma(\alpha(m-1)+\kappa)}, \\ E'_{\alpha,\kappa}(z) &= 1 + \sum_{m=2}^{\infty} \frac{m\Gamma(\kappa)}{\Gamma(\alpha(m-1)+\kappa)} z^{m-1}, \\ E'_{\alpha,\kappa}(z) &= 1 + \sum_{m=2}^{\infty} A_m z^{m-1}. \end{aligned}$$

Here, $A_m = \frac{m\Gamma(\kappa)}{\Gamma(\alpha(m-1)+\kappa)}$. By taking the same computations as in Theorem 2, we get the proof. □

Theorem 4. *If $\alpha \geq 1$ and $\kappa \geq 1$, then $z \to E_{\alpha,\kappa}(z)$ is in $\mathcal{K}$ with respect to the function $-\log(1-z)$.*

Proof. Set

$$E_{\alpha,\kappa}(z) = z + \sum_{m=2}^{\infty} a_{m-1} z^m,$$

and we have $a_{m-1} > 0$ for all $m \geq 2$ and $a_1 = \frac{\Gamma(\kappa)}{\Gamma(\alpha+\kappa)} \leq 1$. For the proof of this result, we use Lemma 3. Therefore, we have to show that $\{ma_{m-1}\}_{m\geq 2}$ is decreasing. Now,

$$\begin{aligned} ma_{m-1} - (m+1)a_m &= \Gamma(\kappa)\left[\frac{m}{\Gamma(\alpha(m-1)+\kappa)} - \frac{m+1}{\Gamma(\alpha m+\kappa)}\right], \\ &= \Gamma(\kappa)\left[\frac{m\Gamma(\alpha m+\kappa) - (m+1)\Gamma(\alpha(m-1)+\kappa)}{\Gamma(\alpha(m-1)+\kappa)\Gamma(\alpha m+\kappa)}\right] > 0. \end{aligned}$$

By restricting parameters, we note that $ma_{m-1} - (m+1)a_m > 0$ for all $m \geq 2$. Thus, $\{ma_{m-1}\}_{m\geq 2}$ is a decreasing sequence—hence the result. □

Theorem 5. *If $\alpha \geq 1$ and $\kappa \geq 1$, then $z \to zE_{\alpha,\kappa}(z^2)$ is in $\mathcal{K}$ respect to the function $\frac{1}{2}\log\left(\frac{1+z}{1-z}\right)$.*

Proof. Set
$$zE_{\alpha,\kappa}\left(z^2\right) = z + \sum_{m=2}^{\infty} A_{2m-1} z^{2m-1}.$$

Here, $A_{2m-1} = a_{m-1} = \frac{\Gamma(\kappa)}{\Gamma(\alpha(m-1)+\kappa)}$ for all $m \geq 2$. In addition, it is clear that $a_1 \leq 1$. Mainly, we have to show that $\{(2m-1)\, a_{m-1}\}_{m\geq 2}$ is decreasing. Now,

$$\begin{aligned}(2m-1)\, a_{m-1} - (2m+1)\, a_m &= \Gamma(\kappa)\left[\frac{(2m-1)}{\Gamma(\alpha(m-1)+\kappa)} - \frac{(2m+1)}{\Gamma(\alpha m+\kappa)}\right], \\ = \Gamma(\kappa)&\left[\frac{(2m-1)\,\Gamma(\alpha m+\kappa) - (2m+1)\,\Gamma(\alpha(m-1)+\kappa)}{\Gamma(\alpha(m-1)+\kappa)\,\Gamma(\alpha m+\kappa)}\right] > 0.\end{aligned}$$

By using conditions on parameters, we observe that $(2m-1)\, a_{m-1} - (2m+1)\, a_m > 0$ for all $m \geq 2$. Thus, $\{(2m-1)\, a_{m-1}\}_{m\geq 2}$ is a decreasing sequence. By applying Lemma 3, we have the required result. □

Theorem 6. *If $\alpha \geq 1$ and $\kappa \geq 3.214319744$, then $E_{\alpha,\kappa}(z) \in \mathcal{S}^*$ in $\mathcal{D}$.*

Proof. Let $p(z) = \frac{zE'_{\alpha,\kappa}(z)}{E_{\alpha,\kappa}(z)}$, $z \in \mathcal{D}$. Then, the function p is analytic in $\mathcal{D}$ with $p(0) = 1$. To prove $E_{\alpha,\kappa}(z)$ is starlike in $\mathcal{D}$, we just prove that $Rep(z) > 0$ in $z \in \mathcal{D}$. For this, it is enough to show $|p(z) - 1| < 1$ for $z \in \mathcal{D}$. By using the inequalities

$$\begin{aligned}\frac{\Gamma(\kappa)}{\Gamma(\alpha m+\kappa)} &\leq \frac{1}{(\kappa)_m},\ \alpha \geq 1, \kappa \geq 1,\ m \in \mathbb{N}, \\ m\,(\kappa)_m &\leq 2^{m-1}\kappa\,(\kappa+1)_{m-1}, \kappa \geq 1,\ m \in \mathbb{N},\end{aligned}$$

we have

$$\begin{aligned}\left|E'_{\alpha,\kappa}(z) - \frac{E_{\alpha,\kappa}(z)}{z}\right| &= \left|\sum_{m=1}^{\infty}\frac{\Gamma(\kappa)}{\Gamma(\alpha m+\kappa)} m z^m\right| \\ &\leq \sum_{m=1}^{\infty}\frac{2^{m-1}}{\kappa\,(\kappa+1)^{m-1}} \\ &\leq \frac{1}{\kappa}\sum_{m=1}^{\infty}\left(\frac{2}{\kappa+1}\right)^{m-1} \\ &= \frac{\kappa+1}{\kappa(\kappa-1)},\quad (\kappa > 1).\end{aligned} \tag{4}$$

Furthermore, using reverse triangle inequality and the inequality $(\kappa)^m \leq (\kappa)_m$, we obtain

$$\begin{aligned}\left|\frac{E_{\alpha,\kappa}(z)}{z}\right| &= \left|1 + \sum_{m=1}^{\infty}\frac{\Gamma(\kappa)}{\Gamma(\alpha m+\kappa)} z^m\right| \\ &\geq 1 - \sum_{m=1}^{\infty}\frac{\Gamma(\kappa)}{\Gamma(\alpha m+\kappa)} \\ &\geq 1 - \sum_{m=1}^{\infty}\frac{1}{(\kappa)_m} \\ &\geq 1 - \frac{1}{\kappa}\sum_{m=1}^{\infty}\left(\frac{1}{\kappa+1}\right)^{m-1} \\ &= \frac{\kappa^2-\kappa-1}{\kappa^2}\quad (\kappa > 0).\end{aligned} \tag{5}$$

By combining (4) and (5), we get

$$\left|\frac{zE'_{\alpha,\kappa}(z)}{E_{\alpha,\kappa}(z)}-1\right|\leq\frac{\kappa(\kappa+1)}{(\kappa-1)(\kappa^2-\kappa-1)}. \tag{6}$$

Therefore, $E_{\alpha,\kappa}(z)\in\mathcal{S}^*$ in $\mathcal{D}$ if $\frac{\kappa(\kappa+1)}{(\kappa-1)(\kappa^2-\kappa-1)}\leq 1$. In other words, we have to show that $\kappa^3-3\kappa^2-\kappa+1\geq 0$. The inequality is satisfied for $\kappa\geq 3.214319744$. Hence, $E_{\alpha,\kappa}(z)$ is starlike in $\mathcal{D}$. □

Remark 1. *Recently, Bansal and Prajpat [11] proved that* $E_{\alpha,\kappa}(z)$ *is starlike, if* $\alpha\geq 1$ *and* $\kappa\geq(3+\sqrt{17})/2\approx 3.56155281$*. The above result improves the result in [11].*

Theorem 7. *If* $\alpha\geq 1$ *and* $\kappa\geq 3.56155281$*, then* $E_{\alpha,\kappa}(z)\in\mathcal{C}$ *in* $\mathcal{D}$*.*

Proof. Let $p(z)=1+\frac{zE''_{\alpha,\kappa}(z)}{E'_{\alpha,\kappa}(z)}$, $z\in\mathcal{D}$. Then, $p(z)$ is analytic in $\mathcal{D}$ with $p(0)=1$. To show that $E_{\alpha,\kappa}(z)$ is convex in $\mathcal{D}$, it is enough to prove that $|p(z)-1|<1$, $z\in\mathcal{D}$. By using the inequalities

$$\begin{aligned}\frac{\Gamma(\kappa)}{\Gamma(\alpha m+\kappa)} &\leq \frac{1}{(\kappa)_m},\ \alpha\geq 1,\kappa\geq 1,\ m\in\mathbb{N},\\ 2m(m+1)(\kappa)_m &\leq 4^{m-1}\kappa(\kappa+1)_{m-1},\kappa\geq 1,\ m\in\mathbb{N},\end{aligned}$$

we have

$$\begin{aligned}\left|zE''_{\alpha,\kappa}(z)\right| &= \left|\sum_{m=1}^{\infty}\frac{\Gamma(\kappa)}{\Gamma(\alpha m+\kappa)}m(m+1)z^m\right|\\ &\leq \sum_{m=1}^{\infty}\frac{4^{m-1}}{2\kappa(\kappa+1)^{m-1}}\\ &\leq \frac{2}{\kappa}\sum_{m=1}^{\infty}\left(\frac{4}{\kappa+1}\right)^{m-1}\\ &= \frac{2(\kappa+1)}{\kappa(\kappa-3)},\quad(\kappa>3).\end{aligned} \tag{7}$$

Furthermore, using the inequality $m(\kappa)^m\leq 2^{m-1}(\kappa)_m$, then we have

$$\begin{aligned}\left|E'_{\alpha,\kappa}(z)\right| &= \left|1+\sum_{m=1}^{\infty}(m+1)\frac{\Gamma(\kappa)}{\Gamma(\alpha m+\kappa)}z^m\right|\\ &\geq 1-\sum_{m=1}^{\infty}(m+1)\frac{\Gamma(\kappa)}{\Gamma(\alpha m+\kappa)}\\ &\geq 1-\sum_{m=1}^{\infty}\frac{1}{(\kappa)_m}\\ &\geq 1-\frac{2}{\kappa}\sum_{m=1}^{\infty}\left(\frac{2}{\kappa+1}\right)^{m-1}\\ &= \frac{\kappa^2-3\kappa-2}{\kappa(\kappa-1)}\quad(\kappa>0).\end{aligned} \tag{8}$$

From (7) and (8), we get

$$\left|\frac{zE''_{\alpha,\kappa}(z)}{E'_{\alpha,\kappa}(z)}\right|\leq\frac{2(\kappa^2-1)}{(\kappa-1)(\kappa^2-3\kappa-2)}. \tag{9}$$

This implies that $E_{\alpha,\kappa}(z) \in \mathcal{C}$ in $\mathcal{D}$ if $\frac{2(\kappa^2-1)}{(\kappa-1)(\kappa^2-3\kappa-2)} \leq 1$. To prove our result, we have to show that $\kappa^3 - 6\kappa^2 + 7\kappa + 6 \geq 0$. The inequality is satisfied for $\kappa \geq 3.5615528$. Hence, $E_{\alpha,\kappa}(z)$ is convex in $\mathcal{D}$. □

Consider the integral operator $\mathcal{F}_\gamma : \mathcal{D} \to \mathbb{C}$, where $\gamma \in \mathbb{C}, \gamma \neq 0$,

$$\mathcal{F}_\gamma(z) = \left\{ \gamma \int_0^z t^{\gamma-2} E_{\alpha,\kappa}(t)\, dt \right\}, \quad z \in \mathcal{D}.$$

Here, $\mathcal{F}_\gamma \in \mathcal{A}$. We prove that $\mathcal{F}_\gamma \in \mathcal{S}$ in $\mathcal{D}$.

Theorem 8. *Let $M \in \mathbb{R}^+$ such that $|E_{\alpha,\kappa}(z)| \leq M$ in $\mathcal{D}$. If*

$$|\gamma - 1| + \frac{\kappa(\kappa+1)}{(\kappa-1)(\kappa^2-\kappa-1)} + \frac{M}{|\gamma|} \leq 1,$$

then $\mathcal{F}_\gamma \in \mathcal{S}$ in $\mathcal{D}$.

Proof. A calculation gives

$$\frac{z\mathcal{F}''_\gamma(z)}{\mathcal{F}'_\gamma(z)} = \frac{zE'_{\alpha,\kappa}(z)}{E_{\alpha,\kappa}(z)} + \frac{z^{\gamma-1}}{\gamma} E_{\alpha,\kappa}(z) + \gamma - 2, \quad z \in \mathcal{D}.$$

Since $E_{\alpha,\kappa}(z) \in \mathcal{A}$, then by Schwarz Lemma, triangle inequality and (6), we obtain

$$\begin{aligned}
\left(1-|z|^2\right) \frac{z\mathcal{F}''_\gamma(z)}{\mathcal{F}'_\gamma(z)} &\leq \left(1-|z|^2\right)\left[|\gamma-1| + \left|\frac{zE'_{\alpha,\kappa}(z)}{E_{\alpha,\kappa}(z)} - 1\right| + \frac{|z|^{\gamma-1}}{|\gamma|}\left|\frac{E_{\alpha,\kappa}(z)}{z}\right|\right] \\
&\leq \left(1-|z|^2\right)\left[|\gamma-1| + \frac{\kappa(\kappa+1)}{(\kappa-1)(\kappa^2-\kappa-1)} + \frac{M}{|\gamma|}\right].
\end{aligned}$$

By using Lemma 2, $\mathcal{F}_\gamma \in \mathcal{S}$ in $\mathcal{D}$. □

Theorem 9. *Let $\alpha \geq 1$, $\mu \in [0,1)$ and $z \in \mathcal{D}$.*
(i) If $\kappa > \frac{(11-3\mu)+\sqrt{\mu^2-12\mu+17}}{2(1-\mu)}$, then $E_{\alpha,\kappa}(z) \in \mathcal{K}\left(\frac{1+\mu}{2}\right)$.
(ii) If $\preceq < 1 - \frac{\{(\kappa+2)(\kappa+\alpha_0)(\kappa+\alpha_0-1)+(\kappa+1)\}}{\kappa(\kappa+1)(\kappa+\alpha_0)(\kappa+\alpha_0-1)}$, then $\frac{E_{\alpha,\kappa}(z)}{z} \in \mathcal{P}(\mu)$.
(iii) If $(1-\mu)\kappa^3 + (2\mu-3)\kappa^2 - \kappa + (1-\mu) > 0$, then $E_{\alpha,\kappa}(z) \in \mathcal{S}^*(\mu)$.
(iv) If $(1-\mu)\kappa^3 + (6\mu-8)\kappa^2 + (7-7\mu)\kappa + (8-6\mu) > 0$, then $E_{\alpha,\kappa}(z) \in \mathcal{C}(\mu)$.

Proof. (i) Using (7) and Lemma 1, we get

$$\left|zE''_{\alpha,\kappa}(z)\right| \leq \frac{2(\kappa+1)}{\kappa(\kappa-3)} < \frac{1-\mu}{4},$$

where $0 \leq \mu < 1 - \frac{8(\kappa+1)}{\kappa(\kappa-3)}$ and $\kappa > \frac{(11-3\mu)+\sqrt{\mu^2-12\mu+17}}{2(1-\mu)}$. This shows that $E_{\alpha,\kappa}(z) \in \mathcal{K}\left(\frac{1+\mu}{2}\right)$.

(ii) To prove $\frac{E_{\alpha,\kappa}(z)}{z} \in \mathcal{P}(\mu)$, we have to show that $|g(z)-1| < 1$, where $g(z) = \frac{\{E_{\alpha,\kappa}(z)/z\}-\mu}{(1-\mu)}$. By using triangle inequality with

$$\begin{aligned}
\frac{\Gamma(\kappa)}{\Gamma(\alpha m+\kappa)} &\leq \frac{1}{(\kappa)_m}, \quad m \in \mathbb{N}, \\
(\kappa)_m &> \kappa(\kappa+\alpha_0)^{m-1}, \quad (\kappa > 0;\ m \in \mathbb{N}\backslash\{1,2\}),
\end{aligned}$$

(see [16]), where

$$\alpha_0 \approx 1.302775637...$$

is the largest root of the equation

$$\alpha^2 + \alpha - 3 = 0,$$

we have

$$\begin{aligned} |g(z)-1| &= \left| \frac{1}{(1-\mu)} \sum_{m=1}^{\infty} \frac{\Gamma(\kappa)}{\Gamma(\alpha m + \kappa)} z^m \right| \\ &\leq \frac{1}{(1-\mu)} \sum_{m=1}^{\infty} \frac{1}{(\kappa)_m} \\ &\leq \frac{1}{(1-\mu)} \left\{ \frac{1}{\kappa} + \frac{1}{\kappa(\kappa+1)} + \sum_{m=3}^{\infty} \frac{1}{\kappa(\kappa+\alpha_0)^{m-1}} \right\} \\ &= \frac{1}{(1-\mu)} \frac{\{(\kappa+2)(\kappa+\alpha_0)(\kappa+\alpha_0-1)+(\kappa+1)\}}{\kappa(\kappa+1)(\kappa+\alpha_0)(\kappa+\alpha_0-1)}. \end{aligned}$$

This implies that $\frac{E_{\alpha,\kappa}(z)}{z} \in \mathcal{P}(\mu)$, for $0 < \mu < 1 - \frac{\{(\kappa+2)(\kappa+\alpha_0)(\kappa+\alpha_0-1)+(\kappa+1)\}}{\kappa(\kappa+1)(\kappa+\alpha_0)(\kappa+\alpha_0-1)}$.

(iii) We use the inequality $\left| \frac{zE'_{\alpha,\kappa}(z)}{E_{\alpha,\kappa}(z)} - 1 \right| < 1 - \mu$ to show the starlikeness of order μ for the function $E_{\alpha,\kappa}(z)$. By using (4) and (5), we have

$$\left| \frac{zE'_{\alpha,\kappa}(z)}{E_{\alpha,\kappa}(z)} - 1 \right| \leq \frac{\kappa(\kappa+1)}{(\kappa-1)(\kappa^2-\kappa-1)} < 1 - \mu.$$

This implies that

$$\mu < 1 - \frac{\kappa(\kappa+1)}{(\kappa-1)(\kappa^2-\kappa-1)}.$$

This completes the proof.

(iv) We use the inequality $\left| \frac{zE''_{\alpha,\kappa}(z)}{E'_{\alpha,\kappa}(z)} \right| < 1 - \mu$ to show that $E_{\alpha,\kappa}(z) \in \mathcal{C}(\mu)$. By using (7) and (8), we have

$$\left| \frac{zE''_{\alpha,\kappa}(z)}{E'_{\alpha,\kappa}(z)} \right| \leq \frac{2(\kappa^2-1)}{(\kappa-3)(\kappa^2-3\kappa-2)} < 1 - \mu.$$

This implies that

$$\mu < 1 - \frac{2(\kappa^2-1)}{(\kappa-3)(\kappa^2-3\kappa-2)},$$

hence the result. □

Substituting $\mu = 0$ in Theorem 9, we obtained the following results.

Corollary 1. *Let $\alpha \geq 1, z \in \mathcal{D}$.*

(i) If $\kappa > \frac{11+\sqrt{17}}{2}$, then $E_{\alpha,\kappa}(z) \in \mathcal{K}\left(\frac{1}{2}\right)$.

(ii) If $\frac{\{(\kappa+2)(\kappa+\alpha_0)(\kappa+\alpha_0-1)+(\kappa+1)\}}{\kappa(\kappa+1)(\kappa+\alpha_0)(\kappa+\alpha_0-1)} < 1$, then $\frac{E_{\alpha,\kappa}(z)}{z} \in \mathcal{P}$.

(iii) If $\kappa^3 - 3\kappa^2 - \kappa + 1 > 0$, then $E_{\alpha,\kappa}(z) \in \mathcal{S}^*$.

(iv) If $\kappa^3 - 8\kappa^2 + 7\kappa + 8 > 0$, then $E_{\alpha,\kappa}(z) \in \mathcal{C}$.

Remark 2. *It is clear that $E_{\alpha,\kappa}(z) \in \mathcal{K}\left(\frac{1}{2}\right)$ when $\alpha \geq 1$, $\kappa > 7.56155$ and $E_{\alpha,\kappa}(z) \in \mathcal{C}$ when $\alpha \geq 1$, $\kappa > 6.796963$. It concludes that our results improve the results of ([17], corollary 2.1).*

3. Hardy Space of Mittag–Leffler Function

Consider the class $\mathcal{H}$ of analytic functions in $\mathcal{D} = \{z : |z| < 1\}$ and $\mathcal{H}^{\infty}$ denote the space bounded functions on $\mathcal{H}$. Let $g \in \mathcal{H}$, set

$$M_q(r,g) = \begin{cases} \left(\frac{1}{2\pi} \int\limits_0^{2\pi} \left| g\left(re^{i\theta}\right) \right|^q d\theta \right)^{1/q}, & 0 < q < \infty, \\ \max\left\{ |g(z)| : |z| \le r \right\}, & q = \infty. \end{cases}$$

If $M_q(r,g)$ is bounded for $r \in [0,1)$, then $g \in \mathcal{H}^q$. It is clear that

$$\mathcal{H}^{\infty} \subset \mathcal{H}^p \subset \mathcal{H}^q,\ 0 < p < q < \infty.$$

For some details, see [18]. It is also known [18] that, if $Re\left(g'(z)\right) > 0$ in $\mathcal{D}$, then

$$\begin{cases} g' \in \mathcal{H}^p, & p < 1, \\ g \in \mathcal{H}^{p/(1-p)}, & 0 < p < 1. \end{cases}$$

Hardy spaces of certain special functions are studied in [10,19,20].

Lemma 6 ([21]). *$\mathcal{P}_0(\mu) * \mathcal{P}_0(\eta) \subset \mathcal{P}_0(\gamma)$, where $\gamma = 1 - 2(1-\mu)(1-\eta)$ and $\mu,\ \eta < 1$. The value γ can not be improved.*

Lemma 7 ([22]). *For $\mu,\ \eta < 1$ and $\gamma = 1 - 2(1-\mu)(1-\eta)$, we have $\mathcal{R}_0(\mu) * \mathcal{R}_0(\eta) \subset \mathcal{R}_0(\gamma)$, or equivalently $\mathcal{P}_0(\mu) * \mathcal{P}_0(\eta) \subset \mathcal{P}_0(\gamma)$.*

Lemma 8 ([23]). *If the function g, convex of order μ, where $\mu \in [0,1)$, is not of the form*

$$g(z) = \begin{cases} l + dz\left(1 - ze^{i\varsigma}\right)^{2\mu-1}, & \mu \neq 1/2, \\ l + d\log\left(1 - ze^{i\varsigma}\right), & \mu = 1/2, \end{cases}$$

for $d,\ l \in \mathbb{C}$, and $\varsigma \in \mathbb{R}$, then the following statements are true:

(i) There exist $\delta = \delta(g) > 0$ such that $g' \in \mathcal{H}^{\delta + 1/[2(1-\mu)]}$.

(ii) If $\mu \in [0,1/2)$, then there exists $\tau = \tau(g) > 0$ such that $g \in \mathcal{H}^{\tau + 1/(1-2\mu)}$.

(iii) If $\mu \ge 1/2$, then $g \in \mathcal{H}^{\infty}$.

Theorem 10. *Let $\mu \in [0,1)$, $(1-\mu)\kappa^3 + (6\mu - 8)\kappa^2 + (7 - 7\mu)\kappa + (8 - 6\mu) > 0$.*

(i) If $\mu \in [0,1/2)$, then $E_{\alpha,\kappa}(z) \in \mathcal{H}^{1/(1-2\mu)}$.

(ii) If $\mu \ge 1/2$, then $E_{\alpha,\kappa}(z) \in \mathcal{H}^{\infty}$.

Proof. By using the definition of the hypergeometric function

$${}_2F_1(a,b,c;z) = \sum_{m=0}^{\infty} \frac{(a)^m (b)^m}{(c)^m} \frac{z^m}{m!},$$

we have

$$\begin{aligned} l + \frac{dz}{\left(1 - ze^{i\varsigma}\right)^{1-2\mu}} &= l + dz_2F_1\left(1, 1-2\alpha, 1; ze^{i\varsigma}\right), \\ &= l + d\sum_{m=0}^{\infty} \frac{(1-2\alpha)_m}{m!} e^{i\varsigma m} z^{m+1}, \end{aligned}$$

for $l,\ d\in\mathbb{C}$, $\mu\neq 1/2$ and for real ς. On the other hand,

$$\begin{aligned} l+d\log\left(1-ze^{i\gamma}\right) &= l-dz_2F_1\left(1,1,2;ze^{i\varsigma}\right), \\ &= l-d\sum_{m=0}^{\infty}\frac{1}{m+1}e^{i\varsigma m}z^{m+1}. \end{aligned}$$

Therefore, the function $E_{\alpha,\kappa}(z)$ is not of the form of $l+dz\left(1-ze^{i\gamma}\right)^{2\mu-1}$ (for $\mu\neq 1/2$) and $l+d\log\left(1-ze^{i\gamma}\right)$ (for $\mu=1/2$). We know that, by part (iv) of Theorem 9, $E_{\alpha,\kappa}(z)\in\mathcal{C}(\mu)$. Therefore, by using Lemma 8, we have the required result. □

Theorem 11. *Let* $\frac{\{(\kappa+2)(\kappa+\alpha_0)(\kappa+\alpha_0-1)+(\kappa+1)\}}{\kappa(\kappa+1)(\kappa+\alpha_0)(\kappa+\alpha_0-1)}<1$ *and* $f\in\mathcal{D}$. *Then, convolution* $E_{\alpha,\kappa}*f$ *is in* $\mathcal{H}^\infty\cap\mathcal{R}$.

Proof. Let $h(z)=E_{\alpha,\kappa}(z)*g(z)$. Then, $h'(z)=\frac{E_{\alpha,\kappa}(z)}{z}*g'(z)$. Using the Corollary 1 part *ii*, we have $\frac{E_{\alpha,\kappa}(z)}{z}\in\mathcal{P}$. As $g\in\mathcal{R}$; therefore, by using Lemma 6 $h\in\mathcal{R}$. Now, the function $\frac{E_{\alpha,\kappa}(z)}{z}$ is complete; therefore, $h(z)$ is complete. This implies that $h(z)$ is bounded. Thus, we have the required result. □

Theorem 12. *Let* $\preceq\ <1-\frac{\{(\kappa+2)(\kappa+\alpha_0)(\kappa+\alpha_0-1)+(\kappa+1)\}}{\kappa(\kappa+1)(\kappa+\alpha_0)(\kappa+\alpha_0-1)}$, $\mu\in[0,1)$ *and* $z\in\mathcal{D}$. *If* $g\in\mathcal{P}(\eta)$, *then* $E_{\alpha,\kappa}(z)*g\in\mathcal{R}(\gamma)$, *where* $\gamma=1-2(1-\mu)(1-\eta)$.

Proof. Let $h(z)=E_{\alpha,\kappa}(z)*g(z)$. Then, it is clear that $h'(z)=\frac{E_{\alpha,\kappa}(z)}{z}*g'(z)$. Using Theorem 9 part (ii), we have $\frac{E_{\alpha,\kappa}(z)}{z}\in\mathcal{P}(\mu)$. As $g\in\mathcal{R}$, therefore, by using Lemma 6 and the fact that $g'\in\mathcal{P}(\eta)$, we have $h'(z)\in\mathcal{P}(\gamma)$, where $.\gamma=1-2(1-\mu)(1-\eta)$. Consequently, $h\in\mathcal{R}(\gamma)$. □

Corollary 2. *Let* $\mu\in[0,1)$ *and* $\preceq\ <1-\frac{\{(\kappa+2)(\kappa+\alpha_0)(\kappa+\alpha_0-1)+(\kappa+1)\}}{\kappa(\kappa+1)(\kappa+\alpha_0)(\kappa+\alpha_0-1)}$. *If* $g\in\mathcal{R}(\eta)$, $\eta=(1-2\mu)/(2-2\mu)$, *then* $E_{\alpha,\kappa}(z)*g\in\mathcal{R}(0)$.

Corollary 3. *Let* $\mu\in[0,1)$ *and* $\frac{\{(\kappa+2)(\kappa+\alpha_0)(\kappa+\alpha_0-1)+(\kappa+1)\}}{\kappa(\kappa+1)(\kappa+\alpha_0)(\kappa+\alpha_0-1)}<1$. *If* $g\in\mathcal{R}(1/2)$, *then* $E_{\alpha,\kappa}(z)*g\in\mathcal{R}(0)$.

4. Applications

Now, we present some applications of the above theorems. It is clear that

$$E_{1,2}(z)=e^z-1,\ E_{1,3}(z)=\frac{2e^z-z-1}{z},\ E_{1,4}(z)=\frac{6e^z-3z^2,-6z-6}{z^2}.$$

From Theorem 9, we get the following:

Corollary 4. (i) *If* $0\leq\mu<\mu_0$, *where* $\mu_0\approx 0.26759$, *then* $E_{1,2}(z)\in\mathcal{P}(\mu)$.
(ii) *If* $0\leq\mu<\mu_1$, *where* $\mu_1\approx 0.55988$, *then* $E_{1,3}(z)\in\mathcal{P}(\mu)$.
(iii) *If* $0\leq\mu<\mu_2$, *where* $\mu_2\approx 0.68904$, *then* $E_{1,4}(z)\in\mathcal{P}(\mu)$.

Corollary 5. *If* $0\leq\mu<\mu_3$, *where* $\mu_3\approx 0.39393$, *then* $E_{1,4}(z)\in\mathcal{S}^*(\mu)$.

Corollary 6. (i) *Let* $0\leq\mu<\mu_4$, *where* $\mu_4\approx 0.2675930$. *If* $g\in\mathcal{R}(\eta)$, $\eta=(1-2\mu)/(2-2\mu)$, *then* $E_{1,2}(z)*g\in\mathcal{R}(0)$.
(ii) *Let* $0\leq\mu<\mu_5$, *where* $\mu_5\approx 0.55987780$. *If* $g\in\mathcal{R}(\eta)$, $\eta=(1-2\mu)/(2-2\mu)$, *then* $E_{1,3}(z)*g\in\mathcal{R}(0)$.
(iii) *Let* $0\leq\mu<\mu_6$, *where* $\mu_6\approx 0.68904320$. *If* $g\in\mathcal{R}(\eta)$, $\eta=(1-2\mu)/(2-2\mu)$, *then* $E_{1,4}(z)*g\in\mathcal{R}(0)$.

5. Conclusions

In this paper, we have studied certain geometric properties of Mittag-Leffler functions such as starlikeness, convexity and close-to-convexity. We have also found the Hardy spaces of Mittag-Leffler functions. Further, we have improved some results in the literature.

Author Contributions: Conceptualization, S.N. and M.R..; Formal analysis, S. N. and M.R.; Funding acquisition, J.L.L.; Investigation, S.N.; Methodology, S.N. and M.R.; Supervision, M.R.; Validation, M.A.; Visualization, M.A..;Writing—original draft, S.N.;Writing—review and editing, J.L.L.

Funding: The work presented here is supported by the Natural Science Foundation of Jiangsu Province under Grant No. BK20151304 and the National Natural Science Foundation of China under Grant No. 11571299.

Acknowledgments: The authors are grateful to the referees for their valuable comments and suggestions which improve the presentation of paper.

Conflicts of Interest: The authors declare no conflict of interest.

References

1. Mittag-Leffler, G.M. Sur la nouvelle fonction $E_\alpha x$. *C. R. Acad. Sci. Paris* **1903**, *137*, 554–558.
2. Wiman, A. Über den fundamental satz in der theorie der funcktionen $E_\alpha x$. *Acta Math.* **1905**, *29*, 191–201. [CrossRef]
3. Gorenflo, R.; Kilbas, A.A.; Mainardi, F.; Rogosin, S.V. Mittag–Leffler functions, Related Topics and Applications. In *Springer Monographs in Mathematics*; Springer: Heidelberg, Germany, 2014.
4. Agarwal, R.P. À propos d'une note de M. Pierre Humbert. *C. R. Acad. Sci. Paris* **1953**, *236*, 2031–2032.
5. Humbert, P. Quelques resultants retifs a la fonction de Mittag–Leffler. *C. R. Acad. Sci.* **1953**, *236*, 1467–1468.
6. Lavault, C. Fractional calculus and generalized Mittag–Leffler type functions. *arXiv* **2017**, arXiv:1703.01912.
7. Srivastava, H.M.; Bansal, D. Close-to-convexity of a certain family of q-Mittag–Leffler functions. *J. Nonlinear Var. Anal.* **2017**, *1*, 61–69.
8. Srivastava, H.M.; Kilicman, A.; Abdulnaby, Z.E.; Ibrahim, R.W. Generalized convolution properties based on the modified Mittag–Leffler function. J. *Non-Linear Sci. Appl.* **2017**, *10*, 4284–4294. [CrossRef]
9. Srivastava, H.M.; Tomovski, Z. Fractional calculus with an itegral operator containing a generalized Mittag–Leffler function in the kernal. *Appl. Math. Comp.* **2009**, *211*, 198–210. [CrossRef]
10. Baricz, Á. Bessel transforms and Hardy space of generalized Bessel functions. *Mathematica* **2006**, *48*, 127–136.
11. Bansal, D.; Prajapat, J.K. Certain geometric properties of the Mittag–Leffler functions. *Complex Var. Elliptic Equ.* **2016**, *61*, 338–350. [CrossRef]
12. Owa, S.; Nunokawa, M.; Saitoh, H.; Srivastava, H.M. Close-to-convexity, starlikeness, and convexity of certain analytic functions. *Appl. Math. Lett.* **2002**, *15*, 63–69. [CrossRef]
13. Pescar, V. A new generalization of Ahlfors's and Becker's criterion of univalence. *Bull. Malaysian Math. Soc. Ser. 2* **1996**, *19*, 53–54.
14. Ozaki, S. On the theory of multivalent functions. *Sci. Rep. Tokyo Bunrika Daigaku* **1935**, *2*, 167–188.
15. Fejer, L. Untersuchungen über Potenzreihen mit mehrfach monotoner Koeffizientenfolge. *Acta Litt. Sci.* **1936**, *8*, 89–115.
16. Baricz, Á.; Ponnusamy, S. Starlikeness and convexity of generalized Bessel functions. *Integral Transforms Spec. Funct.* **2010**, *21*, 641–653. [CrossRef]
17. Răducanu, D. Geometric properties of Mittag–Leffler functions. In *Models and Theories in Social Systems*; Springer: Berlin, Germany, 2019; pp 403–415.
18. Duren, P.L. *Theory of H^p Spaces*; Academic Press: New York, NY, USA; London, UK, 1970.
19. Ponnusamy, S. The Hardy space of hypergeometric functions. *Complex Var. Elliptic Equ.* **1996**, *29*, 83–96. [CrossRef]
20. Yağmur, N.; Orhan, H. Hardy space of generalized Struve functions. *Complex Var. Elliptic Equ.* **2014**, *59*, 929–936. [CrossRef]
21. Stankiewicz, J.; Stankiewicz, Z. Some applications of Hadamard convolution in the theory of functions. *Annales Universitatis Mariae Curie-Sklodowska Lodowska A* **1986**, *40*, 251–265.

22. Ponnusamy, S. Inclusion theorems for convolution product of second order polylogarithms and functions with the derivative in a halfplane. *Rocky Mt. J. Math.* **1998**, *28*, 695–733. [CrossRef]
23. Eenigenburg, P.J.; Keogh, F.R. The Hardy class of some univalent functions and their derivatives. *Mich. Math. J.* **1970**, *17*, 335–346. [CrossRef]

© 2019 by the authors. Licensee MDPI, Basel, Switzerland. This article is an open access article distributed under the terms and conditions of the Creative Commons Attribution (CC BY) license (http://creativecommons.org/licenses/by/4.0/).

symmetry

MDPI

Article

Some Generating Functions for q-Polynomials

Howard S. Cohl [1,*], Roberto S. Costas-Santos [2,*] and Tanay V. Wakhare [3]

1. Applied and Computational Mathematics Division, National Institute of Standards and Technology, Mission Viejo, CA 92694, USA
2. Departamento de Física y Matemáticas, Facultad de Ciencias, Universidad de Alcalá, Alcalá de Henares, 28871 Madrid, Spain
3. Department of Mathematics, University of Maryland, College Park, MD 20742, USA; twakhare@gmail.com

* Correspondence: howard.cohl@nist.gov (H.S.C.); rscosa@gmail.com (R.S.C.-S.)

Received: 14 November 2018; Accepted: 13 December 2018; Published: 16 December 2018

Abstract: Demonstrating the striking symmetry between calculus and q-calculus, we obtain q-analogues of the Bateman, Pasternack, Sylvester, and Cesàro polynomials. Using these, we also obtain q-analogues for some of their generating functions. Our q-generating functions are given in terms of the basic hypergeometric series ${}_4\phi_5$, ${}_5\phi_5$, ${}_4\phi_3$, ${}_3\phi_2$, ${}_2\phi_1$, and q-Pochhammer symbols. Starting with our q-generating functions, we are also able to find some new classical generating functions for the Pasternack and Bateman polynomials.

Keywords: basic hypergeometric functions; generating functions; q-polynomials

MSC: 33D45; 33C20

1. Introduction

We will adopt the following notations for sets: $\mathbb{N} := \{1,2,3,\ldots\}$, $\mathbb{N}_0 := \{0,1,2,\ldots\}$, and $\mathbb{C}$ is the set of complex numbers. We also adopt the conventions that an empty sum vanishes and the empty product is unity.

The generalized hypergeometric series ${}_rF_s$ is given by [1] (1.4.1)

$$ {}_rF_s\left(\begin{matrix} a_1,\ldots,a_r \\ b_1,\ldots,b_s \end{matrix};z\right) = \sum_{n=0}^{\infty} \frac{(a_1)_n\cdots(a_r)_n}{(b_1)_n\cdots(b_s)_n}\frac{z^n}{n!}, $$

where $|z|<1$, the Pochhammer symbol (rising factorial) is defined by $(a)_n := (a)(a+1)\cdots(a+n-1)$, and the (b_i) are such that these denominator factors never vanish. A q-analogue of the hypergeometric series ${}_rF_s$ is the basic hypergeometric series [2]

$$ {}_r\phi_s\left(\begin{matrix} a_1,a_2,\ldots,a_r \\ b_1,b_2,\ldots,b_s \end{matrix}\,\middle|\, q,z\right) = \sum_{n=0}^{\infty} \frac{(a_1;q)_n(a_2;q)_n\cdots(a_r;q)_n}{(b_1;q)_n(b_2;q)_n\cdots(b_s;q)_n}\left[(-1)^n q^{\frac{n(n-1)}{2}}\right]^{1+s-r}\frac{z^n}{(q;q)_n}, \quad (1) $$

where $q \neq 0$ when $r > s+1$. We refer to [2] for convergence properties of the series, however note that (b_i) is such that the denominator factors never vanish. This occurs when $b_i = q^{-m}$ for some $m \in \mathbb{N}_0$. Further note the important limit [1] (p. 15)

$$ \lim_{q\uparrow 1} {}_r\phi_s\left(\begin{matrix} q^{a_1},q^{a_2},\ldots,q^{a_r} \\ q^{b_1},q^{b_2},\ldots,q^{b_s} \end{matrix}\,\middle|\, q,(q-1)^{1+s-r}z\right) = {}_rF_s\left(\begin{matrix} a_1,a_2,\ldots,a_r \\ b_1,b_2,\ldots,b_s \end{matrix};z\right). \quad (2) $$

The q-Pochhammer symbol with non-negative integer subscript is defined by

$$(a;q)_0 = 1, \quad (a;q)_n = (1-a)(1-q)\cdots(1-aq^{n-1}), \quad n \in \mathbb{N}.$$

Some useful properties of the q-Pochhammer symbols that we will take advantage of with $n \in \mathbb{N}_0$ include [1] (1.8.18)

$$(q^{-n};q)_k = \frac{(q;q)_n}{(q;q)_{n-k}}(-1)^k q^{\frac{k(k-1)}{2}-nk}, \qquad k = 0,1,\ldots,n \tag{3}$$

and [1] (1.8.21-22)

$$(a;q)_{2n} = (a^{\frac{1}{2}};q)_n(-a^{\frac{1}{2}};q)_n((aq)^{\frac{1}{2}};q)_n(-(aq)^{\frac{1}{2}};q)_n. \tag{4}$$

Furthermore, one has [1] (p. 12)

$$(a;q)_\infty := \prod_{k=1}^{\infty}(1-aq^{k-1}),$$

and a generalization of the q-Pochhammer symbol for arbitrary $\lambda \in \mathbb{C}$ is given by [1] (1.8.9)

$$(a;q)_\lambda := \frac{(a;q)_\infty}{(aq^\lambda;q)_\infty}, \quad |q| < 1, \tag{5}$$

where the principal value of q^λ is taken. Note that, from (5), it follows that, for $a, \alpha, \beta \in \mathbb{C}$,

$$(a;q)_{\alpha+\beta} = (a;q)_\alpha(aq^\alpha;q)_\beta.$$

We need to define some other q-analogues, such as the q-analogue of a (real) number $[a]_q$, and the q-factorial $[n]_q!$. For the q-number one has [1] (1.8.1)

$$[a]_q := \frac{1-q^a}{1-q},$$

where $q \neq 0$, $q \neq 1$, and the q-factorial [2] (1.2.44)

$$[0]_q! = 1, \qquad [n]_q! = \prod_{k=1}^{n}[k]_q, \quad n \in \mathbb{N}.$$

We also will need a q-analogue of the *binomial theorem* [2] (1.3.2)

$${}_1\phi_0\left(\begin{array}{c} a \\ - \end{array} \middle| q, z\right) = \sum_{k=0}^{\infty}\frac{(a;q)_k}{(q;q)_k}z^k = \frac{(az;q)_\infty}{(z;q)_\infty}, \qquad |z| < 1,\ |q| < 1. \tag{6}$$

The q-binomial coefficient is defined for $a, b \in \mathbb{C}$, [2] (I.40)

$$\begin{bmatrix} a \\ b \end{bmatrix}_q := \frac{(q^{b+1};q)_\infty(q^{a-b+1};q)_\infty}{(q;q)_\infty(q^{a+1};q)_\infty},$$

which specializes if $n \in \mathbb{N}_0$, $k = 0,1,\ldots,n$, to [2] (I.39)

$$\begin{bmatrix} n \\ k \end{bmatrix}_q = \frac{(q;q)_n}{(q;q)_k(q;q)_{n-k}}.$$

Mathematical and Physical Applications

H. Bateman introduced in [3], the hypergeometric polynomial which we refer to in this paper as the Bateman polynomials. The goal of his research in the direction of these polynomials was in trying to better understand some radiation and conduction problems in which one requires the inverse of the Laplace transform. Our generalized q-generating functions for the Bateman and Pasternack (generalized Bateman) polynomials may very well be useful to the study of q-analogues of these radiation and conduction problems. Sylvester (1879) [4] investigated his polynomials and showed that the numbers $\varphi_n(\frac{1}{4})$ can be used for the computation of the numbers of different terms in the determinant of a skew-symmetric matrix of degree $2n$. Similarly, $\varphi_n(\frac{1}{8})$ is significant for the computation of the number of different terms in a determinant of degree $4n$, which is skew-symmetric with respect to both diagonals (see [5] (pp. 255–256)). Cesàro polynomials $g_n^{(s)}(z)$ are in fact the sth mean of the first n partial sums of $1+x+x^2+\ldots$ (see [6] (p. 185)). Our generalizations should apply in q-generalizations of all of these problems, such as for the q-Laplace transform and for q-Bernstein polynomials.

2. The Bateman, Sylvester, Pasternack, and Cesàro Polynomials

H. Bateman introduced in [3], the generalized hypergeometric polynomial:

$$\mathscr{Z}_n(z) = {}_2F_2\left(\begin{matrix}-n, n+1\\ 1,1\end{matrix}\,\middle|\, z\right).$$

By using [7] (Theorem 48), we obtain the generating function

$$\sum_{n=0}^{\infty} \mathscr{Z}_n(z)\, t^n = \frac{1}{1-t}\, {}_1F_1\left(\begin{matrix}\frac{1}{2}\\ 1\end{matrix}\,\middle|\, \frac{-4zt}{(1-t)^2}\right).$$

Using the above information, the Bateman polynomials are defined as [8] (p. 25)

$$\mathscr{B}_n(z) = {}_3F_2\left(\begin{matrix}-n, n+1, \frac{z+1}{2}\\ 1,1\end{matrix}\,\middle|\, 1\right).$$

He also obtained the generating functions:

$$\sum_{n=0}^{\infty} \mathscr{B}_n(z)\, t^n = \frac{1}{1-t}\, {}_2F_1\left(\begin{matrix}\frac{1}{2}, \frac{z+1}{2}\\ 1\end{matrix}\,\middle|\, \frac{-4t}{(1-t)^2}\right),$$

$$\sum_{n=0}^{\infty} \left(\mathscr{B}_n(z-2) - \mathscr{B}_n(z)\right) t^n = \frac{2t}{(1-t)^3}\, {}_2F_1\left(\begin{matrix}\frac{3}{2}, \frac{z+1}{2}\\ 2\end{matrix}\,\middle|\, \frac{-4t}{(1-t)^2}\right).$$

Lemma 1. *Let $t \in \mathbb{C}$. Then the following relation holds:*

$$\sum_{n=0}^{\infty} \mathscr{B}_m(-2n-1)\frac{t^n}{n!} = e^t \mathscr{Z}_m(-t). \tag{7}$$

The Bateman polynomial $\mathscr{B}_n$ was generalized by Pasternack in [9]. He defines the polynomial $\mathscr{B}_n^m$ as

$$\mathscr{B}_n^m(z) = {}_3F_2\left(\begin{matrix}-n, n+1, \frac{z+m+1}{2}\\ 1, m+1\end{matrix}\,\middle|\, 1\right)$$

for $m \in \mathbb{C}\backslash\{-1\}$. These polynomials reduce to the Bateman polynomials when $m = 0$. Further information regarding such polynomials and their connection with (classical) orthogonal polynomials can be found in [10]. Indeed, we can write the Pasternack polynomials in terms of the continuous Hahn polynomials as follows [10] (p. 893):

$$\mathscr{B}_n^m(z) = \frac{1}{i^n(m+1)_n}\, p_n\left(\frac{-iz}{2}; \frac{1+m}{2}, \frac{1-m}{2}, \frac{1-m}{2}, \frac{1+m}{2}\right).$$

We also consider the Sylvester polynomials, defined as (see [6] (p. 185))

$$\varphi_n(z) = \frac{z^n}{n!}\,{}_2F_0\left(\left.\begin{matrix}-n, z\\ -\end{matrix}\right| -\frac{1}{z}\right).$$

Notice that we also can write the Sylvester polynomials in terms of (classical) orthogonal polynomials [1] (p. 48)

$$\varphi_n(z) = (-1)^n L_n^{(-z-n)}(x) = \frac{z^n}{n!}\, C_n(-z; z).$$

Here $L_n^{(\alpha)}$ and C_n represent the Laguerre and Charlier polynomials. It is also known that the Sylvester polynomials satisfy the generating functions

$$\sum_{n=0}^{\infty} \varphi_n(z)\, t^n = \frac{e^{zt}}{(1-t)^z},$$

$$\sum_{n=0}^{\infty} (\lambda)_n\, \varphi_n(z)\, t^n = \frac{1}{(1-zt)^{\lambda}}\,{}_2F_0\left(\left.\begin{matrix}\lambda, z\\ -\end{matrix}\right| \frac{t}{1-zt}\right).$$

The Cesàro polynomials are defined as [6] (p. 449)

$$g_n^{(s)}(z) = \frac{(1+s)_n}{n!}\,{}_2F_1\left(\left.\begin{matrix}-n, 1\\ -s-n\end{matrix}\right| z\right).$$

Observe that this family can be written in terms of Jacobi polynomials [6] (p. 449) as

$$g_n^{(s)}(z) = P_n^{(s+1,-s-n-1)}(2z-1). \tag{8}$$

Furthermore, they satisfy the generating functions: [11] (4.2)

$$\sum_{n=0}^{\infty} g_n^{(s)}(z)\, t^n = (1-t)^{-s-1}(1-zt)^{-1}, \tag{9}$$

$$\sum_{n=0}^{\infty} \binom{n+\ell}{\ell} g_{n+\ell}^{(s)}(z)\, t^n = (1-t)^{-s-1-\ell}(1-zt)^{-1} g_\ell^{(s)}\left(\frac{z(1-t)}{1-zt}\right). \tag{10}$$

The aim of this paper is to obtain the q-analogue of all these families of polynomials as well as the q-analogues of the generating functions stated above. The structure of this paper is as follows. In Section 2, we give some preliminaries on q-calculus and we define some q-analogues of the Bateman, Sylvester, Pasternack, and Cesàro polynomials. In Section 3, we state and prove most (see Remark 1 below) of the q-analogues of the generating functions associated with the q-Bateman, q-Sylvester, q-Pasternack, and q-Cesàro polynomials.

3. The q-Analogues of the Bateman, Sylvester, Pasternack, and Cesàro Polynomials

Taking into account the q-definitions in Section 1 and the polynomials introduced therein, we define the q-Bateman polynomial as

$$\mathscr{Z}_n(z;q) = {}_2\phi_2\left(\begin{matrix} q^{-n}, q^{n+1} \\ q, q \end{matrix} \middle| q, q^n z\right),$$

$$\mathscr{B}_n(z;q) = {}_3\phi_2\left(\begin{matrix} q^{-n}, q^{n+1}, q^{\frac{1+z}{2}} \\ q, q \end{matrix} \middle| q, q^n\right),$$

define the q-Pasternack polynomial as

$$\mathscr{B}_n^m(z;q) = {}_3\phi_2\left(\begin{matrix} q^{-n}, q^{n+1}, q^{\frac{1+z+m}{2}} \\ q, q^{m+1} \end{matrix} \middle| q, q^n\right),$$

define the q-Sylvester polynomial as

$$\varphi_n(z;q) = \frac{z^n}{(q;q)_n}\, {}_2\phi_0\left(\begin{matrix} q^{-n}, q^z \\ - \end{matrix} \middle| q, q^n z^{-1}\right),$$

and define the q-Cesàro polynomial as

$$g_n^{(s)}(z;q) = \frac{(q^{s+1};q)_n}{(q;q)_n}\, {}_2\phi_1\left(\begin{matrix} q^{-n}, q \\ q^{-s-n} \end{matrix} \middle| q, z\right).$$

Lemma 2. *The q-Cesàro polynomial can be written as*

$$g_n^{(s)}(z;q) = \sum_{k=0}^{n} \begin{bmatrix} k+s \\ s \end{bmatrix}_q (zq^s)^{n-k}. \tag{11}$$

4. The Generating Functions

Theorem 1. *Let $q, t, z \in \mathbb{C}$, $|q| < 1$, $|t| < 1$, $|z| < 1$. Then the q-Bateman polynomials satisfy the following generating functions:*

$$\sum_{n=0}^{\infty} \mathscr{B}_n(z;q)\, t^n = \frac{1}{1-t}\, {}_5\phi_5\left(\begin{matrix} -q, q^{\frac{1}{2}}, -q^{\frac{1}{2}}, q^{\frac{z+1}{2}}, 0 \\ q, qt^{\frac{1}{2}}, -qt^{\frac{1}{2}}, (qt)^{\frac{1}{2}}, -(qt)^{\frac{1}{2}} \end{matrix} \middle| q, t\right), \tag{12}$$

$$\sum_{n=0}^{\infty} \mathscr{Z}_n(z;q)\, t^n = \frac{1}{1-t}\, {}_4\phi_5\left(\begin{matrix} -q, q^{\frac{1}{2}}, -q^{\frac{1}{2}}, 0 \\ q, qt^{\frac{1}{2}}, -qt^{\frac{1}{2}}, (qt)^{\frac{1}{2}}, -(qt)^{\frac{1}{2}} \end{matrix} \middle| q, zt\right), \tag{13}$$

$$\sum_{n=0}^{\infty} \Big(\mathscr{B}_n(z-2;q) - \mathscr{B}_n(z;q)\Big)\, t^n = \frac{(1+q)q^{\frac{z-1}{2}}t}{(t;q)_3}\, {}_5\phi_5\left(\begin{matrix} -q^2, q^{\frac{3}{2}}, -q^{\frac{3}{2}}, q^{\frac{z+1}{2}}, 0 \\ q^2, q^2t^{\frac{1}{2}}, -q^2t^{\frac{1}{2}}, (q^3t)^{\frac{1}{2}}, -(q^3t)^{\frac{1}{2}} \end{matrix} \middle| q, qt\right). \tag{14}$$

Proof. Let us start by proving Identity (12). One has

$$\sum_{n=0}^{\infty} \mathscr{B}_n(z;q)\, t^n = \sum_{n=0}^{\infty}\sum_{k=0}^{n} \frac{(q^{-n};q)_k (q^{n+1};q)_k (q^{\frac{z+1}{2}};q)_k}{(q;q)_k (q;q)_k (q;q)_k} (q^k t)^n.$$

Using (3), we obtain

$$\sum_{n=0}^{\infty} \mathscr{B}_n(z;q)\, t^n = \sum_{n=0}^{\infty}\sum_{k=0}^{n} \frac{(q;q)_{n+k} (q^{\frac{z+1}{2}};q)_k}{(q;q)_{n-k} (q;q)_k (q;q)_k (q;q)_k} (-1)^k q^{\frac{k(k-1)}{2}} t^n.$$

Next, we rearrange the double summation and set $n \mapsto n+k$, obtaining

$$\sum_{n=0}^{\infty} \mathscr{B}_n(z;q)\, t^n = \sum_{k=0}^{\infty}\sum_{n=0}^{\infty} \frac{(q;q)_{n+2k}(q^{\frac{z+1}{2}};q)_k}{(q;q)_n(q;q)_k(q;q)_k(q;q)_k}(-1)^k q^{\frac{k(k-1)}{2}} t^{n+k}$$

$$= \sum_{k=0}^{\infty} \frac{(q;q)_{2k}(q^{\frac{z+1}{2}};q)_k}{(q;q)_k(q;q)_k(q;q)_k}(-t)^k q^{\frac{k(k-1)}{2}} \sum_{n=0}^{\infty} \frac{(q^{2k+1};q)_n}{(q;q)_n} t^n.$$

By using (4) and the q-analogue of the binomial theorem (6), we obtain

$$= \sum_{k=0}^{\infty} \frac{(-q;q)_k(q^{\frac{1}{2}};q)_k(-q^{\frac{1}{2}};q)_k(q^{\frac{z+1}{2}};q)_k}{(q;q)_k(q;q)_k}(-t)^k q^{\frac{k(k-1)}{2}} \frac{(q^{2k+1}t;q)_\infty}{(t;q)_\infty}$$

$$= \sum_{k=0}^{\infty} \frac{(-q;q)_k(q^{\frac{1}{2}};q)_k(-q^{\frac{1}{2}};q)_k(q^{\frac{z+1}{2}};q)_k}{(t;q)_{2k+1}(q;q)_k(q;q)_k}(-t)^k q^{\frac{k(k-1)}{2}}$$

$$= \frac{1}{1-t}\sum_{k=0}^{\infty} \frac{(-q;q)_k(q^{\frac{1}{2}};q)_k(-q^{\frac{1}{2}};q)_k(q^{\frac{z+1}{2}};q)_k}{(qt;q)_{2k}(q;q)_k(q;q)_k}(-t)^k q^{\frac{k(k-1)}{2}}$$

$$= \frac{1}{1-t}\sum_{k=0}^{\infty} \frac{(-q;q)_k(q^{\frac{1}{2}};q)_k(-q^{\frac{1}{2}};q)_k(q^{\frac{z+1}{2}};q)_k}{(q;q)_k((qt)^{\frac{1}{2}};q)_k(-(qt)^{\frac{1}{2}};q)_k(qt^{\frac{1}{2}};q)_k(-qt^{\frac{1}{2}};q)_k}(-1)^k q^{\frac{k(k-1)}{2}} \frac{t^k}{(q;q)_k}.$$

Hence, the identity follows. In order to prove Identity (14), one has

$$\sum_{n=0}^{\infty}\left(\mathscr{B}_n(z-2;q) - \mathscr{B}_n(z;q)\right) t^n = \sum_{n=0}^{\infty}\sum_{k=0}^{n} \frac{(q^{-n};q)_k(q^{n+1};q)_k}{(q;q)_k(q;q)_k(q;q)_k} q^{nk}\left((q^{\frac{z-1}{2}};q)_k - (q^{\frac{z+1}{2}};q)_k\right) t^n$$

$$= \sum_{n=0}^{\infty}\sum_{k=0}^{n} \frac{(q^{-n};q)_k(q^{n+1};q)_k}{(q;q)_k(q;q)_k(q;q)_k} q^{nk}(q^{\frac{z+1}{2}};q)_{k-1}(1-q^k)(-q^{\frac{z-1}{2}})t^n$$

$$= \sum_{n=0}^{\infty}\sum_{k=0}^{n} \frac{(q;q)_{n+k}(q^{\frac{z+1}{2}};q)_{k-1}(-1)^{k-1}q^{\frac{z-1}{2}}q^{\frac{k(k-1)}{2}}}{(q;q)_{n-k}(q;q)_{k-1}(q;q)_k(q;q)_k} t^n.$$

Taking into account that the above expression vanishes at $k=0$, we set $k \mapsto k+1$, rearrange the double sum, and set $n \mapsto n+k$, yielding

$$= \sum_{k=0}^{\infty}\sum_{n=0}^{\infty} \frac{(q;q)_{n+2k+1}(q^{\frac{z+1}{2}};q)_k(-1)^k q^{\frac{z-1}{2}} q^{\frac{k(k+1)}{2}}}{(q;q)_{n-1}(q;q)_k(q;q)_{k+1}(q;q)_{k+1}} t^{n+k}.$$

Here, again, the series vanishes at $n=0$, so we set $n \mapsto n+1$. Applying some basic identities of q-Pochhammer symbols, we obtain

$$= (1+q)q^{\frac{z-1}{2}}t\sum_{k=0}^{\infty}\sum_{n=0}^{\infty} \frac{(q^3;q)_{n+2k}(q^{\frac{z+1}{2}};q)_k(-1)^k q^{\frac{k(k-1)}{2}}}{(q;q)_n(q;q)_k(q^2;q)_k(q^2;q)_k}(qt)^k t^n$$

$$= (1+q)q^{\frac{z-1}{2}}t\sum_{k=0}^{\infty} \frac{(q^3;q)_{2k}(q^{\frac{z+1}{2}};q)_k}{(q;q)_k(q^2;q)_k(q^2;q)_k}(-1)^k q^{\frac{k(k-1)}{2}}(qt)^k \sum_{n=0}^{\infty} \frac{(q^{3+2k};q)_n}{(q;q)_n} t^n.$$

Applying again the q-analogue of the binomial theorem (6) and simplifying, we obtain

$$=(1+q)q^{\frac{z-1}{2}}t\sum_{k=0}^{\infty}\frac{(q^3;q)_{2k}(q^{\frac{z+1}{2}};q)_k}{(q;q)_k(q^2;q)_k(q^2;q)_k}(-1)^k q^{\frac{k(k-1)}{2}}(qt)^k\frac{1}{(t;q)_{3+2k}}$$

$$=(1+q)q^{\frac{z-1}{2}}t\sum_{k=0}^{\infty}\frac{(q^{\frac{3}{2}};q)_k(-q^{\frac{3}{2}};q)_k(-q^2;q)_k(q^{\frac{z+1}{2}};q)_k}{(q;q)_k(q^2;q)_k}(-1)^k q^{\frac{k(k-1)}{2}}(qt)^k\frac{1}{(t;q)_{3+2k}}$$

$$=\frac{(1+q)}{(t;q)_3}q^{\frac{z-1}{2}}t\sum_{k=0}^{\infty}\frac{(q^{\frac{3}{2}};q)_k(-q^{\frac{3}{2}};q)_k(-q^2;q)_k(q^{\frac{z+1}{2}};q)_k(-1)^k q^{\frac{k(k-1)}{2}}}{(q;q)_k(q^2;q)_k((tq^3)^{\frac{1}{2}};q)_k(-(tq^3)^{\frac{1}{2}};q)_k(q^2t^{\frac{1}{2}};q)_k(-q^2t^{\frac{1}{2}};q)_k}(qt)^k.$$

Therefore, the identity follows. Let us prove the generating function Equation (14). We have

$$\sum_{n=0}^{\infty}\mathscr{Z}_n(z;q)\,t^n=\sum_{n=0}^{\infty}\sum_{k=0}^{n}\frac{(q^{-n};q)_k(q^{n+1};q)_k(-1)^k q^{\frac{k(k-1)}{2}}}{(q;q)_k(q;q)_k(q;q)_k}z^k q^{nk}t^n$$

$$=\sum_{n=0}^{\infty}\sum_{k=0}^{n}\frac{(q;q)_{n+k}q^{k(k-1)}}{(q;q)_{n-k}(q;q)_k(q;q)_k(q;q)_k}z^k t^n.$$

As in the previous identities, we rearrange the double sums and set $n\mapsto n+k$, obtaining

$$=\sum_{k=0}^{\infty}\sum_{n=0}^{\infty}\frac{(q;q)_{n+2k}q^{k(k-1)}}{(q;q)_n(q;q)_k(q;q)_k(q;q)_k}z^k t^{n+k}=\sum_{k=0}^{\infty}\frac{(q;q)_{2k}q^{k(k-1)}}{(q;q)_k(q;q)_k(q;q)_k}(zt)^k\sum_{n=0}^{\infty}\frac{(q^{2k+1};q)_n}{(q;q)_n}t^n$$

$$=\sum_{k=0}^{\infty}\frac{(q;q)_{2k}q^{k(k-1)}}{(q;q)_k(q;q)_k(q;q)_k}(zt)^k\frac{1}{(t;q)_{2k+1}}=\frac{1}{1-t}\sum_{k=0}^{\infty}\frac{(-q;q)_k(q^{\frac{1}{2}};q)(-q^{\frac{1}{2}};q)q^{k(k-1)}}{(q;q)_k(q;q)_k(tq;q)_{2k}}(zt)^k$$

$$=\frac{1}{1-t}\sum_{k=0}^{\infty}\frac{(-q;q)_k(q^{\frac{1}{2}};q)(-q^{\frac{1}{2}};q)q^{k(k-1)}}{(q;q)_k(q;q)_k((tq)^{\frac{1}{2}};q)_k(-(tq)^{\frac{1}{2}};q)_k(qt^{\frac{1}{2}};q)_k(-qt^{\frac{1}{2}};q)_k}(zt)^k.$$

Hence, the identity follows since $((-1)^k q^{\frac{k(k-1)}{2}})^2=q^{k(k-1)}$. Note that, in order for the generating functions to converge, one must require $|t|<1$. Furthermore, one must also have that the denominator parameters of the basic hypergeometric series must not be equal to a factor of q^{-m} for some $m\in\mathbb{N}_0$. This requires that $|t|<|q|^{-1}$, which is greater than unity since $|q|<1$, so $|t|<1$ suffices. Since $|t|<1$, then $|z|<1$ as well. This completes the proof. □

Theorem 2. *Let $j,m\in\mathbb{N}_0$, $q,t,\lambda\in\mathbb{C}$, $|q|<1$, $|t|<1$. Then the following identities hold:*

$$\sum_{n=0}^{\infty}\mathscr{B}_m^j(-2n-j-1;q)\frac{(q^\lambda;q)_n}{(q;q)_n}t^n=\frac{1}{(t;q)_\lambda}{}_4\phi_3\left(\begin{matrix}q^{-m},q^{m+1},q^\lambda,0\\ q,q^{j+1},qt^{-1}\end{matrix}\middle|\,q,q^m\right),\tag{15}$$

$$\sum_{n=0}^{\infty}\mathscr{B}_m^j(-2n-j-1;q)\,t^n=\frac{1}{1-t}{}_3\phi_2\left(\begin{matrix}q^{-m},q^{m+1},0\\ q^{j+1},qt^{-1}\end{matrix}\middle|\,q,q^m\right),\tag{16}$$

$$\sum_{n=0}^{\infty}\mathscr{B}_m(-2n-1;q)\frac{t^n}{(q;q)_n}=\frac{1}{(t;q)_\infty}{}_4\phi_3\left(\begin{matrix}q^{-m},q^{m+1},0,0\\ q,q,qt^{-1}\end{matrix}\middle|\,q,q^m\right),\tag{17}$$

where the principal value of q^λ is taken.

Proof. Let us prove (15). Setting $z = -2n-1-j$ in the q-Pasternack polynomial, and using their basic hypergeometric expansion, we have

$$\begin{aligned}\sum_{n=0}^{\infty} \mathscr{B}_m^j(-2n-1-j;q)\,\frac{(q^\lambda;q)_n}{(q;q)_n}\,t^n &= \sum_{n=0}^{\infty}\sum_{k=0}^{n} \frac{(q^{-m};q)_k(q^{m+1};q)_k(q^{-n};q)_k}{(q;q)_k(q;q)_k(q^{j+1};q)_k}\,q^{nk}t^n\,\frac{(q^\lambda;q)_n}{(q;q)_n}\\
&= \sum_{n=0}^{\infty}\frac{(q^\lambda;q)_n t^n}{(q;q)_n}\sum_{k=0}^{\min(m,n)}\frac{(q^{-m};q)_k(q^{m+1};q)_k(q^{-n};q)_k q^{mk}}{(q;q)_k(q;q)_k(q^{j+1};q)_k}\\
&= \sum_{k=0}^{m}\frac{(q^{-m};q)_k(q^{m+1};q)_k(q^\lambda;q)_k(tq^m)^k}{(q;q)_k(q;q)_k(q;q)_k(q^{j+1};q)_k}\sum_{n=0}^{\infty}\frac{(q^{-n-k};q)_k(q^{\lambda+k};q)_n t^n}{(q^{k+1};q)_n}\\
&= \sum_{k=0}^{m}\frac{(q^{-m};q)_k(q^{m+1};q)_k(q^\lambda;q)_k(-t)^k q^{mk+\binom{k}{2}-k^2}}{(q;q)_k(q;q)_k(q^{j+1};q)_k}\sum_{n=0}^{\infty}\frac{(q^{\lambda+k};q)_n t^n q^{-nk}}{(q;q)_n}\\
&= \sum_{k=0}^{m}\frac{(q^{-m};q)_k(q^{m+1};q)_k(q^\lambda;q)_k(-t)^k q^{mk+\binom{k}{2}-k^2}}{(q;q)_k(q;q)_k(q^{j+1};q)_k}\,{}_1\phi_0\left(\begin{matrix} q^{\lambda+k}\\ -\end{matrix}\,\middle|\, q, tq^{-k}\right)\\
&= \frac{1}{(t;q)_\lambda}\,{}_4\phi_3\left(\begin{matrix} q^{-m}, q^{m+1}, q^\lambda, 0\\ q, q^{j+1}, qt^{-1}\end{matrix}\,\middle|\, q, t\right),\end{aligned}$$

where we have used (5), (6) [1] (1.8.6), which completes the proof of (15). Observe that, if we set $\lambda = 1$ in (15), we obtain (16). Since $|q| < 1$, taking the limit $\lambda \to \infty$ yields (17). Note that, in order for the generating functions to converge, one must require $|t| < 1$. This completes the proof. □

Setting $t \mapsto t(1-q)$ and taking the $q \uparrow 1$ limit of (17) produces Lemma 1 since

$$\lim_{q\uparrow 1}\frac{(1-q)^n}{(q;q)_n} = \frac{1}{n!}, \qquad \lim_{q\uparrow 1}\frac{1}{(t(1-q);q)_\infty} = e^t,$$

$$\lim_{q\uparrow 1}{}_4\phi_3\left(\begin{matrix} q^{-m}, q^{m+1}, 0, 0\\ q, q, \frac{q}{(1-q)t}\end{matrix}\,\middle|\, q, q^m\right) = {}_2F_2\left(\begin{matrix} -m, m+1\\ 1, 1\end{matrix}\,\middle|\, -t\right),$$

which follows easily by expanding the denominator factor involving t and using (2).

In fact, we are now able to obtain new classical generating functions for the Pasternack and Bateman polynomials by taking the $q \uparrow 1$ limit in (15), (16).

Corollary 1. *Let $j, m \in \mathbb{N}_0$, $t, \lambda \in \mathbb{C}$. Then the following identities hold:*

$$\sum_{n=0}^{\infty}\mathscr{B}_m^j(-2n-1-j)\,\frac{(\lambda)_n}{n!}\,t^n = \frac{1}{(1-t)^\lambda}\,{}_3F_2\left(\begin{matrix} -m, m+1, \lambda\\ 1, j+1\end{matrix}; \frac{-t}{1-t}\right), \tag{18}$$

$$\sum_{n=0}^{\infty}\mathscr{B}_m(-2n-1)\,\frac{(\lambda)_n}{n!}\,t^n = \frac{1}{(1-t)^\lambda}\,{}_3F_2\left(\begin{matrix} -m, m+1, \lambda\\ 1, 1\end{matrix}; \frac{-t}{1-t}\right), \tag{19}$$

$$\sum_{n=0}^{\infty}\mathscr{B}_m^j(-2n-1-j)\,t^n = \frac{1}{1-t}\,{}_2F_1\left(\begin{matrix} -m, m+1\\ j+1\end{matrix}; \frac{-t}{1-t}\right), \tag{20}$$

$$\sum_{n=0}^{\infty}\mathscr{B}_m(-2n-1)\,t^n = \frac{1}{1-t}\,P_m\left(\frac{1+t}{1-t}\right), \tag{21}$$

where $P_m(x)$ is the Legendre polynomial [1] (Section 9.8.3).

Proof. In (15), take the $q \uparrow 1$ limit. Note that

$$(t;q)_\lambda = \frac{(t;q)_\infty}{(tq^\lambda;q)_\infty} = {}_1\phi_0\left(\begin{matrix} q^{-\lambda} \\ - \end{matrix} \middle| q, tq^\lambda\right).$$

Thus, by using (2),

$$\lim_{q\uparrow 1}(t;q)_\lambda = {}_1F_0\left(\begin{matrix} -\lambda \\ - \end{matrix} \middle| t\right) = (1-t)^\lambda.$$

Hence by expanding the denominator factor involving t using (2) in (15) produces (18). Setting $j = 0$ and $\lambda = 1$ in (18), produces (19), (20), respectively. Setting $j = 0$ in (20) produces (21), by noting [1] (9.8.62). □

Theorem 3. *Let $m \in \mathbb{N}_0$, $q, t, z \in \mathbb{C}$, $|q| < 1$, $|t| < 1$. Then the q-Pasternack polynomials satisfy the following generating function:*

$$\sum_{n=0}^{\infty} \mathscr{B}_n^m(z;q)\, t^n = \frac{1}{1-t}\, {}_5\phi_5\left(\begin{matrix} -q, q^{\frac{1}{2}}, -q^{\frac{1}{2}}, q^{\frac{z+m+1}{2}}, 0 \\ q^{m+1}, qt^{\frac{1}{2}}, -qt^{\frac{1}{2}}, (qt)^{\frac{1}{2}}, -(qt)^{\frac{1}{2}} \end{matrix} \middle| q, t\right). \tag{22}$$

Proof. Taking into account the expression of the basic hypergeometric series for these polynomials, we have

$$\begin{aligned}
\sum_{n=0}^{\infty} \mathscr{B}_n^m(z;q)\, t^n &= \sum_{n=0}^{\infty}\sum_{k=0}^{n} \frac{(q^{-n};q)_k (q^{n+1};q)_k (q^{\frac{z+m+1}{2}};q)_k}{(q;q)_k (q;q)_k (q^{m+1};q)_k} (q^k t)^n \\
&= \sum_{n=0}^{\infty}\sum_{k=0}^{n} \frac{(q;q)_{n+k} (q^{\frac{z+m+1}{2}};q)_k}{(q;q)_{n-k}(q;q)_k(q;q)_k(q^{m+1};q)_k} (-1)^k q^{\frac{k(k-1)}{2}} t^n \\
&= \sum_{k=0}^{\infty}\sum_{n=0}^{\infty} \frac{(q;q)_{n+2k} (q^{\frac{z+m+1}{2}};q)_k}{(q;q)_{n}(q;q)_k(q;q)_k(q^{m+1};q)_k} (-1)^k q^{\frac{k(k-1)}{2}} t^{n+k} \\
&= \sum_{k=0}^{\infty} \frac{(q;q)_{2k} (q^{\frac{z+m+1}{2}};q)_k}{(q;q)_k(q;q)_k(q^{m+1};q)_k} (-1)^k q^{\frac{k(k-1)}{2}} t^k \frac{(q^{2k+1}t;q)_\infty}{(t;q)_\infty} \\
&= \frac{1}{1-t}\sum_{k=0}^{\infty} \frac{(q;q)_{2k} (q^{\frac{z+m+1}{2}};q)_k}{(q;q)_k(q;q)_k(q^{m+1};q)_k(qt;q)_{2k}} (-1)^k q^{\frac{k(k-1)}{2}} t^k \\
&= \frac{1}{1-t}\sum_{k=0}^{\infty} \frac{(q^{\frac{1}{2}};q)_k(-q^{\frac{1}{2}};q)_k, (-q;q)_k (q^{\frac{z+m+1}{2}};q)_k\, (-1)^k q^{\frac{k(k-1)}{2}}}{(q;q)_k(q^{m+1};q)_k((qt)^{\frac{1}{2}};q)_k(-(qt)^{\frac{1}{2}};q)_k(qt^{\frac{1}{2}};q)_k(-qt^{\frac{1}{2}};q)_k} t^k,
\end{aligned}$$

which completes the proof. Note that upon comparison with (1), one requires the vanishing numerator element in the basic hypergeometric series due to the factor $(-1)^k q^{\frac{k(k-1)}{2}}$ in the sum, so it is not of type ${}_4\phi_5$. □

Observe that we obtain (12) by setting $m = 0$ in (22).

Theorem 4. *Let $q, t, z, \lambda \in \mathbb{C}$, $|q| < 1$, $|t| < 1$. Then the q-Sylvester polynomials satisfy the following generating functions:*

$$\sum_{n=0}^{\infty}(q^{\lambda};q)_n\,\varphi_n(z;q)\,t^n = \frac{1}{(zt;q)_{\lambda}}\,{}_2\phi_1\left(\begin{matrix} q^{\lambda}, q^{z} \\ ztq^{\lambda}\end{matrix}\,\middle|\, q, t\right), \tag{23}$$

$$\sum_{n=0}^{\infty}\varphi_n(z;q)\,t^n = \frac{1}{(t;q)_z(zt;q)_{\infty}}, \tag{24}$$

where the principal values of q^z and q^{λ} are taken.

Proof. Let us prove the generating function (23) by using an analogous method as before, namely

$$\begin{aligned}\sum_{n=0}^{\infty}(q^{\lambda};q)_n\,\varphi_n(z;q)\,t^n &= \sum_{n=0}^{\infty}\sum_{k=0}^{n}\frac{(q^{-n};q)_k(q^z;q)_k(-1)^k q^{-\frac{k(k-1)}{2}}(q^{\lambda};q)_n}{(q;q)_n(q;q)_k}\,q^{nk}z^{n-k}t^n \\ &= \sum_{n=0}^{\infty}\sum_{k=0}^{n}\frac{(q^z;q)_k(q^{\lambda};q)_n}{(q;q)_{n-k}(q;q)_k}\,z^{n-k}t^n = \sum_{k=0}^{\infty}\sum_{n=0}^{\infty}\frac{(q^z;q)_k(q^{\lambda};q)_{n+k}}{(q;q)_n(q;q)_k}\,z^n t^{n+k} \\ &= \sum_{k=0}^{\infty}\frac{(q^z;q)_k(q^{\lambda};q)_k}{(q;q)_k}\,t^k\sum_{n=0}^{\infty}\frac{(q^{\lambda+k};q)_n(zt)^n}{(q;q)_n} = \sum_{k=0}^{\infty}\frac{(q^z;q)_k(q^{\lambda};q)_k}{(q;q)_k(zt;q)_{\lambda+k}}\,t^k \\ &= \frac{1}{(zt;q)_{\lambda}}\sum_{k=0}^{\infty}\frac{(q^z;q)_k(q^{\lambda};q)_k}{(q;q)_k(ztq^{\lambda};q)_k}\,t^k.\end{aligned}$$

Since $|q| < 1$, (24) follows from taking $\lambda \to \infty$ and applying the q-binomial theorem. □

Now we find the q-analogue of the first generating function for the q-Cesàro polynomials (9).

Theorem 5. *Let $t, z, s \in \mathbb{C}$, $|t| < 1$, $|z| < 1$. Then the q-Cesàro polynomials satisfy the following generating function:*

$$\sum_{n=0}^{\infty} g_n^{(s)}(z;q)\,t^n = \frac{1}{(1-tzq^s)(t;q)_{s+1}}.$$

Proof. Let us prove this by using (11) and some basic properties of the q-Pochhammer symbols and the q-binomial coefficient. One has

$$\begin{aligned}\sum_{n=0}^{\infty} g_n^{(s)}(z;q)\,t^n &= \sum_{n=0}^{\infty}\sum_{k=0}^{n}\begin{bmatrix} k+s \\ s\end{bmatrix}_q (zq^s)^{n-k}t^n = \sum_{n=0}^{\infty}\sum_{k=0}^{n}\begin{bmatrix} n-k+s \\ s\end{bmatrix}_q (zq^s)^k t^n \\ &= \sum_{k=0}^{\infty}(tzq^s)^k\sum_{n=0}^{\infty}\begin{bmatrix} n+s \\ s\end{bmatrix}_q t^n = \frac{1}{(1-tzq^s)(t;q)_{s+1}},\end{aligned}$$

where we have used the geometric series, (5), and (6), which completes the proof. □

The demonstration that we obtain (9) upon taking the limit $q \uparrow 1$ follows by using (5) and then (6).

Remark 1. *We were unable to find the q-analogue of the second generating function (10) for the q-Cesàro polynomials. The method which Agarwal and Manocha [11] used to obtain (10) does not seem to* *straightforwardly generate a corresponding q-analogue. Furthermore, using (8), one can see that (10) is equivalent to the following generating function for Jacobi polynomials [12] (3.15)*

$$\sum_{n=0}^{\infty}\binom{m+n}{m}P_{m+n}^{(\alpha,\beta-n)}(x)t^n=(1-t)^{\beta}(1-\tfrac{1}{2}(x+1)t)^{-\alpha-\beta-m-1}P_m^{(\alpha,\beta)}\left(\frac{x-\frac{1}{2}(x+1)t}{1-\frac{1}{2}(x+1)t}\right). \tag{25}$$

Unfortunately, this formula does not seem amenable to a natural q-analogue. Note that (25) *is given with a misprint in [6] (p. 165, Problem 9(ii)).*

Remark 2. *It has been mentioned by a referee that Theorems 4, 5 can be derived from the results contained in [13]. However, it is not clear to the authors how to go from the q-Bernoulli polynomials to the q-Sylvester and q-Cesàro polynomials. Moreover, the generating functions for q-Sylvester and q-Cesàro polynomials do not look similar to the generating functions given in [13].*

5. Conclusions

In this paper, we introduced several q-polynomials and derived q-analogues of most of the known generating functions for these polynomials. In particular, this was accomplished for the Bateman, Sylvester, Pasternack, and Cesàro polynomials. In Corollary 1, we also were able to find new classical generating functions, by taking $q \uparrow 1$ limits of the q-generating functions we obtained. We were unable to find a q-analogue for the classical generating function for q-Cesàro polynomials (10) (see Remark 1). This would be an interesting project for the future. It would be interesting to see if it is possible to use q-calculus to obtain q-analogues of the results obtained in [3].

Remark 3. *Note that we recently discovered that the Ph.D. thesis of Mohammad Asif [14] (Chapter 4), under the direction of Prof. Mumtaz Ahmad Khan, contains some of the material that appears in this manuscript. Asif treats both q-Bateman polynomials, the q-Pasternack polynomials, and the q-Cesàro polynomials, all of which are defined in precisely the same way, although Asif uses different notations to display these polynomials. Asif also treats q-Shively pseudo–Laguerre polynomials and q-Gottlieb polynomials. Asif does not treat the q-Sylvester polynomials. It should be noted, however, that Asif arrives at the wrong conclusions for the lower parameters in Theorem 1 and in* (17)*. His notation may or may not be at fault in his representation of Theorem 5. He does find the correct result for Theorem 3.*

Author Contributions: H.S.C., R.S.C.-S. and T.V.W. conceived the mathematics; H.S.C., R.S.C.-S. and T.V.W. wrote the paper.

Funding: R. S. Costas-Santos acknowledges that this research was funded by Dirección General de Investigación, Ministerio de Economía y Competitividad of Spain, grant MTM2015-65888-C4-2-P.

Conflicts of Interest: The authors declare no conflict of interest.

References

1. Koekoek, R.; Lesky, P.A.; Swarttouw, R.F. *Hypergeometric Orthogonal Polynomials and Their q-Analogues*; Springer Monographs in Mathematics; With a foreword by Tom H. Koornwinder; Springer-Verlag: Berlin, Germany, 2010.
2. Gasper, G.; Rahman, M. *Basic Hypergeometric Series*, 2nd ed.; Encyclopedia of Mathematics and its Applications; With a foreword by Richard Askey; Cambridge University Press: Cambridge, UK, 2004.
3. Bateman, H. Two systems of polynomials for the solution of Laplace's integral equation. *Duke Math. J.* **1936**, *2*, 569–577. [CrossRef]
4. Sylvester, J.J. Sur la valeur moyenne des coefficients dans le développement d'un déterminant gauche ou symétrique d'un ordre infiniment grand et sur les déterminants doublement gauches. *C. R. de l'Académie des Sci.* **1879**, *89*, 24–26.
5. Erdélyi, A.; Magnus, W.; Oberhettinger, F.; Tricomi, F.G. *Higher Transcendental Functions*; Robert E. Krieger Publishing Co. Inc.: Melbourne, Australia, 1981.

6. Srivastava, H.M.; Manocha, H.L. *A Treatise on Generating Functions*; Ellis Horwood Series: Mathematics and its Applications; Ellis Horwood Ltd.: Chichester, UK, 1984; p. 569.
7. Rainville, E.D. *Special Functions*; The Macmillan Co.: New York, NY, USA, 1960.
8. Bateman, H. Some Properties of a certain Set of Polynomials. *Tohoku Math. J. First Ser.* **1933**, *37*, 23–38.
9. Pasternack, S. A generalization of the polynomial $F_n(x)$. *Lond. Edinb. Dublin Philos. Mag. J. Sci. Ser. 7* **1939**, *28*, 209–226.
10. Koelink, H.T. On Jacobi and continuous Hahn polynomials. *Proc. Am. Math. Soc.* **1996**, *124*, 887–898. [CrossRef]
11. Agarwal, A.K.; Manocha, H.L. On some new generating functions. *Matematicki Vesnik* **1980**, *4*, 395–402.
12. Srivastava, H.M.; Lavoie, J.L.; Tremblay, R. The Rodrigues Type Representations for a Certain Class of Special Functions. *Annali di Matematica Pura ed Applicata* **1979**, *119*, 9–24. [CrossRef]
13. Kim, D.S.; Kim, T.K. q-Bernoulli polynomials and q-umbral calculus. *Sci. China Math.* **2014**, *57*, 1867–1874. [CrossRef]
14. Asif, M. On Some Problems in Special Functions. Ph.D. Thesis, Aligarh Muslim University, Aligarh, India, 2010.

© 2018 by the authors. Licensee MDPI, Basel, Switzerland. This article is an open access article distributed under the terms and conditions of the Creative Commons Attribution (CC BY) license (http://creativecommons.org/licenses/by/4.0/).

symmetry

MDPI

Article

A New Representation for Srivastava's λ-Generalized Hurwitz-Lerch Zeta Functions

Asifa Tassaddiq

College of Computer and Information Sciences, Majmaah University, Majmaah 11952, Saudi Arabia; a.tassaddiq@mu.edu.sa; Tel.: +966-59-370-9784

Received: 1 November 2018; Accepted: 22 November 2018; Published: 8 December 2018

Abstract: Taking inspiration principally from some of the latest research, we develop a new series representation for the λ-generalized Hurwitz-Lerch zeta functions. This representation led to important new results. The Fourier transform played a foundational role in this work. The duality property of the Fourier transform became significant for checking the consistency of the results. Some known data has been verified as special cases of the results obtained in this investigation.

Keywords: Hurwitz-Lerch zeta function; generalized functions; analytic number theory; λ-generalized Hurwitz-Lerch zeta functions; derivative properties; series representation

---

1. Introduction

The Hurwitz-Lerch zeta function has always been remained a focal point for numerous investigators because of its influence on analytic number theory and further practical disciplines. Recently, Srivastava [1] offered a substantially innovative class of Hurwitz-Lerch zeta functions, namely, λ-generalized Hurwitz-Lerch zeta functions. The exploration of its diverse forms has garnered notable concern, and numerous papers have consequently been presented on this subject. Jankov et al. [2] and Srivastava et al. [3] have offered inequalities by considering diverse cases of these functions. Srivastava et al. [4], have presented a nonlinear operator connected to λ-generalized Hurwitz-Lerch zeta functions, in order to investigate the inclusion properties of the definite subclass of a special type of meromorphic functions. Srivastava and Gaboury [5] have considered new expansion formulas for such functions (see, for related data, [6,7]; see also more systematically supplementary revisions cited in these publications). Luo and Raina [8] have discussed an interesting series representation. They also acquired some new inequalities comprising Srivastava's λ-generalized Hurwitz-Lerch zeta functions.

By taking inspiration from all these outcomes, in our current investigation, we consistently present all the special cases of this newly concentrated family of Srivastava's λ-generalized Hurwitz-Lerch zeta functions in the form of a table. On the one hand, we take account of extended Fermi-Dirac and Bose-Einstein functions defined by Srivastava et al. [9], and on the other, we focus on the close relationship of these functions with the family of zeta and related functions. The purpose of this analysis is to discover some fascinating innovative outcomes for Srivastava's λ-generalized Hurwitz-Lerch zeta functions and their different cases by succeeding the methodology of Chaudhry & Qadir [10], Tassaddiq & Qadir [11,12], Tassaddiq [13], Lail & Qadir [14], and Tassaddiq [15]. In these articles [10–15], the authors have investigated new representations for gamma, generalized gamma, extended Fermi-Dirac and Bose-Einstein functions, and Hypergeometric functions, respectively, in terms of complex delta functions. More recently, Tassaddiq [16] has obtained some new results for Srivastava's λ-generalized Hurwitz-Lerch zeta functions by using its Mellin transform representation.

In the present work, we acquire a different representation for the recently introduced family of the λ-generalized Hurwitz-Lerch zeta functions in terms of complex delta functions. We validate this over the space of entire test functions denoted by **Z**. In the usual sense, we can think of a function being defined in the form of an integral or a series of some variables, or in terms of elementary functions. Nevertheless, it requires consideration as an object in itself, characterized by an integral or a series. This is the only possibility to study the function further than its original domain of description. This is necessary for diverse applications of any function. This concern comes to be principally significant while talking about the concept of higher transcendental functions. Such functions have different series, asymptotic, and integral representations to express functions in diverse domains and to give more simple proofs of its properties when compared to others. Therefore, our new representation is a powerful modeling tool that generalizes the domain of the λ-generalized Hurwitz-Lerch zeta functions from complex numbers to complex functions. It applies to functionals that depend on functions, rather than functions that depend on numbers. Since the methodology used is new, therefore each general result in this paper has the capacity to obtain similar new results for well-studied functions. It provides a computational technique to evaluate integrals of the products of these functions. The stability of the results is confirmed by means of classical methods. In any case, this investigation evidence is meaningful for delivering substantial and innovative results. The approach used is simple and interesting.

Next, we will present the basic definitions and preliminaries by dividing this section into two sections, namely (Section 2.1) and (Section 2.2). In Section 2.1, we discuss preliminaries related to Srivastava's λ-generalized Hurwitz-Lerch zeta functions, while in Section 2.2, we discuss basic preliminaries relevant with distributions (generalized functions) that are necessary to understand the results presented in this paper. The organization of the ensuing sections of this paper is as follows: We present a new representation of the λ-generalized Hurwitz-Lerch zeta functions in Section 3. We achieve analogous outcomes for new associated functions. We discuss the convergence and consequences of new representation in Section 4. We present the Fourier transform representation in Section 5. We check the validity of the results achieved by new representation in Section 5. We summarize our present analysis in the last Section 6. Some interesting new formulae created by giving variations to different parameters are presented in Appendix A.

2. Materials and Methods

2.1. Srivastava's λ-Generalized Hurwitz-Lerch Zeta Functions

Consider the ordinary symbolizations

$$\boldsymbol{N} := \{1,2,\ldots\}; \boldsymbol{N}_0 := \boldsymbol{N} \cup \{0\}; \boldsymbol{Z}^- := \{-1,-2,\ldots\}; \boldsymbol{Z}_0^- := \boldsymbol{Z}^- \cup \{0\} \tag{1}$$

where $\boldsymbol{Z}^-$ is the set of negative integers. The symbols $\boldsymbol{R}, \boldsymbol{R}^+$, and $\boldsymbol{C}$ symbolize the sets of real, positive real, and complex numbers, individually throughout the paper.

The standard Fox-Wright function is an extension for the generalized hypergeometric function that is defined by ([8] (p. 2219) Equation (1)) (see also [3], (p. 516), Equation (1)) and [17] p. (493), Equation (2))

$$\begin{gathered} {}_p\Psi^*{}_q\left[\begin{array}{ccc} (\lambda_1,\rho_1),\ldots, & (\lambda_p,\rho_p) \\ (\mu_1,\sigma_1),\ldots, & (\mu_q,\sigma_q) \end{array};z\right] = \sum_{\chi=0}^{\infty} \frac{([\lambda_p])_{\rho_p\chi}}{([\mu_q])_{\sigma_q\chi}}\frac{z^\chi}{\chi!} \\ (\lambda j,\mu_k \in \boldsymbol{C}\mathbf{and}\rho_j,\sigma_k \in \boldsymbol{R}_+ (j=1,\ldots,p;k=1,\ldots,q)). \end{gathered} \tag{2}$$

Pochammar symbols $([\lambda_p])_{\rho_p\chi} := [\lambda_1]_{\rho_p}\chi \ldots [\lambda_p]_{\rho_p\chi}$ are the shifted factorial, defined in terms of the basic gamma function as follows:

$$\begin{gathered} (\lambda)_\rho = \frac{\Gamma(\lambda+\rho)}{\Gamma(\lambda)} = \begin{cases} 1(\rho=0,\rho\in\mathbb{C}\setminus\{0\}) \\ \lambda(\lambda+1)\ldots(\lambda+\chi-1)(\rho=\chi\in\mathbb{N};\lambda\in\mathbb{C}), \end{cases} \\ \Delta := \sum_{j=1}^{q}\sigma_j - \sum_{j=1}^{p}\rho_j\mathbf{and}\nabla := \left(\prod_{j=1}^{p}\rho_j^{-\rho_j}\right)\cdot\left(\prod_{j=1}^{q}\sigma_j^{\sigma_j}\right). \end{gathered} \tag{3}$$

The series given by (2) converges in the complete complex z-plane for $\Delta > -1$; and if $\Delta = 0$, the series (2) converges for specific values of $|z| < \nabla$. For more a comprehensive exchange of such functions, we refer the interested reader to see the references [18–23].

Srivastava's λ-generalized Hurwitz-Lerch zeta function as presented by ([1], p. 1487, Equation (4))

$$\begin{gathered}\Phi^{(\rho_1,\ldots,\rho_p,\sigma_1,\ldots,\sigma_q)}_{\lambda_1,\ldots,\lambda_p,\mu_1,\ldots,\mu_q}(z,s,a;b,\lambda) := \\ \frac{1}{\Gamma(s)}\int_0^\infty t^{s-1} exp\left(-at-\frac{b}{t^\lambda}\right) p^{\Psi^*} q\left[\begin{matrix}(\lambda_1,\rho_1),\ldots, & (\lambda_p,\rho_p) \\ (\mu_1,\sigma_1),\ldots, & (\mu_q,\sigma_q)\end{matrix}; ze^{-t}\right] dt \\ (min[\Re(a),\Re(s)] > 0; \Re(b) \geqq 0; \lambda \geqq 0)\end{gathered} \tag{4}$$

are central for this research paper. Luo and Raina obtained the following series representation ([8], p. 2221, Equation (6))

$$\begin{gathered}\Phi^{(\rho_1,\ldots,\rho_p,\sigma_1,\ldots,\sigma_q)}_{\lambda_1,\ldots,\lambda_p,\mu_1,\ldots,\mu_q}(z,s,a;b,\lambda) = \frac{1}{\lambda\Gamma(s)}\sum_{\chi=0}^{\infty}\frac{([\lambda_p])_{\rho_p\chi}}{([\mu_q])_{\sigma_q\chi}}\frac{1}{\chi!}Z^{\frac{s}{\lambda}}_{\frac{1}{\lambda}}(a+\chi)^{\lambda} b\frac{z^\chi}{(\chi+a)^s} \\ (\ \lambda j \in R(j=1,\ldots,p) \text{and} \mu_j \in R\backslash Z-0(j=1,\ldots,q); \rho j > 0(j,\ldots,p); \sigma j > 0(j=1,\ldots,q); 1+\Delta \geq 0\ )\end{gathered} \tag{5}$$

so that, obviously, one can get the following association with extended Hurwitz-Lerch zeta functions ([17], p. 503, Equation (6.2)) (see also [3,24])

$$\Phi^{(\rho_1,\ldots,\rho_p,\sigma_1,\ldots,\sigma_q)}_{\lambda_1,\ldots,\lambda_p,\mu_1,\ldots,\mu_q}(z,s,a;0,\lambda) = \Phi^{(\rho_1,\ldots,\rho_p,\sigma_1,\ldots,\sigma_q)}_{\lambda_1,\ldots,\lambda_p,\mu_1,\ldots,\mu_q}(z,s,a) = e^b\Phi^{(\rho_1,\ldots,\rho_p,\sigma_1,\ldots,\sigma_q)}_{\lambda_1,\ldots,\lambda_p,\mu_1,\ldots,\mu_q}(z,s,a;b,0). \tag{6}$$

By making use of Equations (4)–(6), we list all the items in the subsequent table that are straightforward to achieve in view of different values of the parameters as specified column and row wise on the next page.

Now if we go through the previous research, we notice that the different cases of λ-generalized Hurwitz-Lerch zeta functions specified in the third column and second row, explicitly $\Theta^\lambda_\mu(\mp z,s,a;b)$, have been defined and explored by [25], (p. 90), Equation (1.6), and [26]. Some of its most interesting versions were studied and considered by [27]. The original class of zeta functions specifically and explicitly is: Hurwitz-Lerch zeta function $\Phi(\pm z,s,a)$, [28], (p. 27), Equation (1.11), extended Fermi-Dirac $\Theta_a(x;s)$, [9], (p. 9), Equation (3.14), extended Bose- Einstein $\Psi_a(x;s)$, [9], (p. 115), Equation (4.4), Fermi-Dirac $F_s(x)$, [9], p. 109, Equation (1.12)], Bose-Einstein $B_s(x)$, [9], (p. 109), Equation (1.12), Polylogarithm $\phi(z,s)$, [28], (Chapter 1], Hurwitz zeta $\zeta(s,a)$ [28], (Chapter 1), and Riemann zeta functions $\zeta(s)$, [28] (Chapter 1), respectively are listed in the last column of Table 1. Two of the items in the first row specifically $\Phi^{(\rho_1,\ldots,\rho_p,\sigma_1,\ldots,\sigma_q)}_{\lambda_1,\ldots,\lambda_p,\mu_1,\ldots,\mu_q}(\pm z,s,a)$ are defined by [1], (p. 1486), Equation 1.11 (see also [17]) and $\Phi^*_\mu(\pm z,s,a)$ defined by [29], p. 100, Equation (1.5). The extended Riemann zeta $\zeta_b(s)$ [30], (p. 308) and Hurwitz zeta functions $\zeta_b(s,a)$ [30], (p. 308) are noticeable in the last two rows. For additional comprehensive study of zeta and related functions, we refer the reader to [1–32] and related discussions therein.

Table 1. Different Special Cases of λ-Generalized Hurwitz-Lerch Zeta Functions.

$min[\Re(a),\Re(s)]>0;\Re(b)\geq 0;\lambda\geq 0;$ $\rho=\rho_1,\ldots\rho_p;,\sigma=\sigma_1,\ldots,\sigma_q;\lambda^*=\lambda_1,\ldots,\lambda_p;\mu=\mu_1,\ldots,\mu_q$		$(p-1=q=0;\lambda_1=\mu;\rho_1=1)$					$(p-1=q=0;\lambda_1=\mu;\rho_1=1)$	
			$\lambda=1$	$\mu=1$	$\lambda=\mu=1$	$b=0$	$b=0$	$\mu=1;b=0$
λ-Generalized Hurwitz-Lerch Zeta Functions	$\Phi^{(\rho;\sigma)}_{\lambda^*,\mu}(\pm z,s,a;b,\lambda)$ [1], (p. 1487), Equation (1.14)	$\Theta^{\lambda}_{\mu}(\mp z,s,a;b)$ [25], (p. 90), Equation (6) and [26]	$\Phi^*_{\mu}(\pm z,s,a,b)$	$\Phi_b(\pm z,s,a,\lambda)$	$\Phi_b(\pm z,s,a)$	$\Phi^{(\rho;\sigma)}_{\lambda^*,\mu}(\pm z,s,a)$ ([1], p. 1486, Equation (1.11)) & [17]	$\Phi^*_{\mu}(\pm z,s,a)$ ([29], p. 100, Equation (1.5))	$\Phi(\pm z,s,a)$ ([28], p. 27, Equation (1.11))
$\Phi^{(\rho;\sigma)}_{\lambda^*,\mu}(\pm e^{-x},s,a;b,\lambda)$ λ-Generalized Extended Fermi-Dirac and Extended Bose-Einstein Functions	$\Theta^{(\rho;\sigma)}_{\lambda^*,\mu}(x,s,a;b,\lambda)$	$\Theta^{\lambda}_{\mu}(x,s,a;b)$	$\Theta^*_{\mu}(x,s,a,b)$	$\Theta_b(x,s,a;\lambda)$	$\Theta_b(x,s,a)$	$\Theta^{(\rho;\sigma)}_{\lambda^*,\mu}(x,s,a)$	$\Theta^*_{\mu}(x,s,a)$ ([27], p. 12, Equation (45))	$\Theta_a(x;s)$ ([9], p. 9, Equation (3.14))
	$\Psi^{(\rho;\sigma)}_{\lambda^*,\mu}(x,s,a;b,\lambda)$	$\Psi^{\lambda}_{\mu}(x,s,a;b)$	$\Psi^*_{\mu}(x,s,a,b)$	$\Psi_b(x,s,a,\lambda)$	$\Psi_b(x,s,a)$	$\Psi^{(\rho;\sigma)}_{\lambda^*,\mu}(x,s,a)$	$\Psi^*_{\mu}(x,s,a)$ ([27], p. 12, Equation (45))	$\Psi_a(x;s)$ [9], p. 115, Equation (4.4)
$\Phi^{(\rho;\sigma)}_{\lambda^*,\mu}(\pm z,s,1;b,\lambda)$ λ-Generalized Polylogarithm Functions	$\phi^{(\rho;\sigma)}_{\lambda^*,\mu}(\pm z,s;b,\lambda)$	$\phi^{\lambda}_{\mu}(\mp z,s,a;b)$	$\phi^*_{\mu}(z,s,b)$	$\phi_b(z,s,\lambda)$	$\phi_b(z,s)$	$\phi^{(\rho;\sigma)}_{\lambda^*,\mu}(z,s)$	$\phi^*_{\mu}(z,s)$ ([27], p. 12, Equation (42))	$\phi(z,s)$ [28], (Chapter 1)
$\Phi^{(\rho;\sigma)}_{\lambda^*,\mu}(\pm e^{-x},s+1,1;b,\lambda)$ λ-Generalized Fermi-Dirac and Bose Einstein Functions	$F^{(\rho;\sigma)}_{\lambda^*,\mu}(x,s;b,\lambda)$	$F^{\lambda}_{\mu}(x,s,a;b)$	$F^*_{\mu}(x,s,b)$	$F_b(x,s,\lambda)$	$F_b(x,s)$	$F^{(\rho;\sigma)}_{\lambda^*,\mu}(x,s)$	$F^*_{\mu}(x,s)$ ([27], p. 12, Equation (45))	$F_s(x)$ ([9], p. 109, Equation (1.12)
	$B^{(\rho;\sigma)}_{\lambda^*,\mu}(x,s;b,\lambda)$	$B^{\lambda}_{\mu}(x,s,a;b)$	$B^*_{\mu}(x,s,b)$	$B_b(x,s,\lambda)$	$B_b(x,s)$	$B^{(\rho;\sigma)}_{\lambda^*,\mu}(x,s)$	$B^*_{\mu}(x,s)$ ([27], p. 12, Equation (45))	$B_s(x)$ ([9], p. 109, Equation (1.12))
$\Phi^{(\rho;\sigma)}_{\lambda^*,\mu}(\pm 1,s,a;b,\lambda)$ λ-Generalized Hurwitz zeta Functions	$\zeta^{(\rho;\sigma)}_{\lambda^*,\mu}(s,a;b,\lambda)$	$\zeta^{\lambda}_{\mu}(s,a;b)$	$\zeta^*_{\mu}(s,a,b)$	$\zeta_b(s,a,\lambda)$	$\zeta_b(s,a)$ [30], p. 308	$\zeta^{(\rho;\sigma)}_{\lambda^*,\mu}(s,a)$	$\zeta^*_{\mu}(s,a)$ [27]	$\zeta(s,a)$ [28], (Chapter 1)
$\Phi^{(\rho;\sigma)}_{\lambda^*,\mu}(\pm 1,s,1;b,\lambda)$ λ-Generalized Riemann Zeta Functions	$\zeta^{(\rho;\sigma)}_{\lambda^*,\mu}(s)$	$\zeta^{\lambda}_{\mu}(s;b)$	$\zeta^*_{\mu}(s,b)$	$\zeta_b(s,\lambda)$	$\zeta_b(s)$ [30], p. 308	$\zeta^{(\rho;\sigma)}_{\lambda^*,\mu}(s)$	$\zeta^*_{\mu}(s)$ [27]	$\zeta(s)$ [28], (Chapter 1)

2.2. Distributions and Test Functions

Continuous linear functionals that act on some space of test functions are commonly known as generalized functions (or distributions). These are the elements of the corresponding dual space of test functions. A review of such elements is significant, because they not only have locally integrable functions, but also consist of additional objects that are not regular distributions. Consequently, several actions such as integration, differentiation, and limits that are defined for functions can be applied to functionals. A delta functional commonly used in singular distribution is defined by

$$\langle \delta(u-a), \varphi(t) \rangle = \varphi(a)(\forall \varphi \in D, a \in R), \tag{7}$$

where for a non-zero a, $\delta(-u) = \delta(u); \delta(au) = \frac{\delta(t)}{|a|}$.

A multi-volume presentation [33] (Vol. I-V) by Gelfand and Shilov is a great treatise on such functions. The commonly used spaces of test functions are the spaces of compact support functions denoted by D, and the space of rapidly decaying functions denoted by S, that also have derivatives of all orders. The spaces D' and S' are the dual spaces of D and S. Spaces S and S' are closed under the Fourier transform, but D and $D\prime$ are not. The Fourier transform of the elements of $D\prime$ are continuous linear functionals acting on the elements of z that comprises of entire functions such that their Fourier transforms are in D [34]. The entire function $\varphi \epsilon z$ does not vanish on some interval $a < u < b$, but vanishes universally. Accordingly

$$z' \supset S' \supset S \supset z; D \cap z \equiv 0; D' \supset S' \supset S \supset D. \tag{8}$$

The elements of Z consist of entire analytic functions satisfying the following set of inequalities

$$|s^q \varphi(s)| \leq C_q e^{a|\tau|}; (q = 0, 1, 2, \dots.) \tag{9}$$

where the constants a and C_q may depend on φ. By ([33], Vol 1, p. 169, Equation (8)), we take the Fourier transform of exponential function

$$F\left[e^{\alpha t}; \omega\right] = 2\pi\delta(\omega - i\alpha) \tag{10}$$

as an example of distribution that is an element of z' and for $\forall g \in z'$ ([33], (p. 159), Equation (4)), see also ([34], p. 201, Equation (9))

$$g(s+b) = \sum_{r=0}^{\infty} g^{(r)}(s)\frac{b^r}{r!}. \tag{11}$$

So that we have the following basic identity

$$\delta(s+b) = \sum_{r=0}^{\infty} \delta^{(r)}(s)\frac{b^r}{r!}; \ \langle \delta^{(r)}(s), \varphi(s) \rangle = (-1)^r \varphi^{(r)}(0). \tag{12}$$

For an additional extensive study of these spaces, we refer the reader to [33] (Vol. I–V), [34,35] and the related bibliography therein.

Throughout this investigation, conditions on the parameters will be considered standard as given in (1)–(6) unless otherwise stated.

3. Results

New Series Representation of the λ-Generalized Hurwitz-Lerch Zeta Functions

Theorem 1. *λ-generalized Hurwitz-Lerch zeta functions have the following representation*

$$\begin{aligned}&\Gamma(s)\Phi^{(\rho_1,\ldots,\rho_p,\sigma_1,\ldots,\sigma_q)}_{\lambda_1,\ldots,\lambda_p,\mu_1,\ldots,\mu_q}(z,s,a;b,\lambda)\\&=2\pi\sum_{\chi,\xi,\psi=0}^{\infty}\frac{([\lambda_p])_{\rho_p\chi}}{([\mu_q])_{\sigma_q\chi}}\frac{(z)^{\chi}}{\chi!}\frac{(-(\chi+a))^{\xi}}{\xi!}\frac{(-b)^{\psi}}{\psi!}\delta(s+\xi-\lambda\psi).\end{aligned}\tag{13}$$

Proof: Let us first replace $t=e^y$ and $s=\sigma+i\tau$ in Equation (4), then we get

$$\begin{aligned}&\Phi^{(\rho_1,\ldots,\rho_p,\sigma_1,\ldots,\sigma_q)}_{\lambda_1,\ldots,\lambda_p,\mu_1,\ldots,\mu_q}(z,s,a)=\\&=\frac{1}{\Gamma(s)}\int_{-\infty}^{\infty}e^{y(\sigma+i\tau)}exp\left(-ae^y-\frac{b}{e^{\lambda y}}\right){}_p\Psi^*{}_q\left[\begin{array}{ccc}(\lambda_1,\rho_1),\ldots, & (\lambda_p,\rho_p) & ;z\,exp(-e^y)\\(\mu_1,\sigma_1),\ldots, & (\mu_q,\sigma_q) & \end{array}\right]dt,\\&(min[\Re(a),\Re(s)]>0).\end{aligned}\tag{14}$$

Now, writing the series form of the Fox-Wright function

$${}_p\Psi^*{}_q\left[\begin{array}{ccc}(\lambda_1,\rho_1),\ldots, & (\lambda_p,\rho_p) & ;zexp(-e^y)\\(\mu_1,\sigma_1),\ldots, & (\mu_q,\sigma_q) & \end{array}\right]=\sum_{\chi=0}^{\infty}\frac{([\lambda_p])_{\rho_p\chi}}{([\mu_q])_{\sigma_q\chi}}\frac{z^{\chi}}{\chi!}exp(-\chi e^y)\tag{15}$$

and then collecting and expanding the exponential terms

$$e^{\sigma y}exp\left(-(a+\chi)e^y-\frac{b}{e^{\lambda y}}\right)=\sum_{\xi,\psi=0}^{\infty}\frac{([\lambda_p])_{\rho_p\chi}}{([\mu_q])_{\sigma_q\chi}}\frac{(z)^{\chi}}{\chi!}\frac{(-(\chi+a))^{\xi}}{\xi!}\frac{(-b)^{\psi}}{\psi!},\tag{16}$$

we get

$$\begin{aligned}&\Phi^{(\rho_1,\ldots,\rho_p,\sigma_1,\ldots,\sigma_q)}_{\lambda_1,\ldots,\lambda_p,\mu_1,\ldots,\mu_q}(z,s,a;b,\lambda)\\&=\sum_{\chi,\xi,\psi=0}^{\infty}\frac{([\lambda_p])_{\rho_p\chi}}{([\mu_q])_{\sigma_q\chi}}\frac{(z)^{\chi}}{\chi!}\frac{(-(\chi+a))^{\xi}}{\xi!}\frac{(-b)^{\psi}}{\psi!}\int_{-\infty}^{\infty}e^{i\tau y}e^{(\sigma+\xi-\psi\lambda)y}dy.\end{aligned}\tag{17}$$

The order of summation and integration is interchangeable due to uniform convergence of the integral. By using Equation (10), we get

$$\begin{aligned}\int_{-\infty}^{\infty}e^{i\tau y}e^{(\sigma+\xi-\psi\lambda)y}dy&=F\left[e^{(\sigma+\xi-\psi\lambda)y};\tau\right]=2\pi\delta(\tau-i(\sigma+\xi-\psi\lambda))\\&=2\pi\delta\left[\tfrac{1}{i}(i\tau-(\sigma+\xi-\psi\lambda))\right]\\&=2\pi|i|\delta(\sigma+i\tau+\xi-\psi\lambda)=2\pi\delta(s+\xi-\psi\lambda).\end{aligned}\tag{18}$$

The above Equations (17) and (18) lead to the required result. □

Remark 1. *We can get analogous outcomes for further associated functions as enumerated row-wise in Table 1, in view of altered parameter values in the form of following corollaries.*

Corollary 1. *λ-Generalized Extended Fermi-Dirac functions have the following representation*

$$\begin{aligned}&\Gamma(s)\Theta^{(\rho_1,\ldots,\rho_p,\sigma_1,\ldots,\sigma_q)}_{\lambda_1,\ldots,\lambda_p,\mu_1,\ldots,\mu_q}(x,s,a;b,\lambda)\\&=2\pi\sum_{\chi,\xi,\psi=0}^{\infty}\frac{([\lambda_p])_{\rho_p\chi}}{([\mu_q])_{\sigma_q\chi}}\frac{(-e^{-x})^{\chi}}{\chi!}\frac{(-(\chi+a))^{\xi}}{\xi!}\frac{(-b)^{\psi}}{\psi!}\delta(s+\xi-\lambda\psi)\end{aligned}\tag{19}$$

and λ*-Generalized Extended Bose-Einstein functions have the following representation*

$$\begin{aligned}&\Gamma(s)\Psi^{(\rho_1,\ldots,\rho_p,\sigma_1,\ldots,\sigma_q)}_{\lambda_1,\ldots,\lambda_p,\mu_1,\ldots,\mu_q}(x,s,a;b,\lambda)\\&=2\pi\sum_{\chi,\xi,\psi=0}^{\infty}\frac{([\lambda_p])_{\rho_p\chi}}{([\mu_q])_{\sigma_q\chi}}\frac{(e^{-x})^{\chi}}{\chi!}\frac{(-(\chi+a))^{\xi}}{\xi!}\frac{(-b)^{\psi}}{\psi!}\delta(s+\xi-\lambda\psi).\end{aligned} \tag{20}$$

Proof. This holds by simply replacing $z \longrightarrow \pm e^{-x}$ on both sides of (13) and by means of the corresponding item specified in column 2 and row 2 of Table 1. □

Corollary 2. λ*-Generalized Fermi-Dirac functions have the following representation*

$$\begin{aligned}&\Gamma(s)F^{(\rho_1,\ldots,\rho_p,\sigma_1,\ldots,\sigma_q)}_{\lambda_1,\ldots,\lambda_p,\mu_1,\ldots,\mu_q}(x,s;b,\lambda)\\&=2\pi\sum_{\chi,\xi,\psi=0}^{\infty}\frac{([\lambda_p])_{\rho_p\chi}}{([\mu_q])_{\sigma_q\chi}}\frac{(-e^{-x})^{\chi}}{\chi!}\frac{(-(\chi+1))^{\xi}}{\xi!}\frac{(-b)^{\psi}}{\psi!}\delta(s+\xi-\lambda\psi)\end{aligned} \tag{21}$$

and λ*-Generalized Extended Bose-Einstein functions have the following representation*

$$\begin{aligned}&\Gamma(s)B^{(\rho_1,\ldots,\rho_p,\sigma_1,\ldots,\sigma_q)}_{\lambda_1,\ldots,\lambda_p,\mu_1,\ldots,\mu_q}(x,s;b,\lambda)\\&=2\pi\sum_{\chi,\xi,\psi=0}^{\infty}\frac{([\lambda_p])_{\rho_p\chi}}{([\mu_q])_{\sigma_q\chi}}\frac{(e^{-x})^{\chi}}{\chi!}\frac{(-(\chi+a))^{\xi}}{\xi!}\frac{(-b)^{\psi}}{\psi!}\delta(s+\xi-\lambda\psi).\end{aligned} \tag{22}$$

Proof. Both results hold by simply replacing $z \longrightarrow \pm e^{-x}; a \longrightarrow 1$ on both sides of (13) and in view of defined item from Table 1 reliable on these parameter values. □

Corollary 3. λ*-Generalized Polylogarithm functions has the following representation*

$$\begin{aligned}&\Gamma(s)\Phi^{(\rho_1,\ldots,\rho_p,\sigma_1,\ldots,\sigma_q)}_{\lambda_1,\ldots,\lambda_p,\mu_1,\ldots,\mu_q}(z,s;b,\lambda)\\&=2\pi z\sum_{\chi,\xi,\psi=0}^{\infty}\frac{([\lambda_p])_{\rho_p\chi}}{([\mu_q])_{\sigma_q\chi}}\frac{(z)^{\chi}}{\chi!}\frac{(-(\chi+1))^{\xi}}{\xi!}\frac{(-b)^{\psi}}{\psi!}\delta(s+\xi-\lambda\psi).\end{aligned} \tag{23}$$

Proof. This holds by simply replacing $a \longrightarrow 1$ on both sides of (13) and using the precise element from Table 1 equivalent to these constraint values. □

Corollary 4. λ*-Generalized Hurwitz zeta functions has the following representation*

$$\Gamma(s)\zeta^{(\rho_1,\ldots,\rho_p,\sigma_1,\ldots,\sigma_q)}_{\lambda_1,\ldots,\lambda_p,\mu_1,\ldots,\mu_q}(s,a;b,\lambda)=2\pi\sum_{\chi,\xi,\psi=0}^{\infty}\frac{([\lambda_p])_{\rho_p\chi}}{([\mu_q])_{\sigma_q\chi}}\frac{1}{\chi!}\frac{(-(\chi+a))^{\xi}}{\xi!}\frac{(-b)^{\psi}}{\psi!}\delta(s+\xi-\lambda\psi). \tag{24}$$

Proof. This holds by simply replacing $z \longrightarrow 1$ on both sides of (13) and, in view of particular items from Table 1, stable with these parameter values. □

Corollary 5. λ*-Generalized Riemann zeta functions has the following representation*

$$\Gamma(s)\zeta^{(\rho_1,\ldots,\rho_p,\sigma_1,\ldots,\sigma_q)}_{\lambda_1,\ldots,\lambda_p,\mu_1,\ldots,\mu_q}(s;b,\lambda)=2\pi\sum_{\chi,\xi,\psi=0}^{\infty}\frac{([\lambda_p])_{\rho_p\chi}}{([\mu_q])_{\sigma_q\chi}}\frac{1}{\chi!}\frac{(-(\chi+1))^{\xi}}{\xi!}\frac{(-b)^{\psi}}{\psi!}\delta(s+\xi-\lambda\psi). \tag{25}$$

Proof. This holds by simply replacing $z \longrightarrow 1; a \longrightarrow 1$ on both sides of (13) and, in view of certain components from Table 1, is firm with these considered values. □

Remark 2. *We can get similar representations for other special cases of these functions by considering different parameter variations in view of Table 1 column-wise.*

By putting $\lambda = 0$ in the above results (13), and in view of the relation (6), we get the following new results:

$$\Gamma(s)\Phi^{(\rho_1,\ldots,\rho_p,\sigma_1,\ldots,\sigma_q)}_{\lambda_1,\ldots,\lambda_p,\mu_1,\ldots,\mu_q}(z,s,a) = 2\pi e^b \sum_{\chi,\xi=0}^{\infty} \frac{([\lambda_p])_{\rho_p\chi}}{([\mu_q])_{\sigma_q\chi}} \frac{(z)^\chi}{\chi!} \frac{(-(\chi+a))^\xi}{\xi!} \delta(s+\xi). \tag{26}$$

Next by considering $b = 0$ in (26), we get the following results:

$$\Gamma(s)\Phi^{(\rho_1,\ldots,\rho_p,\sigma_1,\ldots,\sigma_q)}_{\lambda_1,\ldots,\lambda_p,\mu_1,\ldots,\mu_q}(z,s,a) = 2\pi \sum_{\chi,\xi=0}^{\infty} \frac{([\lambda_p])_{\rho_p\chi}}{([\mu_q])_{\sigma_q\chi}} \frac{(z)^\chi}{\chi!} \frac{(-(\chi+a))^\xi}{\xi!} \delta(s+\xi). \tag{27}$$

Considering, $p - 1 = q = 0$; $(\lambda_1 = \mu; \rho_1 = 1)$, the above Equation (13) would reduce immediately to the following form

$$\Gamma(s)\Theta^\lambda_\mu(z,s,a;b) = 2\pi \sum_{\chi,\xi,\psi=0}^{\infty} (\mu)_\chi \frac{(z)^\chi}{\chi!} \frac{(-(\chi+a))^\xi}{\xi!} \frac{(-b)^\psi}{\psi!} \delta(s+\xi-\lambda\psi). \tag{28}$$

Next, specifying $\mu = 1$ in (28), one can get the following new result as special case

$$\Gamma(s)\Theta(z,s,a;b,\lambda) = 2\pi \sum_{\chi,\xi,\psi=0}^{\infty} (z)^\chi \frac{(-(\chi+a))^\xi}{\xi!} \frac{(-b)^\psi}{\psi!} \delta(s+\xi-\lambda\psi). \tag{29}$$

Next, again by giving variations to different parameters, we can get similar representations for other special cases of these functions.

By putting $\lambda = 0$ in the above result (29), we get the following new result

$$\Gamma(s)\Phi(z,s,a;b) = 2\pi e^b \sum_{\chi,\xi=0}^{\infty} (z)^\chi \frac{(-(\chi+a))^\xi}{\xi!} \delta(s+\xi). \tag{30}$$

By putting $b = 0$ in (30), we get the following results for the original family of Hurwitz-Lerch zeta function and its special cases ([13], Chapter 4):

$$\Gamma(s)\Phi(z,s,a) = 2\pi \sum_{\chi,\xi=0}^{\infty} (z)^\chi \frac{(-(\chi+a))^\xi}{\xi!} \delta(s+\xi), \tag{31}$$

$$\Gamma(s)\phi(z,s) = 2\pi z \sum_{\chi,\xi=0}^{\infty} (z)^\chi \frac{(-(\chi+1))^\xi}{\xi!} \delta(s+\xi), \tag{32}$$

$$\Gamma(s)\zeta(s,a) = 2\pi \sum_{\chi,\xi=0}^{\infty} \frac{(-(\chi+a))^\xi}{\xi!} \delta(s+\xi), \tag{33}$$

$$\Gamma(s)\zeta(s) = 2\pi \sum_{\chi,\xi=0}^{\infty} \frac{(-(\chi+1))^\xi}{\xi!} \delta(s+\xi). \tag{34}$$

Remark 3. *Note that we have obtained a demonstration given in the form of complex delta functions that is only meaningful in the sense of distributions once defined as an inner product with some suitable function. For example, divide both sides of (34) in the usual sense*

$$1 = \frac{\sum_{\chi,\xi=0}^{\infty} \frac{(-(\chi+1))^{\xi}}{\xi!} \delta(s+\xi)}{\Gamma(s)\zeta(s)}. \tag{35}$$

In addition, we get

$$1 = \sum_{\chi,\xi=0}^{\infty} \frac{(-(\chi+1))^{\xi}}{\xi!} \frac{1}{\Gamma(-\xi)\zeta(-\xi)}, \tag{36}$$

where the product $\Gamma(-\xi)\zeta(-\xi)$ contributes only for even values of ξ, because zeros of zeta cancel the poles of gamma functions while for other values of ξ, the right-hand side sum will vanish due to $\Gamma(-\xi)$ in the reciprocal. Therefore, we get

$$1 = \sum_{\chi,\xi=1}^{\infty} \frac{(\chi)^{2\xi}}{(2\xi)!} + 0 \Longrightarrow 1 = \sum_{\chi=0}^{\infty} cosh(\chi) \Longrightarrow 1 = \infty \tag{37}$$

that leads to an obvious contradiction.

Meanwhile, if we consider the inner product

$$\langle \Gamma(s)\zeta(s), \frac{1}{\Gamma(s)\zeta(s)} \rangle = \sum_{\chi,\xi=0}^{\infty} \frac{(-(\chi+1))^{\xi}}{\xi!} \langle \delta(s+\xi), \frac{1}{\Gamma(s)\zeta(s)} \rangle \tag{38}$$

then we get

$$\int_{s\in\mathbb{C}} 1ds = \sum_{\chi,\xi=0}^{\infty} \frac{(-(\chi+1))^{\xi}}{\xi!} \frac{1}{\Gamma(-\xi)\zeta(-\xi)}. \tag{39}$$

Due to the reason as stated above we get

$$\int_{s\in\mathbb{C}} 1ds = \sum_{\chi,\xi=1}^{\infty} \frac{(\chi)^{2\xi}}{(2\xi)!} + 0, \tag{40}$$

$$\int_{s\in\mathbb{C}} 1ds = \int_{-\infty}^{+\infty} 1ds = \sum_{\chi=0}^{\infty} cosh(\chi), \tag{41}$$

and both sides diverge. Therefore, we need to be very rigorous in selecting a class of functions for which this representation is meaningful or convergent.

4. Convergence and Applications of New Series Representation

The representation of the λ -generalized Hurwitz-Lerch zeta function

$$\Phi^{(\rho_1,\ldots,\rho_p,\sigma_1,\ldots,\sigma_q)}_{\lambda_1,\ldots,\lambda_p,\mu_1,\ldots,\mu_q}(z,s,a;b,\lambda)$$

and related functions is attained in the form of the series of delta function that is defined simply if converges as distributions or generalized functions. Therefore, these new representations are well defined for the functions for which these infinite series converge. Meanwhile, the complex delta function acts as a continuous linear functional on the space **Z**. Hence, it is straightforward that the series of delta functions are obviously the continuous linear functionals acting on the space **Z**. (The results may also be true for some larger spaces, but here in our present investigation, we are just restricting to $\boldsymbol{Z}$). Therefore, $\forall\boldsymbol{\Lambda}(s)\epsilon\boldsymbol{Z}$, we get from (13)

$$\begin{aligned}
&\langle\boldsymbol{\Gamma}(s)\Phi^{(\rho_1,\ldots,\rho_p,\sigma_1,\ldots,\sigma_q)}_{\lambda_1,\ldots,\lambda_p,\mu_1,\ldots,\mu_q}(z,s,a;b,\lambda),\boldsymbol{\Lambda}(s)\rangle = \\
2\pi &\sum_{\chi,\xi,\psi=0}^{\infty}\frac{([\lambda_p])_{\rho_p\chi}}{([\mu_q])_{\sigma_q\chi}}\frac{(z)^{\chi}}{\chi!}\frac{(-(\chi+a)^{\xi}}{\xi!}\frac{(-b)^{\psi}}{\psi!}\langle\delta(s+\xi-\lambda\psi),\boldsymbol{\Lambda}(s)\rangle;(\forall\boldsymbol{\Lambda}(s)\epsilon\boldsymbol{Z}) \\
&=\sum_{\chi,\xi,\psi=0}^{\infty}\frac{([\lambda_p])_{\rho_p\chi}}{([\mu_q])_{\sigma_q\chi}}\frac{(z)^{\chi}}{\chi!}\frac{(-(\chi+a))^{\xi}}{\xi!}\frac{(-b)^{\psi}}{\psi!}\boldsymbol{\Lambda}(\lambda\psi-\xi).
\end{aligned}\tag{42}$$

Here, in the above equation, we have used the shifting property of delta functions as follows

$$\langle\delta(s+\xi-\lambda\psi),\boldsymbol{\Lambda}(s)\rangle=\boldsymbol{\Lambda}(\lambda\psi-\xi),\tag{43}$$

which being the elements of space $\boldsymbol{Z}$ are slowly increasing (bounded by a polynomial) test functions and note that sum over the coefficients is

$$\begin{aligned}
\textbf{sumoverthecoefficients} &= \sum_{\chi,\xi,\psi=0}^{\infty}\frac{([\lambda_p])_{\rho_p\chi}}{([\mu_q])_{\sigma_q\chi}}\frac{(z)^{\chi}}{\chi!}\frac{(-(\chi+a))^{\xi}}{\xi!}\frac{(-b)^{\psi}}{\psi!} \\
&= exp(-a-b)p^{\Psi^*}q\left[\begin{array}{ll}(\lambda_1,\rho_1),\ldots, & (\lambda_p,\rho_p) \\ (\mu_1,\sigma_1),\ldots, & (\mu_q,\sigma_q)\end{array};\frac{z}{e}\right],
\end{aligned}\tag{44}$$

which is finite and well defined. Therefore, by using the famous Abel convergent test or by ([35], Proposition 1, p. 46), it is obvious that new series given by (13) converges for $\forall\boldsymbol{\Lambda}(s)\epsilon\boldsymbol{Z}$, which leads to a similar fact for its special and other related cases given in Section 3 and Appendix A.

As already mentioned, in our present investigation, we proved the convergence for slowly increasing functions, but it can now be observed that the series converges for a larger space of functions. Therefore, the condition is necessary and not sufficient, that means for $\forall\boldsymbol{\Lambda}(s)\epsilon\boldsymbol{Z}$, the series is convergent but if the series is convergent, then $\boldsymbol{\Lambda}(s)$ may belong to some other large space for which delta function is meaningful.

Next, by using the new representation of $\Phi^{(\rho_1,\ldots,\rho_p,\sigma_1,\ldots,\sigma_q)}_{\lambda_1,\ldots,\lambda_p,\mu_1,\ldots,\mu_q}(z,s,a;b,\lambda)$, we can find some new integral formulae and verify them by using classical Fourier transform. First, we consider a simple example of a specific set of functions

$$\boldsymbol{\Lambda}(s)=\omega^{s\beta}(\omega\neq 0;s\in C;\beta\epsilon\mathbb{R}).\tag{45}$$

By taking the inner product of these functions with (13) and using the basic (shift) property of delta functions, we get

$$\begin{aligned}
&\int_{s\in\mathbb{C}}\omega^{s\beta}\boldsymbol{\Gamma}(s)\Phi^{(\rho_1,\ldots,\rho_p,\sigma_1,\ldots,\sigma_q)}_{\lambda_1,\ldots,\lambda_p,\mu_1,\ldots,\mu_q}(z,s,a;b,\lambda)ds = \\
&2\pi\sum_{\chi,\xi,\psi=0}^{\infty}\frac{([\lambda_p])_{\rho_p\chi}}{([\mu_q])_{\sigma_q\chi}}\frac{(z)^{\chi}}{\chi!}\frac{(-(\chi+a))^{\xi}}{\xi!}\frac{(-b)^{\psi}}{\psi!}\omega^{(\lambda\psi-\xi)\beta} \\
&=2\pi exp\Big(-a\omega^{-\beta}-\frac{b}{\omega^{-\lambda\beta}}\Big)p^{\Psi^*}q\left[\begin{array}{ll}(\lambda_1,\rho_1),\ldots, & (\lambda_p,\rho_p) \\ (\mu_1,\sigma_1),\ldots, & (\mu_q,\sigma_q)\end{array};z.exp(-\omega^{-\beta})\right].
\end{aligned}\tag{46}$$

Similarly, by considering the action of $\boldsymbol{\Lambda}(s)$ for representations (19)–(34), we can get the following new results:

$$\begin{aligned}
&\int_{s\in\mathbb{C}}\omega^{s\beta}\boldsymbol{\Gamma}(s)\boldsymbol{\Theta}^{(\rho_1,\ldots,\rho_p,\sigma_1,\ldots,\sigma_q)}_{\lambda_1,\ldots,\lambda_p,\mu_1,\ldots,\mu_q}(x,s,a;b,\lambda)ds \\
&=2\pi exp\Big(-a\omega^{-\beta}-\frac{b}{\omega^{-\lambda\beta}}\Big)p^{\Psi^*}q\left[\begin{array}{ll}(\lambda_1,\rho_1),\ldots, & (\lambda_p,\rho_p) \\ (\mu_1,\sigma_1),\ldots, & (\mu_q,\sigma_q)\end{array};-exp(-x-\omega^{-\beta})\right];
\end{aligned}\tag{47}$$

$$\int_{s\in\mathbb{C}} \omega^{s\beta}\Gamma(s)\Psi^{(\rho_1,\ldots,\rho_p,\sigma_1,\ldots,\sigma_q)}_{\lambda_1,\ldots,\lambda_p,\mu_1,\ldots,\mu_q}(x,s,a;b,\lambda)ds = 2\pi exp\left(-a\omega^{-\beta}-\frac{b}{\omega^{-\lambda\beta}}\right)p^{\Psi^*}q\left[\begin{array}{ccc}(\lambda_1,\rho_1),\ldots, & (\lambda_p,\rho_p) & ;exp(-x-\omega^{-\beta})\\ (\mu_1,\sigma_1),\ldots, & (\mu_q,\sigma_q) & \end{array}\right]; \tag{48}$$

$$\int_{s\in\mathbb{C}} \omega^{s\beta}\Gamma(s)F^{(\rho_1,\ldots,\rho_p,\sigma_1,\ldots,\sigma_q)}_{\lambda_1,\ldots,\lambda_p,\mu_1,\ldots,\mu_q}(x,s;b,\lambda)ds = 2\pi exp\left(-\omega^{-\beta}-\frac{b}{\omega^{-\lambda\beta}}\right)p^{\Psi^*}q\left[\begin{array}{ccc}(\lambda_1,\rho_1),\ldots, & (\lambda_p,\rho_p) & ;-exp(-x-\omega^{-\beta})\\ (\mu_1,\sigma_1),\ldots, & (\mu_q,\sigma_q) & \end{array}\right]; \tag{49}$$

$$\int_{s\in\mathbb{C}} \omega^{s\beta}\Gamma(s)B^{(\rho_1,\ldots,\rho_p,\sigma_1,\ldots,\sigma_q)}_{\lambda_1,\ldots,\lambda_p,\mu_1,\ldots,\mu_q}(x,s;b,\lambda)ds = 2\pi exp\left(-\omega^{-\beta}-\frac{b}{\omega^{-\lambda\beta}}\right)p^{\Psi^*}q\left[\begin{array}{ccc}(\lambda_1,\rho_1),\ldots, & (\lambda_p,\rho_p) & ;exp(-x-\omega^{-\beta})\\ (\mu_1,\sigma_1),\ldots, & (\mu_q,\sigma_q) & \end{array}\right]; \tag{50}$$

$$\int_{s\in\mathbb{C}} \omega^{s\beta}\Gamma(s)\Phi^{(\rho_1,\ldots,\rho_p,\sigma_1,\ldots,\sigma_q)}_{\lambda_1,\ldots,\lambda_p,\mu_1,\ldots,\mu_q}(z,s;b,\lambda)ds = 2\pi z.exp\left(-\omega^{-\beta}-\frac{b}{\omega^{-\lambda\beta}}\right)p^{\Psi^*}q\left[\begin{array}{ccc}(\lambda_1,\rho_1),\ldots, & (\lambda_p,\rho_p) & ;zexp(-\omega^{-\beta})\\ (\mu_1,\sigma_1),\ldots, & (\mu_q,\sigma_q) & \end{array}\right]; \tag{51}$$

$$\int_{s\in\mathbb{C}} \omega^{s\beta}\Gamma(s)\zeta^{(\rho_1,\ldots,\rho_p,\sigma_1,\ldots,\sigma_q)}_{\lambda_1,\ldots,\lambda_p,\mu_1,\ldots,\mu_q}(s,a;b;\lambda)ds = 2\pi exp\left(-a\omega^{-\beta}-\frac{b}{\omega^{-\lambda\beta}}\right)p^{\Psi^*}q\left[\begin{array}{ccc}(\lambda_1,\rho_1),\ldots, & (\lambda_p,\rho_p) & ;exp(-\omega^{-\beta})\\ (\mu_1,\sigma_1),\ldots, & (\mu_q,\sigma_q) & \end{array}\right]; \tag{52}$$

$$\int_{s\in\mathbb{C}} \omega^{s\beta}\Gamma(s)\zeta^{(\rho_1,\ldots,\rho_p,\sigma_1,\ldots,\sigma_q)}_{\lambda_1,\ldots,\lambda_p,\mu_1,\ldots,\mu_q}(s;b,\lambda)ds = 2\pi exp\left(-\omega^{-\beta}-\frac{b}{\omega^{-\lambda\beta}}\right)p^{\Psi^*}q\left[\begin{array}{ccc}(\lambda_1,\rho_1),\ldots, & (\lambda_p,\rho_p) & ;exp(-\omega^{-\beta})\\ (\mu_1,\sigma_1),\ldots, & (\mu_q,\sigma_q) & \end{array}\right]. \tag{53}$$

By putting $b = 0$, in (46), we get the following new results: (and if we put $\lambda = 0$, we get e^b times the following results (54)):

$$\int_{s\in\mathbb{C}} \omega^{s\beta}\Gamma(s)\Phi^{(\rho_1,\ldots,\rho_p,\sigma_1,\ldots,\sigma_q)}_{\lambda_1,\ldots,\lambda_p,\mu_1,\ldots,\mu_q}(z,s,a)ds = \sum_{\chi,\xi=0}^{\infty}\frac{(z)^{\chi}}{\chi!}\frac{(-(\chi+a))^{\xi}}{\xi!}\omega^{(\lambda\psi-\xi)\beta} = 2\pi exp(-a\omega^{-\beta})p^{\Psi^*}q\left[\begin{array}{ccc}(\lambda_1,\rho_1),\ldots, & (\lambda_p,\rho_p) & ;z.exp(-\omega^{-\beta})\\ (\mu_1,\sigma_1),\ldots, & (\mu_q,\sigma_q) & \end{array}\right]. \tag{54}$$

By considering $p-1=q=0$ ($\lambda_1 = \mu$; $\rho_1 = 1$), $b \neq 0$ in Equation (46) we can get the following

$$\int_{s\in\mathbb{C}} \omega^{s\beta}\Gamma(s)\Theta^{\lambda}_{\mu}(z,s,a;b)ds = \frac{2\pi exp\left(-(a-1)\omega^{-\beta}-\frac{b}{\omega^{-\lambda\beta}}\right)}{(exp(\omega^{-\beta})-z)^{\mu}}. \tag{55}$$

Taking $b = 0$ in the above results (55) leads to the following new result:

$$\int_{s\in\mathbb{C}} \omega^{s\beta}\Gamma(s)\Phi^{*}_{\mu}(z,s,a)ds = \frac{2\pi exp(-(a-1)\omega^{-\beta})}{(exp(\omega^{-\beta})-z)^{\mu}}. \tag{56}$$

By considering other parametric values as, $p-1=q=0$; $\lambda_1 = \mu$; $\rho_1 = 1$; $b \neq 0$; $\lambda = \mu = 1$ the above result (54) shrinks instantly to the subsequent result:

$$\int_{s\in\mathbb{C}} \omega^{s\beta}\Gamma(s)\Phi_b(z,s,a)ds = \frac{2\pi exp\left(-(a-1)\omega^{-\beta}-\frac{b}{\omega^{-\lambda\beta}}\right)}{(exp(\omega^{-\beta})-z)}. \tag{57}$$

Next, by putting $b = 0$ in the above Equations (57), we get the following [13], (Chapter 4):

$$\int_{s\in\mathbb{C}} \omega^{s\beta}\Gamma(s)\Phi(z,s,a)ds = \frac{exp(-(a-1)\omega^{-\beta})}{(exp(\omega^{-\beta})-z)}. \tag{58}$$

Remark 4. *Results obtained in this section give insights for further new results. For example, consider* $\omega = \frac{1}{e}$*, then we get the Laplace transform of the* λ*-generalized Hurwitz-Lerch zeta functions and the related family of functions. Before going on further with this new representation, we consider the consistency of the new results in the subsequent section.*

5. Fourier Transform Representation

The main purpose of this section is to verify the consistency of the results obtained by the new series representation with the classical Fourier transform representation. Different transform representations have always been of interest for such functions.

By replacing $t = e^y$ and $s = \sigma + i\tau$ in Equation (4), the Fourier transform representation of λ-generalized Hurwitz-Lerch zeta functions is

$$\begin{aligned}&\Gamma(s)\Phi^{(\rho_1,\ldots,\rho_p,\sigma_1,\ldots,\sigma_q)}_{\lambda_1,\ldots,\lambda_p,\mu_1,\ldots,\mu_q}(z,s,a;b,\lambda)\\ &= \sqrt{2\pi}\mathcal{F}\left[e^{\sigma y}exp\left(-ae^y-\tfrac{b}{e^{\lambda y}}\right)p^{\Psi^*}q\left[\begin{matrix}(\lambda_1,\rho_1),\ldots, & (\lambda_p,\rho_p)\\ (\mu_1,\sigma_1),\ldots, & (\mu_q,\sigma_q)\end{matrix};zexp(-e^y)\right];\tau\right]\\ &(min[\Re(a),\Re(s)]>0;\Re(b)\geqq 0;\lambda\geqq 0).\end{aligned} \tag{59}$$

Similarly for the λ-generalized extended Fermi-Dirac functions

$$\begin{aligned}&\Gamma(s)\Theta^{(\rho_1,\ldots,\rho_p,\sigma_1,\ldots,\sigma_q)}_{\lambda_1,\ldots,\lambda_p,\mu_1,\ldots,\mu_q}(x,s,a;b,\lambda)\\ &= \sqrt{2\pi}\mathcal{F}\left[e^{\sigma y}exp\left(-ae^y-\tfrac{b}{e^{\lambda y}}\right)p^{\Psi^*}q\left[\begin{matrix}(\lambda_1,\rho_1),\ldots, & (\lambda_p,\rho_p)\\ (\mu_1,\sigma_1),\ldots, & (\mu_q,\sigma_q)\end{matrix};-e^{-x}exp(-e^y)\right];\tau\right]\\ &(min[\Re(a),\Re(s)]>0;\Re(b)\geqq 0;\lambda\geqq 0)\end{aligned} \tag{60}$$

and Extended Bose-Einstein Functions

$$\begin{aligned}&\Gamma(s)\Psi^{(\rho_1,\ldots,\rho_p,\sigma_1,\ldots,\sigma_q)}_{\lambda_1,\ldots,\lambda_p,\mu_1,\ldots,\mu_q}(x,s,a;b,\lambda)\\ &= \sqrt{2\pi}\mathcal{F}\left[e^{\sigma y}exp\left(-ae^y-\tfrac{b}{e^{\lambda y}}\right)p^{\Psi^*}q\left[\begin{matrix}(\lambda_1,\rho_1),\ldots, & (\lambda_p,\rho_p)\\ (\mu_1,\sigma_1),\ldots, & (\mu_q,\sigma_q)\end{matrix};e^{-x}exp(-e^y)\right];\tau\right]\\ &(min[\Re(a),\Re(s)]>0;\Re(b)\geqq 0;\lambda\geqq 0).\end{aligned} \tag{61}$$

For λ-generalized Fermi-Dirac functions

$$\begin{aligned}&\Gamma(s)F^{(\rho_1,\ldots,\rho_p,\sigma_1,\ldots,\sigma_q)}_{\lambda_1,\ldots,\lambda_p,\mu_1,\ldots,\mu_q}(x,s;b,\lambda)\\ &= \sqrt{2\pi}\mathcal{F}\left[e^{\sigma y}exp\left(-e^y-\tfrac{b}{e^{\lambda y}}\right)p^{\Psi^*}q\left[\begin{matrix}(\lambda_1,\rho_1),\ldots, & (\lambda_p,\rho_p)\\ (\mu_1,\sigma_1),\ldots, & (\mu_q,\sigma_q)\end{matrix};-e^{-x}exp(-e^y)\right];\tau\right]\\ &(min[\Re(a),\Re(s)]>0;\Re(b)\geqq 0;\lambda\geqq 0)\end{aligned} \tag{62}$$

and Bose-Einstein functions

$$\begin{aligned}&\Gamma(s)B^{(\rho_1,\ldots,\rho_p,\sigma_1,\ldots,\sigma_q)}_{\lambda_1,\ldots,\lambda_p,\mu_1,\ldots,\mu_q}(x,s;b,\lambda)\\&=\sqrt{2\pi}\mathcal{F}\left[e^{\sigma y}exp\left(-e^y-\tfrac{b}{e^{\lambda y}}\right)p^{\Psi^*}q\left[\begin{array}{ll}(\lambda_1,\rho_1),\ldots, & (\lambda_p,\rho_p)\\(\mu_1,\sigma_1),\ldots, & (\mu_q,\sigma_q)\end{array};e^{-x}exp(-e^y)\right];\tau\right]\\&(min[\Re(a),\Re(s)]>0;\Re(b)\geqq 0;\lambda\geqq 0).\end{aligned}\tag{63}$$

For λ-generalized Polylogarithm functions

$$\begin{aligned}&\Gamma(s)\Phi^{(\rho_1,\ldots,\rho_p,\sigma_1,\ldots,\sigma_q)}_{\lambda_1,\ldots,\lambda_p,\mu_1,\ldots,\mu_q}(z,s;b,\lambda)\\&=\sqrt{2\pi}\mathcal{F}\left[ze^{\sigma y}exp\left(-e^y-\tfrac{b}{e^{\lambda y}}\right)p^{\Psi^*}q\left[\begin{array}{ll}(\lambda_1,\rho_1),\ldots, & (\lambda_p,\rho_p)\\(\mu_1,\sigma_1),\ldots, & (\mu_q,\sigma_q)\end{array};zexp(-e^y)\right];\tau\right]\\&(min[\Re(a),\Re(s)]>0;\Re(b)\geqq 0;\lambda\geqq 0)\end{aligned}\tag{64}$$

For λ-generalized Hurwitz zeta functions

$$\begin{aligned}&\Gamma(s)\zeta^{(\rho_1,\ldots,\rho_p,\sigma_1,\ldots,\sigma_q)}_{\lambda_1,\ldots,\lambda_p,\mu_1,\ldots,\mu_q}(s,a;b,\lambda)\\&=\sqrt{2\pi}\mathcal{F}\left[e^{\sigma y}exp\left(-ae^y-\tfrac{b}{e^{\lambda y}}\right)p^{\Psi^*}q\left[\begin{array}{ll}(\lambda_1,\rho_1),\ldots, & (\lambda_p,\rho_p)\\(\mu_1,\sigma_1),\ldots, & (\mu_q,\sigma_q)\end{array};exp(-e^y)\right];\tau\right]\\&(min[\Re(a),\Re(s)]>0;\Re(b)\geqq 0;\lambda\geqq 0).\end{aligned}\tag{65}$$

For λ-generalized Riemann zeta functions

$$\begin{aligned}&\Gamma(s)\zeta^{(\rho_1,\ldots,\rho_p,\sigma_1,\ldots,\sigma_q)}_{\lambda_1,\ldots,\lambda_p,\mu_1,\ldots,\mu_q}(s;b,\lambda)\\&=\sqrt{2\pi}\mathcal{F}\left[e^{\sigma y}exp\left(-e^y-\tfrac{b}{e^{\lambda y}}\right)p^{\Psi^*}q\left[\begin{array}{ll}(\lambda_1,\rho_1),\ldots, & (\lambda_p,\rho_p)\\(\mu_1,\sigma_1),\ldots, & (\mu_q,\sigma_q)\end{array};exp(-e^y)\right];\tau\right]\\&(min[\Re(a),\Re(s)]>0;\Re(b)\geqq 0;\lambda\geqq 0).\end{aligned}\tag{66}$$

Similarly, by giving variations to different parameters, we can get similar representations for other special cases of these functions in consideration of Table 1.

6. Verification of the Results Obtained by New Representation

For the Fourier transform of any function $f(t)$, duality property holds as

$$\mathcal{F}\left[\sqrt{2\pi}\mathcal{F}[f(t);\tau];\beta\right]=2\pi f(-\beta).\tag{67}$$

Hence, from (59)–(66), by applying (67), we obtain the following

$$\begin{aligned}&\mathcal{F}\left\{\Gamma(\sigma+i\tau)\Phi^{(\rho_1,\ldots,\rho_p,\sigma_1,\ldots,\sigma_q)}_{\lambda_1,\ldots,\lambda_p,\mu_1,\ldots,\mu_q}(z,\sigma+i\tau,a;b,\lambda);\beta\right\}=\\&\mathcal{F}\left\{\sqrt{2\pi}\mathcal{F}\left\{e^{\sigma y}e^{-ae^y-\frac{b}{e^{\lambda y}}}p^{\Psi^*}q\left[\begin{array}{ll}(\lambda_1,\rho_1),\ldots, & (\lambda_p,\rho_p)\\(\mu_1,\sigma_1),\ldots, & (\mu_q,\sigma_q)\end{array};ze^{-e^y}\right];\tau\right\};\beta\right\}=f(-\beta)\\&=2\pi e^{-\sigma\beta}exp\left(-ae^{-\beta}-\tfrac{b}{e^{-\lambda\beta}}\right)p^{\Psi^*}q\left[\begin{array}{ll}(\lambda_1,\rho_1),\ldots, & (\lambda_p,\rho_p)\\(\mu_1,\sigma_1),\ldots, & (\mu_q,\sigma_q)\end{array};zexp(-e^{-\beta})\right]\\&(min[\Re(a),\Re(s)]>0;\Re(b)\geqq 0;\lambda\geqq 0),\end{aligned}\tag{68}$$

$$\begin{aligned}&Or\int_{-\infty}^{+\infty}e^{i\tau\beta}\Gamma(\sigma+i\tau)\Phi^{(\rho_1,\ldots,\rho_p,\sigma_1,\ldots,\sigma_q)}_{\lambda_1,\ldots,\lambda_p,\mu_1,\ldots,\mu_q}(z,\sigma+i\tau,a;b,\lambda)d\tau\\&=2\pi e^{-\sigma\beta}exp\left(-ae^{-\beta}-\tfrac{b}{e^{-\lambda\beta}}\right)p^{\Psi^*}q\left[\begin{array}{ll}(\lambda_1,\rho_1),\ldots, & (\lambda_p,\rho_p)\\(\mu_1,\sigma_1),\ldots, & (\mu_q,\sigma_q)\end{array};zexp(-e^{-\beta})\right],\end{aligned}\tag{69}$$

which is the special case of our main result (46) for $w = e; s = \sigma + i\tau$ and verifies that results obtained by the new representation are consistent with the classical results.

If we put $\beta = 0$ in the above equation (69), we get the following integral:

$$\begin{aligned} &\int_{-\infty}^{+\infty} \Gamma(\sigma + i\tau)\Phi_{\lambda_1,\ldots,\lambda_p,\mu_1,\ldots,\mu_q}^{(\rho_1,\ldots,\rho_p,\sigma_1,\ldots,\sigma_q)}(z, \sigma + i\tau, a; b, \lambda)d\tau \\ &= 2\pi e^{-a-b} {}_p\Psi^*{}_q\left[\begin{array}{ccc} (\lambda_1, \rho_1), \ldots, & (\lambda_p, \rho_p) & ; \frac{z}{e} \\ (\mu_1, \sigma_1), \ldots, & (\mu_q, \sigma_q) & \end{array}\right], \end{aligned} \tag{70}$$

which is also a specific case of our main result (46). It shows that our new representation produces new results that cannot be found by other methods, but special cases of our obtained results are consistent with the classical results.

Similarly, by considering different parametric values in the above equations and as given in Table 1 in Section 2, we can get the following list of integrals: (one can also note that results obtained by new representation are not only more general than the results obtained by Fourier transform representation but also consistent with the special cases of these results)

$$\begin{aligned} &\int_{-\infty}^{+\infty} e^{i\tau\beta}\Gamma(\sigma + i\tau)\Psi_{\lambda_1,\ldots,\lambda_p,\mu_1,\ldots,\mu_q}^{(\rho_1,\ldots,\rho_p,\sigma_1,\ldots,\sigma_q)}(x, \sigma + i\tau, a; b, \lambda)d\tau \\ &= 2\pi e^{-\sigma\beta} exp\left(-ae^{-\beta} - \frac{b}{e^{-\lambda\beta}}\right) {}_p\Psi^*{}_q\left[\begin{array}{ccc} (\lambda_1, \rho_1), \ldots, & (\lambda_p, \rho_p) & ; e^{-x} exp(-e^{-\beta}) \\ (\mu_1, \sigma_1), \ldots, & (\mu_q, \sigma_q) & \end{array}\right]; \end{aligned} \tag{71}$$

$$\begin{aligned} &\int_{-\infty}^{+\infty} e^{i\tau\beta}\Gamma(\sigma + i\tau)\Theta_{\lambda_1,\ldots,\lambda_p,\mu_1,\ldots,\mu_q}^{(\rho_1,\ldots,\rho_p,\sigma_1,\ldots,\sigma_q)}(x, \sigma + i\tau, a; b, \lambda)d\tau \\ &= 2\pi e^{-\sigma\beta} exp\left(-ae^{-\beta} - \frac{b}{e^{-\lambda\beta}}\right) {}_p\Psi^*{}_q\left[\begin{array}{ccc} (\lambda_1, \rho_1), \ldots, & (\lambda_p, \rho_p) & ; -e^{-x} exp(-e^{-\beta}) \\ (\mu_1, \sigma_1), \ldots, & (\mu_q, \sigma_q) & \end{array}\right]; \end{aligned} \tag{72}$$

$$\begin{aligned} &\int_{-\infty}^{+\infty} e^{i\tau\beta}\Gamma(\sigma + i\tau)B_{\lambda_1,\ldots,\lambda_p,\mu_1,\ldots,\mu_q}^{(\rho_1,\ldots,\rho_p,\sigma_1,\ldots,\sigma_q)}(x, \sigma + i\tau; b, \lambda)d\tau \\ &= 2\pi e^{-\sigma\beta} exp\left(-e^{-\beta} - \frac{b}{e^{-\lambda\beta}}\right) {}_p\Psi^*{}_q\left[\begin{array}{ccc} (\lambda_1, \rho_1), \ldots, & (\lambda_p, \rho_p) & ; e^{-x} exp(-e^{-\beta}) \\ (\mu_1, \sigma_1), \ldots, & (\mu_q, \sigma_q) & \end{array}\right]; \end{aligned} \tag{73}$$

$$\begin{aligned} &\int_{-\infty}^{+\infty} e^{i\tau\beta}\Gamma(\sigma + i\tau)F_{\lambda_1,\ldots,\lambda_p,\mu_1,\ldots,\mu_q}^{(\rho_1,\ldots,\rho_p,\sigma_1,\ldots,\sigma_q)}(x, \sigma + i\tau; b, \lambda)d\tau \\ &= 2\pi e^{-\sigma\beta} exp\left(-e^{-\beta} - \frac{b}{e^{-\lambda\beta}}\right) {}_p\Psi^*{}_q\left[\begin{array}{ccc} (\lambda_1, \rho_1), \ldots, & (\lambda_p, \rho_p) & ; -e^{-x} exp(-e^{-\beta}) \\ (\mu_1, \sigma_1), \ldots, & (\mu_q, \sigma_q) & \end{array}\right]; \end{aligned} \tag{74}$$

$$\begin{aligned} &\int_{-\infty}^{+\infty} e^{i\tau\beta}\Gamma(\sigma + i\tau)\phi_{\lambda_1,\ldots,\lambda_p,\mu_1,\ldots,\mu_q}^{(\rho_1,\ldots,\rho_p,\sigma_1,\ldots,\sigma_q)}(z, \sigma + i\tau; b, \lambda)d\tau \\ &= 2\pi z e^{-\sigma\beta} exp\left(-e^{-\beta} - \frac{b}{e^{-\lambda\beta}}\right) {}_p\Psi^*{}_q\left[\begin{array}{ccc} (\lambda_1, \rho_1), \ldots, & (\lambda_p, \rho_p) & ; z exp(-e^{-\beta}) \\ (\mu_1, \sigma_1), \ldots, & (\mu_q, \sigma_q) & \end{array}\right]; \end{aligned} \tag{75}$$

$$\begin{aligned} &\int_{-\infty}^{+\infty} e^{i\tau\beta}\Gamma(\sigma + i\tau)\zeta_{\lambda_1,\ldots,\lambda_p,\mu_1,\ldots,\mu_q}^{(\rho_1,\ldots,\rho_p,\sigma_1,\ldots,\sigma_q)}(\sigma + i\tau, a; b, \lambda)d\tau \\ &= 2\pi e^{-\sigma\beta} exp\left(-ae^{-\beta} - \frac{b}{e^{-\lambda\beta}}\right) {}_p\Psi^*{}_q\left[\begin{array}{ccc} (\lambda_1, \rho_1), \ldots, & (\lambda_p, \rho_p) & ; exp(-e^{-\beta}) \\ (\mu_1, \sigma_1), \ldots, & (\mu_q, \sigma_q) & \end{array}\right]; \end{aligned} \tag{76}$$

$$\begin{aligned} &\int_{-\infty}^{+\infty} e^{i\tau\beta}\Gamma(\sigma + i\tau)\zeta_{\lambda_1,\ldots,\lambda_p,\mu_1,\ldots,\mu_q}^{(\rho_1,\ldots,\rho_p,\sigma_1,\ldots,\sigma_q)}(\sigma + i\tau; b, \lambda)d\tau \\ &= 2\pi e^{-\sigma\beta} exp\left(-ae^{-\beta} - \frac{b}{e^{-\lambda\beta}}\right) {}_p\Psi^*{}_q\left[\begin{array}{ccc} (\lambda_1, \rho_1), \ldots, & (\lambda_p, \rho_p) & ; exp(-e^{-\beta}) \\ (\mu_1, \sigma_1), \ldots, & (\mu_q, \sigma_q) & \end{array}\right]; \end{aligned} \tag{77}$$

$$\begin{aligned} &\int_{-\infty}^{+\infty} e^{i\tau\beta}\Gamma(\sigma + i\tau)\Phi_{\lambda_1,\ldots,\lambda_p,\mu_1,\ldots,\mu_q}^{(\rho_1,\ldots,\rho_p,\sigma_1,\ldots,\sigma_q)}(z, \sigma + i\tau, a)d\tau \\ &= 2\pi e^{-\sigma\beta} exp(-ae^{-\beta} - b) {}_p\Psi^*{}_q\left[\begin{array}{ccc} (\lambda_1, \rho_1), \ldots, & (\lambda_p, \rho_p) & ; z exp(-e^{-\beta}) \\ (\mu_1, \sigma_1), \ldots, & (\mu_q, \sigma_q) & \end{array}\right]. \end{aligned} \tag{78}$$

For $\beta = 0$, the above results (68)–(78) yield some interesting and simple integral formulae. To confirm the consistency of the results obtained by new representation, it can be noted that the results obtained in this Section (68)–(78) can be generated as special cases of (46)–(48) for $\omega = e, s = \sigma + i\tau$ and vice versa. These are straightforward to obtain by using a basic fact of the Fourier transform and therefore to test the consistency of new representations as they become more important.

7. Discussion and Future Directions

The confluence of distributions (generalized functions) with classical integral transformations has become a remarkably influential tool in the theory of partial differential equations. It has solved various physical and engineering problems that cannot be solved by using classical methods. In this paper, we obtained a new representation for the newly defined family of the λ-generalized Hurwitz-Lerch zeta functions in terms of complex delta functions such that the definition of these functions is formalized over the space of entire test functions denoted by $\boldsymbol{Z}$. This is significant for advancing the foundations of distributional (generalized function) concepts for such higher transcendental functions and enhancing their applications to solve real-world problems. The Riemann hypothesis is a famous and unsolved problem at present in analytic number theory [31]. It states that "all the non-trivial zeros of the zeta function lie on the real line $s = 1/2''$. These zeros appear symmetrically as complex conjugates on this line. The integrals of the zeta function and its generalizations are essential in the investigation of Riemann hypothesis and for the study of zeta functions. Such integrals are also important for the study of distributions in statistical inference and reliability theory [1,26,32]. By using this new definition of the λ-generalized Hurwitz-Lerch zeta functions, one can find such integrals in a simple and uniform way.

λ-generalized Hurwitz-Lerch zeta functions systematically oversimplify the functions of the zeta family and provide understanding for some other potential new members of this family that are not found in the literature. This element is very useful for achieving new results from one main result. Our main result generates at once significant new results for a class of well-studied functions by applying the methodology of this paper. The Fermi-Dirac and Bose-Einstein functions arose in the distribution functions for quantum statistics that deals with two particular kinds of spin symmetry, namely, bosons and fermions. Their close connection considered in this investigation with the λ-generalized Hurwitz-Lerch zeta functions have provided some significant new results for these functions that directly develop the future applications of these representations in quantum physics and related fields. The technique to obtain the results by using new representation explores a required simplicity that is always desirous. These are some straightforward examples. It is expected that the approach developed in this investigation will be doubtlessly significant for further exploration of these higher transcendental functions in future research.

Acknowledgments: The author would like to thank Deanship of Scientific Research at Majmaah University for supporting this work under project number No.1440-15.The author is also thankful to the anonymous reviewers for their useful comments. They really improved the quality of this manuscript.

Conflicts of Interest: The authors declare no conflict of interest.

Appendix A New Results by Considering Special Cases for Section 2 in View of Table 1

$$\Gamma(s)\Theta^{(\rho_1,\ldots,\rho_p,\sigma_1,\ldots,\sigma_q)}_{\lambda_1,\ldots,\lambda_p,\mu_1,\ldots,\mu_q}(x,s,a) = 2\pi e^b \sum_{\chi,\xi=0}^{\infty} \frac{([\lambda_p])_{\rho_p\chi}}{([\mu_q])_{\sigma_q\chi}} \frac{(-e^{-x})^\chi}{\chi!} \frac{(-(\chi+a))^\xi}{\xi!}\delta(s+\xi); \quad \text{(A1)}$$

$$\Gamma(s)\Psi^{(\rho_1,\ldots,\rho_p,\sigma_1,\ldots,\sigma_q)}_{\lambda_1,\ldots,\lambda_p,\mu_1,\ldots,\mu_q}(x,s,a) = 2\pi e^b \sum_{\chi,\xi=0}^{\infty} \frac{([\lambda_p])_{\rho_p\chi}}{([\mu_q])_{\sigma_q\chi}} \frac{(e^{-x})^\chi}{\chi!} \frac{(-(\chi+a))^\xi}{\xi!}\delta(s+\xi); \quad \text{(A2)}$$

$$\Gamma(s)B_{\lambda_1,\ldots,\lambda_p,\mu_1,\ldots,\mu_q}^{(\rho_1,\ldots,\rho_p,\sigma_1,\ldots,\sigma_q)}(x,s) = 2\pi e^b \sum_{\chi,\xi=0}^{\infty} \frac{([\lambda_p])_{\rho_p\chi}}{([\mu_q])_{\sigma_q\chi}} \frac{(e^{-x})^\chi}{\chi!} \frac{(-(\chi+1))^\xi}{\xi!} \delta(s+\xi); \tag{A3}$$

$$\Gamma(s)F_{\lambda_1,\ldots,\lambda_p,\mu_1,\ldots,\mu_q}^{(\rho_1,\ldots,\rho_p,\sigma_1,\ldots,\sigma_q)}(x,s) = 2\pi e^b \sum_{\chi,\xi=0}^{\infty} \frac{([\lambda_p])_{\rho_p\chi}}{([\mu_q])_{\sigma_q\chi}} \frac{(-e^{-x})^\chi}{\chi!} \frac{(-(\chi+1))^\xi}{\xi!} \delta(s+\xi); \tag{A4}$$

$$\Gamma(s)\phi_{\lambda_1,\ldots,\lambda_p,\mu_1,\ldots,\mu_q}^{(\rho_1,\ldots,\rho_p,\sigma_1,\ldots,\sigma_q)}(z,s) = 2\pi z e^b \sum_{\chi,\xi=0}^{\infty} \frac{([\lambda_p])_{\rho_p\chi}}{([\mu_q])_{\sigma_q\chi}} \frac{(z)^\chi}{\chi!} \frac{(-(\chi+1))^\xi}{\xi!} \delta(s+\xi); \tag{A5}$$

$$\Gamma(s)\zeta_{\lambda_1,\ldots,\lambda_p,\mu_1,\ldots,\mu_q}^{(\rho_1,\ldots,\rho_p,\sigma_1,\ldots,\sigma_q)}(s,a) = 2\pi e^b \sum_{\chi,\xi=0}^{\infty} \frac{([\lambda_p])_{\rho_p\chi}}{([\mu_q])_{\sigma_q\chi}} \frac{1}{\chi!} \frac{(-(\chi+a))^\xi}{\xi!} \delta(s+\xi); \tag{A6}$$

$$\Gamma(s)\zeta_{\lambda_1,\ldots,\lambda_p,\mu_1,\ldots,\mu_q}^{(\rho_1,\ldots,\rho_p,\sigma_1,\ldots,\sigma_q)}(s) = 2\pi e^b \sum_{\chi,\xi=0}^{\infty} \frac{([\lambda_p])_{\rho_p\chi}}{([\mu_q])_{\sigma_q\chi}} \frac{1}{\chi!} \frac{(-(\chi+1))^\xi}{\xi!} \delta(s+\xi); \tag{A7}$$

$$\Gamma(s)\Theta_{\lambda_1,\ldots,\lambda_p,\mu_1,\ldots,\mu_q}^{(\rho_1,\ldots,\rho_p,\sigma_1,\ldots,\sigma_q)}(x,s,a) = 2\pi \sum_{\chi,\xi=0}^{\infty} \frac{([\lambda_p])_{\rho_p\chi}}{([\mu_q])_{\sigma_q\chi}} \frac{(-e^{-x})^\chi}{\chi!} \frac{(-(\chi+a))^\xi}{\xi!} \delta(s+\xi) \tag{A8}$$

$$\Gamma(s)\Psi_{\lambda_1,\ldots,\lambda_p,\mu_1,\ldots,\mu_q}^{(\rho_1,\ldots,\rho_p,\sigma_1,\ldots,\sigma_q)}(x,s,a) = 2\pi \sum_{\chi,\xi=0}^{\infty} \frac{([\lambda_p])_{\rho_p\chi}}{([\mu_q])_{\sigma_q\chi}} \frac{(e^{-x})^\chi}{\chi!} \frac{(-(\chi+a))^\xi}{\xi!} \delta(s+\xi) \tag{A9}$$

$$\Gamma(s)B_{\lambda_1,\ldots,\lambda_p,\mu_1,\ldots,\mu_q}^{(\rho_1,\ldots,\rho_p,\sigma_1,\ldots,\sigma_q)}(x,s) = 2\pi \sum_{\chi,\xi=0}^{\infty} \frac{([\lambda_p])_{\rho_p\chi}}{([\mu_q])_{\sigma_q\chi}} \frac{(e^{-x})^\chi}{\chi!} \frac{(-(\chi+1))^\xi}{\xi!} \delta(s+\xi) \tag{A10}$$

$$\Gamma(s)F_{\lambda_1,\ldots,\lambda_p,\mu_1,\ldots,\mu_q}^{(\rho_1,\ldots,\rho_p,\sigma_1,\ldots,\sigma_q)}(x,s) = 2\pi \sum_{\chi,\xi=0}^{\infty} \frac{([\lambda_p])_{\rho_p\chi}}{([\mu_q])_{\sigma_q\chi}} \frac{(-e^{-x})^\chi}{\chi!} \frac{(-(\chi+1))^\xi}{\xi!} \delta(s+\xi) \tag{A11}$$

$$\Gamma(s)\phi_{\lambda_1,\ldots,\lambda_p,\mu_1,\ldots,\mu_q}^{(\rho_1,\ldots,\rho_p,\sigma_1,\ldots,\sigma_q)}(z,s) = 2\pi z \sum_{\chi,\xi=0}^{\infty} \frac{([\lambda_p])_{\rho_p\chi}}{([\mu_q])_{\sigma_q\chi}} \frac{(z)^\chi}{\chi!} \frac{(-(\chi+1))^\xi}{\xi!} \delta(s+\xi) \tag{A12}$$

$$\Gamma(s)\zeta_{\lambda_1,\ldots,\lambda_p,\mu_1,\ldots,\mu_q}^{(\rho_1,\ldots,\rho_p,\sigma_1,\ldots,\sigma_q)}(s,a) = 2\pi \sum_{\chi,\xi=0}^{\infty} \frac{([\lambda_p])_{\rho_p\chi}}{([\mu_q])_{\sigma_q\chi}} \frac{1}{\chi!} \frac{(-(\chi+a))^\xi}{\xi!} \delta(s+\xi) \tag{A13}$$

$$\Gamma(s)\zeta_{\lambda_1,\ldots,\lambda_p,\mu_1,\ldots,\mu_q}^{(\rho_1,\ldots,\rho_p,\sigma_1,\ldots,\sigma_q)}(s) = 2\pi \sum_{\chi,\xi=0}^{\infty} \frac{([\lambda_p])_{\rho_p\chi}}{([\mu_q])_{\sigma_q\chi}} \frac{1}{\chi!} \frac{(-(\chi+1))^\xi}{\xi!} \delta(s+\xi) \tag{A14}$$

$$\Gamma(s)\Theta_\mu^\lambda(x,s,a;b) = 2\pi \sum_{\chi,\xi,\psi=0}^{\infty} (\mu)_\chi \frac{(-e^{-x})^\chi}{\chi!} \frac{(-(\chi+a))^\xi}{\xi!} \frac{(-b)^\psi}{\psi!} \delta(s+\xi-\lambda\psi) \tag{A15}$$

$$\Gamma(s)\Psi_\mu^\lambda(x,s,a;b) = 2\pi \sum_{\chi,\xi,\psi=0}^{\infty} (\mu)_\chi \frac{(e^{-x})^\chi}{\chi!} \frac{(-(\chi+a))^\xi}{\xi!} \frac{(-b)^\psi}{\psi!} \delta(s+\xi-\lambda\psi) \tag{A16}$$

$$\Gamma(s)B_\mu^\lambda(x,s;b) = 2\pi \sum_{\chi,\xi,\psi=0}^{\infty} (\mu)_\chi \frac{(e^{-x})^\chi}{\chi!} \frac{(-(\chi+1))^\xi}{\xi!} \frac{(-b)^\psi}{\psi!} \delta(s+\xi-\lambda\psi) \tag{A17}$$

$$\Gamma(s)F_\mu^\lambda(x,s;b) = 2\pi \sum_{\chi,\xi,\psi=0}^{\infty} (\mu)_\chi \frac{(-e^{-x})^\chi}{\chi!} \frac{(-(\chi+1))^\xi}{\xi!} \frac{(-b)^\psi}{\psi!} \delta(s+\xi-\lambda\psi) \tag{A18}$$

$$\Gamma(s)\phi_\mu^\lambda(z,s;b) = 2\pi z \sum_{\chi,\xi,\psi=0}^{\infty} (\mu)_\chi \frac{(z)^\chi}{\chi!} \frac{(-(\chi+1))^\xi}{\xi!} \frac{(-b)^\psi}{\psi!} \delta(s+\xi-\lambda\psi) \tag{A19}$$

$$\Gamma(s)\zeta_{\mu}^{\lambda}(s,a;b,a) = 2\pi \sum_{\chi,\xi,\psi=0}^{\infty} (\mu)_{\chi} \frac{(e^{-x})^{\chi}}{\chi!} \frac{(-(\chi+a))^{\xi}}{\xi!} \frac{(-b)^{\psi}}{\psi!} \delta(s+\xi-\lambda\psi) \tag{A20}$$

$$\Gamma(s)\zeta_{\mu}^{\lambda}(s;b) = 2\pi \sum_{\chi,\xi,\psi=0}^{\infty} (\mu)_{\chi} \frac{(e^{-x})^{\chi}}{\chi!} \frac{(-(\chi+1))^{\xi}}{\xi!} \frac{(-b)^{\psi}}{\psi!} \delta(s+\xi-\lambda\psi) \tag{A21}$$

$$\Gamma(s)\Phi_{\mu}^{*}(z,s,a,b) = 2\pi e^{b} \sum_{\chi,\xi=0}^{\infty} (\mu)_{\chi} \frac{(z)^{\chi}}{\chi!} \frac{(-(\chi+a))^{\xi}}{\xi!} \delta(s+\xi) \tag{A22}$$

$$\Gamma(s)\Theta_{\mu}^{*}(x,s,a,b) = 2\pi e^{b} \sum_{\chi,\xi=0}^{\infty} (\mu)_{\chi} \frac{(-e^{-x})^{\chi}}{\chi!} \frac{(-(\chi+a))^{\xi}}{\xi!} \delta(s+\xi) \tag{A23}$$

$$\Gamma(s)\Psi_{\mu}^{*}(x,s,a,b) = 2\pi e^{b} \sum_{\chi,\xi=0}^{\infty} (\mu)_{\chi} \frac{(e^{-x})^{\chi}}{\chi!} \frac{(-(\chi+a))^{\xi}}{\xi!} \delta(s+\xi) \tag{A24}$$

$$\Gamma(s)B_{\mu}^{*}(x,s,b) = 2\pi e^{b} \sum_{\chi,\xi=0}^{\infty} (\mu)_{\chi} \frac{(e^{-x})^{\chi}}{\chi!} \frac{(-(\chi+1))^{\xi}}{\xi!} \delta(s+\xi) \tag{A25}$$

$$\Gamma(s)F_{\mu}^{*}(x,s,b) = 2\pi e^{b} \sum_{\chi,\xi=0}^{\infty} (\mu)_{\chi} \frac{(e^{-x})^{\chi}}{\chi!} \frac{(-(\chi+1))^{\xi}}{\xi!} \delta(s+\xi) \tag{A26}$$

$$\Gamma(s)\phi_{\mu}^{*}(z,s;b,\lambda) = 2\pi e^{b} z \sum_{\chi,\xi=0}^{\infty} (\mu)_{\chi} \frac{(z)^{\chi}}{\chi!} \frac{(-(\chi+1))^{\xi}}{\xi!} \delta(s+\xi) \tag{A27}$$

$$\Gamma(s)\zeta_{\mu}^{*}(s,a;b) = 2\pi e^{b} \sum_{\chi,\xi=0}^{\infty} (\mu)_{\chi} \frac{1}{\chi!} \frac{(-(\chi+a))^{\xi}}{\xi!} \delta(s+\xi) \tag{A28}$$

$$\Gamma(s)\zeta_{\mu}^{*}(s;b) = 2\pi e^{b} \sum_{\chi,\xi=0}^{\infty} (\mu)_{\chi} \frac{1}{\chi!} \frac{(-(\chi+1))^{\xi}}{\xi!} \delta(s+\xi) \tag{A29}$$

$$\Gamma(s)\Phi_{\mu}^{*}(z,s,a) = 2\pi \sum_{\chi,\xi=0}^{\infty} (\mu)_{\chi} \frac{(z)^{\chi}}{\chi!} \frac{(-(\chi+a))^{\xi}}{\xi!} \delta(s+\xi) \tag{A30}$$

$$\Gamma(s)\Theta_{\mu}^{*}(x,s,a) = 2\pi \sum_{\chi,\xi=0}^{\infty} (\mu)_{\chi} \frac{(-e^{-x})^{\chi}}{\chi!} \frac{(-(\chi+a))^{\xi}}{\xi!} \delta(s+\xi) \tag{A31}$$

$$\Gamma(s)\Psi_{\mu}^{*}(x,s,a) = 2\pi \sum_{\chi,\xi=0}^{\infty} (\mu)_{\chi} \frac{(e^{-x})^{\chi}}{\chi!} \frac{(-(\chi+a))^{\xi}}{\xi!} \delta(s+\xi) \tag{A32}$$

$$\Gamma(s)B_{\mu}^{*}(x,s) = 2\pi \sum_{\chi,\xi=0}^{\infty} (\mu)_{\chi} \frac{(e^{-x})^{\chi}}{\chi!} \frac{(-(\chi+a))^{\xi}}{\xi!} \delta(s+\xi) \tag{A33}$$

$$\Gamma(s)F_{\mu}^{*}(x,s) = 2\pi \sum_{\chi,\xi=0}^{\infty} (\mu)_{\chi} \frac{(-e^{-x})^{\chi}}{\chi!} \frac{(-(\chi+a))^{\xi}}{\xi!} \delta(s+\xi) \tag{A34}$$

$$\Gamma(s)\phi_{\mu}^{*}(z,s;b,\lambda) = 2\pi z \sum_{\chi,\xi=0}^{\infty} (\mu)_{\chi} \frac{(z)^{\chi}}{\chi!} \frac{(-(\chi+a))^{\xi}}{\xi!} \delta(s+\xi) \tag{A35}$$

$$\Gamma(s)\zeta_{\mu}^{*}(s,a) = 2\pi \sum_{\chi,\xi=0}^{\infty} (\mu)_{\chi} \frac{1}{\chi!} \frac{(-(\chi+a))^{\xi}}{\xi!} \delta(s+\xi) \tag{A36}$$

$$\Gamma(s)\zeta_{\mu}^{*}(s) = 2\pi \sum_{\chi,\xi=0}^{\infty} (\mu)_{\chi} \frac{1}{\chi!} \frac{(-(\chi+1))^{\xi}}{\xi!} \delta(s+\xi) \tag{A37}$$

$$\Gamma(s)\Theta(x,s,a;b,\lambda) = 2\pi \sum_{\chi,\xi,\psi=0}^{\infty} (-e^{-x})^{\chi} \frac{(-(\chi+a))^{\xi}}{\xi!} \frac{(-b)^{\psi}}{\psi!} \delta(s+\xi-\lambda\psi) \tag{A38}$$

$$\Gamma(s)\Psi(x,s,a;b,\lambda) = 2\pi \sum_{\chi,\xi,\psi=0}^{\infty} (e^{-x})^{\chi} \frac{(-(\chi+a))^{\xi}}{\xi!} \frac{(-b)^{\psi}}{\psi!} \delta(s+\xi-\lambda\psi) \tag{A39}$$

$$\Gamma(s)B(x,s;b,\lambda) = 2\pi \sum_{\chi,\xi,\psi=0}^{\infty} (e^{-x})^{\chi} \frac{(-(\chi+1))^{\xi}}{\xi!} \frac{(-b)^{\psi}}{\psi!} \delta(s+\xi-\lambda\psi) \tag{A40}$$

$$\Gamma(s)F(x,s;b,\lambda) = 2\pi \sum_{\chi,\xi,\psi=0}^{\infty} (-e^{-x})^{\chi} \frac{(-(\chi+1))^{\xi}}{\xi!} \frac{(-b)^{\psi}}{\psi!} \delta(s+\xi-\lambda\psi) \tag{A41}$$

$$\Gamma(s)\phi(z,s;b,\lambda) = 2\pi z \sum_{\chi,\xi,\psi=0}^{\infty} (z)^{\chi} \frac{(-(\chi+1))^{\xi}}{\xi!} \frac{(-b)^{\psi}}{\psi!} \delta(s+\xi-\lambda\psi) \tag{A42}$$

$$\Gamma(s)\zeta(s,a;b,\lambda) = 2\pi \sum_{\chi,\xi,\psi=0}^{\infty} \frac{(-(\chi+a))^{\xi}}{\xi!} \frac{(-b)^{\psi}}{\psi!} \delta(s+\xi-\lambda\psi) \tag{A43}$$

$$\Gamma(s)\zeta(s;b,\lambda) = 2\pi \sum_{\chi,\xi,\psi=0}^{\infty} \frac{(-(\chi+1))^{\xi}}{\xi!} \frac{(-b)^{\psi}}{\psi!} \delta(s+\xi-\lambda\psi) \tag{A44}$$

$$\Gamma(s)\Phi(z,s,a;b) = 2\pi e^{b} \sum_{\chi,\xi=0}^{\infty} (z)^{\chi} \frac{(-(\chi+a))^{\xi}}{\xi!} \delta(s+\xi) \tag{A45}$$

$$\Gamma(s)\Theta(x,s,a;b) = 2\pi e^{b} \sum_{\chi,\xi=0}^{\infty} (-e^{-x})^{\chi} \frac{(-(\chi+a))^{\xi}}{\xi!} \delta(s+\xi) \tag{A46}$$

$$\Gamma(s)\Psi(x,s,a;b) = 2\pi e^{b} \sum_{\chi,\xi=0}^{\infty} (e^{-x})^{\chi} \frac{(-(\chi+a))^{\xi}}{\xi!} \delta(s+\xi) \tag{A47}$$

$$\Gamma(s)F(x,s;b) = 2\pi e^{b} \sum_{\chi,\xi=0}^{\infty} (-e^{-x})^{\chi} \frac{(-(\chi+1))^{\xi}}{\xi!} \delta(s+\xi) \tag{A48}$$

$$\Gamma(s)B(x,s;b) = 2\pi e^{b} \sum_{\chi,\xi=0}^{\infty} (e^{-x})^{\chi} \frac{(-(\chi+1))^{\xi}}{\xi!} \delta(s+\xi) \tag{A49}$$

$$\phi(z,s;b) = 2\pi z e^{b} \sum_{\chi,\xi=0}^{\infty} (z)^{\chi} \frac{(-(\chi+1))^{\xi}}{\xi!} \delta(s+\xi) \tag{A50}$$

$$\Gamma(s)\zeta(s,a;b) = 2\pi e^{b} \sum_{\chi,\xi=0}^{\infty} \frac{(-(\chi+a))^{\xi}}{\xi!} \delta(s+\xi) \tag{A51}$$

$$\Gamma(s)\zeta(s;b) = 2\pi e^{b} \sum_{\chi,\xi=0}^{\infty} \frac{(-(\chi+1))^{\xi}}{\xi!} \delta(s+\xi) \tag{A52}$$

References

1. Srivastava, H.M. A new family of the λ-generalized Hurwitz-Lerch Zeta functions with applications. *Appl. Math. Inf. Sci.* **2014**, *8*, 1485–1500. [CrossRef]
2. Jankov, D.; Pogany, T.K.; Saxena, R.K. An extended general Hurwitz-Lerch Zeta function as a Mathieu (a, λ)-series. *Appl. Math. Lett.* **2011**, *24*, 1473–1476. [CrossRef]
3. Srivastava, H.M.; Jankov, D.; Pogany, T.K.; Saxena, R.K. Two-sided inequalities for the extended Hurwitz-Lerch Zeta function. *Comput. Math. Appl.* **2011**, *62*, 516–522. [CrossRef]

4. Srivastava, H.M.; Ghanim, F.; El-Ashwah, R.M. Inclusion properties of certain subclass of univalent meromorphic functions defined by a linear operator associated with the λ-generalized Hurwitz-Lerch zeta function. *Asian-Eur. J. Math.* **2017**, *3*, 34–50.
5. Srivastava, H.M.; Gaboury, S. New expansion formulas for a family of the λ-generalized Hurwitz-Lerch zeta functions. *Int. J. Math. Math. Sci.* **2014**, *2014*, 131067. [CrossRef] [PubMed]
6. Srivastava, H.M.; Gaboury, S.; Ghanim, F. Certain subclasses of meromorphically univalent functions defined by a linear operator associated with the λ-generalized Hurwitz-Lerch zeta function. *Integral Transforms Spec. Funct.* **2015**, *26*, 258–272. [CrossRef]
7. Srivastava, H.M.; Gaboury, S.; Ghanim, F. Some further properties of a linear operator associated with the λ-generalized Hurwitz-Lerch zeta function related to the class of meromorphically univalent functions. *Appl. Math. Comput.* **2015**, *259*, 1019–1029.
8. Luo, M.-J.; Raina, R.K. New Results for Srivastava's λ-Generalized Hurwitz-Lerch Zeta Function. *Filomat* **2017**, *31*, 2219–2229. [CrossRef]
9. Srivastava, H.M.; Chaudhry, M.A.; Qadir, A.; Tassaddiq, A. Some extensions of the Fermi-Dirac and Bose-Einstein functions with applications to zeta and related functions. *Russ. J. Math. Phys.* **2011**, *18*, 107–121. [CrossRef]
10. Chaudhry, M.A.; Qadir, A. Fourier transform and distributional representation of Gamma function leading to some new identities. *Int. J. Math. Math. Sci.* **2004**, *37*, 2091–2096. [CrossRef]
11. Tassaddiq, A.; Qadir, A. Fourier transform and distributional representation of the generalized gamma function with some applications. *Appl. Math. Comput.* **2011**, *218*, 1084–1088. [CrossRef]
12. Tassaddiq, A.; Qadir, A. Fourier transform representation of the extended Fermi-Dirac and Bose-Einstein functions with applications to the family of the zeta and related functions. *Integral Transforms Spec. Funct.* **2011**, *22*, 453–466. [CrossRef]
13. Tassaddiq, A. Some Representations of the Extended Fermi-Dirac and BoseEinstein Functions with Applications. Ph.D. Dissertation, National University of Sciences and Technology Islamabad, Islamabad, Pakistan, 2011.
14. Al-Lail, M.H.; Qadir, A.A. Fourier transform representation of the generalized hypergeometric functions with applications to the confluent and gauss hypergeometric functions. *Appl. Math. Comput.* **2015**, *263*, 392–397. [CrossRef]
15. Tassaddiq, A. A New Representation of the Extended Fermi-Dirac and Bose-Einstein Functions. *Int. J. Math. Appl.* **2017**, *5*, 435–446.
16. Tassaddiq, A. Some new results involving the Srivastava λ-generalized Hurwitz-Lerch zeta functions. *Integral Transforms Spec. Funct.* **2018**, under review.
17. Srivastava, H.M.; Saxena, R.K.; Pogany, T.K.; Saxena, R. Integral and computational representations of the extended Hurwitz-Lerch Zeta function. *Integral Transforms Spec. Funct.* **2011**, *22*, 487–506. [CrossRef]
18. Kilbas, A.A.; Saigo, M. *H-Transforms: Theory and Applications*; Chapman and Hall (CRC Press Company): Boca Raton, FL, USA; London, UK; New York, NY, USA; Washington, DC, USA, 2004.
19. Kilbas, A.A.; Srivastava, H.M.; Trujillo, J.J. *Theory and Applications of Fractional Differential Equations*; North-Holland Mathematical Studies; Elsevier (North-Holland) Science Publishers: Amsterdam, The Netherlands; London, UK; New York, NY, USA, 2006; Volume 204.
20. Mathai, A.M.; Saxena, R.K. *The H-Functions with Applications in Statistics and Other Disciplines*; Wiley Eastern Limited: New Delhi, India, 1978.
21. Mathai, A.M.; Saxena, R.K. *Generalized Hypergeometric Functions with Applications in Statistics and Physical Sciences*; Lecture Notes in Mathematics; Springer: Berlin/Heidelberg, Germany; New York, NY, USA, 1973.
22. Mathai, A.M.; Saxena, R.K.; Haubold, H.J. *The H-Function: Theory and Applications*; Springer: New York, NY, USA; Dordrecht, The Netherlands; Heidelberg, Germany; London, UK, 2010.
23. Olver, F.W.J.; Lozier, D.W.; Boisvert, R.F.; Clark, C.W. (Eds.) *NIST Handbook of Mathematical Functions*; [With 1 CD-ROM (Windows, Macintosh and UNIX)]; U.S. Department of Commerce, National Institute of Standards and Technology: Washington, DC, USA, 2010.
24. Srivastava, H.M. Some generalizations and basic (or q-) extensions of the Bernoulli, Euler and Genocchipolynomials. *Appl. Math. Inform. Sci.* **2011**, *5*, 390–444.
25. Raina, R.K.; Chhajed, P.K. Certain results involving a class of functions associated with the Hurwitz zeta function. *Acta Math. Univ. Comenianae* **2004**, *73*, 89–100.

26. Srivastava, H.M.; Luo, M.-J.; Raina, R.K. New results involving a class of generalized Hurwitz-Lerch zeta functions and their applications. *Turk. J. Anal. Number Theory* **2013**, *1*, 26–35. [CrossRef]
27. Bayad, A.; Chikhi, J. Reduction and duality of the generalized Hurwitz-Lerch zetas. *Fixed Point Theory Appl.* **2013**, *2013*, 82. [CrossRef]
28. Erd´elyi, A.; Magnus, W.; Oberhettinger, F.; Tricomi, F.G. *Higher Transcendental Functions*; McGraw-Hill Book Company: New York, NY, USA; Toronto, ON, Canada; London, UK, 1953.
29. Goyal, S.P.; Laddha, R.K. On the generalized Zeta function and the generalized Lambert function. *Ganita Sandesh* **1997**, *11*, 99–108.
30. Chaudhry, M.A.; Zubair, S.M. *On a Class of Incomplete Gamma Functions with Applications*; Chapman and Hall (CRC Press Company): Boca Raton, FL, USA; London, UK; New York, NY, USA; Washington, DC, USA, 2001.
31. Apostol, T.M. *Introduction to Analytic Number Theory*; Springer: Berlin, Germany; New York, NY, USA; Heidelberg, Germany, 1976.
32. Saxena, R.K.; Pogany, T.K.; Saxena, R.; Jankov, D. On generalized Hurwitz-Lerch Zeta distributions occurring in statistical inference. *Acta Univ. Sapientiae Math.* **2011**, *3*, 43–59.
33. Gel'fand, I.M.; Shilov, G.E. *Generalized Functions: Properties and Operations*; Academic Press: New York, NY, USA, 1969; Volume I–V.
34. Zamanian, A.H. *Distribution Theory and Transform Analysis*; Dover Publications: New York, NY, USA, 1987.
35. Richards, I.; Youn, H. *Theory of Distributions: A Non Technical Introduction*; Cambridge University Press: Cambridge, MA, USA; London, UK; New York, NY, USA, 2007.

© 2018 by the author. Licensee MDPI, Basel, Switzerland. This article is an open access article distributed under the terms and conditions of the Creative Commons Attribution (CC BY) license (http://creativecommons.org/licenses/by/4.0/).

symmetry

MDPI

Article

On Some Statistical Approximation by (p, q)-Bleimann, Butzer and Hahn Operators

Khursheed J. Ansari [1], Ishfaq Ahmad [1,2], M. Mursaleen [3,*] and Iqtadar Hussain [4]

1 Department of Mathematics, College of Science, King Khalid University, Abha 61413, Saudi Arabia; ansari.jkhursheed@gmail.com (K.J.A.); ishfaq.ahmad@iiu.edu.pk (I.A.)
2 Department of Mathematics and Statistics, International Islamic University, Islamabad 44000, Pakistan
3 Department of Mathematics, Aligarh Muslim University, Aligarh 202002, India
4 Department of Mathematics, Statistics and Physics, Qatar University, Doha 2713, Qatar; iqtadarqau@gmail.com
* Correspondence: mursaleenm@gmail.com; Tel.:+91-9411491600

Received: 29 October 2018; Accepted: 4 December 2018; Published: 7 December 2018

Abstract: In this article, we propose a different generalization of (p, q)-BBH operators and carry statistical approximation properties of the introduced operators towards a function which has to be approximated where (p, q)-integers contains symmetric property. We establish a Korovkin approximation theorem in the statistical sense and obtain the statistical rates of convergence. Furthermore, we also introduce a bivariate extension of proposed operators and carry many statistical approximation results. The extra parameter p plays an important role to symmetrize the q-BBH operators.

Keywords: q–Bleimann–Butzer–Hahn operators; (p, q)-integers; (p, q)-Bernstein operators; (p, q)-Bleimann–Butzer–Hahn operators; modulus of continuity; rate of approximation; K-functional

MSC: 41A10; 41A25; 41A36

---

1. Introduction

The q-analog of Bleiman, Butzer and Hahn operators (BBH) [1] is defined by:

$$L_n^q(f;x) = \frac{1}{\ell_n^q(x)} \sum_{k=0}^{n} f\left(\frac{[k]_q}{[n-k+1]_q q^k}\right) q^{\frac{k(k-1)}{2}} \begin{bmatrix} n \\ k \end{bmatrix}_q x^k, \tag{1}$$

where $\ell_n^q(x) = \prod_{k=0}^{n-1}(1+q^s x)$.

For $q = 1$, the sequence of q-BBH operators (1) reduces to the classical BBH-operators [2] in which authors investigated pointwise convergence properties of the BBH-operators in a compact sub-interval of $\mathbb{R}_+$.

Let H_ω denote the space of all real-valued functions f defined on the semi-axis $\mathbb{R}_+$ [3], where ω is the usual modulus of continuity satisfying

$$|f(x) - f(y)| \leq \omega\left(\left|\frac{x}{1+x} - \frac{y}{1+y}\right|\right)$$

for any $x, y \geq 0$.

Gadjiev and Çakar [3] established the Korovkin type theorem which gives the convergence for the sequence of linear positive operators (LPO) to the functions in H_ω.

Now, we recollect the following theorem:

Theorem 1 ([3]). *Let $\{A_n\}$ be the sequence of LPOs from H_ω into $C_B(\mathbb{R}_+)$ such that*

$$\lim_{n\to\infty}\left\|A_n\left(\left(\frac{t}{1+t}\right)^v;x\right)-\left(\frac{x}{1+x}\right)^v\right\|_{C_B}=0,\ \ v=0,1,2,$$

then, for any function $f \in H_\omega$

$$\lim_{n\to\infty}\|A_n(f)-f\|_{C_B}=0.$$

(p,q)-calculus, also called post-quantum calculus, is a generalization of q-calculus which has lots of applications in quantum physics. In approximation theory, the very first (p,q)-type generalization of Bernstein polynomials was introduced by Mursaleen et al. [4] using (p,q)-calculus and improved the said operators (see Erratum [4]). The theory of semigroups of the linear operators is used in order to prove the existence and uniqueness of a weak solutions of boundary value problems in thermoelasticity of dipolar bodies (see [5,6]).

Recently, a very nice application and usage of extra parameter p has been discussed in [7] in the computer-aided geometric design. In that paper, authors applied these (p,q)-Bernstein bases to construct (p,q)-Bézier curves which are further generalizations of q-Bézier curves [8]. For more results on LPOs and its (p,q)-analogues, one can refer to [9–15].

Now, we provide some notations on (p,q)-calculus.

$[n]_{p,q}$ stands for (p,q)-integers defined as

$$[n]_{p,q}=p^{n-1}+p^{n-2}q+p^{n-3}q^2\cdots+q^{n-1}=\begin{cases}\dfrac{p^n-q^n}{p-q} & (p\neq q\neq 1),\\[2mm] \dfrac{1-q^n}{1-q} & (p=1),\\[2mm] n & (p=q=1),\end{cases} \tag{2}$$

$$(ax+by)^n_{p,q}:=\sum_{j=0}^{n}p^{\frac{(n-j)(n-j-1)}{2}}q^{\frac{j(j-1)}{2}}\begin{bmatrix} n\\ j\end{bmatrix}_{p,q}a^{n-j}b^jx^{n-j}y^j,$$

$$(x+y)^n_{p,q}=(x+y)(px+qy)(p^2x+q^2y)\cdots(p^{n-1}x+q^{n-1}y),$$

$$(1-x)^n_{p,q}=(1-x)(p-qx)(p^2-q^2x)\cdots(p^{n-1}-q^{n-1}x),$$

and the binomial coefficients in (p,q)-calculus are given by

$$\begin{bmatrix} n\\ j\end{bmatrix}_{p,q}=\frac{[n]_{p,q}!}{[j]_{p,q}![n-j]_{p,q}!}.$$

By easy computation, we have the relation given below:

$$q^j[n-j+1]_{p,q}=[n+1]_{p,q}-p^{n-j+1}[j]_{p,q}.$$

Authors suggest the readers [16–19].

The (p,q)-analogue of BBH operators was introduced by Mursaleen et al. in [20] as follows:

$$L_n^{p,q}(f;x) = \frac{1}{\ell_n^{p,q}(x)} \sum_{j=0}^{n} f\left(\frac{p^{n-j+1}[j]_{p,q}}{[n-j+1]_{p,q}q^j}\right) p^{\frac{(n-j)(n-j-1)}{2}} q^{\frac{j(j-1)}{2}} \begin{bmatrix} n \\ j \end{bmatrix}_{p,q} x^j, \tag{3}$$

where $x \geq 0$, $0 < q < p \leq 1$, $\ell_n^{p,q}(x) = \prod_{s=0}^{n-1}(p^s + q^s x)$ and function f is defined on the semi axis $\mathbb{R}_+$. If we put $p = 1$, we get the q-BBH operators (1). In that paper, authors established different approximation properties of the sequence of operators (3).

Theorem 2 ([20]). *Let $p = (p_n)$, $q = (q_n)$ satisfying $\lim_{n\to\infty} p_n = 1$, $\lim_{n\to\infty} q_n = 1$ for $0 < q_n < p_n \leq 1$ and if $L_n^{p_n,q_n}(f;x)$ is defined by Label (3). Then, for any function $f \in H_\omega$,*

$$\lim_{n\to\infty} \|L_n^{p_n,q_n} f - f\|_{C_B} = 0.$$

Mursaleen and Nasiruzzaman constructed bivariate (p,q)-BBH operators [21] and studied many nice properties based on that sequence of operators and also given some generalization of that sequence of bivariate operators introducing one more parameter γ in the operators.

The statistical convergence is another notion of convergence, which was introduced by Fast [22] nearly fifty years ago and now this is a very active area of research. The statistical limit of a sequence is an extension of the idea of limit of sequence in an ordinary sense. The natural density of $K \subset \mathbb{N}$ is defined as:

$$\delta(K) = \lim_n \frac{1}{n}\{k \leq n : k \in K\}$$

whenever the limit exists (see [23,24]). The sequence $x = (x_k)$ is said to be statistically convergent to a number L means if, for every $\epsilon > 0$,

$$\delta\{k : |x_k - L| \geq \epsilon\} = 0,$$

and it is denoted by $st - \lim_k x_k = L$. It can be easily seen that every convergent sequence is statistically convergent but not conversely.

Now, we will state some preliminary results on positive linear operators:

Proposition 1 ([25]). *If L is an operator, linear and positive, then, for every $x \in X$, we have*

$$|Lf| \leq L(|f|). \tag{4}$$

Proposition 2 ([25]). *(Hölder's inequality for LPOs). Let $L : X \to Y$ be an operator, linear and positive, and let $1/p + 1/q = 1$, where $p, q > 1$ are real numbers. Then, for every $f, g \in X$*

$$L(|f \cdot g|) \leq (L(|f|^p))^{\frac{1}{p}} \cdot (L(|g|^q))^{\frac{1}{q}}. \tag{5}$$

Remark 1 ([25]). *A particular case of Proposition 2 is the Cauchy–Schwarz's inequality for LPOs, which is obtained from Hölder's inequality for $p = q = 2$ as:*

$$|L(f \cdot g; x)| \leq \sqrt{L(f^2; x)} \cdot \sqrt{L(g^2; x)}. \tag{6}$$

We have organized the rest of the paper as follows. In Section 2, we have constructed (p,q)-BBH operators and calculated some auxiliary results. In Sections 3 and 4, Korovkin type results and rate of convergence are established in statistical sense, respectively. Section 5 is devoted to the construction of the bivariate (p,q)-BBH operators. In Section 6, we have computed rate of statistical convergence for the bivariate (p,q)-BBH operators.

2. Construction of Operators and Moment Estimation

Ersan and Doğru [26] introduced a generalization of (1) and studied different statistical approximation properties of the operators towards a function f which has to be approximated. Inspired with the work of Ersan and Doğru [26], we construct a (p,q)-analogue generalization of the sequence of operators defined in [26] or, on the other hand, we generalize the operators introduced in [20] as follows:

$\forall x \geq 0,\ 0 < q < p \leq 1$, let us define a sequence of (p,q)-BBH operators as follows:

$$\mathcal{B}_n^{p,q}(f;x) = \frac{q/p}{\zeta_n^{p,q}(x)} \sum_{j=0}^{n} f\left(\frac{p^{n-j+1}[j]_{p,q}}{[n-j+1]_{p,q}q^j}\right) p^{\frac{(n-j)(n-j-1)}{2}} q^{\frac{j(j-1)}{2}} \begin{bmatrix} n \\ j \end{bmatrix}_{p,q} x^j, \tag{7}$$

where

$$\zeta_n^{p,q}(x) = \prod_{s=0}^{n-1}(p^s + q^s x) = \sum_{j=0}^{n} p^{\frac{(n-j)(n-j-1)}{2}} q^{\frac{j(j-1)}{2}} \begin{bmatrix} n \\ j \end{bmatrix}_{p,q} x^j. \tag{8}$$

It is easy to verify that, if $p = q = 1$, the operators turn into the classical BBH operators. The sequence of operators (7) is of course more generalized than (1), and it is more flexible than (1).

We need the following lemma to our main result:

Lemma 1. *Let the sequence of operators be given by (7). Then,*

$$\mathcal{B}_n^{p,q}(1;x) = \frac{q}{p} \tag{9}$$

for any $x \geq 0, 0 < q < p \leq 1$.

Proof. The proof is obvious with the help of the relation (8), so we skip the proof. □

Lemma 2. *Let the sequence of operators be given by (7). Then,*

$$\mathcal{B}_n^{p,q}\left(\frac{t}{1+t};x\right) = \frac{q[n]_{p,q}}{[n+1]_{p,q}} \frac{x}{1+x} \tag{10}$$

for any $x \geq 0, 0 < q < p \leq 1$.

Proof. Let $t = \frac{p^{n-j+1}[j]_{p,q}}{[n-j+1]_{p,q}q^j}$, then $\frac{t}{1+t} = \frac{p^{n-j+1}[j]_{p,q}}{[n+1]_{p,q}}$, so

$$\begin{aligned}
&\mathcal{B}_n^{p,q}\left(\frac{t}{1+t};x\right) \\
&= \frac{q/p}{\zeta_n^{p,q}(x)} \frac{[n]_{p,q}}{[n+1]_{p,q}} \sum_{j=1}^{n} p^{n-j+1} p^{\frac{(n-j)(n-j-1)}{2}} q^{\frac{j(j-1)}{2}} \begin{bmatrix} n-1 \\ j-1 \end{bmatrix}_{p,q} x^j \\
&= \frac{q/p}{\zeta_n^{p,q}(x)} \frac{[n]_{p,q}x}{[n+1]_{p,q}} \sum_{j=0}^{n-1} p^{n-j} p^{\frac{(n-j-2)(n-j-1)}{2}} q^{\frac{j(j+1)}{2}} \begin{bmatrix} n-1 \\ j \end{bmatrix}_{p,q} x^j \\
&= \frac{q/p}{\zeta_n^{p,q}(x)} \frac{p^n[n]_{p,q}x}{[n+1]_{p,q}} \sum_{j=0}^{n-1} p^{\frac{(n-j-1)(n-j-2)}{2}} q^{\frac{j(j-1)}{2}} \begin{bmatrix} n-1 \\ j \end{bmatrix}_{p,q} \left(\frac{qx}{p}\right)^j.
\end{aligned}$$

By using (8), the result can be easily obtained. □

Lemma 3. *Let the sequence of operators be given by* (7)*. Then,*

$$\mathcal{B}_n^{p,q}\left(\left(\frac{t}{1+t}\right)^2;x\right)=\frac{pq^3[n]_{p,q}[n-1]_{p,q}}{[n+1]_{p,q}^2}\frac{x^2}{(1+x)(p+qx)}+\frac{p^nq[n]_{p,q}}{[n+1]_{p,q}^2}\frac{x}{1+x} \tag{11}$$

for any $x\geq 0$, $0<q<p\leq 1$.

Proof. It is easy to verify that

$$[j]_{p,q}=p^{j-1}+q[j-1]_{p,q};\qquad [j]_{p,q}^2=q[j]_{p,q}[j-1]_{p,q}+p^{j-1}[j]_{p,q}. \tag{12}$$

With the help of (12), we can have

$$\begin{aligned}
&\mathcal{B}_n^{p,q}\left(\left(\frac{t}{1+t}\right)^2;x\right)\\
&=\frac{q/p}{\zeta_n^{p,q}(x)}\left\{\frac{q[n]_{p,q}[n-1]_{p,q}}{[n+1]_{p,q}^2}\sum_{j=2}^{n}p^{2n-2j+2}p^{\frac{(n-j)(n-j-1)}{2}}q^{\frac{j(j-1)}{2}}\begin{bmatrix}n-2\\ j-2\end{bmatrix}_{p,q}x^j\right.\\
&\qquad\left.+\frac{[n]_{p,q}}{[n+1]_{p,q}^2}\sum_{j=1}^{n}p^{2n-2j+2}p^{j-1}p^{\frac{(n-j)(n-j-1)}{2}}q^{\frac{j(j-1)}{2}}\begin{bmatrix}n-1\\ j-1\end{bmatrix}_{p,q}x^j\right\}\\
&=\frac{q/p}{\zeta_n^{p,q}(x)}\left\{\frac{q[n]_{p,q}[n-1]_{p,q}}{[n+1]_{p,q}^2}\sum_{j=0}^{n-2}p^{2n-2j-2}p^{\frac{(n-j-2)(n-j-3)}{2}}q^{\frac{(j+2)(j+1)}{2}}\begin{bmatrix}n-2\\ j\end{bmatrix}_{p,q}x^{j+2}\right.\\
&\qquad\left.+\frac{[n]_{p,q}}{[n+1]_{p,q}^2}\sum_{j=0}^{n-1}p^{2n-2j}p^{j}p^{\frac{(n-j-1)(n-j-2)}{2}}q^{\frac{(j+1)j}{2}}\begin{bmatrix}n-1\\ j\end{bmatrix}_{p,q}x^{j+1}\right\}\\
&=\frac{q/p}{\zeta_n^{p,q}(x)}\left\{\frac{p^{2n-2}q^2[n]_{p,q}[n-1]_{p,q}}{[n+1]_{p,q}^2}x^2\sum_{j=0}^{n-2}p^{\frac{(n-j-2)(n-j-3)}{2}}q^{\frac{j(j-1)}{2}}\begin{bmatrix}n-2\\ j\end{bmatrix}_{p,q}\left(\frac{q^2x}{p^2}\right)^j\right.\\
&\qquad\left.+\frac{p^{2n}[n]_{p,q}}{[n+1]_{p,q}^2}x\sum_{j=0}^{n-1}p^{\frac{(n-j-1)(n-j-2)}{2}}q^{\frac{j(j-1)}{2}}\begin{bmatrix}n-1\\ j\end{bmatrix}_{p,q}\left(\frac{qx}{p}\right)^j\right\}.
\end{aligned}$$

Now, using (8), we can get the desired result. □

3. Korovkin Type Statistical Approximation Properties

In this section, we obtain the Korovkin type statistical approximation theorem for our sequence of operators (7). Let us give the following theorem:

Theorem 3. *[3] Let* $\{A_n\}$ *be the sequence of LPOs from* H_ω *into* $C_B(\mathbb{R}_+)$ *such that*

$$st-\lim_n\left\|A_n\left(\left(\frac{t}{1+t}\right)^\nu;x\right)-\left(\frac{x}{1+x}\right)^\nu\right\|_{C_B}=0,\ \ \nu=0,1,2.$$

Then, for any function $f\in H_\omega$,

$$st-\lim_n\|A_n(f)-f\|_{C_B}=0.$$

Let us take $p = (p_n)$ and $q = (q_n)$ such that

$$st - \lim_n p_n = 1, \; st - \lim_n q_n = 1. \tag{13}$$

Theorem 4. *Let $\mathcal{B}_n^{p,q}(f;x)$ be the sequence of operators (7) and the sequences $p = (p_n)$ and $q = (q_n)$ satisfy the assumption (13) for $0 < q_n < p_n \leq 1$. Then, for any function $f \in H_\omega$,*

$$st - \lim_n \|\mathcal{B}_n^{p,q}(f;.) - f\| = 0.$$

Proof. For $\nu = 0$ and using (9), we can have

$$st - \lim_n \|\mathcal{B}_n^{p_n,q_n}(1;x) - 1\| = st - \lim_n \left|\frac{q_n}{p_n} - 1\right|.$$

By (13), the following can be easily verified, which is

$$st - \lim_n \|\mathcal{B}_n^{p_n,q_n}(1;x) - 1\| = 0.$$

For $\nu = 1$ and using (10), we get

$$\left\|\mathcal{B}_n^{p_n,q_n}\left(\frac{t}{1+t};x\right) - \frac{x}{1+x}\right\| \leq \left|q_n \frac{[n]_{p_n,q_n}}{[n+1]_{p_n,q_n}} - 1\right| = 1 - q_n \frac{[n]_{p_n,q_n}}{[n+1]_{p_n,q_n}}.$$

For a given $\epsilon > 0$, let us define the following sets:

$$U = \left\{n : \left\|\mathcal{B}_n^{p_n,q_n}\left(\frac{t}{1+t};x\right) - \frac{x}{1+x}\right\| \geq \epsilon\right\},$$

and

$$U' = \left\{n : 1 - q_n \frac{[n]_{p_n,q_n}}{[n+1]_{p_n,q_n}} \geq \epsilon\right\}.$$

It is easily perceived that $U \subset U'$, so we can write

$$\delta\left\{k \leq n : \left\|\mathcal{B}_n^{p_n,q_n}\left(\frac{t}{1+t};x\right) - \frac{x}{1+x}\right\| \geq \epsilon\right\} \leq \delta\left\{k \leq n : 1 - q_n \frac{[n]_{p_n,q_n}}{[n+1]_{p_n,q_n}} \geq \epsilon\right\}.$$

On using (13), it is clear that

$$st - \lim_n \left(1 - q_n \frac{[n]_{p_n,q_n}}{[n+1]_{p_n,q_n}}\right) = 0.$$

Thus,

$$\delta\left\{k \leq n : 1 - q_n \frac{[n]_{p_n,q_n}}{[n+1]_{p_n,q_n}} \geq \epsilon\right\} = 0;$$

then,

$$st - \lim_n \left\|\mathcal{B}_n^{p_n,q_n}\left(\frac{t}{1+t};x\right) - \frac{x}{1+x}\right\| = 0.$$

Lastly, for $\nu = 2$ and using (11), we obtain

$$\begin{aligned}
&\left\| \mathcal{B}_n^{p_n,q_n}\left(\frac{t}{1+t};x\right)^2 - \left(\frac{x}{1+x}\right)^2 \right\| \\
&= \left\| \frac{p_n q_n^3 [n]_{p_n,q_n}[n-1]_{p_n,q_n}}{[n+1]^2_{p_n,q_n}} \frac{x^2}{(1+x)(p_n+q_n x)} + \frac{p_n^n q_n [n]_{p_n,q_n}}{[n+1]^2_{p_n,q_n}} \frac{x}{1+x} - 1 \right\| \\
&\leq \left| \frac{p_n q_n^2 [n]_{p_n,q_n}[n-1]_{p_n,q_n}}{[n+1]^2_{p_n,q_n}} - 1 \right| + \left| \frac{p_n^n q_n [n]_{p_n,q_n}}{[n+1]^2_{p_n,q_n}} \right|.
\end{aligned} \tag{14}$$

Using $[n+1]_{p_n,q_n} = p_n[n]_{p_n,q_n} + q_n^n$, the following can be easily justified that

$$\frac{[n]_{p_n,q_n}[n-1]_{p_n,q_n}}{[n+1]^2_{p_n,q_n}} = \frac{1}{p_n^3}\left(1 - \frac{2q_n^n + p_n q_n^{n-1}}{[n+1]_{p_n,q_n}} + \frac{q_n^{2n} + p_n q_n^{2n-1}}{[n+1]^2_{p_n,q_n}}\right).$$

Substituting it in (14), we can have

$$\begin{aligned}
&\left\| \mathcal{B}_n^{p_n,q_n}\left(\left(\frac{t}{1+t}\right)^2;x\right) - \left(\frac{x}{1+x}\right)^2 \right\| \\
&\leq \left| \frac{q_n^2}{p_n^2} - 1 \right| + \left| \frac{q_n^2}{p_n^2}\left(\frac{2q_n^n + p_n q_n^{n-1}}{[n+1]_{p_n,q_n}} - \frac{q_n^{2n} + p_n q_n^{2n-1}}{[n+1]^2_{p_n,q_n}}\right) \right| + \left| \frac{p_n^n q_n [n]_{p_n,q_n}}{[n+1]^2_{p_n,q_n}} \right| \\
&\leq \frac{q_n^2}{p_n^2} - 1 + \frac{q_n^2}{p_n^2}\left(\frac{2q_n^n + p_n q_n^{n-1}}{[n+1]_{p_n,q_n}} - \frac{q_n^{2n} + p_n q_n^{2n-1}}{[n+1]^2_{p_n,q_n}}\right) + \frac{p_n^n}{[n+1]_{p_n,q_n}} - \frac{p_n^{2n}}{[n+1]^2_{p_n,q_n}}.
\end{aligned}$$

If we choose $\alpha_n = \frac{q_n^2}{p_n^2} - 1$, $\beta_n = \frac{q_n^2}{p_n^2}\left(\frac{2q_n^n + p_n q_n^{n-1}}{[n+1]_{p_n,q_n}} - \frac{q_n^{2n} + p_n q_n^{2n-1}}{[n+1]^2_{p_n,q_n}}\right)$, and $\gamma_n = \frac{p_n^n}{[n+1]_{p_n,q_n}} - \frac{p_n^{2n}}{[n+1]^2_{p_n,q_n}}$, then by (13), we have

$$st - \lim_n \alpha_n = st - \lim_n \beta_n = st - \lim_n \gamma_n = 0. \tag{15}$$

For any given $\epsilon > 0$, now we define four sets as follows:

$$U = \left\{ n : \left\| \mathcal{B}_n^{p_n,q_n}\left(\left(\frac{t}{1+t}\right)^2;x\right) - \left(\frac{x}{1+x}\right)^2 \right\| \geq \epsilon \right\},$$

$$U_1 = \left\{ n : \alpha_n \geq \frac{\epsilon}{3} \right\},\ U_2 = \left\{ n : \beta_n \geq \frac{\epsilon}{3} \right\},\ U_3 = \left\{ n : \gamma_n \geq \frac{\epsilon}{3} \right\}.$$

It is obvious that $U \subset U_1 \cup U_2 \cup U_3$. Then, we obtain

$$\begin{aligned}
&\delta \left\{ k \leq n : \left\| \mathcal{B}_n^{p_n,q_n}\left(\left(\frac{t}{1+t}\right)^2;x\right) - \left(\frac{x}{1+x}\right)^2 \right\| \geq \epsilon \right\} \\
&\leq \delta\left\{ k \leq n : \alpha_n \geq \frac{\epsilon}{3} \right\} + \delta\left\{ k \leq n : \beta_n \geq \frac{\epsilon}{3} \right\} + \delta\left\{ k \leq n : \gamma_n \geq \frac{\epsilon}{3} \right\}.
\end{aligned}$$

It is clear that the right-hand side of the above inequality is zero by (15); then,

$$st - \lim_n \left\| \mathcal{B}_n^{p_n,q_n}\left(\left(\frac{t}{1+t}\right)^2;x\right) - \left(\frac{x}{1+x}\right)^2 \right\| = 0.$$

Hence, the proof is completed. □

4. Rates of Statistical Convergence

This section is devoted to find rates of statistical convergence of operators (7).

The modulus of continuity for the space of functions $f \in H_\omega$ [1] is defined by

$$\tilde{\omega}(f;\delta) = \sup_{x,t\geq 0,\ \left|\frac{t}{1+t}-\frac{x}{1+x}\right|\leq\delta} |f(t)-f(x)|,$$

where $\tilde{\omega}(f;\delta)$ satisfies the following conditions: $\forall f \in H_\omega(R_+)$

1. $\lim\limits_{\delta\to 0}\tilde{\omega}(f;\delta) = 0,$
2. $|f(t)-f(x)| \leq \tilde{\omega}(f;\delta)\left(\frac{\frac{t}{1+t}-\frac{x}{1+x}}{\delta}+1\right).$

Theorem 5. *Let $p=(p_n)$ and $q=(q_n)$ be the sequences satisfying (13) and $0<q_n<p_n\leq 1$, we have*

$$|\mathcal{B}_n^{p_n,q_n}(f;x)-f(x)| \leq \tilde{\omega}(f;\sqrt{\delta_n(x)})\left(\frac{q_n}{p_n}+1\right),$$

where

$$\begin{aligned}
&\delta_n(x)\\
&= \left(\frac{x}{1+x}\right)^2\left(\frac{q_n^4(1+x)}{p_n+q_nx}\frac{[n]_{p_n,q_n}[n-1]_{p_n,q_n}}{[n+1]^2_{p_n,q_n}} - \frac{2q_n^2[n]_{p_n,q_n}}{p_n[n+1]^2_{p_n,q_n}} + \frac{q_n^2}{p_n^2}\right)\\
&\quad + \frac{p_n^{n-1}q_n^2[n]_{p_n,q_n}}{[n+1]^2_{p_n,q_n}}\frac{x}{1+x}.
\end{aligned} \tag{16}$$

Proof.

$$\begin{aligned}
&|\mathcal{B}_n^{p_n,q_n}(f;x)-f(x)|\\
&\quad\leq \mathcal{B}_n^{p_n,q_n}(|f(t)-f(x)|;x)\\
&\quad\leq \tilde{\omega}(f;\delta)\left\{\mathcal{B}_n^{p_n,q_n}(1;x)+\frac{1}{\delta}\mathcal{B}_n^{p_n,q_n}\left(\left|\frac{t}{1+t}-\frac{x}{1+x}\right|;x\right)\right\}.
\end{aligned}$$

By using the Cauchy–Schwarz inequality (see (6)) and using (9)–(11), we have

$$\begin{aligned}
&|\mathcal{B}_n^{p_n,q_n}(f;x)-f(x)|\\
&\leq \tilde{\omega}(f;\delta_n)\left(\frac{q_n}{p_n}+\frac{1}{\delta_n}\left\{\left(\mathcal{B}_n^{p_n,q_n}\left(\left(\frac{t}{1+t}-\frac{x}{1+x}\right)^2;x\right)\right)\right\}^{\frac{1}{2}}\left(\mathcal{B}_n^{p_n,q_n}(1^2;x)\right)^{\frac{1}{2}}\right)\\
&\leq \tilde{\omega}(f;\delta_n)\left(\frac{q_n}{p_n}+\frac{1}{\delta_n}\left\{\left(\frac{x}{1+x}\right)^2\left(\frac{q_n^4(1+x)}{p_n+q_nx}\frac{[n]_{p_n,q_n}[n-1]_{p_n,q_n}}{[n+1]^2_{p_n,q_n}}\right.\right.\right.\\
&\quad\left.\left.\left.-\frac{2q_n^2[n]_{p_n,q_n}}{p_n[n+1]^2_{p_n,q_n}}+\frac{q_n^2}{p_n^2}\right)+\frac{p_n^{n-1}q_n^2[n]_{p_n,q_n}}{[n+1]^2_{p_n,q_n}}\frac{x}{1+x}\right\}^{\frac{1}{2}}\right).
\end{aligned}$$

Thus, it is obvious that, by choosing δ_n as in (16), the theorem is proved. □

Notice that, by conditions in (13), $st-\lim\limits_n = 0$. Then, we have

$$st-\lim_n \tilde{\omega}(f;\delta_n) = 0.$$

This provides us the pointwise rate of statistical convergence of the sequence of operators $B_n^{p_n,q_n}(f;x)$ to $f(x)$.

Now, we will contribute an estimate related to the rate of approximation by means of Lipschitz type maximal functions.

Lenze [27] introduced a Lipschitz type maximal function as

$$\tilde{f}_\alpha(x) = \sup_{t>0,\, t\neq x} \frac{f(t)-f(x)}{|x-t|^\alpha}.$$

The Lipschitz type maximal function space on $E \subset \mathbb{R}_+$ is defined in [1] as follows:

$$\widetilde{W}_{\alpha,E} = \left\{ f : \sup(1+x)^\alpha \tilde{f}_\alpha(x) \leq M \frac{1}{(1+y)^\alpha}; x \geq 0 \text{ and } y \in E \right\},$$

where function f is bounded and continuous on $\mathbb{R}_+$, $0 < \alpha \leq 1$ and M is a positive constant.

Theorem 6. *If $B_n^{p_n,q_n}(f;x)$ is defined by (7), then $\forall f \in \widetilde{W}_{\alpha,E}$, we have*

$$|\mathcal{B}_n^{p_n,q_n}(f;x) - f(x)| \leq M \left(\rho_n(x)^{\frac{\alpha}{2}} \left(\frac{q_n}{p_n}\right)^{\frac{2-\alpha}{2}} + \frac{2q_n}{p_n} d(x,E) \right),$$

where

$$\rho_n(x) = \left(\frac{x}{1+x}\right)^2 \left(\frac{p_n q_n^3(1+x)}{p_n+q_nx} \frac{[n]_{p_n,q_n}[n-1]_{p_n,q_n}}{[n+1]^2_{p_n,q_n}} - \frac{2q_n[n]_{p_n,q_n}}{p_n[n+1]^2_{p_n,q_n}} + \frac{q_n}{p_n} \right) + \frac{p_n^n q_n [n]_{p_n,q_n}}{[n+1]^2_{p_n,q_n}} \frac{x}{1+x}.$$

Proof. A similar technique used in Theorem 7 in [26] will be taken to provide the proof. Letting $x \geq 0$, $(x, x_0) \in \mathbb{R}_+ \times E$, it is understood that

$$|f - f(x)| \leq |f - f(x_0)| + |f(x_0) - f(x)|.$$

Since $\mathcal{B}_n^{p_n,q_n}(f;x)$ is a linear and positive operator, $f \in \widetilde{W}_{\alpha,E}$, using the previous inequality, we have

$$\begin{aligned} &|\mathcal{B}_n^{p_n,q_n}(f;x) - f(x)| \\ &\leq \mathcal{B}_n^{p_n,q_n}(|f - f(x_0)|;x) + |f(x_0) - f(x)|\mathcal{B}_n^{p_n,q_n}(1;x) \\ &\leq M \left(\mathcal{B}_n^{p_n,q_n}\left(\left| \frac{t}{1+t} - \frac{x_0}{1+x_0} \right|^\alpha ;x \right) + \frac{|x-x_0|^\alpha}{(1+x)^\alpha(1+x_0)^\alpha} \mathcal{B}_n^{p_n,q_n}(1;x) \right). \end{aligned} \tag{17}$$

Consequently, we obtain

$$\begin{aligned} &\mathcal{B}_n^{p_n,q_n}\left(\left| \frac{t}{1+t} - \frac{x_0}{1+x_0} \right|^\alpha ;x \right) \\ &\leq \mathcal{B}_n^{p_n,q_n}\left(\left| \frac{t}{1+t} - \frac{x}{1+x} \right|^\alpha ;x \right) + \frac{|x-x_0|^\alpha}{(1+x)^\alpha(1+x_0)^\alpha} \mathcal{B}_n^{p_n,q_n}(1;x). \end{aligned}$$

Using the Hölder's inequality (see (5)) with $p = \frac{2}{\alpha}$ and $q = \frac{2}{2-\alpha}$ and using (9)–(11), we have

$$\begin{aligned}
&\mathcal{B}_n^{p_n,q_n}\left(\left|\frac{t}{1+t} - \frac{x}{1+x}\right|^{\alpha}; x\right) \\
&\leq \mathcal{B}_n^{p_n,q_n}\left(\left(\frac{t}{1+t} - \frac{x}{1+x}\right)^2; x\right)^{\frac{\alpha}{2}} \left(\mathcal{B}_n^{p_n,q_n}(1^2; x)\right)^{\frac{2-\alpha}{2}} \\
&\quad + \frac{|x - x_0|^{\alpha}}{(1+x)^{\alpha}(1+x_0)^{\alpha}} \mathcal{B}_n^{p_n,q_n}(1; x) \\
&= \rho_n(x)^{\frac{\alpha}{2}} \left(\frac{q_n}{p_n}\right)^{\frac{2-\alpha}{2}} + \frac{q_n}{p_n} \frac{|x - x_0|^{\alpha}}{(1+x)^{\alpha}(1+x_0)^{\alpha}}.
\end{aligned}$$

If the above result is substituted in (17), we will get our desired result. Hence, the theorem is proved. □

Corollary 1. *If $B_n^{p_n,q_n}(f; x)$ is defined by (7) and take $E = \mathbb{R}_+$ implies $d(x, E) = 0$, then a special case of Theorem 6 can be obtained as the following result: $\forall f \in \widetilde{W}_{\alpha,R_+}$*

$$|\mathcal{B}_n^{p_n,q_n}(f; x) - f(x)| \leq M \rho_n(x)^{\frac{\alpha}{2}} \left(\frac{q_n}{p_n}\right)^{\frac{2-\alpha}{2}},$$

where $\rho_n(x)$ is the same as in Theorem 6.

5. Construction of the Bivariate Operators

In this section, we define a bivariate version of operators (7) and study their approximation properties.

For $\mathbb{R}_+^2 = [0, \infty) \times [0, \infty)$, $f : \mathbb{R}_+^2 \to \mathbb{R}$ and $0 < q_{n_1}, q_{n_2} < p_{n_1}, p_{n_2} \leq 1$, let us define the bivariate case of the operators (7) as follows:

$$\begin{aligned}
&\mathcal{B}_{n_1,n_2}^{p_{n_1}p_{n_2},q_{n_1}q_{n_2}}(f; x) \\
&= \frac{q_{n_1}/p_{n_1}}{\zeta_{n_1}^{p_{n_1},q_{n_1}}(x)} \frac{q_{n_2}/p_{n_2}}{\zeta_{n_2}^{p_{n_2},q_{n_2}}(y)} \sum_{j_1=0}^{n_1} \sum_{j_2=0}^{n_2} f\left(\frac{p_{n_1}^{n_1-j_1+1}[j_1]_{p_{n_1},q_{n_1}}}{[n_1 - j_1 + 1]_{p_{n_1},q_{n_1}} q_{n_1}^{j_1}}, \frac{p_{n_2}^{n_2-j_2+1}[j_2]_{p_{n_2},q_{n_2}}}{[n_2 - j_2 + 1]_{p_{n_2},q_{n_2}} q_{n_2}^{j_2}}\right) \\
&\quad \times p_{n_1}^{\frac{(n_1-j_1)(n_1-j_1-1)}{2}} q_{n_1}^{\frac{j_1(j_1-1)}{2}} p_{n_2}^{\frac{(n_2-j_2)(n_2-j_2-1)}{2}} q_{n_2}^{\frac{j_2(j_2-1)}{2}} \begin{bmatrix} n_1 \\ j_1 \end{bmatrix}_{p_{n_1},q_{n_1}} \begin{bmatrix} n_2 \\ j_2 \end{bmatrix}_{p_{n_2},q_{n_2}} x^{j_1} y^{j_2},
\end{aligned} \tag{18}$$

where $\zeta_{n_1}^{p_{n_1},q_{n_1}}(x) = \prod_{s=0}^{n_1-1}(p_{n_1}^s + q_{n_1}^s x)$ and $\zeta_{n_2}^{p_{n_2},q_{n_2}}(y) = \prod_{s=0}^{n_2-1}(p_{n_2}^s + q_{n_2}^s y)$.

For $K = I^2 = [0, \infty) \times [0, \infty)$, the modulus of continuity for bivariate case is defined by

$$|f(s,t) - f(x,y)| \leq \omega_2\left(f : \left|\frac{s}{1+s} - \frac{x}{1+x}\right|, \left|\frac{t}{1+t} - \frac{y}{1+y}\right|\right)$$

for each $f \in H_{\omega_2}$. Details of the modulus of continuity for the bivariate case can be found in [28].

Now, we will investigate Korovkin type approximation properties by using the following test functions:

$$e_0(u,v) = 1,\ e_1(u,v) = \frac{u}{1+u},\ e_2(u,v) = \frac{v}{1+v},\ e_3(u,v) = \left(\frac{u}{1+u}\right)^2 + \left(\frac{v}{1+v}\right)^2.$$

Lemma 4.

1. $\mathcal{B}_{n_1,n_2}^{p_{n_1}p_{n_2},q_{n_1}q_{n_2}}(e_0; x, y) = \frac{q_{n_1}q_{n_2}}{p_{n_1}p_{n_2}}$,

2. $\mathcal{B}_{n_1,n_2}^{p_{n_1}p_{n_2},q_{n_1}q_{n_2}}(e_1;x,y) = \frac{q_{n_1}q_{n_2}[n_1]_{p_{n_1},q_{n_1}}}{p_{n_2}[n_1+1]_{p_{n_1},q_{n_1}}}\frac{x}{1+x}$,
3. $\mathcal{B}_{n_1,n_2}^{p_{n_1}p_{n_2},q_{n_1}q_{n_2}}(e_2;x,y) = \frac{q_{n_1}q_{n_2}[n_2]_{p_{n_2},q_{n_2}}}{p_{n_1}[n_2+1]_{p_{n_2},q_{n_2}}}\frac{y}{1+y}$,
4. $\mathcal{B}_{n_1,n_2}^{p_{n_1}p_{n_2},q_{n_1}q_{n_2}}(e_3;x,y)$

$$= \frac{q_{n_1}^4 q_{n_2}}{p_{n_2}}\frac{[n_1]_{p_{n_1},q_{n_1}}[n_1-1]_{p_{n_1},q_{n_1}}}{[n_1+1]_{p_{n_1},q_{n_1}}^2}\frac{x^2}{(1+x)(p_{n_1}+q_{n_1}x)} + \frac{p_{n_1}^{n_1-1}q_{n_1}^2 q_{n_2}}{p_{n_2}}\frac{[n_1]_{p_{n_1},q_{n_1}}}{[n_1+1]_{p_{n_1},q_{n_1}}^2}\frac{x}{1+x}$$
$$+ \frac{q_{n_1}q_{n_2}^4}{p_{n_1}}\frac{[n_2]_{p_{n_2},q_{n_2}}[n_2-1]_{p_{n_2},q_{n_2}}}{[n_2+1]_{p_{n_2},q_{n_2}}^2}\frac{y^2}{(1+y)(p_{n_2}+q_{n_2}y)} + \frac{p_{n_2}^{n_2-1}q_{n_1}q_{n_2}^2}{p_{n_1}}\frac{[n_2]_{p_{n_2},q_{n_2}}}{[n_2+1]_{p_{n_2},q_{n_2}}^2}\frac{y}{1+y}.$$

Let (p_{n_1}), (p_{n_2}), (q_{n_1}) and (q_{n_2}) be the sequences that converge statistically to 1 but not convergent in ordinary sense, so it can be written as for $0 < q_{n_1}, q_{n_2} < p_{n_1}, p_{n_2} \leq 1$,

$$st - \lim_{n_1} p_{n_1} = st - \lim_{n_2} p_{n_2} = st - \lim_{n_1} q_{n_1} = st - \lim_{n_2} q_{n_2} = 1. \tag{19}$$

Now, with condition (19), let us show the statistical convergence of the sequence of bivariate operators (18).

Theorem 7. *Let (p_{n_1}), (p_{n_2}), (q_{n_1}) and (q_{n_2}) be the sequences satisfying the condition (19) and let $\mathcal{B}_{n_1,n_2}^{p_{n_1}p_{n_2},q_{n_1}q_{n_2}}(f;x)$ be the sequence of bivariate positive linear operators acting from $H_{\omega_2}(\mathbb{R}_+^2)$ into $C_B(\mathbb{R}_+)$. Then, for any $f \in H_{\omega_2}$,*

$$st - \lim_{n_1,n_2} \|\mathcal{B}_{n_1,n_2}^{p_{n_1}p_{n_2},q_{n_1}q_{n_2}}(f) - f\| = 0.$$

Proof. Using Lemma 4, the proof can be achieved similarly the proof of Theorem 4. □

6. Rates of Convergence of the Bivariate Operators

For $f \in H_{\omega_2}(\mathbb{R}_+^2)$, modulus of continuity for bivariate case is defined as follows [28]:

$$\tilde{\omega}(f;\delta_1,\delta_2) = \sup_{x,s\geq 0}\left\{|f(s,t)-f(x,y)|; \left|\frac{s}{1+s}-\frac{x}{1+x}\right| \leq \delta_1, \right.$$
$$\left. \left|\frac{t}{1+t}-\frac{y}{1+y}\right| \leq \delta_2, (s,t) \in R_+^2, (x,y) \in R_+^2 \right\}.$$

Here, $\tilde{\omega}(f;\delta_1,\delta_2)$ satisfies the conditions:

$$\tilde{\omega}(f;\delta_1,\delta_2) \to 0 \text{ if } \delta_1 \ \delta_2 \text{ tend to } 0, \text{ and}$$

$$|f(s,t)-f(x,y)| \leq \tilde{\omega}(f;\delta_1,\delta_2)\left(1+\frac{\left|\frac{s}{1+s}-\frac{x}{1+x}\right|}{\delta_1}\right)\left(1+\frac{\left|\frac{t}{1+t}-\frac{y}{1+y}\right|}{\delta_2}\right). \tag{20}$$

Now, we give the rate of the statistical convergence of the bivariate operators (18) by means of modulus of continuity in H_{ω_2}:

Theorem 8. *Let (p_{n_1}), (p_{n_2}), (q_{n_1}) and (q_{n_2}) be the sequences satisfying the condition (19). Then, we have*

$$|\mathcal{B}_{n_1,n_2}^{p_{n_1}p_{n_2},q_{n_1}q_{n_2}}(f;x,y) - f(x,y)| \leq 4\omega\left(f; \sqrt{\delta_{n_1}(x)}\sqrt{\delta_{n_2}(y)}\right), \tag{21}$$

where

$$\delta_{n_1}(x) = \mathcal{B}_{n_1,n_2}^{p_{n_1}p_{n_2},q_{n_1}q_{n_2}}\left(\left(\frac{s}{1+s}-\frac{x}{1+x}\right)^2; x,y\right), \tag{22}$$

$$\delta_{n_2}(y) = \mathcal{B}_{n_1,n_2}^{p_{n_1}p_{n_2},q_{n_1}q_{n_2}}\left(\left(\frac{t}{1+t} - \frac{y}{1+y}\right)^2; x, y\right). \tag{23}$$

Proof. By using (20), we have

$$\begin{aligned}
&|\mathcal{B}_{n_1,n_2}^{p_{n_1}p_{n_2},q_{n_1}q_{n_2}}(f; x, y) - f(x, y)| \leq \tilde{\omega}(f; \delta_1, \delta_2) \\
&\times \left\{\mathcal{B}_{n_1,n_2}^{p_{n_1}p_{n_2},q_{n_1}q_{n_2}}(e_0; x, y) + \frac{1}{\delta_{n_1}}\mathcal{B}_{n_1,n_2}^{p_{n_1}p_{n_2},q_{n_1}q_{n_2}}\left(\left|\frac{s}{1+s} - \frac{x}{1+x}\right|; x, y\right)\right\} \\
&\times \left\{\mathcal{B}_{n_1,n_2}^{p_{n_1}p_{n_2},q_{n_1}q_{n_2}}(e_0; x, y) + \frac{1}{\delta_{n_2}}\mathcal{B}_{n_1,n_2}^{p_{n_1}p_{n_2},q_{n_1}q_{n_2}}\left(\left|\frac{t}{1+t} - \frac{y}{1+y}\right|; x, y\right)\right\}.
\end{aligned} \tag{24}$$

Using Cauchy–Schwarz inequality (see (6)), we have

$$\begin{aligned}
&\mathcal{B}_{n_1,n_2}^{p_{n_1}p_{n_2},q_{n_1}q_{n_2}}\left(\left|\frac{s}{1+s} - \frac{x}{1+x}\right|; x, y\right) \\
&\leq \left\{\mathcal{B}_{n_1,n_2}^{p_{n_1}p_{n_2},q_{n_1}q_{n_2}}\left(\left(\frac{s}{1+s} - \frac{x}{1+x}\right)^2; x, y\right)\right\}^{\frac{1}{2}} \{\mathcal{B}_{n_1,n_2}^{p_{n_1}p_{n_2},q_{n_1}q_{n_2}}(e_0^2; x, y)\}^{\frac{1}{2}},
\end{aligned}$$

and

$$\begin{aligned}
&\mathcal{B}_{n_1,n_2}^{p_{n_1}p_{n_2},q_{n_1}q_{n_2}}\left(\left|\frac{t}{1+t} - \frac{y}{1+y}\right|; x, y\right) \\
&\leq \left\{\mathcal{B}_{n_1,n_2}^{p_{n_1}p_{n_2},q_{n_1}q_{n_2}}\left(\left(\frac{t}{1+t} - \frac{y}{1+y}\right)^2; x, y\right)\right\}^{\frac{1}{2}} \{\mathcal{B}_{n_1,n_2}^{p_{n_1}p_{n_2},q_{n_1}q_{n_2}}(e_0^2; x, y)\}^{\frac{1}{2}}.
\end{aligned}$$

Putting above inequalities in (24) and choosing $\delta_{n_1}(x)$ and $\delta_{n_2}(y)$ as in (22) and (23), respectively, we get our desired result (21). The theorem is completed. □

In the end, we will present the rates of statistical convergence of the bivariate operators (18) by means of Lipschitz type maximal functions.

Let us give the Lipschitz type maximal function space for the bivariate case on $E \times E \subset \mathbb{R}_+ \times \mathbb{R}_+$ as

$$\begin{aligned}
\widetilde{W}_{\alpha_1,\alpha_2,E^2} = \Big\{ f : &\sup(1+s)^{\alpha_1}(1+t)^{\alpha_2}\tilde{f}_{\alpha_1,\alpha_2}(x, y) \leq M\frac{1}{(1+x)^{\alpha_1}}\frac{1}{(1+y)^{\alpha_2}}; \\
&x, y \geq 0, (s, t) \in E^2 \Big\}.
\end{aligned} \tag{25}$$

Here, f is a bounded and continuous function in $\mathbb{R}_+$, M is a positive constant and $0 \leq \alpha_1, \alpha_2 \leq 1$, and then let us define $\tilde{f}_{\alpha_1,\alpha_2}$ as follows:

$$\tilde{f}_{\alpha_1,\alpha_2}(x, y) = \sup_{s,t>0} \frac{|f(s,t) - f(x,y)|}{|s-x|^{\alpha_1}|t-y|^{\alpha_2}}.$$

Theorem 9. *Let* (p_{n_1}), (p_{n_2}), (q_{n_1}) *and* (q_{n_2}) *be the sequences satisfying the condition* (19). *Then, we have*

$$|\mathcal{B}_{n_1,n_2}^{p_{n_1}p_{n_2},q_{n_1}q_{n_2}}(f;x,y)-f(x,y)| \leq M\left(\frac{q_{n_1}q_{n_2}}{p_{n_1}p_{n_2}}\right)$$
$$\times\left\{\delta_{n_1}(x)^{\frac{\alpha_1}{2}}\delta_{n_2}(y)^{\frac{\alpha_2}{2}}\left(\frac{q_{n_1}q_{n_2}}{p_{n_1}p_{n_2}}\right)^{1-\frac{\alpha_1+\alpha_2}{2}}+\delta_{n_1}(x)^{\frac{\alpha_1}{2}}d(y,E)^{\alpha_2}\left(\frac{q_{n_1}q_{n_2}}{p_{n_1}p_{n_2}}\right)^{-\frac{\alpha_1}{2}}\right.$$
$$\left.+\delta_{n_2}(y)^{\frac{\alpha_2}{2}}d(x,E)^{\alpha_1}\left(\frac{q_{n_1}q_{n_2}}{p_{n_1}p_{n_2}}\right)^{-\frac{\alpha_2}{2}}+2d(x,E)^{\alpha_1}d(y,E)^{\alpha_2}\right\},$$

where $0<\alpha_1,\alpha_2\leq 1$, $d(x,E)=\inf\{|x-y|:y\in E\}$, $\delta_{n_1}(x)$ *and* $\delta_{n_2}(y)$ *are defined as in* (22) *and* (23), *respectively.*

Proof. Let $x,y\geq 0$ and $(x_0,y_0)\in E^2$. Then, we can write

$$|f(s,t)-f(x,y)|\leq|f(s,t)-f(x_0,y_0)|+|f(x_0,y_0)-f(x,y)|.$$

Applying the positive linear operators $\mathcal{B}_{n_1,n_2}^{p_{n_1}p_{n_2},q_{n_1}q_{n_2}}(f;x)$ on both the sides of the above inequality and using (25), we obtain

$$\begin{aligned}
&|\mathcal{B}_{n_1,n_2}^{p_{n_1}p_{n_2},q_{n_1}q_{n_2}}(f;x,y)-f(x,y)|\\
&\leq \mathcal{B}_{n_1,n_2}^{p_{n_1}p_{n_2},q_{n_1}q_{n_2}}(|f(s,t)-f(x_0,y_0)|;x,y)\\
&\quad+|f(x_0,y_0)-f(x,y)|\mathcal{B}_{n_1,n_2}^{p_{n_1}p_{n_2},q_{n_1}q_{n_2}}(e_0;x,y)\\
&\leq M\mathcal{B}_{n_1,n_2}^{p_{n_1}p_{n_2},q_{n_1}q_{n_2}}\left(\left|\frac{s}{1+s}-\frac{x_0}{1+x_0}\right|^{\alpha_1}\left|\frac{t}{1+t}-\frac{y_0}{1+y_0}\right|^{\alpha_2};x,y\right)\\
&\quad+M\left|\frac{x}{1+x}-\frac{x_0}{1+x_0}\right|^{\alpha_1}\left|\frac{y}{1+y}-\frac{y_0}{1+y_0}\right|^{\alpha_2}\mathcal{B}_{n_1,n_2}^{p_{n_1}p_{n_2},q_{n_1}q_{n_2}}(e_0;x,y).
\end{aligned}\tag{26}$$

It is known that $(u+v)^\alpha\leq u^\alpha+v^\alpha$ and $0\leq\alpha\leq 1$, so it can be written as

$$\begin{aligned}
\left|\frac{s}{1+s}-\frac{x_0}{1+x_0}\right|^{\alpha_1}&\leq\left|\frac{s}{1+s}-\frac{x}{1+x}\right|^{\alpha_1}+\left|\frac{x}{1+x}-\frac{x_0}{1+x_0}\right|^{\alpha_1}\\
\left|\frac{t}{1+t}-\frac{y_0}{1+y_0}\right|^{\alpha_2}&\leq\left|\frac{t}{1+t}-\frac{y}{1+y}\right|^{\alpha_2}+\left|\frac{y}{1+y}-\frac{y_0}{1+y_0}\right|^{\alpha_2}.
\end{aligned}$$

By using the above inequalities in (26), we have

$$\begin{aligned}
&|\mathcal{B}_{n_1,n_2}^{p_{n_1}p_{n_2},q_{n_1}q_{n_2}}(f;x,y)-f(x,y)|\\
&\leq \mathcal{B}_{n_1,n_2}^{p_{n_1}p_{n_2},q_{n_1}q_{n_2}}\left(\left|\frac{s}{1+s}-\frac{x}{1+x}\right|^{\alpha_1}\left|\frac{t}{1+t}-\frac{y}{1+y}\right|^{\alpha_2};x,y\right)\\
&\quad+\left|\frac{y}{1+y}-\frac{y_0}{1+y_0}\right|^{\alpha_2}\mathcal{B}_{n_1,n_2}^{p_{n_1}p_{n_2},q_{n_1}q_{n_2}}\left(\left|\frac{s}{1+s}-\frac{x}{1+x}\right|^{\alpha_1};x,y\right)\\
&\quad+\left|\frac{x}{1+x}-\frac{x_0}{1+x_0}\right|^{\alpha_1}\mathcal{B}_{n_1,n_2}^{p_{n_1}p_{n_2},q_{n_1}q_{n_2}}\left(\left|\frac{t}{1+t}-\frac{y}{1+y}\right|^{\alpha_2};x,y\right)\\
&\quad+\left|\frac{x}{1+x}-\frac{x_0}{1+x_0}\right|^{\alpha_1}\left|\frac{y}{1+y}-\frac{y_0}{1+y_0}\right|^{\alpha_2}\mathcal{B}_{n_1,n_2}^{p_{n_1}p_{n_2},q_{n_1}q_{n_2}}(e_0;x,y).
\end{aligned}$$

Using Hölder's inequality with $p_1 = \frac{2}{\alpha_1}$, $p_2 = \frac{2}{\alpha_2}$, $q_1 = \frac{2}{2-\alpha_1}$, $q_2 = \frac{2}{2-\alpha_2}$ (see (5)), we obtain

$$\begin{aligned}
&\mathcal{B}_{n_1,n_2}^{p_{n_1}p_{n_2},q_{n_1}q_{n_2}}\left(\left|\frac{s}{1+s}-\frac{x}{1+x}\right|^{\alpha_1}\left|\frac{t}{1+t}-\frac{y}{1+y}\right|^{\alpha_2};x,y\right)\\
&= \mathcal{B}_{n_1,n_2}^{p_{n_1}p_{n_2},q_{n_1}q_{n_2}}\left(\left|\frac{s}{1+s}-\frac{x}{1+x}\right|^{\alpha_1};x,y\right)\mathcal{B}_{n_1,n_2}^{p_{n_1}p_{n_2},q_{n_1}q_{n_2}}\left(\left|\frac{t}{1+t}-\frac{y}{1+y}\right|^{\alpha_2};x,y\right)\\
&\leq \left(\mathcal{B}_{n_1,n_2}^{p_{n_1}p_{n_2},q_{n_1}q_{n_2}}\left(\left(\frac{s}{1+s}-\frac{x}{1+x}\right)^2;x,y\right)\right)^{\frac{\alpha_1}{2}}\left(\mathcal{B}_{n_1,n_2}^{p_{n_1}p_{n_2},q_{n_1}q_{n_2}}(e_0^2;x,y)\right)^{\frac{2-\alpha_1}{2}}\\
&\times\left(\mathcal{B}_{n_1,n_2}^{p_{n_1}p_{n_2},q_{n_1}q_{n_2}}\left(\left(\frac{t}{1+t}-\frac{y}{1+y}\right)^2;x,y\right)\right)^{\frac{\alpha_2}{2}}\mathcal{B}_{n_1,n_2}^{p_{n_1}p_{n_2},q_{n_1}q_{n_2}}(e_0^2;x,y)^{\frac{2-\alpha}{2}}.
\end{aligned}$$

If we use the above inequality in (26), we get our desired result. Thus, the proof is completed. □

Corollary 2. *If we take $E = [0,\infty)$, then because of $d(x,E) = 0$ and $d(y,E) = 0$, we have*

$$|\mathcal{B}_{n_1,n_2}^{p_{n_1}p_{n_2},q_{n_1}q_{n_2}}(f;x,y) - f(x,y)| \leq M\left(\frac{q_{n_1}q_{n_2}}{p_{n_1}p_{n_2}}\right)^{2-\frac{\alpha_1+\alpha_2}{2}}\delta_{n_1}(x)^{\frac{\alpha_1}{2}}\delta_{n_2}(y)^{\frac{\alpha_2}{2}},$$

where $\delta_{n_1}(x)$ and $\delta_{n_2}(y)$ are same as defined as in (22) and (23), respectively.

7. Conclusions

In this paper, we have constructed (p,q)-BBH operators and calculated some auxiliary results for these newly defined operators. We also established Korovkin type results and rate of convergence in a statistical sense. Furthermore, we constructed the bivariate (p,q)-BBH operators and computed rate of statistical convergence for the bivariate (p,q)-BBH operators. Our results are more general than the results for BBH and q-BBH operators.

Author Contributions: Investigation, I.A.and I.H.; Supervision, M.M.; Writing—original draft, K.J.A.

Funding: The authors extend their appreciation to "the Deanship of Scientific Research at King Khalid University" for funding this work through research groups program under Grant No. G.R.P-36-39.

Acknowledgments: We are thankful to the learned referees whose suggestions improved the paper in its present form.

Conflicts of Interest: The authors declare no conflict of interest.

References

1. Aral, A.; Doğru, O. Bleimann Butzer and Hahn operators based on q-integers. *J. Inequal. Appl.* **2007**, *2007*, 79410. [CrossRef]
2. Bleimann, G.; Butzer, P.L.; Hahn, L. A Bernstein-type operator approximating continuous functions on the semi-axis. *Indag. Math.* **1980**, *42*, 255–262. [CrossRef]
3. Gadjiev, A.D.; Çakar, Ö. On uniform approximation by Bleimann, Butzer and Hahn operators on all positive semi-axis. *Trans. Acad. Sci. Azerb. Ser. Phys. Tech. Math. Sci.* **1999**, *19*, 21–26.
4. Mursaleen, M.; Ansari, K.J.; Khan, A. On (p,q)-analogue of Bernstein operators. *Appl. Math. Comput.* **2015**, *266*, 874–882; Erratum in **2015**, *266*, 874–882. [CrossRef]
5. Marin, M. An evolutionary equation in thermoelasticity of dipolar bodies. *J. Math. Phys.* **1999**, *40*, 1391–1399. [CrossRef]
6. Marin, M.; Stan, G. Weak solutions in Elasticity of dipolar bodies with stretch. *Carpath. J. Math.* **2013**, *29*, 33–40.
7. Khalid Khan, D.K. Lobiyal, Bézier curves based on Lupaş (p,q)-analogue of Bernstein functions in CAGD. *J. Comput. Appl. Math.* **2017**, *317*, 458–477. [CrossRef]

8. Srivastava, H.M.; Gupta, V. Rate of convergence for the Bézier variant of the Bleimann–Butzer–Hahn operators. *Appl. Math. Lett.* **2005**, *18*, 849–857. [CrossRef]
9. Ansari, K.J.; Karaisa, A. On the approximation by Chlodowsky type generalization of (p,q)-Bernstein operators. *Int. J. Nonlinear Anal. Appl.* **2017**, 8, 181–200.
10. Dattoli, G.; Lorenzutta, S.; Cesarano, C. Bernstein polynomials and operational methods. *J. Comp. Anal. Appl.* **2016**, *8*, 369–377.
11. Kadak, U.; Mishra, V.N.; Pandey, S. Chlodowsky type generalization of (p,q)-Szász operators involving Brenke type polynomials. *Rev. Real Acad. Cienc. Exact. Fís. Nat. Ser. A Mat.* **2018**, *112*, 1443–1462. [CrossRef]
12. Khan, A.; Sharma, V. Statistical approximation by (p,q)-analogue of Bernstein-Stancu operators. *Ajerbaijan J. Math.* **2018**, *8*, 100–121.
13. Mursaleen, M.; Ansari, K.J.; Khan, A. Some approximation results by (p,q)-analogue of Bernstein-Stancu operators. *Appl. Math. Comput.* **2015**, 264, 392-402; Corrigendum in **2015**, *269*, 744–746.
14. Mursaleen, M.; Alotaibi, A.; Ansari, K.J. On a Kantorovich variant of (p,q)-Szász-Mirakjan operators. *J. Funct. Spaces* **2016**, *2016*, 1035253. [CrossRef]
15. Acar, T. (p,q)-Generalization of Szász-Mirakyan operators. *Math. Meth. Appl. Sci.* **2016**, *39*, 2685–2695. [CrossRef]
16. Hounkonnou, M.N.; Désiré, J.; Kyemba, B. $\mathcal{R}(p,q)$-calculus: Dierentiation and integration. *SUT J. Math.* **2013**, *49*, 145–167.
17. Sadjang, P.N. On the fundamental theorem of (p,q)-calculus and some (p,q)-Taylor formulas. *Result. Math.* **2018**, *73*, 39. [CrossRef]
18. Sahai, V.; Yadav, S. Representations of two parameter quantum algebras and p,q-special functions. *J. Math. Anal. Appl.* **2007**, *335*, 268–279. [CrossRef]
19. Victor, K.; Pokman, C. *Quantum Calculus*; Springer: New York, NY, USA; Berlin/Heidelberg, Germany, 2002.
20. Mursaleen, M.; Nasiruzzaman, M.; Khan, A.; Ansari, K.J. Some approximation results on Bleimann–Butzer–Hahn operators defined by (p,q)-integers. *Filomat* **2016**, *30*, 639–648. [CrossRef]
21. Mursaleen, M. Nasiruzzaman, Some approximation properties of bivariate Bleimann–Butzer–Hahn operators based on (p,q)-integers. *Boll. Unione Mat. Ital.* **2017**, *10*, 271–289. [CrossRef]
22. Fast, H. Sur la convergence statistique. *Colloq. Math.* **1951**, *2*, 241–244. [CrossRef]
23. Niven, I.; Zuckerman, H.S.; Montgomery, H. *An Introduction to the Theory of Numbers*, 5th ed.; Wiley: New York, NY, USA, 1991.
24. Mursaleen, M.; Nasiruzzaman, M.; Ansari, K.J.; Alotaibi, A. Generalized (p,q)-Bleimann–Butzer–Hahn operators and some approximation results. *J. Ineq. Appl.* **2017**, *2017*, 310. [CrossRef] [PubMed]
25. Ansari, K.J. Approximation Properties of Some Operators and Thier q-Analogues. Ph.D. Thesis, Aligarh Muslim University, Aligarh, India, 2015.
26. Ersan, S. Approximation properties of bivariate generalization of Bleimann, Butzer and Hahn operators based on the q-integers. In Proceedings of the 12th WSEAS International Conference on Applied Mathematics, Cairo, Egypt, 29–31 December 2007; pp. 122–127.
27. Lenze, B. Bernstein-Baskakov-Kantorovich operators and Lipschitz-type maximal functions. *Approx. Theory* **1990**, *58*, 469–496.
28. Anastassiou, G.A.; Gal, S.G. *Approximation Theory: Moduli of Continuity and Global Smoothness Preservation*; Birkhauser: Boston, MA, USA, 2000.

© 2018 by the authors. Licensee MDPI, Basel, Switzerland. This article is an open access article distributed under the terms and conditions of the Creative Commons Attribution (CC BY) license (http://creativecommons.org/licenses/by/4.0/).

symmetry

MDPI

Article

Optimizing a Password Hashing Function with Hardware-Accelerated Symmetric Encryption

Rafael Álvarez [1,*], Alicia Andrade [2] and Antonio Zamora [3]

1 Departamento de Ciencia de la Computación e Inteligencia Artificial (DCCIA), Universidad de Alicante, 03690 Alicante, Spain

2 Fac. Ingeniería, Ciencias Físicas y Matemática, Universidad Central, Quito 170129, Ecuador; aandrade@uce.edu.ec

3 Departamento de Ciencia de la Computación e Inteligencia Artificial (DCCIA), Universidad de Alicante, 03690 Alicante, Spain; zamora@dccia.ua.es

* Correspondence: ralvarez@dccia.ua.es

Received: 2 November 2018; Accepted: 22 November 2018; Published: 3 December 2018

Abstract: Password-based key derivation functions (PBKDFs) are commonly used to transform user passwords into keys for symmetric encryption, as well as for user authentication, password hashing, and preventing attacks based on custom hardware. We propose two optimized alternatives that enhance the performance of a previously published PBKDF. This design is based on (1) employing a symmetric cipher, the Advanced Encryption Standard (AES), as a pseudo-random generator and (2) taking advantage of the support for the hardware acceleration for AES that is available on many common platforms in order to mitigate common attacks to password-based user authentication systems. We also analyze their security characteristics, establishing that they are equivalent to the security of the core primitive (AES), and we compare their performance with well-known PBKDF algorithms, such as Scrypt and Argon2, with favorable results.

Keywords: symmetric; encryption; password; hash; cryptography; PBKDF

1. Introduction

Key derivation functions are employed to obtain one or more keys from a master secret. This is especially useful in the case of user passwords, which can be of arbitrary length and are unsuitable to be used directly as fixed-size cipher keys, so, there must be a process for converting passwords into secret keys. This process is performed by password-based key derivation functions (PBKDFs). PBKDFs are also called password hashing functions, and they are commonly employed in user authentication since they have certain advantages over other password processing methods: they are capable of accepting a salt, preventing precalculated table attacks; they are one-way functions (much as ordinary cryptographic hash functions), so the hashed password database cannot be reversed if it is stolen; and they can usually be parameterized in terms of temporal and memory cost to prevent attacks based on massively parallel hardware, like general-purpose graphical processing units (GPGPU) or custom hardware. Password hashing is a field of active research (see [1,2]), with several recent publications [3–10] that improve the current industry standard (PBKDF2, see [11]). Besides password hashing and key derivation, PBKDFs have found applications in the field of cryptocurrencies and blockchain algorithms, where they are used as proof-of-work functions for such designs (see [12]).

Symmetric encryption (see [13]) is a type of cryptography that employs the same (or easily derivable from one another) keys for encryption and decryption, hence the establishment of a symmetric process from *cleartext* to *ciphertext* and, back again, from *ciphertext* to *cleartext*. There are two basic kinds of symmetric cryptosystems: block ciphers and stream ciphers; they differ in that block ciphers have no internal state and usually process data in blocks, while stream ciphers do have an internal state and process data element by element (an element is usually a bit or a byte of data). Nevertheless, most block ciphers can be run in operation modes that make them work as stream ciphers; such is the case in our proposal, where we employ the current advanced encryption standard (AES, [14]) in counter (CTR) mode to work as a stream cipher in the role of a pseudo-random number generator (PRNG); the use of AES as a PRNG has been proposed by the United States National Institute of Standards and Technology (NIST, see [15]). Besides having been independently tested for almost two decades and considered secure by the community, AES has the advantage of being accelerated in hardware on most common modern processors, like those found on laptop, desktop or server machines we use nowadays.

The main contributions of this paper are two different optimizations of a previously proposed PBKDF (see [3]) that favorably compare in performance to the original version and widely employed PBKDFs, such as Scrypt [7] and Argon2 [5]. This is significant for user authentication applications that are based on passwords, as well as in blockchain applications where a proof-of-work algorithm is required. Taking advantage of the fact that hardware acceleration support is available for AES, our proposed PBKDF design reduces the performance advantage of attackers employing GPGPU or custom hardware since its main core primitive is also run on hardware.

The rest of the paper is structured as follows: a short review on related work is included in Section 2, the proposed optimized algorithms are described in Section 3, the results obtained by our studies are presented in Section 4, the significance of the results are discussed in Section 5, while the testing methodology is detailed in Section 6, followed by some conclusions and future lines of work in Section 7.

2. Related Work

There is abundant recent literature on the connection between symmetry and cryptography. Chang et al. proposed a mobility network authentication scheme based on elliptic-curve cryptography (see [16]) that ensures anonymity, security, and convenience. Hung et al. designed a lattice-based revocable certificateless signature scheme (see [17]) that aims to resist cryptanalysis, even in the post-quantum era. Sakalauskas et al. improved upon an asymmetric cipher based on the matrix power function (see [18]) to avoid a successful discrete logarithm attack against the original version. Ramadan et al. published a survey of public key infrastructure (PKI)-based security for mobile systems (see [19]) covering aspects such as authentication, key agreement, and privacy.

Qiao et al. described a black-box traceable ciphertext-policy attribute-based encryption (CP-ABE, see [20,21]) that is scalable and efficient and, therefore, better suited for cryptographic cloud storage. Zhu et al. cryptanalyzed an image encryption algorithm based on a chaos s-box (see [22]), proposing an improved version with better security and performance. Park et al. described the use cases, challenges, and solutions involved in the application of blockchain-based security technologies to cloud computing (see [23]).

Chang et al. proposed password-authenticated key exchange and protected password change protocols that do not involve symmetric or asymmetric cryptosystems (see [24]), basing the security on the computational Diffie–Hellman assumption in the random oracle model. Nam et al. presented a provably secure three-party password-only authenticated key exchange protocol (see [25]) that can run in only two rounds of communication.

Regarding PBKDF and password hashing functions, the finalists to the Password Hashing Competition (*Argon2* [5], *Catena* [6], *Makwa* [8], *Lyra2* [9], and *Yescrypt* [10]) are highly relevant functions related to our proposed optimized PBKDF. Also, there is previous work by the authors in this field, including the original version of the proposal ([3]) and others (see [4,26]).

3. Description

In the following, we describe the parameters and elements, as well as the design, of the original PBKDF function and the proposed optimized versions. There are three variants of our proposal:

the original version (*AESCTR-o*, published in [3]); the intermediate optimization step (*AESCTR-i*); and the final optimization (*AESCTR-f*). We use a pseudocode notation loosely based on the Go language to describe the initialization and output stages of the proposal.

3.1. Parameters

Our proposal and most PBKDF designs share a very similar set of parameters, mainly including the user password (*pass[]*) and random salt (*salt[]*) byte strings to be hashed, the length of the output hash to be generated (*plen*), and some kinds of cost parameters. Those PBKDFs that have been designed to slow down attackers employing GPGPU or specialized hardware usually involve two cost parameters: a time cost (*ptime*) and a memory cost (*pmem*) that, due to its nature, tends to influence the time cost as well. These parameters are the same as in the original version (see [3] for more details).

Most of the variables of the algorithm are unchanged as well; *M[]* is the main memory buffer that is parameterized by *plen* and *pmem*, while *out[]* constitutes the output hash of the function. Also, *M64[]* and *out64[]* are employed in the final optimization (*AESCTR-f*) to perform 64-bit native operations for performance reasons.

The algorithm employs SHA3-256 as a secure cryptographic hash function (see [27]) during the initial seeding phase, and AES-128 (see [14]) is used in CTR mode as a pseudo-random generator; both of these could be swapped for different, equivalent primitives were it necessary in the future.

3.2. Initialization

The initialization stage is unchanged from the original version of the algorithm (see [3]) and is reproduced here in Figure 1. We have added comments to detail the seeding and buffer initialization steps.

```
// Generate a 256-bit hash (seed) from the password and salt
fH := sha3.New256()
fH.Write(pass)
fH.Write(0)
fH.Write(salt)
seed := fH.Sum(nil)

// Build an AES-128-CTR instance from the seed value
blk := aes.NewCipher(seed[0:16])
fC := cipher.NewCTR(blk, seed[16:32])

// Allocate and fill output buffer
out := make([]byte, plen)
fC.XORKeyStream(out, out)

// Allocate and fill memory buffer
M := make([]byte, pmem*plen)
fC.XORKeyStream(M, M)
```

Figure 1. Initialization stage pseudocode (common to all variants).

3.3. Output

The intermediate optimization attempts to improve the performance of the original version (see [3]) by avoiding AES encryption steps inside the loops and just encrypting the *out[]* buffer as the last step. For this reason, there is a second inner loop that generates a new index so that a different row from *M[]* can be processed into *out[]*. These modifications are shown in Figure 2 (pseudocode) and in Figure 3 (flow diagram).

```
for t := 0; t < ptime; t++ {
for m := 0; m < pmem; m++ {
i := (int(out[0:8])%pmem) * plen
for o := 0; o < len(out); o++ {
M[i+o] -= out[o]
}
i = (i*i) % pmem
for o := 0; o < len(out); o++ {
out[o] -= (M[i+o] ^ out[o])
}
}
}
fC.XORKeyStream(out, out)
```

Figure 2. Intermediate (*AESCTR-i*) output stage pseudocode.

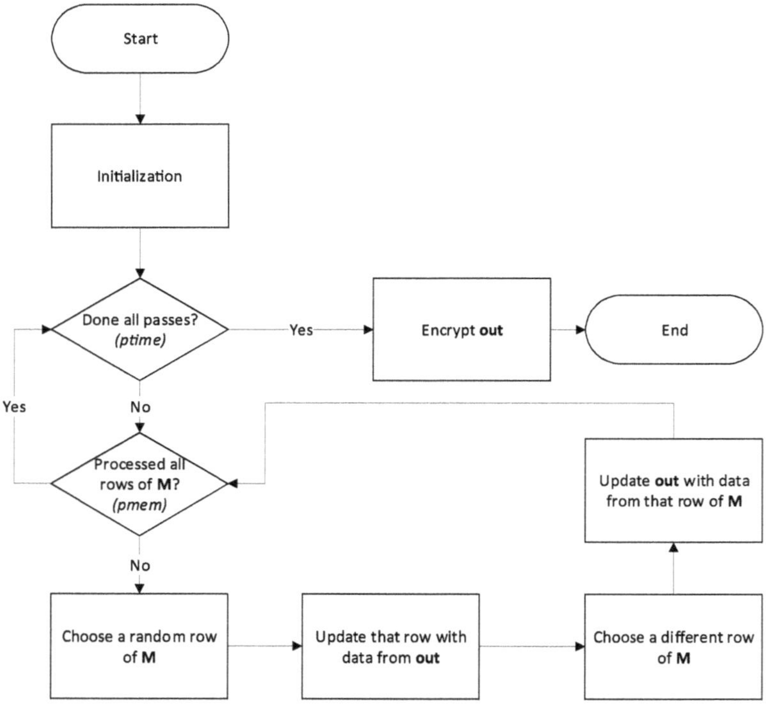

Figure 3. Flow diagram for the intermediate (*AESCTR-i*) output stage.

The final optimization further improves the performance by avoiding writing back to *M[]* and having a second inner loop. Also, memory access and operations are performed in 64-bit. The differences between the final and intermediate optimizations are shown in Figure 4 (pseudocode) and in Figure 5 (flow diagram).

```
for t := 0; t < ptime; t++ {
for m := 0; m < pmem; m++ {
i := (int(out[0:8])%pmem) * plen/8
for o := 0; o < len(out); o++ {
out64[o] -= (M64[i+o] ^ out64[o])
}
}
}
fC.XORKeyStream(out, out)
```

Figure 4. Final (*AESCTR-f*) output stage pseudocode.

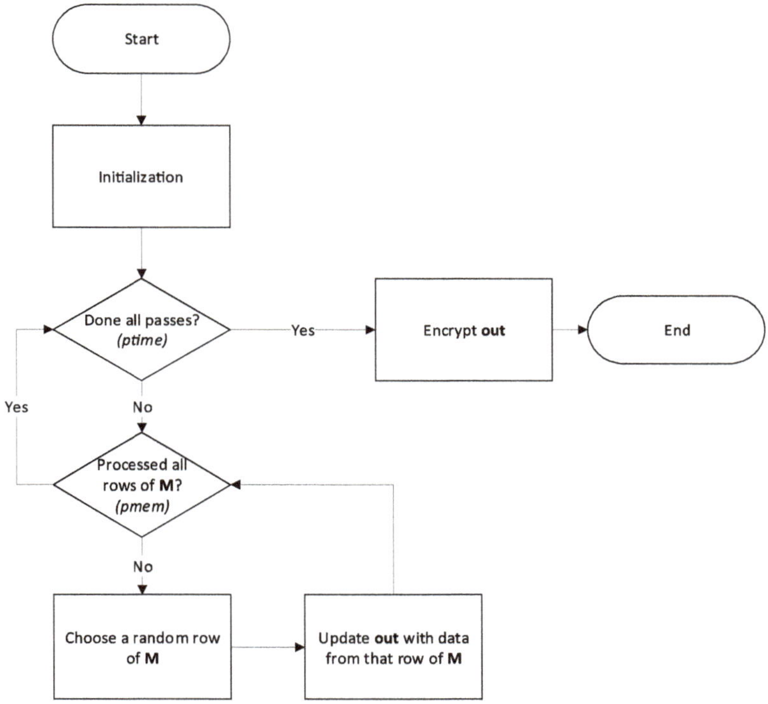

Figure 5. Flow diagram for the final (*AESCTR-f*) output stage.

4. Results

Regarding performance, we benchmarked in the following the optimized proposals, together with the original version ([3]), when modulating the time (*ptime*) and space (*pmem*) complexity parameters. The testing methodology is detailed in Section 6.

In Figure 6, we present the computational cost (execution time in logarithmic scale) of all three variants of the proposal as the *pmem* parameter modulation ranges from 2^8 to 2^{23} entries of 32 bytes (a memory usage varying from 8 KB to 256 MB) and with *ptime* = 1. We can see that the intermediate optimization is a significant improvement over the original function, but it is clearly overtaken by the final optimization, which is about 5 times faster than the original version.

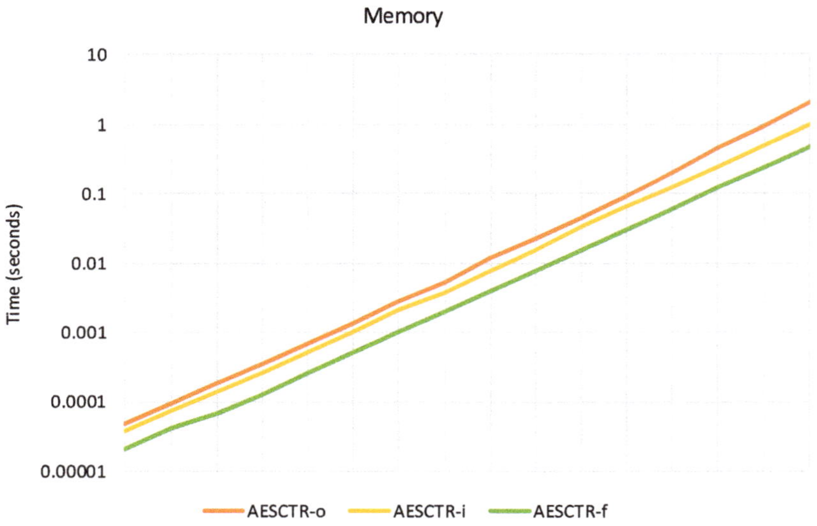

Figure 6. Performance while modulating the spatial cost parameter (*pmem*).

Figure 7 shows the behavior of the proposed optimizations when the time parameter is modulated. In this test, *ptime* values range from 1 to 2^{15} passes and memory usage (*pmem*) is kept constant at 2^8 entries of *plen* bytes (8 KB of memory). In this case, the difference in performance of the final optimization (*AESCTR-f*) is more pronounced than in the case of the memory parameter.

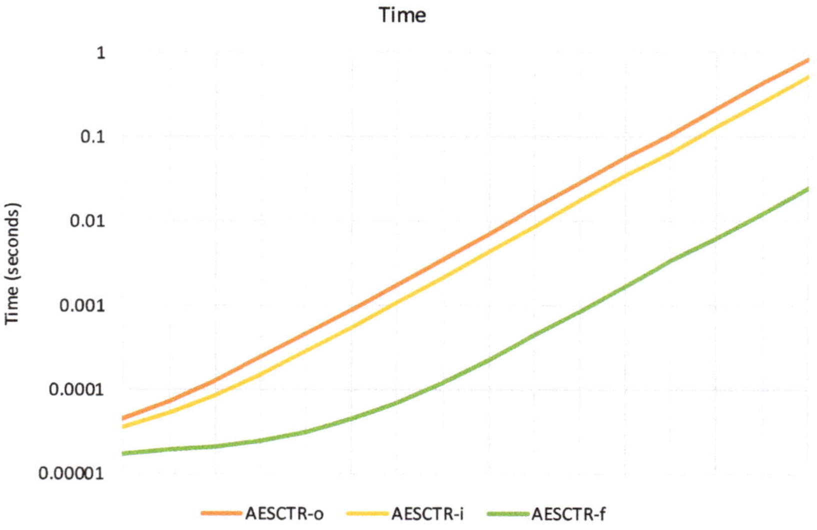

Figure 7. Performance while modulating the temporal cost parameter (*ptime*).

Figure 8 represents the execution time (in logarithmic scale) when both parameters, *pmem* and *ptime*, are simultaneously modulated in a double *for* loop. The outer loop is *pmem*, corresponding to the number of entries in the main memory buffer, *M[]*, going from 2^8 to 2^{15}, and the inner loop corresponds to *ptime*, ranging from 1 to 2^7; the maximum amount of memory usage is 128 MB and occurs when pmem = 2^{15} and ptime = 2^7. The sawtooth shapes are expected in a double-loop arrangement. In this combined benchmark, the performance gain achieved with the final optimization is readily apparent.

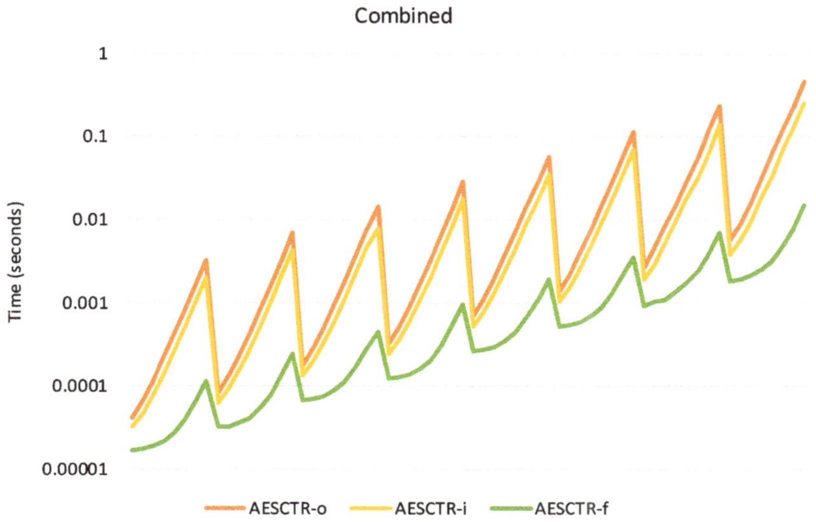

Figure 8. Performance while simultaneously modulating both *ptime* and *pmem* cost parameters.

5. Discussion

In the following, we discuss the security characteristics of our proposal and compare all three variants to Scrypt and Argon2 in terms of performance.

5.1. Comparison with Scrypt and Argon2

Scrypt (see [7]) is a PBKDF that was designed by Colin Percival in 2009. It has been employed for many services and applications, acting as a de facto standard in recent years. It has also been employed as a proof-of-work algorithm in some blockchain implementations.

Argon2 (see [5]) became the Password Hashing Competition (PHC) winner in 2015, overcoming finalists such as Catena [6], Lyra2 [9], Yescrypt [10], and Makwa [8]. It provides very interesting options, such as data-dependent or data-independent memory access, and is gaining traction in many new secure authentication implementations.

As shown in Figure 9, for an equal amount of memory, Scrypt is slower than all three variants of the proposed AES-CTR algorithm, while Argon2 is a very efficient algorithm, but the final optimization (*AESCTR-f*) is slightly faster than Argon2. Moreover, the *speedup* between our proposal and Argon2 increases with memory usage, as shown in Table 1.

It is interesting to note that while Argon2 has been implemented using Intel's SSE4 instruction set and our proposal takes advantage of the native AES instructions, Scrypt does not benefit directly from the hardware acceleration available in modern processors. More details regarding the testing methodology are included in Section 6.

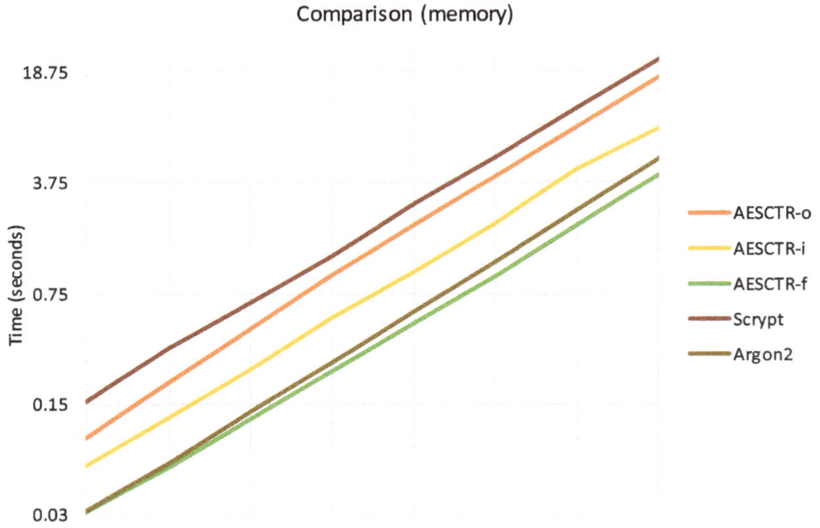

Figure 9. Comparison with Scrypt and Argon2.

Table 1. Speedup between the final optimization and Argon2.

pmem	AESCTR-f (s)	Argon2 (s)	Speedup
8	0.03	0.03	1.04
9	0.06	0.06	1.07
10	0.12	0.13	1.11
11	0.24	0.28	1.16
12	0.49	0.58	1.19
13	0.98	1.21	1.24
14	2.02	2.51	1.25
15	4.15	5.27	1.27

5.2. *Security*

This design is based on two different cryptographic primitives: a pseudo-random generator that provides the initial contents of the memory and output buffers, and a hash function that processes the user password and salt and produces a seed for this pseudo-random generator. We employed a symmetric cipher, AES-128 [14] in CTR mode, as the pseudo-random generator and SHA3-256 [27] as the hash function that provides the seed (128-bit key and initialization vector) for AES.

Both of these cryptographic primitives are well-known, independently tested current standards. Nevertheless, if they were to be deemed insecure in the future, they could be replaced by different, size-compatible, secure alternatives.

The proposed optimized design should be on par with the original version and these primitives at a minimum-security level of 128 bits against brute force attacks since the output of the PBKDF (in all three variants) comes directly from the output of the AES-128 symmetric cipher.

6. Methods

We employed the Go programming language (version 1.11.1, see [28]) for the implementation of all tested algorithms; Go is an excellent choice for cryptography testing since it is a very efficient, compiled language and includes most security standards and algorithms in its standard library. All benchmarks were run on the same computer, a desktop PC with an Intel i7 CPU (6950X, 3.5

GHz and with AES Native Instructions support) and 32 GB of RAM, running Microsoft Windows 10 (1803 release). The length for passwords, salts, and output was chosen as 32 bytes (256 bits), and all tests were run 100 times to avoid the interference from external processes as much as possible.

In the case of the comparison benchmarks, the official implementations available in the golang.org/x/crypto packages were used for Scrypt and Argon2 testing. It should be noted that this implementation of Argon2 supports hardware acceleration via SSE4 instructions and that the recommended parameters (three passes) and a single thread were used. To ensure fair testing, an equal amount of RAM usage was chosen for each algorithm in all comparison tests.

7. Conclusions

We optimized a previously published password-based key derivation function that employs the Advanced Encryption Standard (AES) in counter mode as a core primitive, proposing two new algorithms based on the original design: a more conservative optimization and a fully optimized one. The design philosophy is based on taking advantage of the custom AES instructions available on most modern processors that enable hardware support to defend against brute force password cracking attacks mounted on specialized hardware or general-purpose graphical processing units.

We have also analyzed the performance of all three variants in comparison with Scrypt and Argon2, which are the current industry standards in terms of password hashing functions. The final optimization version of the algorithm (*AESCTR-f*) is faster than Argon2 for an equal amount of memory usage, showing that AES can be an excellent candidate for the design of password hashing functions. Moreover, since the design is based primarily on AES in counter mode as a pseudo-random generator and the final output is directly encrypted with AES, we can establish that our proposal is equivalent, in terms of security, to this extensively analyzed encryption standard.

For future research, there are several possible interesting topics, such as server-side ROM, client-independent update, server relief, parallelism, or specialized implementations, among others.

Server-side ROM is an extra security measure that involves employing a very big random file on the server as part of the hashing process. In this way, an attacker would have to produce the same random file, and the large amount of memory required would further deter the use of specialized hardware. Our proposal can be adapted to accept such a file as part of the algorithm without compromising its performance or security.

Client-independent update is the capability of changing PBKDF parameters without the need for the user to reenter the password. This is a convenient feature in password authentication since the server can increase the security of the PBKDF as necessary but without any friction for the end users. This can be performed without modification to the proposed algorithm by multiple-step processing, but other optimized methods might be possible.

Server relief implies delegating part of the PBKDF computation to the end user so that the server is less impacted by computational requirements of password-authenticating a large number of users simultaneously. It is, in essence, a way of increasing the parallelism of the server and reducing the advantage that attackers might have by using general-purpose graphical processing units or other specialized hardware. This usually involves some kind of protocol in order to share the computational load between the server and the user node in a secure manner.

Multiple-thread parallelism can also be studied and incorporated by modifying the proposed final optimization (*AESCTR-f*). This can be useful in situations where server relief is not possible or when further parallelism is desired. Since Argon2 allows for multiple-thread parallelism, a comparison between the parallel performance scalability of the final optimization and Argon2 might be possible.

Specialized implementation on hardware platforms, such as general-purpose graphic processing units or field programmable gate arrays, could be very useful for further performance testing and optimization of the proposed PBKDF algorithm.

Author Contributions: All three authors contributed equally in the conceptualization, validation, research and writing of this paper.

Funding: Research partially supported by the Spanish Government under Project Grant TEC2014-54110-R (CASUS).

Conflicts of Interest: the authors declare no conflict of interest.

References

1. Hellman, M. A Cryptanalytic Time-memory Trade-off. *IEEE Trans. Inf. Theory* **2006**, *26*, 401–406, doi:10.1109/TIT.1980.1056220. [CrossRef]
2. Provos, N.; Mazieres, D. A Future-Adaptable Password Scheme. In Proceedings of the 1999 USENIX Annual Technical Conference, FREENIX Track, Berkeley, CA, USA, 23–26 August 1999; pp. 81–91.
3. Álvarez-Sánchez, R.; Andrade-Bazurto, A.; Santos-González, I.; Zamora-Gómez, A. AES-CTR as a Password-Hashing Function. In Proceedings of the International Joint Conference SOCO'17-CISIS'17-ICEUTE'17, León, Spain, 6–8 September 2017; Pérez García, H., Alfonso-Cendón, J., Sánchez González, L., Quintián, H., Corchado, E., Eds.; Springer International Publishing: Cham, Switzerland, 2018; pp. 610–617, ISBN 978-3-319-67180-2._59. [CrossRef]
4. Álvarez, R.; Zamora, A. Using Spritz as a Password-Based Key Derivation Function. In Proceedings of the International Joint Conference SOCO'16-CISIS'16-ICEUTE'16, San Sebastián, Spain, 19–21 October 2016.
5. Biryukov, A.; Dinu, D.; Khovratovich, D. Argon2: New Generation of Memory-Hard Functions for Password Hashing and Other Applications. In Proceedings of the IEEE 2016 IEEE European Symposium on Security and Privacy, Saarbrucken, Germany, 21–24 March 2016; pp. 292–302, ISBN 978-1-5090-1751-5. [CrossRef]
6. Forler, C.; Lucks, S.; Wenzel, J. The Catena Password-Scrambling Framework. 2015. Available online: https://password-hashing.net/submissions/specs/Catena-v5.pdf (accessed on 20 November 2018).
7. Percival, C. Stronger Key Derivation via Sequential Memory-Hard Functions. 2009. Available online: http://www.bsdcan.org/2009/schedule/attachments/87_scrypt.pdf (accessed on 20 November 2018).
8. Pornin, T. The Makwa Password Hashing Function. 2015. Available online: http://www.bolet.org/makwa/makwa-spec-20150422.pdf (accessed on 20 November 2018).
9. Simplício, M.A., Jr.; Almeida, L.C.; Andrade, E.R.; dos Santos, P.C.; Barreto, P.S. Lyra2: Password Hashing Scheme with improved security against time-memory trade-offs. *IEEE Trans. Comput.* **2016**, *65*, 3096–3108, doi:10.1109/TC.2016.2516011. [CrossRef]
10. Peslyak, A. yescrypt—A Password Hashing Competition Submission. 2015. Available online: https://password-hashing.net/submissions/specs/yescrypt-v2.pdf (accessed on 20 November 2018).
11. Moriarty, K.; Kaliski, B.; Rusch, A. *PKCS# 5: Password-Based Cryptography Specification Version 2.1*; Technical Report; IETF: Fremont, CA, USA, 2017; doi:10.17487/RFC8018.
12. Biryukov, A.; Dinu, D.; Khovratovich, D. Fast and Tradeoff-Resilient Memory-Hard Functions for Cryptocurrencies and Password Hashing. *IACR Cryptol. ePrint Arch.* **2015**, *2015*, 430:1–430:15.
13. Ferguson, N.; Schneier, B.; Kohno, T. *Cryptography Engineering: Design Principles and Practical Applications*; Wiley Publishing: Hoboken, NJ, USA, 2010, ISBN 978-0-470-47424-2.
14. Daemen, J.; Rijmen, V. AES Proposal: Rijndael. 1999. Available online: http://www.cs.miami.edu/home/burt/learning/Csc688.012/rijndael/rijndael_doc_V2.pdf (accessed on 20 November 2018).
15. Keller, S.S. NIST-Recommended Random Number Generator Based on ANSI X9.31 Appendix A.2.4 Using the 3-Key Triple DES and AES Algorithms. 2005. Available online: http://citeseerx.ist.psu.edu/viewdoc/download?doi=10.1.1.210.70&rep=rep1&type=pdf (accessed on 20 November 2018).
16. Chang, Y.F.; Tai, W.L.; Hsu, M.H. A Secure Mobility Network Authentication Scheme Ensuring User Anonymity. *Symmetry* **2017**, *9*, 307, doi:10.3390/sym9120307. [CrossRef]
17. Hung, Y.H.; Tseng, Y.M.; Huang, S.S. Lattice-Based Revocable Certificateless Signature. *Symmetry* **2017**, *9*, 242, doi:10.3390/sym9100242. [CrossRef]
18. Sakalauskas, E.; Mihalkovich, A.; Venčkauskas, A. Improved Asymmetric Cipher Based on Matrix Power Function with Provable Security. *Symmetry* **2017**, *9*, 9, doi:10.3390/sym9010009. [CrossRef]
19. Ramadan, M.; Du, G.; Li, F.; Xu, C. A Survey of Public Key Infrastructure-Based Security for Mobile Communication Systems. *Symmetry* **2016**, *8*, 85, doi:10.3390/sym8090085. [CrossRef]
20. Qiao, H.; Ba, H.; Zhou, H.; Wang, Z.; Ren, J.; Hu, Y. Practical, Provably Secure, and Black-Box Traceable CP-ABE for Cryptographic Cloud Storage. *Symmetry* **2018**, *10*, 482, doi:10.3390/sym10100482. [CrossRef]

21. Ba, H.; Zhou, H.; Mei, S.; Qiao, H.; Hong, T.; Wang, Z.; Ren, J. Astrape: An Efficient Concurrent Cloud Attestation with Ciphertext-Policy Attribute-Based Encryption. *Symmetry* **2018**, *10*, 425, doi:10.3390/sym10100425. [CrossRef]
22. Zhu, C.; Wang, G.; Sun, K. Cryptanalysis and Improvement on an Image Encryption Algorithm Design Using a Novel Chaos Based S-Box. *Symmetry* **2018**, *10*, 399, doi:10.3390/sym10090399. [CrossRef]
23. Park, J.H.; Park, J.H. Blockchain Security in Cloud Computing: Use Cases, Challenges, and Solutions. *Symmetry* **2017**, *9*, 164, doi:10.3390/sym9080164. [CrossRef]
24. Chang, T.Y.; Hwang, M.S.; Yang, C.C. Password Authenticated Key Exchange and Protected Password Change Protocols. *Symmetry* **2017**, *9*, 134, doi:10.3390/sym9080134. [CrossRef]
25. Nam, J.; Choo, K.K.R.; Han, S.; Paik, J.; Won, D. Two-Round Password-Only Authenticated Key Exchange in the Three-Party Setting. *Symmetry* **2015**, *7*, 105–124, doi:10.3390/sym7010105. [CrossRef]
26. Alvarez, R.; Caballero-Gil, C.; Santonja, J.; Zamora, A. Algorithms for Lightweight Key Exchange. *Sensors* **2017**, *17*, 1517, doi:10.3390/s17071517. [CrossRef] [PubMed]
27. Bertoni, G.; Daemen, J.; Peeters, M.; Van Assche, G. Cryptographic Sponge Functions. 2011. Available online: https://keccak.team/files/CSF-0.1.pdf (accessed on 20 November 2018).
28. The Go Programming Language. Available online: http://www.golang.org (accessed on 20 November 2018).

© 2018 by the authors. Licensee MDPI, Basel, Switzerland. This article is an open access article distributed under the terms and conditions of the Creative Commons Attribution (CC BY) license (http://creativecommons.org/licenses/by/4.0/).

symmetry

MDPI

Article

Complex Spirals and Pseudo-Chebyshev Polynomials of Fractional Degree

Paolo Emilio Ricci

International Telematic University UniNettuno, Corso Vittorio Emanuele II, 39, 00186 Roma, Italy; paoloemilioricci@gmail.com

Received: 9 November 2018; Accepted: 23 November 2018; Published: 28 November 2018

Abstract: The complex Bernoulli spiral is connected to Grandi curves and Chebyshev polynomials. In this framework, pseudo-Chebyshev polynomials are introduced, and some of their properties are borrowed to form classical trigonometric identities; in particular, a set of orthogonal pseudo-Chebyshev polynomials of half-integer degree is derived.

Keywords: Bernoulli spiral; Grandi curves; Chebyshev polynomials; pseudo-Chebyshev polynomials; orthogonality property

MSC: 33C45; 12E10; 42C05; 14H50

---

1. Introduction

The purpose of this article is to emphasize some simple connections among mathematical objects apparently of different types as the Bernoulli spirals, the Grandi (rhodonea) curves, and the first and second kind Chebyshev polynomials. Namely, by considering polar coordinates and the complex form of the Bernoulli spiral, a straightforward connection between the real and imaginary part of the Bernoulli spiral with the Grandi curves follows. Even the Chebyshev polynomials come out immediately.

Since the rhodonea curves exist even for a fractional index, it is possible to define an extension of the first and second kind Chebyshev polynomials to the case of rational degree. Actually, the resulting functions are not polynomials, but irrational functions. However, several properties of these functions can be derived from their trigonometric definition, by using standard identities of circular functions. In particular, for the function of half-integer degree, $T_{n+1/2}$ and $U_{n+1/2}$, the orthogonality property still holds, in the interval $(-1, 1)$, with respect to the same weight function of their polynomial counterparts.

The second section of the article is devoted to recalling the most simple examples of spirals, including the Archimedes, Bernoulli, Fermat, and other spirals, which can be derived by using an analogy with Cartesian coordinates. Namely, the above- mentioned spirals, considered in the plane (ρ, θ), correspond to elementary curves in the plane (x, y), which are, respectively, straight lines, exponential, and power functions. This is, possibly, the motivation for the frequent occurrence of spirals or Grandi curves in natural forms (see, e.g., [1,2]).

2. Spirals

The Archimedes spiral [3] (Figure 1) has the polar equation:

$$\rho = a\,\theta\,, \qquad (a > 0, \quad \theta \in \mathbf{R})\,. \tag{1}$$

If $\theta > 0$, the spiral turns counter-clockwise, if $\theta < 0$, the spiral turns clockwise.

Bernoulli's (logarithmic) spiral [4] (Figure 1) has the polar equation:

$$\rho = a\,b^{\theta}\,, \qquad (a, b \in \mathbf{R}^{+})\,,$$
$$\theta = \log_b \left(\frac{\rho}{a}\right)\,. \tag{2}$$

Varying the parameters a and b, one gets different types of spirals.

The coefficient a changes the size, and the term b controls it to be "narrow" and in what direction it wraps itself.

Since a and b are positive constants, some interesting cases are possible. The most studied logarithmic spiral is called harmonic, as the distance between coils is in the harmonic progression whose ratio is $\phi = \frac{\sqrt{5}-1}{2}$, that is, the "golden ratio" relevant to the unit segment.

The logarithmic spiral was discovered by René Descartes in 1638 and studied by Jakob Bernoulli (1654–1705).

Pierre Varignon (1654–1722) called it an equiangular spiral because:

1. The angle between the tangent at a point and the polar radius passing through that point is constant.
2. The angle of inclination with respect to the concentric circles with the center in the origin is also constant.

It is an example of a fractal. As it is written on J. Bernoulli's tomb: Eadem mutata resurgo, but the spiral represented there is of the Archimedes type.

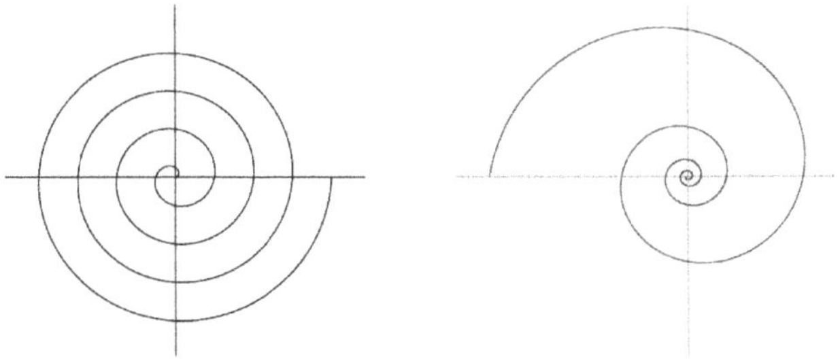

Figure 1. Archimedes' vs. Bernoulli's spiral.

Fermat's spiral (or parabolic) (Figure 2) has the polar equation:

$$\rho = \pm a\,\theta^{1/2}\,. \tag{3}$$

Fermat's (parabolic) spiral suggests the possibility of introducing intermediate graphs between Archimedes' and Bernoulli's spirals.

In fact, in the plane (θ, ρ), the graph of Archimedes' spiral is a straight line, while the Bernoulli spiral has an exponential graph, and the Fermat spiral a parabolic graph.

Then, putting:

$$\rho = a\,\theta^{m/n}\,, \qquad (m, n \text{ positive integers}\,,\ n \neq 0)\,, \tag{4}$$

one gets a family of spirals at varying m and n.

Notice that, if $m > n$, the exponent being greater than one, the coils of the spiral are widening (Figure 2), while if $m < n$, the exponent is less than one, and therefore, the coils of the spiral are shrinking (as in Fermat's case).

Another possibility is to assume $\theta^{m/n}$, where $m/n < 0$ (in this case, the coils are wrapped around the origin (Figure 3) or to use a graph with horizontal asymptotes, in order to get an asymptotic spiral.

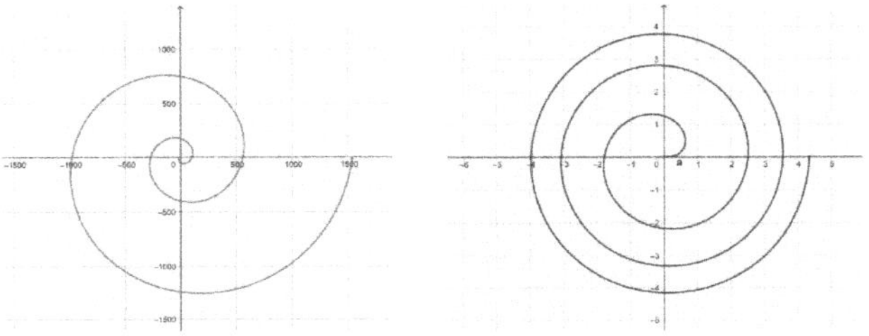

Figure 2. Spiral, $\rho = \theta^{3/2}$, and Fermat spiral, $\rho = \theta^{1/2}$.

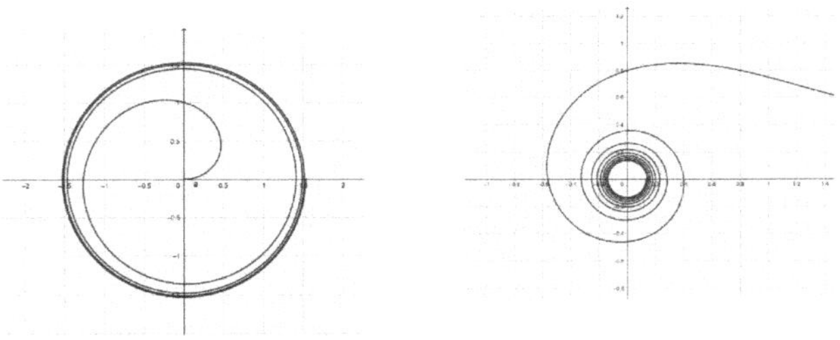

Figure 3. Spiral, $\rho = \theta^{-1/2}$, and asymptotic spiral, $\rho = \arctan(\theta)$.

In what follows, we consider a "canonical form" of the Bernoulli spirals assuming $a = 1\,, b = e^n\,,$ that is the simplified polar equation:

$$\rho = e^{n\,\theta}\,, \qquad (n \in \mathbf{N})\,. \tag{5}$$

3. The Complex Bernoulli Spiral

We now introduce the complex case, putting:

$$\rho = \Re\rho + i\,\Im\rho\,, \tag{6}$$

and considering a Bernoulli spiral of the type:

$$\rho = e^{i\,n\,\theta} = \cos n\theta + i \sin n\theta\,. \tag{7}$$

Therefore, we have:

$$\rho_1 = \Re\rho = \cos n\theta\,, \qquad \rho_2 = \Im\rho = \sin n\theta\,. \tag{8}$$

3.1. Rhodonea Curves

The curves defined in polar coordinates by:

$$\rho_1 = \Re\rho = \cos n\theta\,, \tag{9}$$

are called rhodonea curves or Grandi roses (examples in Figure 4), by the name of Luigi Guido Grandi (1671–1742), who communicated his discovery in a letter to Leibniz in 1713 [5].

Curves of the type:

$$\rho_2 = \Im\rho = \sin n\theta\,, \tag{10}$$

are essentially equivalent to the preceding ones, up to a rotation of $\frac{\pi}{2n}$ radians.

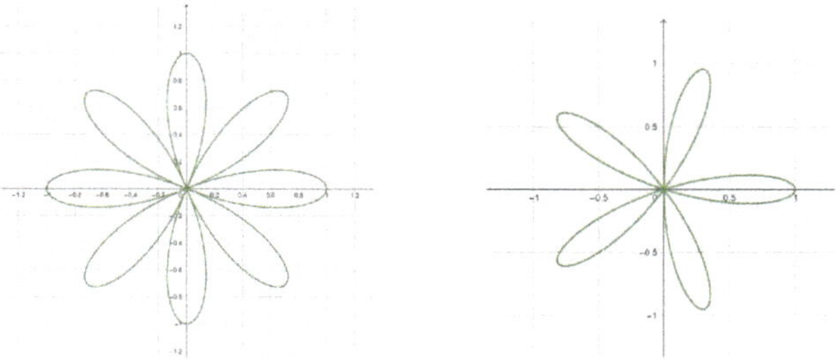

Figure 4. Rhodonea, $\rho = \cos(4\theta)$, and rhodonea, $\rho = \cos(5\theta)$.

3.2. Chebyshev Polynomials

The Chebyshev polynomials of the first and second kind were introduced by Pafnuty L. Chebyshev (1821–1894). They can be derived as the real and imaginary part of the exponential function $e^{i\,n\,\theta} = (\cos\theta + i\sin\theta)^n$ (see Equation (7)), putting $x = \cos\theta$, and using the Euler formula (see [6] for details).

The first kind Chebyshev polynomials are important in approximation theory and Gaussian quadrature rules. In fact, by using their roots—called Chebyshev nodes—the resulting interpolation polynomial minimizes the Runge phenomenon. Furthermore, the relevant approximation is the best approximation to a continuous function under the maximum norm.

Linked with these polynomials are also the Chebyshev polynomials of the second kind, which appear in computing the powers of 2×2 non-singular matrices [7]. Generalizations of such polynomials have been also introduced, in particular for computing powers of higher order matrices (see, e.g., [8,9]).

An excellent book is [10]. The importance of these polynomial sets in applications is shown in [11].

Recently, the Chebyshev polynomials of the first and second kind have been used in order to represent the real and imaginary part of complex Appell polynomials [12,13].

The connection of the second kind Chebyshev polynomials with classical polynomials of number theory has been recently underlined by Kim T., Kim D.S. et al. (see, e.g., [14–16]).

Definition of Chebyshev polynomials of the first kind:

$$\begin{aligned} &T_0(x) = 1\,, \\ &T_1(x) = x\,, \\ &T_{n+1}(x) = 2x\,T_n(x) - T_{n-1}(x)\,, \\ &T_n(x) = \cos(n\,\arccos(x)) \quad \Leftrightarrow \quad T_n(\cos\theta) = \cos(n\,\theta)\,. \end{aligned} \tag{11}$$

Definition of Chebyshev polynomials of the second kind:

$$\begin{aligned} &U_0(x) = 1\,, \\ &U_1(x) = 2x\,, \\ &U_{n+1}(x) = 2x\,U_n(x) - U_{n-1}(x)\,, \\ &\sqrt{1-x^2}\,U_{n-1}(x) = \sin\left(n\,\arccos(x)\right) \quad \Leftrightarrow \quad U_{n-1}(\cos\theta) = \frac{\sin[n\,\theta]}{\sin\theta}\,. \end{aligned} \tag{12}$$

As a consequence of the above considerations, it is possible to note the connection of the rhodonea curves with the first kind Chebyshev polynomials.

In fact, the rhodonea curve $\rho = \cos(n\,\theta)$ (Figure 4) can be interpreted as $T_n(\cos\theta)$ (Equation (11)).

In a similar way, the second kind Chebyshev polynomials have as graphical images the roses of the type $U_n(\cos\theta)$ corresponding to $\sin((n-1)\theta)/\sin(\theta)$ (examples in Figure 5).

Note that, in both cases, the rhodonea curve has n petals if n is odd and $2n$ petals if n is even.

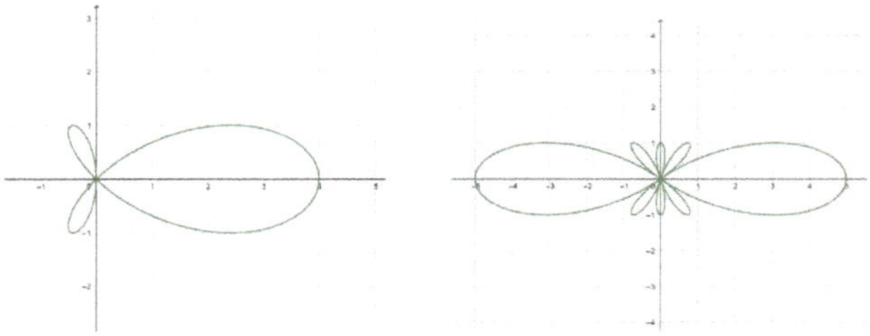

Figure 5. U_4 *rose*, $\rho = \sin(3\theta)/\sin(\theta)$, and U_5 *rose*, $\rho = \sin(4\theta)/\sin(\theta)$.

4. Pseudo-Chebyshev Polynomials

The rhodonea curves exist even for rational values of the index n (see, e.g., [17]). This allows us to consider the sets of first and second kind pseudo-Chebyshev polynomials (graphical examples in Figures 6 and 7). The prefix "pseudo" is used because actually, they are not polynomials, but irrational functions, as it is seen in what follows.

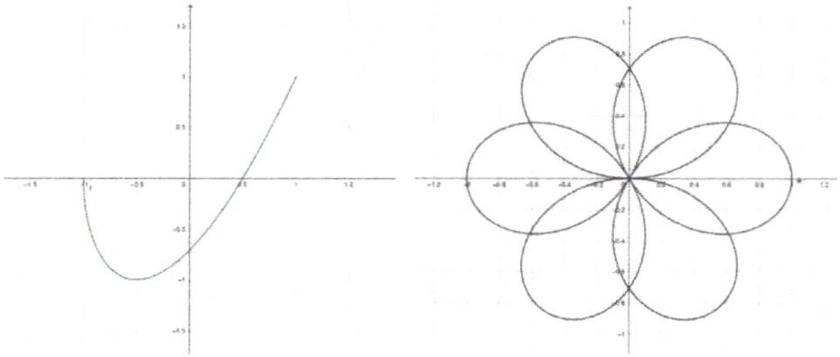

Figure 6. Pseudo $T_{3/2} = \cos(1.5\arccos(x))$, and rhodonea, $\rho = \cos(1.5\,\theta)$.

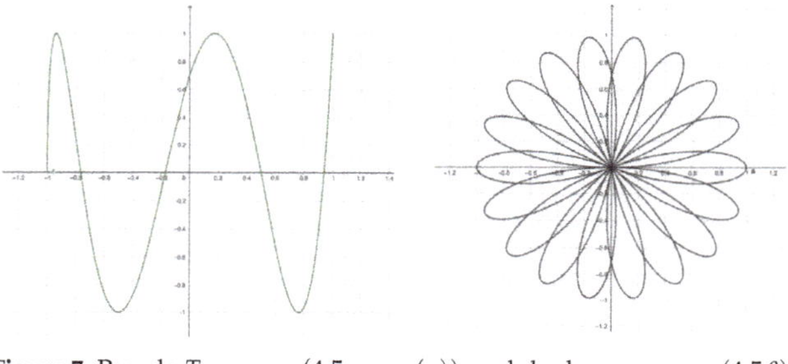

Figure 7. Pseudo $T_{9/2} = \cos(4.5\arccos(x))$, and rhodonea, $\rho = \cos(4.5\,\theta)$.

We start assuming the degree in the form $n+\frac{1}{2}$, that is a semi-integer number. This seems to be the most interesting case, since the resulting functions $T_{n+1/2}$ and $U_{n+1/2}$ are proven to be orthogonal, in the interval $(-1,1)$, with respect to the same, corresponding weights, of the first and second kind Chebyshev polynomials.

We put, by definition:

$$T_{k+\frac{1}{2}}(x) = \cos\left((k+\tfrac{1}{2})\arccos(x)\right) \tag{13}$$

$$\sqrt{1-x^2}\, U_{k-\frac{1}{2}}(x) = \sin\left((k+\tfrac{1}{2}) \arccos(x) \right) . \tag{14}$$

Remark 1. *It is worth noting that the third and fourth kind Chebyshev polynomials $V_n(x)$ and $W_n(x)$ (see, e.g., [18]) have a similar definition, but they do not coincide with the pseudo-Chebyshev, since actually they are true polynomials, and satisfy orthogonality properties with respect to different weights (see Figures 8 and 9).*

The third and fourth kind Chebyshev polynomials have been studied and applied by several scholars (see, e.g., [18–20]), because they are useful in quadrature rules, when the singularities occur only at one of the end points ($+1$ or -1) (see [10]). Furthermore, recently, they have been applied in numerical analysis for solving high odd-order boundary value problems with homogeneous or nonhomogeneous boundary conditions [19].

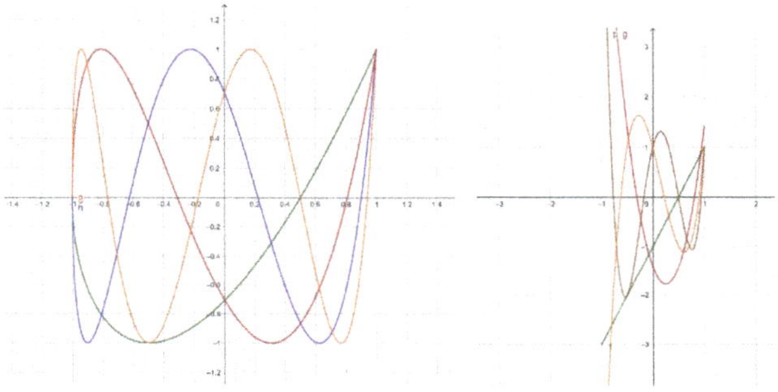

Figure 8. Pseudo $T_{n/2}$, $n = 3,5,7,9$, and third kind $V_k(x)$ $k = 1,2,3,4$.

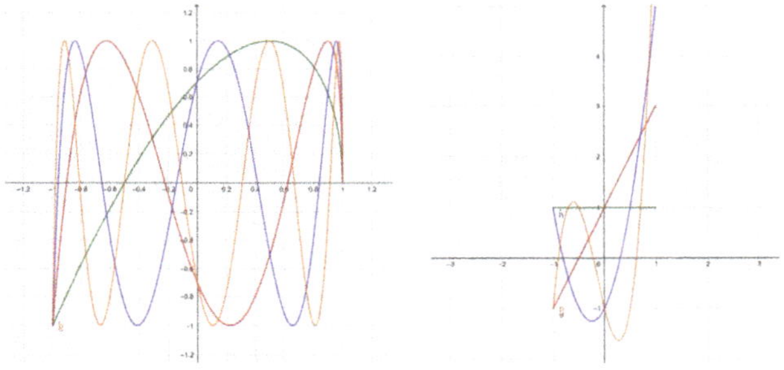

Figure 9. Pseudo $U_{n/2}$, $n = 1,3,5,7$, and fourth kind $W_k(x)$ $k = 0,1,2,3$.

4.1. The Case of Half-Integer Degree

In particular, we have:

$$T_{1/2}(x) = \cos\left(\tfrac{1}{2} \arccos(x)\right) = \sqrt{\frac{1+x}{2}}, \tag{15}$$

$$\sqrt{1-x^2}\, U_{-1/2}(x) = \sin\left(\tfrac{1}{2} \arccos(x)\right) = \sqrt{\frac{1-x}{2}}. \tag{16}$$

Therefore, we find:

$$
\begin{aligned}
T_{3/2}(x) &= \cos\left(\tfrac{3}{2}\arccos(x)\right) = \cos\left(\arccos(x) + \tfrac{1}{2}\arccos(x)\right) \\
&= \cos(\arccos(x))\cos\left(\tfrac{1}{2}\arccos(x)\right) - \sin(\arccos(x))\sin\left(\tfrac{1}{2}\arccos(x)\right) \\
&= x\sqrt{\frac{1+x}{2}} - \sqrt{1-x^2}\sqrt{\frac{1-x}{2}},
\end{aligned} \tag{17}
$$

$$
\begin{aligned}
T_{5/2}(x) &= \cos\left(\tfrac{5}{2}\arccos(x)\right)\cos\left(2\arccos(x) + \tfrac{1}{2}\arccos(x)\right) \\
&= \cos(2\arccos(x))\cos\left(\tfrac{1}{2}\arccos(x)\right) - \sin(2\arccos(x))\sin\left(\tfrac{1}{2}\arccos(x)\right) \\
&= T_2(x)\sqrt{\frac{1+x}{2}} - \sqrt{1-x^2}\,U_1(x)\sqrt{\frac{1-x}{2}}.
\end{aligned} \tag{18}
$$

4.2. Recurrence Relations

We have, in general:

$$
\begin{aligned}
T_{n+1/2}(x) &= \cos\left(n\arccos(x) + \tfrac{1}{2}\arccos(x)\right) \\
&= T_n(x)\sqrt{\frac{1+x}{2}} - \sqrt{1-x^2}\,U_{n-1}(x)\sqrt{\frac{1-x}{2}},
\end{aligned}
$$

that is:

$$
T_{n+1/2}(x) = T_n(x)\,T_{1/2}(x) - \left(1-x^2\right)\,U_{n-1}(x)\,U_{-1/2}(x). \tag{19}
$$

In a similar way, for the second kind, we find:

$$
U_{n+1/2}(x) = U_{n-1}(x)\,T_{1/2}(x) + U_{-1/2}(x)\,T_n(x). \tag{20}
$$

Remark 2. *Note that the number of rose petals of the curves* $\rho = \cos(\frac{n}{2}\theta)$, $n = 1,3,5,\dots$ *is given by the sequence* $\{2,6,10,14,18,22,\dots\}$, *which appears in the Encyclopedia of Integer Sequences [21] at A016825: positive integers congruent to* $2 \mod 4 : a(n) = 4n+2$, *for* $n \geq 0$.

4.3. More General Formulas

By using cosine addition formulas, putting:

$$
\frac{m}{n} = \frac{p}{q} + \frac{r}{s}, \tag{21}
$$

we find:

$$
T_{m/n}(x) = T_{p/q}(x)\,T_{r/s}(x) - \left(1-x^2\right)\,U_{(p/q)-1}(x)\,U_{(r/s)-1}(x), \tag{22}
$$

and by using the sine addition formulas:

$$
U_{m/n}(x) = U_{(p/q)-1}(x)\,T_{r/s}(x) + U_{(r/s)-1}(x)\,T_{p/q}(x). \tag{23}
$$

Particular Results

$$T_1(x) = T_{1/3}(x)\, T_{2/3}(x) - (1-x^2)\; U_{-2/3}(x)\, U_{-1/3}(x)\,. \tag{24}$$

$$T_1(x) = \cos\left[3 \cdot \tfrac{1}{3} \arccos(x)\right] = 4\, T_{1/3}^3(x) - 3\, T_{1/3}(x)\,. \tag{25}$$

$$T_2(x) = \cos\left[3 \cdot \tfrac{2}{3} \arccos(x)\right] = 4\, T_{2/3}^3(x) - 3\, T_{2/3}(x)\,. \tag{26}$$

$$\begin{aligned} T_{2/3}(x) &= \cos\left[2 \cdot \tfrac{1}{3} \arccos(x)\right] = 1 - 2\,\sin^2\left[\tfrac{1}{3} \arccos(x)\right] \\ &= 1 - 2\,(1-x^2)\, U_{-2/3}(x)\,. \end{aligned} \tag{27}$$

$$\begin{aligned} U_{-1/3}(x) &= \frac{\sin\left[2 \cdot \tfrac{1}{3} \arccos(x)\right]}{\sqrt{1-x^2}} \\ = \frac{2}{\sqrt{1-x^2}}\, \sin\left[\frac{1}{3}\arccos(x)\right]\, \cos\left[\frac{1}{3}\arccos(x)\right] &= 2\, U_{-2/3}(x)\, T_{1/3}(x)\,. \end{aligned} \tag{28}$$

$$U_{-2/3}(x) = \frac{\sin\left[\tfrac{1}{3}\arccos(x)\right]}{\sqrt{1-x^2}} = \sqrt{\frac{1 - T_{1/3}^2(x)}{1-x^2}}\,. \tag{29}$$

Combining the above equations, we find:

$$\begin{aligned} T_1(x) &= T_{1/3}(x)\, T_{2/3}(x) - 2\, T_{1/3}(x)\, \left(1 - T_{1/3}^2(x)\right) \\ &= T_{1/3}(x)\left(2\, T_{1/3}^2(x) + T_{2/3}(x) - 2\right)\,. \end{aligned} \tag{30}$$

4.4. Orthogonality for Half-Integer Degree

Theorem 1. *The Chebyshev functions $T_{m/2}(x)$ satisfy the orthogonality property:*

$$\int_{-1}^{1} T_{m/2}(x)\, T_{n/2}(x) \frac{1}{\sqrt{1-x^2}}\, dx = 0\,, \qquad (m \neq n)\,, \tag{31}$$

where m,n are positive odd numbers such that $m + n = 2k$, $k = 2, 3, 4, \ldots,$

$$\int_{-1}^{1} T_{m/2}^2(x) \frac{1}{\sqrt{1-x^2}}\, dx = \frac{\pi}{2}\,. \tag{32}$$

Proof. As a consequence of Werner formulas, we have, under the above conditions,

$$\begin{aligned} &\int_{-1}^{1} \cos(m/2 \arccos(x))\, \cos(n/2 \arccos(x)) \frac{1}{\sqrt{1-x^2}}\, dx = (\text{putting } x = \cos(2t)\,) \\ &= 2\int_{0}^{\pi/2} \cos(mt)\cos(nt)\, dt = 0\,, \end{aligned} \tag{33}$$

and:

$$\int_{-1}^{1} \cos^2(m/2\arccos(x))\frac{1}{\sqrt{1-x^2}}\,dx = 2\int_0^{\pi/2} \cos^2(mt)\,dt = \frac{\pi}{2}\,. \tag{34}$$

□

Theorem 2. *The Chebyshev functions $U_{m/2}(x)$ satisfy the orthogonality property:*

$$\int_{-1}^{1} U_{m/2}(x)\,U_{n/2}(x)\sqrt{1-x^2}\,dx = 0\,, \qquad (m \neq n)\,, \tag{35}$$

where m, n are positive odd numbers such that $m + n = 2k$, $k = 2, 3, 4, \ldots$,

$$\int_{-1}^{1} U_{m/2}^2(x)\,\sqrt{1-x^2}\,dx = \frac{\pi}{2}\,. \tag{36}$$

Proof. We have, under the above conditions,

$$\begin{aligned}\int_{-1}^{1} \sin(m/2\arccos(x))\,\sin(n/2\arccos(x))\,\sqrt{1-x^2}\,dx &= (\text{putting } x = \cos(2t)\,) \\ &= 2\int_0^{\pi/2} \sin(mt)\sin(nt)\,dt = 0\,,\end{aligned} \tag{37}$$

and:

$$\int_{-1}^{1} \sin^2(m/2\arccos(x))\,\sqrt{1-x^2}\,dx = 2\int_0^{\pi/2} \sin^2(mt)\,dt = \frac{\pi}{2}\,. \tag{38}$$

□

5. Conclusions

The complex form of the Bernoulli spiral, by using Euler formulas, allows us to emphasize connections with Grandi (rhodonea) curves. The rhodonea with the fractional index can be viewed as an extension of first and second kind Chebyshev polynomials to irrational functions. The properties of these "pseudo-Chebyshev functions" are borrowed from classical trigonometric identities. In particular, in the case of half-integer degree, the corresponding functions satisfy the same orthogonality property of the corresponding Chebyshev polynomials of integer degree.

Funding: This research received no external funding.

Acknowledgments: Dedicated to Hari M. Srivastava with deep admiration.

Conflicts of Interest: The author declares no conflict of interest.

References

1. Bini, D.; Cherubini, C.; Filippi, S.; Gizzi, A.; Ricci, P.E. On the universality of spiral waves. *Commun. Comput. Phys.* **2010**, *8*, 610–622.
2. Gielis, J. *The Geometrical Beauty of Plants*; Atlantis Press: Paris, France, 2017.
3. Heath, T.L. *The Works of Archimedes*; Google Books; Cambridge University Press: Cambridge, UK, 1897.
4. Archibald, R.C. Notes on the logarithmic spiral, golden section and the Fibonacci series. In *Dynamic Symmetry*; Hambidge, J., Ed.; Yale University Press: New Haven, CT, USA, 1920; pp. 16–18.
5. Tenca, L. *Guido Grandi Matematico Cremonense*; Istituto lombardo di Scienze e Lettere: Milano, Italy, 1950.
6. Rivlin, T.J. *The Chebyshev Polynomials*; Wiley: New York, NY, USA, 1990.

7. Ricci, P.E. Alcune osservazioni sulle potenze delle matrici del secondo ordine e sui polinomi di Tchebycheff di seconda specie. *Atti della Accademia delle Scienze di Torino* **1975**, *109*, 405–410.
8. Ricci, P.E. Sulle potenze di una matrice. *Rend. Mat.* **1976**, *9*, 179–194.
9. Ricci, P.E. I polinomi di Tchebycheff in più variabili. *Rend. Mat.* **1978**, *11*, 295–327.
10. Mason, J.C.; Handscomb, D.C. *Chebyshev Polynomials*; Chapman and Hall: New York, NY, USA; CRC: Boca Raton, FL, USA, 2003.
11. Boyd, J.P. *Chebyshev and Fourier Spectral Methods*, 2nd ed.; Dover: Mineola, NY, USA, 2001.
12. Srivastava, H.M.; Ricci, P.E.; Natalini, P. A Family of Complex Appell Polynomial Sets. *Real Acad. Sci. Exact. Fis Nat. Ser. A Math.* **2018**, submitted.
13. Srivastava, H.M.; Manocha, H.L. *A Treatise on Generating Functions*; Halsted Press (Ellis Horwood Limited, Chichester), John Wiley and Sons: New York, NY, USA; Chichester, UK; Brisbane, Australia; Toronto, ON, Canada, 1984.
14. Kim, T.; Kim, D.S.; Dolgy, D.V.; Park, J.-W. Sums of finite products of Chebyshev polynomials of the second kind and of Fibonacci polynomials. *J. Inequal. Appl.* **2018**. [CrossRef] [PubMed]
15. Kim, T.; Kim, D.S.; Kwon, J.; Dolgy, D.V. Expressing Sums of Finite Products of Chebyshev Polynomials of the Second Kind and of Fibonacci Polynomials by Several Orthogonal Polynomials. *Mathematics* **2018**, *6*, 210. [CrossRef]
16. Kim, D.S.; Dolgy, D.V.; Kim, T.; Rim, S.-H. Identities involving Bernoulli and Euler polynomials arising from Chebyshev polynomials. *Proc. Jangjeon Math. Soc.* **2012**, *15*, 361–370.
17. Hall, L. Trochoids, Roses, and Thorns-Beyond the Spirograph. *Coll. Math. J.* **1992**, *23*, 20–35.
18. Aghigh, K.; Masjed-Jamei, M.; Dehghan, M. A survey on third and fourth kind of Chebyshev polynomials and their applications. *Appl. Math. Comput.* **2008**, *199*, 2–12. [CrossRef]
19. Doha, E.H.; Abd-Elhameed, W.M.; Alsuyuti, M.M. On using third and fourth kinds Chebyshev polynomials for solving the integrated forms of high odd-order linear boundary value problems. *J. Egypt. Math. Soc.* **2014**, *23*, 397–405. [CrossRef]
20. Kim, T.; Kim, D.S.; Dolgy, D.V.; Kwon, J. Sums of finite products of Chebyshev polynomials of the third and fourth kinds. *Adv. Differ. Equ.* **2018**, *2018*, 283. [CrossRef]
21. Sloane, N.J.A.; Plouffe, S. *The Encyclopedia of Integer Sequences*; Academic Press: San Diego, CA, USA, 1995.

© 2018 by the author. Licensee MDPI, Basel, Switzerland. This article is an open access article distributed under the terms and conditions of the Creative Commons Attribution (CC BY) license (http://creativecommons.org/licenses/by/4.0/).

symmetry

MDPI

Article

Generalized Liouville–Caputo Fractional Differential Equations and Inclusions with Nonlocal Generalized Fractional Integral and Multipoint Boundary Conditions

Ahmed Alsaedi [1], Madeaha Alghanmi [1], Bashir Ahmad [1,*] and Sotiris K. Ntouyas [1,2]

1 Nonlinear Analysis and Applied Mathematics (NAAM)-Research Group, Department of Mathematics, Faculty of Science, King Abdulaziz University, P.O. Box 80203, Jeddah 21589, Saudi Arabia; aalsaedi@hotmail.com (A.A.); madeaha@hotmail.com (M.A.)

2 Department of Mathematics, University of Ioannina, 451 10 Ioannina, Greece; sntouyas@uoi.gr

* Correspondence: bashirahmad_qau@yahoo.com

Received: 8 October 2018; Accepted: 20 November 2018; Published: 26 November 2018

Abstract: We develop the existence criteria for solutions of Liouville–Caputo-type generalized fractional differential equations and inclusions equipped with nonlocal generalized fractional integral and multipoint boundary conditions. Modern techniques of functional analysis are employed to derive the main results. Examples illustrating the main results are also presented. It is imperative to mention that our results correspond to the ones for a symmetric second-order nonlocal multipoint integral boundary value problem under suitable conditions (see the last section).

Keywords: differential equation; differential inclusion; Liouville–Caputo-type fractional derivative; fractional integral; existence; fixed point

1. Introduction

Fractional order differential and integral operators extensively appear in the mathematical modeling of various scientific and engineering phenomena. The main advantage for using these operators is their nonlocal nature, which can describe the past history of processes and material involved in the phenomena. Thus, fractional-order models are more realistic and informative than their corresponding integer-order counterparts. Examples include bio-engineering [1], Chaos and fractional dynamics [2], ecology [3], financial economics [4], etc. Widespread applications of methods of fractional calculus in numerous real world phenomena motivated many researchers to develop this important branch of mathematical analysis—for instance, see the texts [5–8].

Fractional differential equations equipped with a variety of boundary conditions have recently been studied by several researchers. In particular, overwhelming interest has been shown in the study of nonlocal nonlinear fractional-order boundary value problems (FBVPs). The concept of nonlocal conditions dates back to the work of Bitsadze and Samarski [9] and these conditions facilitate describing the physical phenomena taking place inside the boundary of the given domain. In computational fluid dynamics (CFD) studies of blood flow problems, it is hard to justify the assumption of a circular cross-section of a blood vessel due to its changing geometry throughout the vessel. This issue has been addressed by the introduction of integral boundary conditions. In addition, integral boundary conditions are used in regularization of ill-posed parabolic backward problems. Moreover, integral boundary conditions play an important role in mathematical models for bacterial self-regularization [10].

On the other hand, multivalued (inclusions) problems are found to be of special significance in studying dynamical systems and stochastic processes. Examples include granular systems [11,12], control problems [13,14], dynamics of wheeled vehicles [15], etc. For more details, see the text [16], which addresses the pressing issues in stochastic processes, queueing networks, optimization and their application in finance, control, climate control, etc. In previous work [17], synchronization processes involving fractional differential inclusions are studied.

The area of investigation for nonlocal nonlinear fractional boundary value problems includes existence and uniqueness of solutions, stability and oscillatory properties, analytic and numerical methods. The literature on the topic is now much enriched and covers fractional order differential equations and inclusions involving Riemann–Liouville, Liouville–Caputo (Caputo), Hadamard type derivatives, etc. For some recent works on the topic, we refer the reader to a series of papers [18–36] and the references cited therein.

In this paper, we introduce and study a new class of boundary value problems of Liouville–Caputo-type generalized fractional differential equations and inclusions (instead of taking the usual Liouville–Caputo fractional order derivative) supplemented with nonlocal generalized fractional integral and multipoint boundary conditions. Precisely, we consider the problems:

$$\begin{cases} {}^{\rho}_{c}D^{\alpha}_{0+}y(t) = f(t, y(t)), \ \ t \in J := [0,T], \\ y(T) = \sum_{i=1}^{m} \sigma_i {}^{\rho}I^{\beta}_{0+}y(\eta_i) + \kappa, \ \ \delta y(0) = \sum_{j=1}^{k} \mu_j y(\xi_j), \\ 0 < \eta_1 < \cdots < \eta_i < \cdots < \eta_m < \xi_1 < \cdots < \xi_j < \cdots < \xi_k < T, \end{cases} \tag{1}$$

and

$$\begin{cases} {}^{\rho}_{c}D^{\alpha}_{0+}y(t) \in F(t, y(t)), \ \ t \in J := [0,T], \\ y(T) = \sum_{i=1}^{m} \sigma_i {}^{\rho}I^{\beta}_{0+}y(\eta_i) + \kappa, \ \ \delta y(0) = \sum_{j=1}^{k} \mu_j y(\xi_j), \\ 0 < \eta_1 < \cdots < \eta_i < \cdots < \eta_m < \xi_1 < \cdots < \xi_j < \cdots < \xi_k < T, \end{cases} \tag{2}$$

where ${}^{\rho}_{c}D^{\alpha}_{0+}$ is the Liouville–Caputo-type generalized fractional derivative of order $1 < \alpha \le 2$, ${}^{\rho}I^{\beta}_{0+}$ is the generalized (Katugampola type) fractional integral of order $\beta > 0, \rho > 0$, $f : J \times \mathbb{R} \to \mathbb{R}$ is a continuous function, $\sigma_i, \mu_j, \kappa \in \mathbb{R}, i = 1, 2, \ldots, m, j = 1, 2, \ldots, k$, $\delta = t^{1-\rho}\frac{d}{dt}$, and $F : J \times \mathbb{R} \to \mathcal{P}(\mathbb{R})$ is a multivalued function ($\mathcal{P}(\mathbb{R})$ is the family of all nonempty subsets of $\mathbb{R}$).

The rest of the paper is arranged as follows: Section 2 contains some preliminary concepts related to our work and a vital lemma associated with the linear variant of the given problem, which is used to convert the given problems into fixed point problems. In Section 3, the existence and uniqueness results for problem (1) are obtained by using a Banach contraction mapping principle, Krasnoselskii's fixed point theorem and Leray–Schauder nonlinear alternative. Existence results for the inclusions problem (2) are studied in Section 4 via Leray–Schauder nonlinear alternative, and Covitz and Nadler fixed point theorem for multi-valued maps. Examples illustrating the obtained results are also included.

2. Preliminaries

Denote by $X^p_c(a,b)$ the space of all complex-valued Lebesgue measurable functions φ on (a,b) equipped with the norm:

$$\|\varphi\|_{X^p_c} = \Big(\int_a^b |x^c \varphi(x)|^p \frac{dx}{x} \Big)^{1/p} < \infty \, c \in \mathbb{R}, 1 \le p \le \infty.$$

Let $L^1(a,b)$ represent the space of all Lebesgue measurable functions ψ on (a,b) endowed with the norm:

$$\|\psi\|_{L^1} = \int_a^b |\psi(x)| dx < \infty.$$

We further recall that $AC^n(J,\mathbb{R}) = \{x : J \to \mathbb{R} : x, x', \ldots, x^{(n-1)} \in C(J,\mathbb{R})$ and $x^{(n-1)}$ is absolutely continuous $\}$. For $0 \le \epsilon < 1$, we define $C_{\epsilon,\rho}(J,\mathbb{R}) = \{f : J \to \mathbb{R} : (t^{\rho} - a^{\rho})^{\epsilon} f(t) \in C(J,\mathbb{R})\}$

endowed with the norm $\|f\|_{C_{\epsilon,\rho}} = \|(t^\rho - a^\rho)^\epsilon f(t)\|_C$. Moreover, we define the class of functions f that have absolutely continuous δ^{n-1}-derivative, denoted by $AC^n_\delta(J,\mathbb{R})$, as follows: $AC^n_\delta(J,\mathbb{R}) = \Big\{f : J \to \mathbb{R} : \delta^{n-1} f \in AC(J,\mathbb{R}), \delta = t^{1-\rho}\frac{d}{dt}\Big\}$, which is equipped with the norm $\|f\|_{C^n_\delta} = \sum_{k=0}^{n-1} \|\delta^k f\|_C$. More generally, the space of functions endowed with the norm $\|f\|_{C^n_{\delta,\epsilon}} = \sum_{k=0}^{n-1} \|\delta^k f\|_C + \|\delta^n f\|_{C_{\epsilon,\rho}}$ is defined by

$$C^n_{\delta,\epsilon}(J,\mathbb{R}) = \Big\{f : J \to \mathbb{R} : \delta^{n-1} f \in C(J,\mathbb{R}), \delta^n f \in C_{\epsilon,\rho}(J,\mathbb{R}), \delta = t^{1-\rho}\frac{d}{dt}\Big\}.$$

Notice that $C^n_{\delta,0} = C^n_\delta$.

Definition 1 ([37]). *For $-\infty < a < t < b < \infty$, the left-sided and right-sided generalized fractional integrals of $f \in X^p_c(a,b)$ of order $\alpha > 0$ and $\rho > 0$ are respectively defined by*

$$({}^\rho I^\alpha_{a+} f)(t) = \frac{\rho^{1-\alpha}}{\Gamma(\alpha)} \int_a^t \frac{s^{\rho-1}}{(t^\rho - s^\rho)^{1-\alpha}} f(s)ds, \tag{3}$$

$$({}^\rho I^\alpha_{b-} f)(t) = \frac{\rho^{1-\alpha}}{\Gamma(\alpha)} \int_t^b \frac{s^{\rho-1}}{(s^\rho - t^\rho)^{1-\alpha}} f(s)ds. \tag{4}$$

Definition 2 ([38]). *For $0 \le a < x < b < \infty$, the generalized fractional derivatives, associated with the generalized fractional integrals (3) and (4), are respectively defined by*

$$\begin{aligned} ({}^\rho D^\alpha_{a+} f)(t) &= \Big(t^{1-\rho}\frac{d}{dt}\Big)^n ({}^\rho I^{n-\alpha}_{a+} f)(t) \\ &= \frac{\rho^{\alpha-n+1}}{\Gamma(n-\alpha)} \Big(t^{1-\rho}\frac{d}{dt}\Big)^n \int_a^t \frac{s^{\rho-1}}{(t^\rho - s^\rho)^{\alpha-n+1}} f(s)ds, \end{aligned} \tag{5}$$

$$\begin{aligned} ({}^\rho D^\alpha_{b-} f)(t) &= \Big(-t^{1-\rho}\frac{d}{dt}\Big)^n ({}^\rho I^{n-\alpha}_{b-} f)(t) \\ &= \frac{\rho^{\alpha-n+1}}{\Gamma(n-\alpha)} \Big(-t^{1-\rho}\frac{d}{dt}\Big)^n \int_t^b \frac{s^{\rho-1}}{(s^\rho - t^\rho)^{\alpha-n+1}} f(s)ds, \end{aligned} \tag{6}$$

if the integrals exist.

Definition 3 ([39]). *The left-sided and right-sided Liouville–Caputo-type generalized fractional derivatives of $f \in AC^n_\delta[a,b]$ of order $\alpha \ge 0$ are respectively defined via the above generalized fractional derivatives as*

$${}^\rho_c D^\alpha_{a+} f(x) = {}^\rho D^\alpha_{a+}\Big[f(t) - \sum_{k=0}^{n-1} \frac{\delta^k f(a)}{k!}\Big(\frac{t^\rho - a^\rho}{\rho}\Big)^k\Big](x), \quad \delta = x^{1-\rho}\frac{d}{dx}, \tag{7}$$

$${}^\rho_c D^\alpha_{b-} f(x) = {}^\rho D^\alpha_{b-}\Big[f(t) - \sum_{k=0}^{n-1} \frac{(-1)^k \delta^k f(b)}{k!}\Big(\frac{b^\rho - t^\rho}{\rho}\Big)^k\Big](x), \quad \delta = x^{1-\rho}\frac{d}{dx}, \tag{8}$$

where $n = [\alpha] + 1$.

Lemma 1 ([39]). *Let $\alpha \ge 0, n = [\alpha] + 1$ and $f \in AC^n_\delta[a,b]$, where $0 < a < b < \infty$. Then,*
1. if $\alpha \notin \mathbb{N}$,

$${}^\rho_c D^\alpha_{a+} f(t) = \frac{1}{\Gamma(n-\alpha)} \int_a^t \Big(\frac{t^\rho - s^\rho}{\rho}\Big)^{n-\alpha-1} \frac{(\delta^n f)(s)ds}{s^{1-\rho}} = {}^\rho I^{n-\alpha}_{a+}(\delta^n f)(t), \tag{9}$$

$${}^\rho_c D^\alpha_{b-} f(t) = \frac{1}{\Gamma(n-\alpha)} \int_t^b \Big(\frac{s^\rho - t^\rho}{\rho}\Big)^{n-\alpha-1} \frac{(-1)^n(\delta^n f)(s)ds}{s^{1-\rho}} = {}^\rho I^{n-\alpha}_{b-}(\delta^n f)(t). \tag{10}$$

2. *If* $\alpha \in \mathbb{N}$,

$$ {}^{\rho}_{c}D^{\alpha}_{a^+}f = \delta^n f, \quad {}^{\rho}_{c}D^{\alpha}_{b^-}f = (-1)^n\delta^n f. \tag{11}$$

Lemma 2 ([39]). *Let* $f \in AC^n_\delta[a,b]$ *or* $C^n_\delta[a,b]$ *and* $\alpha \in \mathbb{R}$. *Then,*

$$ {}^{\rho}I^{\alpha}_{a^+}\,{}^{\rho}_{c}D^{\alpha}_{a^+}f(x) = f(x) - \sum_{k=0}^{n-1}\frac{(\delta^k f)(a)}{k!}\Big(\frac{x^\rho - a^\rho}{\rho}\Big)^k,$$

$$ {}^{\rho}I^{\alpha}_{b^-}\,{}^{\rho}_{c}D^{\alpha}_{b^-}f(x) = f(x) - \sum_{k=0}^{n-1}\frac{(-1)^k(\delta^k f)(a)}{k!}\Big(\frac{b^\rho - x^\rho}{\rho}\Big)^k.$$

In particular, for $0 < \alpha \le 1$, *we have*

$$ {}^{\rho}I^{\alpha}_{a^+}\,{}^{\rho}_{c}D^{\alpha}_{a^+}f(x) = f(x) - f(a), \; {}^{\rho}I^{\alpha}_{b^-}\,{}^{\rho}_{c}D^{\alpha}_{b^-}f(x) = f(x) - f(b).$$

For computational convenience, we introduce the notations:

$$ A_1 = 1 - \sum_{j=1}^{k}\mu_j\frac{\xi_j^\rho}{\rho}, \quad A_2 = \sum_{j=1}^{k}\mu_j, \tag{12}$$

$$ B_1 = \frac{T^\rho}{\rho} - \sum_{i=1}^{m}\sigma_i\frac{\eta_i^{\rho(\beta+1)}}{\rho^{\beta+1}\Gamma(\beta+2)}, \quad B_2 = 1 - \sum_{i=1}^{m}\sigma_i\frac{\eta_i^{\rho\beta}}{\rho^{\beta}\Gamma(\beta+1)}, \tag{13}$$

$$ \Omega = A_1B_2 + B_1A_2. \tag{14}$$

The following lemma, related to the linear variant of problem (1), plays a key role in converting the given problem into a fixed point problem.

Lemma 3. *Let* $h \in C(0,T) \cap L(0,T)$, $y \in AC^2_\delta(J)$ *and* $\Omega \neq 0$. *Then, the solution of the boundary value problem (BVP):*

$$ \begin{cases} {}^{\rho}_{c}D^{\alpha}_{0^+}y(t) = h(t), \; t \in J := [0,T], \\ y(T) = \sum_{i=1}^{m}\sigma_i\,{}^{\rho}I^{\beta}_{0^+}y(\eta_i) + \kappa, \quad \delta y(0) = \sum_{j=1}^{k}\mu_j y(\xi_j), \\ 0 < \eta_1 < \cdots < \eta_i < \cdots < \eta_m < \xi_1 < \cdots < \xi_j < \cdots < \xi_k < T, \end{cases} \tag{15}$$

is given by

$$ \begin{aligned} y(t) = {} & {}^{\rho}I^{\alpha}_{0^+}h(t) + \tfrac{1}{\Omega}\Big\{ -B_1\sum_{j=1}^{k}\mu_j\,{}^{\rho}I^{\alpha}_{0^+}h(\xi_j) + A_1\Big[\sum_{i=1}^{m}\sigma_i\,{}^{\rho}I^{\alpha+\beta}_{0^+}h(\eta_i) - {}^{\rho}I^{\alpha}_{0^+}h(T) + \kappa\Big]\Big\} \\ & + \tfrac{t^\rho}{\rho\Omega}\Big\{ B_2\sum_{j=1}^{k}\mu_j\,{}^{\rho}I^{\alpha}_{0^+}h(\xi_j) + A_2\Big[\sum_{i=1}^{m}\sigma_i\,{}^{\rho}I^{\alpha+\beta}_{0^+}h(\eta_i) - {}^{\rho}I^{\alpha}_{0^+}h(T) + \kappa\Big]\Big\}. \end{aligned} \tag{16}$$

Proof. Applying ${}^{\rho}I^{\alpha}_{0^+}$ on the fractional differential equation in (15) and using Lemma 2, the solution of fractional differential equation in (15) for $t \in J$ is

$$ y(t) = {}^{\rho}I^{\alpha}_{0^+}h(t) + c_1 + c_2\frac{t^\rho}{\rho} = \frac{\rho^{1-\alpha}}{\Gamma(\alpha)}\int_0^t s^{\rho-1}(t^\rho - s^\rho)^{\alpha-1}h(s)ds + c_1 + c_2\frac{t^\rho}{\rho}, \tag{17}$$

for some $c_1, c_2 \in \mathbb{R}$. Taking $\delta - derivative$ of (17), we get

$$ \delta y(t) = {}^{\rho}I^{\alpha-1}_{0^+}h(t) + c_2 = \frac{\rho^{2-\alpha}}{\Gamma(\alpha-1)}\int_0^t s^{\rho-1}(t^\rho - s^\rho)^{\alpha-2}h(s)ds + c_2. \tag{18}$$

Using the boundary condition $\delta y(0) = \sum_{j=1}^{k} \mu_j y(\xi_j)$ in (18), we get

$$c_2 = \sum_{j=1}^{k} \mu_j {}^{\rho}I_{0+}^{\alpha} h(\xi_j) + c_1 \sum_{j=1}^{k} \mu_j + c_2 \sum_{j=1}^{k} \mu_j \frac{\xi_j^{\rho}}{\rho},$$

which, on account of (12), takes the form:

$$A_1 c_2 - A_2 c_1 = \sum_{j=1}^{k} \mu_j {}^{\rho}I_{0+}^{\alpha} h(\xi_j). \tag{19}$$

Applying the generalized integral operator ${}^{\rho}I_{0+}^{\beta}$ on (17), we get

$${}^{\rho}I_{0+}^{\beta} y(t) = {}^{\rho}I_{0+}^{\alpha+\beta} h(t) + c_1 \frac{t^{\rho\beta}}{\rho^{\beta}\Gamma(\beta+1)} + c_2 \frac{t^{\rho(\beta+1)}}{\rho^{\beta+1}\Gamma(\beta+2)}, \tag{20}$$

which, together with the boundary condition $y(T) = \sum_{i=1}^{m} \sigma_i \, {}^{\rho}I_{0+}^{\beta} y(\eta_i) + \kappa$, yields

$$\begin{aligned} {}^{\rho}I_{0+}^{\alpha} h(T) + c_1 + c_2 \frac{T^{\rho}}{\rho} &= \sum_{i=1}^{m} \sigma_i \, {}^{\rho}I_{0+}^{\alpha+\beta} h(\eta_i) + \sum_{i=1}^{m} \sigma_i c_1 \frac{\eta_i^{\rho\beta}}{\rho^{\beta}\Gamma(\beta+1)} \\ &+ \sum_{i=1}^{m} \sigma_i c_2 \frac{\eta_i^{\rho(\beta+1)}}{\rho^{\beta+1}\Gamma(\beta+2)} + \kappa. \end{aligned} \tag{21}$$

Using the notations (13) in (21), we obtain

$$B_1 c_2 + B_2 c_1 = \sum_{i=1}^{m} \sigma_i \, {}^{\rho}I_{0+}^{\alpha+\beta} h(\eta_i) - {}^{\rho}I_{0+}^{\alpha} h(T) + \kappa. \tag{22}$$

Solving the system of Equations (19) and (22) for c_1 and c_2, we find that

$$c_1 = \frac{1}{\Omega}\Big\{ - B_1 \sum_{j=1}^{k} \mu_j {}^{\rho}I_{0+}^{\alpha} h(\xi_j) + A_1 \Big[\sum_{i=1}^{m} \sigma_i \, {}^{\rho}I_{0+}^{\alpha+\beta} h(\eta_i) - {}^{\rho}I_{0+}^{\alpha} h(T) + \kappa \Big] \Big\}. \tag{23}$$

and

$$c_2 = \frac{1}{\Omega}\Big\{ B_2 \sum_{j=1}^{k} \mu_j {}^{\rho}I_{0+}^{\alpha} h(\xi_j) + A_2 \Big[\sum_{i=1}^{m} \sigma_i \, {}^{\rho}I_{0+}^{\alpha+\beta} h(\eta_i) - {}^{\rho}I^{\alpha} h(T) + \kappa \Big] \Big\}. \tag{24}$$

Substituting the values of c_1 and c_2 in (17), we get Equation (16). The converse follows by direct computation. The proof is completed. □

3. Main Results for the Problem (1)

By Lemma 3, we define an operator $\mathcal{G} : \mathcal{C} \to \mathcal{C}$ ($\mathcal{C} = C(J, \mathbb{R})$) associated with problem (1) as

$$\begin{aligned} \mathcal{G}y(t) &= {}^{\rho}I_{0+}^{\alpha} f(t, y(t)) + \frac{1}{\Omega}\Big\{ - B_1 \sum_{j=1}^{k} \mu_j {}^{\rho}I_{0+}^{\alpha} f(\xi_j, y(\xi_j)) + A_1 \Big[\sum_{i=1}^{m} \sigma_i \, {}^{\rho}I_{0+}^{\alpha+\beta} f(\eta_i, y(\eta_i)) \\ &\quad - {}^{\rho}I_{0+}^{\alpha} f(T, y(T)) + \kappa \Big] \Big\} + \frac{t^{\rho}}{\rho\Omega}\Big\{ B_2 \sum_{j=1}^{k} \mu_j {}^{\rho}I_{0+}^{\alpha} f(\xi_j, y(\xi_j)) \\ &\quad + A_2 \Big[\sum_{i=1}^{m} \sigma_i \, {}^{\rho}I_{0+}^{\alpha+\beta} f(\eta_i, y(\eta_i)) - {}^{\rho}I_{0+}^{\alpha} f(T, y(t)) + \kappa \Big] \Big\}. \end{aligned} \tag{25}$$

In the following, for brevity, we use the notations:

$$\begin{aligned}\Lambda &= \frac{T^{\rho\alpha}}{\rho^{\alpha}\Gamma(\alpha+1)}+\frac{1}{|\Omega|}\Big\{|B_1|\sum_{j=1}^{k}|\mu_j|\frac{\xi_j^{\rho\alpha}}{\rho^{\alpha}\Gamma(\alpha+1)}+|A_1|\Big[\sum_{i=1}^{m}|\sigma_i|\frac{\eta_i^{\rho(\alpha+\beta)}}{\rho^{\alpha+\beta}\Gamma(\alpha+\beta+1)}+\frac{T^{\rho\alpha}}{\rho^{\alpha}\Gamma(\alpha+1)}\Big]\Big\}\\ &\quad+\frac{T^{\rho}}{\rho|\Omega|}\Big\{|B_2|\sum_{j=1}^{k}|\mu_j|\frac{\xi_j^{\rho\alpha}}{\rho^{\alpha}\Gamma(\alpha+1)}+|A_2|\Big[\sum_{i=1}^{m}|\sigma_i|\frac{\eta_i^{\rho(\alpha+\beta)}}{\rho^{\alpha+\beta}\Gamma(\alpha+\beta+1)}+\frac{T^{\rho\alpha}}{\rho^{\alpha}\Gamma(\alpha+1)}\Big]\Big\}. \qquad (26)\end{aligned}$$

In the first result, we establish the existence of solutions for problem (1) via Leray–Schauder nonlinear alternative [40].

Theorem 1. *Suppose that the following conditions hold:*

(H_1) *For a function $\phi \in L^1([0,T],\mathbb{R}^+)$, and a nondecreasing function $\psi:\mathbb{R}^+\to\mathbb{R}^+$ such that $|f(t,y)|\le \phi(t)\psi(\|y\|)$, $\forall (t,y)\in[0,T]\times\mathbb{R}$;*

(H_2) *there exists a positive constant $\mathcal{M}$ such that*

$$\frac{\mathcal{M}}{\psi(\mathcal{M})\Lambda_1+\dfrac{|\kappa|(\rho|A_1|+T^{\rho}|A_2|)}{\rho|\Omega|}}>1,$$

where

$$\begin{aligned}\Lambda_1 &= {}^{\rho}I_{0^+}^{\alpha}\phi(T)+\frac{1}{|\Omega|}\Big\{|B_1|\sum_{j=1}^{k}|\mu_j|{}^{\rho}I_{0^+}^{\alpha}\phi(\xi_j)+|A_1|\Big[\sum_{i=1}^{m}|\sigma_i|\,{}^{\rho}I_{0^+}^{\alpha+\beta}\phi(\eta_i)+{}^{\rho}I_{0^+}^{\alpha}\phi(T)\Big]\Big\}\\ &\quad+\frac{T^{\rho}}{\rho|\Omega|}\Big\{|B_2|\sum_{j=1}^{k}|\mu_j|{}^{\rho}I_{0^+}^{\alpha}\phi(\xi_j)+|A_2|\Big[\sum_{i=1}^{m}|\sigma_i|\,{}^{\rho}I_{0^+}^{\alpha+\beta}\phi(\eta_i)+{}^{\rho}I_{0^+}^{\alpha}\phi(T)\Big]\Big\}. \qquad (27)\end{aligned}$$

Then, there exists at least one solution for problem (1) on $[0,T]$.

Proof. Firstly, we show that the operator $\mathcal{G}:\mathcal{C}\to\mathcal{C}$ defined by (25) is continuous and completely continuous.

Step 1: $\mathcal{G}$ is continuous.

Let $\{y_n\}$ be a sequence such that $y_n\to y$ in $\mathcal{C}$. Then,

$$\begin{aligned}|\mathcal{G}(y_n)(t)-\mathcal{G}(y)(t)| &\le {}^{\rho}I_{0^+}^{\alpha}|f(t,y_n(t))-f(t,y(t))|+\frac{1}{|\Omega|}\Bigg\{|B_1|\sum_{j=1}^{k}|\mu_j|{}^{\rho}I_{0^+}^{\alpha}|f(\xi_j,y_n(\xi_j))-f(\xi_j,y(\xi_j))|\\ &\quad+|A_1|\Big[\sum_{i=1}^{m}|\sigma_i|\,{}^{\rho}I_{0^+}^{\alpha+\beta}|f(\eta_i,y_n(\eta_i))-f(\eta_i,y(\eta_i))|+{}^{\rho}I_{0^+}^{\alpha}|f(T,y_n(T))-f(T,y(T))|\Big]\Bigg\}\\ &\quad+\frac{t^{\rho}}{\rho|\Omega|}\Big\{|B_2|\sum_{j=1}^{k}|\mu_j|{}^{\rho}I_{0^+}^{\alpha}|f(\xi_j,y_n(\xi_j))-f(\xi_j,y(\xi_j))|\\ &\quad+|A_2|\Big[\sum_{i=1}^{m}|\sigma_i|\,{}^{\rho}I_{0^+}^{\alpha+\beta}|f(\eta_i,y_n(\eta_i))-f(\eta_i,y(\eta))|+{}^{\rho}I_{0^+}^{\alpha}|f(T,y_n(T))-f(T,y(T))|\Big]\Big\}\\ &\le \Lambda\|f(\cdot,y_n)-f(\cdot,y)\|.\end{aligned}$$

In view of continuity of f, it follows from the above inequality that

$$\|\mathcal{G}(y_n)-\mathcal{G}(y)\|\le\Lambda\|f(\cdot,y_n)-f(\cdot,y)\|\to 0, \text{ as } n\to\infty.$$

Step 2: $\mathcal{G}$ maps bounded sets into bounded sets in $\mathcal{C}$.

For a positive number r, it will be shown that there exists a positive constant ℓ such that $\|\mathcal{G}(y)\| \leq \ell$ for any $y \in B_r = \{y \in C : \|y\| \leq r\}$. By (H_1), for each $t \in J$, we have

$$\begin{aligned}
|\mathcal{G}(y)(t)| \leq\ & {}^{\rho}I^{\alpha}_{0^+}|f(t,y(t))| + \frac{1}{|\Omega|}\Big\{|B_1|\sum_{j=1}^{k}|\mu_j|{}^{\rho}I^{\alpha}_{0^+}|f(\xi_j,y(\xi_j))| + |A_1|\Big[\sum_{i=1}^{m}|\sigma_i|\,{}^{\rho}I^{\alpha+\beta}_{0^+}|f(\eta_i,y(\eta_i))| \\
& + {}^{\rho}I^{\alpha}_{0^+}|f(T,y(T))| + |\kappa|\Big]\Big\} + \frac{t^{\rho}}{\rho|\Omega|}\Big\{|B_2|\sum_{j=1}^{k}|\mu_j|{}^{\rho}I^{\alpha}_{0^+}|f(\xi_j,y(\xi_j))| \\
& + |A_2|\Big[\sum_{i=1}^{m}|\sigma_i|\,{}^{\rho}I^{\alpha+\beta}_{0^+}|f(\eta_i,y(\eta_i))| + {}^{\rho}I^{\alpha}_{0^+}|f(T,y(T))| + |\kappa|\Big]\Big\} \\
\leq\ & {}^{\rho}I^{\alpha}_{0^+}\phi(T)\psi(\|y\|) + \frac{1}{|\Omega|}\Bigg\{|B_1|\sum_{j=1}^{k}|\mu_j|{}^{\rho}I^{\alpha}_{0^+}\phi(\xi_j)\psi(\|y\|) + |A_1|\Big[\sum_{i=1}^{m}|\sigma_i|\,{}^{\rho}I^{\alpha+\beta}_{0^+}\phi(\eta_i)\psi(\|y\|) \\
& + {}^{\rho}I^{\alpha}_{0^+}\phi(T)\psi(\|y\|) + |\kappa|\Big]\Big\} + \frac{T^{\rho}}{\rho|\Omega|}\Big\{|B_2|\sum_{j=1}^{k}|\mu_j|{}^{\rho}I^{\alpha}_{0^+}\phi(\xi_j)\psi(\|y\|) \\
& + |A_2|\Big[\sum_{i=1}^{m}|\sigma_i|\,{}^{\rho}I^{\alpha+\beta}_{0^+}\phi(\eta_i)\psi(\|y\|) + {}^{\rho}I^{\alpha}_{0^+}\phi(T)\psi(\|y\|) + |\kappa|\Big]\Big\} \\
\leq\ & \psi(\|y\|)\Big({}^{\rho}I^{\alpha}_{0^+}\phi(T) + \frac{1}{|\Omega|}\Big\{|B_1|\sum_{j=1}^{k}|\mu_j|{}^{\rho}I^{\alpha}_{0^+}\phi(\xi_j) + |A_1|\Big[\sum_{i=1}^{m}|\sigma_i|\,{}^{\rho}I^{\alpha+\beta}_{0^+}\phi(\eta_i) \\
& + {}^{\rho}I^{\alpha}_{0^+}\phi(T)\Big]\Big\} + \frac{T^{\rho}}{\rho|\Omega|}\Big\{|B_2|\sum_{j=1}^{k}|\mu_j|{}^{\rho}I^{\alpha}_{0^+}\phi(\xi_j) + |A_2|\Big[\sum_{i=1}^{m}|\sigma_i|\,{}^{\rho}I^{\alpha+\beta}_{0^+}\phi(\eta_i) + {}^{\rho}I^{\alpha}_{0^+}\phi(T)\Big]\Big\}\Big) \\
& + \frac{|\kappa|(\rho|A_1| + T^{\rho}|A_2|)}{\rho|\Omega|}. \\
\leq\ & \psi(\|r\|)\Big({}^{\rho}I^{\alpha}_{0^+}\phi(T) + \frac{1}{|\Omega|}\Big\{|B_1|\sum_{j=1}^{k}|\mu_j|{}^{\rho}I^{\alpha}_{0^+}\phi(\xi_j) + |A_1|\Big[\sum_{i=1}^{m}|\sigma_i|\,{}^{\rho}I^{\alpha+\beta}_{0^+}\phi(\eta_i) \\
& + {}^{\rho}I^{\alpha}_{0^+}\phi(T)\Big]\Big\} + \frac{T^{\rho}}{\rho|\Omega|}\Big\{|B_2|\sum_{j=1}^{k}|\mu_j|{}^{\rho}I^{\alpha}_{0^+}\phi(\xi_j) + |A_2|\Big[\sum_{i=1}^{m}|\sigma_i|\,{}^{\rho}I^{\alpha+\beta}_{0^+}\phi(\eta_i) + {}^{\rho}I^{\alpha}_{0^+}\phi(T)\Big]\Big\}\Big) \\
& + \frac{|\kappa|(\rho|A_1| + T^{\rho}|A_2|)}{\rho|\Omega|} := \ell.
\end{aligned}$$

Step 3: $\mathcal{G}$ maps bounded sets into equicontinuous sets of C.

Let B_r be a bounded set of C as in Step 2, Then, for $t_1, t_2 \in (0, T]$ with $t_1 < t_2$, and $y \in B_r$, we have

$$\begin{aligned}
& |\mathcal{G}(y)(t_2) - \mathcal{G}(y)(t_1)| \\
\leq\ & \left|{}^{\rho}I^{\alpha}_{0^+}f(t_2,y(t_2)) - {}^{\rho}I^{\alpha}_{0^+}f(t_1,y(t_1))\right| + \frac{|t_2^{\rho} - t_1^{\rho}|}{\rho|\Omega|}\Big\{|B_2|\sum_{j=1}^{k}|\mu_j|{}^{\rho}I^{\alpha}_{0^+}|f(\xi_j,y(\xi_j))| \\
& + |A_2|\Big[\sum_{i=1}^{m}|\sigma_i|\,{}^{\rho}I^{\alpha+\beta}_{0^+}|f(\eta_i,y(\eta_i))| + {}^{\rho}I^{\alpha}_{0^+}|f(T,y(T))| + |\kappa|\Big]\Big\} \\
\leq\ & \frac{\rho^{1-\alpha}\psi(r)}{\Gamma(\alpha)}\left|\int_0^{t_1}\Big[\frac{s^{\rho-1}}{(t_2^{\rho} - s^{\rho})^{1-\alpha}} - \frac{s^{\rho-1}}{(t_1^{\rho} - s^{\rho})^{1-\alpha}}\Big]\phi(s)ds + \int_{t_1}^{t_2}\frac{s^{\rho-1}}{(t_2^{\rho} - s^{\rho})^{1-\alpha}}\phi(s)ds\right| \\
& + \frac{|t_2^{\rho} - t_1^{\rho}|}{\rho|\Omega|}\Big\{\psi(r)\Big(|B_2|\sum_{j=1}^{k}|\mu_j|{}^{\rho}I^{\alpha}_{0^+}\phi(\xi_j) + |A_2|\Big[\sum_{i=1}^{m}|\sigma_i|\,{}^{\rho}I^{\alpha+\beta}_{0^+}\phi(\eta_i) + {}^{\rho}I^{\alpha}_{0^+}\phi(T)\Big]\Big) + |A_2||\kappa|\Big\} \\
\to\ & 0 \text{ as } t_2 \to t_1,
\end{aligned}$$

independently of $y \in B_r$. In view of steps 1–3, it follows by the Arzelá–Ascoli theorem that the operator $\mathcal{G} : C \longrightarrow C$ is completely continuous.

Step 4: We show that there exists an open set $V \subseteq C$ with $y \neq \lambda\mathcal{G}(y)$ for $\lambda \in (0,1)$ and $y \in \partial V$.

Let $y \in \mathcal{C}$ be a solution of $y = \lambda \mathcal{G} y$ for $\lambda \in [0,1]$. Then, for $t \in [0,T]$, we have

$$\begin{aligned}
|y(t)| &= |\lambda(\mathcal{G}y)(t)| \\
&\le {}^{\rho}I_{0+}^{\alpha}|f(t,y(t))| + \frac{1}{|\Omega|}\Big\{|B_1|\sum_{j=1}^{k}|\mu_j|^{\rho}I_{0+}^{\alpha}|f(\xi_j,y(\xi_j))| + |A_1|\Big[\sum_{i=1}^{m}|\sigma_i|\,{}^{\rho}I_{0+}^{\alpha+\beta}|f(\eta_i,y(\eta_i))| \\
&\quad + {}^{\rho}I_{0+}^{\alpha}|f(T,y(T))| + |\kappa|\Big]\Big\} + \frac{t^{\rho}}{\rho|\Omega|}\Big\{|B_2|\sum_{j=1}^{k}|\mu_j|^{\rho}I_{0+}^{\alpha}|f(\xi_j,y(\xi_j))| \\
&\quad + |A_2|\Big[\sum_{i=1}^{m}|\sigma_i|\,{}^{\rho}I_{0+}^{\alpha+\beta}|f(\eta_i,y(\eta_i))| + {}^{\rho}I_{0+}^{\alpha}|f(T,y(T))| + |\kappa|\Big]\Big\} \\
&\le \psi(\|y\|)\Big({}^{\rho}I_{0+}^{\alpha}|\phi(T) + \frac{1}{|\Omega|}\Big\{|B_1|\sum_{j=1}^{k}|\mu_j|^{\rho}I_{0+}^{\alpha}\phi(\xi_j) + |A_1|\Big[\sum_{i=1}^{m}|\sigma_i|\,{}^{\rho}I_{0+}^{\alpha+\beta}\phi(\eta_i) \\
&\quad + {}^{\rho}I_{0+}^{\alpha}\phi(T)\Big]\Big\} + \frac{T^{\rho}}{\rho|\Omega|}\Big\{|B_2|\sum_{j=1}^{k}|\mu_j|^{\rho}I_{0+}^{\alpha}\phi(\xi_j) + |A_2|\Big[\sum_{i=1}^{m}|\sigma_i|\,{}^{\rho}I_{0+}^{\alpha+\beta}\phi(\eta_i) + {}^{\rho}I_{0+}^{\alpha}\phi(T)\Big]\Big\}\Big) \\
&\quad + \frac{|\kappa|(\rho|A_1| + T^{\rho}|A_2|)}{\rho|\Omega|},
\end{aligned}$$

which, on taking the norm for $t \in J$, implies that

$$\frac{\|y\|}{\psi(\|y\|)\Lambda_1 + \dfrac{|\kappa|(\rho|A_1| + T^{\rho}|A_2|)}{\rho|\Omega|}} \le 1.$$

By the assumption (H_2), we can find a positive number $\mathcal{M}$ such that $\|y\| \neq \mathcal{M}$. Introduce $V = \{y \in \mathcal{C} : \|y\| < \mathcal{M}\}$ and observe that the operator $\mathcal{G} : \overline{V} \to \mathcal{C}$ is continuous and completely continuous. By the definition of V, there does not exist any $y \in \partial V$ satisfying $y = \lambda\mathcal{G}(y)$ for some $\lambda \in (0,1)$. Hence, we deduce by the nonlinear alternative of Leray–Schauder type [40] that $\mathcal{G}$ has a fixed point $y \in \overline{V}$ that is indeed a solution of the problem (1). This completes the proof. □

In the next result, we prove the existence of solutions for problem (1) by applying Krasnoselskii's fixed point theorem [41].

Theorem 2. *Let $f : [0,T] \times \mathbb{R} \to \mathbb{R}$ be a continuous function such that the following assumptions hold:*

(H_3) $|f(t,x) - f(t,y)| \le L\|x - y\|, \quad \forall t \in [0,T],\ L > 0,\ x,y \in \mathbb{R};$
(H_4) $|f(t,y)| \le \Phi(t),\ \forall (t,y) \in [0,T] \times \mathbb{R}$, *and* $\Phi \in C([0,T],\mathbb{R}^+)$.

Then, problem (1) has at least one solution on $[0,T]$, provided that

$$L\left(\frac{T^{\rho}}{\rho|\Omega|}\left\{|B_2|\sum_{j=1}^{k}|\mu_j|\frac{\xi_j^{\rho\alpha}}{\rho^{\alpha}\Gamma(\alpha+1)} + |A_2|\left[\sum_{i=1}^{m}|\sigma_i|\frac{\eta_i^{\rho(\alpha+\beta)}}{\rho^{\alpha+\beta}\Gamma(\alpha+\beta+1)} + \frac{T^{\rho\alpha}}{\rho^{\alpha}\Gamma(\alpha+1)}\right]\right\}\right) < 1. \tag{28}$$

Proof. Let us fix $\bar{r} \ge \|\Phi\|\Lambda + \frac{|\kappa|(\rho|A_1| + T^{\rho}|A_2|)}{\rho|\Omega|}$, where $\|\Phi\| = \sup_{t\in J}|\Phi(t)|$ and consider $B_{\bar{r}} = \{y \in \mathcal{C} : \|y\| \le \bar{r}\}$. Let us split the operator $\mathcal{G} : \mathcal{C} \to \mathcal{C}$ defined by (25) on $B_{\bar{r}}$ as $\mathcal{G} = \mathcal{A} + \mathcal{B}$, where $\mathcal{A}$ and $\mathcal{B}$ are given by

$$\mathcal{A}(t) = {}^{\rho}I_{0+}^{\alpha}f(t,y(t)) + \frac{1}{\Omega}\Big\{-B_1\sum_{j=1}^{k}\mu_j{}^{\rho}I_{0+}^{\alpha}f(\xi_j,y(\xi_j)) + A_1\Big[\sum_{i=1}^{m}\sigma_i\,{}^{\rho}I_{0+}^{\alpha+\beta}f(\eta_i,y(\eta_i)) - {}^{\rho}I_{0+}^{\alpha}f(T,y(T)) + \kappa\Big]\Big\},$$

and

$$\mathcal{B}(t) = \frac{t^{\rho}}{\rho\Omega}\Big\{B_2\sum_{j=1}^{k}\mu_j{}^{\rho}I_{0+}^{\alpha}f(\xi_j,y(\xi_j)) + A_2\Big[\sum_{i=1}^{m}\sigma_i\,{}^{\rho}I_{0+}^{\alpha+\beta}f(\eta_i,y(\eta_j)) - {}^{\rho}I_{0+}^{\alpha}f(T,y(t)) + \kappa\Big]\Big\}.$$

For $x, y \in B_{\bar{r}}$, we find that

$$\begin{aligned}
\|\mathcal{A}x+\mathcal{B}y\| &\le \sup_{t\in J}\Bigg\{{}^{\rho}I_{0^+}^{\alpha}|f(t,x(t))|+\frac{1}{\Omega}\Big\{|B_1|\sum_{j=1}^{k}|\mu_j|^{\rho}I_{0^+}^{\alpha}f(\xi_j,x(\xi_j))+A_1\Big[\sum_{i=1}^{m}|\sigma_i|\,{}^{\rho}I_{0^+}^{\alpha+\beta}|f(\eta_i,x(\eta_i))| \\
&\quad +{}^{\rho}I_{0^+}^{\alpha}|f(T,x(T))|+|\kappa|\Big]\Big\}+\frac{t^{\rho}}{\rho|\Omega|}\Big\{|B_2|\sum_{j=1}^{k}|\mu_j|^{\rho}I_{0^+}^{\alpha}|f(\xi_j,y(\xi_j))| \\
&\quad +|A_2|\Big[\sum_{i=1}^{m}|\sigma_i|\,{}^{\rho}I_{0^+}^{\alpha+\beta}|f(\eta_i,y(\eta_j))|+{}^{\rho}I_{0^+}^{\alpha}|f(T,y(T))|+|\kappa|\Big]\Big\}\Bigg\} \\
&\le \|\Phi\|\Bigg\{\frac{T^{\rho\alpha}}{\rho^{\alpha}\Gamma(\alpha+1)}+\frac{1}{|\Omega|}\Big\{|B_1|\sum_{j=1}^{k}|\mu_j|\frac{\xi_j^{\rho\alpha}}{\rho^{\alpha}\Gamma(\alpha+1)} \\
&\quad +|A_1|\Big[\sum_{i=1}^{m}|\sigma_i|\frac{\eta_i^{\rho(\alpha+\beta)}}{\rho^{\alpha+\beta}\Gamma(\alpha+\beta+1)}+\frac{T^{\rho\alpha}}{\rho^{\alpha}\Gamma(\alpha+1)}\Big]\Big\}+\frac{T^{\rho}}{\rho|\Omega|}\Big\{|B_2|\sum_{j=1}^{k}|\mu_j|\frac{\xi_j^{\rho\alpha}}{\rho^{\alpha}\Gamma(\alpha+1)} \\
&\quad +|A_2|\Big[\sum_{i=1}^{m}|\sigma_i|\frac{\eta_i^{\rho(\alpha+\beta)}}{\rho^{\alpha+\beta}\Gamma(\alpha+\beta+1)}+\frac{T^{\rho\alpha}}{\rho^{\alpha}\Gamma(\alpha+1)}\Big]\Big\}\Bigg\}+\frac{|\kappa|(\rho|A_1|+T^{\rho}|A_2|)}{\rho|\Omega|} \\
&\le \|\Phi\|\Lambda+\frac{|\kappa|(\rho|A_1|+T^{\rho}|A_2|)}{\rho|\Omega|}<\bar{r}.
\end{aligned}$$

Thus, $\mathcal{A}x+\mathcal{B}y \in B_{\bar{r}}$. Now, for $x, y \in B_{\bar{r}}$ and for each $t \in J$, we obtain

$$\begin{aligned}
\|\mathcal{B}x-\mathcal{B}y\| &\le \sup_{t\in J}\Big\{\frac{t^{\rho}}{\rho|\Omega|}\Big\{|B_2|\sum_{j=1}^{k}|\mu_j|^{\rho}I_{0^+}^{\alpha}|f(\xi_j,x(\xi_j))-f(\xi_j,y(\xi_j))| \\
&\quad +|A_2|\Big[\sum_{i=1}^{m}|\sigma_i|\,{}^{\rho}I_{0^+}^{\alpha+\beta}|f(\eta_i,x(\eta_i))-f(\eta_i,y(\eta_i))|+{}^{\rho}I_{0^+}^{\alpha}|f(T,x(T))-f(T,y(T))|\Big]\Big\}\Big\} \\
&\le L\Big(\frac{T^{\rho}}{\rho|\Omega|}\Big\{|B_2|\sum_{j=1}^{k}|\mu_j|\frac{\xi_j^{\rho\alpha}}{\rho^{\alpha}\Gamma(\alpha+1)}+|A_2|\Big[\sum_{i=1}^{m}|\sigma_i|\frac{\eta_i^{\rho(\alpha+\beta)}}{\rho^{\alpha+\beta}\Gamma(\alpha+\beta+1)}+\frac{T^{\rho\alpha}}{\rho^{\alpha}\Gamma(\alpha+1)}\Big]\Big\}\Big)\|x-y\|,
\end{aligned}$$

which, together with condition (28), implies that $\mathcal{B}$ is a contraction. Continuity of f implies that the operator $\mathcal{A}$ is continuous. In addition, $\mathcal{A}$ is uniformly bounded on $B_{\bar{r}}$ as

$$\begin{aligned}
\|\mathcal{A}y\| &\le \|\Phi\|\Big(\frac{T^{\rho\alpha}}{\rho^{\alpha}\Gamma(\alpha+1)}+\frac{1}{|\Omega|}\Big\{|B_1|\sum_{j=1}^{k}|\mu_j|\frac{\xi_j^{\rho\alpha}}{\rho^{\alpha}\Gamma(\alpha+1)}+|A_1|\Big[\sum_{i=1}^{m}|\sigma_i|\frac{\eta_i^{\rho(\alpha+\beta)}}{\rho^{\alpha+\beta}\Gamma(\alpha+\beta+1)} \\
&\quad +\frac{T^{\rho\alpha}}{\rho^{\alpha}\Gamma(\alpha+1)}\Big]\Big\}\Big)+\frac{|A_1||\kappa|}{|\Omega|}.
\end{aligned}$$

In order to show the compactness of the operator $\mathcal{A}$, let $\sup_{(t,y)\in J\times B_{\bar{r}}}|f(t,y)| = \bar{f} < \infty$. Consequently, for $t_1, t_2 \in J$, $t_1 < t_2$, we have

$$\begin{aligned}
\|(\mathcal{A}y)(t_2)-(\mathcal{A}y)(t_1)\| &\le \Big\|\frac{\rho^{1-\alpha}}{\Gamma(\alpha)}\Big[\int_0^{t_1}s^{\rho-1}[(t_2^{\rho}-s^{\rho})^{\alpha-1}-(t_1^{\rho}-s^{\rho})^{\alpha-1}]f(s,y(s))ds \\
&\quad +\int_{t_1}^{t_2}s^{\rho-1}(t_2^{\rho}-s^{\rho})^{\alpha-1}f(s,y(s))ds\Big]\Big\| \\
&\le \frac{\bar{f}}{\rho^{\alpha}\Gamma(\alpha+1)}\Big\{2(t_2^{\rho}-t_1^{\rho})^{\alpha}+|t_2^{\rho\alpha}-t_1^{\rho\alpha}|\Big\}.
\end{aligned}$$

As the right-hand side of the above inequality tends to zero independently of $y \in B_{\bar{r}}$ when $t_2 \to t_1$, therefore $\mathcal{A}$ is equicontinuous. Thus, $\mathcal{A}$ is relatively compact on $B_{\bar{r}}$. Hence, the conclusion of Arzelá-Ascoli theorem applies and that $\mathcal{A}$ is compact on $B_{\bar{r}}$. Since all the conditions of Krasnoselskii's fixed point theorem hold true, it follows by Krasnoselskii's fixed point theorem that problem (1) has at least one solution on J. □

Our final result in this section is concerned with the uniqueness of solutions for problem (1) and is based on a Banach fixed point theorem.

Theorem 3. *Assume that $f:[0,T]\times\mathbb{R}\to\mathbb{R}$ is continuous and the condition (H_3) holds. Then, problem (1) has a unique solution on J if*

$$L\Lambda<1, \tag{29}$$

where Λ is defined by (26).

Proof. In view of the condition (29), consider a set $B_{\tilde r}=\{y\in\mathcal{C}:\|y\|\le\tilde r\}$, where

$$\tilde r>\frac{f_0\Lambda+\frac{|\kappa|(\rho|A_1|+T^\rho|A_2|)}{\rho|\Omega|}}{1-L\Lambda},\quad \sup_{t\in[0,T]}|f(t,0)|=f_0$$

and show that $\mathcal{G}B_{\tilde r}\subset B_{\tilde r}$ ($\mathcal{G}$ is defined by (25)). For $y\in B_{\tilde r}$, using (H_3), we get

$$\begin{aligned}
&|\mathcal{G}(y)(t)|\\
\le\ & {}^\rho I^\alpha_{0^+}[|f(t,y(t))-f(t,0)|+|f(t,0)|]+\frac{1}{|\Omega|}\Big\{|B_1|\sum_{j=1}^k|\mu_j|{}^\rho I^\alpha_{0^+}[|f(\xi_j,y(\xi_j))-f(\xi_j,0)|+|f(\xi_j,0)|]\\
&+A_1\Big[\sum_{i=1}^m|\sigma_i|\,{}^\rho I^{\alpha+\beta}_{0^+}[|f(\eta_i,y(\eta_i))-f(\eta_i,0)|+|f(\eta_i,0)|]+{}^\rho I^\alpha_{0^+}[|f(T,y(T))-f(T,0)|+|f(T,0)|]\Big]\\
&+|\kappa|\Big\}+\frac{t^\rho}{\rho|\Omega|}\Big\{|B_2|\sum_{j=1}^k\mu_j{}^\rho I^\alpha_{0^+}[|f(\xi_j,y(\xi_j))-f(\xi_j,0)|+|f(\xi_j,0)|]\\
&+|A_2|\Big[\sum_{i=1}^m\sigma_i\,{}^\rho I^{\alpha+\beta}_{0^+}[|f(\eta_i,y(\eta_i))-f(\eta_i,0)|+|f(\eta_i,0)|]+{}^\rho I^\alpha_{0^+}[|f(T,y(T))-f(T,0)|+|f(T,0)|]\\
&+|\kappa|\Big]\Big\}\\
\le\ & (L\|y\|+f_0)\Bigg[\frac{T^{\rho\alpha}}{\rho^\alpha\Gamma(\alpha+1)}+\frac{1}{|\Omega|}\Big\{|B_1|\sum_{j=1}^k|\mu_j|\frac{\xi_j^{\rho\alpha}}{\rho^\alpha\Gamma(\alpha+1)}+|A_1|\Big[\sum_{i=1}^m|\sigma_i|\frac{\eta_i^{\rho(\alpha+\beta)}}{\rho^{\alpha+\beta}\Gamma(\alpha+\beta+1)}\\
&+\frac{T^{\rho\alpha}}{\rho^\alpha\Gamma(\alpha+1)}\Big]\Big\}+\frac{T^\rho}{\rho|\Omega|}\Big\{|B_2|\sum_{j=1}^k|\mu_j|\frac{\xi_j^{\rho\alpha}}{\rho^\alpha\Gamma(\alpha+1)}+|A_2|\Big[\sum_{i=1}^m|\sigma_i|\frac{\eta_i^{\rho(\alpha+\beta)}}{\rho^{\alpha+\beta}\Gamma(\alpha+\beta+1)}\\
&+\frac{T^{\rho\alpha}}{\rho^\alpha\Gamma(\alpha+1)}\Big]\Big\}\Bigg]+\frac{|\kappa|(\rho|A_1|+T^\rho|A_2|)}{\rho|\Omega|}\\
\le\ & (L\tilde r+f_0)\Lambda+\frac{|\kappa|(\rho|A_1|+T^\rho|A_2|)}{\rho|\Omega|}\le\tilde r,
\end{aligned}$$

which, on taking the norm for $t\in J$, yields $\|\mathcal{G}(y)\|\le\tilde r$. This shows that $\mathcal{G}$ maps $B_{\tilde r}$ into itself. Now, we establish that the operator $\mathcal{G}$ is a contraction. For that, let $y,z\in\mathcal{C}$. Then, we get

$$\begin{aligned}
|\mathcal{G}(y)(t)-\mathcal{G}(z)(t)|\ \le\ & {}^\rho I^\alpha_{0^+}|f(t,y(t))-f(t,z(t))|+\frac{1}{|\Omega|}\Big\{|B_1|\sum_{j=1}^k|\mu_j|{}^\rho I^\alpha_{0^+}|f(\xi_j,y(\xi_j))-f(\xi_j,z(\xi_j))|\\
&+A_1\Big[\sum_{i=1}^m|\sigma_i|\,{}^\rho I^{\alpha+\beta}_{0^+}|f(\eta_i,y(\eta_i))-f(\eta_i,z(\eta_i))|+{}^\rho I^\alpha_{0^+}|f(T,y(T))-f(T,z(T))|\Big]\Big\}\\
&+\frac{t^\rho}{\rho|\Omega|}\Big\{|B_2|\sum_{j=1}^k|\mu_j|{}^\rho I^\alpha_{0^+}|f(\xi_j,y(\xi_j))-f(\xi_j,z(\xi_j))|\\
&+|A_2|\Big[\sum_{i=1}^m|\sigma_i|\,{}^\rho I^{\alpha+\beta}_{0^+}|f(\eta_i,y(\eta_i))-f(\eta_i,z(\eta_i))|+{}^\rho I^\alpha_{0^+}|f(T,y(T))-f(T,z(T))|\Big]\Big\}
\end{aligned}$$

$$\leq \quad L\Lambda\|y-z\|.$$

Consequently, we obtain

$$\|\mathcal{G}(y)-\mathcal{G}(z)\| \leq L\Lambda\|y-z\|,$$

which implies that $\mathcal{G}$ is a contraction by the condition (29). Hence, $\mathcal{G}$ has a unique fixed point by a Banach fixed point theorem. Equivalently, we deduce that problem (1) has a unique solution on J. The proof is completed. □

Example 1. *Consider the following boundary value problem*

$$\begin{cases} {}^{1/3}{}_{c}D^{7/5}_{0^+}y(t) = f(t,y(t)),\ t \in [0,2], \\ y(2) = 1/2\ {}^{1/3}I^{3/5}y(1/4) + 2/3\ {}^{1/3}I^{3/5}y(3/4) + 2/9, \\ \delta y(0) = 2/5\, y(1) + 4/5\, y(3/2), \end{cases} \tag{30}$$

where $\rho = 1/3,\ \alpha = 7/5,\ \sigma_1 = 1/2,\ \sigma_2 = 2/3,\ \beta = 3/5,\ \eta_1 = 1/4,\ \eta_2 = 3/4,\ \mu_1 = 2/5,\ \mu_2 = 4/5,\ \kappa = 2/9,\ \xi_1 = 1,\ \xi_2 = 3/2,\ T = 2$ *and* $f(t,y(t))$ *will be fixed later.*

Using the given data, we find that $|A_1| = 2.947314182$, $|A_2| = 1.2$, $|B_1| = 0.491608875$, $|B_2| = 1.181571585$, $|\Omega| = 4.072393340$, and $\Lambda = 27.12293267$ (A_i, B_i $(i=1,2), \Omega$ and Λ are respectively given by Equations (12), (13), (14) and (26)).

For illustrating Theorem 1, we take

$$f(t,y) = \frac{(1+t)}{60}\Big(\frac{|y|}{|y|+1} + y + \frac{1}{8}\Big). \tag{31}$$

Clearly, $f(t,x)$ is continuous and satisfies the condition (H_1) with $\phi(t) = \frac{(1+t)}{60}$, $\psi(\|y\|) = ||y|| + \frac{9}{8}$. By the condition (H_2), we find that $\mathcal{M} > 2.390158$. Thus, all conditions of Theorem 1 are satisfied and, consequently, there exists at least one solution for problem (30) with $f(t,y(t))$ given by (31) on $[0,2]$.

In order to illustrate Theorem 2, we choose

$$f(t,y) = \frac{\tan^{-1}y + e^{-t}}{4\sqrt{81+\sin t}}. \tag{32}$$

It is easy to check that $f(t,x)$ is continuous and satisfies the conditions (H_3) and (H_4) with $L = 1/36$ and $\Phi(t) = \frac{\pi + 2e^{-t}}{8\sqrt{81+\sin t}}$. In addition,

$$L\Big(\frac{T^{\rho}}{\rho|\Omega|}\Big\{|B_2|\sum_{j=1}^{k}|\mu_j|\frac{\xi_j^{\rho\alpha}}{\rho^{\alpha}\Gamma(\alpha+1)} + |A_2|\Big[\sum_{i=1}^{m}|\sigma_i|\frac{\eta_i^{\rho(\alpha+\beta)}}{\rho^{\alpha+\beta}\Gamma(\alpha+\beta+1)} + \frac{T^{\rho\alpha}}{\rho^{\alpha}\Gamma(\alpha+1)}\Big]\Big\}\Big) \approx 0.420512 < 1.$$

Thus, all of the conditions of Theorem 2 hold true. Thus, by the conclusion of Theorem 2, problem (30) has at least one solution on $[0,2]$.

With $L\Lambda \approx 0.753415 < 1$, one can note that the assumptions of Theorem 3 are also satisfied. Hence, the conclusion of Theorem 3 applies and the problem (30) with $f(t,y)$ given (32) has a unique solution on $[0,2]$.

4. Existence Results for the Problem (2)

This section is devoted to the existence of solutions for problem (2).

Definition 4. *A function $y \in C([0,T],\mathbb{R})$ possessing Liouville–Caputo-type generalized deravative of order α is said to be a solution of the boundary value problem (2) if $y(T) = \sum_{i=1}^{m} \sigma_i {}^{\rho}I_{0+}^{\beta} y(\eta_i) + \kappa$, $\delta y(0) = \sum_{j=1}^{k} \mu_j y(\xi_j)$ and there exists function $v \in L^1([0,1],\mathbb{R})$ such that $v(t) \in F(t, y(t))$ a.e. on $[0,T]$ and*

$$\begin{aligned} y(t) &= {}^{\rho}I_{0+}^{\alpha} v(t) + \frac{1}{\Omega}\Big\{ - B_1 \sum_{j=1}^{k} \mu_j {}^{\rho}I_{0+}^{\alpha} v(\xi_j) + A_1 \Big[\sum_{i=1}^{m} \sigma_i \, {}^{\rho}I_{0+}^{\alpha+\beta} v(\eta_i) - {}^{\rho}I_{0+}^{\alpha} v(T) + \kappa \Big] \Big\} \\ &+ \frac{t^{\rho}}{\rho\Omega}\Big\{ B_2 \sum_{j=1}^{k} \mu_j {}^{\rho}I_{0+}^{\alpha} v(\xi_j) + A_2 \Big[\sum_{i=1}^{m} \sigma_i \, {}^{\rho}I_{0+}^{\alpha+\beta} v(\eta_i) - {}^{\rho}I_{0+}^{\alpha} v(T) + \kappa \Big] \Big\}. \end{aligned} \tag{33}$$

4.1. The Carathéodory Case

Here, we present an existence result for problem (2) when F has convex values and is of the Carathéodory type. The main tool of our study is a nonlinear alternative of Leray–Schauder type [40].

Theorem 4. *Assume that:*

- (C_1) *$F : [0,T] \times \mathbb{R} \to \mathcal{P}_{cp,c}(\mathbb{R})$ is L^1-Carathéodory, where $\mathcal{P}_{cp,c}(\mathbb{R}) = \{Y \in \mathcal{P}(\mathbb{R}) : Y \text{ is compact and convex}\}$;*
- (C_2) *there exists a continuous nondecreasing function $\Psi : [0,\infty) \to (0,\infty)$ and a function $\Phi \in L^1([0,T],\mathbb{R}^+)$ such that*

$$\|F(t,y)\|_{\mathcal{P}} := \sup\{|x| : x \in F(t,y)\} \le \Phi(t)\Psi(\|y\|) \text{ for each } (t,y) \in [0,T] \times \mathbb{R};$$

- (C_3) *there exists a constant $W > 0$ such that*

$$\frac{\|W\|}{\Psi(\|W\|)\Lambda_2 + \dfrac{|\kappa|(\rho|A_1| + T^{\rho}|A_2|)}{\rho|\Omega|}} > 1,$$

where

$$\begin{aligned} \Lambda_2 &= {}^{\rho}I_{0+}^{\alpha} \Phi(T) + \frac{1}{|\Omega|}\Big\{ |B_1| \sum_{j=1}^{k} |\mu_j| {}^{\rho}I_{0+}^{\alpha} \Phi(\xi_j) + |A_1| \Big[\sum_{i=1}^{m} |\sigma_i| \, {}^{\rho}I_{0+}^{\alpha+\beta} \Phi(\eta_i) + {}^{\rho}I_{0+}^{\alpha} \Phi(T) \Big] \Big\} \\ &+ \frac{T^{\rho}}{\rho|\Omega|}\Big\{ |B_2| \sum_{j=1}^{k} |\mu_j| {}^{\rho}I_{0+}^{\alpha} \Phi(\xi_j) + |A_2| \Big[\sum_{i=1}^{m} |\sigma_i| \, {}^{\rho}I_{0+}^{\alpha+\beta} \Phi(\eta_i) + {}^{\rho}I_{0+}^{\alpha} \Phi(T) \Big] \Big\}. \end{aligned} \tag{34}$$

Then, problem (2) has at least one solution on $[0,T]$.

Proof. In order to convert problem (2) into a fixed point problem, we introduce an operator $\mathcal{N} : \mathcal{C} \longrightarrow \mathcal{P}(\mathcal{C})$ by

$$\mathcal{N}(y) = \{h \in \mathcal{C} : h(t) = \mathcal{F}(y)(t)\}, \tag{35}$$

where

$$\begin{aligned} \mathcal{F}(y)(t) = \; & {}^{\rho}I_{0+}^{\alpha} v(t) + \frac{1}{\Omega}\Big\{ - B_1 \sum_{j=1}^{k} \mu_j {}^{\rho}I_{0+}^{\alpha} v(\xi_j) + A_1 \Big[\sum_{i=1}^{m} \sigma_i \, {}^{\rho}I_{0+}^{\alpha+\beta} v(\eta_i) - {}^{\rho}I_{0+}^{\alpha} v(T) + \kappa \Big] \Big\} \\ & + \tfrac{t^{\rho}}{\rho\Omega}\Big\{ B_2 \textstyle\sum_{j=1}^{k} \mu_j {}^{\rho}I_{0+}^{\alpha} v(\xi_j) + A_2 \Big[\sum_{i=1}^{m} \sigma_i \, {}^{\rho}I_{0+}^{\alpha+\beta} v(\eta_i) - {}^{\rho}I_{0+}^{\alpha} v(T) + \kappa \Big] \Big\}, \end{aligned}$$

for $v \in S_{F,y}$. Obviously, the fixed points of the operator $\mathcal{N}$ correspond to solutions of the problem (2).

It will be shown in several steps that the operator $\mathcal{N}$ satisfies the assumptions of the Leray–Schauder nonlinear alternative [40].

Step 1. $\mathcal{N}(y)$ is convex for each $y \in \mathcal{C}$.

This step is obvious since $S_{F,y}$ is convex (F has convex values).

Step 2. $\mathcal{N}$ maps bounded sets (balls) into bounded sets in $\mathcal{C}$.

For a positive number R, let $B_R = \{y \in \mathcal{C} : \|y\| \leq R\}$ be a bounded ball in $\mathcal{C}$. Then, for each $h \in \mathcal{N}(y)$, $y \in B_R$, there exists $v \in S_{F,y}$ such that

$$\begin{aligned} h(t) = & \; {}^{\rho}I^{\alpha}_{0^+}v(t) + \frac{1}{\Omega}\Big\{ - B_1 \sum_{j=1}^{k} \mu_j {}^{\rho}I^{\alpha}_{0^+}v(\xi_j) + A_1\Big[\sum_{i=1}^{m} \sigma_i \, {}^{\rho}I^{\alpha+\beta}_{0^+}v(\eta_i) - {}^{\rho}I^{\alpha}_{0^+}v(T) + \kappa\Big]\Big\} \\ & + \tfrac{t^{\rho}}{\rho\Omega}\Big\{ B_2 \textstyle\sum_{j=1}^{k} \mu_j {}^{\rho}I^{\alpha}_{0^+}v(\xi_j) + A_2\Big[\textstyle\sum_{i=1}^{m} \sigma_i \, {}^{\rho}I^{\alpha+\beta}_{0^+}v(\eta_i) - {}^{\rho}I^{\alpha}_{0^+}v(T) + \kappa\Big]\Big\}. \end{aligned}$$

Then, for $t \in [0, T]$, we have

$$\begin{aligned} |h(t)| \leq & \; {}^{\rho}I^{\alpha}_{0^+}|v(t)| + \frac{1}{|\Omega|}\Big\{ |B_1| \sum_{j=1}^{k} \mu_j {}^{\rho}I^{\alpha}_{0^+}|v(\xi_j)| + |A_1|\Big[\sum_{i=1}^{m} \sigma_i \, {}^{\rho}I^{\alpha+\beta}_{0^+}|v(\eta_i)| + {}^{\rho}I^{\alpha}_{0^+}|v(T)| + |\kappa|\Big]\Big\} \\ & + \frac{t^{\rho}}{\rho|\Omega|}\Big\{ |B_2| \sum_{j=1}^{k} \mu_j {}^{\rho}I^{\alpha}|v(\xi_j)| + |A_2|\Big[\sum_{i=1}^{m} \sigma_i \, {}^{\rho}I^{\alpha+\beta}_{0^+}|v(\eta_i)| + {}^{\rho}I^{\alpha}_{0^+}|v(T)| + |\kappa|\Big]\Big\} \\ \leq & \; \Psi(\|y\|)\Big({}^{\rho}I^{\alpha}_{0^+}\Phi(T) + \frac{1}{|\Omega|}\Big\{ |B_1| \sum_{j=1}^{k} \mu_j {}^{\rho}I^{\alpha}_{0^+}\Phi(\xi_j) + |A_1|\Big[\sum_{i=1}^{m} \sigma_i \, {}^{\rho}I^{\alpha+\beta}_{0^+}|\Phi(\eta_i)| + {}^{\rho}I^{\alpha}_{0^+}\Phi(T)\Big]\Big\} \\ & + \frac{T^{\rho}}{\rho|\Omega|}\Big\{ |B_2| \sum_{j=1}^{k} \mu_j {}^{\rho}I^{\alpha}\Phi(\xi_j) + |A_2|\Big[\sum_{i=1}^{m} \sigma_i \, {}^{\rho}I^{\alpha+\beta}_{0^+}\Phi(\eta_i) + {}^{\rho}I^{\alpha}_{0^+}\Phi(T)\Big]\Big\}\Big) \\ & + \frac{|\kappa|(\rho|A_1| + T^{\rho}|A_2|)}{\rho|\Omega|}. \end{aligned}$$

Thus,

$$\begin{aligned} \|h\| \leq & \; \Psi(R)\Big({}^{\rho}I^{\alpha}_{0^+}\Phi(T) + \frac{1}{|\Omega|}\Big\{ |B_1| \sum_{j=1}^{k} |\mu_j| {}^{\rho}I^{\alpha}_{0^+}\Phi(\xi_j) + |A_1|\Big[\sum_{i=1}^{m} |\sigma_i| \, {}^{\rho}I^{\alpha+\beta}_{0^+}\Phi(\eta_i) + {}^{\rho}I^{\alpha}_{0^+}\Phi(T)\Big]\Big\} \\ & + \frac{T^{\rho}}{\rho|\Omega|}\Big\{ |B_2| \sum_{j=1}^{k} |\mu_j| {}^{\rho}I^{\alpha}_{0^+}\Phi(\xi_j)| + |A_2|\Big[\sum_{i=1}^{m} |\sigma_i| \, {}^{\rho}I^{\alpha+\beta}_{0^+}\Phi(\eta_i) + {}^{\rho}I^{\alpha}_{0^+}\Phi(T)\Big]\Big\}\Big) \\ & + \frac{|\kappa|(\rho|A_1| + T^{\rho}|A_2|)}{\rho|\Omega|} := \tilde{\ell}. \end{aligned}$$

Step 3. $\mathcal{N}$ maps bounded sets into equicontinuous sets of $\mathcal{C}$.

Let $t_1, t_2 \in (0, T]$, $t_1 < t_2$, and let $y \in B_R$. Then,

$$\begin{aligned} & |h(t_2) - h(t_1)| \\ \leq & \; \Big| {}^{\rho}I^{\alpha}_{0^+}v(t_2) - {}^{\rho}I^{\alpha}_{0^+}v(t_1)\Big| + \frac{|t_2^{\rho} - t_1^{\rho}|}{\rho|\Omega|}\Big\{ |B_2| \sum_{j=1}^{k} |\mu_j| {}^{\rho}I^{\alpha}_{0^+}|v(\xi_j| \\ & + |A_2|\Big[\sum_{i=1}^{m} |\sigma_i| \, {}^{\rho}I^{\alpha+\beta}_{0^+}|v(\eta_i)| + {}^{\rho}I^{\alpha}_{0^+}|v(T)| + |\kappa|\Big]\Big\} \\ \leq & \; \frac{\rho^{1-\alpha}\Psi(R)}{\Gamma(\alpha)}\Big| \int_0^{t_1} \Big[\frac{s^{\rho-1}}{(t_2^{\rho} - s^{\rho})^{1-\alpha}} - \frac{s^{\rho-1}}{(t_1^{\rho} - s^{\rho})^{1-\alpha}} \Big]\Phi(s)ds + \int_{t_1}^{t_2} \frac{s^{\rho-1}}{(t_2^{\rho} - s^{\rho})^{1-\alpha}}\Phi(s)ds \Big| \\ & + \frac{|t_2^{\rho} - t_1^{\rho}|}{\rho|\Omega|}\Big\{ \Psi(R)\Big(|B_2| \sum_{j=1}^{k} |\mu_j| {}^{\rho}I^{\alpha}_{0^+}\Phi(\xi_j) + |A_2|\Big[\sum_{i=1}^{m} |\sigma_i| \, {}^{\rho}I^{\alpha+\beta}_{0^+}\Phi(\eta_i) + {}^{\rho}I^{\alpha}_{0^+}\Phi(T)\Big]\Big) + |A_2||\kappa|\Big\} \\ \to & \; 0 \text{ as } t_2 - t_1 \to 0, \end{aligned}$$

independently of $y \in B_R$. In view of the foregoing steps, the Arzelá–Ascoli theorem applies and that the operator $\mathcal{N} : \mathcal{C} \to \mathcal{P}(\mathcal{C})$ is completely continuous.

In our next step, we show that $\mathcal{N}$ is u.s.c. We just need to establish that $\mathcal{N}$ has a closed graph as it is already shown to be completely continuous [42] (Proposition 1.2).

Step 4. $\mathcal{N}$ has a closed graph.

Let $y_n \to y_*, h_n \in \mathcal{N}(y_n)$ and $h_n \to h_*$. Then, we have to show that $h_* \in \mathcal{N}(y_*)$. Associated with $h_n \in \mathcal{N}(y_n)$, we have that $v_n \in S_{F,y_n}$ such that for each $t \in [0,T]$,

$$\begin{aligned} h_n(t) &= {}^\rho I^\alpha v_n(t) + \frac{1}{\Omega}\Big\{ - B_1 \sum_{j=1}^k \mu_j {}^\rho I^\alpha v_n(\xi_j) + A_1\Big[\sum_{i=1}^m \sigma_i \, {}^\rho I_{0^+}^{\alpha+\beta} v_n(\eta_i) - {}^\rho I^\alpha v_n(T) + \kappa\Big]\Big\} \\ &\quad + \frac{t^\rho}{\rho\Omega}\Big\{ B_2 \sum_{j=1}^k \mu_j {}^\rho I_{0^+}^\alpha v_n(\xi_j) + A_2\Big[\sum_{i=1}^m \sigma_i \, {}^\rho I_{0^+}^{\alpha+\beta} v_n(\eta_i) - {}^\rho I_{0^+}^\alpha v_n(T) + \kappa\Big]\Big\}. \end{aligned}$$

Thus, it is sufficient to establish that there exists $v_* \in S_{F,y_*}$ such that for each $t \in [0,T]$,

$$\begin{aligned} h_*(t) &= {}^\rho I_{0^+}^\alpha v_*(t) + \frac{1}{\Omega}\Big\{ - B_1 \sum_{j=1}^k \mu_j {}^\rho I^\alpha v_*(\xi_j) + A_1\Big[\sum_{i=1}^m \sigma_i \, {}^\rho I_{0^+}^{\alpha+\beta} v_*(\eta_i) - {}^\rho I^\alpha v_*(T) + \kappa\Big]\Big\} \\ &\quad + \frac{t^\rho}{\rho\Omega}\Big\{ B_2 \sum_{j=1}^k \mu_j {}^\rho I_{0^+}^\alpha v_*(\xi_j) + A_2\Big[\sum_{i=1}^m \sigma_i \, {}^\rho I_{0^+}^{\alpha+\beta} v_*(\eta_i) - {}^\rho I_{0^+}^\alpha v_*(T) + \kappa\Big]\Big\}. \end{aligned}$$

Next, we introduce the linear operator $\Theta : L^1([0,T],\mathbb{R}) \to \mathcal{C}$ as

$$\begin{aligned} v \mapsto \Theta v(t) &= {}^\rho I_{0^+}^\alpha v(t) + \frac{1}{\Omega}\Big\{ - B_1 \sum_{j=1}^k \mu_j {}^\rho I^\alpha v(\xi_j) + A_1\Big[\sum_{i=1}^m \sigma_i \, {}^\rho I_{0^+}^{\alpha+\beta} v(\eta_i) - {}^\rho I_{0^+}^\alpha v(T) + \kappa\Big]\Big\} \\ &\quad + \frac{t^\rho}{\rho\Omega}\Big\{ B_2 \sum_{j=1}^k \mu_j {}^\rho I^\alpha v(\xi_j) + A_2\Big[\sum_{i=1}^m \sigma_i \, {}^\rho I_{0^+}^{\alpha+\beta} v(\eta_i) - {}^\rho I_{0^+}^\alpha v(T) + \kappa\Big]\Big\}. \end{aligned}$$

Observe that $\|h_n(t) - h_*(t)\| \to 0$ as $n \to \infty$. Therefore, by a closed graph result obtained in [43], $\Theta \circ S_F$ is a closed graph operator. Moreover, we have that $h_n(t) \in \Theta(S_{F,y_n})$. As $y_n \to y_*$, we have that

$$\begin{aligned} h_*(t) &= {}^\rho I_{0^+}^\alpha v_*(t) + \frac{1}{\Omega}\Big\{ - B_1 \sum_{j=1}^k \mu_j {}^\rho I^\alpha v_*(\xi_j) + A_1\Big[\sum_{i=1}^m \sigma_i \, {}^\rho I_{0^+}^{\alpha+\beta} v_*(\eta_i) - {}^\rho I_{0^+}^\alpha v_*(T) + \kappa\Big]\Big\} \\ &\quad + \frac{t^\rho}{\rho\Omega}\Big\{ B_2 \sum_{j=1}^k \mu_j {}^\rho I^\alpha v_*(\xi_j) + A_2\Big[\sum_{i=1}^m \sigma_i \, {}^\rho I_{0^+}^{\alpha+\beta} v_*(\eta_i) - {}^\rho I_{0^+}^\alpha v_*(T) + \kappa\Big]\Big\} \end{aligned}$$

for some $v_* \in S_{F,y_*}$.

Step 5. There exists an open set $\mathcal{U} \subseteq C([0,T],\mathbb{R})$ with $y \notin \lambda\mathcal{N}(y)$ for any $\lambda \in (0,1)$ and all $y \in \partial\mathcal{U}$.

Let $\lambda \in (0,1)$ and $y \in \lambda\mathcal{N}(y)$. Then, we can find $v \in L^1([0,T],\mathbb{R})$ and $v \in S_{F,y}$ such that, for $t \in [0,T]$, we have

$$\begin{aligned} y(t) &= \lambda {}^\rho I_{0^+}^\alpha v(t) + \frac{\lambda}{\Omega}\Big\{ - B_1 \sum_{j=1}^k \mu_j {}^\rho I_{0^+}^\alpha v(\xi_j) + A_1\Big[\sum_{i=1}^m \sigma_i \, {}^\rho I_{0^+}^{\alpha+\beta} v(\eta_i) - {}^\rho I_{0^+}^\alpha v(T) + \kappa\Big]\Big\} \\ &\quad + \lambda\frac{t^\rho}{\rho\Omega}\Big\{ B_2 \sum_{j=1}^k \mu_j {}^\rho I_{0^+}^\alpha v(\xi_j) + A_2\Big[\sum_{i=1}^m \sigma_i \, {}^\rho I_{0^+}^{\alpha+\beta} v(\eta_i) - {}^\rho I_{0^+}^\alpha v(T) + \kappa\Big]\Big\}. \end{aligned}$$

As in Step 2, one can find that

$$\begin{aligned} |y(t)| &\le {}^\rho I_{0^+}^\alpha |v(t)| + \frac{1}{|\Omega|}\Big\{ |B_1| \sum_{j=1}^k |\mu_j| {}^\rho I_{0^+}^\alpha |v(\xi_j)| + |A_1|\Big[\sum_{i=1}^m \sigma_i \, {}^\rho I_{0^+}^{\alpha+\beta} |v(\eta_i)| + {}^\rho I_{0^+}^\alpha |v(T)| + |\kappa|\Big]\Big\} \\ &\quad + \frac{t^\rho}{\rho|\Omega|}\Big\{ |B_2| \sum_{j=1}^k |\mu_j| {}^\rho I_{0^+}^\alpha |v(\xi_j)| + |A_2|\Big[\sum_{i=1}^m |\sigma_i| \, {}^\rho I_{0^+}^{\alpha+\beta} |v(\eta_i)| + {}^\rho I_{0^+}^\alpha |v(T)| + |\kappa|\Big]\Big\} \end{aligned}$$

$$\leq \Psi(y)\Big({}^{\rho}I^{\alpha}\Phi(T) + \frac{1}{|\Omega|}\Big\{|B_1|\sum_{j=1}^{k}|\mu_j|{}^{\rho}I^{\alpha}\Phi(\xi_j) + |A_1|\Big[\sum_{i=1}^{m}|\sigma_i|\,{}^{\rho}I_{0^+}^{\alpha+\beta}|\Phi(\eta_i)| + {}^{\rho}I^{\alpha}|\Phi(T)|\Big]\Big\}$$
$$+\frac{T^{\rho}}{\rho|\Omega|}\Big\{|B_2|\sum_{j=1}^{k}|\mu_j|{}^{\rho}I^{\alpha}\Phi(\xi_j) + |A_2|\Big[\sum_{i=1}^{m}|\sigma_i|\,{}^{\rho}I_{0^+}^{\alpha+\beta}\Phi(\eta_i) + {}^{\rho}I^{\alpha}\Phi(T)\Big]\Big\}\Big)$$
$$+\frac{|\kappa|(\rho|A_1| + T^{\rho}|A_2|)}{\rho|\Omega|},$$

which implies that

$$\frac{\|y\|}{\Psi(\|y\|)\Lambda_2 + \dfrac{|\kappa|(\rho|A_1| + T^{\rho}|A_2|)}{\rho|\Omega|}} \leq 1.$$

By the assumption (C_3), there exists W such that $\|y\| \neq W$. Let us set

$$U = \{y \in C(J, \mathbb{R}) : \|y\| < W\}.$$

Observe that the operator $\mathcal{N} : \overline{\mathcal{U}} \to \mathcal{P}(C(J,\mathbb{R}))$ is a compact multi-valued map, u.s.c. with convex closed values. From the choice of $\mathcal{U}$, there does not exist $y \in \partial U$ satisfying $y \in \lambda\mathcal{N}(y)$ for some $\lambda \in (0,1)$. In consequence, we deduce by the nonlinear alternative of Leray–Schauder type [40] that the operator $\mathcal{N}$ has a fixed point $y \in \overline{\mathcal{U}}$ that is a solution of problem (2). This completes the proof. □

4.2. The Lipschitz Case

Let (X,d) be a metric space induced from the normed space $(X; \|\cdot\|)$. Define $H_d : \mathcal{P}(X) \times \mathcal{P}(X) \to \mathbb{R} \cup \{\infty\}$ as $H_d(P,Q) = \max\{\sup_{p\in P} d(p,Q), \sup_{q\in Q} d(P,q)\}$, where $d(P,q) = \inf_{p\in P} d(p;q)$ and $d(p,Q) = \inf_{q\in Q} d(p;q)$. Then, $(\mathcal{P}_{cl,b}(X), H_d)$ is a metric space [16]. (Here, $\mathcal{P}_{cl,b}(X) = \{Y \in \mathcal{P}(X) : Y \text{ is closed and bounded}\}$).

In the following result, we apply a fixed point theorem (If $N : X \to \mathcal{P}_{cl}(X)$ is a contraction, then $FixN \neq \emptyset$, where $\mathcal{P}_{cl}(X) = \{Y \in \mathcal{P}(X) : Y \text{ is closed}\}$) due to Covitz and Nadler [44].

Theorem 5. *Let the following conditions hold:*

(C_4) $F : [0,T] \times \mathbb{R} \to \mathcal{P}_{cp}(\mathbb{R})$ *is such that* $F(\cdot, y) : [0,T] \to \mathcal{P}_{cp}(\mathbb{R})$ *is measurable for each* $y \in \mathbb{R}$*, where* $\mathcal{P}_{cp}(\mathbb{R}) = \{Y \in \mathcal{P}(\mathbb{R}) : Y \text{ is compact}\}$;

(C_5) $H_d(F(t,y), F(t,\bar{y})) \leq \theta(t)|y - \bar{y}|$ *for almost all* $t \in [0,T]$ *and* $y, \bar{y} \in \mathbb{R}$ *with* $\theta \in C([0,T], \mathbb{R}^+)$ *and* $d(0, F(t,0)) \leq \theta(t)$ *for almost all* $t \in [0,T]$*, where*

Then, problem (2) has at least one solution on $[0,T]$ *if* $\|\theta\|\Lambda < 1$*, i.e.,*

$$\begin{aligned} K := {} & \|\theta\|\Big[\frac{T^{\rho\alpha}}{\rho^{\alpha}\Gamma(\alpha+1)} + \frac{1}{|\Omega|}\Big\{|B_1|\sum_{j=1}^{k}|\mu_j|\frac{\xi_j^{\rho\alpha}}{\rho^{\alpha}\Gamma(\alpha+1)} \\ & +|A_1|\Big[\sum_{i=1}^{m}|\sigma_i|\frac{\eta_i^{\rho(\alpha+\beta)}}{\rho^{\alpha+\beta}\Gamma(\alpha+\beta+1)} + \frac{T^{\rho\alpha}}{\rho^{\alpha}\Gamma(\alpha+1)}\Big]\Big\} \\ & +\frac{T^{\rho}}{\rho|\Omega|}\Big\{|B_2|\sum_{j=1}^{k}|\mu_j|\frac{\xi_j^{\rho\alpha}}{\rho^{\alpha}\Gamma(\alpha+1)} + |A_2|\Big[\sum_{i=1}^{m}|\sigma_i|\frac{\eta_i^{\rho(\alpha+\beta)}}{\rho^{\alpha+\beta}\Gamma(\alpha+\beta+1)} + \frac{T^{\rho\alpha}}{\rho^{\alpha}\Gamma(\alpha+1)}\Big]\Big\}\Big] < 1. \end{aligned} \tag{36}$$

Proof. By the assumption (C_4), it is clear that the set $S_{F,y}$ is nonempty for each $y \in \mathcal{C}$ and thus there exists a measurable selection for F (see Theorem III.6 [45]). Firstly, it will be shown that $\mathcal{N}(y) \in \mathcal{P}_{cl}(\mathcal{C})$ for each $y \in \mathcal{C}$, where the operator $\mathcal{N}$ is defined by (35). Let $\{u_n\}_{n\geq 0} \in \mathcal{F}(y)$ be such that $u_n \to u$ $(n \to \infty)$ in $\mathcal{C}$. Then, $u \in \mathcal{C}$ and we can find $v_n \in S_{F,y_n}$ such that, for each $t \in [0,T]$,

$$u_n(t) = {}^{\rho}I_{0^+}^{\alpha}v_n(t) + \frac{1}{\Omega}\Big\{-B_1\sum_{j=1}^{k}\mu_j{}^{\rho}I^{\alpha}v_n(\xi_j) + A_1\Big[\sum_{i=1}^{m}\sigma_i\,{}^{\rho}I_{\eta_0}^{\alpha+\beta}v_n(\eta_i) - {}^{\rho}I^{\alpha}v_n(T) + \kappa\Big]\Big\}$$

$$+\frac{t^{\rho}}{\rho\Omega}\Big\{B_2\sum_{j=1}^{k}\mu_j{}^{\rho}I_{0^+}^{\alpha}v_n(\xi_j)+A_2\Big[\sum_{i=1}^{m}\sigma_i\,{}^{\rho}I_{0^+}^{\alpha+\beta}v_n(\eta_i)-{}^{\rho}I_{0^+}^{\alpha}v_n(T)+\kappa\Big]\Big\}.$$

Since F has compact values, we pass onto a subsequence (if necessary) such that v_n converges to v in $L^1([0,T],\mathbb{R})$, which implies that $v \in S_{F,y}$ and for each $t \in [0,T]$, we have

$$\begin{aligned}u_n(t)\to u(t) &= {}^{\rho}I_{0^+}^{\alpha}v(t)+\frac{1}{\Omega}\Big\{-B_1\sum_{j=1}^{k}\mu_j{}^{\rho}I_{0^+}^{\alpha}v_n(\xi_j)+A_1\Big[\sum_{i=1}^{m}\sigma_i\,{}^{\rho}I_{0^+}^{\alpha+\beta}v(\eta_i)-{}^{\rho}I_{0^+}^{\alpha}v(T)+\kappa\Big]\Big\}\\ &\quad+\frac{t^{\rho}}{\rho\Omega}\Big\{B_2\sum_{j=1}^{k}\mu_j{}^{\rho}I_{0^+}^{\alpha}v(\xi_j)+A_2\Big[\sum_{i=1}^{m}\sigma_i\,{}^{\rho}I_{0^+}^{\alpha+\beta}v(\eta_i)-{}^{\rho}I_{0^+}^{\alpha}v(T)+\kappa\Big]\Big\}.\end{aligned}$$

Thus, $u \in \mathcal{N}(y)$.

Next, it will be shown that there exists $K < 1$ (defined by (36)) such that

$$H_d(\mathcal{N}(y),\mathcal{N}(\bar{y})) \le K\|y-\bar{y}\| \quad \text{for each } y,\bar{y}\in\mathcal{C}.$$

Let $y,\bar{y} \in \mathcal{C}$ and $h_1 \in \mathcal{F}(y)$. Then, there exists $v_1(t) \in F(t,y(t))$ for each $t \in [0,T]$ and that

$$\begin{aligned}h_1(t) &= {}^{\rho}I_{0^+}^{\alpha}v_1(t)+\frac{1}{\Omega}\Big\{-B_1\sum_{j=1}^{k}\mu_j{}^{\rho}I_{0^+}^{\alpha}v_1(\xi_j)+A_1\Big[\sum_{i=1}^{m}\sigma_i\,{}^{\rho}I_{0^+}^{\alpha+\beta}v_1(\eta_i)-{}^{\rho}I_{0^+}^{\alpha}v_1(T)+\kappa\Big]\Big\}\\ &\quad+\frac{t^{\rho}}{\rho\Omega}\Big\{B_2\sum_{j=1}^{k}\mu_j{}^{\rho}I_{0^+}^{\alpha}v_1(\xi_j)+A_2\Big[\sum_{i=1}^{m}\sigma_i\,{}^{\rho}I_{0^+}^{\alpha+\beta}v_1(\eta_i)-{}^{\rho}I_{0^+}^{\alpha}v_1(T)+\kappa\Big]\Big\}.\end{aligned}$$

By (C_5), $H_d(F(t,y),F(t,\bar{y})) \le \theta(t)|y(t)-\bar{y}(t)|$ and that there exists $w \in F(t,\bar{y}(t))$ satisfying the inequality: $|v_1(t)-w| \le \theta(t)|y(t)-\bar{y}(t)|,\ t\in[0,T]$.

Next, we introduce $\mathcal{S}:[0,T]\to\mathcal{P}(\mathbb{R})$ as

$$\mathcal{S}(t)=\{w\in\mathbb{R}:|v_1(t)-w|\le\theta(t)|y(t)-\bar{y}(t)|\}.$$

By Proposition III.4 [45], the multivalued operator $\mathcal{S}(t)\cap F(t,\bar{y}(t))$ is measurable. Thus, there exists a function $v_2(t)$, which is a measurable selection for $\mathcal{S}$ and that $v_2(t)\in F(t,\bar{y}(t))$. Thus, for each $t\in[0,T]$, we have $|v_1(t)-v_2(t)|\le\theta(t)|y(t)-\bar{y}(t)|$.

Next, we define

$$\begin{aligned}h_2(t) &= {}^{\rho}I_{0^+}^{\alpha}v_2(t)+\frac{1}{\Omega}\Big\{-B_1\sum_{j=1}^{k}\mu_j{}^{\rho}I_{0^+}^{\alpha}v_2(\xi_j)+A_1\Big[\sum_{i=1}^{m}\sigma_i\,{}^{\rho}I_{0^+}^{\alpha+\beta}v_n(\eta_i)-{}^{\rho}I_{0^+}^{\alpha}v_2(T)+\kappa\Big]\Big\}\\ &\quad+\frac{t^{\rho}}{\rho\Omega}\Big\{B_2\sum_{j=1}^{k}\mu_j{}^{\rho}I_{0^+}^{\alpha}v_2(\xi_j)+A_2\Big[\sum_{i=1}^{m}\sigma_i\,{}^{\rho}I_{0^+}^{\alpha+\beta}v_2(\eta_i)-{}^{\rho}I_{0^+}^{\alpha}v_2(T)+\kappa\Big]\Big\}\end{aligned}$$

for each $t \in [0,T]$. Then,

$$
\begin{aligned}
&|h_1(t) - h_2(t)| \\
&\leq {}^{\rho}I^{\alpha}_{0^+}|v_1(t) - v_2(t)| + \frac{1}{|\Omega|}\Big\{|B_1|\sum_{j=1}^{k}|\mu_j|^{\rho}I^{\alpha}_{0^+}|v_1(\xi_j) - v_2(\xi_j)| + |A_1|\Big[\sum_{i=1}^{m}|\sigma_i|\,{}^{\rho}I^{\alpha+\beta}_{0^+}|v_1(\eta_i) - v_2(\eta_i)| \\
&+{}^{\rho}I^{\alpha}_{0^+}|v_1(T) - v_2(T)|\Big]\Big\} + \frac{t^{\rho}}{\rho|\Omega|}\Big\{|B_2|\sum_{j=1}^{k}\mu_j{}^{\rho}I^{\alpha}_{0^+}|v_2(\xi_j) - v_1(\xi_j)| + |A_2|\Big[\sum_{i=1}^{m}\sigma_i\,{}^{\rho}I^{\alpha+\beta}_{0^+}|v_2(\eta_i) - v_1(\eta_i)| \\
&+{}^{\rho}I^{\alpha}_{0^+}|v_2(T) - v_1(T)|\Big]\Big\} \\
&\leq \|\theta\|\Big[\frac{T^{\rho\alpha}}{\rho^{\alpha}\Gamma(\alpha+1)} + \frac{1}{|\Omega|}\Big\{|B_1|\sum_{j=1}^{k}|\mu_j|\frac{\xi_j^{\rho\alpha}}{\rho^{\alpha}\Gamma(\alpha+1)} + |A_1|\Big[\sum_{i=1}^{m}|\sigma_i|\frac{\eta_i^{\rho(\alpha+\beta)}}{\rho^{\alpha+\beta}\Gamma(\alpha+\beta+1)} + \frac{T^{\rho\alpha}}{\rho^{\alpha}\Gamma(\alpha+1)}\Big]\Big\} \\
&+\frac{T^{\rho}}{\rho|\Omega|}\Big\{|B_2|\sum_{j=1}^{k}|\mu_j|\frac{\xi_j^{\rho\alpha}}{\rho^{\alpha}\Gamma(\alpha+1)} + |A_2|\Big[\sum_{i=1}^{m}|\sigma_i|\frac{\eta_i^{\rho(\alpha+\beta)}}{\rho^{\alpha+\beta}\Gamma(\alpha+\beta+1)} + \frac{T^{\rho\alpha}}{\rho^{\alpha}\Gamma(\alpha+1)}\Big]\Big\}\Big]\|y - \bar{y}\|.
\end{aligned}
$$

Hence,

$$
\begin{aligned}
\|h_1 - h_2\| \;\leq\; &\|\theta\|\Big[\frac{T^{\rho\alpha}}{\rho^{\alpha}\Gamma(\alpha+1)} + \frac{1}{|\Omega|}\Big\{|B_1|\sum_{j=1}^{k}|\mu_j|\frac{\xi_j^{\rho\alpha}}{\rho^{\alpha}\Gamma(\alpha+1)} + |A_1|\Big[\sum_{i=1}^{m}|\sigma_i|\frac{\eta_i^{\rho(\alpha+\beta)}}{\rho^{\alpha+\beta}\Gamma(\alpha+\beta+1)} \\
&+\frac{T^{\rho\alpha}}{\rho^{\alpha}\Gamma(\alpha+1)}\Big]\Big\} + \frac{T^{\rho}}{|\rho\Omega|}\Big\{|B_2|\sum_{j=1}^{k}|\mu_j|\frac{\xi_j^{\rho\alpha}}{\rho^{\alpha}\Gamma(\alpha+1)} \\
&+|A_2|\Big[\sum_{i=1}^{m}|\sigma_i|\frac{\eta_i^{\rho(\alpha+\beta)}}{\rho^{\alpha+\beta}\Gamma(\alpha+\beta+1)} + \frac{T^{\rho\alpha}}{\rho^{\alpha}\Gamma(\alpha+1)}\Big]\Big\}\Big]\|y - \bar{y}\|.
\end{aligned}
$$

Analogously, interchanging the roles of y and $\overline{y}$, we find that

$$
\begin{aligned}
H_d(\mathcal{N}(y),\mathcal{N}(\bar{y})) \;\leq\; &\|\theta\|\Big[\frac{T^{\rho\alpha}}{\rho^{\alpha}\Gamma(\alpha+1)} + \frac{1}{|\Omega|}\Big\{|B_1|\sum_{j=1}^{k}|\mu_j|\frac{\xi_j^{\rho\alpha}}{\rho^{\alpha}\Gamma(\alpha+1)} + |A_1|\Big[\sum_{i=1}^{m}|\sigma_i|\frac{\eta_i^{\rho(\alpha+\beta)}}{\rho^{\alpha+\beta}\Gamma(\alpha+\beta+1)} \\
&+\frac{T^{\rho\alpha}}{\rho^{\alpha}\Gamma(\alpha+1)}\Big]\Big\} + \frac{T^{\rho}}{|\rho\Omega|}\Big\{|B_2|\sum_{j=1}^{k}|\mu_j|\frac{\xi_j^{\rho\alpha}}{\rho^{\alpha}\Gamma(\alpha+1)} \\
&+|A_2|\Big[\sum_{i=1}^{m}|\sigma_i|\frac{\eta_i^{\rho(\alpha+\beta)}}{\rho^{\alpha+\beta}\Gamma(\alpha+\beta+1)} + \frac{T^{\rho\alpha}}{\rho^{\alpha}\Gamma(\alpha+1)}\Big]\Big\}\Big]\|y - \bar{y}\|.
\end{aligned}
$$

This shows that $\mathcal{N}$ is a contraction. Therefore, the operator $\mathcal{N}$ has a fixed point y by Covitz and Nadler [44], which corresponds to a solution of problem (2). This completes the proof. □

Example 2. *Consider the following inclusions problem:*

$$
\begin{cases}
{}^{1/3}{}_{c}D^{7/5}_{0^+}y(t) \in F(t,y(t)),\; t \in [0,2], \\
y(2) = 1/2\;{}^{1/3}I^{3/5}y(1/4) + 2/3\;{}^{1/3}I^{3/5}y(3/4) + 2/9, \\
\delta y(0) = 2/5\,y(1) + 4/5\,y(3/2),
\end{cases}
\tag{37}
$$

where $F(t,y(t))$ will be defined later.

The values of $|A_1|$, $|A_2|$, $|B_1|$, $|B_2|$, $|\Omega|$ and Λ are the same as those in Example 1. For illustrating Theorem 4, we take

$$
F(t,y(t)) = \Big[\frac{e^{-t}}{\sqrt{4000+t}}\Big(\sin y + \frac{1}{2}\Big),\; \frac{(1+t)}{60}\Big(\tan^{-1}y + y + \frac{1}{4}\Big)\Big]. \tag{38}
$$

It is easy to check that $F(t, y(t))$ is L^1-Carathéodory. In view of (C_2), we find that $\Phi(t) = \frac{(1+t)}{60}$, $\Psi(\|y\|) = \|y\| + \frac{2\pi+1}{4}$ and the condition (C_3) implies that $W > 3.289470$. Thus, all hypotheses of Theorem 4 hold true and the conclusion of Theorem 4 applies to problem (37) with $F(t, y(t))$ given by (38) on $[0, 2]$.

Now, we illustrate Theorem 5 by considering

$$F(t, y(t)) = \Big[\frac{(t+1)}{4\sqrt{900+t}}\Big(\tan^{-1} y + \sin t\Big), \frac{e^{-t}\cos t}{250}\Big(\frac{|y|}{|y|+1} + \frac{1}{8}\Big)\Big]. \tag{39}$$

Clearly,

$$H_d(F(t, y), F(t, \bar{y})) \leq \frac{(t+1)}{120}\|y - \bar{y}\|.$$

Letting $\theta(t) = \frac{(t+1)}{120}$, we observe that $d(0, F(t, 0)) \leq \theta(t)$ for almost all $t \in [0, 2]$ and that $K \approx 0.6780733168 < 1$ (K is given by (36)). As the assumptions of Theorem 5 hold true, there exists at least one solution for problem (37) with $F(t, y(t))$ given by (39) on $[0, 2]$.

5. Conclusions

We have developed the existence theory for fractional differential equations and inclusions involving the Liouville–Caputo-type generalized derivative, supplemented with nonlocal generalized fractional integral and multipoint boundary conditions. Our results are based on the modern techniques of the functional analysis. In case of a single valued problem (1), we have obtained three results: the first two results deal with the existence of solutions while the third one is concerned with the uniqueness of solutions for the given problem. The first existence result relies on a Leray–Schauder nonlinear alternative, which allows the nonlinearity $f(t, y)$ to behave like $|f(t, y)| \leq \phi(t)\psi(\|y\|)$ (see (H_1)) and the second results, depending on Krasnoselskii's fixed point theorem, handles the nonlinearity $f(t, y)$ of the form described by the conditions (H_3) and (H_4). The third result provides a criterion ensuring a unique solution of the problem at hand by requiring the nonlinear function $f(t, y)$ to satisfy the classical Lipschitz condition and is based on a Banach fixed point theorem. The tools of the fixed point theory chosen for our case are easy to apply and extend the scope of the obtained results in the scenario of simplicity of the assumptions. Again, for the inclusion problem (2), the idea is to assume a simple set of conditions to establish the existence of solutions for problem (2) involving both convex and nonconvex valued maps. As a matter of fact, the fixed point theorems chosen to solve the multivalued problem (2) are standard and popular in view of their applicability. Concerning the choice of the method to solve a given problem, one needs to loook at the set of assumptions satisfied by the single and multivalued maps involved in the problem, which decides the selection of the tool to be employed. As an application of the present work, the generalization of the Feynman and Wiener path integrals developed by Laskin [46], in the context of fractional quantum mechanics and fractional statistical mechanics, can be enhanced further. We emphasize that we obtain new results associated with symmetric solutions of a second-order ordinary differential equation equipped with nonlocal fractional integral and multipoint boundary conditions if we take $0 < \eta_1 < \cdots < \eta_i < \cdots < \eta_m < \xi_1 < \cdots < \xi_j < \cdots < \xi_k < T/2$ and $f(t, x)$ to be symmetric on the interval $[0, T]$ for all $x \in \mathbb{R}$ when $\alpha \to 2^-$ $(\rho = 1)$.

Author Contributions: Formal Analysis, M.A. and B.A.; Investigation, A.A.; Methodology, S.K.N.

Funding: This project was funded by the Deanship of Scientific Research (DSR) at King Abdulaziz University, Jeddah, Saudi Arabia, under Grant No. (RG-1-130-39).

Acknowledgments: This project was funded by the Deanship of Scientific Research (DSR) at King Abdulaziz University, Jeddah, Saudi Arabia, under Grant No. (RG-1-130-39). The authors acknowledge with thanks DSR technical and financial support. The authors thank the editor and the reviewers for their constructive remarks that led to the improvement of the original manuscript.

Conflicts of Interest: The authors declare no conflict of interest.

References

1. Magin, R.L. *Fractional Calculus in Bioengineering*; Begell House: Chicago, IL, USA, 2006.
2. Zaslavsky, G.M. *Hamiltonian Chaos and Fractional Dynamics*; Oxford University Press: New York, USA, 2008.
3. Javidi, M.; Ahmad, B. Dynamic analysis of time fractional order phytoplankton-toxic phytoplankton-zooplankton system. *Ecol. Model.* **2015**, *318*, 8–18. [CrossRef]
4. Fallahgoul, H.A.; Focardi, S.M.; Fabozzi, F.J. *Fractional Calculus and Fractional Processes with Applications to Financial Economics: Theory and Application*; Elsevier/Academic Press: London, UK, 2017.
5. Kilbas, A.A.; Srivastava, H.M.; Trujillo, J.J. North-Holland Mathematics Studies. In *Theory and Applications of Fractional Differential Equations*; Elsevier Science B.V.: Amsterdam, The Netherlands, 2006.
6. Lakshmikantham, V.; Leela, S.; Devi, J.V. *Theory of Fractional Dynamic Systems*; Cambridge Academic Publishers: Cambridge, UK, 2009.
7. Diethelm, K. *The Analysis of Fractional Differential Equations: An Application-Oriented Exposition Using Differential Operators of Caputo Type*; Springer-Verlag: Berlin, Genmany, 2010.
8. Ahmad, B.; Alsaedi, A.; Ntouyas, S.K.; Tariboon, J. *Hadamard-Type Fractional Differential Equations, Inclusions and Inequalities*, Springer: Cham, Switzerland, 2017.
9. Bitsadze, A.; Samarskii, A. On some simple generalizations of linear elliptic boundary problems. *Russ. Acad. Sci. Dokl. Math.* **1969**, *10*, 398–400.
10. Ciegis, R.; Bugajev, A. Numerical approximation of one model of the bacterial self-organization. *Nonlinear Anal. Model. Control* **2012**, *17*, 253–270.
11. Richard, P.; Nicodemi, M.; Delannay, R.; Ribiere, P.; Bideau, D. Slow relaxation and compaction of granular system. *Nat. Mater.* **2005**, *4*, 121–128. [CrossRef] [PubMed]
12. Quezada, J.C.; Sagnol, L.; Chazallon, C. Shear Test on Viscoelastic Granular Material Using Contact Dynamics Simulations. In *EPJ Web of Conferences*, Powders & Grains: Montpellier, France, 2017.
13. Pereira, F.L.; de Sousa, J.B.; de Matos, A.C. An Algorithm for Optimal Control Problems Based on Differential Inclusions, In Proceedings of the 34th Conference on Decision & Control, New Orleans, LA, USA, 13–15 December 1995.
14. Korda, M.; Henrion, D.; Jones, C.N. Convex computation of the maximum controlled invariant set for polynomial control systems. *SIAM J. Control Optim.* **2014**, *52*, 2944–2969. [CrossRef]
15. Bastien, J. Study of a driven and braked wheel using maximal monotone differential inclusions: Applications to the nonlinear dynamics of wheeled vehicles. *Arch. Appl. Mech.* **2014**, *84*, 851–880. [CrossRef]
16. Kisielewicz, M. *Stochastic Differential Inclusions and Applications*; Springer: New York, NY, USA, 2013.
17. Danca, M.F. Synchronization of piecewise continuous systems of fractional order. *Nonlinear Dyn.* **2014**, *78*, 2065–2084. [CrossRef]
18. Ahmad, B.; Nieto, J.J. Riemann-Liouville fractional integro-differential equations with fractional nonlocal integral boundary conditions. *Bound. Val. Prob.* **2011**, *2011*, 36. [CrossRef]
19. Liang, S.; Zhang, J. Existence of multiple positive solutions for m-point fractional boundary value problems on an infinite interval. *Math. Comput. Model.* **2011**, *54*, 1334–1346. [CrossRef]
20. Bai, Z.B.; Sun, W. Existence and multiplicity of positive solutions for singular fractional boundary value problems. *Comput. Math. Appl.* **2012**, *63*, 1369–1381. [CrossRef]
21. Agarwal, R.P.; O'Regan, D.; Stanek, S. Positive solutions for mixed problems of singular fractional differential equations. *Math. Nachr.* **2012**, *285*, 27–41. [CrossRef]
22. Ahmad, B.; Ntouyas, S.K.; Alsaedi, A. A study of nonlinear fractional differential equations of arbitrary order with Riemann–Liouville type multistrip boundary conditions. *Math. Probl. Eng.* **2013**, *2013*, 9. [CrossRef]
23. Ahmad, B.; Ntouyas, S.K. Existence results for higher order fractional differential inclusions with multi-strip fractional integral boundary conditions. *Electron. J. Qual. Theory Differ. Equat.* **2013**, *2013*, 1–19. [CrossRef]
24. O'Regan, D.; Staněk, S. Fractional boundary value problems with singularities in space variables. *Nonlinear Dyn.* **2013**, *71*, 641–652. [CrossRef]
25. Zhai, C.; Xu, L. Properties of positive solutions to a class of four-point boundary value problem of Caputo fractional differential equations with a parameter. *Commun. Nonlinear Sci. Numer. Simul.* **2014**, *19*, 2820–2827. [CrossRef]

26. Graef, J.R.; Kong, L.; Wang, M. Existence and uniqueness of solutions for a fractional boundary value problem on a graph. *Fract. Calc. Appl. Anal.* **2014**, *17*, 499–510. [CrossRef]
27. Wang, G.; Liu, S.; Zhang, L. Eigenvalue problem for nonlinear fractional differential equations with integral boundary conditions. *Abstr. Appl. Anal.* **2014**, *2014*. [CrossRef]
28. Henderson, J.; Kosmatov, N. Eigenvalue comparison for fractional boundary value problems with the Caputo derivative. *Fract. Calc. Appl. Anal.* **2014**, *17*, 872–880. [CrossRef]
29. Zhang, L.; Ahmad, B.; Wang, G. Successive iterations for positive extremal solutions of nonlinear fractional differential equations on a half line. *Bull. Aust. Math. Soc.* **2015**, *91*, 116–128. [CrossRef]
30. Henderson, J.; Luca, R. Nonexistence of positive solutions for a system of coupled fractional boundary value problems. *Bound. Val. Probl.* **2015**, *2015*, 138. [CrossRef]
31. Ntouyas, S.K.; Etemad, S. On the existence of solutions for fractional differential inclusions with sum and integral boundary conditions. *Appl. Math. Comput.* **2015**, *266*, 235–243. [CrossRef]
32. Mei, Z.D.; Peng, J.G.; Gao, J.H. Existence and uniqueness of solutions for nonlinear general fractional differential equations in Banach spaces. *Indagat. Math.* **2015**, *26*, 669–678. [CrossRef]
33. Ahmad, B.; Ntouyas, S.K. Existence results for fractional differential inclusions with Erdelyi-Kober fractional integral conditions. *Ann. Univ. Ovidius Constanta-Seria Mat.* **2017**, *25*, 5–24. [CrossRef]
34. Srivastava, H.M. Remarks on some families of fractional-order differential equations. *Integral Transforms Spec. Funct.* **2017**, *28*, 560–564. [CrossRef]
35. Wang, G.; Pei, K.; Agarwal, R.P.; Zhang, L.; Ahmad, B. Nonlocal Hadamard fractional boundary value problem with Hadamard integral and discrete boundary conditions on a half-line. *J. Comput. Appl. Math.* **2018**, *343*, 230–239. [CrossRef]
36. Ahmad, B.; Luca, R. Existence of solutions for sequential fractional integro-differential equations and inclusions with nonlocal boundary conditions. *Appl. Math. Comput.* **2018**, *339*, 516–534. [CrossRef]
37. Katugampola, U.N. New approach to a generalized fractional integral. *Appl. Math. Comput.* **2015**, *218*, 860–865. [CrossRef]
38. Katugampola, U.N. A new approach to generalized fractional derivatives. *Bull. Math. Anal. Appl.* **2014**, *6*, 1–15.
39. Jarad, F.; Abdeljawad, T.; Baleanu, D. On the generalized fractional derivatives and their Caputo modification. *J. Nonlinear Sci. Appl.* **2017**, *10*, 2607–2619. [CrossRef]
40. Granas, A.; Dugundji, J. *Fixed Point Theory*; Springer-Verlag: New York, NY, USA, 2003.
41. Krasnoselskii, M.A. Two remarks on the method of successive approximations. *Uspekhi Mat. Nauk* **1955**, *10*, 123–127.
42. Deimling, K. *Multivalued Differential Equations*; Walter De Gruyter: New York, NY, USA, 1992.
43. Lasota, A.; Opial, Z. An application of the Kakutani-Ky Fan theorem in the theory of ordinary differential equations. *Bull. Acad. Polon. Sci. Ser. Sci. Math. Astronom. Phys.* **1965**, *13*, 781–786.
44. Covitz, H.; Nadler, S.B. Multivalued contraction mappings in generalized metric spaces. *Israel J. Math.* **1970**, *8*, 5–11. [CrossRef]
45. Castaing, C.; Valadier, M. *Convex Analysis and Measurable Multifunctions*; Springer-Verlag: Berlin/Heidelberg, Germany, 1977.
46. Laskin, N. Fractional quantum mechanics and Levy path integrals. *Phys. Lett. A* **2000**, *268*, 298–305. [CrossRef]

© 2018 by the authors. Licensee MDPI, Basel, Switzerland. This article is an open access article distributed under the terms and conditions of the Creative Commons Attribution (CC BY) license (http://creativecommons.org/licenses/by/4.0/).

symmetry MDPI

Article

Convolution and Partial Sums of Certain Multivalent Analytic Functions Involving Srivastava–Tomovski Generalization of the Mittag–Leffler Function

Yi-Hui Xu [1] and Jin-Lin Liu [2,*]

[1] Department of Mathematics, Suqian College, Suqian 223800, China; yuanziqixu@126.com
[2] Department of Mathematics, Yangzhou University, Yangzhou 225002, China
* Correspondence: jlliu@yzu.edu.cn

Received: 7 October 2018; Accepted: 23 October 2018; Published: 5 November 2018

Abstract: We derive several properties such as convolution and partial sums of multivalent analytic functions associated with an operator involving Srivastava–Tomovski generalization of the Mittag–Leffler function.

Keywords: analytic function; Hadamard product (convolution); partial sum; Srivastava–Tomovski generalization of Mittag–Leffler function; subordination

1. Introduction

The Mittag–Leffler function $E_\alpha(z)$ [1] and its generalization $E_{\alpha,\beta}(z)$ [2] are defined by the following series:

$$E_\alpha(z) = \sum_{n=0}^{\infty} \frac{z^n}{\Gamma(\alpha n + 1)} \quad (z, \alpha \in \mathbb{C};\ \mathrm{Re}(\alpha) > 0) \tag{1}$$

and

$$E_{\alpha,\beta}(z) = \sum_{n=0}^{\infty} \frac{z^n}{\Gamma(\alpha n + \beta)} \quad (z, \alpha, \beta \in \mathbb{C};\ \mathrm{Re}(\alpha) > 0), \tag{2}$$

respectively. It is known that these functions are extensions of exponential, hyperbolic, and trigonometric functions, since

$$E_1(z) = E_{1,1}(z) = e^z,$$
$$E_2(z^2) = E_{2,1}(z^2) = \cosh z$$

and

$$E_2(-z^2) = E_{2,1}(-z^2) = \cos z.$$

The functions $E_\alpha(z)$ and $E_{\alpha,\beta}(z)$ arise naturally in the resolvent of fractional integro-differential and fractional differential equations which are involved in random walks, super-diffusive transport problems, the kinetic equation, Lévy flights, and in the study of complex systems. In particular, the Mittag–Leffler function is an explicit formula for the solution the Riemann–Liouville fractional integrals that was developed by Hille and Tamarkin.

In [3], Srivastava and Tomovski defined a generalized Mittag–Leffler function $E^{\gamma,k}_{\alpha,\beta}(z)$ as follows:

$$E^{\gamma,k}_{\alpha,\beta}(z) = \sum_{n=0}^{\infty} \frac{(\gamma)_{nk} z^n}{\Gamma(\alpha n + \beta) n!}, \tag{3}$$

$$(\alpha, \beta, \gamma, k, z \in \mathbb{C};\quad \mathrm{Re}(\alpha) > \max\{0, \mathrm{Re}(k) - 1\};\quad \mathrm{Re}(k) > 0),$$

where $(x)_n$ is the Pochhammer symbol

$$(x)_n = \frac{\Gamma(x+n)}{\Gamma(x)} = x(x+1)\cdots(x+n-1) \quad (n \in \mathbb{N}; \quad x \in \mathbb{C})$$

and $(x)_0 = 1$. They proved that the function $E^{\gamma,k}_{\alpha,\beta}(z)$ given by (3) is an entire function in the complex plane. Recently, Attiya [4] proved that, if $\mathrm{Re}(\alpha) \geq 0$ with $\mathrm{Re}(k) = 1$ and $\beta \neq 0$, the power series in (3) converges absolutely and analytically in $\mathbb{U} = \{z : |z| < 1\}$ for all $\gamma \in \mathbb{C}$. We call the function $E^{\gamma,k}_{\alpha,\beta}(z)$ the Srivastava–Tomovski generalization of the Mittag–Leffler function.

Let $\mathcal{A}(p)$ be the class of functions of the form

$$f(z) = z^p + \sum_{n=2}^{\infty} a_{n+p-1} z^{n+p-1} \quad (p \in \mathbb{N}) \tag{4}$$

which are analytic in $\mathbb{U}$. For $p = 1$, we write $\mathcal{A} := \mathcal{A}(1)$. The Hadamard product (or convolution) of two functions

$$f_j(z) = z^p + \sum_{n=2}^{\infty} a_{n+p-1,j} z^{n+p-1} \in \mathcal{A}(p) \quad (j = 1, 2)$$

is given by

$$(f_1 * f_2)(z) = z^p + \sum_{n=2}^{\infty} a_{n+p-1,1} a_{n+p-1,2} z^{n+p-1} = (f_2 * f_1)(z).$$

Let $\mathcal{P}$ denote the class of functions φ with $\varphi(0) = 1$. Suppose that f and g are analytic in $\mathbb{U}$. If there exists a Schwarz function w such that $f(z) = g(w(z))$ for $z \in \mathbb{U}$, then we say that the function f is subordinate to g and write $f(z) \prec g(z)$ for $z \in \mathbb{U}$. Furthermore, if g is univalent in $\mathbb{U}$, then the following equivalence holds true:

$$f(z) \prec g(z) \quad (z \in \mathbb{U}) \Leftrightarrow f(0) = g(0) \quad \text{and} \quad f(\mathbb{U}) \subset g(\mathbb{U}).$$

Throughout this paper, we assume that

$$\alpha, \beta, \gamma, k \in \mathbb{C}; \quad \mathrm{Re}(\alpha) > \max\{0, \mathrm{Re}(k) - 1\} \quad \text{and} \quad \mathrm{Re}(k) > 0.$$

We define the function $Q^{\gamma,k}_{\alpha,\beta}(z) \in \mathcal{A}(p)$ associated with the Srivastava–Tomovski generalization of the Mittag–Leffler function by

$$Q^{\gamma,k}_{\alpha,\beta}(z) = \frac{\Gamma(\alpha+\beta)}{(\gamma)_k} z^{p-1} \left(E^{\gamma,k}_{\alpha,\beta}(z) - \frac{1}{\Gamma(\beta)} \right) \quad (z \in \mathbb{U}). \tag{5}$$

For $f \in \mathcal{A}(p)$, we introduce a new operator $H^{\gamma,k}_{\alpha,\beta} : \mathcal{A}(p) \to \mathcal{A}(p)$ by

$$\begin{aligned} H^{\gamma,k}_{\alpha,\beta} f(z) &= Q^{\gamma,k}_{\alpha,\beta}(z) * f(z) \\ &= z^p + \sum_{n=2}^{\infty} \frac{\Gamma(\gamma+nk)\Gamma(\alpha+\beta)}{\Gamma(\gamma+k)\Gamma(\alpha n+\beta) n!} a_{n+p-1} z^{n+p-1}. \end{aligned} \tag{6}$$

Note that $H^{1,1}_{0,\beta} f(z) = f(z)$. From (6), we easily have the following identity:

$$z \left(H^{\gamma,k}_{\alpha,\beta} f(z) \right)' = \left(\frac{\gamma}{k} + 1 \right) H^{\gamma+1,k}_{\alpha,\beta} f(z) - \left(\frac{\gamma}{k} + 1 - p \right) H^{\gamma,k}_{\alpha,\beta} f(z). \tag{7}$$

It is noteworthy to mention that the Fox–Wright hypergeometric function ${}_q\Psi_s$ is more general than many of the extensions of the Mittag–Leffler function.

Now, we introduce a new subclass of $\mathcal{A}(p)$ by using the operator $H^{\gamma,k}_{\alpha,\beta}$.

Definition 1. *A function $f \in \mathcal{A}(p)$ is said to be in $\Omega_{\alpha,\beta}^{\gamma,k}(\lambda;\varphi)$ if it satisfies the first-order differential subordination:*

$$(1-\lambda)z^{-p}H_{\alpha,\beta}^{\gamma,k}f(z)+\frac{\lambda}{p}z^{-p+1}\left(H_{\alpha,\beta}^{\gamma,k}f(z)\right)'\prec\varphi(z), \tag{8}$$

where $\lambda \in \mathbb{C}$ and $\varphi \in \mathcal{P}$.

Lemma 1. *([5]). Let $g(z)=1+\sum_{n=m}^{\infty}b_nz^n$ $(m \in \mathbb{N})$ be analytic in $\mathbb{U}$. If $\mathrm{Re}(g(z))>0$ $(z \in \mathbb{U})$, then*

$$\mathrm{Re}\,(g(z)) \geq \frac{1-|z|^m}{1+|z|^m} \quad (z \in \mathbb{U}).$$

The study of the Mittag–Leffler function is an interesting topic in Geometric Function Theory. Many properties of the Mittag–Leffler function and the generalized Mittag–Leffler function can be found, e.g., in [6–22]. In this paper we shall make a further contribution to the subject by showing some interesting properties such as convolution and partial sums for functions in the class $\Omega_{\alpha,\beta}^{\gamma,k}(\lambda;\varphi)$.

2. Properties of the Class $\Omega_{\alpha,\beta}^{\gamma,k}(\lambda;\varphi)$

Theorem 1. *Let $\lambda \geq 0$ and*

$$f_j(z)=z^p+\sum_{n=2}^{\infty}a_{n+p-1,j}z^{n+p-1}\in\Omega_{\alpha,\beta}^{\gamma,k}(\lambda;\varphi_j) \quad (j=1,2), \tag{9}$$

where

$$\varphi_j(z)=\frac{1+A_jz}{1+B_jz} \quad \text{and} \quad -1 \leq B_j < A_j \leq 1. \tag{10}$$

If $f \in \mathcal{A}(p)$ is defined by

$$H_{\alpha,\beta}^{\gamma,k}f(z)=\left(H_{\alpha,\beta}^{\gamma,k}f_1(z)\right)*\left(H_{\alpha,\beta}^{\gamma,k}f_2(z)\right), \tag{11}$$

then $f \in \Omega_{\alpha,\beta}^{\gamma,k}(\lambda;\varphi)$, where

$$\varphi(z)=\rho+(1-\rho)\frac{1+z}{1-z} \tag{12}$$

and ρ is given by

$$\rho=\begin{cases} 1-\frac{4(A_1-B_1)(A_2-B_2)}{(1-B_1)(1-B_2)}\left(1-\frac{p}{\lambda}\int_0^1\frac{t^{\frac{p}{\lambda}-1}}{1+t}dt\right) & (\lambda>0), \\ 1-\frac{2(A_1-B_1)(A_2-B_2)}{(1-B_1)(1-B_2)} & (\lambda=0). \end{cases} \tag{13}$$

The bound ρ is sharp when $B_1=B_2=-1$.

Proof. We consider the case when $\lambda>0$. Since $f_j \in \Omega_{\alpha,\beta}^{\gamma,k}(\lambda;\varphi_j)$, it follows that

$$\begin{aligned} p_j(z) &= (1-\lambda)z^{-p}H_{\alpha,\beta}^{\gamma,k}f_j(z)+\frac{\lambda}{p}z^{-p+1}\left(H_{\alpha,\beta}^{\gamma,k}f_j(z)\right)' \\ &\prec \frac{1+A_jz}{1+B_jz} \quad (j=1,2) \end{aligned} \tag{14}$$

and

$$H_{\alpha,\beta}^{\gamma,k} f_j(z) = \frac{p}{\lambda} z^{-\frac{p(1-\lambda)}{\lambda}} \int_0^z t^{\frac{p}{\lambda}-1} p_j(t)dt$$
$$= \frac{p}{\lambda} z^p \int_0^1 t^{\frac{p}{\lambda}-1} p_j(tz)dt \quad (j = 1,2). \tag{15}$$

Now, if $f \in \mathcal{A}(p)$ is defined by (11), we find from (14) that

$$H_{\alpha,\beta}^{\gamma,k} f(z) = \left(H_{\alpha,\beta}^{\gamma,k} f_1(z)\right) * \left(H_{\alpha,\beta}^{\gamma,k} f_2(z)\right)$$
$$= \left(\frac{p}{\lambda} z^p \int_0^1 t^{\frac{p}{\lambda}-1} p_1(tz)dt\right) * \left(\frac{p}{\lambda} z^p \int_0^1 t^{\frac{p}{\lambda}-1} p_2(tz)dt\right)$$
$$= \frac{p}{\lambda} z^p \int_0^1 t^{\frac{p}{\lambda}-1} p_0(tz)dt, \tag{16}$$

where

$$p_0(z) = \frac{p}{\lambda} \int_0^1 t^{\frac{p}{\lambda}-1} (p_1 * p_2)(tz)dt. \tag{17}$$

Further, by using (14) and the Herglotz theorem, we see that

$$\mathrm{Re}\left\{\left(\frac{p_1(z) - \rho_1}{1 - \rho_1}\right) * \left(\frac{1}{2} + \frac{p_2(z) - \rho_2}{2(1 - \rho_2)}\right)\right\} > 0 \quad (z \in \mathbb{U}),$$

which leads to

$$\mathrm{Re}\{(p_1 * p_2)(z)\} > \rho_0 = 1 - 2(1 - \rho_1)(1 - \rho_2) \quad (z \in \mathbb{U}),$$

where

$$0 \le \rho_j = \frac{1 - A_j}{1 - B_j} < 1 \quad (j = 1,2).$$

Moreover, according to Lemma, we have

$$\mathrm{Re}\{(p_1 * p_2)(z)\} \ge \rho_0 + (1 - \rho_0)\frac{1 - |z|}{1 + |z|} \quad (z \in \mathbb{U}). \tag{18}$$

Thus, it follows from (16) to (18) that

$$\mathrm{Re}\left\{(1 - \lambda)z^{-p} H_{\alpha,\beta}^{\gamma,k} f(z) + \frac{\lambda}{p} z^{-p+1} \left(H_{\alpha,\beta}^{\gamma,k} f(z)\right)'\right\} = \mathrm{Re}\{p_0(z)\}$$
$$= \frac{p}{\lambda} \int_0^1 t^{\frac{p}{\lambda}-1} \mathrm{Re}\{(p_1 * p_2)(tz)\}dt$$
$$\ge \frac{p}{\lambda} \int_0^1 t^{\frac{p}{\lambda}-1} \left(\rho_0 + (1 - \rho_0)\frac{1 - |z|t}{1 + |z|t}\right) dt$$
$$> \rho_0 + \frac{p(1 - \rho_0)}{\lambda} \int_0^1 t^{\frac{p}{\lambda}-1} \frac{1 - t}{1 + t} dt$$
$$= 1 - 4(1 - \rho_1)(1 - \rho_2) \left(1 - \frac{p}{\lambda} \int_0^1 \frac{t^{\frac{p}{\lambda}-1}}{1 + t} dt\right)$$
$$= \rho,$$

which proves that $f \in \Omega_{\alpha,\beta}^{\gamma,k}(\lambda; \varphi)$ for the function φ given by (12).

In order to show that the bound ρ is sharp, we take the functions $f_j \in \mathcal{A}(p)$ $(j = 1,2)$ defined by

$$H_{\alpha,\beta}^{\gamma,k} f_j(z) = \frac{p}{\lambda} z^{-\frac{p(1-\lambda)}{\lambda}} \int_0^z t^{\frac{p}{\lambda}-1} \left(\frac{1 + A_j t}{1 - t}\right) dt \quad (j = 1,2), \tag{19}$$

for which we have

$$p_j(z) = (1-\lambda)z^{-p}H_{\alpha,\beta}^{\gamma,k}f_j(z) + \frac{\lambda}{p}z^{-p+1}\left(H_{\alpha,\beta}^{\gamma,k}f_j(z)\right)' = \frac{1+A_jz}{1-z} \quad (j=1,2)$$

and

$$(p_1 * p_2)(z) = \frac{1+A_1z}{1-z} * \frac{1+A_2z}{1-z} = 1-(1+A_1)(1+A_2) + \frac{(1+A_1)(1+A_2)}{1-z}.$$

Hence, for the function f given by (11), we have

$$(1-\lambda)z^{-p}H_{\alpha,\beta}^{\gamma,k}f(z) + \frac{\lambda}{p}z^{-p+1}\left(H_{\alpha,\beta}^{\gamma,k}f(z)\right)'$$
$$= \frac{p}{\lambda}\int_0^1 t^{\frac{p}{\lambda}-1}\left(1-(1+A_1)(1+A_2) + \frac{(1+A_1)(1+A_2}{1-tz}\right)dt$$
$$\to \rho \quad (as \quad z \to -1),$$

which shows that the number ρ is the best possible when $B_1 = B_2 = -1$.

For the case when $\lambda = 0$, the proof of Theorem 1 is simple, and we choose to omit the details involved. Now the proof of Theorem 1 is completed. □

Theorem 2. *Let α, β, γ, k, and λ be positive real numbers. Let $f(z) = z^p + \sum_{n=2}^{\infty} a_{n+p-1}z^{n+p-1} \in \mathcal{A}(p)$, $s_1(z) = z^p$, and $s_m(z) = z^p + \sum_{n=2}^{m} a_{n+p-1}z^{n+p-1}$ $(m \geq 2)$. Suppose that*

$$\sum_{n=2}^{\infty} c_n|a_{n+p-1}| \leq 1, \tag{20}$$

where

$$c_n = \frac{1-B}{A-B} \cdot \frac{\Gamma(\gamma+nk)\Gamma(\alpha+\beta)}{\Gamma(\beta+n\alpha)\Gamma(\gamma+k)n!}\left(1+\frac{\lambda}{p}(n-1)\right) \tag{21}$$

and $-1 \leq B < A \leq 1$.

(i) If $-1 \leq B \leq 0$, then $f \in \Omega_{\alpha,\beta}^{\gamma,k}\left(\lambda; \frac{1+Az}{1+Bz}\right)$.

(ii) If $\{c_n\}_1^{\infty}$ is nondecreasing, then

$$\mathrm{Re}\left\{\frac{f(z)}{s_m(z)}\right\} > 1 - \frac{1}{c_{m+1}} \tag{22}$$

and

$$\mathrm{Re}\left\{\frac{s_m(z)}{f(z)}\right\} > \frac{c_{m+1}}{1+c_{m+1}} \tag{23}$$

for $z \in \mathbb{U}$. The estimates in (22) and (23) are sharp for each $m \in \mathbb{N}$.

Proof. From the assumptions of Theorem 2, we have $c_n > 0$ $(n \in \mathbb{N})$. Let

$$J(z) = (1-\lambda)z^{-p}H_{\alpha,\beta}^{\gamma,k}f(z) + \frac{\lambda}{p}z^{-p+1}\left(H_{\alpha,\beta}^{\gamma,k}f(z)\right)'$$
$$= 1 + \sum_{n=2}^{\infty} \frac{\Gamma(\gamma+nk)\Gamma(\alpha+\beta)}{\Gamma(\beta+n\alpha)\Gamma(\gamma+k)n!}\left(1+\frac{\lambda}{p}(n-1)\right)a_{n+p-1}z^{n-1}. \tag{24}$$

(i) For $-1 \leq B \leq 0$ and $z \in \mathbb{U}$, it follows from (20), (21), and (24), that

$$\left|\frac{J(z)-1}{A-BJ(z)}\right| = \left|\frac{\sum_{n=2}^{\infty} \frac{\Gamma(\gamma+nk)\Gamma(\alpha+\beta)}{\Gamma(\beta+n\alpha)\Gamma(\gamma+k)n!}\left(1+\frac{\lambda}{p}(n-1)\right)a_{n+p-1}z^{n-1}}{A-B-B\sum_{n=2}^{\infty}\frac{\Gamma(\gamma+nk)\Gamma(\alpha+\beta)}{\Gamma(\beta+n\alpha)\Gamma(\gamma+k)n!}\left(1+\frac{\lambda}{p}(n-1)\right)a_{n+p-1}z^{n-1}}\right| \leq \frac{\sum_{n=2}^{\infty} c_n|a_{n+p-1}|}{1-B+B\sum_{n=2}^{\infty}c_n|a_{n+p-1}|} \leq 1,$$

which implies that

$$(1-\lambda)z^{-p}H_{\alpha,\beta}^{\gamma,k}f(z)+\frac{\lambda}{p}z^{-p+1}\left(H_{\alpha,\beta}^{\gamma,k}f(z)\right)' \prec \frac{1+Az}{1+Bz}.$$

Hence, $f \in \Omega_{\alpha,\beta}^{\gamma,k}\left(\lambda; \frac{1+Az}{1+Bz}\right)$.

(ii) Under the hypothesis in part (ii) of Theorem 2, we can see from (21) that $c_{n+1} > c_n > 1$ $(n \in \mathbb{N})$. Therefore, we have

$$\sum_{n=2}^{m}|a_{n+p-1}| + c_{m+1}\sum_{n=m+1}^{\infty}|a_{n+p-1}| \leq \sum_{n=2}^{\infty}c_n|a_{n+p-1}| \leq 1. \tag{25}$$

Upon setting

$$p_1(z) = c_{m+1}\left\{\frac{f(z)}{s_m(z)} - \left(1-\frac{1}{c_{m+1}}\right)\right\} = 1 + \frac{c_{m+1}\sum_{n=m+1}^{\infty}a_{n+p-1}z^{n-1}}{1+\sum_{n=2}^{\infty}a_{n+p-1}z^{n-1}},$$

and applying (25), we find that

$$\left|\frac{p_1(z)-1}{p_1(z)+1}\right| \leq \frac{c_{m+1}\sum_{n=m+1}^{\infty}|a_{n+p-1}|}{2-2\sum_{n=2}^{m}|a_{n+p-1}| - c_{m+1}\sum_{n=m+1}^{\infty}|a_{n+p-1}|} \leq 1 \quad (z \in \mathbb{U}),$$

which readily yields (22).

If we take

$$f(z) = z^p - \frac{z^{m+p}}{c_{m+1}}, \tag{26}$$

then

$$\frac{f(z)}{s_m(z)} = 1 - \frac{z^m}{c_{m+1}} \to 1 - \frac{1}{c_{m+1}} \quad \text{and} \quad z \to 1^-,$$

which shows that the bound in (22) is the best possible for each $m \in \mathbb{N}$.

Similarly, if we put

$$p_2(z) = (1+c_{m+1})\left(\frac{s_m(z)}{f(z)} - \frac{c_{m+1}}{1+c_{m+1}}\right),$$

then we can deduce that

$$\left|\frac{p_2(z)-1}{p_2(z)+1}\right| \leq \frac{(1+c_{m+1})\sum_{n=m+1}^{\infty}|a_{n+p-1}|}{2-2\sum_{n=2}^{m}|a_{n+p-1}| - (c_{m+1}-1)\sum_{n=m+1}^{\infty}|a_{n+p-1}|} \leq 1 \quad (z \in \mathbb{U}),$$

which yields (23).

The bound in (23) is sharp for each $m \in \mathbb{N}$, with the extremal function f given by (26). The proof of Theorem 2 is thus completed. □

Author Contributions: All authors contributed equally.

Funding: This research is supported by National Natural Science Foundation of China (Grant No. 11571299) and Natural Science Foundation of Jiangsu Gaoxiao (Grant No. 17KJB110019).

Acknowledgments: The authors would like to express sincere thanks to the referees for careful reading and suggestions which helped us to improve the paper.

Conflicts of Interest: The authors declare no conflict of interest.

References

1. Mittag-Leffler, G.M. Sur la nouvelle fonction $E(x)$. *C. R. Acad. Sci. Paris* **1903**, *137*, 554–558.
2. Wiman, A. Über den Fundamental satz in der Theorie der Funcktionen $E(x)$. *Acta Math.* **1905**, *29*, 191–201. [CrossRef]
3. Srivastava, H.M.; Tomovski, Ž. Fractional calculus with an integral operator containing a generalized Mittag-Leffler function in the kernal. *Appl. Math. Comput.* **2009**, *211*, 198–210.
4. Attiya, A.A. Some applications of Mittag-Leffler function in the unit disk. *Filomat* **2016**, *30*, 2075–2081. [CrossRef]
5. MacGregor, T.H. Functions whose derivative has a positive real part. *Trans. Am. Math. Soc.* **1962**, *104*, 532–537. [CrossRef]
6. Tomovski, Z. Generalized Cauchy type problems for nonlinear fractional differential equations with composite fractional derivative operator. *Nonlinear Anal.* **2012**, *75*, 3364–3384. [CrossRef]
7. Tomovski, Z.; Hilfer, R.; Srivastava, H.M. Fractional and operational calculus with generalized fractional derivative operators and Mittag-Leffler type functions. *Integr. Transf. Spec. Funct.* **2010**, *21*, 797–814. [CrossRef]
8. Bansal, D.; Prajapat, J.K. Certain geometric properties of the Mittag-Leffler functions. *Complex Var. Elliptic Eq.* **2016**, *61*, 338–350. [CrossRef]
9. Grag, M.; Manohar, P.; Kalla, S.L. A Mittag-Leffler-type function of two variables. *Integral Transf. Spec. Funct.* **2013**, *24*, 934–944. [CrossRef]
10. Srivastava, H.M.; Bansal, D. Close-to-convexity of a certain family of q-Mittag-Leffler functions. *J. Nonlinear Var. Anal.* **2017**, *1*, 61–69.
11. Srivastava, H.M.; Frasin, B.A.; Pescar, V. Univalence of integral operators involving Mittag-Leffler functions. *Appl. Math. Inf. Sci.* **2017**, *11*, 635–641. [CrossRef]
12. Liu, J.-L. Notes on Jung-Kim-Srivastava integral operator. *J. Math. Anal. Appl.* **2004**, *294*, 96–103. [CrossRef]
13. Assante, D.; Cesarano, C.; Fornaro, C.; Vazquez, L. Higher order and fractional diffusive equations. *J. Eng. Sci. Technol. Rev.* **2015**, *8*, 202–204. [CrossRef]
14. Cesarano, C.; Fornaro, C.; Vazquez, L. A note on a special class of hermite polynomials. *Int. J. Pure Appl. Math.* **2015**, *98*, 261–273. [CrossRef]
15. Kapoor, G.P.; Mishra, A.K. Coefficient estimates for inverses of starlike functions of positive order. *J. Math. Anal. Appl.* **2007**, *329*, 922–934. [CrossRef]
16. Ma, W.C.; Minda, D. A unified treatment of some special classes of univalent functions. In Proceedings of the Conference on Complex Analysis, Tianjin, China, 18–23 August 1992; pp. 157–169.
17. Marin, M.; Florea, O. On temporal behavior of solutions in thermoelasticity of porous micropolar bodies. *An. St. Univ. Ovidius Constanta-Seria Math.* **2014**, *22*, 169–188.
18. Miller, S.S.; Mocanu, P.T. Differential subordinations and univalent functions. *Mich. Math. J.* **1981**, *28*, 157–171. [CrossRef]
19. Nishiwaki, J.; Owa, S. Coefficient inequalities for analytic functions. *Int. J. Math. Math. Sci.* **2002**, *29*, 285–290. [CrossRef]
20. Ruscheweyh, S. *Convolutions in Geometric Function Theory*; Les Presses de 1'Université de Montréal: Montréal, QC, Canada, 1982.

21. Seoudy, T.M.; Aouf, M.K. Coefficient estimates of new classes of q-starlike and q-convex functions of complex order. *J. Math. Inequal.* **2016**, *10*, 135–145. [CrossRef]
22. Srivastava, H.M. Some Fox-Wright generalized hypergeometric functions and associated families of convolution operators. *Appl. Anal. Discret. Math.* **2007**, *1*, 56–71.

© 2018 by the authors. Licensee MDPI, Basel, Switzerland. This article is an open access article distributed under the terms and conditions of the Creative Commons Attribution (CC BY) license (http://creativecommons.org/licenses/by/4.0/).

symmetry

MDPI

Article

Modified Kudryashov Method to Solve Generalized Kuramoto-Sivashinsky Equation

Adem Kilicman [1,*] and Rathinavel Silambarasan [2]

1 Department of Mathematics, Faculty of Science, Universiti Putra Malaysia, UPM Serdang 43400, Malaysia
2 Department of Information Technology, School of Information Technology and Engineering, Vellore Institute of Technology, Vellore 632014, India; silambu_vel@yahoo.co.in
* Correspondence: akilic@upm.edu.my

Received: 21 September 2018; Accepted: 17 October 2018; Published: 21 October 2018

check for updates

Abstract: The generalized Kuramoto–Sivashinsky equation is investigated using the modified Kudryashov method for the new exact solutions. The modified Kudryashov method converts the given nonlinear partial differential equation to algebraic equations, as a result of various steps, which upon solving the so-obtained equation systems yields the analytical solution. By this way, various exact solutions including complex structures are found, and their behavior is drawn in the 2D plane by Maple to compare the uniqueness and wave traveling of the solutions.

Keywords: generalized Kuramoto–Sivashinsky equation; modified Kudryashov method; exact solutions; Maple graphs

1. Introduction

In engineering and science, the problems arising from the wave propagation of communication between two (or) more systems such as electromagnetic waves in wireless sensor networks, water flow in dams during an earthquake, stability of the output in electricity current, viscous flows in fluid dynamics, magneto hydro dynamics, turbulence in microtides and other physical phenomena are described by the non-linear evolution equations (NLEE). In modeling such aforesaid media continuously described by the generalized Kuramoto–Sivashinsky equation (GKSE) [1] given by the nonlinear partial differential equation for $u = u(x,t)$ and non-zero constants α, β and γ:

$$u_t + uu_x + \alpha u_{xx} + \beta u_{xxx} + \gamma u_{xxxx} = 0. \tag{1}$$

The GKSE and its solutions play huge roles in flowing in viscous fluids, feedback in the output of self-loop controllers, trajectory systems and gas dynamics. The process of solving NLEE analytically and numerically uses symbolic computation procedures such as exact solution techniques and cardinal function methods such as wavelet transforms, respectively. When $\alpha = \gamma = 1$ and $\beta = 0$, Equation (1) leads to the Kuramoto–Sivashinsky equation (KSE). N. A. Kudryashov solved Equation (1) by the method of Weiss–Tabor–Carnevale and obtained exact solutions in [1]. E. J. Parkes et al. applied the tanh method for Equation (1) by taking $\alpha = \beta = 1$ and solving using the Mathematica package; they also solved Equation (1) by taking $\alpha = -1$ and $\beta = 1$ in [2]. B. Abdel-Hamid in [3] assumed the initial solution as the PDE for u and solved exactly for $\alpha = 1$ and $\beta = 0$ in Equation (1). D. Baldwin et al. [4] applied the tanh and sech methods to Equation (1) with $\alpha = \gamma = 1$ and solved using the Mathematica package. C. Li et al. [5] solved Equation (1) of the form $u_t + \beta u^\alpha u_x + \gamma u^\tau u_{xx} + \delta u_{xxxx} = 0$ using the Bernoulli equation as the auxiliary differential equation. By the simplest equation method, again, N. A. Kudryashov solved Equation (1) by considering $u_x = u^m u_x$ and obtained the solution for general m with some restrictions in [6]. A. H. Khater et al.

in [7] used Chebyshev polynomials and applied the collocation points to solve approximations of Equation (1). M. G. Porshokouhi et al. in [8] solved Equation (1) for different values of constants and approximately solved by the variational iteration method. In [9], C.M. Khalique reduced Equation (1) by Lie symmetry and solved exactly by the simplest equation method with Riccati and Bernoulli equations separately. D. Feng in [10] by taking $\beta = 0$ and $uu_x = \gamma uu_x$ in Equation (1) solved using the Riccati equation as the auxiliary differential equation. M. Lakestani et al. used the B-spline approximation function and solved Equation (1) numerically in [11], where they used tanh exact solutions for error estimations. J. Yang et al. in [12] used the sine-cosine method and dynamic bifurcation method to solve the more generalized GKSE and its related equations to Equation (1). In [13], J. Rashidinia et al. solved Equation (1) by Chebyshev wavelets. O. Acan et al. applied the reduced differential transform method to solve Equation (1) by taking $\beta = 0$ in [14].

For solving the nonlinear partial differential equations, there have been many schemes applied such as the Kudryashov method by M. Foroutan et al. in [15] and K. K. Ali et al. in [16]; the modified Kudryashov method by K. Hosseini et al. in [17,18], D. Kumar et al. in [19], A. K. Joardar et al. in [20] and A.R. Seadawy et al. in [21]; the generalized Kudryashov method by F. Mahmud et al. in [22], S. T. Demiray et al. in [23] and S. Bibi et al. in [24]; the sine-cosine method by K. R. Raslan et al. in [25]; the sine-Gordon method by H. Bulut et al. in [26]; the sinh-Gordon equation expansion method by H. M. Baskonus et al. in [27], Y. Xian-Lin et al. in [28] and A. Esen et al. in [29]; the extended trial equation method by K. A. Gepreel in [30], Y. Pandir et al. in [31] and Y. Gurefe et al. in [32]; the Exp-function method by L.K. Ravi et al. in [33], A. R. Seadawy et al. in [34] and M. Nur Alam et al. in [35]; the Jacobi elliptic function method by S. Liu et al. in [36]; the F-expansion method by A. Ebaid et al. in [37]; and the extended $\left(\frac{G'}{G}\right)$ method by E. M. E. Zayed and S. Al-Joudi et al. in [38].

The GKSE Equation (1) does not have the solution for general α and β; however, for the different values of α and β, the solution exists for (1), which can be found in [1–14]. In this work, we apply the modified Kudryashov method (MKM) to solve the GKSE in which we compute the constants α and β by the MKM. Then, for the each solution, a two-dimensional graph is drawn to show the wave traveling.

2. Analysis of the Modified Kudryashov Method

The modified Kudryashov method involves the following steps in solving the nonlinear partial differential equations (NLPDE) [17–21]:

Step 1. Consider the given NLPDE of the following form $u = u(x,t)$.

$$P\left(u, u_t, u_x, u_{tt}, u_{xx}, u_{xt}, \cdots\right) = 0. \tag{2}$$

Step 2. Apply the wave transformation $u(x,t) = u(\eta)$ in Equation (2), where:

$$\eta = \mu(x - \lambda t). \tag{3}$$

Here, μ is the wave variable and λ is the velocity; both are non-zero constants. Hence, Equation (2) transforms to the following ODE:

$$O\left(u, u^{'}, u^{''}, uu^{'}, \cdots\right) = 0, \tag{4}$$

where the prime represents the derivative with respect to η.

Step 3. Let the initial solution guess of Equation (4) be,

$$u(\eta) = A_0 + \sum_{i=1}^{N} A_i \left[Q(\eta)\right]^i, \tag{5}$$

where N is a non-zero and positive constant calculated by the principle of homogeneous balancing of Equation (4), A_i; $i = 0, 1, 2, \cdots$ are unknowns to be calculated and $Q(\eta)$ is the solution of the following auxiliary ODE:

$$\frac{dQ(\eta)}{d\eta} = Q(\eta)\left[Q(\eta) - 1\right]\ln(a);\ a \neq 1, \tag{6}$$

given by,

$$Q(\eta) = \frac{1}{1 + Da^{\eta}}, \tag{7}$$

where D is the integral constant and we assume $D = 1$.

Step 4. Substituting Equations (5) and (6) in Equation (4) leads to the polynomial in $Q(\eta)^i$; $i = 0, 1, 2, \cdots$. As $Q(\eta)^i \neq 0$, so collecting its coefficients and then equating to zero give the systems of overdetermined algebraic equations, which upon solving give the unknowns of Equations (3) and (5).

Step 5. Finally, substituting the values of Step 4 in Equation (5) and then in Equation (3) gives the solution $u(x, t)$ of Equation (2).

3. MKM Application to Solve the Generalized Kuramoto–Sivashinsky Equation

Applying the wave transformation with Equation (3) to Equation (1) leads to the ODE, and then, integrating once the ODE by taking integration constant to zero transforms to the following ODE:

$$-\lambda u + \frac{u^2}{2} + \alpha\mu u^{(1)} + \beta\mu^2 u^{(2)} + \gamma\mu^3 u^{(3)} = 0, \tag{8}$$

where $u = u(\eta)$ and the superscripts (.) represent the derivatives w. r. t. η. By the homogeneous balancing of Equation (8), $N = 3$, and hence, the initial guess solution of Equation (8) from Equation (5) is given by,

$$u(\eta) = A_0 + A_1 Q(\eta) + A_2\left(Q(\eta)\right)^2 + A_3\left(Q(\eta)\right)^3. \tag{9}$$

Substituting Equations (6) and (9) in Equation (8) results in the sixth order polynomial of $Q(\eta)$. Collecting the coefficients of $\left(Q(\eta)\right)^i$; $i = 0, 1, \cdots, 6$ and equating each coefficient to zero gives the systems of algebraic equations, which upon solving by Maple give the unknowns in Equations (9), (3) and (α, β) in Equation (8). The resulting values are substituted in Equation (9) along with Equations (3) and (7), which give the exact solution of Equation (1) for the specific values of constants α and β. Substituting the α and β values in Equation (1) and the unknowns A_i; $i = 0, 1, 2, 3$ in Equation (9) where $Q(\eta)$ is given by Equation (7) yields the following exact solutions. Let $\delta_1 = \gamma\mu\ln(a)$, $\delta_2 = \gamma\mu^2\ln(a)^2$ and $\delta_3 = \gamma\mu^3\ln(a)^3$ in the following cases.

Case 1. For $\alpha = \delta_2$ and $\beta = 4\delta_1$ in Equation (1), the unknown coefficients are given by,

$$A_0 = A_1 = 0,\ A_2 = 120\delta_3,\ A_3 = -120\delta_3,\ \lambda = 6\delta_3.$$

Therefore, the exact solution of Equation (1) is given by (Figure 1),

$$u_1(x,t) := \frac{120\delta_3 a^{\mu x - 6\delta_3 \mu t}}{\left(1 + a^{\mu x - 6\delta_3 \mu t}\right)^3}. \tag{10}$$

Further, for the same α and β value, the second set of unknown coefficients are given by,

$$A_0 = -12\delta_3,\ A_1 = 0,\ A_2 = 120\delta_3,\ A_3 = -120\delta_3,\ \lambda = -6\delta_3.$$

Therefore, the exact solution of Equation (1) is given by (Figure 1),

$$u_2(x,t) := -\frac{12\delta_3\left(1 + a^{3\mu x}e^{3(6\delta_3\mu \ln(a)t)} + 3a^{2\mu x}e^{2(6\delta_3\mu \ln(a)t)} - 7a^{\mu x}e^{6\delta_3\mu \ln(a)t}\right)}{\left(1 + a^{\mu x}e^{6\delta_3\mu \ln(a)t}\right)^3}. \tag{11}$$

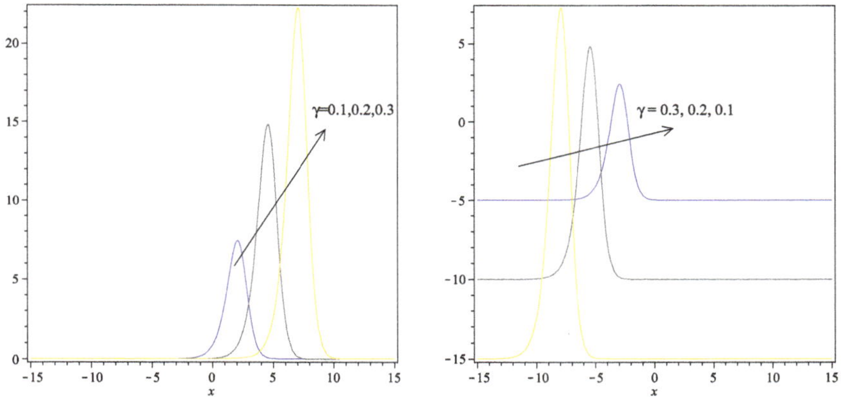

(a) $u_1(x,t)$ (b) $u_2(x,t)$

Figure 1. Solutions in Case 1, Equations (10) and (11), respectively from left to right for $a = 5$, $\mu = 1$ and $t = 1$ in $x \in [-15, 15]$ for different values of γ.

Case 2. For $\alpha = \delta_2$ and $\beta = -4\delta_1$ in Equation (1), the unknown coefficients are given by,

$$A_0 = 0,\ A_1 = -120\delta_3,\ A_2 = 240\delta_3,\ A_3 = -120\delta_3,\ \lambda = -6\delta_3.$$

Therefore, the exact solution of Equation (1) is given by (Figure 2),

$$u_3(x,t) := -\frac{120\delta_3 a^{2(\mu x + 6\delta_3 \mu t)}}{\left(1 + a^{\mu x + 6\delta_3 \mu t}\right)^3}. \tag{12}$$

Further, for the same α and β value, the second set of unknown coefficients are given by,

$$A_0 = 12\delta_3,\ A_1 = -120\delta_3,\ A_2 = 240\delta_3,\ A_3 = -120\delta_3,\ \lambda = 6\delta_3.$$

Therefore, the exact solution of Equation (1) is given by (Figure 2),

$$u_4(x,t) := \frac{12\delta_3\left(a^{3(\mu x - 6\delta_3\mu t)} - 7a^{2(\mu x - 6\delta_3\mu t)} + 3a^{\mu x - 6\delta_3\mu t} + 1\right)}{\left(1 + a^{\mu x - 6\delta_3 \mu t}\right)^3}. \tag{13}$$

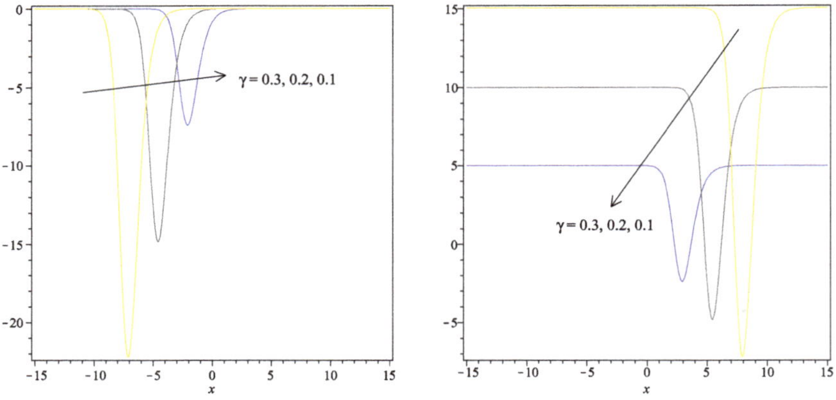

(**a**) $u_3(x,t)$ (**b**) $u_4(x,t)$

Figure 2. Solutions in Case 2, Equations (12) and (13), respectively from left to right for $a = 5$, $\mu = 1$ and $t = 1$ in $x \in [-15, 15]$ for different values of γ.

Case 3. For $\alpha = -19\delta_2$ and $\beta = 0$ in Equation (1), the unknown coefficients are given by,

$$A_0 = -60\delta_3,\ A_1 = 0,\ A_2 = 180\delta_3,\ A_3 = -120\delta_3,\ \lambda = -30\delta_3.$$

Therefore, the exact solution of Equation (1) is given by (Figure 3),

$$u_5(x,t) := -\frac{60\delta_3 e^{2(30\delta_3\mu\ln(a)t)}\left(a^{3\mu x}e^{30\delta_3\mu\ln(a)t} + 3a^{2\mu x}\right)}{\left(1 + a^{\mu x}e^{30\delta_3\mu\ln(a)t}\right)^3}. \tag{14}$$

Further, for the same α and β value, the second set of unknown coefficients are given by,

$$A_0 = A_1 = 0,\ A_2 = 180\delta_3,\ A_3 = -120\delta_3,\ \lambda = 30\delta_3.$$

Therefore, the exact solution of Equation (1) is given by (Figure 3),

$$u_6(x,t) := \frac{60\delta_3\left(1 + 3a^{\mu x - 30\delta_3\mu t}\right)}{\left(1 + a^{\mu x - 30\delta_3\mu t}\right)^3}. \tag{15}$$

Case 4. For $\alpha = 47\delta_2$ and $\beta = 12\delta_1$ in Equation (1), the unknown coefficients are given by,

$$A_0 = A_1 = A_2 = 0,\ A_3 = -120\delta_3,\ \lambda = -60\delta_3.$$

Therefore, the exact solution of Equation (1) is given by (Figure 4),

$$u_7(x,t) := -\frac{120\delta_3}{\left(1 + a^{\mu x + 60\delta_3\mu t}\right)^3}. \tag{16}$$

Further, for the same α and β, the second set of unknown coefficients are given by,

$$A_0 = 120\delta_3,\ A_1 = A_2 = 0,\ A_3 = -120\delta_3,\ \lambda = 60\delta_3.$$

Therefore, the exact solution of Equation (1) is given by (Figure 4),

$$u_8(x,t) := \frac{120\delta_3\left(3a^{\mu x}e^{2(60\delta_3\mu\ln(a)t)} + 3a^{2\mu x}e^{60\delta_3\mu\ln(a)t} + a^{3\mu x}\right)}{\left(a^{\mu x} + e^{60\delta_3\mu\ln(a)t}\right)^3}. \quad (17)$$

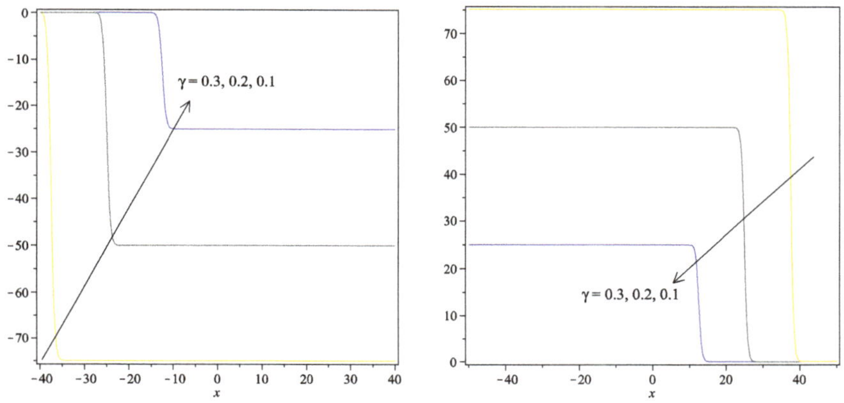

(a) $u_5(x,t)$ (b) $u_6(x,t)$

Figure 3. Solutions in Case 3, Equations (14) and (15), respectively from left to right for $a = 5$, $\mu = 1$ and $t = 1$ in $x \in [-40, 40]$ for $u_5(x,t)$ and in $x \in [-50, 50]$ for $u_6(x,t)$ for different values of γ.

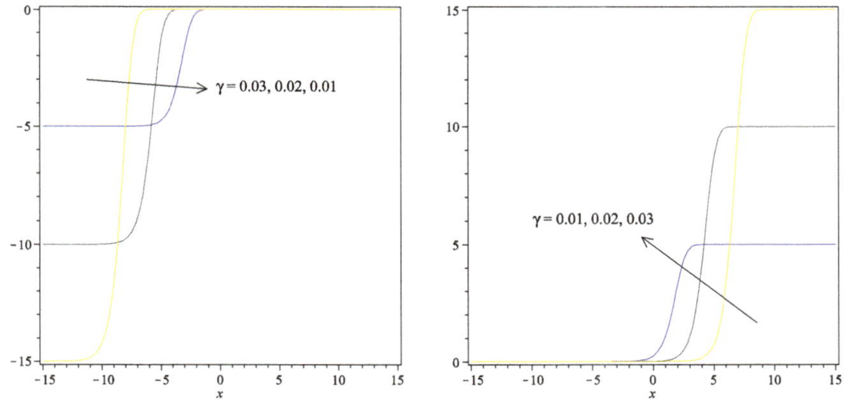

(a) $u_7(x,t)$ (b) $u_8(x,t)$

Figure 4. Solutions in Case 4, Equations (16) and (17), respectively from left to right for $a = 5$, $\mu = 1$ and $t = 1$ in $x \in [-15, 15]$ for different values of γ.

Case 5. For $\alpha = 47\delta_2$ and $\beta = -12\delta_1$ in Equation (1), the unknown coefficients are given by,

$$A_0 = 0,\ A_1 = -360\delta_3,\ A_2 = 360\delta_3,\ A_3 = -120\delta_3,\ \lambda = -60\delta_3.$$

Therefore, the exact solution of Equation (1) is given by (Figure 5),

$$u_9(x,t) := -\frac{120\delta_3\left(3a^{2\mu x}e^{2(60\delta_3\mu\ln(a)t)} + 3a^{\mu x}e^{60\delta_3\mu\ln(a)t} + 1\right)}{\left(1 + a^{\mu x}e^{60\delta_3\mu\ln(a)t}\right)^3}. \quad (18)$$

Further, for the same α and β value, the second set of unknown coefficients are given by,

$$A_0 = 120\delta_3,\ A_1 = -360\delta_3,\ A_2 = 360\delta_3,\ A_3 = -120\delta_3,\ \lambda = 60\delta_3.$$

Therefore, the exact solution of Equation (1) is given by (Figure 5),

$$u_{10}(x,t) := \frac{120\delta_3 a^{3(\mu x - 60\delta_3\mu t)}}{\left(1 + a^{\mu x - 60\delta_3\mu t}\right)^3}. \quad (19)$$

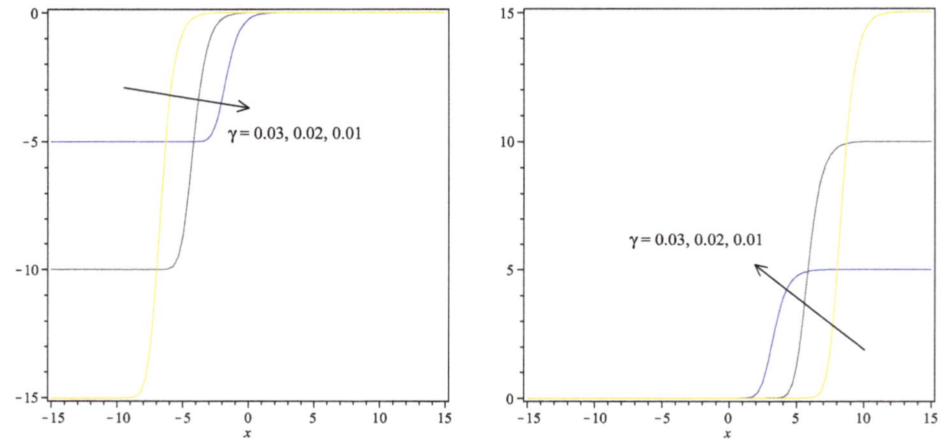

(a) $u_9(x,t)$ **(b)** $u_{10}(x,t)$

Figure 5. Solutions in Case 5, Equations (18) and (19), respectively from left to right for $a = 5$, $\mu = 1$ and $t = 1$ in $x \in [-15, 15]$ for different values of γ.

Case 6. For $\alpha = 73\delta_2$ and $\beta = 16\delta_1$ in Equation (1), the unknown coefficients are given by,

$$A_0 = 180\delta_3,\ A_1 = 0,\ A_2 = -60\delta_3,\ A_3 = -120\delta_3,\ \lambda = 90\delta_3.$$

Therefore, the exact solution of Equation (1) is given by (Figure 6),

$$u_{11}(x,t) := \frac{60\delta_3\left(8a^{\mu x}e^{2(90\delta_3\mu\ln(a)t)} + 9a^{2\mu x}e^{90\delta_3\mu\ln(a)t} + 3a^{3\mu x}\right)}{\left(e^{90\delta_3\mu\ln(a)t} + a^{\mu x}\right)^3}. \quad (20)$$

Further, for the same α and β value, the second set of unknown coefficients are given by,

$$A_0 = A_1 = 0,\ A_2 = -60\delta_3,\ A_3 = -120\delta_3,\ \lambda = -90\delta_3.$$

Therefore, the exact solution of Equation (1) is given by (Figure 6),

$$u_{12}(x,t) := -\frac{60\delta_3\left(3 + a^{\mu x + 90\delta_3\mu t}\right)}{\left(1 + a^{\mu x + 90\delta_3\mu t}\right)^3}. \tag{21}$$

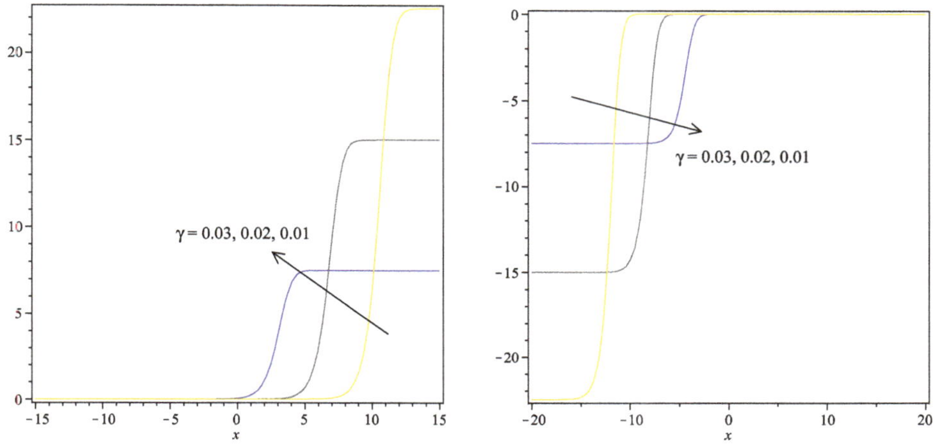

(a) $u_{11}(x,t)$ **(b)** $u_{12}(x,t)$

Figure 6. Solutions in Case 6, Equations (20) and (21), respectively from left to right for $a = 5$, $\mu = 1$ and $t = 1$ in $x \in [-15, 15]$ for $u_{11}(x,t)$ and $x \in [-20, 20]$ for $u_{12}(x,t)$ for different values of γ.

Case 7. For $\alpha = 73\delta_2$ and $\beta = -16\delta_1$ in Equation (1), the unknown coefficients are given by,

$$A_0 = 180\delta_3,\ A_1 = -480\delta_3,\ A_2 = 420\delta_3,\ A_3 = -120\delta_3,\ \lambda = 90\delta_3.$$

Therefore, the exact solution of Equation (1) is given by (Figure 7),

$$u_{13}(x,t) := \frac{60\delta_3\left(a^{2\mu x} e^{90\delta_3\mu \ln(a) t} + 3a^{3\mu x}\right)}{\left(e^{90\delta_3\mu \ln(a) t} + a^{\mu x}\right)^3}. \tag{22}$$

Further, for the same α and β value, the second set of unknown coefficients are given by,

$$A_0 = 0,\ A_1 = -480\delta_3,\ A_2 = 420\delta_3,\ A_3 = -120\delta_3,\ \lambda = -90\delta_3.$$

Therefore, the exact solution of Equation (1) is given by (Figure 7),

$$u_{14}(x,t) := -\frac{60\delta_3\left(8a^{2\mu x} e^{2(90\delta_3\mu \ln(a) t)} + 9a^{\mu x} e^{90\delta_3\mu \ln(a) t} + 3\right)}{\left(1 + a^{\mu x} e^{90\delta_3\mu \ln(a) t}\right)^3}. \tag{23}$$

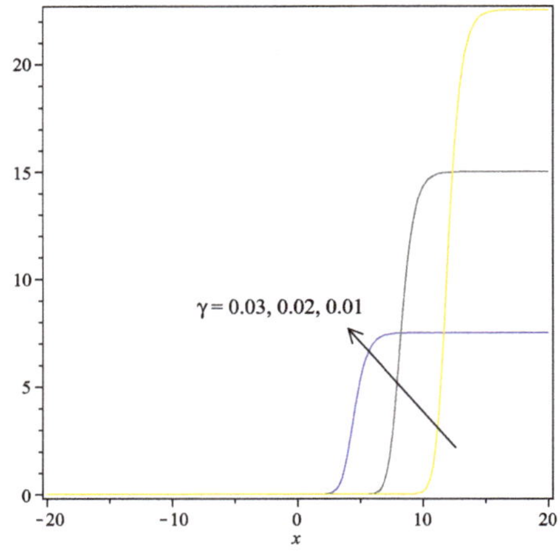

(a) $u_{13}(x,t)$

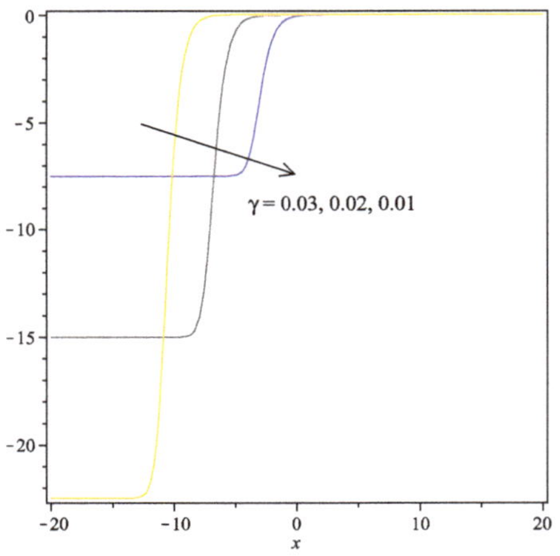

(b) $u_{14}(x,t)$

Figure 7. Solutions in Case 7, Equations (22) and (23), respectively from left to right for $a = 5$, $\mu = 1$ and $t = 1$ in $x \in [-20, 20]$ for different values of γ.

Case 8. For $\alpha = \frac{19}{11}\delta_2$ and $\beta = 0$ in Equation (1), the unknown coefficients are given by,

$$A_0 = \frac{60}{11}\delta_3,\ A_1 = -\frac{720}{11}\delta_3,\ A_2 = 180\delta_3,\ A_3 = -120\delta_3,\ \lambda = \frac{30}{11}\delta_3.$$

Therefore, the exact solution of Equation (1) is given by (Figure 8),

$$u_{15}(x,t) := \frac{60\delta_3 a^{\left(\mu x - \frac{30}{11}\delta_3\mu t\right)}\left(a^{2\left(\mu x - \frac{30}{11}\delta_3\mu t\right)} - 9a^{\left(\mu x - \frac{30}{11}\delta_3\mu t\right)} + 12\right)}{11\left(1 + a^{\left(\mu x - \frac{30}{11}\delta_3\mu t\right)}\right)^3}. \tag{24}$$

Further, for the same α and β value, the second set of unknown coefficients are given by,

$$A_0 = 0,\ A_1 = -\frac{720}{11}\delta_3,\ A_2 = 180\delta_3,\ A_3 = -120\delta_3,\ \lambda = -\frac{30}{11}\delta_3.$$

Therefore, the exact solution of Equation (1) is given by (Figure 8),

$$u_{16}(x,t) := -\frac{60\delta_3\left(1 - 9a^{\left(\mu x + \frac{30}{11}\delta_3\mu t\right)} + 12a^{2\left(\mu x + \frac{30}{11}\delta_3\mu t\right)}\right)}{11\left(1 + a^{\left(\mu x + \frac{30}{11}\delta_3\mu t\right)}\right)^3}. \tag{25}$$

Case 9. For $\alpha = -\delta_2$ and $\beta = 4i\delta_1$ in Equation (1), the unknown coefficients are given by,

$$A_0 = 0,\ A_1 = -60\mu^3\ln(a)^3\left(\gamma - i\gamma\right),\ A_2 = 60(3-i)\delta_3,\ A_3 = -120\delta_3,\ \lambda = 4i\delta_3.$$

Therefore, the exact complex solution of Equation (1) is given by,

$$u_{17}(x,t) := \frac{60\delta_3 a^{\mu x - 4i\delta_3\mu t}\left(i + 1 + (i-1)\,a^{\mu x - 4i\delta_3\mu t}\right)}{\left(1 + a^{\mu x - 4i\delta_3\mu t}\right)^3}. \tag{26}$$

The 2D graph of real and imaginary parts of $u_{17}(x,t)$ are drawn in Figure 9.

Further, for the same α and β value, the second set of unknown coefficients are given by,

$$A_0 = -8i\delta_3,\ A_1 = -60\mu^3 \ln(a)^3 (\gamma - i\gamma),\ A_2 = 60(3-i)\delta_3,\ A_3 = -120\delta_3,\ \lambda = -4i\delta_3.$$

Therefore, the exact complex solution of Equation (1) is given by,

$$u_{18}(x,t) := -\frac{8\delta_3}{(1+a^{\mu x+4i\delta_3\mu t})^3}\left[i(1+a^{3(\mu x+4i\delta_3\mu t)}) + \left(\frac{15-9i}{2}\right)a^{2(\mu x+4i\delta_3\mu t)} - \left(\frac{15+9i}{2}\right)a^{\mu x+4i\delta_3\mu t}\right]. \quad (27)$$

where $i = \sqrt{-1}$. The 2D graphs of the real and imaginary parts of $u_{18}(x,t)$ are drawn in Figure 10.

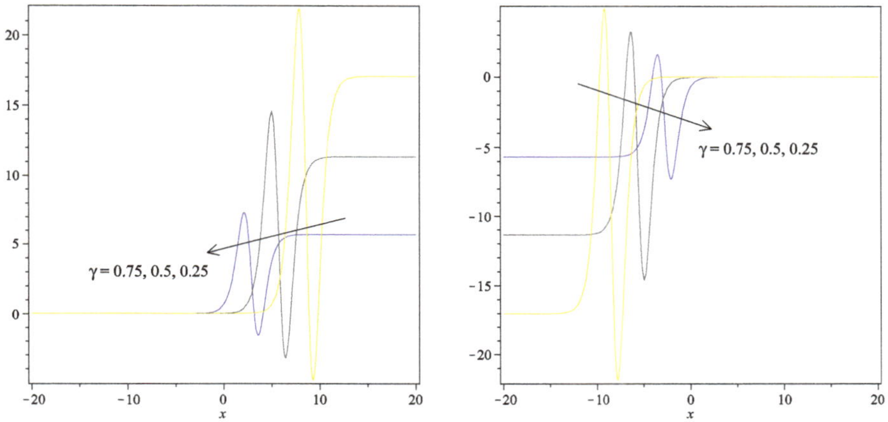

(a) $u_{15}(x,t)$ (b) $u_{16}(x,t)$

Figure 8. Solutions in Case 8, Equations (24) and (25), respectively from left to right for $a = 5$, $\mu = 1$ and $t = 1$ in $x \in [-20, 20]$ for different values of γ.

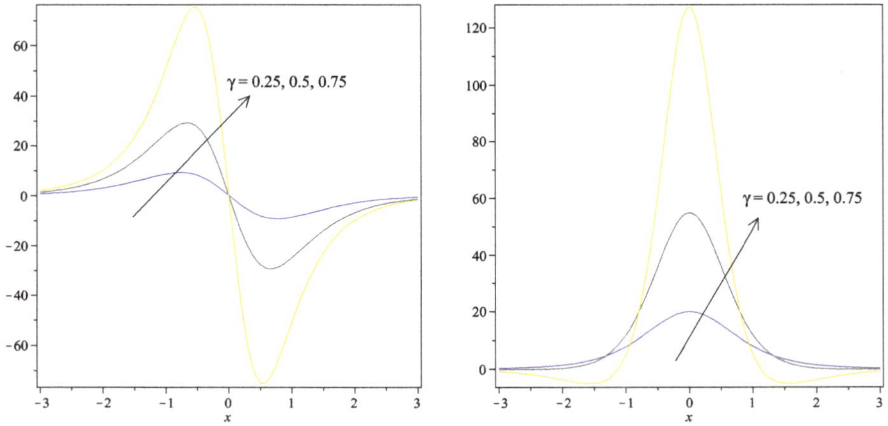

(a) Real part of $u_{17}(x,t)$ (b) Imaginary part of $u_{17}(x,t)$

Figure 9. Real and imaginary part of the solution in Case 9, Equation (26), respectively from left to right for $a = 5$, $\mu = 1$ and $t = 1$ in $x \in [-3, 3]$ for different values of γ.

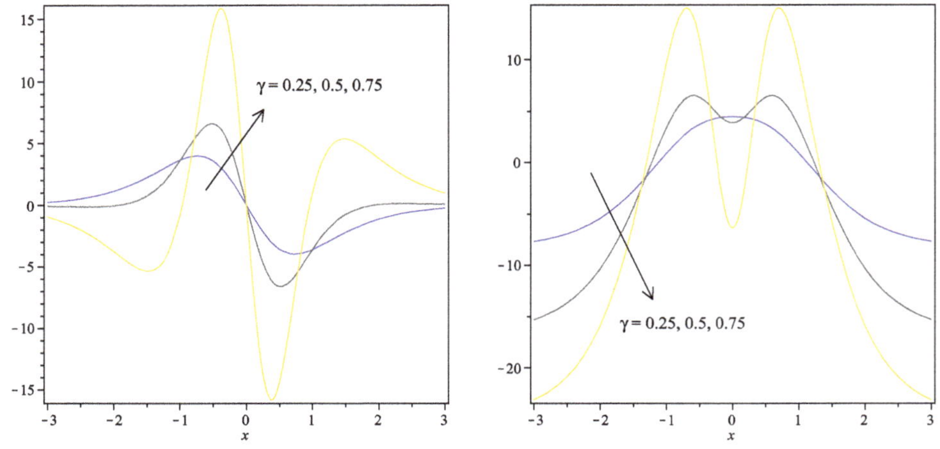

(a) Real part of $u_{18}(x,t)$ **(b)** Imaginary part of $u_{18}(x,t)$

Figure 10. Real and imaginary part of the solution in Case 9, Equation (27), respectively from left to right for $a = 5$, $\mu = 1$ and $t = 1$ in $x \in [-3, 3]$ for different values of γ.

Case 10. For $\alpha = -\delta_2$ and $\beta = -4i\delta_1$ in Equation (1), the unknown coefficients are given by,

$$A_0 = 0,\ A_1 = -60\mu^3 \ln(a)^3(\gamma + i\gamma),\ A_2 = 60(3+i)\delta_3,\ A_3 = -120\delta_3,\ \lambda = -4i\delta_3.$$

Therefore, the exact complex solution of Equation (1) is given by,

$$u_{19}(x,t) := -\frac{60\delta_3 a^{\mu x + 4i\delta_3\mu t}\left(i - 1 + (i+1)a^{\mu x + 4i\delta_3\mu t}\right)}{\left(1 + a^{\mu x + 4i\delta_3\mu t}\right)^3}. \tag{28}$$

The 2D graphs of real and imaginary parts of $u_{19}(x,t)$ are drawn in Figure 11.

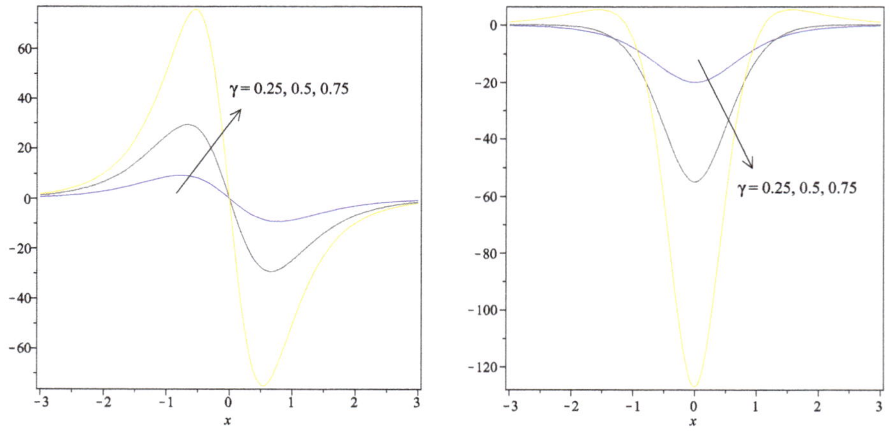

(a) Real part of $u_{19}(x,t)$ **(b)** Imaginary part of $u_{19}(x,t)$

Figure 11. Real and imaginary part of the solution in Case 10, Equation (28), respectively from left to right for $a = 5$, $\mu = 1$ and $t = 1$ in $x \in [-3, 3]$ for different values of γ.

Further, for the same α and β value, the second set of unknown coefficients are given by,

$$A_0 = 8i\delta_3,\ A_1 = -60\mu^3 \ln(a)^3(\gamma + i\gamma),\ A_2 = 60(3+i)\delta_3,\ A_3 = -120\delta_3,\ \lambda = 4i\delta_3.$$

Therefore, the exact complex solution of Equation (1) is given by,

$$u_{20}(x,t) := \frac{8\delta_3}{\left(1+a^{\mu x - 4i\delta_3 \mu t}\right)^3}\left[i(1+a^{3(\mu x-4i\delta_3\mu t)}) - \left(\frac{15+9i}{2}\right)a^{2(\mu x-4i\delta_3\mu t)} + \left(\frac{15-9i}{2}\right)a^{\mu x-4i\delta_3\mu t}\right]. \quad (29)$$

where $i = \sqrt{-1}$. The 2D graphs of the real and imaginary parts of $u_{20}(x,t)$ are drawn in Figure 12.

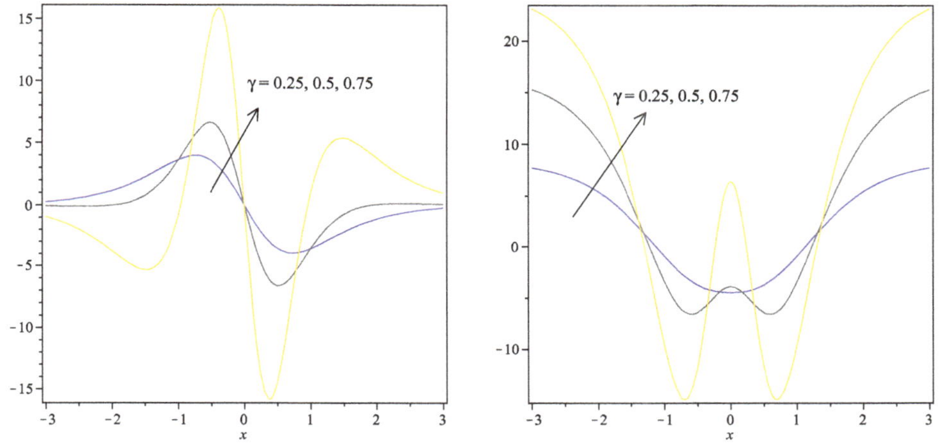

(a) Real part of $u_{20}(x,t)$ **(b)** Imaginary part of $u_{20}(x,t)$

Figure 12. Real and imaginary part of the solution in Case 10, Equation (29), respectively from left to right for $a = 7$, $\mu = 1$ and $t = 1$ in $x \in [-3,3]$ for different values of γ.

4. Conclusions

In this work, the generalized Kuramoto–Sivashinsky equation is solved, and the exact solutions have been found. The aforesaid GKSE has solutions for the different values of α and β, which we obtained by the application of the modified Kudryashov method, and we found 10 classes of (α, β) pairs and their corresponding two distinct exact solutions for each pair of Equation (1) in Cases 1–10. The two-dimensional simulations of the solutions in Figures 1–12 show their behavioral pattern and wave train traveling for different values of γ. However, the wave structures vary when the values of a, μ, t and the domain changes in the 2D plane. The solutions found in this work will be useful in studying electromagnetic waves, fluid flows and the areas where GKSE plays a vital role. All the solutions are validated in the Maple computer algebra system by substituting them in the original equation. Our new solutions are compared with the previous solutions of GKSE in Appendices A and B.

Author Contributions: Both authors contributed equally. Both authors read and approved the final manuscript.

Funding: The research is partially supported by the University Putra Malaysia research grant having vot number UPM-GPB/2017/9543000.

Conflicts of Interest: The authors declare no conflict of interest.

Appendix A. GKSE in the Previous Studies

N.A. Kudryashov in [6] solved for the exact solution of Equation (1). Based on the homogeneous balancing, he has taken the following initial solution.

$$u(\eta) = A_0 + A_1 g(\eta) + A_2 g(\eta)^2 + A_3 g(\eta)^3.$$

where $g(\eta)$ is the solution of $\frac{dg(\eta)}{d\eta} = b - g(\eta)^2$, and obtained the following values.

1.

$$A_0 = -\frac{\beta^3}{576\gamma^2},\ A_1 = \frac{5\beta^2}{4\gamma},\ A_2 = -15\beta,\ A_3 = 120\gamma,\ \alpha = \frac{47\beta^2}{144\gamma},\ b = \frac{\beta^2}{576\gamma^2},\ C_0 = -\frac{5\beta^3}{144\gamma^2}.$$

2.

$$A_0 = \frac{30\beta^3}{128\gamma^2},\ A_1 = -\frac{30\beta^2}{16\gamma},\ A_2 = -30\beta,\ A_3 = 120\gamma,\ \alpha = \frac{\beta^2}{16\gamma},\ b = \frac{\beta^2}{64\gamma^2},\ C_0 = \frac{3\beta^3}{32\gamma^2}.$$

In the same work, he solved Equation (1) with the auxiliary equations $\left(\frac{dg(z)}{dz}\right)^2 + 4g(z)^3 - ag(z)^2 - 2bg(z) + d = 0$ and $\frac{d^2g(z)}{dz^2} + 6g(z)^2 - ag(z) - b = 0$ and obtained other values for unknowns.

C.M. Khalique in [9] solved Equation (1) by taking the Bernoulli equation $\frac{dh(\eta)}{d\eta} = ah(\eta) + bh(\eta)^2$ and Riccati equation $\frac{dh(\eta)}{d\eta} = ah(\eta)^2 + bh(\eta) + c$ as the auxiliary ODE and obtained the following values respectively by using each ODE. For both the auxiliary equation the constant values are $a = 1,\ b = 3$ and $c = 1$:

1.

$$A_0 = \nu - 6a^3\gamma,\ A_1 = -120a^2b\gamma,\ A_2 = 240ab^2\gamma,\ A_3 = -120b^3\gamma,\ \alpha = a^2\gamma,\ \beta = 4a\gamma.$$

2.

$$A_0 = -990\gamma + 60\gamma k + \nu,\ A_1 = 60\gamma + 180\gamma k,\ A_2 = 60\gamma k,\ A_3 = -120\gamma,\ \alpha = 365\gamma,\ \beta = -36\gamma - 4\gamma k.$$

While comparing the above values, our solutions of Equation (1) in this work are new to the surveyed literature.

Appendix B. Studying GKSE by GKM and SGEEM

1. For solving Equation (1) by the generalized Kudryashov method [22–24], the homogeneous balancing of Equation (8) gives $N = M + 3$, which has infinite solutions. For the value $M = 1$, this gives $N = 4$. Therefore,

$$u(\eta) = \frac{A_0 + A_1 Q(\eta) + A_2 (Q(\eta))^2 + A_3 (Q(\eta))^3 + A_4 (Q(\eta))^4}{B_0 + B_1 Q(\eta)}.$$

   where $Q(\eta)$ is the solution of $\frac{dQ(\eta)}{d\eta} = Q(\eta)(Q(\eta) - 1)$, Applying these equations to Equation (8) leads to the polynomial in $Q(\eta)$ and its powers. Collecting the coefficients of $(Q(\eta))^i;\ i = 0, 1, 2, \cdots$ and attempting to solve the overdetermined equations results in the continuous execution of Maple. Hence, we conclude that Equation (1) cannot be solved by the generalized Kudryashov method.
2. Next, for solving Equation (1) by the sine-Gordon equation expansion method [26], the homogeneous balancing is the same as the MKM given by $N = 3$. Thus,

$$u(\eta) = A_0 + A_1 \tanh(\eta) + B_1 \mathrm{sech}(\eta) + A_2 \tanh^2(\eta) + B_2 \tanh(\eta)\mathrm{sech}(\eta) + A_3 \tanh^3(\eta) + B_3 \tanh^2(\eta)\mathrm{sech}(\eta).$$

Substituting the above equation $u(\eta)$ in Equation (8) and following the steps in [26] lead to the polynomials in $\sin(w)$, $\cos(w)$, their products and powers. Collecting the coefficients, equating them to zero and solving in Maple result in the continuous execution. Thus, we conclude that Equation (1) cannot be solved by the sine-Gordon equation expansion method either.

References

1. Kudryashov, N.A. Exact solutions of the generalized Kuramoto-Sivashinsky equation. *Phys. Lett. A* **1990**, *147*, 287–290. [CrossRef]
2. Parkes, E.J.; Duffy, B.R. An automated tanh-function method for finding solitary wave solutions to non-linear evolution equations. *Comput. Phys. Commun.* **1996**, *98*, 288–300. [CrossRef]
3. Hamid, B.A. An exact solution to the Kuramoto-Sivashinsky equation. *Phys. Lett. A* **1999**, *263*, 338–340. [CrossRef]
4. Baldwin, D.; Göktas, Ü.; Hereman, W.; Hong, L.; Martino, R.S.; Miller, J.C. Symbolic computation of exact solutions expressible in hyperbolic and elliptic functions for nonlinear PDEs. *J. Symb. Comput.* **2004**, *37*, 669–705. [CrossRef]
5. Li, C.; Chen, G.; Zhao, S. Exact travelling wave solutions to the generalized Kuramoto-Sivashinsky equation. *Latin Am. Appl. Res.* **2004**, *34*, 65–68.
6. Kudryashov, N.A. Solitary and Periodic Solutions of the Generalized Kuramoto-Sivashinsky Equation. *Regul. Chaotic Dyn.* **2008**, *13*, 234–238. [CrossRef]
7. Khater, A.H.; Temsah, R.S. Numerical solutions of the generalized Kuramoto-Sivashinsky equation by Chebyshev spectral collocation methods. *Comput. Math. Appl.* **2008**, *56*, 1465–1472. [CrossRef]
8. Porshokouhi, M.G.; Ghanbari, B. Application of He's variational iteration method for solution of the family of Kuramoto-Sivashinsky equations. *J. King Saud Univ.-Sci.* **2011**, *23*, 407–411. [CrossRef]
9. Khalique, C.M. Exact Solutions of the Generalized Kuramoto-Sivashinsky Equation. *Caspian J. Math. Sci.* **2012**, *1*, 109–116.
10. Feng, D. Exact Solutions of Kuramoto-Sivashinsky Equation. *Int. J. Educ. Manag. Eng.* **2012**, *6*, 61–66. [CrossRef]
11. Lakestani, M.; Dehghan, M. Numerical solutions of the generalized Kuramoto-Sivashinsky equation using B-spline functions. *Appl. Math. Model.* **2012**, *36*, 605–617. [CrossRef]
12. Yang, J.; Lu, X.; Tang, X. Exact travelling wave solutions for the generalized Kuramoto-Sivashinsky equation. *J. Math. Sci. Adv. Appl.* **2015**, *31*, 1–13.
13. Rashidinia, J.; Jokar, M. Polynomial scaling functions for numerical solution of generalized Kuramoto-Sivashinsky equation. *Appl. Anal.* **2015**, *10*, 1–10. [CrossRef]
14. Acan, O.; Keskin, Y. Approximate solution of Kuramoto-Sivashinsky equation using reduced differential transform method. *AIP Conf. Proc.* **2015**, *1648*, 470003-1–470003-4.
15. Foroutan, M.; Manafian, J.; Taghipour-Farshi, H. Exact solutions for Fitzhugh-Nagumo model of nerve excitation via Kudryashov method. *Opt. Quantum Electron.* **2017**, *49*, 352. [CrossRef]
16. Ali, K.K.; Nuruddeen, R.I.; Hadhoud, A.R. New exact solitary wave solutions for the extended (3 + 1)-dimensional Jimbo-Miwa equations. *Results Phys.* **2018**, *9*, 12–16. [CrossRef]
17. Hosseini, K.; Mayeli, P.; Ansari, R. Modified Kudryashov method for solving the conformable time-fractional Klein-Gordon equations with quadratic and cubic nonlinearities. *Optik* **2017**, *130*, 737–742. [CrossRef]
18. Hosseini, K.; Ansari, R. New exact solutions of nonlinear conformable time-fractional Boussinesq equations using the modified Kudryashov method. *Waves Random Complex Med.* **2017**, *27*, 628–636. [CrossRef]
19. Kumar, D.; Seadawy, A.R.; Joardar, A.K. Modified Kudryashov method via new exact solutions for some conformable fractional differential equations arising in mathematical biology. *Chin. J. Phys.* **2018**, *56*, 75–85. [CrossRef]
20. Joardar, A.K.; Kumar, D.; al Woadud, K.M.A. New exact solutions of the combined and double combined sinh-cosh-Gordon equations via modified Kudryashov method. *Int. J. Phys. Res.* **2018**, *1*, 25–30. [CrossRef]
21. Seadawy, A.R.; Kumar, D.; Hosseini, K.; Samadani, F. The system of equations for the ion sound and Langmuir waves and its new exact solutions, *Results Phys.* **2018**, *9*, 1631–1634.
22. Mahmud, F.; Samsuzzoha, M.; Akbar, M.A. The generalized Kudryashov method to obtain exact traveling wave solutions of the PHI-four equation and the Fisher equation. *Results Phys.* **2017**, *7*, 4296–4302. [CrossRef]

23. Demiray, S.T.; Pandir, Y.; Bulut, H. Generalized Kudryashov Method for Time-Fractional Differential Equations. *Abstract Appl. Anal.* **2014**, *2014*, 901540.
24. Bibi, S.; Ahmed, N.; Khan, U.; Mohyud-Din, S.T. Some new exact solitary wave solutions of the van der Waals model arising in nature. *Results Phys.* **2018**, *9*, 648–655. [CrossRef]
25. Raslan, K.R.; L-Danaf, T.S.E.; Ali, K.K. New exact solutions of coupled generalized regularized long wave equations. *J. Egypt. Math. Soc.* **2017**, *25*, 400–405. [CrossRef]
26. Bulut, H.; Sulaiman, T.A.; Baskonus, H.M.; Sandulyak, A.A. New solitary and optical wave structures to the (1 + 1)-dimensional combined KdV-mKdV equation. *Optik* **2017**, *135*, 327–336. [CrossRef]
27. Baskonus, H.M.; Sulaiman, T.A.; Bulut, H. Dark, bright and other optical solitons to the decoupled nonlinear Schrödinger equation arising in dual-core optical fibers. *Opt. Quantum Electron.* **2018**, *50*, 165. [CrossRef]
28. Yang, X.-L.; Tang, J.-S. Travelling Wave Solutions for Konopelchenko-Dubrovsky Equation Using an Extended sinh-Gordon Equation Expansion Method. *Commun. Theor. Phys.* **2008**, *50*, 1047–1051.
29. Esen, A.; Sulaiman, T.A.; Bulut, H.; Baskonus, H.M. Optical solitons to the space-time fractional (1+1)-dimensional coupled nonlinear Schrödinger equation. *Optik* **2018**, *167*, 150–156. [CrossRef]
30. Gepreel, K.A. Extended trial equation method for nonlinear coupled Schrodinger Boussinesq partial differential equations. *J. Egypt. Math. Soc.* **2016**, *24*, 381–391. [CrossRef]
31. Pandir, Y.; Gurefe, Y.; Misirli, E. The Extended Trial Equation Method for Some Time Fractional Differential Equations. *Discret. Dyn. Nat. Soc.* **2013**, *2013*, 491359. [CrossRef]
32. Gurefe, Y.; Misirli, E.; Sonmezoglu, A.; Ekici, M. Extended trial equation method to generalized nonlinear partial differential equations. *Appl. Math. Comput.* **2013**, *219*, 5253–5260. [CrossRef]
33. Ravi, L.K.; Ray, S.S.; Sahoo, S. New exact solutions of coupled Boussinesq-Burgers equations by Exp-function method. *J. Ocean Eng. Sci.* **2017**, *2*, 34–46. [CrossRef]
34. Seadawy, A.R.; Lu, D.; Khater, M.M.A. Solitary wave solutions for the generalized Zakharov-Kuznetsov-Benjamin-Bona-Mahony nonlinear evolution equation. *J. Ocean Eng. Sci.* **2017**, *2*, 137–142. [CrossRef]
35. Alam, M.N.; Alam, M.M. An analytical method for solving exact solutions of a nonlinear evolution equation describing the dynamics of ionic currents along microtubules. *J. Taibah Univ. Sci.* **2017**, *11*, 939–948. [CrossRef]
36. Liu, S.; Fu, Z.; Liu, S.; Zhao, Q. Jacobi elliptic function expansion method and periodic wave solutions of nonlinear wave equations. *Phys. Lett. A* **2001**, *289*, 69–74. [CrossRef]
37. Ebaid, A.; Aly, E.H. Exact solutions for the transformed reduced Ostrovsky equation via the F-expansion method in terms of Weierstrass-elliptic and Jacobian-elliptic functions. *Wave Motion* **2012**, *49*, 296–308. [CrossRef]
38. Zayed, E.M.E.; Al-Joudi, S. Applications of an Extended G'/G-Expansion Method to Find Exact Solutions of Nonlinear PDEs in Mathematical Physics. *Math. Probl. Eng.* **2010**, *2010*, 768573. [CrossRef]

© 2018 by the authors. Licensee MDPI, Basel, Switzerland. This article is an open access article distributed under the terms and conditions of the Creative Commons Attribution (CC BY) license (http://creativecommons.org/licenses/by/4.0/).

symmetry

MDPI

Article

On the Existence of the Solutions of a Fredholm Integral Equation with a Modified Argument in Hölder Spaces

Merve Temizer Ersoy * and Hasan Furkan

Institute of Science and Technology, Kahramanmaraş Sütçü İmam University, Kahramanmaraş 46100, Turkey; hasanfurkan@gmail.com

* Correspondence: mervetemizer@hotmail.com

Received: 29 September 2018; Accepted: 16 October 2018; Published: 19 October 2018

Abstract: This article concerns the entity of solutions of a quadratic integral equation of the Fredholm type with an altered argument, $x(t) = p(t) + x(t) \int_0^1 k(t,\tau)(Tx)(\tau)d\tau$, where p, k are given functions, T is the given operator satisfying conditions specified later and x is an unknown function. Through the classical Schauder fixed point theorem and a new conclusion about the relative compactness in Hölder spaces, we obtain the existence of solutions under certain assumptions. Our work is more general than the previous works in the Conclusion section. At the end, we introduce several tangible examples where our entity result can be adopted.

Keywords: Fredholm integral equation; Schauder fixed point theorem; Hölder condition

---

1. Introduction

The work of differential equations, with an altered argument being latest, has continued for decades. For more data and consequences related to these equations, see [1–3]. These topics have linear modifications of their arguments and have been worked on by the authors in the papers [1–15]. Integral equations of course stem from several applications in specification numerous real-world problems (see [16,17] and the references therein). Quadratic integral equations arise naturally in applications to real-world problems. For example, problems in the kinetic theory of gases and in the theory of radiative transfer lead to the quadratic integral equation:

$$x(t) = 1 + tx(t) \int_0^1 \frac{\Phi(\tau)}{t+\tau} x(\tau)d\tau,$$

(see [18–21]). The integral equations of a similar form have been examined by several authors [22–28]. Furthermore, some studies using similar techniques have been dedicated to a micropolar porous body and vibrations in thermoelasticity [29,30].

Very recently, J. Banaś and R. Nalepa et al. [4] studied the following equation:

$$x(t) = p(t) + x(t) \int_a^b k(t,\tau)x(\tau)d\tau. \quad (1)$$

Further, J. Caballero, M. Darwish and K. Sadarangani et al. [5] studied the following equation:

$$x(t) = p(t) + x(t) \int_0^1 k(t,\tau)x(r(\tau))d\tau. \quad (2)$$

Furthermore, J. Cabelloro Mena, R. Nalepa and K. Sadarangani et al. [6] studied the following equation:

$$x(t) = p(t) + x(t) \int_0^1 k(t,\tau) \left\{ \max_{\eta \in [0, r(\tau)]} |x(\eta)| \right\} d\tau. \tag{3}$$

The purpose of this paper is to examine the existence of solutions of the following integral equation of the Fredholm type with a changed argument,

$$x(t) = p(t) + x(t) \int_0^1 k(t,\tau)(Tx)(\tau) d\tau, \quad t \in I = [0,1]. \tag{4}$$

Equation (4) is more general than many equations considered up to now and includes (1), (2) and (3) as special cases. Notice that Equation (1) in [4] for $a = 0$ and $b = 1$ is a particular case of (4) with $(Tx)(\tau) = x(\tau)$. Furthermore, it should be noted that Equation (4) is more general than Equation (2) considered in [5]. If we take $(Tx)(\tau) = x\left(r(\tau)\right)$, then the equation:

$$x(t) = p(t) + x(t) \int_0^1 k(t,\tau) x(r(\tau)) d\tau$$

is obtained from Equation (4). Further, notice that Equation (3) in [6] is a particular case of (4), for $(Tx)(\tau) = \max_{\eta \in [0, r(\tau)]} |x(\eta)|$, where the function $r : [0,1] \to [0,1]$ is continuous and nondecreasing.

Compared to the previous works [4–6], we have further generalized the new assumptions in finding the solutions of (1), (2) and (3).

Our solutions substitute for the spaces of functions satisfactory the Hölder condition, and this is a source of the originality of the article. To do this, we will use a recent consequence about the classical Schauder fixed point theorem and the relative compactness in Hölder spaces.

2. Preliminaries and Notations

In this section, we present definitions, notations and theorems that are used along this paper. The following known definitions are available in [4,5,31,32].

Let $[a,b]$ be a closed interval in $\mathbb{R}$; by $C[a,b]$, we indicate the space of continuous functions defined on $[a,b]$ equipped with the supremum norm, i.e.,

$$\|x\|_\infty = \sup\left\{|x(t)| : t \in [a,b]\right\}$$

for $x \in C[a,b]$. For a fixed α with $0 < \alpha \leq 1$, by $H_\alpha[a,b]$, we will indicate the spaces of the real functions x defined on $[a,b]$ and satisfying the Hölder condition, that is those functions x for which there exists a constant H_x^α such that:

$$|x(t) - x(s)| \leq H_x^\alpha |t-s|^\alpha \tag{5}$$

for all $t, s \in [a,b]$. It is well proven that $H^\alpha[a,b]$ is a linear subspace of $C[a,b]$. Furthermore, for $x \in H^\alpha[a,b]$, by H_x^α, we will indicate the least possible stable value for which Inequality (5) is satisfied. Rather, we put:

$$H_x^\alpha = \sup\left\{ \frac{|x(t) - x(s)|}{|t-s|^\alpha} : t, s \in [a,b] \text{ and } t \neq s \right\}. \tag{6}$$

The space $H_\alpha[a,b]$ with $0 < \alpha \leq 1$ may be equipped with the norm:

$$\|x\|_\alpha = |x(a)| + H_x^\alpha$$

for $x \in H_\alpha[a,b]$. Here, H_x^α is defined by (6). In [4], the authors demonstrated that $(H_\alpha[a,b], \|\cdot\|_\alpha)$ with $0 < \alpha \leq 1$ is a Banach space.

Lemma 1. *For $0 < \alpha \leq 1$ and $x \in H_\alpha[a,b]$, we have:*

$$\|x\|_\infty \leq \max\left(1, (b-a)^\alpha\right) \|x\|_\alpha.$$

In particular, the inequality $\|x\|_\infty \leq \|x\|_\alpha$ is satisfied for $a = 0$ and $b = 1$ [4].

Lemma 2. *For $0 < \alpha < \beta \leq 1$, we have:*

$$H_\beta[a,b] \subset H_\alpha[a,b] \subset C[a,b].$$

Furthermore, for $x \in H_\beta[a,b]$, we have:

$$\|x\|_\alpha \leq \max\left(1, (b-a)^{\beta-\alpha}\right) \|x\|_\beta.$$

Particularly, the inequality $\|x\|_\infty \leq \|x\|_\alpha \leq \|x\|_\beta$ is satisfied for $a = 0$ and $b = 1$, [4].

Lemma 3. *Let us assume that $0 < \alpha < \beta \leq 1$ and E is a bounded subset in $H_\beta[a,b]$, then E is a relatively compact subset in $H_\alpha[a,b]$ [5].*

Lemma 4. *Assume that $0 < \alpha < \beta \leq 1$, and by B_r^β, we indicate the ball centered at θ and radius r in the space $H_\beta[a,b]$, i.e., $B_r^\beta = \{x \in H_\beta[a,b] : \|x\|_\beta \leq r\}$. B_r^β is a closed subset of $H_\alpha[a,b]$ [5].*

Corollary 1. *Assume that $0 < \alpha < \beta \leq 1$ and B_r^β is a relatively compact subset in $H_\alpha[a,b]$ and a closed subset of $H_\alpha[a,b]$, then B_r^β is a compact subset in the space $H_\alpha[a,b]$, [5].*

Now let us give the following theorem, which is the base tool used in our study.

Theorem 1 (Schauder's fixed point theorem). *Let E be a nonempty, compact subset of a Banach space $(X, \|\cdot\|)$, convex, and let $T : E \to E$ be a continuity mapping. Then, T has at least one fixed point in E [7].*

3. Main Result

Now, we are ready to give the main result of the paper. In this section, we introduce the following sufficient conditions for the main theorem in our study, and we will prove the solvability of Equation (4) in Hölder spaces.

Hereafter, we suppose unless stated otherwise that α and β are arbitrarily fixed numbers such that $0 < \alpha < \beta \leq 1$.

Theorem 2. *Assume that the following Conditions (i)–(iv) are satisfied:*

(i) $p \in H_\beta[0,1]$.

(ii) $k : [0,1] \times [0,1] \to \mathbb{R}$ *is a continuous function such that there exists a constant $k_\beta > 0$ such that:*

$$|k(t,\tau) - k(s,\tau)| \leq k_\beta |t-s|^\beta,$$

for any $t, s, \tau \in [0,1]$.

(iii) The operator $T : H_\beta[0,1] \to C[0,1]$ is continuous on $H_\beta[0,1]$ with respect to the norm $\|\cdot\|_\alpha$, and there exists a function $f : \mathbb{R}_+ \to \mathbb{R}_+$, which is non-decreasing such that the inequality holds:

$$\|Tx\|_\infty \leq f\left(\|x\|_\beta\right),$$

for any $x \in H_\beta[0,1]$.

(iv) There exists a positive solution r_0 of the inequality:

$$\|p\|_\beta + (2K + k_\beta) r f(r) \le r,$$

where the constant K is defined by:

$$\sup\left\{\int_0^1 |k(t,\tau)|d\tau : t \in [0,1]\right\} \le K.$$

Then, Equation (4) has at least one solution $x = x(t)$ belonging to space $H_\alpha[0,1]$.

Proof. Now, let us consider $x \in H_\beta[0,1]$ and the operator F defined on the space $H_\beta[0,1]$ by the formula:

$$(Fx)(t) = p(t) + x(t)\int_0^1 k(t,\tau)(Tx)(\tau)d\tau,$$

for $t \in [0,1]$. Then, for arbitrarily fixed $t, s \in [0,1], (t \neq s)$, in view of our assumptions, we get:

$$\begin{aligned}
& (Fx)(t) - (Fx)(s) \\
= \; & p(t) + x(t)\int_0^1 k(t,\tau)(Tx)(\tau)d\tau - p(s) - x(s)\int_0^1 k(s,\tau)(Tx)(\tau)d\tau \\
= \; & p(t) - p(s) + x(t)\int_0^1 k(t,\tau)(Tx)(\tau)d\tau - x(s)\int_0^1 k(s,\tau)(Tx)(\tau)d\tau \\
& + x(s)\int_0^1 k(t,\tau)(Tx)(\tau)d\tau - x(s)\int_0^1 k(t,\tau)(Tx)(\tau)d\tau \\
= \; & p(t) - p(s) + (x(t) - x(s))\int_0^1 k(t,\tau)(Tx)(\tau)d\tau \\
& + x(s)\int_0^1 (k(t,\tau) - k(s,\tau))(Tx)(\tau)d\tau
\end{aligned}$$

and:

$$\begin{aligned}
& \frac{|(Fx)(t) - (Fx)(s)|}{|t-s|^\beta} \\
\le \; & \frac{|p(t) - p(s)|}{|t-s|^\beta} + \tfrac{|x(t)-x(s)|}{|t-s|^\beta}\int_0^1 |k(t,\tau)||(Tx)(\tau)|d\tau \\
& + \frac{|x(s)|}{|t-s|^\beta}\int_0^1 |k(t,\tau) - k(s,\tau)||(Tx)(\tau)|d\tau \\
\le \; & H_p^\beta + \|x\|_\beta \|Tx\|_\infty \int_0^1 |k(t,\tau)|d\tau \\
& + |x(s)|\int_0^1 \frac{|k(t,\tau) - k(s,\tau)|}{|t-s|^\beta}|(Tx)(\tau)|d\tau \\
\le \; & H_p^\beta + \|x\|_\beta \|Tx\|_\infty K + \|x\|_\infty \|Tx\|_\infty \int_0^1 k_\beta \frac{|t-s|^\beta}{|t-s|^\beta}d\tau \\
\le \; & H_p^\beta + \|x\|_\beta \|Tx\|_\infty K + \|x\|_\beta \|Tx\|_\infty k_\beta \\
\le \; & H_p^\beta + \|x\|_\beta f\left(\|x\|_\beta\right) K + \|x\|_\beta f\left(\|x\|_\beta\right) k_\beta \\
= \; & H_p^\beta + (K + k_\beta)\|x\|_\beta f\left(\|x\|_\beta\right).
\end{aligned} \tag{7}$$

This demonstrates that the operator F maps $H_\beta[0,1]$ into itself.

Besides, for any $x \in H_\beta[0,1]$, we get:

$$\begin{aligned} |(Fx)(0)| &\leq |p(0)| + |x(0)| \int_0^1 |k(0,\tau)||(Tx)(\tau)|d\tau \\ &\leq |p(0)| + \|x\|_\infty \|Tx\|_\infty K \\ &\leq |p(0)| + \|x\|_\beta \|Tx\|_\infty K \\ &\leq |p(0)| + \|x\|_\beta f\left(\|x\|_\beta\right) K. \end{aligned} \tag{8}$$

By the inequalities (7) and (8), we derive that:

$$\|Fx\|_\beta \leq \|p\|_\beta + (2K + k_\beta)\|x\|_\beta f\left(\|x\|_\beta\right). \tag{9}$$

Since positive number r_0 is the solution of the inequality given in Hypothesis (iv), from (9) and function $f : \mathbb{R}_+ \to \mathbb{R}_+$, which is non-decreasing, we conclude that the inequality:

$$\|Fx\|_\beta \leq \|p\|_\beta + (2K + k_\beta) r_0 f(r_0) \leq r_0 \tag{10}$$

holds. As a result, it follows that F transform the ball:

$$B_{r_0}^\beta = \{x \in H_\beta[0,1] : \|x\|_\beta \leq r_0\}$$

into itself. That is, $F : B_{r_0}^\beta \to B_{r_0}^\beta$. Thus, we have that the set $B_{r_0}^\beta$ is relatively compact in $H_\alpha[0,1]$ for any $0 < \alpha < \beta \leq 1$. Furthermore, $B_{r_0}^\beta$ is a compact subset in $H_\alpha[0,1]$.

In the sequel, we will demonstrate that the operator F is continuous on $B_{r_0}^\beta$ with respect to the norm $\|\cdot\|_\alpha$, where $0 < \alpha < \beta \leq 1$.

Let $y \in B_{r_0}^\beta$ be an arbitrary point in $B_{r_0}^\beta$. Then, we get:

$$\begin{aligned} & (Fx)(t) - (Fy)(t) - ((Fx)(s) - (Fy)(s)) \\ = \;& p(t) + x(t) \int_0^1 k(t,\tau)(Tx)(\tau)d\tau \\ & -p(t) - y(t) \int_0^1 k(t,\tau)(Ty)(\tau)d\tau \\ & -p(s) - x(s) \int_0^1 k(s,\tau)(Tx)(\tau)d\tau \\ & +p(s) + y(s) \int_0^1 k(s,\tau)(Ty)(\tau)d\tau \end{aligned} \tag{11}$$

for any $x \in B_{r_0}^\beta$ and $t, s \in [0,1]$. Equality (11) can be rewritten as:

$$\begin{aligned} & (Fx)(t) - (Fy)(t) - ((Fx)(s) - (Fy)(s)) \\ = \;& x(t) \int_0^1 k(t,\tau)(Tx)(\tau)d\tau - y(t) \int_0^1 k(t,\tau)(Tx)(\tau)d\tau \\ & +y(t) \int_0^1 k(t,\tau)(Tx)(\tau)d\tau - y(t) \int_0^1 k(t,\tau)(Ty)(\tau)d\tau \\ & -x(s) \int_0^1 k(s,\tau)(Tx)(\tau)d\tau + y(s) \int_0^1 k(s,\tau)(Tx)(\tau)d\tau \\ & -y(s) \int_0^1 k(s,\tau)(Tx)(\tau)d\tau + y(s) \int_0^1 k(s,\tau)(Ty)(\tau)d\tau. \end{aligned} \tag{12}$$

By (12), we have:

$$\begin{aligned} & (Fx)(t) - (Fy)(t) - ((Fx)(s) - (Fy)(s)) \\ = \;& (x(t) - y(t)) \int_0^1 k(t,\tau)(Tx)(\tau)d\tau \\ & +y(t) \int_0^1 k(t,\tau)((Tx)(\tau) - (Ty)(\tau))d\tau \\ & -(x(s) - y(s)) \int_0^1 k(s,\tau)(Tx)(\tau)d\tau \\ & -y(s) \int_0^1 k(s,\tau)((Tx)(\tau) - (Ty)(\tau))d\tau. \end{aligned} \tag{13}$$

(13) yields the following equality:

$$\begin{array}{rl} & ((Fx)(t)-(Fy)(t))-((Fx)(s)-(Fy)(s)) \\ = & [(x(t)-y(t))-(x(s)-y(s))]\int_0^1 k(t,\tau)(Tx)(\tau)d\tau \\ & +(x(s)-y(s))\int_0^1 (k(t,\tau)-k(s,\tau))(Tx)(\tau)d\tau \\ & +(y(t)-y(s))\int_0^1 k(t,\tau)((Tx)(\tau)-(Ty)(\tau))d\tau \\ & +y(s)\int_0^1 (k(t,\tau)-k(s,\tau))((Tx)(\tau)-(Ty)(\tau))d\tau. \end{array} \tag{14}$$

Hence, taking into account (14), we can write:

$$\begin{array}{rl} & \frac{|(Fx)(t)-(Fy)(t)-((Fx)(s)-(Fy)(s))|}{|t-s|^\alpha} \\ \leq & \frac{|(x(t)-y(t))-(x(s)-y(s))|}{|t-s|^\alpha}\int_0^1 |k(t,\tau)||(Tx)(\tau)|d\tau \\ & +\frac{|x(s)-y(s)|}{|t-s|^\alpha}\int_0^1 |k(t,\tau)-k(s,\tau)||(Tx)(\tau)|d\tau \\ & +\frac{|y(t)-y(s)|}{|t-s|^\alpha}\int_0^1 |k(t,\tau)||(Tx)(\tau)-(Ty)(\tau)|d\tau \\ & +\frac{|y(s)|}{|t-s|^\alpha}\int_0^1 |k(t,\tau)-k(s,\tau)||(Tx)(\tau)-(Ty)(\tau)|d\tau \\ \leq & \|x-y\|_\alpha\|Tx\|_\infty K+\|x-y\|_\infty\|Tx\|_\infty\int_0^1 k_\beta|t-s|^{\beta-\alpha}d\tau \\ & +\|y\|_\alpha\|Tx-Ty\|_\infty K+\|y\|_\infty\|Tx-Ty\|_\infty\int_0^1 k_\beta|t-s|^{\beta-\alpha}d\tau \\ \leq & K\|x-y\|_\alpha\|Tx\|_\infty+k_\beta\|x-y\|_\alpha\|Tx\|_\infty \\ & +K\|y\|_\alpha\|Tx-Ty\|_\infty+k_\beta\|y\|_\alpha\|Tx-Ty\|_\infty \\ \leq & Kf\left(\|x\|_\beta\right)\|x-y\|_\alpha+k_\beta f\left(\|x\|_\beta\right)\|x-y\|_\alpha \\ & +K\|y\|_\alpha\|Tx-Ty\|_\infty+k_\beta\|y\|_\alpha\|Tx-Ty\|_\infty \\ = & (K+k_\beta)f\left(\|x\|_\beta\right)\|x-y\|_\alpha+(K+k_\beta)\|y\|_\alpha\|Tx-Ty\|_\infty \end{array} \tag{15}$$

for all $t,s\in[0,1]$ with $t\neq s$. Besides, for $x,y\in B^\beta_{r_0}$, we obtain the following inequality:

$$\begin{array}{rl} & |(Fx)(0)-(Fy)(0)| \\ = & \Big|p(0)+x(0)\int_0^1 k(0,\tau)(Tx)(\tau)d\tau \\ & -p(0)-y(0)\int_0^1 k(0,\tau)(Ty)(\tau)d\tau\Big| \\ = & \Big|x(0)\int_0^1 k(0,\tau)(Tx)(\tau)d\tau-y(0)\int_0^1 k(0,\tau)(Tx)(\tau)d\tau \\ & +y(0)\int_0^1 k(0,\tau)(Tx)(\tau)d\tau-y(0)\int_0^1 k(0,\tau)(Ty)(\tau)d\tau\Big| \\ = & \Big|(x(0)-y(0))\int_0^1 k(0,\tau)(Tx)(\tau)d\tau \\ & +y(0)\int_0^1 k(0,\tau)((Tx)(\tau)-(Ty)(\tau))d\tau\Big| \\ \leq & \|x-y\|_\infty K\|Tx\|_\infty+\|y\|_\infty K\|Tx-Ty\|_\infty \\ \leq & K\|x-y\|_\alpha\|Tx\|_\infty+K\|y\|_\alpha\|Tx-Ty\|_\infty \\ \leq & Kf\left(\|x\|_\beta\right)\|x-y\|_\alpha+K\|y\|_\alpha\|Tx-Ty\|_\infty. \end{array} \tag{16}$$

From (15) and (16), we have that:

$$\begin{array}{rl} & \|Fx-Fy\|_\alpha \\ = & |(Fx-Fy)(0)|+H^\alpha_{Fx-Fy} \\ = & |(Fx)(0)-(Fy)(0)| \\ & +\sup\left\{\frac{|(Fx)(t)-(Fy)(t)-((Fx)(s)-(Fy)(s))|}{|t-s|^\alpha}:t,s\in[0,1]\text{ and }t\neq s\right\} \\ \leq & (2K+k_\beta)f\left(\|x\|_\beta\right)\|x-y\|_\alpha+(2K+k_\beta)\|y\|_\alpha\|Tx-Ty\|_\infty. \end{array} \tag{17}$$

Moreover, since $\|y\|_\alpha \le \|y\|_\beta \le r_0$ and $f\left(\|x\|_\beta\right) \le f(r_0)$, we derive from (17) that the following inequality holds:

$$\|Fx - Fy\|_\alpha \le (2K + k_\beta) f(r_0)\|x - y\|_\alpha + (2K + k_\beta) r_0 \|Tx - Ty\|_\infty. \tag{18}$$

Since the operator $T : H_\beta[0,1] \to C[0,1]$ is continuous on $H_\beta[0,1]$ with respect to the norm $\|\cdot\|_\alpha$, it is also continuous at the point $y \in B^\beta_{r_0}$. Let us take an arbitrary $\varepsilon > 0$. Since the operator T is continuous at the point $y \in B^\beta_{r_0}$, there exists $\delta > 0$ such that the inequality:

$$\|Tx - Ty\|_\infty < \frac{\varepsilon}{2(2K + k_\beta) r_0}$$

is satisfied for all $x \in B^\beta_{r_0}$, where $\|x - y\|_\alpha < \delta$ and:

$$0 < \delta < \frac{\varepsilon}{2(2K + k_\beta) f(r_0)}.$$

Then, taking into account (18), we derive the following inequality:

$$\begin{aligned} \|Fx - Fy\|_\alpha &\le (2K + k_\beta) f(r_0)\|x - y\|_\alpha + (2K + k_\beta) r_0 \|Tx - Ty\|_\infty \\ &< \frac{\varepsilon}{2} + \frac{\varepsilon}{2} \\ &= \varepsilon. \end{aligned}$$

As a result, we infer that the operator F is continuous at the point $y \in B^\beta_{r_0}$. Because y was chosen arbitrarily, we deduce that F is continuous on $B^\beta_{r_0}$ with respect to the norm $\|\cdot\|_\alpha$. As $B^\beta_{r_0}$ is compact in $H_\alpha[0,1]$, from the classical Schauder fixed point theorem, we get the desired result. □

4. Examples

In this part, we conclude the article by presenting two examples that illustrate the generality and efficiency of our results.

Example 1. *Let us consider the following quadratic integral equation:*

$$x(t) = \sqrt[6]{q \cos t + \hat{q}} + x(t) \int_0^1 \sqrt[5]{mt^2 + \tau} \sin x^2(\tau) d\tau, \ \ t \in I = [0,1]. \tag{19}$$

Here, $q, \hat{q}$ and m are the suitable nonnegative constants to be determined such that Conditions (i)–(iv) of Theorem 2 hold.

Set $p(t) = \sqrt[6]{q \cos t + \hat{q}}$ and $k(t,\tau) = \sqrt[5]{mt^2 + \tau}$ for all $t, \tau \in [0,1]$.

It is easily seen that:

$$\begin{aligned} |p(t) - p(s)| &= \left|\sqrt[6]{q \cos t + \hat{q}} - \sqrt[6]{q \cos s + \hat{q}}\right| \\ &\le \left|\sqrt[6]{q \cos t + \hat{q} - q \cos s - \hat{q}}\right| \\ &\le \sqrt[6]{q |\cos t - \cos s|} \\ &\le \sqrt[6]{2q \left|\sin\left(\frac{t+s}{2}\right)\right| \left|\sin\left(\frac{t-s}{2}\right)\right|} \\ &\le \sqrt[6]{q}\sqrt[6]{|t-s|} \\ &= \sqrt[6]{q}|t-s|^{\frac{1}{6}} \end{aligned}$$

for all $t, s \in [0,1]$. *This says that* $p \in H_{\frac{1}{6}}[0,1]$ *and, moreover,* $H_p^{\frac{1}{6}} = \sqrt[6]{q}$. *Therefore, we can take the constants* α *and* β *as* $0 < \alpha < \frac{1}{6}$ *and* $\beta = \frac{1}{6}$. *Therefore, Assumption (i) of Theorem 2 holds. Note that:*

$$\begin{aligned} \|p\|_{\frac{1}{6}} &= |p(0)| + \sup\left\{ \frac{|p(t) - p(s)|}{|t-s|^{\frac{1}{6}}} : t, s \in [0,1] \text{ and } t \neq s \right\} \\ &= |p(0)| + H_p^{\frac{1}{6}} = \sqrt[6]{q + \hat{q}} + \sqrt[6]{q}. \end{aligned}$$

Further, we have:

$$\begin{aligned} |k(t,\tau) - k(s,\tau)| &= \left| \sqrt[5]{mt^2 + \tau} - \sqrt[5]{ms^2 + \tau} \right| \\ &\leq \left| \sqrt[5]{m(t^2 - s^2)} \right| \\ &= \sqrt[5]{m} \sqrt[5]{|(t^2 - s^2)|} \\ &= \sqrt[5]{m} \sqrt[5]{|t-s|} \sqrt[5]{|t+s|} \\ &\leq \sqrt[5]{2m} |t-s|^{\frac{1}{6}} |t-s|^{\frac{1}{30}} \\ &\leq \sqrt[5]{2m} |t-s|^{\frac{1}{6}} \end{aligned}$$

for all $t, s, \tau \in [0,1]$. *Assumption (ii) of Theorem 2 holds with* $k_\beta = k_{\frac{1}{6}} = \sqrt[5]{2m}$.

Since $(Tx)(\tau) = \sin x^2(\tau)$ *and:*

$$\left| \sin x^2(\tau) \right| \leq \left| x^2(\tau) \right| = |x(\tau)||x(\tau)| \leq \|x\|_\infty^2 \leq \|x\|_\beta^2$$

for all $x \in H_\beta[0,1]$ *and* $\tau \in [0,1]$, *the inequality:*

$$\|Tx\|_\infty = \sup_{\tau \in [0,1]} \left| \sin x^2(\tau) \right| \leq \|x\|_\beta^2$$

holds. Therefore, we can choose the function $f : \mathbb{R}_+ \to \mathbb{R}_+$ *as* $f(x) = x^2$. *This function is non-decreasing and satisfies the inequality in Assumption (iii).*

We will show that the operator $T : H_\beta[0,1] \to C[0,1]$ *is continuous on* $H_\beta[0,1]$ *with respect to the norm* $\|\cdot\|_\alpha$. *Let us take* $x, y \in H_\beta[0,1]$ *and* $\tau \in [0,1]$.

It is clear that:

$$\begin{aligned} \left| \sin x^2(\tau) - \sin y^2(\tau) \right| &\leq \left| x^2(\tau) - y^2(\tau) \right| \\ &= |x(\tau) - y(\tau)|\,|x(\tau) + y(\tau)| \\ &= |x(\tau) - y(\tau)|\,|x(\tau) - y(\tau) + 2y(\tau)| \\ &\leq |x(\tau) - y(\tau)|\,(|x(\tau) - y(\tau)| + 2\,|y(\tau)|) \\ &\leq \|x - y\|_\infty (\|x - y\|_\infty + 2\|y\|_\infty) \\ &\leq \|x - y\|_\alpha\,(\|x - y\|_\alpha + 2\|y\|_\alpha) \end{aligned}$$

and:

$$\|Tx - Ty\|_\infty \leq \sup_{\tau \in [0,1]} \left| \sin x^2(\tau) - \sin y^2(\tau) \right| \leq \|x - y\|_\alpha (\|x - y\|_\alpha + 2\|y\|_\alpha).$$

Now, we will show that T is continuous at the point $y \in H_\beta[0,1]$ with respect to the norm $\|\cdot\|_\alpha$. Let us take an arbitrary $\varepsilon > 0$. Then, there exists the positive number δ such that $\|x-y\|_\alpha < \delta$ and the inequality:

$$\|Tx - Ty\|_\infty < \delta\,(\delta + 2\|y\|_\alpha) < \varepsilon$$

is satisfied for all $x \in H_\beta[0,1]$, where $0 < \delta < \sqrt{\|y\|_\alpha^2 + \varepsilon} - \|y\|_\alpha$. Therefore, we can choose the positive number δ as $\delta = \frac{1}{2}\sqrt{\|y\|_\alpha^2 + \varepsilon} - \|y\|_\alpha$. As a result, we infer that the operator T is continuous at the point $y \in B_{r_0}^\beta$. Since y was chosen arbitrarily, we deduce that T is continuous on $H_\beta[0,1]$ with respect to the norm $\|\cdot\|_\alpha$.

Further, we can calculate that:

$$\begin{aligned}
\sup\left\{\int_0^1 |k(t,\tau)|d\tau : t \in [0,1]\right\} &= \sup\left\{\int_0^1 \left|\sqrt[5]{mt^2+\tau}\right| d\tau : t \in [0,1]\right\} \\
&= \sup\left\{\frac{5}{6}\left(\sqrt[5]{(mt^2+1)^6} - \sqrt[5]{(mt^2)^6}\right) : t \in [0,1]\right\} \\
&\le \sup\left\{\frac{5}{6}\sqrt[5]{(mt^2+1)^6} : t \in [0,1]\right\} \\
&= \frac{5}{6}\sqrt[5]{(m+1)^6} \\
&\le \sqrt[5]{(m+1)^6} \\
&= K.
\end{aligned}$$

In this case, the inequality appearing in assumption (vi) of Theorem 2 takes the following form:

$$\|p\|_{\frac{1}{6}} + (2K + k_\beta) r f(r) \le r$$

which is equivalent to:

$$\sqrt[6]{q} + \sqrt[6]{q+\hat{q}} + \left(2\sqrt[5]{(m+1)^6} + \sqrt[5]{2m}\right) r r^2 \le r. \tag{20}$$

There exists a positive number r_0 satisfying (20) provided that the constants $q, \hat{q}$ and m are chosen as suitable.

For example, if one chose $q = \frac{1}{10^{18}}, \hat{q} = 0$ and $m = \frac{1}{2^{16}}$, then the inequality:

$$\frac{2}{10^3} + \left(2\sqrt[5]{\left(\frac{1}{2^{16}}+1\right)^6} + 0.125\right) r^3 \le r$$

holds for $r = r_0 = \frac{1}{10}$. Therefore, using Theorem 2, we infer that there is at least one solution x of Equation (19) in the space $H_\alpha[0,1]$ with $0 < \alpha < \frac{1}{6}$.

Example 2. *Let us consider the following quadratic integral equation:*

$$x(t) = \frac{1}{10^6}\arctan\sqrt[3]{t + \ln q} + x(t)\int_0^1 \sqrt[3]{m \sin t + \tau}\sqrt{|x(\tau)|}d\tau, \ \ t \in I = [0,1], \tag{21}$$

where q and m are the suitable positive constants to be selected for which Conditions (i)–(iv) of Theorem 2 hold.

Set $p(t) = \frac{1}{10^6}\arctan\sqrt[3]{t+\ln q}$ and $k(t,\tau) = \sqrt[3]{m\sin t + \tau}$ for all $t, \tau \in [0,1]$.

It is obvious that the inequality:

$$
\begin{aligned}
|p(t)-p(s)| &= \left|\frac{1}{10^6}\arctan\sqrt[3]{t+\ln q}-\frac{1}{10^6}\arctan\sqrt[3]{s+\ln q}\right| \\
&\le \left|\frac{1}{10^6}\arctan\left(\sqrt[3]{t+\ln q}-\sqrt[3]{s+\ln q}\right)\right| \\
&= \frac{1}{10^6}\left|\sqrt[3]{t+\ln q}-\sqrt[3]{s+\ln q}\right| \\
&\le \frac{1}{10^6}\left|\sqrt[3]{t+\ln q-s-\ln q}\right| \\
&\le \frac{1}{10^6}\sqrt[3]{|t-s|} \\
&= \frac{1}{10^6}|t-s|^{\frac{1}{3}}
\end{aligned}
$$

holds for all $t,s\in[0,1]$. Therefore, $p\in H_{\frac{1}{3}}[0,1]$ and $H_p^{\frac{1}{3}}=\frac{1}{10^6}$. Hence, the constants α and β can be taken as $0<\alpha<\frac{1}{3}$ and $\beta=\frac{1}{3}$.

Therefore, Assumption (i) of Theorem 2 is satisfied. Note that:

$$
\begin{aligned}
\|p\|_{\frac{1}{3}} &= |p(0)|+\sup\left\{\frac{|p(t)-p(s)|}{|t-s|^{\frac{1}{3}}} : t,s\in[0,1] \text{ and } t\ne s\right\} \\
&= |p(0)|+H_p^{\frac{1}{3}}=\frac{1}{10^6}\left|\arctan\sqrt[3]{\ln q}\right|+\frac{1}{10^6}.
\end{aligned}
$$

Further, we have:

$$
\begin{aligned}
|k(t,\tau)-k(s,\tau)| &= \left|\sqrt[3]{m\sin t+\tau}-\sqrt[3]{m\sin s+\tau}\right| \\
&\le \left|\sqrt[3]{m(\sin t-\sin s)}\right| \\
&\le \sqrt[3]{2m\left|\cos\left(\frac{t+s}{2}\right)\right|\left|\sin\left(\frac{t-s}{2}\right)\right|} \\
&\le \sqrt[3]{m}|t-s|^{\frac{1}{3}}
\end{aligned}
$$

for all $t,s,\tau\in[0,1]$. Assumption (ii) of Theorem 2 is satisfied with $k_\beta=k_{\frac{1}{3}}=\sqrt[3]{m}$.

Since $(Tx)(\tau)=\sqrt{|x(\tau)|}$ and:

$$
\sqrt{|x(\tau)|}\le\sqrt{\|x\|_\infty}\le\sqrt{\|x\|_\beta}
$$

for all $x\in H_\beta[0,1]$ and $\tau\in[0,1]$, the inequality:

$$
\|Tx\|_\infty=\sup_{\tau\in[0,1]}\left|\sqrt{|x(\tau)|}\right|\le\sqrt{\|x\|_\beta}
$$

holds. Therefore, we can choose the function $f:\mathbb{R}_+\to\mathbb{R}_+$ as $f(x)=\sqrt{x}$. This function is non-decreasing and satisfies the inequality in Assumption (iii).

We will show that the operator $T: H_\beta[0,1]\to C[0,1]$ is continuous on $H_\beta[0,1]$ with respect to the norm $\|\cdot\|_\alpha$. Let us take $x,y\in H_\beta[0,1]$ and $\tau\in[0,1]$. It is certain that:

$$
\left|\sqrt{|x(\tau)|}-\sqrt{|y(\tau)|}\right|\le\sqrt{|x(\tau)-y(\tau)|}\le\sqrt{\|x-y\|_\infty}\le\sqrt{\|x-y\|_\alpha}
$$

and:

$$\|Tx - Ty\|_\infty \leq \sup_{\tau \in [0,1]} \left| \sqrt{|x(\tau)|} - \sqrt{|y(\tau)|} \right| \leq \sqrt{\|x - y\|_\alpha}.$$

Now, we will show that T is continuous at the point $y \in H_\beta[0,1]$ with respect to the norm $\|\cdot\|_\alpha$. Let us take an arbitrary $\varepsilon > 0$. Then, there exists the positive number δ such that $\|x - y\|_\alpha < \delta$ and the inequality:

$$\|Tx - Ty\|_\infty \leq \sqrt{\|x - y\|_\alpha} < \varepsilon$$

is satisfied for all $x \in H_\beta[0,1]$. Here, we can choose the positive number δ as $\delta = \varepsilon^2$.

As a result, we infer that the operator T is continuous at the point $y \in B^\beta_{r_0}$. Since y was chosen arbitrarily, we deduce that T is continuous on $H_\beta[0,1]$ with respect to the norm $\|\cdot\|_\alpha$.

Further, we can calculate that:

$$\begin{aligned}
& \sup\left\{\int_0^1 |k(t,\tau)|d\tau : t \in [0,1]\right\} \\
= & \sup\left\{\int_0^1 \left|\sqrt[3]{m \sin t + \tau}\right| d\tau : t \in [0,1]\right\} \\
= & \sup\left\{\frac{1}{3}\left(\sqrt[3]{(m \sin t + 1)^4} - \sqrt[3]{(m \sin t)^4}\right) : t \in [0,1]\right\} \\
\leq & \sup\left\{\frac{1}{3}\sqrt[3]{(m \sin t + 1)^4} : t \in [0,1]\right\} \\
\leq & \frac{1}{3}\sqrt[3]{(m+1)^4} \\
\leq & \sqrt[3]{(m+1)^4} \\
= & K.
\end{aligned}$$

In this case, the inequality appearing in Assumption (vi) of Theorem 2 takes the following form:

$$\|p\|_{\frac{1}{3}} + (2K + k_\beta) r f(r) \leq r$$

which is equivalent to:

$$\frac{1}{10^6}\left(\left|\arctan \sqrt[3]{\ln q}\right| + 1\right) + \left(2\sqrt[3]{(m+1)^4} + \sqrt[3]{m}\right) r\sqrt{r} \leq r. \tag{22}$$

There exists a positive number r_0 satisfying (22) for chosen suitable constants q and m.

For example, if one chooses $q = 1$ and $m = \frac{1}{5^{12}}$, then the inequality:

$$\frac{1}{10^6} + \left(2\sqrt[3]{\left(1 + \frac{1}{5^{12}}\right)^4} + 0.0016\right) r^{\frac{3}{2}} \leq r$$

holds for $r = r_0 = \frac{1}{10^4}$. Therefore, using Theorem 2, we infer that there is at least one solution x of Equation (21) in the space $H_\alpha[0,1]$ with $0 < \alpha < \frac{1}{3}$.

5. Conclusions

In this paper, we have investigated the existence of solutions of the integral Equation (4). It should be noted that Equation (4) is more general than many equations considered up to now. For example, it includes the equations examined in previous studies [4–6]. That is, if we take the operator T as

$(Tx)(\tau) = x(\tau)$, we obtain the integral Equation (1) in [4] with $a = 0$ and $b = 1$. On the other hand, if we take $(Tx)(\tau) = x(r(\tau))$, we have the integral Equation (2) in [5]. Further, if we take $(Tx)(\tau) = \max_{\eta \in [0,r(\tau)]} |x(\eta)|$, we have the integral Equation (3) in [6].

Author Contributions: Writing, original draft, M.T.E. and H.F. Writing, review and editing, M.T.E. and H.F.

Acknowledgments: M.T.E. was supported by the Scientific and Technological Research Council of Turkey (TUBITAK Programme, 2228-B).

Conflicts of Interest: The authors declare no conflict of interest.

References

1. Kulenovic, M.R.S. Oscillation of the Euler differential equation with delay. *Czech. Math. J.* **1995**, *45*, 1–16.
2. Mureşan, V. On a class of Volterra integral equations with deviating argument. *Stud. Univ. Babes-Bolyai Math.* **1999**, *44*, 47–54.
3. Mureşan, V. Volterra integral equations with iterations of linear modification of the argument. *Novi Sad J. Math.* **2003**, *33*, 1–10.
4. Banaś, J.; Nalepa, R. On the space of functions with growths tempered by a modulus of continuity and its applications. *J. Funct. Spaces Appl.* **2013**. [CrossRef]
5. Caballero, J.; Abdalla, M.; Sadarangani, K. Solvability of a quadratic integral equation of fredholm type in Hölder spaces. *Electron. J. Differ. Equ.* **2014**, *31*, 1–10. [CrossRef]
6. Caballero, M.J.; Nalepa, R.; Sadarangani, K. Solvability of a quadratic integral equation of Fredholm type with Supremum in Hölder Spaces. *J. Funct. Spaces Appl.* **2014**. [CrossRef]
7. Schauder, J. Der Fixpunktsatz in Funktionalriiumen. *Stud. Math.* **1930**, *2*, 171–180. [CrossRef]
8. López, B.; Harjani, J.; Sadaragani, K. Existence of positive solutions in the space of Lipschitz functions to a class of fractional differential equations of arbitrary order. *Racsam* **2018**, *112*, 1281–1294. [CrossRef]
9. Bacoţiu, C. Volterra-Fredholm nonlinear systems with modified argument via weakly Picard operators theory. *Carpathian J. Math.* **2008**, *24*, 1–19.
10. Benchohra, M.; Darwish, M.A. On unique solvability of quadratic integral equations with linear modification of the argument. *Miskolc Math. Notes* **2009**, *10*, 3–10.
11. Dobriţoiu, M. Analysis of a nonlinear integral equation with modified argument from physics. *Int. J. Math. Models Methods Appl. Sci.* **2008**, *2*, 403–412.
12. Kato, T.; Mcleod, J.B. The functional-differential equation $y'(x) = ay(\lambda x) + by(x)$. *Bull. Am. Math. Soc.* **1971**, *77*, 891–937.
13. Lauran, M. Existence results for some differential equations with deviating argument. *Filomat* **2011**, *25*, 21–31. [CrossRef]
14. Mureşan, V. A functional-integral equation with linear modification of the argument via weakly Picard operators. *Fixed Point Theory* **2008**, *9*, 189–197.
15. Mureşan, V. A Fredholm-Volterra integro-differential equation with linear modification of the argument. *J. Appl. Math.* **2010**, *3*, 147–158.
16. Agarwal, R.P.; O'Regan, D. *Infinite Interval Problems for Differential, Difference and Integral Equations*; Springer: Dordrecht, The Netherlands, 2001; ISBN 978-94-015-9171-3.
17. Agarwal, R.P.; O'Regan, D.; Wong, P.J.Y. *Positive Solutions of Differential, Difference and Integral Equations*; Springer: Dordrecht, The Netherlands, 1999.
18. Case, K.M.; Zweifel, P.F. *Linear Transport Theory*; Addison Wesley: Reading, MA, USA, 1967.
19. Chandrasekhar, S. *Radiative Transfer*; Dover Publications: New York, NY, USA, 1960.
20. Hu, S.; Khavani, M.; Zhuang, W. Integral equations arising in the kinetic theory of gases. *J. Appl. Anal.* **1989**, *34*, 261–266. [CrossRef]
21. Kelly, C.T. Approximation of solutions of some quadratic integral equations in transport theory. *J. Integral Equ.* **1982**, *4*, 221–237.
22. Banas, J.; Lecko, M.; El-Sayed, W.G. Existence theorems of some quadratic integral equation. *J. Math. Anal. Appl.* **1998**, *222*, 276–285. [CrossRef]
23. Banas, J.; Caballero, J.; Rocha, J.; Sadarangani, K. Monotonic solutions of a class of quadratic integral equations of Volterra type. *Comput. Math. Appl.* **2005**, *49*, 943–952. [CrossRef]

24. Caballero, J.; Rocha, J.; Sadarangani, K. On monotonic solutions of an integral equation of Volterra type. *J. Comput. Appl. Math.* **2005**, *174*, 119–133. [CrossRef]
25. Darwish, M.A. On solvability of some quadratic functional-integral equation in Banach algebras. *Commun. Appl. Anal.* **2007**, *11*, 441–450.
26. Darwish, M.A.; Ntouyas, S.K. On a quadratic fractional Hammerstein-Volterra integral equations with linear modification of the argument. *Nonlinear Anal. Theory Methods Appl.* **2011**, *74*, 3510–3517. [CrossRef]
27. Darwish, M.A. On quadratic integral equation of fractional orders. *J. Math. Anal. Appl.* **2005**, *311*, 112–119. [CrossRef]
28. Agarwal, R.P.; Banas, J.; Banas, K.; O'Regan, D. Solvability of a quadratic Hammerstein integral equation in the class of functions having limits at infinity. *J. Integral Equ. Appl.* **2011**, *23*, 157–181. [CrossRef]
29. Marin, M. An approach of a heat-flux dependent theory for micropolar porous media. *Meccanica* **2016**, *51*, 1127–1133. [CrossRef]
30. Marin, M. Some estimates on vibrations in thermoelasticity of dipolar bodies. *J. Vib. Control* **2010**, *16*, 33–47. [CrossRef]
31. Caballero, J.; Darwish, M.A.; Sadarangani, K. Positive Solutions in the Space of Lipschitz Functions for Fractional Boundary Value Problems with Integral Boundary Conditions. *Mediterr. J. Math.* **2017**. [CrossRef]
32. Cabrera, I.; Harjani, J.; Sadarangani, K. Existence and Uniqueness of Solutions for a Boundary Value Problem of Fractional Type with Nonlocal Integral Boundary Conditions in Hölder Spaces. *Mediterr. J. Math.* **2018**. [CrossRef]

© 2018 by the authors. Licensee MDPI, Basel, Switzerland. This article is an open access article distributed under the terms and conditions of the Creative Commons Attribution (CC BY) license (http://creativecommons.org/licenses/by/4.0/).

symmetry

MDPI

Article

A Class of Nonlinear Boundary Value Problems for an Arbitrary Fractional-Order Differential Equation with the Riemann-Stieltjes Functional Integral and Infinite-Point Boundary Conditions

Hari M. Srivastava [1,2,*], Ahmed M. A. El-Sayed [3] and Fatma M. Gaafar [4]

1 Department of Mathematics and Statistics, University of Victoria, Victoria, BC V8W 3R4, Canada
2 Department of Medical Research, China Medical University Hospital, China Medical University, Taichung 40402, Taiwan
3 Department of Mathematics, Faculty of Science, Alexandria University, Alexandria 21500, Egypt; amasayed@alexu.edu.eg
4 Department of Mathematics, Faculty of Science, Damanhour University, Damanhour 22516, Egypt; fatmagaafar2@yahoo.com
* Correspondence: harimsri@math.uvic.ca

Received: 7 September 2018; Accepted: 14 October 2018; Published: 16 October 2018

Abstract: In this paper, we investigate the existence of an absolute continuous solution to a class of first-order nonlinear differential equation with integral boundary conditions (BCs) or with infinite-point BCs. The Liouville-Caputo fractional derivative is involved in the nonlinear function. We first consider the existence of a solution for the first-order nonlinear differential equation with m-point nonlocal BCs. The existence of solutions of our problems is investigated by applying the properties of the Riemann sum for continuous functions. Several examples are given in order to illustrate our results.

Keywords: nonlinear boundary value problems; fractional-order differential equations; Riemann-Stieltjes functional integral; Liouville-Caputo fractional derivative; infinite-point boundary conditions; advanced and deviated arguments; existence of at least one solution

MSC: primary 26A33, 34B18, 34K37; secondary 34A08, 34B10

---

1. Introduction

Our objective in this article is to investigate the existence of absolute continuous solutions of the nonlocal first-order boundary value problem (BVP) with the nonlinear function involving the Liouville-Caputo fractional derivative:

$$\frac{dx}{dt} = f\big(t, D^{\alpha}x(t)\big) \quad \text{a.e.} \qquad (0 < t < 1;\ 0 < \alpha \leqq 1), \tag{1}$$

together with either the Riemann-Stieltjes functional integral boundary condition (with the advanced or deviated argument ϕ) given by

$$\int_0^1 x\big(\phi(s)\big)\, dg(s) = x_0 \qquad (g : [0,1] \to [0,1];\ g(s) \geqq 0) \tag{2}$$

or the infinite-point boundary conditions given by

$$\sum_{k=1}^{\infty} a_k\, x(\phi(\tau_k)) = x_0 \qquad (a_k > 0;\ \tau_k \in (0,1);\ \phi(\tau_k) \leqq \tau_k), \tag{3}$$

where $g : [0,1] \to [0,1]$ is an increasing function, $\alpha \in (0,1]$ and D^{α} denotes the Liouville-Caputo fractional derivative of order α. The integral in (2) is the Riemann-Stieltjes type with respect to $g(s)$. In the case when $g(s) = s$, the Riemann-Stieltjes integral in the boundary condition given by (2) reduces to the relatively more familiar Riemann integral.

In the case when $\alpha = 1$, the BVP (1) becomes the implicit differential problem given by

$$\frac{dx}{dt} = f\left(t, \frac{dx}{dt}\right) \quad \text{a.e.} \qquad (0 < t < 1)$$

under the Riemann-Stieltjes functional integral BC (2) or infinite-point BCs (3).

Our results in this article are based upon *Kolmogorov's Compactness Criterion* (see [1]) and upon Schauder's Fixed Point Theorem (see [2]).

Nonlinear BVPs with nonlocal multi-point BCs have received a lot of attention in recent years. In fact, various conditions are obtained for the existence of solutions by (for example) Alvan et al. [3], Benchohra et al. [4], Boucherif [5], El-Sayed and Bin-Taher [6], Gao and Han [7], Hamani et al. [8] and Nieto et al. [9] (see also the references to the related earlier works which are cited in each of these investigations).

BVPs with integral BCs arise naturally in semiconductor problems [10], thermal conduction problems [11], hydrodynamic problems [12], population dynamics model [13], and so on (see also [14]). Recently, these BVPs were extensively studied by (among others) Akcan and Çetin [15], Boucherif [16], Benchohra et al. [17], Chalishajar and Kumar [18], Dou et al. [19], Li and Zhang [20], Liu et al. [21], Song et al. [22], Tokmagambetov and Torebek [23], Wang et al. [24] and Yang and Qin [25] (see also the references to the related earlier works which are cited in each of these investigations).

The study of BVPs involving infinite-point BCs has become attractive recently. In the year 2011, Gao and Han [7] firstly studied the solutions to thefractional-order differential equation problem with infinite-point BCs. Ever since then, many significant and interesting cases of BVPs of fractional order were considered with infinite-point BCs by (for example) Ge et al. [26], Guo et al. [27], Hu and Zhang [28], Li et al. [29], Liu et al. [30], Zhang and Zhong [31] and Zhang [32] (see also to the references cited therein). In the year 2016, Xu and Yang [33] proposed a generalization of the PID controller and studied two kinds of fractional-order differential equations arising in control theory together with the infinite point boundary conditions. Their results can describe the corresponding control system accurately and also provide a platform for the understanding of our environment. However, investigations on the infinite-point BVPs for differential equations of fractional or integer order have gradually aroused people's attentions and interests, but such investigations are still not too many.

Motivated by the above-mentioned developments and results, we consider the BVP given by (1) and (2) or by (1) and (3). In each case, we determine sufficient conditions on f guaranteeing that the problem (1) under the Riemann-Stieltjes functional integral BC (2) or the problem (1) under infinite-point BC (3) has a solution. We first find the solutions of the problem (1) with the m-point BCs given by

$$\sum_{k=1}^{m} a_k\, x(\phi(\tau_k)) = x_0 \qquad (a_k \neq 0;\ 0 < \tau_k < 1), \tag{4}$$

and then, by using the properties of the Riemann sum for continuous functions, we investigate the solutions of the BVP given by (1) and (2) as well as the BVP given by Equations (1) and (3). The solutions of our problems in the Carathéodory sense are given under some weak conditions on f, which are sufficiently general and easy to check.

Our work has the following salient features. Firstly, a unified investigation involving both the Riemann-Stieltjes integral as well as infinite points is presented here in the BCs of the BVP (1). Secondly, to the best of our knowledge, most (if not all) of the earlier works dealt with the Riemann-Stieltjes integral BCs or infinite-point BCs as separate cases. Here, if we have a way of getting the continuous solution of the m-point BVP, we can (in a simple way) get a solution to the BVP with the Riemann-Stieltjes integral or infinite points in the BCs.

2. Preliminaries

Let $C(I)$ be the space of continuous functions defined on I with the norm given by

$$||x|| = \sup_{t\in I} |x(t)|,$$

and $AC[0,1]$ be the space of all absolutely continuous functions on $[0,1]$.

In addition, let $L_1(I)$ denote the class of the Lebesgue-integrable functions on the interval $I = [0,1]$ with the norm given by

$$||y||_{L_1} = \int_0^1 y(\xi)d\xi.$$

Definition 1. *The Riemann-Liouville fractional integral of the function $f \in L_1[0,T]$ of order $\beta > 0$ is defined by (see [34,35])*

$$I^\beta f(t) = \frac{1}{\Gamma(\beta)} \int_0^t (t-s)^{\beta-1} f(s)\, ds.$$

Definition 2. *The Caputo (or, more precisely, the Liouville-Caputo) fractional derivative of $f(t)$ of order α $(0 < \alpha \leqq 1)$ is defined as follows (see [34,35])*

$$D^\alpha f(t) = I^{1-\alpha} \frac{d}{dt}\{f(t)\} = \frac{1}{\Gamma(1-\alpha)} \int_0^t (t-s)^{-\alpha} \frac{d}{ds}\{f(s)\}\, ds.$$

3. Existence of Solutions to (1) with the *m*-Point BCs (4)

Definition 3. *A function x is called a solution of problem (1) with the m-point BCs (4) if $x \in AC[0,1]$ and satisfies (1) and (4).*

We make several assumptions as detailed below:

(i) The function $f : [0,1] \times \mathbb{R} \to \mathbb{R}$ is a Carathéodory function, that is, it possesses the following properties:

 (a) For each $t \in [0,1]$, $f(t,\cdot)$ is continuous;

 (b) For each $u \in \mathbb{R}$, $f(\cdot,u)$ is measurable.

(ii) The function $\phi : [0,1] \to [0,1]$ is continuous and advanced, $\phi(t) \geqq t$, or continuous and deviated, $\phi(t) \leqq t$.

(iii) There exists an integrable function $a \in L_1[0,1]$ and a constant $b > 0$ such that

$$|f(t,u)| \leqq a(t) + b\,|u| \qquad \text{for each } t \in [0,1] \text{ and } u \in \mathbb{R}$$

Lemma 1. *The boundary value problem given by (1) and (4) is equivalent to the following integral equation:*

$$x(t) = A\left(x_0 - \sum_{k=1}^m a_k \int_0^{\phi(\tau_k)} y(\xi)\, d\xi\right) + \int_0^t y(\xi)\, d\xi, \tag{5}$$

where $y(t)$ is the solution of fractional-order integral equation given by

$$y(t) = f(t, I^{1-\alpha}y(t)) \qquad (t \in [0,1]) \tag{6}$$

and

$$A = \left(\sum_{k=1}^{m} a_k\right)^{-1}.$$

Proof. We begin by considering the problem (1) with the m-point BCs in (4). If we put $y(t) = x'(t)$ in (1), then Definition 2 implies that

$$y(t) = f(t, I^{1-\alpha}y(t)).$$

We also have

$$x(t) = x(0) + I^1 y(t). \tag{7}$$

We now use the nonlocal condition (4) in order to compute the constant $x(0)$. Indeed, upon setting $t = \phi(\tau_k) \in (0,1)$ in Equation (7), we get

$$x(\phi(\tau_k)) = \int_0^{\phi(\tau_k)} y(\xi)\,d\xi + x(0),$$

so that we have

$$\sum_{k=1}^{m} a_k\, x(\phi(\tau_k)) = \sum_{k=1}^{m} a_k \int_0^{\phi(\tau_k)} y(\xi)\,d\xi + x(0) \sum_{k=1}^{m} a_k.$$

From Equation (4), we find that

$$x_0 = \sum_{k=1}^{m} a_k \int_0^{\phi(\tau_k)} y(\xi)\,d\xi + x(0) \sum_{k=1}^{m} a_k,$$

which yields

$$x(0) = A\left(x_0 - \sum_{k=1}^{m} a_k \int_0^{\phi(\tau_k)} y(\xi)\,d\xi\right).$$

Substituting this last evaluation in Equation (7), we obtain formula (5).

Finally, in order to complete the proof of the above Lemma, we show that Equation (5) satisfies problem (1) together with the m-point BCs in (4). In fact, from (5), we obtain

$$D^{\alpha}x(t) = I^{1-\alpha}\,\frac{d}{dt}\{x(t)\} = I^{1-\alpha}y(t).$$

In addition, upon differentiating (5) with respect to t, we have

$$\frac{dx}{dt} = y(t) = f(t, I^{1-\alpha}y(t)) = f(t, D^{\alpha}x(t)).$$

Again, from (5), we have

$$\sum_{k=1}^{m} a_k\, x(\phi(\tau_k)) = x_0 - \sum_{k=1}^{m} a_k \int_0^{\phi(\tau_k)} y(\xi)\,d\xi + \sum_{k=1}^{m} a_k \int_0^{\phi(\tau_k)} y(\xi)\,d\xi = x_0.$$

This proves the equivalence between the nonlocal problem given by (1) and (2) and the integral Equation (5). □

For the problem (1) with the m-point BCs (4), we prove Theorem 1 below.

Theorem 1. *Suppose that the assumptions* (i) *to* (iii) *are satisfied. If*

$$\frac{b}{\Gamma(2-\alpha)} < 1,$$

then the fractional-order integral equation (6) *has a solution* $y \in L_1[0,1]$.

Suppose also that the coefficients a_k *satisfy the following inequality*:

$$\sum_{k=1}^{m} a_k \neq 0.$$

Then the problem (1) *together with the m-point BCs in* (4) *has at least one solution* $x \in AC[0,1]$ *given by* (5).

Proof. Let us define the operator T associated with Equation (6) by

$$(Ty)(t) = f\big(t, I^{1-\alpha}y(t)\big).$$

In addition, for a positive number r, let

$$B_r = \{y : y \in L_1(I) \quad \text{and} \quad ||y||_{L_1} \leqq r\} \subset L_1[0,1],$$

where

$$r \geqq \frac{||a||_{L_1}}{1 - \frac{b}{\Gamma(2-\alpha)}}.$$

Clearly, B_r is nonempty, closed, convex and bounded.

From the assumption (i), we can deduce that the operator T is continuous.

Suppose that y is an arbitrary element in B_r. We will show that $TB_r \subset r$. Indeed, from (6) and the assumptions (i) and (iii), we get

$$\begin{aligned}
||Ty||_{L_1} &= \int_0^1 |Ty(t)|\, dt \\
&\leqq \int_0^1 |a(t)| dt + b \int_0^1 \int_0^t \frac{(t-\xi)^{-\alpha}}{\Gamma(1-\alpha)}\, |y(\xi)|\, d\xi\, dt \\
&\leqq ||a||_{L_1} + b \int_0^1 \int_\xi^1 \frac{(t-\xi)^{-\alpha}}{\Gamma(1-\alpha)}\, dt\, |y(\xi)|\, d\xi \\
&\leqq ||a||_{L_1} + \frac{b}{\Gamma(2-\alpha)}\, ||y||_{L_1} \leqq ||a||_{L_1} + \frac{b}{\Gamma(2-\alpha)}\, r \leqq r,
\end{aligned}$$

which implies that $TB_r \subset B_r$.

We will now show that T is a compact operator. In fact, if we let Ω be a bounded subset of B_r, then $T(\Omega)$ is clearly seen to be bounded in $L_1[0,1]$, that is, the first condition of Kolmogorov's Compactness Criterion (see [1]) is satisfied.

We next prove that

$$(Ty)_h \to Ty \text{ uniformly in } L_1[0,1] \qquad (h \to 0),$$

where

$$(Ty)_h(t) = \frac{1}{h} \int_t^{t+h} (Ty)(\xi)\, d\xi.$$

For each $y \in \Omega$, we thus find that

$$\begin{aligned}||(Ty)_h - Ty||_{L_1} &= \int_0^1 |(Ty)_h(t) - (Ty)(t)|\, dt \\ &= \int_0^1 |\frac{1}{h}\int_t^{t+h} (Ty)(\xi)\, d\xi - (Ty)(t)|\, dt \\ &\leqq \int_0^1 \left(\frac{1}{h}\int_t^{t+h} |(Ty)(\xi) - (Ty)(t)|\, d\xi\right) dt \\ &\leqq \int_0^1 \frac{1}{h}\int_t^{t+h} |f(\xi, I^{1-\alpha}y(\xi)) - f(t, I^{1-\alpha}y(t))|\, d\xi\, dt.\end{aligned}$$

By the assumptions (i) and (iii), $y \in \Omega$ implies that $f \in L_1[0,1]$, so it follows that (see [36])

$$\frac{1}{h}\int_t^{t+h} |f(\xi, I^{1-\alpha}y(\xi)) - f(t, I^{1-\alpha}y(t))|\, d\xi \to 0 \qquad (h \to 0) \text{ a.e.} \qquad (t \in [0,1]).$$

Then, by Kolmogorov's Compactness Criterion (see [1]), we find that $T(\Omega)$ is relatively compact, that is, T is a compact operator.

As a consequence of Schauder's Fixed Point Theorem (see [2]), the operator T has a fixed point in B_r. This proves the existence of the solution $y \in L_1[0,1]$ of Equation (6). Consequently, based on the above Lemma, problem (1) together with the m-point BCs (4) possess a solution $x \in AC(0,1)$.

Now, from Equation (5), we have

$$x(0) = \lim_{t\to 0+} x(t) = A\, x_0 - A \sum_{k=1}^{m} a_k \int_0^{\phi(\tau_k)} y(\xi)\, d\xi$$

and

$$x(1) = \lim_{t\to 1-} x(t) = A\, x_0 - A \sum_{k=1}^{m} a_k \int_0^{\phi(\tau_k)} y(\xi)\, d\xi + \int_0^1 y(\xi)\, d\xi,$$

from which we deduce that Equation (5) has a solution $x \in AC[0,1]$.

Consequently, the nonlocal problem given by (1) and (4) has a solution $x \in AC[0,1]$ given by (5). □

4. Riemann-Stieltjes Functional Integral BCs

Let $x \in AC[0,1]$ be a solution of the problem (1) with the m-point BCs in (4). Then, we have the following theorem.

Theorem 2. *Suppose that the assumptions (i) to (iii) are satisfied. If*

$$\frac{b}{\Gamma(2-\alpha)} < 1$$

and $g : [0,1] \to [0,1]$ is an increasing function, then there exists a solution $x \in AC[0,1]$ of the following problem:

$$x'(t) = f(t, D^{\alpha}x(t)) \text{ a.e.} \qquad (t \in (0,1);\ \alpha \in (0,1]),$$

together with the Riemann-Stieltjes functional integral condition:

$$\int_0^1 x(\phi(s))\, dg(s) = x_0,$$

which is represented by

$$x(t) = [g(1) - g(0)]^{-1} x_0 - [g(1) - g(0)]^{-1} \cdot \int_0^1 \int_0^{\phi(s)} y(\xi)\, d\xi\, dg(s) + \int_0^t y(\xi)\, d\xi. \tag{8}$$

Proof. Let

$$a_k = g(t_k) - g(t_{k-1}) \qquad (\tau_k \in (t_{k-1}, t_k);\ 0 \leqq t_0 < t_1 < t_2, \cdots < t_n \leqq 1).$$

Then, the multi-point nonlocal condition (4) becomes

$$\sum_{k=1}^{m} [g(t_k) - g(t_{k-1})]\, x(\phi(\tau_k)) = x_0.$$

From the continuity of the solution x of the multi-point nonlocal problem given by (1) and (4), we can get

$$\lim_{m\to\infty} \sum_{k=1}^{m} [g(t_k) - g(t_{k-1})]\, x(\phi(\tau_k)) = \int_0^1 x(\phi(s))\, dg(s).$$

Furthermore, the multi-point nonlocal boundary condition (4) can be transformed into the following Riemann-Stieltjes functional integral form:

$$\int_0^1 x(\phi(s))\, dg(s) = x_0.$$

In addition, from the functional integral Equation (5), we have

$$\begin{aligned} \lim_{m\to\infty} x(t) &= [g(1) - g(0)]^{-1} x_0 - [g(1) - g(0)]^{-1} \\ &\quad \cdot \lim_{m\to\infty} \sum_{k=1}^{m} [g(t_k) - g(t_{k-1})] \int_0^{\phi(\tau_k)} y(\xi) d\xi + \int_0^t y(\xi) d\xi \\ &= [g(1) - g(0)]^{-1} x_0 - [g(1) - g(0)]^{-1} \\ &\quad \cdot \int_0^1 \int_0^{\phi(s)} y(\xi)\, d\xi\, dg(s) + \int_0^t y(\xi) d\xi. \end{aligned}$$

Hence, the continuous solution of the first-order nonlinear differential Equation (1) with the Riemann-Stieltjes functional integral condition (2) is given by (8). □

We would like to provide two examples of the first order BVP (1) with the Riemann-Stieltjes functional integral boundary condition (2) (with the advanced or deviated argument ϕ) whose solutions are ensured by Theorem 2.

Example 1. *Let the nonlinear function $f(t,u)$ in (1) be given by*

$$f(t,u) = \cos\left(3(t+1)\right) + \frac{1}{5}\left(t^3 \sin u + e^{-t}\, u\right).$$

It is clear that the assumptions (i) and (iii) of Theorem 2 are fulfilled with

$$a(t) = \cos\left(3(t+1)\right) \in L_1[0,1] \qquad \text{and} \qquad b = \frac{2}{5}.$$

Let the fractional order in (1) be $\alpha = \frac{1}{2}$. Then

$$\frac{b}{\Gamma(2-\alpha)} \approx 0.4503338 < 1.$$

In this case, the first-order BVP (1) *has the following form:*

$$\frac{dx}{dt} = \cos\left(3(t+1)\right) + \frac{1}{5}\left[t^3 \sin D^{1/2}x(t) + e^{-t} D^{1/2}x(t)\right]. \tag{9}$$

Let the function $g : [0,1] \to [0,1]$ *be defined by the formula*

$$g(t) = t \ln(1+t).$$

If $\beta \in (0,1)$, *we consider the advanced function* $\phi(t) = t^{\beta}$. *Then, the integral condition* (2) *assumes the following form:*

$$\int_0^1 x(t^{\beta})\, d\big(t \ln(1+t)\big) = x_0. \tag{10}$$

Thus, clearly, one can obtain the existence of a solution of (9) *and* (10).

Example 2. *Let* $f(t,u)$, $g(t)$, α *and* β *be as in Example* (1) *and consider the deviated function* $\phi(t) = \beta t$. *Then the functional integral condition* (2) *becomes*

$$\int_0^1 x(\beta t)\, d\big(t \ln(1+t)\big) = x_0. \tag{11}$$

Therefore, we can obtain the existence of a solution of (9) *and* (11).

We now consider another Riemann-Stieltjes nonlocal integral boundary condition.

Corollary 1. *Let the assumptions of Theorem* 2 *be satisfied. Then there exists a solution* $x \in AC[0,1]$ *of the following problem:*

$$x'(t) = f\big(t, D^{\alpha}x(t)\big) \ \text{a.e.} \qquad \big(t \in (0,\,1);\ \alpha \in (0,1]\big),$$

together with the Riemann-Stieltjes nonlocal integral condition given by

$$\int_c^d x(\phi(s))\, dg(s) = x_0 \qquad (0 < c < d < 1),$$

which is represented by

$$x(t) = [g(d) - g(c)]^{-1}\, x_0 - [g(d) - g(c)]^{-1} \cdot \int_c^d \int_0^{\phi(s)} y(\xi)\, d\xi\, dg(s) + \int_0^t y(\xi)\, d\xi.$$

Proof. The proof of the above corollary is similar to that of Theorem 2. Here, in this case, we let

$$a_k = g(t_k) - g(t_{k-1}) \qquad \big(\tau_k \in (t_{k-1}, t_k);\ 0 < c \leqq t_0 < t_1 < t_2, \cdots < t_n \leqq d < 1\big).$$

□

5. Infinite-Point Boundary Conditions

Let $x \in AC[0,1]$ be the solution of the nonlocal problem given by (1) and (4). Then, we have the following theorem.

Theorem 3. *Let the assumptions* (i) *and* (iii) *be satisfied and let*

$$\phi(\tau_k) \leqq \tau_k \qquad \text{and} \qquad \frac{b}{\Gamma(2-\alpha)} < 1.$$

Suppose also that the following series:

$$\sum_{k=1}^{\infty} a_k = B^{-1}$$

is convergent. Then there exists a solution $x \in AC[0,1]$ of the nonlocal problem (1) and (3) given by the following integral equation:

$$x(t) = B\,x_0 - B \sum_{k=1}^{\infty} a_k \int_0^{\phi(\tau_k)} y(\xi)\,d\xi + \int_0^t y(\xi)\,d\xi \tag{12}$$

for every solution y of the functional equation (6).

Proof. Let $x \in AC[0,1]$ be a solution of the infinite point BVP (1) and (4) given by (5). Since

$$\left|a_k\,x(\phi(\tau_k))\right| \leqq a_k\,||x|| \qquad \text{and} \qquad \left|a_k \int_0^{\phi(\tau_k)} y(\xi)\,d\xi\right| \leqq a_k\,||y||_{L_1},$$

by the comparison test, the series in (3) and

$$\sum_{k=1}^{\infty} a_k \int_0^{\phi(\tau_k)} y(\xi)\,d\xi$$

are convergent. Thus, by taking the limit as $m \to \infty$ in (5), we obtain

$$x(t) = B\,x_0 - B \sum_{k=1}^{\infty} a_k \int_0^{\phi(\tau_k)} y(\xi)\,d\xi + \int_0^t y(\xi)\,d\xi,$$

which, for every solution y of the functional Equation (6), satisfies the differential Equation (1). Furthermore, from (12), we have

$$\begin{aligned}\sum_{k=1}^{\infty} a_k\,x(\phi(\tau_k)) &= B^{-1}\,B\,x_0 - B^{-1}\,B \sum_{k=1}^{\infty} a_k \int_0^{\phi(\tau_k)} y(\xi)\,d\xi \\ &\quad + \sum_{k=1}^{\infty} a_k \int_0^{\phi(\tau_k)} y(\xi)\,d\xi = x_0.\end{aligned} \tag{13}$$

This proves that the solution of the integral equation (12) satisfies the problem given by (1) under infinite-point BCs (3). □

6. Further Illustrative Examples

In this section, we consider the following examples with a view to illustrating some of our main results.

Example 3. *Consider the following infinite-point BVP:*

$$\frac{dx}{dt} = \frac{\ln\left(1 + |D^{2/3}x(t)|\right)}{2+t^2} + t^3 e^{-t^2} \quad a.e. \qquad (0 < t < 1)$$

together with

$$\sum_{k=1}^{\infty} \frac{1}{k^3}\,x\left(\frac{k^2-1}{k^2} - \sigma \sin^2\left(\sqrt{\frac{k^2-1}{k^2}}\right)\right) \qquad (0 \leqq \sigma \leqq 1).$$

If we set

$$f(t,u) = \frac{\ln\left(1+|u(t)|\right)}{2+t^2} + t^3 e^{-t^2},$$

then

$$|f(t,u)| \leqq t^3 e^{-t^2} + \frac{1}{3}\,|u|.$$

We also set

$$a(t) = t^3 e^{-t^2} \in L_1[0,1] \qquad \text{and} \qquad b = \frac{1}{3}.$$

Thus, clearly, assumptions (i) and (iii) are satisfied.

On the other hand, we have

$$\alpha = \frac{2}{3} \quad \text{so that} \quad \frac{b}{\Gamma(2-\alpha)} \approx 0.3731148 < 1.$$

Now, if we let

$$\phi(\tau_k) = \tau_k - \lambda \sin^2\left(\sqrt{\tau_k}\right) \qquad \text{and} \qquad \tau_k = \frac{k^2-1}{k^2} \in (0,1),$$

then

$$\phi(\tau_k) \leqq \tau_k.$$

In addition, the following series:

$$\sum_{k=1}^{\infty} a_k = \sum_{k=1}^{\infty} \frac{1}{k^3}$$

is convergent. Therefore, by appealing to Theorem 3, the given infinite-point BVP has an absolute continuous solution.

Example 4. *Consider the following infinite-point implicit BVP:*

$$\frac{dx}{dt} = \frac{[x'(t)]^3}{2(1+|x'(t)|^2)} + \frac{1}{4\pi}\sin\left(\pi\, x'(t)\right) + \cos t^3 + 3 \quad \text{a.e.} \qquad \left(0<t<1;\ x'(t) := \frac{dx}{dt}\right)$$

together with

$$\sum_{k=1}^{\infty} 10\left(\frac{3}{4}\right)^k x\left(\frac{1}{k^3} - \lambda \exp\left(-\frac{1}{k^3}\right)\right) \qquad (0 \leqq \lambda \leqq 1).$$

If we set

$$f(t,u) = \frac{u^3}{2(1+|u|^2)} + \frac{1}{4\pi}\sin\left(\pi\, u\right) + \cos t^3 + 3,$$

then

$$|f(t,u)| \leqq \cos t^3 + 3 + \frac{3}{4}\,|u|,$$

Now, putting

$$a(t) = \cos t^3 + 3 \in L_1[0,1] \ \text{ and } \ b = \frac{3}{4},$$

the assumptions (i) and (iii) hold true.

We have

$$\alpha = 1 \quad \text{so that} \quad \frac{b}{\Gamma(2-\alpha)} = \frac{3}{4} < 1.$$

On the other hand, if we let

$$\phi(\tau_k) = \tau_k - \lambda \exp\left(-\tau_k\right) \qquad \text{and} \qquad \tau_k = \frac{1}{k^3} \in (0,1),$$

then

$$\phi(\tau_k) \leqq \tau_k.$$

We also see that the following series:

$$\sum_{k=1}^{\infty} a_k = 10 \sum_{k=1}^{\infty} \left(\frac{2}{3}\right)^{k-1}$$

is convergent. Therefore, by applying Theorem 3, the given infinite-point implicit BVP has an absolute continuous solution in $[0,1]$.

7. Conclusions

In our present investigation, we have considered the existence of an absolute continuous solution to a class of first-order nonlinear differential equation with integral boundary conditions (BCs) or with infinite-point BCs (see Theorems 2 and 3 and the above Corollary). We have demonstrated that, if we can get the continuous solutions to BVPs with m-point BCs, we can easily get the solutions to these problems with integral BCs or with infinite-point BCs. Several examples have also been given in order to illustrate some of our main results. We note that the fractional differential Equation (1) involves the ordinary derivative $\frac{dx}{dt}$ of order 1 on its left-hand side. In the foreseeable future, we propose to investigate the possibility of extending our results to such other higher-order derivatives as

$$\frac{d^2x}{dt^2}, \quad \frac{d^3x}{dt^3}, \quad \frac{d^4x}{dt^4}, \quad \cdots,$$

occurring on the left-hand side of the fractional differential Equation (1), involving the Liouville-Caputo fractional derivatives together with integral BCs and/or the infinite-point BCs.

Author Contributions: All authors contributed equally.

Funding: This research received no external funding.

Conflicts of Interest: The authors declare no conflict of interest.

References

1. Dugundji, J.; Granas, A. *Fixed Point Theory*; Monografie Mathematyczne: Warsaw, Poland, 1982.
2. Deimling, K. *Nonlinear Functional Analysis*; Springer: Berlin/Heidelberg, Germany; New York, NY, USA, 1985.
3. Alvan, M.; Darzi, R.; Mahmoodi, A. Existence results for a new class of boundary value problems of nonlinear fractional differential equations. *Mathematics* **2016**, *4*, 13. [CrossRef]
4. Benchohra, M.; Hamani, S.; Ntouyas, S. Boundary value problems for differential equations with fractional order and nonlocal conditions. *Nonlinear Anal.* **2009**, *71*, 2391–2396. [CrossRef]
5. Boucherif, A. First-order differential inclusions with nonlocal initial conditions. *Appl. Math. Lett.* **2002**, *15*, 409–414. [CrossRef]
6. El-Sayed, A.M.A.; Bin-Taher, E.O. Positive solutions for a nonlocal multi-point boundary-value problem of fractional and second order. *Electron. J. Differ. Equ.* **2013**, *2013*, 64.
7. Gao, H.; Han, X. Existence of positive solutions for fractional differential equation with nonlocal boundary condition. *Int. J. Differ. Equ.* **2011**, *2011*, 328394. [CrossRef]
8. Hamani, S.; Benchora, M.; Graef, J.R. Existence results for boundary-value problems with nonlinear fractional differential inclusions and integral conditions. *Electron. J. Q. Theory Differ. Equ.* **2013**, *20*, 1–16.
9. Nieto, J.J.; Ouahab, A.; Venktesh, V. Implicit fractional differential equations via the Liouville-Caputo derivative. *Mathematics* **2015**, *3*, 398–411. [CrossRef]

10. Ionkin, N.I. Solution of a boundary value problem in heat conduction theory with nonlocal boundary conditions. *Differ. Equ.* **1977**, *13*, 294–304.
11. Cannon, J.R. The solution of the heat equation subject to the specification of energy. *Q. Appl. Math.* **1963**, *21*, 155–160. [CrossRef]
12. Chegis, R.Y. Numerical solution of a heat conduction problem with an integral boundary condition. *Lietuvos Matematikos Rinkinys* **1984**, *24*, 209–215.
13. Wang, Y.; Liu, L.; Zhang, X.; Wu, Y. Positive solutions of an abstract fractional semipositone differential system model for bioprocesses of HIV infection. *Appl. Math. Comput.* **2015**, *258*, 312–324. [CrossRef]
14. Cattani, C.; Srivastava, H.M.; Yang, X.-J. (Eds.) *Fractional Dynamics*; Emerging Science Publishers: Berlin, Germany; Warsaw, Poland, 2015.
15. Akcan, U.; Çetin, E. The lower and upper solution method for three-point boundary value problems with integral boundary conditions on a half-line. *Filomat* **2018**, *32*, 341–353. [CrossRef]
16. Boucherif, A. Second-order boundary value problems with integral boundary conditions. *Nonlinear Anal.* **2009**, *70*, 364–371. [CrossRef]
17. Benchohra, M.; Nieto, J.J.; Ouahab, A. Second-order boundary value problem with integral boundary conditions. *Bound. Value Prob.* **2011**, *2011*, 260309. [CrossRef]
18. Chalishajar, D.; Kumar, A. Existence, uniqueness and Ulam's stability of solutions for a coupled system of fractional differential equations with integral boundary conditions. *Mathematics* **2018**, *6*, 96. [CrossRef]
19. Dou, J.; Zhou, D.; Pang, H. Existence and multiplicity of positive solutions to a fourth-order impulsive integral boundary value problem with deviating argument. *Bound. Value Prob.* **2016**, *2016*, 166. [CrossRef]
20. Li, Y.; Zhang, H. Solvability for system of nonlinear singular differential equations with integral boundary conditions. *Bound. Value Prob.* **2014**, *2014*, 158. [CrossRef]
21. Liu, L.; Hao, X.; Wu, Y. Positive solutions for singular second order differential equations with integral boundary conditions. *Math. Comput. Model.* **2013**, *57*, 836–847. [CrossRef]
22. Song, G.; Zhao, Y.; Sun, X. Integral boundary value problems for first order impulsive integro-differential equations of mixed type. *J. Comput. Appl. Math.* **2011**, *235*, 2928–2935. [CrossRef]
23. Tokmagambetov, N.; Torebek, B.T. Well-posed problems for the fractional laplace equation with integral boundary conditions. *Electron. J. Differ. Equ.* **2018**, *2018*, 90.
24. Wang, X.; Wang, L.; Zeng, Q. Fractional differential equations with integral boundary conditions. *J. Nonlinear Sci. Appl.* **2015**, *8*, 309–314. [CrossRef]
25. Yang, W.; Qin, Y. Positive solutions for nonlinear Caputo type fractional q-difference equations with integral boundary conditions. *Mathematics* **2016**, *4*, 63. [CrossRef]
26. Ge, F.; Zhou, H.; Kou, C. Existence of solutions for a coupled fractional differential equations with infinitely many points boundary conditions at resonance on an unbounded domain. *Differ. Equ. Dyn. Syst.* **2016**, *24*, 1–17. [CrossRef]
27. Guo, L.; Liu, L.; Wu, Y. Existence of positive solutions for singular fractional differential equations with infinite-point boundary conditions. *Nonlinear Anal.* **2016**, *21*, 635–650. [CrossRef]
28. Hu, L.; Zhang, S. Existence results for a coupled system of fractional differential equations with p-Laplacian operator and infinite-point boundary conditions. *Bound. Value Prob.* **2017**, *2017*, 88. [CrossRef]
29. Li, B.; Sun, S.; Sun, Y. Existence of solutions for fractional Langevin equation with infinite-point boundary conditions. *J. Appl. Math. Comput.* **2017**, *53*, 683–692. [CrossRef]
30. Liu, S.; Liu, J.; Dai, Q.; Li, H. Uniqueness results for nonlinear fractional differential equations with infinite-point integral boundary conditions. *J. Nonlinear Sci. Appl.* **2017**, *10*, 1281–1288. [CrossRef]
31. Zhang, X. Positive solutions for a class of singular fractional differential equation with infinite-point boundary value conditions. *Appl. Math. Lett.* **2015**, *39*, 22–27. [CrossRef]
32. Zhong, Q.; Zhang, X. Positive solution for higher-order singular infinite-point fractional differential equation with p-Laplacian. *Adv. Differ. Equ.* **2016**, *2016*, 11. [CrossRef]
33. Xu, B.; Yang, Y. Eigenvalue intervals for infinite-point fractional boundary value problem and application in systems theory. *Int. J. Circuits Syst. Signal Process.* **2016**, *10*, 215–224.
34. Kilbas, A.A.; Srivastava, H.M.; Trujillo, J.J. *Theory and Applications of Fractional Differential Equations*; North-Holland Mathematical Studies; Elsevier (North-Holland) Science Publishers: Amsterdam, The Netherlands; London, UK; New York, NY, USA, 2006; Volume 204.

35. Srivastava, H.M.; Saad, K.M. Some new models of the time-fractional gas dynamics equation. *Adv. Math. Models Appl.* **2018**, *3*, 5–17.
36. Swartz, C. *Measure, Integration and Functional Spaces*; World Scientific Publishing Company: Singapore, 1994.

© 2018 by the authors. Licensee MDPI, Basel, Switzerland. This article is an open access article distributed under the terms and conditions of the Creative Commons Attribution (CC BY) license (http://creativecommons.org/licenses/by/4.0/).

symmetry

MDPI

Article

Third-Order Hankel Determinant for Certain Class of Analytic Functions Related with Exponential Function

Hai-Yan Zhang, Huo Tang * and Xiao-Meng Niu

School of Mathematics and Statistics, Chifeng University, Chifeng 024000, China; cfxyzhhy@163.com (H.-Y.Z.); ndnxm@126.com (X.-M.N.)

* Correspondence: thth2009@163.com

Received: 30 July 2018; Accepted: 9 October 2018; Published: 15 October 2018

Abstract: Let S_l^* denote the class of analytic functions f in the open unit disk $\mathbb{D} = \{z : |z| < 1\}$ normalized by $f(0) = f'(0) - 1 = 0$, which is subordinate to exponential function, $\frac{zf'(z)}{f(z)} \prec e^z$ $(z \in \mathbb{D})$. In this paper, we aim to investigate the third-order Hankel determinant $H_3(1)$ for this function class S_l^* associated with exponential function and obtain the upper bound of the determinant $H_3(1)$. Meanwhile, we give two examples to illustrate the results obtained.

Keywords: analytic function; Hankel determinant; exponential function; upper bound

MSC: 30C45; 30C50; 30C80

1. Introduction

Let $\mathcal{S}$ denote the class of functions f which are analytic and univalent in the open unit disk $\mathbb{D} = \{z : |z| < 1\}$ of the form

$$f(z) = z + \sum_{n=2}^{\infty} a_n z^n \quad (z \in \mathbb{D}). \tag{1}$$

Assume that $\mathcal{P}$ denote the class of analytic functions p normalized by

$$p(z) = 1 + c_1 z + c_2 z^2 + c_3 z^3 + \cdots$$

and satisfying the condition Re $p(z) > 0$ $(z \in \mathbb{D})$.

It is easy to see that, if $p(z) \in \mathcal{P}$, then exists a Schwarz function $\omega(z)$ with $\omega(0) = 0$ and $|\omega(z)| < 1$, such that (see [1])

$$p(z) = \frac{1 + w(z)}{1 - w(z)} \quad (z \in \mathbb{D}).$$

Now, we start with recalling the definition of subordination.

Suppose that f and g are two analytic functions in $\mathbb{D}$. Then, we say that the function g is subordinate to the function f, and we write

$$g(z) \prec f(z) \quad (z \in \mathbb{D}),$$

if there exists a Schwarz function $\omega(z)$ with $\omega(0) = 0$ and $|\omega(z)| < 1$, such that (see [2])

$$g(z) = f(\omega(z)) \quad (z \in \mathbb{D}).$$

Recently, Mendiratta et al. in [3] introduced the following subclass S_l^* of analytic functions associated with exponential function.

Definition 1. *(see [3]). A function $f \in \mathcal{S}$ is said to be in the class S_l^*, if it satisfies the following condition:*

$$\frac{zf'(z)}{f(z)} \prec e^z \ (z \in \mathbb{D}). \tag{2}$$

We easily observe that, $f \in S_l^*$, if and only if

$$\left|\log \frac{zf'(z)}{f(z)}\right| < 1 \ \ (z \in \mathbb{D}). \tag{3}$$

In fact, if we choose $f(z) = z + \frac{1}{4}z^2$, then, from Equation (3), we can sketch the figure of the function class S_l^* (see Figure 1).

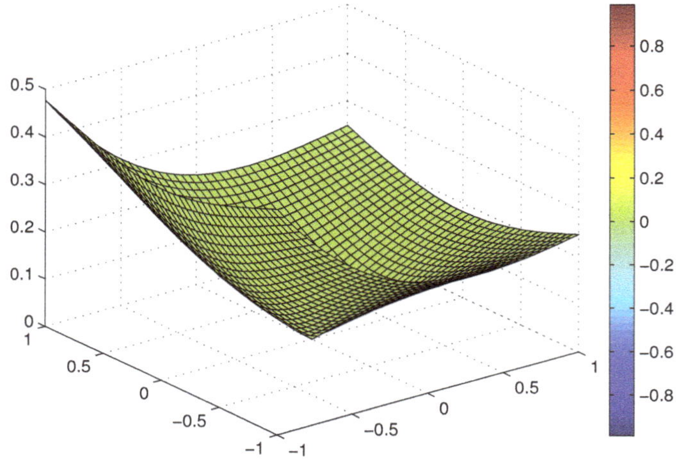

Figure 1. the figure of the function class S_l^* for $f(z) = z + \frac{1}{4}z^2$.

The qth Hankel determinant for $q \geq 1$ and $n \geq 1$ is stated by Noonan and Thomas [4] as

$$H_q(n) = \begin{vmatrix} a_n & a_{n+1} & \cdots & a_{n+q-1} \\ a_{n+1} & a_{n+2} & \cdots & a_{n+q} \\ \vdots & \vdots & & \vdots \\ a_{n+q-1} & a_{n+q} & \cdots & a_{n+2q-2} \end{vmatrix} \quad (a_1 = 1).$$

This determinant has been considered by several authors, for example, Noor [5] determined the rate of growth of $H_q(n)$ as $n \to \infty$ for functions $f(z)$ given by Equation (1) with bounded boundary and Ehrenborg [6] studied the Hankel determinant of exponential polynomials.

In particular, we have

$$H_2(1) = \begin{vmatrix} a_1 & a_2 \\ a_2 & a_3 \end{vmatrix} = a_3 - a_2^2 \ \ (a_1 = 1, \ n = 1, \ q = 2),$$

$$H_2(2) = \begin{vmatrix} a_2 & a_3 \\ a_3 & a_4 \end{vmatrix} = a_2a_4 - a_3^2 \quad (n = 2,\ q = 2),$$

and

$$H_3(1) = \begin{vmatrix} a_1 & a_2 & a_3 \\ a_2 & a_3 & a_4 \\ a_3 & a_4 & a_5 \end{vmatrix} \quad (n = 1,\ q = 3).$$

Since $f \in \mathcal{S},\ a_1 = 1$, thus

$$H_3(1) = a_3(a_2a_4 - a_3^2) - a_4(a_4 - a_2a_3) + a_5(a_3 - a_2^2).$$

We note that $H_2(1)$ is the well-known Fekete-Szego functional (see, for instance, [7–12]).

In recent years, many authors studied the second-order Hankel determinant $H_2(2)$ and the third-order Hankel determinant $H_3(1)$ for various classes of functions, the interested readers can see, for example, [13–22]. We note that, they discussed the determinants $H_2(2)$ and $H_3(1)$ based on the function classes, which are all subordinate to a certain function $\frac{1+Az}{1+Bz}$ $(-1 \leq B < A \leq 1;$ $z \in \mathbb{D})$. Until now, very few researchers have studied the above determinants for the function class, subordinated to e^z $(z \in \mathbb{D})$. So, in this paper, we aim to investigate the third-order Hankel determinant $H_3(1)$ for the function class S_l^*, which is associated with exponential function, and obtain the upper bound of the above determinant.

2. Main Results

In order to prove our desired results, we shall require the following lemmas.

Lemma 1. *(see [23]). If $p(z) \in \mathcal{P}$, then exists some $x,\ z$ with $|x| \leq 1,\ |z| \leq 1$, such that*

$$2c_2 = c_1^2 + x(4 - c_1^2),$$

$$4c_3 = c_1^3 + 2c_1x(4 - c_1^2) - (4 - c_1^2)c_1x^2 + 2(4 - c_1^2)(1 - |x|^2)z.$$

Lemma 2. *(see [24]). Let $p(z) \in \mathcal{P}$, then*

$$|c_n| \leq 2,\ n = 1, 2, \cdots.$$

Lemma 3. *(see [3]). If the function $f(z) \in S_l^*$ and of the form Equation (1), then*

$$|a_2| \leq 1,\ \ |a_3| \leq \frac{3}{4},\ \ |a_4| \leq \frac{17}{36},\ \ |a_5| \leq 1. \tag{4}$$

We now state and prove the main results of our present investigation.

Theorem 1. *If the function $f(z) \in S_l^*$ and of the form Equation (1), then we have*

$$|a_3 - a_2^2| \leq \frac{1}{2}. \tag{5}$$

Proof. Since $f(z) \in S_l^*$, according to the definition of subordination, then there exists a Schwarz function $\omega(z)$ with $\omega(0) = 0$ and $|\omega(z)| < 1$, such that

$$\frac{zf'(z)}{f(z)} = e^{\omega(z)}.$$

Now

$$\begin{aligned} \tfrac{zf'(z)}{f(z)} &= \tfrac{z+\sum_{n=2}^{\infty} na_n z^n}{z+\sum_{n=2}^{\infty} a_n z^n} \\ &= (1+\sum_{n=2}^{\infty} na_n z^{n-1})[1 - a_2 z + (a_2^2 - a_3)z^2 - (a_2^3 - 2a_2 a_3 + a_4)z^3 + \cdots] \\ &= 1 + a_2 z + (2a_3 - a_2^2)z^2 + (a_2^3 - 3a_2 a_3 + 3a_4)z^3 + \cdots . \end{aligned} \tag{6}$$

Define a function

$$p(z) = \frac{1+\omega(z)}{1-\omega(z)} = 1 + c_1 z + c_2 z^2 + \cdots .$$

Then, we notice that $p(z) \in \mathcal{P}$ and

$$\omega(z) = \frac{p(z)-1}{1+p(z)} = \frac{c_1 z + c_2 z^2 + c_3 z^3 + \cdots}{2 + c_1 z + c_2 z^2 + c_3 z^3 + \cdots}.$$

On the other hand,

$$\begin{aligned} e^{\omega(z)} &= 1 + \omega(z) + \tfrac{\omega(z)^2}{2!} + \tfrac{\omega(z)^3}{3!} + \cdots \\ &= 1 + \tfrac{c_1 z + c_2 z^2 + c_3 z^3 + \cdots}{2 + c_1 z + c_2 z^2 + c_3 z^3 + \cdots} + \tfrac{1}{2}(\tfrac{c_1 z + c_2 z^2 + c_3 z^3 + \cdots}{2 + c_1 z + c_2 z^2 + c_3 z^3 + \cdots})^2 + \tfrac{1}{6}(\tfrac{c_1 z + c_2 z^2 + c_3 z^3 + \cdots}{2 + c_1 z + c_2 z^2 + c_3 z^3 + \cdots})^3 + \cdots \\ &= 1 + \tfrac{1}{2}(c_1 z + c_2 z^2 + c_3 z^3 + \cdots)[1 - \tfrac{c_1 z}{2} + (\tfrac{c_1^2}{4} - \tfrac{c_2}{2})z^2 - (\tfrac{c_1^3}{8} - \tfrac{c_1 c_2}{2} + \tfrac{c_3}{2})z^3 + \cdots] \\ &+ \tfrac{1}{8}(c_1 z + c_2 z^2 + c_3 z^3 + \cdots)^2[1 - \tfrac{c_1 z}{2} + (\tfrac{c_1^2}{4} - \tfrac{c_2}{2})z^2 - (\tfrac{c_1^3}{8} - \tfrac{c_1 c_2}{2} + \tfrac{c_3}{2})z^3 + \cdots]^2 \\ &+ \tfrac{1}{48}(c_1 z + c_2 z^2 + c_3 z^3 + \cdots)^3[1 - \tfrac{c_1 z}{2} + (\tfrac{c_1^2}{4} - \tfrac{c_2}{2})z^2 - (\tfrac{c_1^3}{8} - \tfrac{c_1 c_2}{2} + \tfrac{c_3}{2})z^3 + \cdots]^3 + \cdots \\ &= 1 + \tfrac{1}{2}c_1 z + (\tfrac{c_2}{2} - \tfrac{c_1^2}{8})z^2 + (\tfrac{c_1^3}{48} - \tfrac{c_1 c_2}{4} + \tfrac{c_3}{2})z^3 + \cdots . \end{aligned} \tag{7}$$

On comparing the coefficients of z, z^2, z^3 between the Equations (6) and (7), we obtain

$$a_2 = \frac{c_1}{2},\ a_3 = \frac{c_2}{4} + \frac{c_1^2}{16},\ a_4 = \frac{c_3}{6} + \frac{c_1 c_2}{24} - \frac{c_1^3}{288}. \tag{8}$$

So,

$$|a_3 - a_2^2| = \left|\frac{c_2}{4} + \frac{c_1^2}{16} - \frac{c_1^2}{4}\right| = \left|\frac{c_2}{4} - \frac{3c_1^2}{16}\right|.$$

Using Lemma 1, we thus know that

$$|a_3 - a_2^2| = \left|\frac{x(4-c_1^2)}{8} - \frac{c_1^2}{16}\right|.$$

Letting $|x| = t \in [0,1]$, $c_1 = c \in [0,2]$ and applying the triangle inequality, the above equation reduces to

$$|a_3 - a_2^2| \le \frac{t(4-c^2)}{8} + \frac{c^2}{16}.$$

Suppose that

$$F(c,t) := \frac{t(4-c^2)}{8} + \frac{c^2}{16},$$

then we get

$$\frac{\partial F}{\partial t} = \frac{4-c^2}{8} \geq 0,$$

which shows that $F(c,t)$ is an increasing function on the closed interval [0,1] about t. Therefore, the function $F(c,t)$ can get the maximum value at $t = 1$, that is

$$\max F(c,t) = F(c,1) = \frac{(4-c^2)}{8} + \frac{c^2}{16}.$$

Next, let

$$G(c) := \frac{(4-c^2)}{8} + \frac{c^2}{16} = \frac{1}{2} - \frac{c^2}{16}.$$

Then, we easily find the function $G(c)$ have a maximum value at $c = 0$, also which is

$$|a_3 - a_2^2| \leq G(0) = \frac{1}{2}.$$

The proof of Theorem 1 is thus completed. □

Theorem 2. *If the function $f(z) \in S_l^*$ and of the form Equation (1), then we have*

$$|a_2a_3 - a_4| \leq \frac{896\sqrt{2}+385}{3087}. \tag{9}$$

Proof. From the Equation (8), we have

$$\begin{aligned}|a_2a_3 - a_4| &= |\tfrac{c_1c_2}{8} + \tfrac{c_1^3}{32} - \tfrac{c_3}{6} - \tfrac{c_1c_2}{24} + \tfrac{c_1^3}{288}| \\ &= |\tfrac{c_1c_2}{12} - \tfrac{c_3}{6} + \tfrac{5c_1^3}{144}|.\end{aligned}$$

Again, by applying Lemma 1, we get

$$|a_2a_3 - a_4| = \left|\frac{(4-c_1^2)c_1x^2}{24} - \frac{(4-c_1^2)c_1x}{24} - \frac{(4-c_1^2)(1-|x|^2)z}{12} + \frac{5c_1^3}{144}\right|.$$

Assume that $|x| = t \in [0,1]$, $c_1 = c \in [0,2]$. Then, using the triangle inequality, we deduce that

$$|a_2a_3 - a_4| \leq \frac{(4-c^2)ct^2}{24} + \frac{(4-c^2)ct}{24} + \frac{(4-c^2)}{12} + \frac{5c^3}{144}.$$

Setting

$$F(c,t) := \frac{(4-c^2)ct^2}{24} + \frac{(4-c^2)ct}{24} + \frac{(4-c^2)}{12} + \frac{5c^3}{144}.$$

Hence, we have

$$\frac{\partial F}{\partial t} = \frac{(4-c^2)ct}{12} + \frac{(4-c^2)c}{24} \geq 0,$$

namely, that $F(c,t)$ is an increasing function on the closed interval [0,1] about t. This implies that the maximum value of $F(c,t)$ occurs at $t = 1$, which is

$$\max F(c,t) = F(c,1) = \frac{(4-c^2)c}{24} + \frac{(4-c^2)c}{24} + \frac{(4-c^2)}{12} + \frac{5c^3}{144}.$$

Now define

$$G(c) := \frac{(4-c^2)c}{24} + \frac{(4-c^2)c}{24} + \frac{(4-c^2)}{12} + \frac{5c^3}{144},$$

then

$$G'(c) = \frac{(4-c^2)}{12} - \frac{c^2}{6} - \frac{c}{6} + \frac{15c^2}{144}.$$

Let $G'(c) = 0$, then the root is $c = r = \frac{-4+8\sqrt{2}}{7}$. And so the function $G(c)$ have a maximum value attained at $c = r = \frac{-4+8\sqrt{2}}{7}$, also which is

$$|a_2a_3 - a_4| \le G(r) = \frac{896\sqrt{2}+385}{3087}.$$

The proof of Theorem 2 is completed. □

Theorem 3. *If the function $f(z) \in S_l^*$ and of the form Equation (1), then we have*

$$|a_2a_4 - a_3^2| \le \frac{7}{12}. \tag{10}$$

Proof. Suppose that $f(z) \in S_l^*$, then from Equation (8), we have

$$\begin{aligned} |a_2a_4 - a_3^2| &= |\tfrac{c_1c_3}{12} + \tfrac{c_1^2c_2}{48} - \tfrac{c_1^4}{576} - (\tfrac{c_2}{4} + \tfrac{c_1^2}{16})^2| \\ &= |\tfrac{c_1c_3}{12} - \tfrac{c_1^2c_2}{96} - \tfrac{c_1^4}{576} - \tfrac{c_2^2}{16} - \tfrac{c_1^4}{256}|. \end{aligned}$$

In view of Lemma 1, we thus obtain

$$\begin{aligned} |a_2a_4 - a_3^2| &= \left| \tfrac{c_1c_3}{12} + \tfrac{c_1^2c_2}{48} - \tfrac{c_1^4}{576} - (\tfrac{c_2}{4} + \tfrac{c_1^2}{16})^2 \right| \\ &= \left| \tfrac{xc_1^2(4-c_1^2)}{192} - \tfrac{x^2c_1^2(4-c_1^2)}{48} - \tfrac{x^2(4-c_1^2)^2}{64} - \tfrac{c_1(4-c_1^2)(1-|x|^2)z}{24} - \tfrac{c_1^4}{256} \right|. \end{aligned}$$

Also, let $|x| = t \in [0,1]$, $c_1 = c \in [0,2]$. Then, using the triangle inequality, we get

$$|a_2a_4 - a_3^2| \le \frac{tc^2(4-c^2)}{192} + \frac{t^2c^2(4-c^2)}{48} + \frac{t^2(4-c^2)^2}{64} + \frac{(4-c^2)}{12} + \frac{c^4}{256}.$$

Assume that

$$F(c,t) := \frac{tc^2(4-c^2)}{192} + \frac{t^2c^2(4-c^2)}{48} + \frac{t^2(4-c^2)^2}{64} + \frac{(4-c^2)}{12} + \frac{c^4}{256},$$

thus, we have

$$\frac{\partial F}{\partial t} = \frac{c^2(4-c^2)}{192} + \frac{tc^2(4-c^2)}{24} + \frac{t(4-c^2)^2}{32} \ge 0,$$

which implies that $F(c,t)$ increases on the closed interval [0,1] about t. That is, that $F(c,t)$ have a maximum value at $t = 1$, which is

$$\max F(c,t) = F(c,1) = \frac{5c^2(4-c^2)}{192} + \frac{(4-c^2)^2}{64} + \frac{(4-c^2)}{12} + \frac{c^4}{256}.$$

Taking

$$G(c) := \frac{5c^2(4-c^2)}{192} + \frac{(4-c^2)^2}{64} + \frac{(4-c^2)}{12} + \frac{c^4}{256},$$

then we have

$$G'(c) = \frac{5c(4-c^2)}{96} - \frac{c(4-c^2)}{16} - \frac{c}{6} - \frac{5c^3}{96} + \frac{c^3}{64}.$$

If $G'(c) = 0$, then the root is $c = 0$. After a simple calculation, we can deduce that $G''(0) < 0$, which means that the function $G(c)$ can take the maximum value at $c = 0$, also which is

$$|a_2a_4 - a_3^2| \leq G(0) = \frac{7}{12},$$

and so we complete the proof of Theorem 3. □

Theorem 4. *If the function $f(z) \in S_l^*$ and of the form Equation (1), then we have*

$$|H_3(1)| \leq \frac{165,095 + 60,928\sqrt{2}}{444,528} \approx 0.565. \tag{11}$$

Proof. Because

$$H_3(1) = a_3(a_2a_4 - a_3^2) - a_4(a_4 - a_2a_3) + a_5(a_3 - a_2^2),$$

so, by applying the triangle inequality, we obtain

$$|H_3(1)| \leq |a_3||a_2a_4 - a_3^2| + |a_4||a_4 - a_2a_3| + |a_5||a_3 - a_2^2|. \tag{12}$$

Next, substituting Equations (4), (5), (8) and (10) into (12), we easily get the desired assertion Equation (11).

Finally, we give two examples to illustrate the results obtained. □

Example 1. *If we choose the function $f(z) = e^z - 1 = z + \sum_{n=2}^{\infty} \frac{z^n}{n!} \in S_l^*$, then we have*

$$\begin{aligned}|H_3(1)| &\leq |a_3||a_2a_4 - a_3^2| + |a_4||a_4 - a_2a_3| + |a_5||a_3 - a_2^2| \\ &= \frac{1}{3!} \times |\frac{1}{2!} \times \frac{1}{4!} - \frac{1}{3!} \times \frac{1}{3!}| + \frac{1}{4!} \times |\frac{1}{4!} - \frac{1}{2!} \times \frac{1}{3!}| + \frac{1}{5!} \times |\frac{1}{3!} - \frac{1}{2!} \times \frac{1}{2!}| \\ &\approx 0.004 < 0.565.\end{aligned}$$

Example 2. *If we put the function $f(z) = -\log(1 - z) = z + \sum_{n=2}^{\infty} \frac{z^n}{n} \in S_l^*$, then we get*

$$\begin{aligned}|H_3(1)| &\leq |a_3||a_2a_4 - a_3^2| + |a_4||a_4 - a_2a_3| + |a_5||a_3 - a_2^2| \\ &= \frac{1}{3} \times |\frac{1}{2} \times \frac{1}{4} - \frac{1}{3} \times \frac{1}{3}| + \frac{1}{4} \times |\frac{1}{4} - \frac{1}{2} \times \frac{1}{3}| + \frac{1}{5} \times |\frac{1}{3} - \frac{1}{2} \times \frac{1}{2}| \\ &\approx 0.042 < 0.565.\end{aligned}$$

3. Conclusions

In this paper, we mainly investigate the third-order Hankel determinant $H_3(1)$ for the function class S_l^*, which is subordinate to exponential function, and obtain the upper bound of the above determinant. The results obtained generalize and unify the theories of Hankel determinants in geometric function theory.

Author Contributions: All of the authors in this paper investigated this research. H.-Y.Z. wrote and reviewed the original draft, X.-M.N. sketched the figure and H.T. wrote, reviewed and edited this research.

Funding: The second author was supported by the Natural Science Foundation of the People's Republic of China under Grants 11561001 and 11271045, the Program for Young Talents of Science and Technology in Universities of Inner Mongolia Autonomous Region under Grant NJYT-18-A14, the Natural Science Foundation of Inner Mongolia of the People's Republic of China under Grant 2018MS01026, and the Higher School Foundation of Inner Mongolia of the People's Republic of China under Grants NJZY17300 and NJZY17301.

Acknowledgments: The authors thank the reviewers for their useful suggestions and valuable comments.

Conflicts of Interest: The authors declare no conflict of interest.

References

1. Srivastava, H.M.; Owa, S. (Eds.) *Current Topics in Analytic Function Theory*; World Scientific Publishing Company: London, UK, 1992.
2. Miller, S.S.; Mocanu, P.T. *Differential Subordinations: Theory and Applications, Series on Monographs and Textbooks in Pure and Applied Mathematics*; Marcel Dekker Incorporated: New York, NY, USA, 2000; p. 225.
3. Mendiratta, R.; Nagpal, S.; Ravichandran, V. On a subclass of strongly starlike functions associated with exponential function. *Bull. Malays. Math. Sci. Soc.* **2015**, *38*, 365–386. [CrossRef]
4. Noonan, J.W.; Thomas, D.K. On the second Hankel determinant of areally mean p-valent functions. *Trans. Am. Math. Soc.* **1976**, *223*, 337–346.
5. Noor, K.I. Hankel determinant problem for the class of functions with bounded boundary rotation. *Rev. Roum. Math. Pure Appl.* **1983**, *28*, 731–739.
6. Ehrenborg, R. The Hankel determinant of exponential polynomials. *Am. Math. Mon.* **2000**, *107*, 557–560. [CrossRef]
7. Fekete, M.; Szegö, G. Eine benberkung uber ungerada schlichte funktionen. *J. Lond. Math. Soc.* **1933**, *8*, 85–89. [CrossRef]
8. Koepf, W. On the Fekete-Szego problem for close-to-convex functions. *Proc. Am. Math. Soc.* **1987**, *101*, 89–95.
9. Koepf, W. On the Fekete-Szego problem for close-to-convex functions II. *Arch. Math.* **1987**, *49*, 420–433. [CrossRef]
10. Srivastava, H.M.; Hussain, S.; Raziq, A.; Raza, M. The Fekete-Szego functional for a subclass of analytic functions associated with quasi-subordination. *Carpath. J. Math.* **2018**, *34*, 103–113.
11. Srivastava, H.M.; Mishra, A.K.; Das, M.K. The Fekete-Szego problem for a subclass of close-to-convex functions. *Complex Var. Theory Appl.* **2001**, *44*, 145–163. [CrossRef]
12. Tang, H.; Srivastava, H.M.; Sivasubramanian, S.; Gurusamy, P. The Fekete-Szego functional problems for some classes of m-fold symmetric bi-univalent functions. *J. Math. Inequal.* **2016**, *10*, 1063–1092. [CrossRef]
13. Babalola, K.O. On $H_3(1)$ Hankel determinant for some classes of univalent functions. *Inequal. Theory Appl.* **2010**, *6*, 1–7.
14. Bansal, D. Upper bound of second Hankel determinant for a new class of analytic functions. *Appl. Math. Lett.* **2013**, *26*, 103–107. [CrossRef]
15. Bansal, D.; Maharana, S.; Prajapat, J.K. Third order Hankel determinant for certain univalent functions. *J. Korean Math. Soc.* **2015**, *52*, 1139–1148. [CrossRef]
16. Caglar, M.; Deniz, E.; Srivastava, H.M. Second Hankel determinant for certain subclasses of bi-univalent functions. *Turkish J. Math.* **2017**, *41*, 694–706. [CrossRef]
17. Janteng, A.; Halim, S.; Darus, M. Coefficient inequality for a function whose derivative has a positive real part. *J. Inequal. Pure Appl. Math.* **2006**, *7*, 1–5.
18. Janteng, A.; Halim, S.A.; Darus, M. Hankel determinant for starlike and convex functions. *Int. J. Math. Anal.* **2007**, *13*, 619–625.
19. Lee, S.K.; Ravichandran, V.; Subramaniam, S. Bounds for the second Hankel determinant of certain univalent functions. *J. Inequal. Appl.* **2013**, *281*, 1–17. [CrossRef]
20. Raza, M.; Malik, S.N. Upper bound of the third Hankel determinant for a class of analytic functions related with lemniscate of bernoulli. *J. Inequal. Appl.* **2013**, *1*, 1–8. [CrossRef]
21. Srivastava, H.M.; Altinkaya, S.; Yalcin, S. Hankel determinant for a subclass of bi-univalent functions defined by using a symmetric q-derivative operator. *Filomat* **2018**, *32*, 503–516. [CrossRef]
22. Zhang, H.-Y.; Tang, H.; Ma, L.-N. Upper bound of third Hankel determinant for a class of analytic functions. *Pure Appl. Math.* **2017**, *33*, 211–220. (In Chinese)
23. Libera, R.J.; Zlotkiewicz, E.J. Coefficient bounds for the inverse of a function with derivative in P. *Proc. Am. Math. Soc.* **1983**, *87*, 251–257. [CrossRef]
24. Pommerenke, C. Univalent functions. *J. Math. Soc. Jpn.* **1975**, *49*, 759–780.

© 2018 by the authors. Licensee MDPI, Basel, Switzerland. This article is an open access article distributed under the terms and conditions of the Creative Commons Attribution (CC BY) license (http://creativecommons.org/licenses/by/4.0/).

symmetry

MDPI

Article

Geometric Properties of Lommel Functions of the First Kind

Young Jae Sim [1], Oh Sang Kwon [1] and Nak Eun Cho [2,*]

1 Department of Mathematics, Kyungsung University, Busan 48434, Korea; yjsim@ks.ac.kr (Y.J.S.); oskwon@ks.ac.kr (O.S.K.)
2 Department of Applied Mathematics, Pukyong National University, Busan 48513, Korea
* Correspondence: necho@pknu.ac.kr

Received: 25 July 2018; Accepted: 18 September 2018; Published: 1 October 2018

Abstract: In the present paper, we find sufficient conditions for starlikeness and convexity of normalized Lommel functions of the first kind using the admissible function methods. Additionally, we investigate some inclusion relationships for various classes associated with the Lommel functions. The functions belonging to these classes are related to the starlike functions, convex functions, close-to-convex functions and quasi-convex functions.

Keywords: Lommel functions; univalent functions; starlike functions; convex functions; inclusion relationships

1. Introduction

Let $\mathcal{A}$ denote the family of functions f of the form:

$$f(z) = z + \sum_{n=2}^{\infty} a_n z^n$$

which are analytic in the open unit disk $\mathbb{D}$ and satisfy the usual normalization condition $f(0) = f'(0) - 1 = 0$. Let $\mathcal{S}$ denote the subclass of $\mathcal{A}$ which are univalent in $\mathbb{D}$. Also let $\mathcal{S}^*(\alpha)$ and $\mathcal{C}(\alpha)$ denote the subclasses of $\mathcal{A}$ consisting of functions which are starlike of order α and convex of order α in $\mathbb{D}$, respectively. Analytically, these classes are characterized by the equivalence:

$$f \in \mathcal{S}^*(\alpha) \Longleftrightarrow \Re\left\{\frac{zf'(z)}{f(z)}\right\} > \alpha \quad (0 \leq \alpha < 1, z \in \mathbb{D})$$

and

$$f \in \mathcal{C}(\alpha) \Longleftrightarrow \Re\left\{1 + \frac{zf''(z)}{f'(z)}\right\} > \alpha \quad (0 \leq \alpha < 1, z \in \mathbb{D}).$$

For convenience, let $\mathcal{S}^*(0) = \mathcal{S}^*$ and $\mathcal{C}(0) = \mathcal{C}$ which are the classes of starlike functions and convex functions, respectively. Furthermore, let $\mathcal{C}(\beta, \alpha)$ and $\mathcal{C}^*(\beta, \alpha)$ be the subclasses of $\mathcal{A}$ defined by

$$\mathcal{C}(\beta, \alpha) = \left\{f \in \mathcal{A} : \exists g \in \mathcal{S}^*(\alpha) \quad \text{s.t.} \quad \Re\left\{\frac{zf'(z)}{g(z)}\right\} > \beta \quad (0 \leq \alpha, \beta < 1; z \in \mathbb{D})\right\}$$

and

$$\mathcal{C}^*(\beta, \alpha) = \left\{f \in \mathcal{A} : \exists g \in \mathcal{K}(\alpha) \quad \text{s.t.} \quad \Re\left\{\frac{(zf'(z))'}{g'(z)}\right\} > \beta \quad (0 \leq \alpha, \beta < 1; z \in \mathbb{D})\right\},$$

respectively. The functions in the classes $\mathcal{C}(\beta,\alpha)$ and $\mathcal{C}^*(\beta,\alpha)$ are known as close-to-convex functions and quasi-convex functions, respectively.

The Lommel function of the first kind $s_{\mu,\nu}$ which is expressed in terms of a hypergeometric series

$$s_{\mu,\nu}(z) = \frac{z^{\mu+1}}{(\mu-\nu+1)(\mu+\nu+1)}\,{}_1F_2\left(1;\frac{\mu-\nu+3}{2},\frac{\mu+\nu+3}{2};-\frac{z^2}{4}\right),$$

where $\mu \pm \nu$ are not negative odd integers, is a particular solution of the following inhomogeneous Bessel differential equation [1]:

$$z^2w''(z) + zw'(z) + (z^2-\nu^2)w(z) = z^{\mu+1}.$$

It is observed that the function $s_{\mu,\nu}$ does not belong to the class $\mathcal{A}$. Recently, Yağmur [2] and Baricz et al. [3] considered the following function $h_{\mu,\nu}$ defined by:

$$h_{\mu,\nu}(z) = (\mu-\nu+1)(\mu+\nu+1)z^{(1-\mu)/2}s_{\mu,\nu}(\sqrt{z})$$

and they obtained some geometric properties of the function $h_{\mu,\nu}$. For another interesting properties of Lommel function, we can refer to [4,5].

The above function $h_{\mu,\nu}$ belongs to $\mathcal{A}$ and is expressed by:

$$h_{\mu,\nu}(z) = \sum_{n=1}^{\infty}\frac{(-1/4)^{n-1}}{\left(\frac{\mu-\nu+3}{2}\right)_{n-1}\left(\frac{\mu+\nu+3}{2}\right)_{n-1}}z^n \quad ((-\mu\pm\nu-3)/2 \notin \mathbb{N} := \{1,2,\cdots\}), \tag{1}$$

where $(\lambda)_n$ is the Pochhammer symbol which defined in terms of Euler's gamma function such that $(\lambda)_n = \Gamma(\lambda+n)/\Gamma(\lambda) = \lambda(\lambda+1)\cdots(\lambda+n-1)$.

Corresponding to the function $h_{\mu,\nu}$ defined by (1), we consider a linear operator $L_{\mu,\nu} : \mathcal{A} \to \mathcal{A}$ defined by:

$$L_{\mu,\nu}f(z) = h_{\mu,\nu}(z) * f(z) \quad ((-\mu\pm\nu-3)/2 \notin \mathbb{N}, z \in \mathbb{D}, f \in \mathcal{A}), \tag{2}$$

in terms of the Hadamard product (or convolution) $*$. Then it can be easily observed from (1) and (2) that the following relation holds:

$$z(L_{\mu+1,\nu+1}f(z))' = \left(\frac{\mu+\nu+3}{2}\right)L_{\mu,\nu}f(z) - \left(\frac{\mu+\nu+1}{2}\right)L_{\mu+1,\nu+1}f(z). \tag{3}$$

In a few years ago, many authors introduced new subclasses of univalent (or multivalent) functions by using several linear operators and found many properties of them [6–13]. In [14,15], various inclusion relationships associated with several subclasses of analytic functions were investigated.

Motivated by their works, by using the linear operator $L_{\mu,\nu}$, we define new subclasses of $\mathcal{A}$ as follows:

$$\mathcal{S}^*_{\mu,\nu}(\alpha) := \left\{f \in \mathcal{A} : \Re\left\{\frac{z(L_{\mu,\nu}f(z))'}{L_{\mu,\nu}f(z)}\right\} > \alpha \ (0 \le \alpha < 1; z \in \mathbb{D})\right\},$$

$$\mathcal{K}_{\mu,\nu}(\alpha) := \left\{f \in \mathcal{A} : \Re\left\{1+\frac{z(L_{\mu,\nu}f(z))''}{(L_{\mu,\nu}f(z))'}\right\} > \alpha \ (0 \le \alpha < 1; z \in \mathbb{D})\right\},$$

$$\mathcal{C}_{\mu,\nu}(\beta,\alpha) := \left\{f \in \mathcal{A} : \exists g \in \mathcal{S}^*_{\mu,\nu}(\alpha) \text{ s.t. } \Re\left\{\frac{z(L_{\mu,\nu}f(z))'}{L_{\mu,\nu}g(z)}\right\} > \beta \ (0 \le \alpha,\beta < 1; z \in \mathbb{D})\right\}$$

and

$$\mathcal{C}^*_{\mu,\nu}(\beta,\alpha) := \left\{f \in \mathcal{A} : \exists g \in \mathcal{K}_{\mu,\nu}(\alpha) \text{ s.t. } \Re\left\{\frac{(z(L_{\mu,\nu}f(z))')'}{(L_{\mu,\nu}g(z))'}\right\} > \beta \ (0 \le \alpha,\beta < 1; z \in \mathbb{D})\right\}.$$

Here, we note that a function f belongs to the class $\mathcal{S}^*_{\mu,\nu}(\alpha)$ ($\mathcal{K}_{\mu,\nu}(\alpha)$, $\mathcal{C}_{\mu,\nu}(\beta,\alpha)$ and $\mathcal{C}^*_{\mu,\nu}(\beta,\alpha)$) is equivalent to that the function $L_{\mu,\nu}f(z)$ belongs to the class $\mathcal{S}^*(\alpha)$ ($\mathcal{K}(\alpha)$, $\mathcal{C}(\beta,\alpha)$ and $\mathcal{C}^*(\beta,\alpha)$, respectively). Further, from the linearity of the operator $L_{\mu,\nu}$, the following relations hold:

$$f(z) \in \mathcal{K}_{\mu,\nu}(\alpha) \Longleftrightarrow zf'(z) \in \mathcal{S}^*_{\mu,\nu}(\alpha) \tag{4}$$

and

$$f(z) \in \mathcal{C}^*_{\mu,\nu}(\beta,\alpha) \Longleftrightarrow zf'(z) \in \mathcal{C}_{\mu,\nu}(\beta,\alpha). \tag{5}$$

In the present paper some geometric properties of the normalized Lommel function of the first kind are obtained by applying the method of admissible function. In Section 2, we find some sufficient conditions for starlikeness and convexity for the function $h_{\mu,\nu}$. In Section 3, we investigate some inclusion relationships for the classes $\mathcal{S}^*_{\mu,\nu}(\alpha)$, $\mathcal{K}_{\mu,\nu}(\alpha)$, $\mathcal{C}_{\mu,\nu}(\beta,\alpha)$ and $\mathcal{C}^*_{\mu,\nu}(\beta,\alpha)$ which are related to the function $h_{\mu,\nu}$.

The following lemmas will be used for the proof of our results.

Lemma 1. ([16] Miller and Mocanu) *Let Ω be a set in the complex plane $\mathbb{C}$ and let b be a complex number such that $\Re(b) > 0$. Suppose that the function $\psi : \mathbb{C}^3 \times \mathbb{D} \to \mathbb{C}$ satisfies the condition*

$$\psi(\mathrm{i}\rho, \sigma, a + \mathrm{i}b; z) \notin \Omega$$

for all real $\rho, \sigma, a, b \in \mathbb{R}$ with $\sigma \le -|b - \mathrm{i}\rho|^2/(2\Re(b))$, $\sigma + a \le 0$ and $z \in \mathbb{D}$. If the function $p(z)$ defined by $p(z) = b + b_1 z + b_2 z^2 + \ldots$ is analytic in $\mathbb{D}$ and if

$$\psi(p(z), zp'(z), z^2 p''(z); z) \in \Omega,$$

then $\Re\{p(z)\} > 0$ in $\mathbb{D}$.

Lemma 2. ([17] Miller and Mocanu) *Let $u = u_1 + \mathrm{i}u_2$, $v = v_1 + \mathrm{i}v_2$ with $u_1, u_2, v_1, v_2 \in \mathbb{R}$ and $\Delta \subset \mathbb{C}^2$. Suppose that $\Phi : \Delta \to \mathbb{C}$ satisfies the following conditions*

1. *$\Phi(u, v)$ is continuous in Δ;*
2. *$(1, 0) \in \Delta$ and $\Re\{\Phi(1, 0)\} > 0$;*
3. *$\Re\{\Phi(\mathrm{i}u_2, v_1)\} \le 0$ for all $(\mathrm{i}u_2, v_1) \in \Delta$ such that $v_1 \le -(1 + u_2^2)/2$.*

Let p be an analytic function in $\mathbb{D}$ such that $p(0) = 1$ and $(p(z), zp'(z)) \in \Delta$ for all $z \in \mathbb{D}$. If $\Re\{\Phi(p(z), zp'(z))\} > 0$ in $\mathbb{D}$, then $\Re\{p(z)\} > 0$ in $\mathbb{D}$.

For analytic functions f and g, we say that f is subordinate to g, denoted by $f \prec g$, if there is an analytic function $\omega : \mathbb{D} \to \mathbb{D}$ with $|\omega(z)| \le |z|$ such that $f(z) = g(\omega(z))$. Further, if g is univalent, then the definition of subordination $f \prec g$ can be simplified into the conditions $f(0) = g(0)$ and $f(\mathbb{D}) \subseteq g(\mathbb{D})$ (See [18], p. 36).

Lemma 3. ([19] Eenigenburg et al.) *Let h be convex univalent in $\mathbb{D}$ and w be analytic in $\mathbb{D}$ with $\Re\{w(z)\} \ge 0$ in $\mathbb{D}$. If q is analytic in $\mathbb{D}$ and $q(0) = h(0)$, then the subordination*

$$q(z) + w(z) z q'(z) \prec h(z) \quad (z \in \mathbb{D})$$

implies that

$$q(z) \prec h(z) \quad (z \in \mathbb{D}).$$

Lemma 4. ([2] Yağmur) *If $\mu > -1$, $\nu \in \mathbb{R}$ where $\mu \pm \nu$ are not negative odd integers, and*

$$(\mu+1)[(\mu+1)(\mu+3)-\nu^2] \geq \frac{1}{8},$$

then $\Re\{h_{\mu,\nu}(z)/z\} > 0$ *in* $\mathbb{D}$.

2. Sufficient Conditions for Starlikeness and Convexity

We find some sufficient conditions for starlikeness and convexity of the function $h_{\mu,\nu}$ given by (1).

Theorem 1. *Let* μ *and* ν *be real numbers such that* $\mu \pm \nu$ *are not negative odd integers,* $\mu > 2$,

$$(\mu+1)[(\mu+1)(\mu+3)-\nu^2] \geq \frac{1}{8} \tag{6}$$

and

$$-\frac{1}{2}(\mu-2)+\frac{1}{96}(\mu-2)^{-1}-\frac{1}{4}\left((\mu-1)^2-\nu^2\right) \leq 0. \tag{7}$$

Then the function $h_{\mu,\nu}$ *is a starlike univalent function in* $\mathbb{D}$.

Proof. Since

$$h_{\mu,\nu}(z) = (\mu-\nu+1)(\mu+\nu+1)z^{(1-\mu)/2}s_{\mu,\nu}(\sqrt{z})$$

and the function $s_{\mu,\nu}$ satisfies the inhomogeneous differential equation

$$z^2 s''_{\mu,\nu}(z) + z s'_{\mu,\nu}(z) + (z^2-\nu^2)s_{\mu,\nu}(z) = z^{\mu+1},$$

we have

$$z^2 h''_{\mu,\nu}(z) + \mu z h'_{\mu,\nu}(z) + \frac{1}{4}\left(z+(\mu-1)^2-\nu^2\right)h_{\mu,\nu}(z) - \left(\frac{\mu-\nu+1}{2}\right)\left(\frac{\mu+\nu+1}{2}\right)z = 0. \tag{8}$$

Set

$$p(z) = \frac{z h'_{\mu,\nu}(z)}{h_{\mu,\nu}(z)}. \tag{9}$$

From (6) and Lemma 4, $\Re\{h_{\mu,\nu}(z)/z\} > 0$ for all $z \in \mathbb{D}$ and this implies that $h_{\mu,\nu}(z) \neq 0$ holds for all $z \in \mathbb{D} \setminus \{0\}$. Therefore p is analytic in $\mathbb{D}$ and $p(0) = 1$. Furthermore, by (8) and (9), we have the following equation

$$\left[zp'(z) + p(z)^2 + (\mu-1)p(z) + \frac{1}{4}\left(z+(\mu-1)^2-\nu^2\right)\right]h_{\mu,\nu}(z) = \left(\frac{\mu-\nu+1}{2}\right)\left(\frac{\mu+\nu+1}{2}\right)z.$$

Now, we put

$$\tilde{p}(z) = zp'(z) + p(z)^2 + (\mu-1)p(z) + \frac{1}{4}\left(z+(\mu-1)^2-\nu^2\right).$$

Then we have

$$\tilde{p}(z)h_{\mu,\nu}(z) = \left(\frac{\mu-\nu+1}{2}\right)\left(\frac{\mu+\nu+1}{2}\right)z.$$

Differentiating the above equation and multiplying by z, we get

$$[z\tilde{p}'(z) + (p(z)-1)\tilde{p}(z)]h_{\mu,\nu}(z) = 0.$$

Since $z\tilde{p}'(z) + (p(z)-1)\tilde{p}(z) = 0$ at $z = 0$ and $h_{\mu,\nu}(z) \neq 0$ for all $z \in \mathbb{D} \setminus \{0\}$, we have

$$z\tilde{p}'(z) + (p(z)-1)\tilde{p}(z) = 0$$

in $\mathbb{D}$, or equivalently,

$$\begin{aligned} &p(z)^3 + (\mu-2)p(z)^2 + z^2p''(z) + 3zp'(z)p(z) + (\mu-1)zp'(z) \\ &+\frac{1}{4}\left(z + (\mu-1)(\mu-5) - \nu^2\right)p(z) - \frac{1}{4}\left((\mu-1)^2 - \nu^2\right) = 0 \end{aligned} \tag{10}$$

in $\mathbb{D}$. Now, let $\Omega = \{0\}$ and define a function $\psi : \mathbb{C}^3 \times \mathbb{D} \to \mathbb{C}$ by

$$\begin{aligned} &\psi(r,s,t;z) \\ &= r^3 + (\mu-2)r^2 + t + 3rs + (\mu-1)s + \frac{1}{4}(z + (\mu-1)(\mu-5) - \nu^2)r - \frac{1}{4}((\mu-1)^2 - \nu^2). \end{aligned}$$

Then the Equation (10) can be rewritten as

$$\psi(p(z), zp'(z), z^2p''(z); z) \in \Omega.$$

Moreover it holds that

$$\begin{aligned} &\Re\{\psi(\rho\mathrm{i}, \sigma, a + \mathrm{i}b; z)\} \\ &= -(\mu-2)\rho^2 + a + (\mu-1)\sigma + \frac{1}{4}\Re\left\{\left(z + (\mu-1)(\mu-5) - \nu^2\right)\mathrm{i}\rho\right\} - \frac{1}{4}\left((\mu-1)^2 - \nu^2\right) \\ &< -\frac{1}{2}(\mu-2)(1 + 3|\rho|^2) + \frac{1}{4}|\rho| - \frac{1}{4}\left((\mu-1)^2 - \nu^2\right), \end{aligned} \tag{11}$$

for $z \in \mathbb{D}$ and $\rho, \sigma, a, b \in \mathbb{R}$ with $\sigma \leq -(1+\rho^2)/2$ and $\sigma + a \leq 0$. Define a function $g : [0, \infty) \to \mathbb{R}$ by

$$g(\rho) = -\frac{1}{2}(\mu-2)(1 + 3\rho^2) + \frac{1}{4}\rho.$$

Then, $g'(\rho) = 0$ occurs when $\rho = \rho^* := 1/(12(\mu-2)) > 0$ and $g''(\rho^*) = -3(\mu-2) < 0$. Therefore, the function g has its maximum

$$g(\rho^*) = -\frac{1}{2}(\mu-2) + \frac{1}{96}(\mu-2)^{-1}$$

on the half interval $[0, \infty)$. Hence from (7) and (11) we have

$$\begin{aligned} &\Re\{\psi(\rho\mathrm{i}, \sigma, a + \mathrm{i}b; z)\} \\ &< g(\rho) - \frac{1}{4}\left((\mu-1)^2 - \nu^2\right) \\ &\leq -\frac{1}{2}(\mu-2) + \frac{1}{96}(\mu-2)^{-1} - \frac{1}{4}\left((\mu-1)^2 - \nu^2\right) \\ &\leq 0, \end{aligned}$$

for all $z \in \mathbb{D}$ and all $\rho, \sigma, a, b \in \mathbb{R}$ with $\sigma \leq -(1+\rho^2)/2$ and $\sigma + a \leq 0$. By Lemma 1, we have $\Re\{p(z)\} > 0$ in $\mathbb{D}$ which shows that $h_{\mu,\nu}$ is starlike in $\mathbb{D}$. □

Example 1. *We note that $\mu = 5/2$ and $\nu = 1/2$ satisfy the condition of Theorem 1. Therefore the function*

$$h_{5/2,1/2}(z) = 12\left(\frac{z + 2\cos\sqrt{z} - 2}{z}\right) \tag{12}$$

is starlike in $\mathbb{D}$.

Theorem 2. *Let μ and ν be real numbers such that $\mu \pm \nu$ are not negative odd integers, $\mu > 2$,*

$$(\mu+1)[(\mu+1)(\mu+3)-\nu^2] \geq \frac{1}{8}$$

and

$$\begin{cases} \frac{1}{96}(\mu-2)^{-1}+(\mu-2)-\frac{1}{4}\left((\mu-1)^2-\nu^2\right) \leq 0, & \text{if } \mu \leq \frac{25}{12}, \\ -\frac{1}{2}\mu+\frac{5}{4}-\frac{1}{4}\left((\mu-1)^2-\nu^2\right) \leq 0, & \text{if } \mu > \frac{25}{12}. \end{cases} \tag{13}$$

Then the function $h_{\mu,\nu}$ is a convex univalent function in $\mathbb{D}$.

Proof. First of all, we observe that the condition (13) implies (7) in Theorem 1. To see this, we assume that the inequality (13) holds. For the case $2 < \mu \leq 25/12$, from the inequality $-(\mu-2)/2 \leq \mu-2$, we can easily obtain the inequality (7). For the case $\mu > 25/12$, it is sufficient to check the following inequality holds:

$$-\frac{1}{2}(\mu-2)^2+\frac{1}{96} \leq -\frac{1}{2}\mu(\mu-2)+\frac{5}{4}(\mu-2).$$

And the above inequality is true for $\mu > 25/12$, since

$$-\frac{1}{2}\mu(\mu-2)+\frac{5}{4}(\mu-2)+\frac{1}{2}(\mu-2)^2-\frac{1}{96}=\frac{1}{4}\left(\mu-\frac{49}{24}\right)>\frac{1}{96}.$$

Therefore the function $h_{\mu,\nu}$ is starlike univalent, hence $h'_{\mu,\nu}(z) \neq 0$ in $\mathbb{D}$. Now, set

$$p(z)=1+\frac{zh''_{\mu,\nu}(z)}{h'_{\mu,\nu}(z)} \quad (z \in \mathbb{D}).$$

Since $h'_{\mu,\nu}(z) \neq 0$ in $\mathbb{D}$, p is analytic in $\mathbb{D}$ with $p(0)=1$. And we have

$$zh''_{\mu,\nu}(z)=(p(z)-1)h'_{\mu,\nu}(z) \tag{14}$$

and

$$2zh''_{\mu,\nu}(z)+z^2h^{(3)}_{\mu,\nu}(z)=[zp'(z)+p(z)^2-p(z)]h'_{\mu,\nu}(z). \tag{15}$$

Furthermore, from (8), we have

$$(p(z)+\mu-1)zh'_{\mu,\nu}(z)+\frac{1}{4}\left(z+(\mu-1)^2-\nu^2\right)h_{\mu,\nu}(z)-\left(\frac{\mu-\nu+1}{2}\right)\left(\frac{\mu+\nu+1}{2}\right)z=0. \tag{16}$$

Differentiating (16) and multiplying by z, we get

$$\begin{aligned} &z^2p'(z)h'_{\mu,\nu}(z)+(p(z)+\mu-1)zh'_{\mu,\nu}(z)+(p(z)+\mu-1)z^2h''_{\mu,\nu}(z) \\ &+\frac{1}{4}zh_{\mu,\nu}(z)+\frac{1}{4}\left(z+(\mu-1)^2-\nu^2\right)zh'_{\mu,\nu}(z) \\ &-\left(\frac{\mu-\nu+1}{2}\right)\left(\frac{\mu+\nu+1}{2}\right)z=0. \end{aligned} \tag{17}$$

Substituting (17) into (16), we obtain

$$\begin{aligned} &(p(z)+\mu-1)z^2h''_{\mu,\nu}(z)+[zp'(z)+\frac{1}{4}\left(z+(\mu-1)^2-\nu^2\right)]zh'_{\mu,\nu}(z) \\ &-\frac{1}{4}\left((\mu-1)^2-\nu^2\right)h_{\mu,\nu}(z)=0. \end{aligned} \tag{18}$$

Differentiating (18) and using the equalities (14) and (15), we get

$$z^2p''(z) + 3zp'(z)p(z) + (\mu-1)zp'(z) + p(z)^3 + (\mu-2)p(z)^2 \\ + \frac{1}{4}\left(z + (\mu-1)(\mu-5) - \nu^2\right)p(z) + \frac{1}{4}\left(z - (\mu-1)^2 + \nu^2\right) = 0. \tag{19}$$

Now, let $\Omega = \{0\}$ and define a function $\psi : \mathbb{C}^3 \times \mathbb{D} \to \mathbb{C}$ by

$$\psi(r,s,t;z) \\ = t + 3rs + (\mu-1)s + r^3 + (\mu-2)r^2 + \frac{1}{4}\left(z + (\mu-1)(\mu-5) - \nu^2\right)r + \frac{1}{4}\left(z - (\mu-1)^2 + \nu^2\right).$$

Then, (19) becomes

$$\psi(p(z), zp'(z), z^2p''(z); z) \in \Omega.$$

And simple calculations give us that

$$\begin{aligned} &\Re\{\psi(\mathrm{i}\rho,\sigma,a+\mathrm{i}b;z)\} \\ &= a + (\mu-1)\sigma - (\mu-2)\rho^2 + \frac{1}{4}\Re\{\mathrm{i}\rho z\} + \frac{1}{4}\Re\left\{z - (\mu-1)^2 + \nu^2\right\} \\ &\le -(\mu-2)\rho^2 + (\mu-2)\sigma + \frac{1}{4}\Re\{(1+\mathrm{i}\rho)z\} - \frac{1}{4}\left((\mu-1)^2 - \nu^2\right) \\ &< -(\mu-2)\rho^2 - \frac{1}{2}(\mu-2)(1+\rho^2) + \frac{1}{4}\sqrt{1+\rho^2} - \frac{1}{4}\left((\mu-1)^2 - \nu^2\right) \\ &= -\frac{3}{2}(\mu-2)u^2 + (\mu-2) + \frac{1}{4}u - \frac{1}{4}\left((\mu-1)^2 - \nu^2\right), \end{aligned} \tag{20}$$

for all $z \in \mathbb{D}$ and all $\rho, \sigma, a, b, u \in \mathbb{R}$ with $\sigma \le -(1+\rho^2)/2$, $\sigma + a \le 0$ and $u = \sqrt{1+\rho^2}$. Define a function $g : [1,\infty) \to \mathbb{R}$ by

$$g(u) = -\frac{3}{2}(\mu-2)u^2 + \frac{1}{4}u + \mu - 2.$$

Then, by putting $u^* = 1/(12(\mu-2)) > 0$, we have $g'(u^*) = 0$. Moreover it holds that $g''(u) = -3(\mu-2) < 0$ for all $u \in [1,\infty)$. Therefore $u = u^*$ gives the maximum value for g when $\mu \le 25/12$. On the other hand, when $\mu > 25/12$ the function g is maximized by setting $u = 1$. Hence, for the case $\mu \le 25/12$, it follows from (13) and (20) that

$$\begin{aligned} &\Re\{\psi(\mathrm{i}\rho,\sigma,a+\mathrm{i}b;z)\} \\ &< g(u^*) - \frac{1}{4}\left((\mu-1)^2 - \nu^2\right) \\ &\le \frac{1}{96}(\mu-2)^{-1} + (\mu-2) - \frac{1}{4}\left((\mu-1)^2 - \nu^2\right) \\ &\le 0, \end{aligned}$$

for all $z \in \mathbb{D}$ and all $\rho, \sigma, a, b \in \mathbb{R}$ with $\sigma \le -(1+\rho^2)/2$ and $\sigma + a \le 0$. Similarly, for the case $\mu > 25/12$, we obtain

$$\begin{aligned} &\Re\{\psi(\mathrm{i}\rho,\sigma,a+\mathrm{i}b;z)\} \\ &< g(1) - \frac{1}{4}\left((\mu-1)^2 - \nu^2\right) \\ &\le -\frac{1}{2}\mu + \frac{5}{4} - \frac{1}{4}\left((\mu-1)^2 - \nu^2\right) \\ &\le 0, \end{aligned}$$

for all $z \in \mathbb{D}$ and all $\rho, \sigma, a, b \in \mathbb{R}$ with $\sigma \le -(1+\rho^2)/2$ and $\sigma + a \le 0$. By Lemma 1, we thus have $\Re\{p(z)\} > 0$ in $\mathbb{D}$ which shows that $h_{\mu,\nu}$ is convex in $\mathbb{D}$. □

Example 2. *We note that* $\mu = 5/2$ *and* $\nu = 1/2$ *satisfy the condition of Theorem 2. Therefore the function* $h_{5/2,1/2}$ *given by (12) is convex in* $\mathbb{D}$.

3. Inclusion Relationships

Now, we investigate some inclusion relationships for the classes $\mathcal{S}^*_{\mu,\nu}(\alpha)$, $\mathcal{K}_{\mu,\nu}(\alpha)$, $\mathcal{C}_{\mu,\nu}(\beta,\alpha)$ and $\mathcal{C}^*_{\mu,\nu}(\beta,\alpha)$. We begin by proving our first inclusion relationship for the class $\mathcal{S}^*_{\mu,\nu}(\alpha)$.

Theorem 3. *Let* μ, ν *and* α *be real numbers such that* $\mu \pm \nu$ *are not negative odd integers,* $0 \le \alpha < 1$ *and* $2\alpha + \mu + \nu + 1 \ge 0$. *Then*

$$\mathcal{S}^*_{\mu,\nu}(\alpha) \subset \mathcal{S}^*_{\mu+1,\nu+1}(\alpha).$$

Proof. Let $f \in \mathcal{S}^*_{\mu,\nu}(\alpha)$ and define a function $\phi : \mathbb{C} \to \mathbb{C}$ by

$$\phi(z) = \frac{1}{1-\alpha}\left(\frac{z(L_{\mu+1,\nu+1}f(z))'}{L_{\mu+1,\nu+1}f(z)} - \alpha\right). \tag{21}$$

Then ϕ is analytic in $\mathbb{D}$ and $\phi(0) = 1$. From the equality (3), we get

$$\left(\frac{\mu+\nu+3}{2}\right)\frac{L_{\mu,\nu}f(z)}{L_{\mu+1,\nu+1}f(z)} = \frac{z(L_{\mu+1,\nu+1}f(z))'}{L_{\mu+1,\nu+1}f(z)} + \frac{\mu+\nu+1}{2}. \tag{22}$$

By combining (21) and (22), we obtain

$$\frac{L_{\mu,\nu}f(z)}{L_{\mu+1,\nu+1}f(z)} = \frac{2}{\mu+\nu+3}\left[(1-\alpha)\phi(z) + \alpha + \frac{\mu+\nu+1}{2}\right]. \tag{23}$$

Now, by applying the logarithmic differentiation on both sides of (23) and multiplying the resulting equation by z, we have

$$\frac{z(L_{\mu,\nu}f(z))'}{L_{\mu,\nu}f(z)} = \frac{z(L_{\mu+1,\nu+1}f(z))'}{L_{\mu+1,\nu+1}f(z)} + \frac{(1-\alpha)z\phi'(z)}{(1-\alpha)\phi(z) + \alpha + \frac{\mu+\nu+1}{2}}$$

which, in view of (21), yields

$$\frac{1}{1-\alpha}\left(\frac{z(L_{\mu,\nu}f(z))'}{L_{\mu,\nu}f(z)} - \alpha\right) = \phi(z) + \frac{z\phi'(z)}{(1-\alpha)\phi(z) + \alpha + \frac{\mu+\nu+1}{2}}. \tag{24}$$

Now, we define a function $\Phi : \mathbb{C}^2 \to \mathbb{C}$ by

$$\Phi(u,v) = u + \frac{v}{(1-\alpha)u + \alpha + \frac{\mu+\nu+1}{2}}.$$

Observe that Φ is continuous on

$$\Delta := \left(\mathbb{C} \setminus \left\{\frac{\alpha + \frac{\mu+\nu+1}{2}}{\alpha - 1}\right\}\right) \times \mathbb{C},$$

$(1,0) \in \Delta$ and $\Re\{\Phi(1,0)\} > 0$. Since $f \in \mathcal{S}^*_{\mu,\nu}(\alpha)$, it follows from (24) that $\Re\{\Phi(\phi(z), z\phi'(z), z^2\phi''(z))\} > 0$ for all $z \in \mathbb{D}$. Also, for $(iu_2, v_1) \in \Delta$ with u_2, $v_1 \in \mathbb{R}$ such that $v_1 \le -(1+u_2^2)/2$, we have

$$\begin{aligned}\Re\left\{\Phi(\mathrm{i}u_2, v_1)\right\} &= \Re\left\{\mathrm{i}u_2 + \frac{v_1}{\mathrm{i}(1-\alpha)u_2 + \alpha + \frac{\mu+\nu+1}{2}}\right\}\\ &= \frac{v_1\left(\alpha + \frac{\mu+\nu+1}{2}\right)}{(1-\alpha)^2 u_2^2 + \left(\alpha + \frac{\mu+\nu+1}{2}\right)^2}\\ &\leq -\frac{1}{2}\left(1+u_2^2\right)\frac{\alpha + \frac{\mu+\nu+1}{2}}{(1-\alpha)^2 u_2^2 + \left(\alpha + \frac{\mu+\nu+1}{2}\right)^2}\\ &< 0\end{aligned}$$

which shows that $\Re\left\{\Phi(\mathrm{i}u_2, v_1)\right\} < 0$. Therefore, by Lemma 2, we have

$$\Re\left\{\phi(z)\right\} > 0 \ (z \in \mathbb{D}).$$

Thus, by making use of (21), we find that $f \in \mathcal{S}^*_{\mu+1,\nu+1}(\alpha)$. This completes the proof of Theorem 3. □

Theorem 4. *Let μ, ν and α be real numbers such that $\mu \pm \nu$ are not negative odd integers, $0 \leq \alpha < 1$ and $2\alpha + \mu + \nu + 1 \geq 0$. Then*

$$\mathcal{K}_{\mu,\nu}(\alpha) \subset \mathcal{K}_{\mu+1,\nu+1}(\alpha).$$

Proof. By applying (4) and Theorem 3, we observe that

$$\begin{aligned}f \in \mathcal{K}_{\mu,\nu}(\alpha) &\Longleftrightarrow zf' \in \mathcal{S}^*_{\mu,\nu}(\alpha)\\ &\Longrightarrow zf' \in \mathcal{S}^*_{\mu+1,\nu+1}(\alpha)\\ &\Longleftrightarrow f \in \mathcal{K}_{\mu+1,\nu+1}(\alpha)\end{aligned}$$

which proves Theorem 4. □

Theorem 5. *Let μ, ν, α and β be real numbers such that $\mu \pm \nu$ are not negative odd integers, $0 \leq \alpha < 1$, $0 \leq \beta < 1$ and $2\alpha + \mu + \nu + 1 \geq 0$. Then*

$$\mathcal{C}_{\mu,\nu}(\beta, \alpha) \subset \mathcal{C}_{\mu+1,\nu+1}(\beta, \alpha).$$

Proof. Let $f \in \mathcal{C}_{\mu,\nu}(\beta, \alpha)$. Then there exists a function $g \in \mathcal{S}^*_{\mu,\nu}(\alpha)$ such that

$$\Re\left\{\frac{z(L_{\mu,\nu}f(z))'}{L_{\mu,\nu}g(z)}\right\} > \beta. \tag{25}$$

Define a function $\phi : \mathbb{D} \to \mathbb{C}$ by

$$\phi(z) = \frac{1}{1-\beta}\left(\frac{z(L_{\mu+1,\nu+1}f(z))'}{L_{\mu+1,\nu+1}g(z)} - \beta\right). \tag{26}$$

Then, ϕ is analytic in $\mathbb{D}$ with $\phi(0) = 1$. Using the identity (3), we also have

$$
\begin{aligned}
\frac{z(L_{\mu,\nu}f(z))'}{L_{\mu,\nu}g(z)} &= \frac{L_{\mu,\nu}(zf'(z))}{L_{\mu,\nu}g(z)} \\
&= \frac{z(L_{\mu+1,\nu+1}(zf'(z)))' + \left(\frac{\mu+\nu+1}{2}\right) L_{\mu+1,\nu+1}(zf'(z))}{z(L_{\mu+1,\nu+1}g(z))' + \left(\frac{\mu+\nu+1}{2}\right) L_{\mu+1,\nu+1}g(z)} \\
&= \frac{\frac{z(L_{\mu+1,\nu+1}(zf'(z)))'}{L_{\mu+1,\nu+1}g(z)} + \left(\frac{\mu+\nu+1}{2}\right) \frac{L_{\mu+1,\nu+1}(zf'(z))}{L_{\mu+1,\nu+1}g(z)}}{\frac{z(L_{\mu+1,\nu+1}g(z))'}{L_{\mu+1,\nu+1}g(z)} + \frac{\mu+\nu+1}{2}}.
\end{aligned} \tag{27}
$$

Now we define a function $q : \mathbb{D} \to \mathbb{C}$ by

$$
q(z) = \frac{1}{1-\alpha}\left(\frac{z(L_{\mu+1,\nu+1}g(z))'}{L_{\mu+1,\nu+1}g(z)} - \alpha\right). \tag{28}
$$

Since $g \in \mathcal{S}^*_{\mu,\nu}(\alpha)$, by Theorem 3, we have $g \in \mathcal{S}^*_{\mu+1,\nu+1}(\alpha)$ and therefore we get $\Re\{q(z)\} > 0$ in $\mathbb{D}$. Upon substituting from (26) and (28) into (27), we have

$$
\frac{z(L_{\mu,\nu}f(z))'}{L_{\mu,\nu}g(z)} = \frac{\frac{z(L_{\mu+1,\nu+1}(zf'(z)))'}{L_{\mu+1,\nu+1}g(z)} + \left(\frac{\mu+\nu+1}{2}\right)((1-\beta)\phi(z)+\beta)}{(1-\alpha)q(z)+\alpha+\frac{\mu+\nu+1}{2}}. \tag{29}
$$

By logarithmically differentiating both sides of (26) with respect to z, we have

$$
\frac{z(L_{\mu+1,\nu+1}(zf'(z)))'}{L_{\mu+1,\nu+1}g(z)} = ((1-\beta)\phi(z)+\beta)((1-\alpha)q(z)+\alpha) + (1-\beta)z\phi'(z)
$$

which, in conjunction with (29), yields

$$
\frac{1}{1-\beta}\left(\frac{z(L_{\mu,\nu}f(z))'}{L_{\mu,\nu}g(z)} - \beta\right) = \phi(z) + \frac{z\phi'(z)}{(1-\alpha)q(z)+\alpha+\frac{\mu+\nu+1}{2}}.
$$

Put

$$
\omega(z) = \frac{1}{(1-\alpha)q(z)+\alpha+\frac{\mu+\nu+1}{2}}.
$$

Then, ω is analytic in $\mathbb{D}$ and, from the inequality (25), we have

$$
\Re\{\phi(z) + \omega(z)z\phi'(z)\} > 0
$$

in $\mathbb{D}$. Using the fact that $\Re\{q(z)\} > 0$ in $\mathbb{D}$ and the inequality $2\alpha+\mu+\nu+1 \geq 0$, we have $\Re\{\omega(z)\} > 0$ in $\mathbb{D}$. Applying Lemma 3 with $h(z) = (1+z)/(1-z)$, we have $\Re\{\phi(z)\} > 0$ in $\mathbb{D}$. Thus, by making use of (26), we get $f \in \mathcal{C}_{\mu+1,\nu+1}(\beta,\alpha)$. This completes the proof of Theorem 5. □

Finally, we state the inclusion relationship for the class $\mathcal{C}^*_{\mu,\nu}(\beta,\alpha)$.

Theorem 6. *Let μ, ν, α and β be real numbers such that $\mu \pm \nu$ are not negative odd integers, $0 \leq \alpha < 1$, $0 \leq \beta < 1$ and $2\alpha+\mu+\nu+1 \geq 0$. Then*

$$
\mathcal{C}^*_{\mu,\nu}(\beta,\alpha) \subset \mathcal{C}^*_{\mu+1,\nu+1}(\beta,\alpha).
$$

Proof. By applying (5) and Theorem 5, we observe that

$$
\begin{aligned}
f \in \mathcal{C}^*_{\mu,\nu}(\beta,\alpha) &\Longleftrightarrow zf'(z) \in \mathcal{C}_{\mu,\nu}(\beta,\alpha) \\
&\Longrightarrow zf'(z) \in \mathcal{C}_{\mu+1,\nu+1}(\beta,\alpha) \\
&\Longleftrightarrow f \in \mathcal{C}^*_{\mu+1,\nu+1}(\beta,\alpha)
\end{aligned}
$$

which proves Theorem 6. □

Author Contributions: All of the authors in this paper investigated this research. Y.J.S. and O.S.K. wrote and reviewed the original draft. And N.E.C. wrote, reviewed and edited this research.

Funding: The first author was supported by the National Research Foundation of Korea(NRF) grant funded by the Korea government(MSIP; Ministry of Science, ICT & Future Planning) (No. NRF-2017R1C1B5076778). The third author was supported by the Basic Science Research Program through the National Research Foundation of Korea (NRF) funded by the Ministry of Education, Science and Technology (No. 2016R1D1A1A09916450).

Conflicts of Interest: The authors declare no conflict of interest.

References

1. Lommel, E. Ueber eine mit den Besselschen Functionen verwandte Function. *Math. Ann.* **1876**, *9*, 425–444. [CrossRef]
2. Yağmur, N. Hardy space of Lommel functions. *Bull. Korean Math. Soc.* **2015**, *52*, 1035–1046. [CrossRef]
3. Baricz, Á.; Dimitrov, D.K.; Orhan, H.; Yağmur, N. Radii of starlikeness of some special functions. *Proc. Am. Math. Soc.* **2016**, *144*, 3355–3367. [CrossRef]
4. Baricz, Á.; Koumandos, S. Turán type inequalities for some Lommel function of the first kind. *Proc. Edinb. Math. Soc.* **2016**, *59*, 569–579. [CrossRef]
5. Baricz, Á.; Szász, R. Close-to-convexity of some special functions and their derivatives. *Bull. Malays. Math. Sci. Soc.* **2016**, *39*, 427–437. [CrossRef]
6. Bernardi, S.D. Convex and starlike univalent functions. *Trans. Am. Math. Soc.* **1969**, *35*, 429–446. [CrossRef]
7. Carlson, B.C.; Shaffer, D.B. Starlike and prestarlike hypergeometric functions. *SIAM J. Math. Anal.* **1984**, *159*, 737–745. [CrossRef]
8. Libera, R.J. Some classes of regular univalent functions. *Proc. Am. Math. Soc.* **1965**, *16*, 755–758. [CrossRef]
9. Liu, J.L. The Noor integral and strongly starlike functions. *J. Math. Anal. Appl.* **2001**, *261*, 441–447. [CrossRef]
10. Noor, K.I. On new classes of integral operators. *J. Nat. Geometry* **1999**, *16*, 71–80.
11. Noor, K.I.; Alkhorasani, H.A. Properties of close-to-convexity preserved by some integral operators. *J. Math. Anal. Appl.* **1985**, *112*, 509–516. [CrossRef]
12. Ruscheweyh, S. New criteria for univalent functions. *Proc. Am. Math. Soc.* **1975**, *49*, 109–115. [CrossRef]
13. Srivastava, H.M.; Owa, S. Some characterization and distortion theorems involving fractional calculus, generalized hypergeometric functions, Hadamard products, linear operators, and certain subclasses of analytic functions. *Nagoya Math. J.* **1987**, *106*, 1–28. [CrossRef]
14. Cho, N.E.; Kwon, O.S.; Srivastava, H.M. Inclusion relationships and argument properties for certain subclasses of multivalent functions associated with a family of linear operators. *J. Math. Anal. Appl.* **2004**, *292*, 470–483. [CrossRef]
15. Choi, J.H.; Saigo, M.; Srivastava, H.M. Some inclusion properties of a certain family of integral operators. *J. Math. Anal. Appl.* **2002**, *276*, 432–445. [CrossRef]
16. Miller, S.S.; Mocanu, P.T. Differential subordinations and inequalities in the complex plane. *J. Differ. Equ.* **1987**, *67*, 199–211. [CrossRef]
17. Miller, S.S.; Mocanu, P.T. Second order differential inequalities in the complex plane. *J. Math. Anal. Appl.* **1978**, *65*, 289–305. [CrossRef]
18. Pommerenke, C. *Univalent Functions*; Vandenhoeck and Ruprecht: Göttingen, Germany, 1975; ISBN 3-525-40133-7.
19. Eenigenburg, P.; Miller, S.S.; Mocanu, P.T.; Reade, M.O. On a Briot-Bouquet differential subordination. *Rev. Roum. Math. Pures Appl.* **1984**, *29*, 567–573.

© 2018 by the authors. Licensee MDPI, Basel, Switzerland. This article is an open access article distributed under the terms and conditions of the Creative Commons Attribution (CC BY) license (http://creativecommons.org/licenses/by/4.0/).

symmetry

MDPI

Article

Symmetric Identities for (P, Q)-Analogue of Tangent Zeta Function

Cheon Seoung Ryoo

Department of Mathematics, Hannam University, Daejeon 34430, Korea; ryoocs@hnu.kr

Received: 25 August 2018; Accepted: 10 September 2018; Published: 11 September 2018

Abstract: The goal of this paper is to define the (p, q)-analogue of tangent numbers and polynomials by generalizing the tangent numbers and polynomials and Carlitz-type q-tangent numbers and polynomials. We get some explicit formulas and properties in conjunction with (p, q)-analogue of tangent numbers and polynomials. We give some new symmetric identities for (p, q)-analogue of tangent polynomials by using (p, q)-tangent zeta function. Finally, we investigate the distribution and symmetry of the zero of (p, q)-analogue of tangent polynomials with numerical methods.

Keywords: tangent numbers; tangent polynomials; Carlitz-type q-tangent numbers; Carlitz-type q-tangent polynomials; (p, q)-analogue of tangent numbers and polynomials; (p, q)-analogue of tangent zeta function; symmetric identities; zeros

MSC: 11B68; 11S40; 11S80

---

1. Introduction

The field of the special polynomials such as tangent polynomials, Bernoulli polynomials, Euler polynomials, and Genocchi polynomials is an expanding area in mathematics (see [1–16]). Many generalizations of these polynomials have been studied (see [1,3–9,11–18]). Srivastava [14] developed some properties and q-extensions of the Euler polynomials, Bernoulli polynomials, and Genocchi polynomials. Choi, Anderson and Srivastava have discussed q-extension of the Riemann zeta function and related functions (see [5,17]). Dattoli, Migliorati and Srivastava derived a generalization of the classical polynomials (see [6]).

It is the purpose of this paper to introduce and investigate a new some generalizations of the Carlitz-type q-tangent numbers and polynomials, q-tangent zeta function, Hurwiz q-tangent zeta function. We call them Carlitz-type (p, q)-tangent numbers and polynomials, (p, q)-tangent zeta function, and Hurwitz (p, q)-tangent zeta function. The structure of the paper is as follows: In Section 2 we define Carlitz-type (p, q)-tangent numbers and polynomials and derive some of their properties involving elementary properties, distribution relation, property of complement, and so on. In Section 3, by using the Carlitz-type (p, q)-tangent numbers and polynomials, (p, q)-tangent zeta function and Hurwitz (p, q)-tangent zeta function are defined. We also contains some connection formulae between the Carlitz-type (p, q)-tangent numbers and polynomials and the (p, q)-tangent zeta function, Hurwitz (p, q)-tangent zeta function. In Section 4 we give several symmetric identities about (p, q)-tangent zeta function and Carlitz-type (p, q)-tangent polynomials and numbers. In the following Section, we investigate the distribution and symmetry of the zero of Carlitz-type (p, q)-tangent polynomials using a computer. Our paper ends with Section 6, where the conclusions and future developments of this work are presented. The following notations will be used throughout this paper.

- $\mathbb{N}$ denotes the set of natural numbers.
- $\mathbb{Z}_0^- = \{0, -1, -2, -2, \ldots\}$ denotes the set of nonpositive integers.
- $\mathbb{R}$ denotes the set of real numbers.
- $\mathbb{C}$ denotes the set of complex numbers.

We remember that the classical tangent numbers T_n and tangent polynomials $T_n(x)$ are defined by the following generating functions (see [19])

$$\frac{2}{e^{2t}+1} = \sum_{n=0}^{\infty} T_n \frac{t^n}{n!}, \quad (|2t| < \pi), \tag{1}$$

and

$$\left(\frac{2}{e^{2t}+1}\right) e^{xt} = \sum_{n=0}^{\infty} T_n(x) \frac{t^n}{n!}, \quad (|2t| < \pi). \tag{2}$$

respectively. Some interesting properties of basic extensions and generalizations of the tangent numbers and polynomials have been worked out in [11,12,18–20]. The (p,q)-number is defined as

$$[n]_{p,q} = \frac{p^n - q^n}{p-q} = p^{n-1} + p^{n-2}q + p^{n-3}q^2 + \cdots + p^2q^{n-3} + pq^{n-2} + q^{n-1}.$$

It is clear that (p,q)-number contains symmetric property, and this number is q-number when $p = 1$. In particular, we can see $\lim_{q\to 1}[n]_{p,q} = n$ with $p = 1$. Since $[n]_{p,q} = p^{n-1}[n]_{\frac{q}{p}}$, we observe that (p,q)-numbers and p-numbers are different. In other words, by substituting q by $\frac{q}{p}$ in the definition q-number, we cannot have (p,q)-number. Duran, Acikgoz and Araci [7] introduced the (p,q)-analogues of Euler polynomials, Bernoulli polynomials, and Genocchi polynomials. Araci, Duran, Acikgoz and Srivastava developed some properties and relations between the divided differences and (p,q)-derivative operator (see [1]). The (p,q)-analogues of tangent polynomials were described in [20]. By using (p,q)-number, we construct the Carlitz-type (p,q)-tangent polynomials and numbers, which generalized the previously known tangent polynomials and numbers, including the Carlitz-type q-tangent polynomials and numbers. We begin by recalling here the Carlitz-type q-tangent numbers and polynomials (see [18]).

Definition 1. *For any complex x we define the Carlitz-type q-tangent polynomials, $T_{n,q}(x)$, by the equation*

$$F_q(t,x) = \sum_{n=0}^{\infty} T_{n,q}(x)\frac{t^n}{n!} = [2]_q \sum_{m=0}^{\infty}(-1)^m q^m e^{[2m+x]_q t}. \tag{3}$$

The numbers $T_{n,q}(0)$ are called the Carlitz-type q-tangent numbers and are denoted by $T_{n,q}$. Based on this idea, we generalize the Carlitz-type q-tangent number $T_{n,q}$ and q-tangent polynomials $T_{n,q}(x)$. It follows that we define the following (p,q)-analogues of the the Carlitz-type q-tangent number $T_{n,q}$ and q-tangent polynomials $T_{n,q}(x)$. In the next section we define the (p,q)-analogue of tangent numbers and polynomials. After that we will obtain some their properties.

2. (p,q)-Analogue of Tangent Numbers and Polynomials

Firstly, we construct (p,q)-analogue of tangent numbers and polynomials and derive some of their relevant properties.

Definition 2. *For $0 < q < p \leq 1$, the Carlitz-type (p,q)-tangent numbers $T_{n,p,q}$ and polynomials $T_{n,p,q}(x)$ are defined by means of the generating functions*

$$F_{p,q}(t) = \sum_{n=0}^{\infty} T_{n,p,q}\frac{t^n}{n!} = [2]_q \sum_{m=0}^{\infty}(-1)^m q^m e^{[2m]_{p,q}t}, \tag{4}$$

and

$$F_{p,q}(t,x) = \sum_{n=0}^{\infty} T_{n,p,q}(x)\frac{t^n}{n!} = [2]_q \sum_{m=0}^{\infty} (-1)^m q^m e^{[2m+x]_{p,q}t}, \tag{5}$$

respectively.

Setting $p = 1$ in (4) and (5), we can obtain the corresponding definitions for the Carlitz-type q-tangent numbers $T_{n,q}$ and q-tangent polynomials $T_{n,q}(x)$ respectively. Obviously, if we put $p = 1$, then we have

$$T_{n,p,q}(x) = T_{n,q}(x), \quad T_{n,p,q} = T_{n,q}.$$

Putting $p = 1$, we have

$$\lim_{q\to 1} T_{n,p,q}(x) = T_n(x), \quad \lim_{q\to 1} T_{n,p,q} = T_n.$$

Theorem 1. *For $n \in \mathbb{N} \cup \{0\}$, one has*

$$T_{n,p,q} = [2]_q \left(\frac{1}{p-q}\right)^n \sum_{l=0}^{n} \binom{n}{l} (-1)^l \frac{1}{1+q^{2l+1}p^{2(n-l)}}. \tag{6}$$

Proof. By (4), we have

$$\begin{aligned}\sum_{n=0}^{\infty} T_{n,p,q}\frac{t^n}{n!} &= [2]_q \sum_{m=0}^{\infty} (-1)^m q^m e^{[2m]_{p,q}t}\\ &= \sum_{n=0}^{\infty} \left([2]_q \left(\frac{1}{p-q}\right)^n \sum_{l=0}^{n} \binom{n}{l} (-1)^l \frac{1}{1+q^{2l+1}p^{2(n-l)}}\right)\frac{t^n}{n!}.\end{aligned}$$

Equating the coefficients of $\frac{t^n}{n!}$, we arrive at the desired result (6). □

If we put $p = 1$ in Theorem 1, we obtain (cf. [18])

$$T_{n,q} = [2]_q \left(\frac{1}{1-q}\right)^n \sum_{l=0}^{n} \binom{n}{l} (-1)^l \frac{1}{1+q^{2l+1}}.$$

Next, we construct the Carlitz-type (h,p,q)-tangent polynomials $T^{(h)}_{n,p,q}(x)$. Define the Carlitz-type (h,p,q)-tangent polynomials $T^{(h)}_{n,p,q}(x)$ by

$$T^{(h)}_{n,p,q}(x) = [2]_q \sum_{m=0}^{\infty} (-1)^m q^m p^{hm} [2m+x]^n_{p,q}. \tag{7}$$

Theorem 2. *For $n \in \mathbb{N} \cup \{0\}$, one has*

$$\begin{aligned}T_{n,p,q}(x) &= [2]_q \left(\frac{1}{p-q}\right)^n \sum_{l=0}^{n} \binom{n}{l} (-1)^l q^{xl} p^{(n-l)x} \frac{1}{1+q^{2l+1}p^{2(n-l)+h}}\\ &= [2]_q \sum_{m=0}^{\infty} (-1)^m q^m [2m+x]^n_{p,q}.\end{aligned}$$

Proof. By (5), we obtain

$$T_{n,p,q}(x) = [2]_q \left(\frac{1}{p-q}\right)^n \sum_{l=0}^{n} \binom{n}{l} (-1)^l q^{xl} p^{(n-l)x} \frac{1}{1+q^{2l+1}p^{2(n-l)}}. \tag{8}$$

Again, by using (5) and (8), we obtain

$$\begin{aligned}&\sum_{n=0}^{\infty} T_{n,p,q}(x)\frac{t^n}{n!}\\&=\sum_{n=0}^{\infty}\left([2]_q\left(\frac{1}{p-q}\right)^n\sum_{l=0}^{n}\binom{n}{l}(-1)^l q^{xl}p^{(n-l)x}\frac{1}{1+q^{2l+1}p^{2(n-l)}}\right)\frac{t^n}{n!}\\&=[2]_q\sum_{m=0}^{\infty}(-1)^m q^m e^{[2m+x]_{p,q}t}.\end{aligned}\tag{9}$$

Since $[x+2y]_{p,q}=p^{2y}[x]_{p,q}+q^x[2y]_{p,q}$, we have

$$T_{n,p,q}(x)=[2]_q\sum_{l=0}^{n}\binom{n}{l}[x]_{p,q}^{n-l}q^{xl}\sum_{k=0}^{l}\binom{l}{k}(-1)^k\left(\frac{1}{p-q}\right)^l\frac{1}{1+q^{2k+1}p^{2(n-k)}}.\tag{10}$$

By using (9) and (10), (p,q)-number, and the power series expansion of e^{xt}, we give Theorem 2. □

Furthermore, by (7) and Theorem 2, we have

$$T_{n,p,q}(x)=\sum_{l=0}^{n}\binom{n}{l}[x]_{p,q}^{n-l}q^{xl}T_{l,p,q}^{(2n-2l)},$$

$$T_{n,p,q}(x+y)=\sum_{l=0}^{n}\binom{n}{l}p^{xl}q^{y(n-l)}[y]_{p,q}^{l}T_{n-l,p,q}^{(2l)}.$$

From (4) and (5), we can derive the following properties of the Carlitz-type tangent numbers $T_{n,p,q}$ and polynomials $T_{n,p,q}(x)$. So, we choose to omit the details involved.

Proposition 1. *For any positive integer n, one has*

(1) $T_{n,p,q}(x)=\dfrac{[2]_q}{[2]_{q^m}}[m]_{p,q}^n\sum_{a=0}^{m-1}(-1)^a q^a T_{n,p^m,q^m}\left(\frac{2a+x}{m}\right),(m=odd).$

(2) $T_{n,p^{-1},q^{-1}}(2-x)=(-1)^n p^n q^n T_{n,p,q}(x).$

Theorem 3. *For $n\in\mathbb{N}\cup\{0\}$, one has*

$$qT_{n,p,q}(2)+T_{n,p,q}=\begin{cases}[2]_q, & \text{if } n=0,\\ 0, & \text{if } n\neq 0.\end{cases}$$

Theorem 4. *If n is a positive integer, then we have*

$$\sum_{l=0}^{n-1}(-1)^l q^l[2l]_{p,q}^m=\frac{(-1)^{n+1}q^nT_{m,p,q}(2n)+T_{m,p,q}}{[2]_q}.$$

Proof. By (4) and (5), we get

$$-[2]_q\sum_{l=0}^{\infty}(-1)^{l+n}q^{l+n}e^{[2l+2n]_{p,q}t}+[2]_q\sum_{l=0}^{\infty}(-1)^l q^l e^{[2l]_{p,q}t}=[2]_q\sum_{l=0}^{n-1}(-1)^l q^l e^{[2l]_{p,q}t}.\tag{11}$$

Hence, by (4), (5) and (11), we have

$$\begin{aligned}&(-1)^{n+1}q^{n}\sum_{m=0}^{\infty}T_{m,p,q}(2n)\frac{t^{m}}{m!}+\sum_{m=0}^{\infty}T_{m,p,q}\frac{t^{m}}{m!}\\&=\sum_{m=0}^{\infty}\left([2]_{q}\sum_{l=0}^{n-1}(-1)^{l}q^{l}[2l]_{p,q}^{m}\right)\frac{t^{m}}{m!}.\end{aligned}$$

Equating coefficients of $\frac{t^m}{m!}$ gives Theorem 4. □

3. (p,q)-Analogue of Tangent Zeta Function

Using Carlitz-type (p,q)-tangent numbers and polynomials, we define the (p,q)-tangent zeta function and Hurwitz (p,q)-tangent zeta function. These functions have the values of the Carlitz-type (p,q)-tangent numbers $T_{n,p,q}$, and polynomials $T_{n,p,q}(x)$ at negative integers, respectively. From (4), we note that

$$\begin{aligned}\left.\frac{d^{k}}{dt^{k}}F_{p,q}(t)\right|_{t=0}&=[2]_{q}\sum_{m=0}^{\infty}(-1)^{n}q^{m}[2m]_{p,q}^{k}\\&=T_{k,p,q},\ (k\in\mathbb{N}).\end{aligned}$$

From the above equation, we construct new (p,q)-tangent zeta function as follows:

Definition 3. *We define the (p,q)-tangent zeta function for $s\in\mathbb{C}$ with $Re(s)>0$ by*

$$\zeta_{p,q}(s)=[2]_{q}\sum_{n=1}^{\infty}\frac{(-1)^{n}q^{n}}{[2n]_{p,q}^{s}}. \tag{12}$$

Notice that $\zeta_{p,q}(s)$ is a meromorphic function on $\mathbb{C}$(cf.7). Remark that, if $p=1, q\to 1$, then $\zeta_{p,q}(s)=\zeta_{T}(s)$ which is the tangent zeta function (see [19]). The relationship between the $\zeta_{p,q}(s)$ and the $T_{k,p,q}$ is given explicitly by the following theorem.

Theorem 5. *Let $k\in\mathbb{N}$. We have*

$$\zeta_{p,q}(-k)=T_{k,p,q}.$$

Please note that $\zeta_{p,q}(s)$ function interpolates $T_{k,p,q}$ numbers at non-negative integers. Similarly, by using Equation (5), we get

$$\left.\frac{d^{k}}{dt^{k}}F_{p,q}(t,x)\right|_{t=0}=[2]_{q}\sum_{m=0}^{\infty}(-1)^{m}q^{m}[2m+x]_{p,q}^{k} \tag{13}$$

and

$$\left.\left(\frac{d}{dt}\right)^{k}\left(\sum_{n=0}^{\infty}T_{n,p,q}(x)\frac{t^{n}}{n!}\right)\right|_{t=0}=T_{k,p,q}(x),\ \text{for } k\in\mathbb{N}. \tag{14}$$

Furthermore, by (13) and (14), we are ready to construct the Hurwitz (p,q)-tangent zeta function.

Definition 4. *For $s\in\mathbb{C}$ with $Re(s)>0$ and $x\notin\mathbb{Z}_0^-$, we define*

$$\zeta_{p,q}(s,x)=[2]_{q}\sum_{n=0}^{\infty}\frac{(-1)^{n}q^{n}}{[2n+x]_{p,q}^{s}}. \tag{15}$$

Obverse that the function $\zeta_{p,q}(s,x)$ is a meromorphic function on $\mathbb{C}$. We note that, if $p=1$ and $q\to 1$, then $\zeta_{p,q}(s,x)=\zeta_{T}(s,x)$ which is the Hurwitz tangent zeta function (see [19]). The function

$\zeta_{p,q}(-k,x)$ interpolates the numbers $T_{k,p,q}(x)$ at non-negative integers. Substituting $s=-k$ with $k \in \mathbb{N}$ into (15), and using Theorem 2, we easily arrive at the following theorem.

Theorem 6. *Let $k \in \mathbb{N}$. One has*

$$\zeta_{p,q}(-k,x) = T_{k,p,q}(x).$$

4. Some Symmetric Properties About (P,Q)-Analogue of Tangent Zeta Function

Our main objective in this section is to obtain some symmetric properties about (p,q)-tangent zeta function. In particular, some of these symmetric identities are also related to the Carlitz-type (p,q)-tangent polynomials and the alternate power sums. To end this section, we focus on some symmetric identities containing the Carlitz-type (p,q)-tangent zeta function and the alternate power sums.

Theorem 7. *Let w_1 and w_2 be positive odd integers. Then we have*

$$\begin{aligned}&[2]_{q^{w_1}}[w_1]^s_{p,q}\sum_{i=0}^{w_2-1}(-1)^i q^{w_1 i}\zeta_{p^{w_2},q^{w_2}}\left(s, w_1x+\frac{2w_1 i}{w_2}\right)\\&=[2]_{q^{w_2}}[w_2]^s_{p,q}\sum_{j=0}^{w_1-1}(-1)^j q^{w_2 j}\zeta_{p^{w_1},q^{w_1}}\left(s, w_2x+\frac{2w_2 j}{w_1}\right).\end{aligned}$$

Proof. For any $x,y\in\mathbb{C}$, we observe that $[xy]_{p,q}=[x]_{p^y,q^y}[y]_{p,q}$. By substituting $w_1x+\frac{2w_1i}{w_2}$ for x in Definition 4, replace p by p^{w_2} and replace q by q^{w_2}, respectively, we derive

$$\begin{aligned}\zeta_{p^{w_2},q^{w_2}}\left(s, w_1x+\frac{2w_1 i}{w_2}\right)&=[2]_{q^{w_2}}\sum_{n=0}^{\infty}\frac{(-1)^n q^{w_2 n}}{[w_1x+\frac{2w_1 i}{w_2}+2n]^s_{p^{w_2},q^{w_2}}}\\&=[2]_{q^{w_2}}[w_2]^s_{p,q}\sum_{n=0}^{\infty}\frac{(-1)^n q^{w_2 n}}{[w_1w_2x+2w_1i+2w_2n]^s_{p,q}}.\end{aligned}$$

Since for any non-negative integer m and positive odd integer w_1, there exist unique non-negative integer r such that $m=w_1r+j$ with $0\le j\le w_1-1$. Thus, this can be written as

$$\begin{aligned}&\zeta_{p^{w_2},q^{w_2}}\left(s, w_1x+\frac{2w_1 i}{w_2}\right)\\&=[2]_{q^{w_2}}[w_2]^s_{p,q}\sum_{\substack{w_1r+j=0\\0\le j\le w_1-1}}^{\infty}\frac{(-1)^{w_1r+j}q^{w_2(w_1r+j)}}{[2w_2(w_1r+j)+w_1w_2x+2w_1i]^s_{p,q}}\\&=[2]_{q^{w_2}}[w_2]^s_{p,q}\sum_{j=0}^{w_1-1}\sum_{r=0}^{\infty}\frac{(-1)^{w_1r+j}q^{w_2(w_1r+j)}}{[w_1w_2(2r+x)+2w_1i+2w_2j]^s_{p,q}}.\end{aligned}$$

It follows from the above equation that

$$\begin{aligned}&[2]_{q^{w_1}}[w_1]^s_{p,q}\sum_{i=0}^{w_2-1}(-1)^i q^{w_1 i}\zeta_{p^{w_2},q^{w_2}}\left(s, w_1x+\frac{2w_1 i}{w_2}\right)\\&=[2]_{q^{w_1}}[2]_{q^{w_2}}[w_1]^s_{p,q}[w_2]^s_{p,q}\\&\qquad\times\sum_{i=0}^{w_2-1}\sum_{j=0}^{w_1-1}\sum_{r=0}^{\infty}\frac{(-1)^{r+i+j}q^{(w_1w_2r+w_1i+w_2j)}}{[w_1w_2(2r+x)+2w_1i+2w_2j]^s_q}.\end{aligned}\tag{16}$$

From the similar method, we can have that

$$\zeta_{p^{w_1},q^{w_1}}\left(s, w_2x + \frac{2w_2j}{w_1}\right) = [2]_{q^{w_1}} \sum_{n=0}^{\infty} \frac{(-1)^n q^{w_1 n}}{[w_2x + \frac{2w_2j}{w_1} + 2n]^s_{p^{w_1},q^{w_1}}}$$
$$= [2]_{q^{w_1}} [w_1]^s_{p,q} \sum_{n=0}^{\infty} \frac{(-1)^n q^{w_1 n}}{[w_1w_2x + 2w_2j + 2w_1n]^s_{p,q}}.$$

After some calculations in the above, we have

$$\begin{aligned}
&[2]_{q^{w_2}} [w_2]^s_{p,q} \sum_{j=0}^{w_1-1} (-1)^j q^{w_2 j} \zeta^{(h)}_{p^{w_1},q^{w_1}}\left(s, w_2x + \frac{2w_2j}{w_1}\right) \\
&= [2]_{q^{w_1}} [2]_{q^{w_2}} [w_1]^s_{p,q} [w_2]^s_{p,q} \\
&\quad \times \sum_{i=0}^{w_2-1} \sum_{j=0}^{w_1-1} \sum_{r=0}^{\infty} \frac{(-1)^{r+i+j} q^{(w_1w_2r + w_1i + w_2j)}}{[w_1w_2(2r+x) + 2w_1i + 2w_2j]^s_{p,q}}.
\end{aligned} \tag{17}$$

Thus, from (16) and (17), we obtain the result. □

Corollary 1. *For $s \in \mathbb{C}$ with $Re(s) > 0$, we have*

$$\zeta_{p,q}(s, w_1x) = [w_1]^{-s}_{p,q} \sum_{j=0}^{w_1-1} (-1)^j q^j \zeta_{p^{w_1},q^{w_1}}\left(s, \frac{x+2j}{w_1}\right).$$

Proof. Let $w_2 = 1$ in Theorem 7. Then we immediately get the result. □

Next, we also derive some symmetric identities for Carlitz-type (p, q)-tangent polynomials by using (p, q)-tangent zeta function.

Theorem 8. *Let w_1 and w_2 be any positive odd integers. The following multiplication formula holds true for the Carlitz-type (p, q)-tangent polynomials:*

$$[2]_{q^{w_1}} [w_2]^n_{p,q} \sum_{i=0}^{w_2-1} (-1)^i q^{w_1 i} T_{n,p^{w_2},q^{w_2}}\left(w_1x + \frac{2w_1i}{w_2}\right)$$
$$= [2]_{q^{w_2}} [w_1]^n_{p,q} \sum_{j=0}^{w_1-1} (-1)^j q^{w_2 j} T_{n,p^{w_1},q^{w_1}}\left(w_2x + \frac{2w_2j}{w_1}\right).$$

Proof. By substituting $T_{n,p,q}(x)$ for $\zeta_{p,q}(s, x)$ in Theorem 7, and using Theorem 6, we can find that

$$\begin{aligned}
&[2]_{q^{w_1}} [w_1]^{-n}_{p,q} \sum_{i=0}^{w_2-1} (-1)^i q^{w_1 i} \zeta_{p^{w_2},q^{w_2}}\left(-n, w_1x + \frac{2w_1i}{w_2}\right) \\
&= [2]_{q^{w_1}} [w_1]^{-n}_{p,q} \sum_{i=0}^{w_2-1} (-1)^i q^{w_1 i} T_{n,p^{w_2},q^{w_2}}\left(w_1x + \frac{2w_1i}{w_2}\right),
\end{aligned} \tag{18}$$

and

$$\begin{aligned}
&[2]_{q^{w_2}} [w_2]^{-n}_{p,q} \sum_{j=0}^{w_1-1} (-1)^j q^{w_2 j} \zeta_{p^{w_1},q^{w_1}}\left(-n, w_2x + \frac{2w_2j}{w_1}\right) \\
&= [2]_{q^{w_2}} [w_2]^{-n}_{p,q} \sum_{j=0}^{w_1-1} (-1)^j q^{w_2 j} T_{n,p^{w_1},q^{w_1}}\left(w_2x + \frac{2w_2j}{w_1}\right).
\end{aligned} \tag{19}$$

Thus, by (18) and (19), this concludes our proof. □

Considering $w_1 = 1$ in the Theorem 8, we obtain as below equation.

$$T_{n,p,q}(x) = \frac{[2]_q}{[2]_{q^{w_2}}}[w_2]^n_{p,q} \sum_{j=1}^{w_2-1} (-1)^j q^j T_{n,p^{w_2},q^{w_2}}\left(\frac{x+2j}{w_2}\right).$$

Furthermore, by applying the addition theorem for the Carlitz-type (h,p,q)-tangent polynomials $T^{(h)}_{n,p,q}(x)$, we can obtain the following theorem.

Theorem 9. *Let w_1 and w_2 be any positive odd integers. Then one has*

$$\begin{aligned}
&[2]_{q^{w_2}} \sum_{l=0}^{n} \binom{n}{l} [w_2]^l_q [w_1]^{n-l}_{p,q} p^{w_1 w_2 x l} T^{(2l)}_{n-l,p^{w_1},q^{w_1}}(w_2 x) \mathcal{T}_{n,l,p^{w_2},q^{w_2}}(w_1) \\
&= [2]_{q^{w_1}} \sum_{l=0}^{n} \binom{n}{l} [w_1]^l_{p,q} [w_2]^{n-l}_{p,q} p^{w_1 w_2 x l} T^{(2l)}_{n-l,p^{w_2},q^{w_2}}(w_1 x) \mathcal{T}_{n,l,p^{w_1},q^{w_1}}(w_2).
\end{aligned}$$

Proof. From Theorem 8, we have

$$\begin{aligned}
&[2]_{q^{w_1}} [w_2]^n_{p,q} \sum_{i=0}^{w_2-1} (-1)^i q^{w_1 i} T_{n,p^{w_2},q^{w_2}}\left(w_1 x + \frac{2w_1 i}{w_2}\right) \\
&= [2]_{q^{w_1}} [w_2]^n_{p,q} \sum_{i=0}^{w_2-1} (-1)^i q^{w_1 i} \sum_{l=0}^{n} \binom{n}{l} q^{2w_1(n-l)i} p^{w_1 w_2 x l} \\
&\qquad \times T^{(2l)}_{n-l,p^{w_2},q^{w_2}}(w_1 x) \left(\frac{[w_1]_{p,q}}{[w_2]_{p,q}}\right)^l [2i]^l_{p^{w_1},q^{w_1}} \\
&= [2]_{q^{w_1}} [w_2]^n_{p,q} \sum_{l=0}^{n} \binom{n}{l} \left(\frac{[w_1]_{p,q}}{[w_2]_{p,q}}\right)^l p^{w_1 w_2 x l} T^{(2l)}_{n-l,p^{w_2},q^{w_2}}(w_1 x) \\
&\qquad \times \sum_{i=0}^{w_2-1} (-1)^i q^{w_1 i} q^{2(n-l)w_1 i} [2i]^l_{p^{w_1},q^{w_1}}.
\end{aligned}$$

Therefore, we obtain that

$$\begin{aligned}
&[2]_{q^{w_1}} [w_2]^n_{p,q} \sum_{i=0}^{w_2-1} (-1)^i q^{w_1 i} T_{n,p^{w_2},q^{w_2}}\left(w_1 x + \frac{2w_1 i}{w_2}\right) \\
&= [2]_{q^{w_1}} \sum_{l=0}^{n} \binom{n}{l} [w_1]^l_{p,q} [w_2]^{n-l}_{p,q} p^{w_1 w_2 x l} T^{(2l)}_{n-l,p^{w_2},q^{w_2}}(w_1 x) \mathcal{T}_{n,l,p^{w_1},q^{w_1}}(w_2),
\end{aligned} \tag{20}$$

and

$$\begin{aligned}
&[2]_{q^{w_2}} [w_1]^n_{p,q} \sum_{j=0}^{w_1-1} (-1)^j q^{w_2 j} T_{n,p^{w_1},q^{w_1}}\left(w_2 x + \frac{2w_2 j}{w_1}\right) \\
&= [2]_{q^{w_2}} \sum_{l=0}^{n} \binom{n}{l} [w_2]^l_q [w_1]^{n-l}_{p,q} p^{w_1 w_2 x l} T^{(2l)}_{n-l,p^{w_1},q^{w_1}}(w_2 x) \mathcal{T}_{n,l,p^{w_2},q^{w_2}}(w_1).
\end{aligned} \tag{21}$$

where $\mathcal{T}_{n,l,p,q}(k) = \sum_{i=0}^{k-1} (-1)^i q^{(1+2n-2l)i} [2i]^l_{p,q}$ is called as the alternate power sums. Thus, the theorem can be established by (20) and (21). □

5. Zeros of the Carlitz-Type (P, Q)-Tangent Polynomials

The purpose of this section is to support theoretical predictions using numerical experiments and to discover new exciting patterns for zeros of the Carlitz-type (p, q)-tangent polynomials $T_{n,p,q}(x)$. We propose some conjectures by numerical experiments. The first values of the $T_{n,p,q}(x)$ are given by

$$T_{0,p,q}(x) = 1,$$

$$T_{1,p,q}(x) = -\frac{-p^x - p^x q^3 + q^x + p^2 q^{1+x}}{(p-q)(1+p^2 q)(1-q+q^2)},$$

$$\begin{aligned} T_{2,p,q}(x) = {} & \frac{p^{2x} + p^{2+2x}q^3 + p^{2x}q^5 + p^{2+2x}q^8 - 2p^x q^x + q^{2x} - 2p^{4+x}q^{1+x}}{(p-q)^2(1+p^4 q)(1+p^2 q^3)(1-q+q^2-q^3+q^4)} \\ & - \frac{2p^x q^{5+x} - 2p^{4+x}q^{6+x} + p^4 q^{1+2x} + p^2 q^{3+2x} + p^6 q^{4+2x}}{(p-q)^2(1+p^4 q)(1+p^2 q^3)(1-q+q^2-q^3+q^4)}. \end{aligned}$$

Tables 1 and 2 present the numerical results for approximate solutions of real zeros of $T_{n,p,q}(x)$. The numbers of zeros of $T_{n,p,q}(x)$ are tabulated in Table 1 for a fixed $p = \frac{1}{2}$ and $q = \frac{1}{10}$.

Table 1. Numbers of real and complex zeros of $T_{n,p,q}(x), p = \frac{1}{2}, q = \frac{1}{10}$.

Degree n	Real Zeros	Complex Zeros
1	1	0
2	2	0
3	1	2
4	2	2
5	1	4
6	2	4
7	1	6
8	2	6
9	1	8
10	2	8
11	1	10
12	2	10
13	1	12
14	2	12
⋮	⋮	⋮
30	2	28

Table 2. Numerical solutions of $T_{n,p,q}(x) = 0, p = \frac{1}{2}, q = \frac{1}{10}$.

Degree n	x
1	0.0147214
2	−0.0451666, 0.0490316
3	0.0737013
4	−0.0782386, 0.0906197
5	0.102727
6	−0.0935042, 0.111767

The use of computer has made it possible to identify the zeros of the Carlitz-type (p, q)-tangent polynomials $T_{n,p,q}(x)$. The zeros of the Carlitz-type (p, q)-tangent polynomials $T_{n,p,q}(x)$ for $x \in \mathbb{C}$ are plotted in Figure 1.

In Figure 1(top-left), we choose $n = 10, p = 1/2$ and $q = 1/10$. In Figure 1(top-right), we choose $n = 20, p = 1/2$ and $q = 1/10$. In Figure 1(bottom-left), we choose $n = 30, p = 1/2$ and $q = 1/10$. In Figure 1(bottom-right), we choose $n = 40, p = 1/2$ and $q = 1/10$. It is amazing that the structure of the real roots of the Carlitz-type (p,q)-tangent polynomials $T_{n,p,q}(x)$ is regular. Thus, theoretical prediction on the regular structure of the real roots of the Carlitz-type (p,q)-tangent polynomials $T_{n,p,q}(x)$ is await for further study (Table 1). Next, we have obtained the numerical solution satisfying Carlitz-type (p,q)-tangent polynomials $T_{n,p,q}(x) = 0$ for $x \in \mathbb{R}$. The numerical solutions are tabulated in Table 2 for a fixed $p = \frac{1}{2}$ and $q = \frac{1}{10}$ and various value of n.

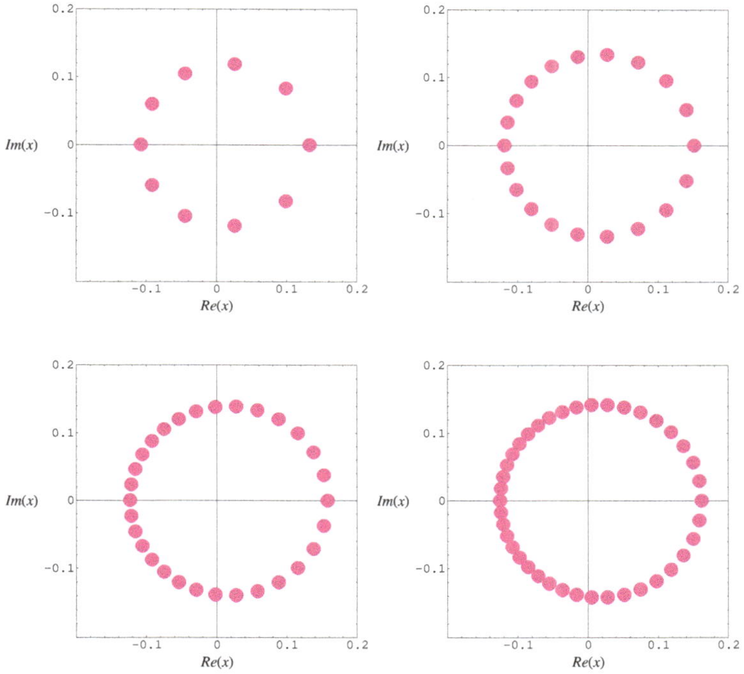

Figure 1. Zeros of $T_{n,p,q}(x)$.

6. Conclusions and Future Developments

This study constructed the Carlitz-type (p,q)-tangent numbers and polynomials. We have derived several formulas for the Carlitz-type (h,q)-tangent numbers and polynomials. Some interesting symmetric identities for Carlitz-type (p,q)-tangent polynomials are also obtained. Moreover, the results of [18] can be derived from ours as special cases when $q = 1$. By numerical experiments, we will make a series of the following conjectures:

Conjecture 1. *Prove or disprove that* $T_{n,p,q}(x), x \in \mathbb{C}$, *has* $Im(x) = 0$ *reflection symmetry analytic complex functions. Furthermore,* $T_{n,p,q}(x)$ *has* $Re(x) = a$ *reflection symmetry for* $a \in \mathbb{R}$.

Many more values of n have been checked. It still remains unknown if the conjecture holds or fails for any value n (see Figure 1).

Conjecture 2. *Prove or disprove that* $T_{n,p,q}(x) = 0$ *has n distinct solutions.*

In the notations: $R_{T_{n,p,q}(x)}$ denotes the number of real zeros of $T_{n,p,q}(x)$ lying on the real plane $Im(x) = 0$ and $C_{T_{n,p,q}(x)}$ denotes the number of complex zeros of $T_{n,p,q}(x)$. Since n is the degree of the polynomial $T_{n,p,q}(x)$, we get $R_{T_{n,p,q}(x)} = n - C_{T_{n,p,q}(x)}$ (see Tables 1 and 2).

Conjecture 3. *Prove or disprove that*

$$R_{T_{n,p,q}(x)} = \begin{cases} 1, & \text{if } n = odd, \\ 2, & \text{if } n = even. \end{cases}$$

We expect that investigations along these directions will lead to a new approach employing numerical method regarding the research of the Carlitz-type (p,q)-tangent polynomials $T_{n,p,q}(x)$ which appear in applied mathematics, and mathematical physics (see [11,18–20]).

Funding: This work was supported by the National Research Foundation of Korea(NRF) grant funded by the Korea government(MEST) (No. 2017R1A2B4006092).

Acknowledgments: The author would like to thank the referees for their valuable comments, which improved the original manuscript in its present form.

Conflicts of Interest: The author declares no conflict of interest.

References

1. Araci, S.; Duran, U.; Acikgoz, M.; Srivastava, H.M. A certain (p,q)-derivative operato rand associated divided differences. *J. Inequal. Appl.* **2016**, *2016*, 301. [CrossRef]
2. Andrews, G.E.; Askey, R.; Roy, R. Special Functions. In *Encyclopedia of Mathematics and Its Applications 71*; Cambridge University Press: Cambridge, UK, 1999.
3. Bayad, A. Identities involving values of Bernstein, q-Bernoulli, and q-Euler polynomials. *Russ. J. Math. Phys.* **2011**, *18*, 133–143. [CrossRef]
4. Carlitz, L. Expansion of q-Bernoulli numbers and polynomials. *Duke Math. J.* **1958**, *25*, 355–364. [CrossRef]
5. Choi, J.; Anderson, P.J.; Srivastava, H.M. Some q-extensions of the Apostal-Bernoulli and the Apostal-Euler polynomials of order n, and the multiple Hurwiz zeta function. *Appl. Math. Comput.* **2008**, *199*, 723–737.
6. Dattoli, G.; Migliorati, M.; Srivastava, H.M. Sheffer polynomials, monomiality principle, algebraic methods and the theory of classical polynomials. *Math. Comput. Model.* **2007**, *45*, 1033–1041. [CrossRef]
7. Duran, U.; Acikgoz, M.; Araci, S. On (p,q)-Bernoulli, (p,q)-Euler and (p,q)-Genocchi polynomials. *J. Comput. Theor. Nanosci.* **2016**, *13*, 7833–7846. [CrossRef]
8. Kupershmidt, B.A. Reflection symmetries of q-Bernoulli polynomials. *J. Nonlinear Math. Phys.* **2005**, *12*, 412–422. [CrossRef]
9. Kurt, V. A further symmetric relation on the analogue of the Apostol-Bernoulli and the analogue of the Apostol-Genocchi polynomials. *Appl. Math. Sci.* **2009**, *3*, 53–56.
10. He, Y. Symmetric identities for Carlitz's q-Bernoulli numbers and polynomials. *Adv. Differ. Equ.* **2013**, *246*. [CrossRef]
11. Ryoo, C.S.; Agarwal, R.P. Some identities involving q-poly-tangent numbers and polynomials and distribution of their zeros. *Adv. Differ. Equ.* **2017**, *213*. [CrossRef]
12. Shin, H.; Zeng, J. The q-tangent and q-secant numbers via continued fractions. *Eur. J. Comb.* **2010**, *31*, 1689–1705. [CrossRef]
13. Simsek, Y. Twisted (h,q)-Bernoulli numbers and polynomials related to twisted (h,q)-zeta function and L-function. *J. Math. Anal. Appl.* **2006**, *324*, 790–804. [CrossRef]
14. Srivastava, H.M. Some generalizations and basic (or q-) extensions of the Bernoulli, Euler and Genocchi Polynomials. *Appl. Math. Inform. Sci.* **2011**, *5*, 390–444.
15. Srivastava, H.M.; Pintér, Á. Remarks on some relationships between the Bernoulli and Euler polynomials. *Appl. Math. Lett.* **2004**, *17*, 375–380. [CrossRef]
16. Srivastava, H.M.; Pintér, Á. Addition theorems for the Appell polynomials and the associated classes of polynomial expansions. *Aequ. Math.* **2013**, *85*, 483–495.

17. Choi, J.; Anderson, P.J.; Srivastava, H.M. Carlitz's q-Bernoulli and q-Euler numbers and polynomials and a class of generalized q-Hurwiz zeta functions. *Appl. Math. Comput.* **2009**, *215*, 1185–1208
18. Hwang, K.W.; Ryoo, C.S. On Carlitz-type q-tangent numbers and polynomials and computation of their zeros. *J. App. Math. Inform.* **2017**, *35*, 495–504. [CrossRef]
19. Ryoo, C.S. A note on the tangent numbers and polynomials. *Adv. Stud. Theor. Phys.* **2013**, *7*, 447–454. [CrossRef]
20. Agarwal, R.P.; Kang, J.Y.; Ryoo, C.S. Some properties of (p,q)-tangent polynomials. *J. Comput. Anal. Appl.* **2018**, *24*, 1439–1454.

© 2018 by the author. Licensee MDPI, Basel, Switzerland. This article is an open access article distributed under the terms and conditions of the Creative Commons Attribution (CC BY) license (http://creativecommons.org/licenses/by/4.0/).

symmetry

MDPI

Article

Sharp Bounds on the Higher Order Schwarzian Derivatives for Janowski Classes

Nak Eun Cho [1], Virendra Kumar [2,*] and V. Ravichandran [3]

1 Department of Applied Mathematics, Pukyong National University, Busan 48513, Korea; necho@pknu.ac.kr
2 Department of Mathematics, Ramanujan college, University of Delhi, H-Block, Kalkaji, New Delhi 110019, India
3 Department of Mathematics, National Institute of Technology, Tiruchirappalli, Tamil Nadu 620015, India; ravic@nitt.edu; vravi68@gmail.com
* Correspondence: vktmaths@yahoo.in

Received: 25 July 2018; Accepted: 14 August 2018; Published: 18 August 2018

Abstract: Higher order Schwarzian derivatives for normalized univalent functions were first considered by Schippers, and those of convex functions were considered by Dorff and Szynal. In the present investigation, higher order Schwarzian derivatives for the Janowski star-like and convex functions are considered, and sharp bounds for the first three consecutive derivatives are investigated. The results obtained in this paper generalize several existing results in this direction.

Keywords: higher order Schwarzian derivatives; Janowski star-like function; Janowski convex function; bound on derivatives

MSC: 30C45; 30C50; 30C80

---

1. Introduction

Let $\mathcal{A}$ be the class of analytic functions defined on the unit disk $\mathbb{D} := \{z \in \mathbb{C} \ : \ |z| < 1\}$ and having the form:

$$f(z) = z + a_2 z^2 + a_3 z^3 + \cdots. \tag{1}$$

The subclass of $\mathcal{A}$ consisting of univalent functions is denoted by $\mathcal{S}$. An analytic function f is subordinate to another analytic function g if there is an analytic function w with $|w(z)| \leq |z|$ and $w(0) = 0$ such that $f(z) = g(w(z))$, and we write $f \prec g$. If g is univalent, then $f \prec g$ if and only if $f(0) = g(0)$ and $f(\mathbb{D}) \subseteq g(\mathbb{D})$. The classes $\mathcal{S}^*$ and $\mathcal{K}$ of star-like and convex functions, respectively, are among the most studied subclasses of $\mathcal{S}$. These classes are defined, respectively, as:

$$\mathcal{S}^* := \left\{ f \in \mathcal{S} : \operatorname{Re}\left(\frac{zf'(z)}{f(z)}\right) > 0,\ z \in \mathbb{D} \right\}$$

and:

$$\mathcal{K} := \left\{ f \in \mathcal{S} : \operatorname{Re}\left(1 + \frac{zf''(z)}{f'(z)}\right) > 0,\ z \in \mathbb{D} \right\}.$$

The Koebe function $k(z) = z/(1-z)^2 \in \mathcal{S}^*$ and $z/(1-z) \in \mathcal{K}$.

General forms of these classes were considered by Janowski [1]. For $-1 \leq B < A \leq 1$, these classes are defined by:

$$\mathcal{S}^*[A,B] := \left\{ f \in \mathcal{S} : \frac{zf'(z)}{f(z)} \prec \frac{1+Az}{1+Bz} \right\} \quad \text{and} \quad \mathcal{K}[A,B] := \left\{ f \in \mathcal{S} : 1 + \frac{zf''(z)}{f'(z)} \prec \frac{1+Az}{1+Bz} \right\}.$$

These classes are called the class of Janowski star-like and Janowski convex functions, respectively. On specializing the parameters A and B, we get several well-known classes such as $\mathcal{S}^* := \mathcal{S}^*[1,-1]$ and $\mathcal{K} := \mathcal{K}[1,-1]$. The functions h_0 and k_0 defined by:

$$h_0(z) = \begin{cases} z(1+Bz)^{\frac{A}{B}-1}, & B \neq 0; \\ ze^{Az}, & B = 0, \end{cases} \tag{2}$$

and:

$$k_0(z) = \begin{cases} \frac{1}{A}[(1+Bz)^{\frac{A}{B}} - 1], & B \neq 0, A \neq 0; \\ \frac{1}{B}\log(1+Bz), & A = 0; \\ \frac{1}{A}[e^{Az} - 1], & B = 0. \end{cases} \tag{3}$$

belong to the classes $\mathcal{S}^*[A,B]$ and $\mathcal{K}[A,B]$ $(-1 \leq B < A \leq 1)$, respectively. In particular, $2z/(2-z) \in \mathcal{S}^*[1/2,-1/2]$ and $4z/(2-z)^2 \in \mathcal{K}[1/2,-1/2]$.

The quantity $a_2^2 - a_3$ is associated with the Schwarzian derivative of function $f \in \mathcal{S}$. Recall that the Schwarzian derivative of a locally univalent function f is defined by:

$$\mathbf{S}(f)(z) := \left(\frac{f''(z)}{f'(z)}\right)' - \frac{1}{2}\left(\frac{f''(z)}{f'(z)}\right)^2$$

which is an important quantity in univalent function theory. For example, the quantity $a_3 - \mu a_2^2 = (f'''(0) - 3\mu(f''(0))^2/2)/6$ is called the Fekete–Szegö functional, and finding the sharp bound on modulus of this quantity is popularly known as the Fekete–Szegö problem. Nehari [2] (see also [3]) proved that the necessary condition for an analytic function f to be in the class $\mathcal{S}$ is $|\mathbf{S}(f)(z)| \leq 6(1-|z|^2)^{-2}$, and the sufficient condition is $|\mathbf{S}(f)(z)| \leq 2(1-|z|^2)^{-2}$. In both directions, the results are the best possible in the sense that the constants two and six cannot be replaced by the smaller numbers. The sharpness of the later condition was verified by Hille [4]. The first inequality is sharp in the case of the Koebe function, whereas the sharpness in the second can be seen in the case of the function $f_0(z) = (1/2)\log((1+z)/(1-z))$. Later, Nehari [5] proved that if f is a convex function, then $|\mathbf{S}(f)(z)| \leq 2(1-|z|^2)^{-2}$.

Aharanov and Harmelin [6] studied the higher order Schwarzian derivatives $\sigma_n(f)$ with invariance under composition on the left by Möbius transformations T, $\sigma_n(T \circ f) = \sigma_n(f)$, and their relation to univalence of the function f. The higher order Schwarzian derivative is defined as follows (see [6,7]):

$$\sigma_3(f) = \mathbf{S}(f)$$

and for any integer $n \geq 4$, it is given by:

$$\sigma_{n+1}(f) = (\sigma_n(f))' - (n-1)\sigma_n(f)\frac{f''}{f'}.$$

In particular,

$$\sigma_4(f) = \frac{f^{(4)}}{f'} - 6\frac{f'''f'}{f'^2} + \left(\frac{f''}{f'}\right)^3$$

and for:

$$\sigma_5(f) = \frac{f^{(5)}}{f'} - 10\frac{f^{(4)}f''}{f'^2} - 6\left(\frac{f'''}{f'}\right)^2 + 48\frac{f'''f'^2}{f'^3} - 36\left(\frac{f''}{f'}\right)^4.$$

Schippers [7] derived the differential equation for the Loewner flow of the Schwarzian derivative of univalent functions and used this to investigate the bounds on the modulus of higher order Schwarzian derivatives. These bounds were shown to be sharp in the case of the Koebe function. He also proved certain two-point distortion theorems for the higher order Schwarzian derivatives in terms of the hyperbolic metric. Later, the higher order Schwarzian derivatives for convex functions were considered by Dorff and Szynal [8]. Since the class $\mathcal{K}$ is linearly invariant (see [9]), so there is no loss in restricting consideration to $\sigma_n(f)(0) =: \mathbf{S}_n$. From the above definition of $\sigma_n(f)$, we see that $\mathbf{S}_3 = \sigma_3(f)(0) = 6(a_3 - a_2^2)$, $\mathbf{S}_4 = \sigma_4(f)(0) = 24(a_4 - 3a_2a_3 + 2a_2^3)$ and $\mathbf{S}_5 = \sigma_5(f)(0) = 24(5a_5 - 20a_2a_4 - 9a_3^2 + 48a_3a_2^2 - 24a_2^4)$. Droff and Szynal proved that $|\mathbf{S}_3| \leq 2$, $|\mathbf{S}_4| \leq 4$ and $|\mathbf{S}_5| \leq 12$ with inequality in the case of the function:

$$f_n(z) = \int_0^z (1 - t^{n-1})^{-\frac{2}{n-1}} dt, \quad n = 3, 4, 5.$$

They also conjectured that the maximal value of $|\mathbf{S}_n|$ for $n = 6, 7, 8, \cdots$ is attained in the case of the function f_n defined above.

In general, it is not so easy for researchers to deal with the higher order Schwarzian derivatives as the methods in geometric function theory known at present time are not substantial enough. However, they have a very important role in geometric/univalent function theory. In particular, Gal [10], by using the powerful method of admissible functions of Miller and Mocanu [11], investigated the geometric criterion of univalence, which combines higher order Schwarzian derivatives with those of the Ruscheweyh and Sălăgean operators. Therefore, it is very natural to consider higher order Schwarzian derivatives for various geometric results of analytic functions. In this direction, Tamanoi [12], investigated various properties of higher Schwarzian derivatives and their relation with combinatorial polynomials. Tamanoi also proved that the higher Schwarzian derivatives are Möbius invariant; see [12] (p. 135 Theorems 3–3(ii)). Later, in 2011, Kim and Sugawa [13] investigated relations between the Aharonov invariants ([13,14]) and Tamanoi's Schwarzian derivatives of higher order and gave a recursive formula for Tamanoi's Schwarzians. In the same paper, they proposed a new definition of invariant Schwarzian derivatives of a non-constant holomorphic function between Riemann surfaces with conformal metrics. In 2011, Kim and Sugawa reviewed the Peschl–Minda derivatives [15,16] and Schwarzian derivatives of higher order due to Aharonov [14], Tamanoi [12] and Kim and Sugawa [13] for a non-constant holomorphic map between Riemann surfaces with conformal metrics. They also proved that the higher-order Schwarzian derivatives of Aharonov and Tamanoi cannot be extended to holomorphic functions between projective Riemann surfaces unlike the classical Schwarzian derivatives. The higher order Schwarzian derivative are useful in the study of the properties of non-linear dynamical system and has been studied extensively by several researchers [17]. For many applications of the higher order Schwarzian derivatives related to the real functions, the reader may refer to [17,18] and the references cited therein. Kwon and Sim [19], in 2017, using the theory of admissible functions investigated some sufficient conditions for normalized analytic functions to be star-like, associated with Tamanoi's Schwarzian derivative of third order.

Motivated by the works of Schippers [7] and Dorff and Szynal [8] and other related works cited above, in this paper, we shall consider the higher order Schwarzian derivatives for Janowski star-like and convex functions. The sharp bound on the first three consecutive Schwarzian derivatives for Janowski star-like and convex functions is investigated. We shall also point out some relevant connections of our results with the existing result. Several examples in support of our main results are also given with explanations. To prove our results, we need the following results:

Let $\mathcal{B}$ be the class of Schwarz functions consisting of analytic functions of the form $w(z) = c_1z + c_2z^2 + c_3z^3 + \cdots$ $(z \in \mathbb{D})$ and satisfying the condition $|w(z)| < 1$ for $z \in \mathbb{D}$. Let $\mathcal{P}$ denote the class of analytic functions of the form $p(z) = 1 + p_1z + p_2z^2 + p_3z^3 + \cdots$ for which $\operatorname{Re} p(z) > 0$ $(z \in \mathbb{D})$. The following correspondence between the classes $\mathcal{B}$ and $\mathcal{P}$ holds:

$$p \in \mathcal{P} \text{ if and only if } w(z) = \frac{p(z) - 1}{p(z) + 1} \in \mathcal{B}. \tag{4}$$

Comparing coefficients in (4), we have:

$$c_1 = \frac{p_1}{2}, \ c_2 = \frac{2p_2 - p_1^2}{4}, \ c_3 = \frac{4p_3 - 4p_1p_2 + p_1^3}{8}, \ c_4 = \frac{8p_4 - 8p_1p_3 - 4p_2^2 + 6p_1^2p_2 - p_1^4}{16}. \quad (5)$$

Consider the functional $\Psi(\mu, \nu) = |c_3 + \mu c_1 c_2 + \nu c_1^3|$ for $w \in \mathcal{B}$ and $\mu, \nu \in \mathbb{R}$.

Lemma 1. *If $w \in \mathcal{B}$ [20] (p. 128 Lemma 2), then for any real numbers μ and ν, we have:*

$$|\Psi(\mu,\nu)| \leq \begin{cases} 1, & (\mu,\nu) \in \Omega_1 \cup \Omega_2 \cup \{(2,1)\}; \\ |\nu|, & (\mu,\nu) \in \bigcup_{k=3}^{7} \Omega_k; \\ \frac{2}{3}(|\mu|+1)\left(\frac{|\mu|+1}{3(|\mu|+\nu+1)}\right)^{1/2}, & (\mu,\nu) \in \Omega_8 \cup \Omega_9; \\ \frac{1}{3}\nu\left(\frac{\mu^2-4}{\mu^2-4\nu}\right)\left(\frac{\mu^2-4}{3(\nu-1)}\right)^{1/2}, & (\mu,\nu) \in \Omega_{10} \cup \Omega_{11} - \{(2,1)\}; \\ \frac{2}{3}(|\mu|-1)\left(\frac{|\mu|-1}{3(|\mu|-\nu-1)}\right)^{1/2}, & (\mu,\nu) \in \Omega_{12}. \end{cases}$$

Here, the symbols Ω_k's are defined as follows:

$$\Omega_1 := \left\{(\mu,\nu) \in \mathbb{R}^2 : |\mu| \leq 1/2, \ |\nu| \leq 1\right\},$$

$$\Omega_2 := \left\{(\mu,\nu) \in \mathbb{R}^2 : \frac{1}{2} \leq |\mu| \leq 2, \ \frac{4}{27}(|\mu|+1)^3 - (|\mu|+1) \leq \nu \leq 1\right\},$$

$$\Omega_3 := \left\{(\mu,\nu) \in \mathbb{R}^2 : |\mu| \leq \frac{1}{2}, \ \nu \leq -1\right\}, \Omega_4 := \left\{(\mu,\nu) \in \mathbb{R}^2 : |\mu| \geq 1/2, \ \nu \leq -\frac{2}{3}(|\mu|+1)\right\},$$

$$\Omega_5 := \left\{(\mu,\nu) \in \mathbb{R}^2 : |\mu| \leq 2, \ \nu \geq 1\right\}, \Omega_6 := \left\{(\mu,\nu) \in \mathbb{R}^2 : 2 \leq |\mu| \leq 4, \ \nu \geq \frac{1}{12}(\mu^2+8)\right\},$$

$$\Omega_7 := \left\{(\mu,\nu) \in \mathbb{R}^2 : |\mu| \geq 4, \ \nu \geq \frac{2}{3}(|\mu|-1)\right\},$$

$$\Omega_8 := \left\{(\mu,\nu) \in \mathbb{R}^2 : \frac{1}{2} \leq |\mu| \leq 2, \ -\frac{2}{3}(|\mu|+1) \leq \nu \leq \frac{4}{27}(|\mu|+1)^3 - (|\mu|+1)\right\},$$

$$\Omega_9 := \left\{(\mu,\nu) \in \mathbb{R}^2 : |\mu| \geq 2, \ -\frac{2}{3}(|\mu|+1) \leq \nu \leq \frac{2|\mu|(|\mu|+1)}{\mu^2+2|\mu|+4}\right\},$$

$$\Omega_{10} := \left\{(\mu,\nu) \in \mathbb{R}^2 : 2 \leq |\mu| \leq 4, \ \frac{2|\mu|(|\mu|+1)}{\mu^2+2|\mu|+4} \leq \nu \leq \frac{1}{12}(\mu^2+8)\right\},$$

$$\Omega_{11} := \left\{(\mu,\nu) \in \mathbb{R}^2 : |\mu| \geq 4, \ \frac{2|\mu|(|\mu|+1)}{\mu^2+2|\mu|+4} \leq \nu \leq \frac{2|\mu|(|\mu|-1)}{\mu^2-2|\mu|+4}\right\},$$

$$\Omega_{12} := \left\{(\mu,\nu) \in \mathbb{R}^2 : |\mu| \geq 4, \ \frac{2|\mu|(|\mu|-1)}{\mu^2-2|\mu|+4} \leq \nu \leq \frac{2}{3}(|\mu|-1)\right\}.$$

The extremal functions, up to rotations, are of the form:

$$w_1(z) = z^3, \ w_2(z) = z, \ w_3(z) = \frac{z(t_1 - z)}{1 - t_1 z}, \ w_4(z) = \frac{z(t_2 + z)}{1 + t_2 z}$$

and $w_5(z) = c_1 z + c_2 z^2 + c_3 z^3 + \cdots$, where the parameters t_1, t_2 and the coefficients c_i are given by:

$$t_1 = \left(\frac{|\mu|+1}{3(|\mu|+\nu+1)}\right)^{1/2}, \ t_2 = \left(\frac{|\mu|-1}{3(|\mu|-\nu-1)}\right)^{1/2}, \ c_1 = \left(\frac{2\nu(\mu^2+2)-3\mu^2}{3(\nu-1)(\mu^2-4\nu)}\right)^{1/2},$$

$$c_2 = (1 - c_1^2)e^{i\theta_0}, \quad c_3 = -c_1 c_2 e^{i\theta_0}, \quad \theta_0 = \pm \arccos\left[\frac{\mu}{2}\left(\frac{\nu(\mu^2+8) - 2(\mu^2+2)}{2\nu(\mu^2+2) - 3\mu^2}\right)^{1/2}\right].$$

Lemma 2. *If $w \in \mathcal{B}$ [21](see also, [22]), then for any complex numbers τ, we have:*

$$|c_2 - \tau c_1^2| \leq \max\{1; |\tau|\}.$$

The result is sharp for the functions $w(z) = z$ or $w(z) = z^2$.

Lemma 3. *Let $\hat{\alpha}, \hat{\beta}, \hat{\gamma}$ and $\hat{a}$ satisfy the inequalities* [23] (p. 506 Lemma 2.1), $0 < \hat{\alpha} < 1, 0 < \hat{a} < 1$ *and:*

$$8\hat{a}(1-\hat{a})[(\hat{\alpha}\hat{\beta} - 2\hat{\gamma})^2 + (\hat{\alpha}(\hat{a}+\hat{\alpha}) - \hat{\beta})^2] + \hat{\alpha}(1-\hat{\alpha})(\hat{\beta} - 2\hat{a}\hat{\alpha})^2 \leq 4\hat{a}\hat{\alpha}^2(1-\hat{\alpha})^2(1-\hat{a}). \tag{6}$$

If $p(z) = 1 + p_1 z + p_2 z^2 + p_3 z^3 + \cdots \in \mathcal{P}$, then:

$$|\hat{\gamma} p_1^4 + \hat{a} p_2^2 + 2\hat{\alpha} p_1 p_3 - (3/2)\hat{\beta} p_1^2 p_2 - p_4| \leq 2.$$

2. Main Results

The following theorem gives the sharp bound on the first three consecutive higher order Schwarzian derivatives for Janowski convex functions. In fact, Theorem 1 is a generalization of the result in [8] (p. 8 Theorem 1) due to Dorff and Szynal.

Theorem 1. *Let $f \in \mathcal{K}[A, B]$. Then, the following implications hold:*

1. *If $-1 \leq B < A \leq 1$, then $|\mathbf{S}_3| \leq (A - B)$.*

2. *(a) If either of the set of conditions:*

$$|3A + B| \leq 1$$

or $1 \leq |3A + B| \leq 4$ *and:*

$$\frac{4}{27}\left(\left|\frac{3A+B}{2}\right| + 1\right)^3 - \left(\left|\frac{3A+B}{2}\right| + 1\right) \leq \frac{A(A+B)}{2} \leq 1$$

hold, then $|\mathbf{S}_4| \leq 2(A - B)$.

(b) If $1 \leq |3A + B| \leq 4$ and:

$$-\frac{2}{3}\left(\left|\frac{3A+B}{2}\right| + 1\right) \leq \frac{A(A+B)}{2} \leq \frac{4}{27}\left(\left|\frac{3A+B}{2}\right| + 1\right)^3 - \left(\left|\frac{3A+B}{2}\right| + 1\right)$$

hold, then:

$$|\mathbf{S}_4| \leq \frac{4(A-B)}{3}\left(\left|\frac{3A+B}{2}\right| + 1\right)\left(\frac{\left|\frac{3A+B}{2}\right| + 1}{3\left(\left|\frac{3A+B}{2}\right| + \frac{A(A+B)}{2} + 1\right)}\right)^{1/2}.$$

3. *If $-1 < B < A < 1$ and:*

$$\left(A^2 - 2A + 2\right)(A-1)^2(A-B)^2(A+B-2)(A+B+2)$$
$$- 36\left(A^2 - 1\right)^2(A+B-2)(A+B) + 24(A+1)(A-1)^2(A-B) \geq 0 \tag{7}$$

hold, then $|\mathbf{S}_5| \leq 6(A - B)$.

All estimates are sharp.

Proof. Let $f \in \mathcal{K}[A,B]$. For such a function f, by definition, we can write:

$$1+\frac{zf''(z)}{f'(z)}=\frac{1+Aw(z)}{1+Bw(z)}, \tag{8}$$

where $w(z)=c_1z+c_2z^2+c_3z^3+\cdots \in \mathcal{B}$. Comparing the coefficients of the like power terms in (8), we have:

$$a_2=\frac{1}{2}(A-B)c_1,\quad a_3=\frac{1}{6}(A-B)\left[(A-2B)c_1^2+c_2\right], \tag{9}$$

$$a_4=\frac{1}{24}(A-B)\left[(A^2-5AB+6B^2)c_1^3+(3A-7B)c_1c_2+2c_3\right] \tag{10}$$

and:

$$a_5=\frac{1}{120}(A-B)[(A^3-24B^3-9A^2B+26B^2A)c_1^4+2(3A^2-17AB+23B^2)c_1^2c_2 \\ +3(A-3B)c_2^2+4(2A-5B)c_1c_3+6c_4]. \tag{11}$$

(1) From (9), we have:

$$\begin{aligned}\mathbf{S}_3 &= 6(a_3-a_2^2)\\ &= (A-B)\left[c_2-\frac{A+B}{2}c_1^2\right].\end{aligned} \tag{12}$$

Now, an application of Lemma 2 in (12) gives the desired estimate on $|\mathbf{S}_3|$. The function for which equality holds is given by (8) with the choice $w(z)=z^2$.

(2) Next, we consider:

$$\begin{aligned}\mathbf{S}_4 &= 24(a_4-3a_2a_3+2a_2^3)\\ &= 2(A-B)\left[c_3+\mu c_1c_2+\nu c_1^3\right]\\ &= 2(A-B)\Upsilon(\mu,\nu),\end{aligned} \tag{13}$$

where $\Upsilon(\mu,\nu):=c_3+\mu c_1c_2+\nu c_1^3$ with $\mu:=-(3A+B)/2$ and $\nu:=A(A+B)/2$.

Assume that Ω_i's are as defined in Lemma 1 with μ and ν as given above. We observe that $\nu=A(A+B)/2\geq -1$ as $AB\geq -2-A^2$, and so, $(\mu,\nu)\notin\Omega_3$. Furthermore, $|\mu|=|-(3A+B)/2|<2$ because $-4<3A+B<4$. Therefore, we can conclude that $(\mu,\nu)\notin\Omega_i$ $(i=6,7,9,10,11,12)$. Moreover, $\nu<1$ as $A^2+AB<2$. This reveals that $(\mu,\nu)\notin\Omega_5$. We now claim that $(\mu,\nu)\notin\Omega_4$. For this, we first assume that $\mu\leq 0$. Then, $|\mu|\geq 1/2$ gives $3A+B\geq 1$. Furthermore, the condition $\nu\leq(-2/3)(|\mu|+1)$ holds if $3A(A+B)/2\leq -2(3A+B+2)$ or equivalently if $-3A(A+B)\geq 2(3A+B+2)=2(3A+B)+4\geq 6$, that is if $-AB\geq 2+A^2$. Clearly, this is false. Similarly, in the case when $\mu\leq 0$, the condition $\nu\leq(-2/3)(|\mu|+1)$ does not hold. Thus, we conclude that our claim is true. Further, if $|3A+B|\leq 1$, then $(\mu,\nu)\in\Omega_1$. Furthermore, if $1\leq|3A+B|\leq 4$ and:

$$\frac{4}{27}\left(\left|-\frac{3A+B}{2}\right|+1\right)^3-\left(\left|-\frac{3A+B}{2}\right|+1\right)\leq\frac{A(A+B)}{2}\leq 1,$$

then $(\mu,\nu)\in\Omega_2$. In view of Lemma 1, we see that if $(\mu,\nu)\in\Omega_1\cup\Omega_2$, then $|\Upsilon(\mu,\nu)|\leq 1$, and hence, $|\mathbf{S}_4|\leq 2(A-B)$. The function for which equality holds is given by (8) with the choice $w(z)=z^3$.

Now, if $1 \leq |3A+B| \leq 4$ and:

$$-\frac{2}{3}\left(\left|\frac{3A+B}{2}\right|+1\right) \leq \frac{A(A+B)}{2} \leq \frac{4}{27}\left(\left|\frac{3A+B}{2}\right|+1\right)^3 - \left(\left|\frac{3A+B}{2}\right|+1\right),$$

then $(\mu,\nu) \in \Omega_8$. Now, an application of Lemma 1, in this case, gives:

$$\begin{aligned}
|\Upsilon(\mu,\nu)| &\leq \frac{2}{3}(|\mu|+1)\left(\frac{|\mu|+1}{3(|\mu|+\nu+1)}\right)^{1/2} \\
&= \frac{2}{3}\left(\left|\frac{3A+B}{2}\right|+1\right)\left(\frac{\left|\frac{3A+B}{2}\right|+1}{3\left(\left|\frac{3A+B}{2}\right|+\frac{A(A+B)}{2}+1\right)}\right)^{1/2}.
\end{aligned}$$

This inequality together with (13) gives the desired bound on $|\mathbf{S}_4|$. To show the sharpness, we consider the function f defined by (8) with the choice of the Schwarz function:

$$w(z) = \frac{z(t_1 - z)}{1 - t_1 z},$$

where:

$$t_1 := \left(\frac{\left|\frac{3A+B}{2}\right|+1}{3\left(\left|\frac{3A+B}{2}\right|+\frac{A(A+B)}{2}+1\right)}\right)^{1/2}.$$

For this Schwarz function w, we see that $c_1 = t_1, c_2 = t_1^2 - 1, c_3 = t_1^3 - t_1$ and:

$$\begin{aligned}
\mathbf{S}_4 &= 2(A-B)\left[c_3 + \left|\frac{3A+B}{2}\right| c_1 c_2 + \frac{A(A+B)}{2} c_1^3\right] \\
&= 2(A-B)\left[t_1^3 - t_1 + \left|\frac{3A+B}{2}\right| t_1(t_1^2-1) + \frac{A(A+B)}{2} t_1^3\right] \\
&= 2(A-B)\left[t_1^3\left(\left|\frac{3A+B}{2}\right| + \frac{A(A+B)}{2}+1\right) - t_1\left(\left|\frac{3A+B}{2}\right|+1\right)\right] \\
&= \frac{4(A-B)}{3}\left(\left|\frac{3A+B}{2}\right|+1\right)\left(\frac{\left|\frac{3A+B}{2}\right|+1}{3\left(\left|\frac{3A+B}{2}\right|+\frac{A(A+B)}{2}+1\right)}\right)^{1/2}.
\end{aligned}$$

This confirms the sharpness of the result.

(3) Now, it remains to find the estimate on $|\mathbf{S}_5|$. Using (5)–(11), we get:

$$\begin{aligned}
\mathbf{S}_5 &= 24(5a_5 - 20a_2a_4 - 9a_3^2 + 48a_2^2a_3 - 24a_2^4) \\
&= 3(A-B)\left[\hat{\gamma}p_1^4 + \hat{a}p_2^2 + 2\hat{\alpha}p_1p_3 - (3/2)\hat{\beta}p_1^2p_2 - p_4\right] \\
&= 3(A-B)\Psi(\hat{\gamma},\hat{a},\hat{\alpha},\hat{\beta}), \qquad (14)
\end{aligned}$$

where $\Psi(\hat{\gamma},\hat{a},\hat{\alpha},\hat{\beta}) := \hat{\gamma}p_1^4 + \hat{a}p_2^2 + 2\hat{\alpha}p_1p_3 - (3/2)\hat{\beta}p_1^2p_2 - p_4$ with the parameters $\hat{\gamma},\hat{a},\hat{\alpha}$ and $\hat{\beta}$ given by:

$$\hat{\gamma} := \frac{(A-1)^2(2-A-B)}{16}, \quad \hat{a} := \frac{2-A-B}{4}, \quad \hat{\alpha} := \frac{1-A}{2}, \quad \hat{\beta} := \frac{(1-A)(3-B-2A)}{6}.$$

Since the case $A=1, B=-1$ was considered by Dorff and Szynal [8], we assume that A and B are constrained as $-1 < B < A < 1$. Under these conditions, it is a simple matter to verify that $0 < \hat{\alpha} < 1$ and $0 < \hat{a} < 1$. Moreover, the condition (6) holds if and only if:

$$\frac{(A-1)^2\left(A^2-2A+2\right)(A-B)^2(A+B-2)(A+B+2)}{1152}+\frac{1}{48}(A-1)^2(A+1)(A-B)$$
$$\geq\frac{1}{32}\left(A^2-1\right)^2(A+B-2)(A+B)$$

or equivalently, if and only if (7) holds. Therefore, in view of Lemma 3, we conclude that if the above condition holds, then we must have $|\Psi(\hat{\gamma},\hat{a},\hat{\alpha},\hat{\beta})|\leq 2$, and thus, from (14), the result follows at once. Equality holds in the case of the function f defined by (8) with the choice of the Schwarz function $w(z)=z^4$. □

Remark 1. *In particular, when $A=1$ and $B=-1$, Theorem 1 reduces to the result in [8] (p. 8 Theorem 1) due to Dorff and Szynal.*

Example 1. *Setting $A=1/2$ and $B=-1$ in Equation (3), we get the function $f_1\in\mathcal{K}$ defined by:*

$$f_1(z)=\frac{2\left(1-\sqrt{1-z}\right)}{\sqrt{1-z}}=z+\frac{3z^2}{4}+\frac{5z^3}{8}+\frac{35z^4}{64}+\frac{63z^5}{128}+\cdots.$$

Here, we see that, $B=-1<1/2=A, a_2=3/4, a_3=5/8, a_4=35/64$ and $|\mathbf{S}_3|=6|a_3-a_2^2|=3/8<3/2=A-B$. This supports Part (1) of Theorem 1. Now, for the function f_1, we see that $3A+B=1/2<1$ and $|\mathbf{S}_4|=24|a_4-3a_2a_3+2a_2^3|=3/8<2(A-B)=3$. This supports Part 2(a) of Theorem 1.

Example 2. *Let $A=1/2$ and $B=-1$ in Equation (3). Then, we get the function $f_2\in\mathcal{K}$ defined by:*

$$f_2(z)=e^z-1=z+\frac{z^2}{2}+\frac{z^3}{6}+\frac{z^4}{24}+\frac{z^5}{120}+\cdots.$$

Here, $1<|3A+B|=3<4, a_2=1/2, a_3=1/6, a_4=1/24$ and $|\mathbf{S}_4|=24|a_4-3a_2a_3+2a_2^3|=1<2(A-B)=2$. This supports Part 2(b) of Theorem 1. For an example satisfying Part (3) of Theorem 1, we set $A=3/4$ and $B=1/4$ in Equation (3), and thus, we get the function f_3 defined by:

$$f_3(z)=z+\frac{z^2}{4}+\frac{z^3}{48}.$$

Here, $a_2=1/4, a_3=1/48, a_4=0, a_5=0$ and $|\mathbf{S}_5|=24|5a_5-20a_2a_4-9a_3^2+48a_2^2a_3-24a_2^4|=27/32<3=6(A-B)$.

Theorem 2. *Let $f\in\mathcal{S}^*[A,B]$. Then, the following inequalities hold:*

1. *If $-1\leq B<A\leq 1$, then $|\mathbf{S}_3|\leq 3(A-B)$.*
2. *(a) If A and B satisfy either:*

$$|B-3A|\leq 1/2 \ \ and \ \ |A(2A-B)|\leq 1$$

or:

$$1/2\leq|B-3A|\leq 1 \ \ and \ \ \frac{4}{27}(|B-3A|+1)^3-(|B-3A|+1)\leq A(2A-B)\leq 1,$$

then $|\mathbf{S}_4|\leq 8(A-B)$.

(b) Let us denote:

$$T(A,B):=-\frac{2}{3}(|B-3A|+1).$$

If either of the following sets of conditions:

$$1/2 \le |B-3A| \le 2,\ \ T(A,B) \le A(2A-B) \le \frac{4}{27}(|B-3A|+1)^3 - (|B-3A|+1),$$

or:

$$|B-3A| \ge 2,\ \ T(A,B) \le A(2A-B) \le \frac{2|B-3A|(|B-3A|+1)}{(B-3A)^2+|B-3A|+4}$$

hold, then:

$$|\mathbf{S}_4| \le \frac{16(A-B)(|B-3A|+1)^{3/2}}{3(3(|B-3A|+A(2A-B)+1))^{1/2}}.$$

(c) Moreover,

$$|\mathbf{S}_4| \le \begin{cases} 8(A-B)|A(2A-B)|, & 2 \le |B-3A| \le 4,\ A(2A-B) \ge ((B-3A)^2+8)/12; \\ 48, & A=1,\ B=-1. \end{cases}$$

3. *If A and B satisfy the conditions* $0 < (A+1)\left[43A^2 - 43A(B-1) + 10B^2 - 23B + 10\right] < 80$, $0 < 13A - 3B + 10 < 20$ *and:*

$$\begin{aligned} &450(13A-3B-10)(13A-3B+10)\left(4A^2-4AB+B^2-1\right)^2 - 50(2A-B-1)(2A-B+1) \\ &\left(8A^2+A(7-9B)+(B-7)B\right)^2 - (13A-3B-10)(13A-3B+10)(1849A^4-172A^3(23B-20) \\ &\quad + A^2\left(2976B^2-5916B+2878\right) - 4A\left(230B^3-798B^2+928B-231\right) \\ &\quad + 100B^4 - 520B^3 + 1177B^2 - 630B + 98)(A-B)^2 \le 0, \end{aligned} \tag{15}$$

then $|\mathbf{S}_5| \le 15(A-B)$.

All estimates are sharp.

Proof. Since $f \in \mathcal{S}^*[A,B]$, it follows that there exists a Schwarz function $w(z) = c_1z + c_2z^2 + c_3z^3 + \cdots$ such that:

$$\frac{zf'(z)}{f(z)} = \frac{1+Aw(z)}{1+Bw(z)}. \tag{16}$$

Comparing the coefficients on both sides of (16), we have:

$$a_2 = (A-B)c_1,\ \ a_3 = \frac{1}{2}(A-B)\left(Ac_1^2 - 2Bc_1^2 + c_2\right), \tag{17}$$

$$a_4 = \frac{1}{6}(A-B)\left[(A-2B)(A-3B)c_1^3 + (3A-7B)c_1c_2 + 2c_3\right] \tag{18}$$

and:

$$\begin{aligned} a_5 = \frac{1}{24}(A-B)[(A^3-9A^2B+26AB^2-24B^3)c_1^4 + \left(6A^2-34AB+46B^2\right)c_1^2c_2 \\ + 4\left(2A-5B\right)c_1c_3 + 3\left(A-3B\right)c_2^2 + 6c_4]. \end{aligned} \tag{19}$$

As in the proof of Theorem 1, using (17)–(19), we get:

$$\mathbf{S}_3 = -3(A-B)\left(Ac_1^2 - c_2\right),\ \ \mathbf{S}_4 = 8(A-B)\left[A(2A-B)c_1^3 + (B-3A)\,c_1c_2 + c_3\right] \tag{20}$$

and:

$$\mathbf{S}_5 = -\frac{3}{16}(A-B)[(A+1)\left(43A^2-43A(B-1)+10B^2-23B+10\right)p_1^4 \\ -4\left(43A^2+A(53-33B)+5B^2-23B+15\right)p_1^2p_2+80(2A-B+1)p_1p_3 \\ +4(13A-3B+10)p_2^2-80p_4]. \quad (21)$$

(1) From (20), we have:

$$\begin{aligned} |\mathbf{S}_3| &= 3(A-B)\left|Ac_1^2-c_2\right| \\ &\leq 3(A-B)\max\{1;|A|\} \\ &= 3(A-B). \end{aligned}$$

This gives the required estimate on $|\mathbf{S}_3|$. The extremal function in this case is given by (16) with the function $w(z)=z^2$.

(2) Again, from (20), we get:

$$\begin{aligned} |\mathbf{S}_4| &= 8(A-B)\left|A(2A-B)c_1^3+(B-3A)c_1c_2+c_3\right| \\ &\leq 8(A-B)\left|c_3+\mu c_1c_2+\nu c_1^3\right|, \end{aligned} \quad (22)$$

where $\mu := B-3A$ and $\nu := A(2A-B)$. Assume that $\Omega_i^\prime s$ are as defined in Lemma 1 with the setting μ and ν mentioned above. In particular, $|\mu| \not> 4$ and $|\mu|=4$ if and only if $A=1$ and $B=-1$. It can be easily verified that for $A=1$ and $B=-1$, $(\mu,\nu)\notin\Omega_{11}\cup\Omega_{12}$. Now, in view of Lemma 1, from (22), we see that if A and B satisfy either:

$$|B-3A| \leq \frac{1}{2} \text{ and } |A(2A-B)| \leq 1$$

or:

$$\frac{1}{2} \leq |B-3A| \leq 1 \text{ and } \frac{4}{27}(|B-3A|+1)^3-(|B-3A|+1) \leq A(2A-B) \leq 1,$$

then $|\mathbf{S}_4| \leq 8(A-B)$. The extremal function in this case is given by (16) with the choice of the function $w(z)=z^3$.

Further, since $\nu > -1$, it follows that $(\mu,\nu)\notin\Omega_3$. Moreover, a computation reveals that $(\mu,\nu)\notin\Omega_4\cup\Omega_5$. Furthermore, it can be verified that $(\mu,\nu)\in\Omega_7$ if and only if $A=1$ and $B=-1$. Now, an application of Lemma 1 gives:

$$|\mathbf{S}_4| \leq \begin{cases} 8(A-B)|A(2A-B)|, & 2\leq|B-3A|\leq 4,\ A(2A-B)\geq((B-3A)^2+8)/12; \\ 48, & A=1,\ B=-1. \end{cases}$$

The extremal function in this case is given by (16) with the choice of the function $w(z)=z$.

Similarly, we can prove that if either of the sets of following conditions:

$$1/2 \leq |B-3A| \leq 2,\ -\frac{2}{3}(|B-3A|+1) \leq A(2A-B) \leq \frac{4}{27}(|B-3A|+1)^3-(|B-3A|+1),$$

or:

$$|B-3A| \geq 2,\ -\frac{2}{3}(|B-3A|+1) \leq A(2A-B) \leq \frac{2|B-3A|(|B-3A|+1)}{(B-3A)^2+|B-3A|+4}$$

hold, then

$$|\mathbf{S}_4| \leq \frac{16(A-B)(|B-3A|+1)^{3/2}}{3(3(|B-3A|+A(2A-B)+1))^{1/2}}.$$

To show the sharpness, we consider the function f defined by (16) with the choice of the Schwarz function $w(z) = z(t_2 - z)/(1 - t_2 z)$, where:

$$t_2 := \left(\frac{|B-3A|+1}{3\left(|B-3A|+A(2A-B)+1\right)}\right)^{1/2}.$$

For the Schwarz function w, given above, we see that $c_1 = t_2, c_2 = t_2^2 - 1, c_3 = t_2^3 - t_2$ and:

$$\begin{aligned}
\mathbf{S}_4 &= 8(A-B)\left(A(2A-B)c_1^3 + |B-3A|c_1c_2 + c_3\right) \\
&= 8(A-B)\left(A(2A-B)t^3 + |B-3A|t_2(t_2^2-1) + t_2^3 - t_2\right) \\
&= 8(A-B)\left([A(2A-B) + |B-3A| + 1]t_2^3 - (|B-3A|+1)t_2\right) \\
&= -\frac{16(A-B)}{3}\frac{(|B-3A|+1)^{3/2}}{(3(|B-3A|+A(2A-B)+1))^{1/2}}.
\end{aligned}$$

This confirms the sharpness of the result.

(3) Finally, it remains to find the estimate on $|\mathbf{S}_5|$. The expression for $\mathbf{S}_5$ given in (21) can be written as:

$$\mathbf{S}_5 = -\frac{15}{2}(A-B)[\hat{\gamma}p_1^4 + \hat{a}p_2^2 + 2\hat{\alpha}p_1p_3 - (3/2)\hat{\beta}p_1^2p_2 - p_4], \tag{23}$$

where:

$$\hat{\gamma} := (A+1)\left(43A^2 - 43A(B-1) + 10B^2 - 23B + 10\right)/80, \quad \hat{a} := (13A - 3B + 10)/20$$

$$\hat{\alpha} := (2A - B + 1)/2, \quad \hat{\beta} := \left(43A^2 + A(53 - 33B) + 5B^2 - 23B + 15\right)/30.$$

In order to apply Lemma 3, we assume that the parameters A and B satisfy the conditions $0 < (A+1)\left[43A^2 - 43A(B-1) + 10B^2 - 23B + 10\right] < 80$ and $0 < (13A - 3B + 10) < 20$ together with the condition:

$$\begin{aligned}
&(13A-3B-10)(13A-3B+10)(A-B)^2(1849A^4 - 172A^3(23B-20) \\
&\quad + A^2(2976B^2 - 5916B + 2878) - 4A(230B^3 - 798B^2 + 928B - 231) \\
&\quad\quad + 100B^4 - 520B^3 + 1177B^2 - 630B + 98) \\
&\quad + 50(2A-B-1)(2A-B+1)\left(8A^2 + A(7-9B) + (B-7)B\right)^2 \\
&\quad\quad \geq 450(13A-3B-10)(13A-3B+10)\left(4A^2 - 4AB + B^2 - 1\right)^2,
\end{aligned}$$

or equivalently, if (15) holds.

Thus, all conditions of Lemma 3 are fulfilled. Therefore, we have $|\mathbf{S}_5| \leq 15(A-B)$. The function f defined by (16) with the choice of the Schwarz function $w(z) = z^4$ shows that the result is sharp. This completes the proof. □

Example 3. *Let $A = 1/2$ and $B = -1$. Then, from Equation (2), we get the function f_4 defined by:*

$$f_4(z) = \frac{z}{(1-z)^{3/2}} = z + \frac{3z^2}{2} + \frac{15z^3}{8} + \frac{35z^4}{16} + \frac{315z^5}{128} + \frac{693z^6}{256} + \cdots.$$

The function $f_4 \in \mathcal{S}^*$ and satisfies the assertion in Part (1) of Theorem 2, as for this function, we have $a_2 = 3/2, a_3 = 15/8, a_4 = 35/16$ and $|\mathbf{S}_3| = 6|a_3 - a_2^2| = 9/4 < 9/2 = 3(A-B)$.

Example 4. *(i) Let $A = 1/2$ and $B = -1$ in Equation (2). Then, we get the function $f_5 \in \mathcal{S}^*$ defined by:*

$$f_5(z) = ze^{z/4} = z + \frac{z^2}{4} + \frac{z^3}{32} + \frac{z^4}{384} + \frac{z^5}{6144} + \cdots.$$

Here, we have $|B - 3A| = 3/4 < 1/2$ and $|A(2A - B)| = 1/8 < 1, a_2 = 1/4, a_3 = 1/32, a_4 = 1/384$ and $|\mathbf{S_4}| = 1/4 < 2 = 8(A - B)$. This verifies the result asserted in Part 2(a) of Theorem 2.

(ii) Let $A = 0$ and $B = -1$. Then, we have:

$$T(A,B) := -\frac{2}{3}(|B - 3A| + 1) = -\frac{4}{3} \quad 1/2 < |B - 3A| = 1 < 2$$

and

$$T(A,B) = -\frac{4}{3} < A(2A - B) = 0 < \frac{4}{27}(|B - 3A| + 1)^3 - (|B - 3A| + 1) = 12.$$

For $A = 0$ and $B = -1$, we get the function f_6 from Equation (2) defined by:

$$f_6(z) = \frac{z}{1 - z} = z + z^2 + z^3 + z^4 + z^5 + \cdots.$$

Now, a computation gives

$$|\mathbf{S_4}| = 0 < \frac{32}{3\sqrt{3}} = \frac{16(A - B)(|B - 3A| + 1)^{3/2}}{3[3(|B - 3A| + A(2A - B) + 1)]^{1/2}}.$$

This confirms the correctness of the assertion in 2(b) of Theorem 2.

(iii) For an example of a function satisfying the assertion in 2(c) of Theorem 2, we take $A = 1$ and $B = -1$ in Equation (2). Then, we get the function f_7 defined by:

$$f_7(z) = \frac{z}{(1 - z)^2} = z + 2z^2 + 3z^3 + 4z^4 + \cdots.$$

For this function, it can be verified that $|S_4| = 48$.

(iv) For $A = -1/4$ and $B = -1/2$, computations show that $0 < 13A - 3B + 10 = 33/4 < 20$:

$$0 < (A + 1)\left[43A^2 - 43A(B - 1) + 10B^2 - 23B + 10\right] = \frac{507}{64} < 80$$

and Inequality (15) becomes $-2777538049/65536 < 0$. Setting $A = -1/4$ and $B = -1/2$ in (2), we get the function f_8 defined by:

$$f_8(z) = z + \frac{z^2}{4} + \frac{5z^4}{128} + \frac{35z^5}{2048} + \cdots.$$

For this function, we see that $\mathbf{S_4} = 0 < 15/4 = 15(A - B)$. This verifies the assertion in Part (3) of Theorem 2.

3. Conclusion

In Theorems 1 and 2, the sharp bounds on the first three consecutive derivatives for Janowski convex and star-like functions are investigated. Examples 1 and 2 support the conclusions of Theorem 1, whereas Examples 3 and 4 validate the assertions in Theorem 2. The results obtained in this paper generalize several existing results in this direction, and they are pointed out. It would be interesting to investigate the estimation on other higher order Schwarzian derivatives and their applications to study the properties of a non-linear dynamical system.

Author Contributions: All authors contributed equally.

Acknowledgments: The authors would like to express their gratitude to the referees for many valuable suggestions regarding a previous version of this paper.

Funding: The first author was supported by the Basic Science Research Program through the National Research Foundation of Korea (NRF) funded by the Ministry of Education, Science and Technology (No. 2016R1D1A1A09916450)

Conflicts of Interest: The authors declare no conflict of interest.

References

1. Janowski, W. Extremal problems for a family of functions with positive real part and for some related families. *Ann. Polon. Math.* **1970/1971**, *23*, 159–177. [CrossRef]
2. Nehari, Z. The Schwarzian derivative and schlicht functions. *Bull. Am. Math. Soc.* **1949**, *55*, 545–551. [CrossRef]
3. Kraus, W. Über den Zusammenhang einiger Characteristiken eines einfach zusammenhängenden Bereiches mit der Kreisabbildung. *Mitt. Math. Sem. Giessen.* **1932**, *21*, 1–28.
4. Hille, E. Remarks on a paper by Zeev Nehari. *Bull. Amer. Math. Soc.* **1949**, *55*, 552–553. [CrossRef]
5. Nehari, Z. A property of convex conformal maps. *J. Anal. Math.* **1976**, *30*, 390–393. [CrossRef]
6. Harmelin, R. Aharonov invariants and univalent functions. *Israel J. Math.* **1982**, *43*, 244–254. [CrossRef]
7. Schippers, E. Distortion theorems for higher order Schwarzian derivatives of univalent functions. *Proc. Am. Math. Soc.* **2000**, *128*, 3241–3249. [CrossRef]
8. Dorff, M.; Szynal, J. Higher order Schwarzian derivatives for convex univalent functions. *Tr. Petrozavodsk. Gos. Univ. Ser. Mat.* **2009**, *15*, 7–11.
9. Koepf, W. Close-to-convex functions and linear-invariant families. *Ann. Acad. Sci. Fenn. Ser. A I Math.* **1983**, *8*, 349–355. [CrossRef]
10. Gal, S.G. Higher order derivatives of Schwarz, Sălăgean and Rushcheweyh types in the geometric theory of complex functions. *Rev. Roumaine Math. Pures Appl.* **2002**, *47*, 33–42.
11. Miller, S.S.; Mocanu, P.T. Second order differential inequalities in the complex plane. *J. Math. Anal. Appl.* **1978**, *65*, 289–305. [CrossRef]
12. Tamanoi, H. Higher Schwarzian operators and combinatorics of the Schwarzian derivative. *Math. Ann.* **1996**, *305*, 127–151. [CrossRef]
13. Kim, S.A.; Sugawa, T. Invariant Schwarzian derivatives of higher order. *Complex Anal. Oper. Theory* **2011**, *5*, 659–670. [CrossRef]
14. Aharonov, D. A necessary and sufficient condition for univalence of a meromorphic function. *Duke Math. J.* **1969**, *36*, 599–604.
15. Kim, S.A.; Sugawa, T. Invariant differential operators associated with a conformal metric. *Michigan Math. J.* **2007**, *55*, 459–479.
16. Schippers, E. The calculus of conformal metrics. *Ann. Acad. Sci. Fenn. Math.* **2007**, *32*, 497–521
17. Hacibekiroğlu, G.; Çãğlar, M.; Polatoğlu, Y. The higher order Schwarzian derivative: Its applications for chaotic behavior and new invariant sufficient condition of chaos. *Nonlinear Anal. Real World Appl.* **2009**, *10*, 1270–1275. [CrossRef]
18. Aguilar, R.M. Higher order Schwarzians for geodesic flows, moment sequences and the radius of adapted complexifications. *Q. J. Math.* **2013**, *64*, 1–36. [CrossRef]
19. Kwon, O.; Sim, Y. Starlikeness and Schwarzian derivatives of higher order of analytic functions. *Commun. Korean Math. Soc.* **2017**, *32*, 93–106. [CrossRef]
20. Prokhorov, D.V.; Szynal, J. Inverse coefficients for (α, β)-convex functions. *Ann. Univ. Mariae Curie-Skłodowska Sect. A* **1981**, *35*, 125–143.
21. Keogh, F.R.; Merkes, E.P. A coefficient inequality for certain classes of analytic functions. *Proc. Am. Math. Soc.* **1969**, 20, 8–12. [CrossRef]
22. Ali, R.M.; Ravichandran, V.; Seenivasagan V. Coefficient bounds for p-valent functions. *Appl. Math. Comput.* **2007**, *187*, 35–46. [CrossRef]
23. Ravichandran, V.; Verma, S. Bound for the fifth coefficient of certain star-like functions. *C. R. Math. Acad. Sci. Paris* **2015**, *353*, 505–510. [CrossRef]

© 2018 by the authors. Licensee MDPI, Basel, Switzerland. This article is an open access article distributed under the terms and conditions of the Creative Commons Attribution (CC BY) license (http://creativecommons.org/licenses/by/4.0/).

symmetry

MDPI

Article

On the Convolution Quadrature Rule for Integral Transforms with Oscillatory Bessel Kernels

Junjie Ma [1,2,*] and Huilan Liu [1,2,*]

1 School of Mathematics and Statistics, Guizhou University, Guiyang 550025, China
2 Guizhou Provincial Key Laboratory of Public Big Data, Guizhou University, Guiyang 550025, China
* Correspondence: jjma@gzu.edu.cn (J.M.); hlliu5@gzu.edu.cn (H.L.)

Received: 13 June 2018; Accepted: 23 June 2018; Published: 25 June 2018

Abstract: Lubich's convolution quadrature rule provides efficient approximations to integrals with special kernels. Particularly, when it is applied to computing highly oscillatory integrals, numerical tests show it does not suffer from fast oscillation. This paper is devoted to studying the convergence property of the convolution quadrature rule for highly oscillatory problems. With the help of operational calculus, the convergence rate of the convolution quadrature rule with respect to the frequency is derived. Furthermore, its application to highly oscillatory integral equations is also investigated. Numerical results are presented to verify the effectiveness of the convolution quadrature rule in solving highly oscillatory problems. It is found from theoretical and numerical results that the convolution quadrature rule for solving highly oscillatory problems is efficient and high-potential.

Keywords: highly oscillatory; convolution quadrature rule; volterra integral equation; Bessel kernel; convergence

1. Introduction

Highly oscillatory integrals (HOI) arise frequently in antenna problems involving Sommerfeld integrals (see [1,2]), computation of mutual impedance between conductors (see [3,4]) and many other oscillatory problems (HOP). Generally, an oscillatory integral can be written as

$$I[f] = \int_0^b f(t)W(t)dt. \tag{1}$$

Here $W(\cdot)$ denotes a highly oscillatory function and $f(\cdot)$ is slowly varied. Due to the high oscillation, classical quadrature rules (e.g., Newton-Cotes and Gauss rules) are often ineffective, and calculation of this class of integrals is deemed to be a challenging problem ([5]).

Past decades witness a rapid development of researches on calculation of HOIs. Based on Filon's idea ([6]), Iserles and Nørsett developed the Filon-type method by approximating the slowly varied function by its Hermite interpolant. Both of theoretical and numerical results manifested that this method enjoyed high-order convergence rates with respect to the frequency ([7]). To get stable and fast algorithms, Domínguez, et al. ([8]), and Xiang, et al. ([9]), proposed the Clenshaw-Curtis-Filon-type method, respectively, which enjoyed extensive applications at present.

Although Filon's methodology leads to many efficient algorithms, most of them suffer to complicate computation of moment integrals. An alternative way to addressing this problem is transforming the integral interval into the complex plane, which derives the numerical steepest descent method. In [10], the complex integration method in the standard case was discussed extensively. In [11], the general form and corresponding error analysis were studied. This method was extended to the case of HOIs on semi-finite intervals in [12]. It is notable that the numerical steepest descent method with Gauss-Laguerre quadrature may not provide satisfactory solutions in practice (see [13]). Therefore, computation of transformed integrals is still an interesting topic.

Another important quadrature rule for HOIs is Levin's method ([14]). It enjoys a wide application for its being free of computing complex moments and less restrictions to integrands. The spirit of this method is transforming the integration problem into an ODE, and solving this equation by collocation methods. It is well-know that implementing of Levin methods comes down to efficient solutions of linear systems, which are often singular and dense. In [15], an SVD solver for the ill-conditioned system was presented. In [16], Olver developed a moment-free method by using the shifted GMRES. Recently, by employing the property of Chebyshev polynomials and preconditioners, a sparse and well-conditioned Levin method was constructed in [17].

There are many other important methods for calculating HOIs, for example, the homotopy perturbation method ([18]), the generalized quadrature rule ([19]), the extrapolation method ([20]). For simplicity, we omit the details. It is quite unexpected that little attention has been paid to the convolution quadrature rule (CQ) for HOIs ([21,22]), especially its asymptotic property in the case of high oscillation. In fact, CQ is well-known for its efficient in evaluating convolution integrals, and oscillatory integrals of convolution-type play significant roles in solving oscillatory and evolutionary problems (see [23,24]). Therefore, it is a meaningful issue to study CQ for HOIs.

Consider integral transforms with Bessel kernels as

$$\int_0^x f(t)J_m(\omega(x-t))dt,\ x\in[0,T], \tag{2}$$

with $m \geq 0$ being an integer and $\omega \gg 1$, and oscillatory Volterra integral equations as

$$\int_0^x J_0(\omega(x-t))u(t)dt = f(x),\ x\in[0,T], \tag{3}$$

where $f(\cdot)$ is sufficiently smooth, $f(0) = 0$, and $u(t)$ is unknown. In this paper, we are devoted to studying convergence property of CQ with respective to the frequency for solving above two problems. The same models have been considered in [25,26], where authors concluded that Filon-type methods enjoyed the property that the higher the oscillation, the better the approximation. In the following, we will find CQ share a similar property as Filon-type methods, and even better when they are applied to solving highly oscillatory integral equations. The remaining parts are organized as follows. In Section 2, we briefly review CQ and give the convergence analysis. A modified rule is also proposed in this part. Then we study CQ for solving Volterra integral equations with highly oscillatory kernels in Section 3. Some numerical experiments are carried out in Section 4 to verify our given results.

2. Convergence of the Convolution Quadrature Rule

In this section, we revisit Lubich's convolution quadrature firstly. Then the convergence property with respect to the frequency is studied. In [21,22], Lubich proposed an algorithm for computing the following integral,

$$I[f,g] = \int_0^x f(t)g(x-t)dt,\ x>0. \tag{4}$$

Let $F(\cdot)$ denote the Laplace transform of $f(\cdot)$ and satisfy

- $F(s)$ is analytic in the region $|\arg(s-c)| < \pi - \varphi, \varphi < \pi/2, c \in \mathbb{R}$;
- there exist constants M and μ, such that $|F(s)| \leq M|s|^{-\mu}$.

By the definition of Laplace transform, it follows that

$$f(t) = \frac{1}{2\pi i}\int_\Gamma F(\lambda)e^{\lambda t}d\lambda, \tag{5}$$

where Γ is a curve locates in the analytic region of $F(s)$ and goes from $\infty \cdot e^{-i(\pi-\varphi)}$ to $\infty \cdot e^{i(\pi-\varphi)}$.

Substituting (5) into (4) gives

$$\int_0^x f(t)g(x-t)dt = \frac{1}{2\pi i}\int_\Gamma F(\lambda)\int_0^x e^{\lambda t}g(x-t)dtd\lambda. \tag{6}$$

Noting that $y(x) = \int_0^x e^{\lambda t}g(x-t)dt$ satisfies the initial value problem

$$\begin{cases} \dfrac{dy}{dx} = \lambda y(x) + g(x), \\ y(0) = 0. \end{cases} \tag{7}$$

Defining the grid $\{t_n := nh, n = 0, 1, ...\}$, we can approximate $y(x)$ by

$$\sum_{j=0}^{k} \alpha_j y_{n+j-k} = h\beta(\lambda y_n + g(nh)), n \geq 0, \tag{8}$$

where $y_{-k} = ... = y_{-1} = 0$, $g(x) = 0(x < 0)$. Multiplying both sides by $\zeta^n (n \geq 0)$ in (8) and summing give

$$(\alpha_0\zeta^k + ... + \alpha_k)\mathbf{y}(\zeta) = \beta(h\lambda\mathbf{y}(\zeta) + h\mathbf{g}(\zeta)), \tag{9}$$

where $\mathbf{y}(\zeta) = \sum_{n=0}^{\infty} y_n\zeta^n$, $\mathbf{g}(\zeta) = \sum_{n=0}^{\infty} g(nh)\zeta^n$. Letting

$$\delta(\zeta) = \frac{\alpha_0\zeta^k + ... + \alpha_k}{\beta}, \tag{10}$$

it follows that

$$\mathbf{y}(\zeta) = \left(\frac{\delta(\zeta)}{h} - \lambda\right)^{-1}\mathbf{g}(\zeta). \tag{11}$$

Since $F(s)$ is analytic in the inside region of the curve Γ, we have, by Cauchy's integral formula (see [27], p. 32),

$$\frac{1}{2\pi i}\int_\Gamma F(\lambda)\left(\frac{\delta(\zeta)}{h} - \lambda\right)^{-1} d\lambda = F\left(\frac{\delta(\zeta)}{h}\right). \tag{12}$$

Therefore, it follows that

$$\begin{aligned} \frac{1}{2\pi i}\int_\Gamma F(\lambda)\mathbf{y}(\zeta)d\lambda =& \frac{1}{2\pi i}\int_\Gamma F(\lambda)\left(\frac{\delta(\zeta)}{h} - \lambda\right)^{-1}\mathbf{g}(\zeta)d\lambda \\ =& F\left(\frac{\delta(\zeta)}{h}\right)\mathbf{g}(\zeta). \end{aligned} \tag{13}$$

Here the coefficient corresponding to ζ^n denotes an approximation to the integral (4) at $x = nh$. Suppose

$$F\left(\frac{\delta(\zeta)}{h}\right) = \sum_{j=0}^{\infty} w_j(h)\zeta^j. \tag{14}$$

Then CQ for (4) is defined as

$$Q_h^{cq}(x) = \sum_{0 \le jh \le x} w_j(h) g(x - jh). \tag{15}$$

By the definition of coefficients of Taylor expansions, we get

$$w_n(h) = \frac{1}{2\pi i} \int_{|z|=\rho} F\left(\frac{\delta(\zeta)}{h}\right) z^{-n-1} dz, \tag{16}$$

where ρ is sufficiently small such that the disc $|z| \le \rho$ falls in the analytic region of $F\left(\delta(\zeta)/h\right)$. Letting $z = \rho e^{i\theta}$, we obtain

$$w_n(h) = \frac{\rho^{-n}}{2\pi} \int_0^{2\pi} F\left(\frac{\delta(\rho e^{i\theta})}{h}\right) e^{-in\theta} d\theta. \tag{17}$$

Discretizing $\int_0^{2\pi} F\left(\frac{\delta(\rho e^{i\theta})}{h}\right) e^{-in\theta} d\theta$ by the composite trapezoid rule gives

$$\begin{aligned} \int_0^{2\pi} F\left(\frac{\delta(\rho e^{i\theta})}{h}\right) e^{-in\theta} d\theta &= \sum_{l=0}^{L-1} \int_{l\frac{2\pi}{L}}^{(l+1)\frac{2\pi}{L}} F\left(\frac{\delta(\rho e^{i\theta})}{h}\right) e^{-in\theta} d\theta \\ &\approx \sum_{l=0}^{L-1} \frac{\pi}{L} \left(F\left(\frac{\delta(\rho e^{il\frac{2\pi}{L}})}{h}\right) e^{-inl\frac{2\pi}{L}} + F\left(\frac{\delta(\rho e^{i(l+1)\frac{2\pi}{L}})}{h}\right) e^{-in(l+1)\frac{2\pi}{L}} \right) \\ &= \frac{2\pi}{L} \sum_{l=0}^{L-1} F\left(\frac{\delta(\rho e^{il\frac{2\pi}{L}})}{h}\right) e^{-inl\frac{2\pi}{L}}. \end{aligned} \tag{18}$$

The last equal sign works due to $F\left(\frac{\delta(\rho e^{i\theta})}{h}\right) e^{-in\theta}$ is $2\pi-$periodic. This leads to

$$w_n(h) \approx \frac{\rho^{-n}}{L} \sum_{l=0}^{L-1} F\left(\frac{\delta(\rho e^{il\frac{2\pi}{L}})}{h}\right) e^{-inl\frac{2\pi}{L}},\ n = 0, 1, ..., N. \tag{19}$$

Its computation complexity is $O(N \log N)$ by FFT. In this paper, we adopt $L = 10\ N$, $\rho^N = \sqrt{\epsilon}, \epsilon = 10^{-16}$ to guarantee a precision of order $O(\sqrt{\epsilon})$ in (19).

Remark 1. *In [28], convolution quadrature weights were rewritten as*

$$w_j(h) = \int_0^\infty g(s) \phi_j(s/h) ds. \tag{20}$$

Here $\phi_j(t) = e^{-t}\frac{t^j}{j!}$ *for backward differentiation formula of order 1 (BDF1), and* $\phi_j(t) = \frac{H_j(\sqrt{2t})}{j!}\left(\frac{t}{2}\right)^{j/2} e^{-3t/2}$ *for BDF2 with* $H_j(\cdot)$ *denoting the jth Hermite polynomial. Recurrence relations of frequently-used bases can be found in Table 1.*

Table 1. Recurrence relations for the CQ basis functions.

Scheme	Initial Basis ($\phi_{-n} = 0, n \geq 1$)	Recurrence for Basis Functions ($j \geq 1$)
BDF1	$\phi_0(t) = e^{-t}$	$j\phi_j(t) - t\phi_{j-1}(t) = 0$
BDF2	$\phi_0(t) = e^{-3t/2}$	$j\phi_j(t) - 2t\phi_{j-1}(t) + t\phi_{j-2}(t) = 0$
BDF3	$\phi_0(t) = e^{-11t/6}$	$j\phi_j(t) - 3t\phi_{j-1}(t) + 3t\phi_{j-2}(t) - t\phi_{j-3}(t) = 0$
BDF4	$\phi_0(t) = e^{-25t/12}$	$j\phi_j(t) - 4t\phi_{j-1}(t) + 6t\phi_{j-2}(t) - 4t\phi_{j-3}(t) + t\phi_{j-4}(t) = 0$

Existing convergence analysis of CQ often restricts to the property with respect to the stepsize. For example, setting

$$\mathcal{F}(s)g(x) = \int_0^x f(t)g(x-t)dt, \tag{21}$$

$$\mathcal{F}(s_h)g(x) = \sum_{0\leq jh\leq x} w_j(h)g(x-jh), \tag{22}$$

the convergence rate of CQ is as follows

Theorem 1. *([21], Theorem 3.1) Suppose that $\delta(\zeta)$ satisfies*

- *$\delta(\zeta)$ is analytic and without zeros in a neighbourhood of the closed unit disc $|\zeta| \leq 1$, with the exception of a zero at $\zeta = 1$;*
- *$|\arg \delta(\zeta)| \leq \pi - \alpha$, for $|\zeta| < 1$, for some $\alpha > \varphi$;*
- *$\dfrac{\delta(e^{-h})}{h} = 1 + O(h^p)$, for some $p \geq 1$.*

Then we have

$$\begin{aligned}|\mathcal{F}(s_h)g(x) - \mathcal{F}(s)g(x)| \leq & Cx^{\mu-1}\{h|g(0)| + ... + h^{p-1}|g^{(p-2)}(0)| \\ & + h^p(|g^{(p-1)}(0)| + x \max_{0\leq t\leq x} |g^{(p)}(t)|)\},\end{aligned} \tag{23}$$

where the constant C does not depend on the stepsize h.

One point which should be remarkable is that A-stable linear multistep methods for solving the initial value problem (7) are more reliable in actual computation. Therefore, in this paper, we make use of BDF2 in numerical experiments.

In the following parts, we study the convergence property of CQ for

$$I[f] = \int_0^b f(t)J_m(\omega t)dt, \ \omega \gg 1. \tag{24}$$

Here $J_m(\cdot)$ denotes the first kind Bessel function of order m with m being a nonnegative integer. Defining $\tilde{f}(s) := f(b-s)$, we transform (1) into

$$I[f] = \int_0^b \tilde{f}(b-s)J_m(\omega s)ds. \tag{25}$$

Here the Laplace transform of $J_m(\omega s)$ can be represented as (see [29], p. 1024)

$$F_{J_m}(s) = \frac{\left(\sqrt{s^2+\omega^2} - s\right)^m}{\omega^m\sqrt{s^2+\omega^2}}. \tag{26}$$

Before elaborating on the convergence property of CQ, the following lemma should be presented first.

Lemma 1. *Suppose $f(\cdot)$ is continuous on the interval S, where S may be a closed interval on the positive real axis or $[a,\infty)$ for some $a \geq 0$. Assume $\int_S |f(t)|dt$, $\int_S |f'(t)|dt$ exist. Then for any $v > 0$ and sufficiently large ω, there exists a constant C, which does not depend on the frequency ω, such that*

$$\left|\int_S f(t) J_m(\omega t) dt\right| \leq C\omega^{-1}. \tag{27}$$

Furthermore, if the interval S does not contain origin, then it follows

$$\left|\int_S f(t) J_m(\omega t) dt\right| \leq C\omega^{-3/2}. \tag{28}$$

This lemma can be proved by integration by parts and we omit the detail. Now the main theorem of this section follows.

Theorem 2. *Suppose $f(\cdot) \in C^2[0,b]$, and $\int_0^b |f(t)|dt$, $\int_0^b |f'(t)|dt$, $\int_0^b |f''(t)|dt$ exist. Then there exists a positive constant C independent of ω, such that, as the frequency ω tends to infinity, CQ satisfies*

$$|Q_h^{cq}(b) - I[f]| \leq C\omega^{-3/2}. \tag{29}$$

Proof. Denote by δ_j the coefficients of $\delta(\zeta) = \sum_{j=0}^{\infty} \delta_j \zeta^j$, and define

$$s_h g(x) := \frac{1}{h} \sum_{0 \leq jh \leq x} \delta_j g(x - jh), \ 0 < x \leq b. \tag{30}$$

Then $I[f]$ can be represented as a Dunford-Taylor integral

$$I[f] = \mathcal{F}(s_h)\tilde{f}(x) = \frac{1}{2\pi i}\int_\Gamma F_{J_m}(\lambda)(s_h - \lambda)^{-1}\tilde{f}(x)d\lambda. \tag{31}$$

By noting

$$F_{J_m}(s) = \int_0^\infty e^{-ts} J_m(\omega t) dt, \tag{32}$$

we have

$$\begin{aligned} \mathcal{F}(s_h)\tilde{f}(x) =& \frac{1}{2\pi i}\int_\Gamma \int_0^\infty e^{-t\lambda} J_m(\omega t)(s_h - \lambda)^{-1}\tilde{f}(x) dt d\lambda \\ =& \int_0^\infty J_m(\omega t) e^{-ts_h}\tilde{f}(x) dt. \end{aligned} \tag{33}$$

This implies

$$\begin{aligned} \mathcal{F}(s_h)\tilde{f}(x) - \mathcal{F}(s)\tilde{f}(x) =& \int_0^x J_m(\omega t)(e^{-ts_h}\tilde{f}(x) - \tilde{f}(x-t))dt \\ &+ \int_x^\infty J_m(\omega t) e^{-ts_h}\tilde{f}(x) dt. \end{aligned} \tag{34}$$

Define $\phi_1(t) := e^{-ts_h}\tilde{f}(x) - \tilde{f}(x-t)$ and $\phi_2(t) := e^{-ts_h}\tilde{f}(x)$. By recurrence relations and derivatives of Bessel functions, we get

$$
\begin{aligned}
\mathcal{F}(s_h)\tilde{f}(x) - \mathcal{F}(s)\tilde{f}(x) =& \phi_1(0) \int_0^x J_m(\omega t)dt + \frac{1}{\omega}(\phi_1(t) - \phi_1(0))J_{m+1}(\omega t)\big|_{t=0}^{x} \\
&- \frac{1}{\omega} \int_0^x \left(\phi_1'(t) - \frac{m+1}{t}(\phi_1(t) - \phi_1(0))\right) J_{m+1}(\omega t)dt \\
&+ \frac{1}{\omega}\phi_2(t)J_{m+1}(\omega t)\big|_{t=x}^{\infty} \\
&- \frac{1}{\omega} \int_x^{\infty} \left(\phi_2'(t) - \frac{m+1}{t}\phi_2(t)\right) J_{m+1}(\omega t)dt.
\end{aligned}
\tag{35}
$$

According to $\phi_1(0) = 0$ and Lemma 1, it follows

$$
|\mathcal{F}(s_h)\tilde{f}(x) - \mathcal{F}(s)\tilde{f}(x)| \leq C\omega^{-3/2}, \tag{36}
$$

where the constant C does not depend on ω. This completes the proof. □

This theorem verifies the accuracy of CQ will increase as the frequency tends to infinity. According to the proof of Theorem 2, we can eliminate low order terms in (35) and get a modified convolution quadrature rule.

- Modified convolution quadrature rule for (1) of the first kind (MCQ1).

$$
Q_h^{mcq1}(b) = \mathcal{F}(s_h)(f(x) - f(b)) \mid_{x=1} + f(b) \int_0^b J_m(\omega t)dt. \tag{37}
$$

According to ([30], p. 681), we have

$$
\int_0^1 t^{\mu} J_{\nu}(\omega t)dt = \frac{2^{\mu}\Gamma(\frac{\mu+\nu+1}{2})}{\omega^{\mu+1}\Gamma(\frac{\nu-\mu+1}{2})} + \omega^{-\mu}(\mu+\nu-1)J_{\nu}(\omega)s_{\mu-1,\nu-1}^{(2)}(\omega) - \omega^{-\mu}J_{\nu-1}(\omega)s_{\mu,\nu}^{(2)}(\omega). \tag{38}
$$

Here $s_{\mu,\nu}^{(2)}(x)$ denotes the Lommel function of the second kind, and can be efficient computed by asymptotic expansions ([31]). This implies

$$
\int_0^b J_m(\omega t)dt = \frac{1}{b^2\omega} + \frac{(m-1)J_m(\omega b)s_{-1,m-1}^{(2)}(\omega b) - J_{m-1}(\omega b)s_{0,m}^{(2)}(\omega b)}{b}. \tag{39}
$$

Analogy to the proof of Theorem 2, we immediately arrive at the following results.

Theorem 3. *Suppose $f(\cdot) \in C^2[0,b]$, and $\int_0^b |f(t)|dt$, $\int_0^b |f'(t)|dt$, $\int_0^b |f''(t)|dt$ exist. Then there exists a positive constant C independent of ω, such that, as the frequency ω tends to infinity, MCQ1 satisfies*

$$
|Q_h^{mcq1}(b) - I[f]| \leq C\omega^{-2}. \tag{40}
$$

Remark 2. *By Lubich's methodology of eliminating low order terms, we also obtain a modified convolution quadrature rule of the second kind for (1) (MCQ2). Although these two modified quadrature rules share the same convergence rates with respect to the stepsize, their convergence properties are quite different in the case of calculation of HOI. We will illustrate this phenomenon in Section 4.*

3. Application to a Volterra Equation

In literature, convolution quadrature rules are important tools for solving Volterra equations with convolution kernels ([32]). Many numerical experiments show they are efficient for solving some highly oscillatory Volterra integral equations (HOVIE). Consider HOVIE (3),

$$\int_0^x J_0(\omega(x-t))u(t)dt = f(x),\ x \in [0,T],$$

where $f(\cdot)$ is sufficiently smooth and $f(0) = 0$. Let

$$I_N := \{x_n : 0 = x_0 < x_1 < ... < x_N = T\}, \tag{41}$$

be a uniform grid on $[0,T]$ with spacing $h := T/N$. This equation arises from acoustic scattering problems (see [33]). By applying CQ to (3), we get

$$\sum_{j=0}^{k} w_{k-j}(h)\tilde{u}_j = f_k,\ k = 1,2,...,N. \tag{42}$$

Here u_j denotes the numerical solution $\tilde{u}(x)$ at x_j and $f_k := f(x_k)$. Once the initial value u_0 is known, the numerical solution at the uniform grid I_N can be obtained by solving the linear system (42), and the numerical solution on $[0,T]$ can be written as

$$\tilde{u}(x) := \sum_{j=0}^{N} \tilde{u}_j \phi_j(x). \tag{43}$$

Following the methodology from [24,26], we establish the convergence analysis by expressing the error function in terms of moments with highly oscillatory kernels. So let us consider some integrals involving Bessel kernels.

Lemma 2. *Define the functional $T_x : C([0,T]) \to R$ as*

$$T_x(f) = \int_0^x J_m(\omega s)f(s)ds, \tag{44}$$

with $x \in [0,T]$ and $f \in C([0,T])$. Then $\infty-$norm of the functional T_x satisfies

$$\|T_x\| \leq C\omega^{-1/2},\ as\ \omega \to \infty, \tag{45}$$

where C is a constant independent of ω.

Proof. It is easy to show

$$\|T_x\| = \int_0^x \mid J_m(\omega s) \mid ds. \tag{46}$$

By the variable transformation $t = \omega s$, we have

$$\|T_x\| = \frac{1}{\omega}\int_0^{\omega x} \mid J_m(t) \mid dt. \tag{47}$$

According to the asymptotic expansions of Bessel functions (see [34], p. 228), there exists $Z > 0$ such that for any $z > Z$, and a constant C^*, we have

$$\mid J_m(z) \mid \leq \frac{C^*}{\sqrt{z}}. \tag{48}$$

This implies

$$\begin{aligned} \|T_x\| =& \frac{1}{\omega}\int_0^Z \mid J_m(t) \mid dt + \frac{1}{\omega}\int_Z^{\omega x} \mid J_m(t) \mid dt \\ \leq& \frac{1}{\omega}\int_0^Z \mid J_m(t) \mid dt + \frac{1}{\omega}\int_Z^{\omega x} \frac{C^*}{\sqrt{z}} dt. \end{aligned} \tag{49}$$

Therefore, there exists a constant C, such that

$$\|T_x\| \leq C\omega^{-1/2}, \text{ as } \omega \to \infty. \tag{50}$$

This completes the proof. □

Lemma 3. *For any integers $\mu, v > 0$ and $x \in (0, T]$, the following integral*

$$I = \omega^{1/2} \int_0^x \frac{J_\mu(\omega t) J_v(\omega(x-t))}{t} dt \tag{51}$$

is uniformly bounded with respect to $\omega > 0$.

Proof. The variable transformation $s = \omega t$ gives

$$I = \omega^{1/2} \int_0^{\omega x} \frac{J_\mu(s) J_v(\omega x - s)}{s} ds. \tag{52}$$

According to ([34], p. 242), we have

$$I = \frac{\omega^{1/2} J_{\mu+v}(\omega x)}{\mu}. \tag{53}$$

By the asymptotic expansion of Bessel functions, we completes the proof. □

Now we get the convergence property of CQ for the numerical solution to HOVIE.

Theorem 4. *Suppose that $f \in C^3[0, T]$, then the convolution quadrature rule for solving (3) introduces a unique numerical solution $\tilde{u}$, and satisfies*

$$|u(x) - \tilde{u}(x)| \leq C\omega^{-1/2}, x \in I_N, \omega \to \infty, \tag{54}$$

where C is a constant independent of ω.

Proof. By noting

$$\mathcal{F}(s_h)\tilde{u}(x) = f(x) \tag{55}$$

and

$$\mathcal{F}(s)u(x) = f(x) \tag{56}$$

with $x \in I_h$, we get the error equation

$$\mathcal{F}(s_h)\epsilon(x) - (\mathcal{F}(s)u(x) - \mathcal{F}(s_h)u(x)) = 0, \tag{57}$$

where $\epsilon(x) := \tilde{u}(x) - u(x)$. By Remark 1 and Lemma 1, we have

$$w_j(h) = \begin{cases} O(\omega^{-1}), \; j = 0, \\ O(\omega^{-3/2}), \; j > 0. \end{cases} \tag{58}$$

Therefore, the remaining work is proving $\mathcal{F}(s)u(x) - \mathcal{F}(s_h)u(x)$ behaves as $O(\omega^{-3/2})$. A similar process to the proof of Theorem 2 gives

$$\begin{aligned} \mathcal{F}(s_h)u(x) - \mathcal{F}(s)u(x) =& \phi_1(0) \int_0^x J_0(\omega t)dt + \frac{1}{\omega}(\phi_1(t) - \phi_1(0))J_1(\omega t)|_0^x \\ & - \frac{1}{\omega} \int_0^x \left(\phi_1'(t) - \frac{1}{t}(\phi_1(t) - \phi_1(0)) \right) J_1(\omega t)dt \\ & + \frac{1}{\omega}\phi_2(t)J_1(\omega t)|_x^\infty - \frac{1}{\omega}\int_x^\infty \left(\phi_2'(t) - \frac{1}{t}\phi_2(t) \right) J_1(\omega t)dt, \end{aligned} \tag{59}$$

where $\phi_1(t) := e^{-ts_h}u(x) - u(x-t)$ and $\phi_2(t) := e^{-ts_h}u(x)$. Consider the integrals

$$\tilde{I} := \int_0^x \left(\phi_1'(t) - \frac{1}{t}(\phi_1(t) - \phi_1(0)) \right) J_1(\omega t)dt + \int_x^\infty \left(\phi_2'(t) - \frac{1}{t}\phi_2(t) \right) J_1(\omega t)dt. \tag{60}$$

By using integration by parts we have

$$\begin{aligned} \tilde{I} =& \frac{1}{\omega} \left(\phi_1'(t) - \frac{\phi_1(t)}{t} \right) J_2(\omega t)|_0^x - \frac{1}{\omega} \int_0^x \frac{t^2\phi_1''(t) - 3t\phi_1'(t) + 3\phi_1(t)}{t^2} J_2(\omega t)dt \\ & + \frac{1}{\omega} \left(\phi_2'(t) - \frac{\phi_2(t)}{t} \right) J_2(\omega t)|_x^\infty - \frac{1}{\omega}\int_x^\infty \left(\phi_2''(t) - \frac{3\phi_2'(t)}{t} + \frac{3\phi_2(t)}{t^2} \right) J_2(\omega t)dt. \end{aligned} \tag{61}$$

According to [35], the exact solution to (3) can be written as

$$u(x) = f'(x) + \omega^2 \int_0^x J_0(\omega s)f(x-s)ds - \omega \int_0^x J_1(\omega s)f'(x-s)ds. \tag{62}$$

Furthermore, we have

$$u'(x) = f''(x) + \omega^2 \int_0^x J_0(\omega s)f'(x-s)ds - \omega J_1(\omega x)f'(0) - \omega \int_0^x J_1(\omega s)f''(x-s)ds, \tag{63}$$

$$u''(x) = f'''(x) + \frac{\omega J_1(\omega x)}{x}f'(0) + \omega \int_0^x \frac{J_1(\omega s)}{s} f''(x-s)ds. \tag{64}$$

A direct calculation implies $\frac{t^2\phi_1''(t) - 3t\phi_1'(t) + 3\phi_1(t)}{t^2} + \frac{\omega J_1(\omega(x-t))}{x-t}f'(0)$ is $O(\omega)$ as ω goes to infinity. By Lemmas 2 and 3, we obtain

$$\left| \int_0^x \frac{t^2\phi_1''(t) - 3t\phi_1'(t) + 3\phi_1(t)}{t^2} J_2(\omega t)dt \right| \sim O(\omega^{1/2}), \omega \to \infty. \tag{65}$$

With the help of Lemma 1, we have $\tilde{I} = O(\omega^{-1/2})$. It follows that

$$| \; \mathcal{F}(s)u(x) - \mathcal{F}(s_h)u(x) \; | = O(\omega^{-3/2}). \tag{66}$$

Combining Equations (58) and (66) gives

$$|u(x) - \tilde{u}(x)| \leq C\omega^{-1/2}, x \in I, \omega \to \infty, \tag{67}$$

where C is a constant independent of ω. This completes the proof. □

4. Numerical Results

In this section, we present some numerical results to verify given estimates in previous sections. All experiments are performed in MATLAB 2013b.

As a first example, we consider the following HOI,

$$I_1 = \int_0^1 J_0(\omega t)\frac{1}{1+25t^2}dt,\ I_2 = \int_0^2 J_1(\omega t)\cos(t)e^{-t}dt. \tag{68}$$

In Figures 1–3, we show the convergence rates of three convolution quadrature rules. Slowly varied lines in these figures manifest that given asymptotic orders in Section 2 are optimal. Absolute errors of these methods are given in Tables 2 and 3. The numerical results illustrate MCQ1 is much more efficient than other two methods in computing HOI.

Table 2. Comparisons of quadrature rules for $\int_0^1 J_0(\omega t)\frac{1}{1+25t^2}dt$.

ω	20	100	200	400	600	800	1000
CQ	9.7×10^{-4}	3.4×10^{-5}	1.1×10^{-5}	9.4×10^{-7}	1.5×10^{-6}	1.3×10^{-6}	1.7×10^{-7}
MCQ1	8.3×10^{-4}	5.0×10^{-6}	6.3×10^{-7}	6.6×10^{-8}	2.1×10^{-8}	1.2×10^{-8}	7.2×10^{-9}
MCQ2	9.4×10^{-4}	3.5×10^{-5}	1.1×10^{-5}	9.4×10^{-7}	1.5×10^{-6}	1.3×10^{-6}	1.7×10^{-7}

Table 3. Comparisons of quadrature rules for $\int_0^2 J_1(\omega t)\cos(t)e^{-t}dt$.

ω	20	100	200	400	600	800	1000
CQ	1.3×10^{-5}	7.8×10^{-6}	1.1×10^{-5}	1.3×10^{-6}	1.4×10^{-6}	1.4×10^{-6}	4.0×10^{-7}
MCQ1	6.3×10^{-6}	9.3×10^{-7}	1.6×10^{-7}	2.5×10^{-8}	1.9×10^{-8}	9.2×10^{-9}	4.6×10^{-9}
MCQ2	1.3×10^{-5}	7.8×10^{-6}	1.1×10^{-5}	1.3×10^{-6}	1.4×10^{-6}	1.4×10^{-6}	4.0×10^{-7}

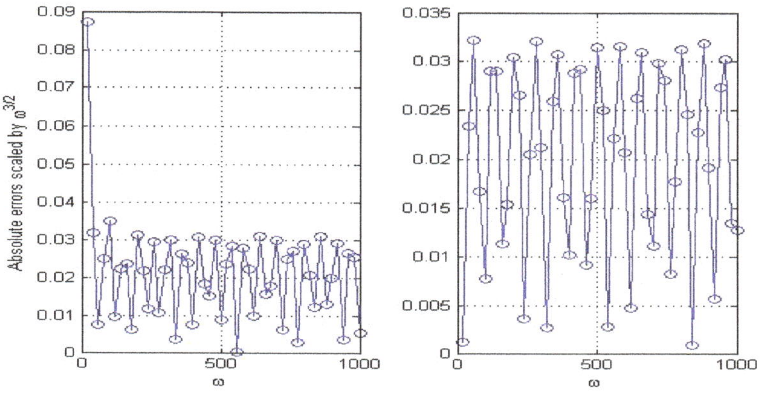

Figure 1. Asymptotic convergence rates of CQ for $\int_0^1 J_0(\omega t)\frac{1}{1+25t^2}dt$ (**left**) and $\int_0^2 J_1(\omega t)\cos(t)e^{-t}dt$ (**right**).

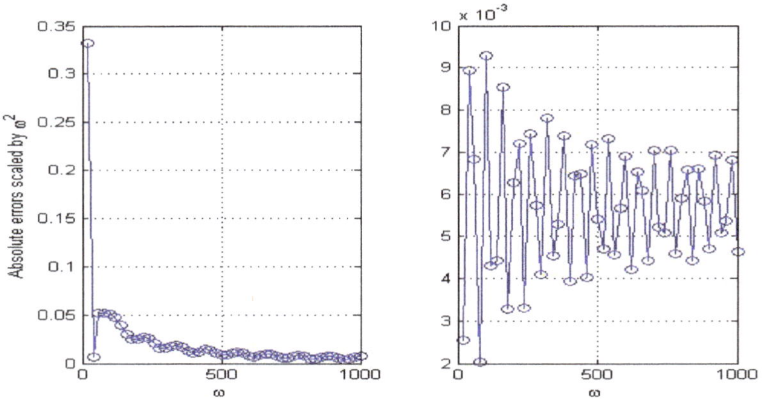

Figure 2. Asymptotic convergence rates of MCQ1 for $\int_0^1 J_0(\omega t)\frac{1}{1+25t^2}dt$ (**left**) and $\int_0^2 J_1(\omega t)\cos(t)e^{-t}dt$ (**right**).

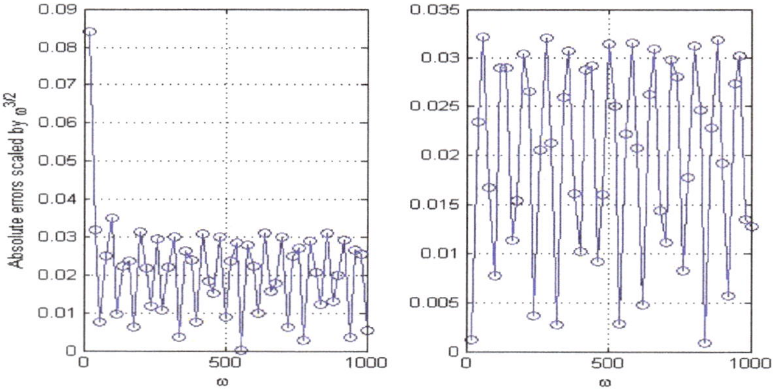

Figure 3. Asymptotic convergence rates of MCQ2 for $\int_0^1 J_0(\omega t)\frac{1}{1+25t^2}dt$ (**left**) and $\int_0^2 J_1(\omega t)\cos(t)e^{-t}dt$ (**right**).

In the second example, we consider application of CQ to solving HOVIE (3). Firstly, by letting $T = 2$ and $N = 20$, we give the computed solution in Table 4 with various ω. Absolute errors listed in this table show CQ shares the same property as Filon methods ([24,26]), that is, the higher the oscillation, the better the approximation. Then we compare these two methods in Figure 4, where we can learn CQ behaves better than Filon methods.

Table 4. CQ for Volterra integral equations with $f(x) = xe^{-x}$.

ω	$x = 0.1$	$x = 0.4$	$x = 0.8$	$x = 1.2$	$x = 1.6$	$x = 1.8$	$x = 2$
10	3.3×10^{-1}	7.6×10^{-2}	4.7×10^{-2}	2.5×10^{-2}	8.0×10^{-3}	1.4×10^{-2}	3.1×10^{-3}
100	9.1×10^{-2}	5.5×10^{-3}	6.3×10^{-4}	7.0×10^{-5}	3.5×10^{-4}	3.6×10^{-5}	2.6×10^{-4}
200	5.0×10^{-2}	1.4×10^{-4}	3.6×10^{-4}	1.1×10^{-4}	1.2×10^{-4}	1.1×10^{-4}	1.8×10^{-5}
500	1.9×10^{-2}	1.9×10^{-4}	1.2×10^{-5}	4.6×10^{-5}	3.7×10^{-5}	2.2×10^{-5}	6.8×10^{-6}
1000	1.0×10^{-2}	5.9×10^{-7}	3.9×10^{-5}	1.2×10^{-5}	8.7×10^{-10}	9.6×10^{-6}	9.2×10^{-6}

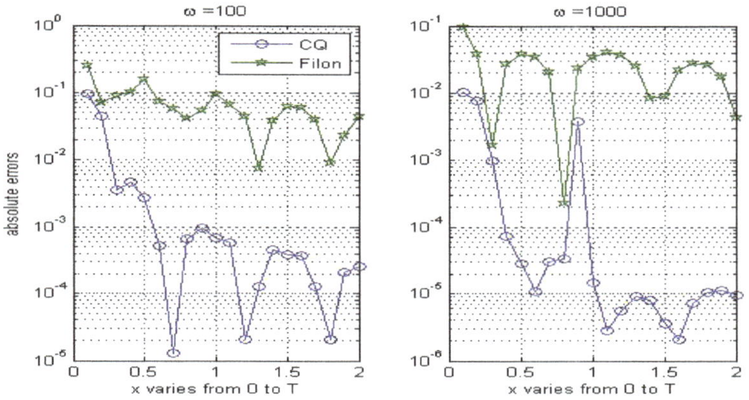

Figure 4. Comparisons between Filon methods and CQ for solving Volterra equations with $f(x) = \frac{x}{1+x^2}$.

5. Conclusions

The above theoretical and numerical results contribute to the study on the convergence property of CQ for solving HOP. The theoretical results in Section 2 reveal the convergence rate of CQ with respect to the frequency, that is, CQ converges in negative powers of ω as ω goes to infinity. Among them, the new modified rule (MCQ1) enjoys the fastest convergence rate. When we apply CQ to solving HOVIE, similar phenomenon is detected and analyzed. The numerical results in Section 4 show given convergence orders in Section 2 are optimal. In addition, this paper merely opens a window to the convergence theory of CQ for HOP, much work on various versions of CQ, such as Runge-Kutta CQ ([36]), Fourier CQ ([37]), and so forth, is needed in the future.

Author Contributions: J.M. and H.L. conceived and designed the experiments; J.M. performed the experiments; H.L. analyzed the data; J.M. contributed reagents/materials/analysis tools; J.M. and H.L. wrote the paper.

Funding: This work is supported by NSF of China (No. 11761020), Scientific Research Foundation for Young Talents of Department of Education of Guizhou Province (No. 2016125), and Science and Technology Foundation of Guizhou Province (No. QKH[2017]5788).

Acknowledgments: The authors thank referees for their helpful suggestions.

Conflicts of Interest: The authors declare no conflict of interest.

Abbreviations

The following abbreviations are used in this manuscript:

BDF	backward differentiation formula
CQ	convolution quadrature rule
FFT	fast Fourier transform
GMRES	generalized minimal residual method
HOI	highly oscillatory integral
HOP	highly oscillatory problem
ODE	ordinary differential equation
HOVIE	highly oscillatory Volterra integral equation
MCQ1	modified convolution quadrature of the first kind
MCQ2	modified convolution quadrature of the second kind

References

1. Michalski, K. Extrapolation methods for sommerfeld integral tails. *IEEE Trans. Antennas Propag.* **1998**, *46*, 1405–1418. [CrossRef]
2. Rakov, V.; Uman, M. Review and evaluation of lightning return stroke models including some aspects of their application. *IEEE Trans. Electromagn. C* **1998**, *40*, 403–426. [CrossRef]
3. Theodoulidis, T. Exact solution of Pollaczek's integral for evaluation of earth-return impedane for underground conductors. *IEEE Trans. Electromagn. C* **2012**, *54*, 806–814. [CrossRef]
4. Ma, J. Fast and high-precision calculation of earth return mutual impedance between conductors over a multilayered soil. *COMPEL* **2018**, *37*, 1214–1227. [CrossRef]
5. Iserles, A. On the numerical quadrature of highly-oscillating integrals I: Fourier transforms. *IMA J. Numer. Anal.* **2004**, *24*, 365–391. [CrossRef]
6. Filon, L.N.G. On a quadrature formula for trigonometric integrals. *Proc. R. Soc. Edinb.* **1928**, *49*, 38–47. [CrossRef]
7. Iserles, A.; Nørsett, S.P. Efficient quadrature of highly oscillatory integrals using derivatives. *Proc. R. Soc. A* **2005**, *461*, 1383–1399. [CrossRef]
8. Domínguez, V.; Graham, I.G.; Smyshlyaev, V.P. Stability and error estimates for Filon-Clenshaw-Curtis rules for highly-oscillatory integrals. *IMA J. Numer. Anal.* **2011**, *31*, 1253–1280. [CrossRef]
9. Xiang, S.; Cho, Y.; Wang, H.; Brunner, H. Clenshaw-Curtis-Filon-type methods for highly oscillatory Bessel transforms and applications. *IMA J. Numer. Anal.* **2011**, *31*, 1281–1314. [CrossRef]
10. Milovanovíc, G. Numerical calculation of integrals involving oscillatory and singular kernels and some applications of quadratures. *Comput. Math. Appl.* **1998**, *36*, 19–39. [CrossRef]
11. Huybrechs, D.; Vandewalle, S. On the evaluation of highly oscillatory integrals by analytic continuation. *SIAM J. Numer. Anal.* **2006**, *44*, 1026–1048. [CrossRef]
12. Majidian, H. Numerical approximation of highly oscillatory integrals on semi-finite intervals by steepest descent method. *Numer. Algorithms* **2013**, *63*, 537–548. [CrossRef]
13. Ma, J. Implementing the complex integral method with the transformed Clenshaw-Curtis quadrature. *Appl. Math. Comput.* **2015**, *250*, 792–797. [CrossRef]
14. Levin, D. Procedures for computing one- and two-dimensional integrals of functions with rapid irregular oscillations. *Math. Comput.* **1982**, *38*, 531–538. [CrossRef]
15. Li, J.; Wang, X.; Wang, T. A universal solution to one-dimensional highly oscillatory integrals. *Sci. China* **2008**, *10*, 1614–1622.
16. Olver, S. Shifted GMRES for oscillatory integrals. *Numer. Math.* **2010**, *114*, 607–628. [CrossRef]
17. Ma, J.; Liu, H. A well-conditioned Levin method for calculation of highly oscillatory integrals and its application. *J. Comput. Appl. Math.* **2018**, *342*, 451–462. [CrossRef]
18. Chen, R.; Xiang, S. Note on the homotopy perturbation method for multivariate vector-value oscillatory integrals. *Appl. Math. Comput.* **2009**, *215*, 78–84. [CrossRef]
19. Evans, G.A.; Chung, K.C. Some theoretical aspects of generalised quadrature rules. *J. Complex.* **2003**, *19*, 272–285. [CrossRef]
20. Sidi, A. Extrapolation method for oscillatory infinite integrals. *Math. Comput.* **1988**, *51*, 249–266. [CrossRef]

21. Lubich, C. Convolution quadrature and discretized operational calculus. I. *Numer. Math.* **1988**, *52*, 129–145. [CrossRef]
22. Lubich, C. Convolution quadrature and discretized operational calculus. II. *Numer. Math.* **1988**, *52*, 413–425. [CrossRef]
23. Lubich, C. On convolution quadrature and Hille-Phillips operational calculus. *Appl. Numer. Math.* **1992**, *9*, 187–199. [CrossRef]
24. Xiang, S.; Brunner, H. Efficient methods for Volterra integral equations with highly oscillatory Bessel kernels. *BIT Numer. Math.* **2013**, *53*, 241–263. [CrossRef]
25. Xiang, S.; Wang, H. Fast integration of highly oscillatory integrals with exotic oscillators. *Math. Comput.* **2010**, *79*, 829–844. [CrossRef]
26. Ma, J.; Xiang, S.; Kang, H. On the convergence rates of Filon methods for the solution of a Volterra integral equation with a highly oscillatory Bessel kernel. *Appl. Math. Lett.* **2013**, *26*, 699–705. [CrossRef]
27. Stein, M.S.; Shakarchi, R. *Complex Analysis*; Princeton University Press: Princeton, NJ, USA, 2003.
28. Davis, P.; Duncan, D. Convolution spline approximations for time domain boundary integral equations. *J. Integral Equ. Appl.* **2014**, *26*, 369–410. [CrossRef]
29. Abramowitz, M.; Stegun, I.A. *Handbook of Mathematical Functions*; Dover Publications: Mineola, NY, USA, 1964.
30. Gradshteyn, I.S.; Ryzhik, I.M. *Table of Integrals, Series, and Products*; Academic Press: Cambridge, MA, USA, 1994.
31. Watson, G.N. *A Treatise on the Theory of Bessel Functions*; Cambridge University Press: Cambridge, UK, 1952.
32. Kauthen, J.P. A survey of singularly perturbed Volterra equations. *Appl. Numer. Math.* **1997**, *24*, 95–114. [CrossRef]
33. Davis, P.; Duncan, D. Stability and convergence of collocation schemes for retarded potential integral equations. *SIAM J. Numer. Anal.* **2004**, *42*, 1167–1188. [CrossRef]
34. Olver, F.J.; Lozier, D.W.; Boisvert, R.F.; Clark, C.W. *NIST Handbook of Mathematical Functions*; Cambridge University Press: Cambridge, UK, 2010.
35. Wang, H.; Xiang, S. Asymptotic expansion and Filon-type methods for a Volterra integral equation with a highly oscillatory kernel. *IMA J. Numer. Anal.* **2011**, *31*, 469–490. [CrossRef]
36. Banjai, L.; Lubich, C.; Melenk, J.M. Runge-Kutta convolution quadrature for operators arising in wave propagation. *Numer. Math.* **2011**, *119*, 1–20. [CrossRef]
37. Xu, K.; Austin, A.P.; Wei, K. A Fast Algorithm for the Convolution of Functions with Compact Support Using Fourier Extensions. *SIAM J. Sci. Comput.* **2017**, *39*, A3089–A3106. [CrossRef]

© 2018 by the authors. Licensee MDPI, Basel, Switzerland. This article is an open access article distributed under the terms and conditions of the Creative Commons Attribution (CC BY) license (http://creativecommons.org/licenses/by/4.0/).